# STATICS AND MECHANICS OF MATERIALS: AN INTEGRATED APPROACH

## Second Edition

### William F. Riley
Distinguished Professor Emeritus
Iowa State University

### Leroy D. Sturges
Aerospace Engineering and Engineering Mechanics
Iowa State University

### Don H. Morris
Engineering Science and Mechanics
Virginia Polytechnic Institute and State University

JOHN WILEY & SONS, INC.

*William F. Riley*
*1925–2000*
*Friend, Colleague, Mentor*

ACQUISITIONS EDITOR   Joseph Hayton
MARKETING MANAGER   Katherine Hepburn
DESIGNER   Dawn L. Stanley

Library of Congress Cataloging-in-Publication Data

Sturges, Leroy D.
    Statics and mechanics of materials : an integrated approach / Leroy D. Sturges, Don H. Morris.—2nd ed.
        p. cm.
    Rev. ed. of: Statics and mechanics of materials / William F. Riley, Leroy D. Sturges, Don H. Morris. c1995.
        ISBN 0-471-43446-9
    1. Strength of materials. 2. Statics. I. Morris, Don H., 1939- II. Riley, William F. (William Franklin), 1925- Statics and mechanics of materials. III. Title.

TA405.R56 2001
620.1'12--dc21

2001046630

95–4072
CIP

# PREFACE

## APPROACH/PHILOSOPHY

The purpose of courses in engineering mechanics is to describe the effects that forces have on bodies and structures. The traditional introduction to mechanics consists of a course in statics followed by a course in mechanics of materials. The principles of statics are used to determine the forces that act on or in a structure, assuming that the structure is perfectly rigid and does not deform. Then, these forces, along with the theory developed in mechanics of materials, are used to determine how the material deforms or reacts.

This book approaches the teaching of mechanics using the just-in-time approach. As soon as the student has studied equilibrium of concurrent force systems, he or she is ready to calculate stretches of wires and rods using a one-dimensional Hooke's law. After studying rigid body equilibrium, the student is ready to calculate stresses and deformation in members such as shafts and beams. When the two subjects are integrated in a unified course in this manner, students can immediately see the use of the principles of statics; they can see the interrelationship of statics and mechanics of materials.

**Free-Body Diagrams.** We strongly feel that a proper free-body diagram is very important in all mechanics courses. It is our approach that, whenever an equation of equilibrium is written, it must be accompanied by a complete, proper free-body diagram. Furthermore, since the primary purpose of a free-body diagram is to show the forces acting on a body, the free-body diagram should not be used for any other purpose. We encourage students to draw separate diagrams to show deformation and compatibility relationships.

**Problem-Solving Procedure.** Students are urged to develop the ability to reduce problems to a series of simpler component problems that can be easily analyzed and combined to give the solution of the initial problem. Along with an effective methodology for problem decomposition and solution, the ability to present results in a clear, logical, and neat manner is emphasized throughout the text.

**Homework Problems.** The illustrative examples and homework problems have been selected with special attention devoted to problems that require an understanding of the principles of mechanics of materials without demanding excessive time for computational work. Over 1100 homework problems are included so that problem assignments may be varied from term to term. The problems in each set represent a considerable range of difficulty and are grouped according to this range of difficulty. Mastery, in general, is not achieved by solving a large number of simple but similar problems. While the solution of simple problems is necessary to build a student's problem-solving skills and confidence, we believe that a student gains mastery of a subject through application of basic theory to the solution of problems that appear somewhat difficult.

**SI versus U.S. Customary Units.** U.S. customary units and SI units are used in approximately equal proportions in the text for both example problems and homework problems. To help the instructor who wants to assign problems of one type or the other, odd-numbered homework problems are in U.S. customary units and even-numbered homework problems are in SI units.

**Answers Provided.** Answers to about half of the homework problems are included at the end of this book. Since the convenient designation of problems for which answers are provided is of great value to those who make up assignment sheets, the problems for which answers are provided are indicated by means of an asterisk (*) after the problem number.

**Emphasis on Fundamentals.** This book is designed to emphasize the required fundamental principles, with numerous applications to demonstrate and develop logical, orderly methods of procedure. Instead of deriving numerous formulas for all types of problems, we have stressed the use of free-body diagrams and the equations of equilibrium, together with the geometry of the deformed body and the observed relations between stress and strain, for the analysis of the force system acting on a body.

**Emphasis on Clarity.** The emphasis is always on keeping the material understandable to the student. Clarity is never sacrificed for the sake of mathematical elegance. Calculus and vector methods are used where necessary and where appropriate. However, if scalar methods are more appropriate and/or are commonly used by practicing engineers, then these methods are generally used in the example problems.

## NEW TO THIS EDITION

In addition to new and revised homework problems, this edition has 27 new example problems including 6 new computational example problems. The review of vector operations, which was previously in an appendix, has been integrated into the appropriate sections of Chapters 2 and 5. The structural applications in Chapter 6 have been rearranged so that the discussion of frames and machines precedes the discussion of trusses. A common mistake that students make when they see trusses before frames is to draw forces along the axis of structural members that are not two-force members. This new arrangement should help students recognize that forces in structures often do not act along the axis of the structural members. Design of ductile and brittle materials (including theories of failure) has been added to Chapter 10, and eccentrically loaded columns have been added to Chapter 11. In addition, several minor sections in Chapters 4 through 9 have been promoted to major sections for better organization of the material.

## ORGANIZATION

After a brief introduction to mechanics in Chapter 1, Chapter 2 describes the characteristics of forces and develops the mathematics necessary to work with concurrent forces. These concepts are immediately used in Chapter 3 to calculate the forces acting on a particle in equilibrium. The basic discussion is completed in Chapter 4, where stress, strain, and the relationship between loads and deformation is presented. Chapter 5 continues the description of forces, develops the concept of equivalent force-couple systems, and explores the effects of forces and couples on rigid bodies. Chapter 6 presents the equilibrium of rigid bodies and its use in several structural applications. The final five chapters consist of standard topics of Mechanics of Materials—torsion of circular shafts (Chapter 7), flexural stresses in beams (Chapter 8), deflections of beams (Chapter 9), combined loadings (Chapter 10), and columns and other compressive members (Chapter 11). Second moments of areas are introduced and developed where needed in Chapters 7 and 8.

## PEDAGOGY

Every chapter opens with a brief introduction and ends with a summary of important concepts covered in the chapter, followed by a set of review problems. All principles are illustrated by one or more example problems and several homework problems. The homework problems are graded in difficulty and are separated into groups of introductory, intermediate, and challenging problems. Several sections include a set of computer problems that require students to analyze how the solution depends on some parameter of the problem. While the computations could be accomplished by the student writing a FORTRAN program, the computations could just as easily be carried out using MathCAD, Mathematica, or a spreadsheet program. The important concept of the computer problems is that they require students to analyze how the solution depends on some parameter of the problem.

**Design.**  Most chapters conclude with a section on design that includes example problems and a set of homework problems. The emphasis in these problems is that there is often more than just one criterion to be satisfied in a design specification. An acceptable design must satisfy all specified criteria. In addition, standard lumber, pipes, beams, and so on come in specific sizes. The student must choose an appropriate structural member from these standard materials. Since each different choice of a beam or a piece of lumber has a different specific weight and affects the overall problem differently, students are also introduced to the idea that design is an iterative process.

## ACKNOWLEDGMENTS

We are grateful for comments and suggestions received from colleagues and from users of the earlier edition of this book. We would like to particularly mention those people who provided feedback on revision plans and parts or all of the new edition. They include Jim Elkins, Oklahoma Christian University; Hamid Garmestani, FAMU-FSU; David S. Hansen, United States Air Force Academy; Takeru Igusa, Johns Hopkins University; Thomas Juliano, New Jersey Institute of Technology; Richard D. Keane, University of Illinois; John S. Klegka, United States Military Academy; Nels Madsen, Auburn University; K.T. Ramesh, Johns Hopkins University; Risa J. Robinson, Rochester Institute of Technology; and Douglas J. Wendel, Snow College. Final judgments concerning organization of material and emphasis of topics, however, were made by the authors. Finally, we thank our families for their constant support of our efforts.

**E-mail address.**  We will be pleased to receive comments from readers and will attempt to acknowledge all such communications. Comments can be sent by e-mail to *sturges@iastate.edu* or to *dhmorris@vt.edu*.

# CONTENTS

## CHAPTER 11: COLUMNS

## APPENDIX A: TABLES OF PROPERTIES

## ANSWERS TO SELECTED PROBLEMS 707

## INDEX 717

# GENERAL PRINCIPLES

## 1-1 INTRODUCTION

*Mechanics* is the branch of the physical sciences that deals with the response of bodies to the action of forces. The subject matter of this field constitutes a large part of our knowledge of the laws governing the behavior of solid bodies as well as the laws governing the behavior of gases and liquids. The laws of mechanics find application in most machines and structures involved in engineering practice. For convenience, the study of mechanics is divided into three parts: namely, the mechanics of rigid bodies, the mechanics of deformable bodies, and the mechanics of fluids.

A study of the mechanics of rigid bodies can be further subdivided into three main divisions: statics, kinematics, and kinetics. *Statics* deals with bodies that are acted upon by balanced forces and hence are at rest or have uniform motion. Such bodies are said to be in "equilibrium." Statics is an important part of the study of mechanics because it provides methods for determining support reactions and relationships between internal force distributions and external loads for stationary structures. Many practical engineering problems involving loads carried by structural components can be solved using relationships developed in statics. The relationships developed in statics between internal force distributions and external loads play an important role in the development of deformable body mechanics.

*Kinematics* deals with the motion of bodies without considering the manner in which the motion is produced. Kinematics is sometimes referred to as the "geometry of motion." Kinematics forms an important part of the study of mechanics, not only because of its application to problems in which forces are involved, but also because of its application to problems that involve only motions of parts of a machine. For many motion problems, the principles of kinematics alone are sufficient for solution of the problem. Such problems are discussed in Kinematics of Machinery books, in which the motion of machine elements such as cam shafts, gears, connecting rods, and quick-return mechanisms is considered.

*Kinetics* deals with bodies that are acted upon by unbalanced forces; hence, they have nonuniform or accelerated motions. A study of kinetics is an important part of the study of mechanics because it provides relationships between the motion of a body and the forces and moments acting upon the body. Kinetic relationships may be obtained by direct application of Newton's laws of motion or by using integrated forms of the equations of motion that result in the principles of work–energy or impulse–momentum. Frequently, the term *dynamics* is used in the technical literature to denote the subdivisions of mechanics with which the idea of motion is most closely associated—namely, kinematics and kinetics.

The branch of mechanics that deals with internal force distributions and the deformations developed in actual engineering structures and machine components when they are subjected to systems of forces is known as *mechanics of deformable bodies.* Books covering this part of mechanics commonly have the title *Mechanics of Materials, Strength of Materials,* or *Mechanics of Deformable Bodies.*

The branch of mechanics that deals with liquids and gases at rest or in motion is known as *fluid mechanics.* Fluids can be classified as compressible or incompressible. A fluid is said to be compressible if the density of the fluid varies with temperature and pressure. If the volume of a fluid remains constant during a change in pressure, the fluid is said to be incompressible. Liquids are considered incompressible for most engineering applications. A subdivision of fluid mechanics that deals with incompressible liquids is commonly known as *hydraulics.*

This integrated book on statics and mechanics of materials will cover all topics normally treated in books dealing with the statics of rigid and deformable bodies. The book will provide the foundation required for follow-on courses in many fields of engineering.

## 1-2 FUNDAMENTAL QUANTITIES OF MECHANICS

The fundamental quantities of mechanics are space, time, mass, and force. Three of the quantities—space, time, and mass—are absolute quantities, meaning that they are independent of each other and cannot be expressed in terms of the other quantities or in simpler terms. The quantity known as a force is not independent of the other three quantities but is related to the mass of the body and to the manner in which the velocity of the body varies with time. The following paragraphs provide a brief description of these and other important concepts.

*Space* is the geometric region in which the physical events of interest in mechanics occur. The region extends without limit in all directions. The measure used to describe the size of a physical system is known as a *length.* The position of a point in space can be determined relative to some reference point by using linear and angular measurements with respect to a coordinate system whose origin is at the reference point. The basic reference system used as an aid in solving mechanics problems is one that is considered fixed in space. Measurements relative to this system are called *absolute.*

*Time* can be defined as the interval between two events. Measurements of this interval are made by making comparisons with some reproducible event such as the time required for the earth to orbit the sun or the time required for the earth to rotate on its axis. Solar time is earth rotation time measured with respect to the sun and is used for navigation on earth and for daily living purposes.

Any device that is used to indicate the passage of time is referred to as a *clock*. Reproducible events commonly used as sensing mechanisms for clocks include the swing of a pendulum, oscillation of a spiral spring and balance wheel, and oscillation of a piezoelectric crystal. The time required for one of these devices to complete one cycle of motion is known as the *period*. The frequency of the motion is the number of cycles occurring in a given unit of time.

*Matter* is any substance that occupies space. A *body* is matter bounded by a closed surface. The property of a body that causes it to resist any change in motion is known as *inertia*. *Mass* is a quantitative measure of inertia. The resistance a body offers to a change in translational motion is independent of the size and shape of the body. It depends only on the mass of the body. The resistance a body offers to a change in rotational motion depends on the distribution of the body's mass. Mass is also a factor in the gravitational attraction between two bodies.

A *force* can be defined as the action of one body upon another body. Our concept of force comes mainly from personal experiences in which we are one of the bodies and tension or compression of our muscles results when we try to pull or push the second body. This is an example of force resulting from direct contact between bodies. A force can also be exerted between bodies that are physically separated. Gravitational forces exerted by the earth on the moon and on artificial satellites to keep them in earth orbit are examples. Since a body cannot exert a force on a second body unless the second body offers a resistance, a force never exists alone. Forces always occur in pairs, and the two forces have equal magnitude and opposite sense. Although a single force never exists, it is usually convenient in mechanics problems to think only of the actions of other bodies on the body in question without taking into account the reactions of the body in question. The external effect of a force on a body is either acceleration of the body or development of resisting forces (reactions) on the body.

A *particle* has mass but no size or shape. When a body (large or small) in a mechanics problem can be treated as a particle, the analysis is greatly simplified, since the mass can be assumed to be concentrated at a point and the concept of rotation is not involved in the solution of the problem.

A *rigid body* can be represented as a collection of particles. The size and shape of the body remain constant at all times and under all conditions of loading. The rigid-body concept represents an idealization of the true situation, since all real bodies will change shape to a certain extent when subjected to a system of forces. Such changes are small for most structural elements and machine parts encountered in engineering practice; therefore, they have only a negligible effect upon the reactions required to maintain equilibrium of the body.

**Newton's Laws** The foundations for studies in engineering mechanics are the laws formulated and published by Sir Isaac Newton in 1687. In a treatise called *The Principia,* Newton stated the basic laws governing the motion of a particle as[1]:

---

[1] As stated in Dr. Ernst Mach, *The Science of Mechanics,* 9th ed. The Open Court Publishing Company, LaSalle, Ill., 1942.

> ### NEWTON'S LAWS OF MOTION
>
> **Law 1:** "Every body perseveres in its state of rest or of uniform motion in a straight line, except insofar as it is compelled to change that state by impressed forces."
>
> **Law 2:** "Change of motion is proportional to the moving force impressed, and takes place in the direction of the straight line in which such force is impressed."
>
> **Law 3:** "Reaction is always equal and opposite to action; that is to say, the actions of two bodies upon each other are always equal and directly opposite."

These laws, which have come to be known as Newton's Laws of Motion, are commonly expressed today as

*Law 1.* In the absence of external forces, a particle originally at rest or moving with a constant velocity will remain at rest or continue to move with a constant velocity along a straight line.

*Law 2.* If an external force acts on a particle, the particle will be accelerated in the direction of the force and the magnitude of the acceleration will be directly proportional to the force and inversely proportional to the mass of the particle.

*Law 3.* For every action there is an equal and opposite reaction. The forces of action and reaction between bodies are equal in magnitude, opposite in direction, and collinear.

Newton's three laws were developed from a study of planetary motion (the motion of particles); therefore, they apply only to the motion of particles. During the eighteenth century, Leonhard Euler (1707–1783) extended Newton's work on particles to rigid-body systems.

The first law of motion is a special case of the second law and covers the case in which the particle is in equilibrium. Thus, the first law provides the foundation for the study of statics. The second law of motion provides the foundation for the study of dynamics. The mathematical statement of the second law, which is widely used in dynamics, is

$$\mathbf{F} = m\mathbf{a} \tag{1-1}$$

where $\mathbf{F}$ is the external force acting on the particle, $m$ is the mass of the particle, and $\mathbf{a}$ is the acceleration of the particle. The third law of motion provides the foundation for an understanding of the concept of a force, since in practical engineering applications the word *action* is taken to mean force. Thus, if one body exerts a force on a second body, the second body exerts an equal and opposite force on the first.

The law that governs the mutual attraction between two bodies was also formulated by Newton and is known as the Law of Gravitation. This law can be expressed mathematically as

$$F = G\frac{m_1 m_2}{r^2} \tag{1-2}$$

TABLE 1-1 Solar System Masses and Distances

Mass

| | | |
|---|---|---|
| of the earth | $= 4.095(10^{23})$ slug | $= 5.976(10^{24})$ kg |
| of the moon | $= 5.037(10^{21})$ slug | $= 7.350(10^{22})$ kg |
| of the sun | $= 1.364(10^{29})$ slug | $= 1.990(10^{30})$ kg |

Mean or average radius

| | | |
|---|---|---|
| of the earth* | $= 3960$ mi | $= 6370$ km |
| of the moon | $= 1080$ mi | $= 1740$ km |
| of the sun | $= 432,000$ mi | $= 696,000$ km |

Mean or average distance from the earth

| | | |
|---|---|---|
| to the moon | $= 239,000$ mi | $= 384,000$ km |
| to the sun | $= 92.96(10^6)$ mi | $= 149.6(10^6)$ km |

*Radius of a sphere of equal volume

| | | |
|---|---|---|
| Polar radius | $= 3950$ mi | $= 6357$ km |
| Equatorial radius | $= 3963$ mi | $= 6378$ km |

where $F$ is the magnitude of the mutual force of attraction between the two bodies, $G$ is the universal gravitational constant, $m_1$ is the mass of one of the bodies, $m_2$ is the mass of the other body, and $r$ is the distance between the centers of mass of the two bodies.

Approximate values of the universal gravitational constant that are suitable for most engineering computations are

$$G = 3.439(10^{-8}) \text{ ft}^3/(\text{slug} \cdot \text{s}^2) \text{ (in the U.S. customary system of units)}$$

$$G = 6.673(10^{-11}) \text{ m}^3/(\text{kg} \cdot \text{s}^2) \text{ (in the SI system of units)}$$

The mutual forces of attraction between the two bodies represent the action of one body on the other; therefore, they obey Newton's third law, which requires that they be equal in magnitude, opposite in direction, and collinear (lie along the line joining the centers of mass of the two bodies). The law of gravitation is very important in all studies involving the motion of planets or artificial satellites.

Some of the quantities and constants that may be of interest in applying the law of universal gravitation are listed in Table 1-1.

**Mass and Weight** The mass $m$ of a body is an absolute quantity that is independent of the position of the body and independent of the surroundings in which the body is placed. The weight $W$ of a body is the gravitational attraction exerted on the body by the planet Earth or by any other massive body, such as the moon. Therefore, the weight of a body depends on the position of the body relative to some other body. Thus, for Eq. (1-2), at the surface of the earth,

$$W = G\frac{m_e m}{r_e^2} = mg \qquad (1\text{-}3)$$

where $m_e$ is the mass of the earth, $r_e$ is the mean radius of the earth, and $g = Gm_e/r_e^2$ is the gravitational acceleration.

Approximate values for the gravitational acceleration that are suitable for most engineering computations are

$$g = 32.2 \text{ ft/s}^2$$
$$g = 9.81 \text{ m/s}^2$$

A source of some confusion arises because the pound is sometimes used as a unit of mass and the kilogram is sometimes used as a unit of force. In grocery stores in Europe, weights of packages are marked in kilograms. In the United States, weights of packages are often marked in both pounds and kilograms. Similarly, a unit of mass called the *pound* or the *pound mass,* which is the mass whose weight is one pound under standard gravitational conditions, is sometimes used.

Throughout this book, without exception, the pound (lb) will be used as the unit of force and the slug will be used as the unit of mass for problems and examples when the U.S. customary system of units is used. Similarly, the newton (N) will be used as the unit of force and the kilogram (kg) will be used as the unit of mass for problems and examples when the SI system of units is used.

## ▎ Example Problem 1-1

A body weighs 250 lb at the earth's surface. Determine

(a) The mass of the body.
(b) The weight of the body 500 miles above the earth's surface.
(c) The weight of the body on the moon's surface.

### SOLUTION

(a) The weight of the body at the earth's surface is given by Eq. (1-3) as

$$W = mg$$

Thus,

$$m = \frac{W}{g} = \frac{250}{32.2} = 7.76 \, \frac{\text{lb} \cdot \text{s}^2}{\text{ft}} = 7.76 \text{ slug} \qquad \textbf{Ans.}$$

(b) The force of attraction between two bodies is given by Eq. (1-2) as

$$W = F = G\frac{m_1 m_2}{r^2} \quad \text{or} \quad Wr^2 = Gm_1 m_2 = \text{Constant}$$

The mean radius of the earth (see Table 1-1) is

$$r_e = 3960 \text{ mi}$$

Thus, for the two positions of the body,

$$Wr_e^2 = W_{500}\,(r_e + 500)^2 = Gm_1 m_2 = \text{Constant}$$

$$W_{500} = \frac{Wr_e^2}{(r_e + 500)^2} = \frac{250(3960)^2}{(3960 + 500)^2} = 197.1 \text{ lb} \qquad \textbf{Ans.}$$

(c) The mean radius and mass of the moon (see Table 1-1) are

$$r_m = 1080 \text{ mi} = 5.702(10^6) \text{ ft}$$
$$m_m = 5.037(10^{21}) \text{ slug}$$

Also,

$$G = 3.439(10^{-8}) \text{ ft}^3/(\text{slug} \cdot \text{s}^2)$$

On the moon's surface, the weight of the body is given by Eq. (1-2) as

$$W = \frac{Gmm_m}{r^2_m} = 3.439(10^{-8})\frac{7.77(5.037)(10^{21})}{[5.702(10^6)]^2} = 41.4 \text{ lb} \ \blacksquare \qquad \textbf{Ans.}$$

## PROBLEMS

### Introductory Problems

**1-1\*** Calculate the mass $m$ of a body that weighs 600 lb at the surface of the earth.

**1-2\*** Calculate the weight $W$ of a body at the surface of the earth if it has a mass $m$ of 675 kg.

**1-3** If a man weighs 180 lb at sea level, determine the weight $W$ of the man
(a) At the top of Mt. McKinley (20,320 ft above sea level).
(b) At the top of Mt. Everest (29,028 ft above sea level).

**1-4\*** Calculate the weight $W$ of a navigation satellite at a distance of 20,200 km above the earth's surface if the satellite weighs 9750 N at the earth's surface.

**1-5** Compute the gravitational force acting between two spheres that are touching each other if each sphere weighs 1125 lb and has a diameter of 20 in.

**1-6** Two spherical bodies have masses of 60 kg and 80 kg, respectively. Determine the gravitational force of attraction between the spheres if the distance from center to center is 600 mm.

### Intermediate Problems

**1-7\*** Determine the weight $W$ of a satellite when it is in orbit 8500 mi above the surface of the earth if the satellite weighs 7600 lb at the earth's surface.

**1-8\*** Determine the weight $W$ of a satellite when it is in orbit $20.2(10^6)$ m above the surface of the earth if the satellite weighs 8450 N at the earth's surface.

**1-9\*** If a woman weighs 135 lb when standing on the surface of the earth, how much would she weigh when standing on the surface of the moon?

**1-10** Determine the weight $W$ of a body that has a mass of 1000 kg
(a) At the surface of the earth.
(b) At the top of Mt. McKinley (6193 m above sea level).
(c) In a satellite at an altitude of 250 km.

**1-11** If a man weighs 210 lb at sea level, determine the weight $W$ of the man
(a) At the top of Mt. Everest (29,028 ft above sea level).
(b) In a satellite at an altitude of 200 mi.

**1-12** A space traveler weighs 800 N on earth. A planet having a mass of $5(10^{25})$ kg and a diameter of $30(10^6)$ m orbits a distant star. Determine the weight $W$ of the traveler on the surface of this planet.

### Challenging Problems

**1-13\*** The planet Jupiter has a mass of $1.302(10^{26})$ slug and a visible diameter (top of the cloud layer) of 88,700 mi. Determine the gravitational acceleration $g$
(a) At a point 100,000 miles above the top of the clouds.
(b) At the top of the cloud layers.

**1-14\*** The planet Saturn has a mass of $5.67(10^{26})$ kg and a visible diameter (top of the cloud layer) of 120,000 km. The weight $W$ of a planetary probe on earth is 4.50 kN. Determine
(a) The weight of the probe when it is 600,000 km above the top of the clouds.
(b) The weight of the probe as it begins its penetration of the cloud layers.

**1-15** The first U.S. satellite, *Explorer 1,* had a mass of approximately 1 slug. Determine the force exerted on the satellite by the earth at the low and high points of its orbit, which were 175 mi and 2200 mi, respectively, above the surface of the earth.

**1-16** A neutron star has a mass of $2(10^{30})$ kg and a diameter of 10 km. Determine the gravitational force of attraction on a 10-kg space probe

(a) When it is 1000 km from the center of the star.
(b) At the instant of impact with the surface of the star.

---

# 1-3 UNITS OF MEASUREMENT

The building blocks of mechanics are the physical quantities used to express the laws of mechanics. Some of these quantities are mass, length, force, and time. Physical quantities are often divided into fundamental quantities and derived quantities. *Fundamental quantities* cannot be defined in terms of other physical quantities. The number of quantities regarded as fundamental is the minimum number needed to give a consistent and complete description of all the physical quantities ordinarily encountered in the subject area. Length and time are examples of quantities viewed as fundamental in mechanics. *Derived quantities* are those whose defining operations are based on measurements of other physical quantities. Area, volume, velocity, and acceleration are examples of derived quantities in mechanics. Some quantities may be viewed as either fundamental or derived; mass and force are examples of such quantities. In the SI system of units, mass is regarded as a fundamental quantity and force is a derived quantity. In the U.S. customary system of units, force is regarded as a fundamental quantity and mass is a derived quantity.

The magnitude of each of the fundamental quantities is defined by an arbitrarily chosen unit or "standard." The familiar yard, foot, and inch, for example, come from the practice of using the human arm, foot, and thumb as length standards. However, for any type of precise calculations, such units of length are unsatisfactory. The first truly international standard of length was a bar of platinum–iridium alloy, called the *standard meter,*[2] which was kept at the International Bureau of Weights and Measures in Sèvres, France. The distance between two fine lines engraved on gold plugs near the ends of the bar is defined to be one meter. Historically, the meter was intended to be one ten-millionth of the distance from the pole to the equator along the meridian line through Paris. Accurate measurements made after the standard meter bar was constructed show that it differs from its intended value by approximately 0.023%.

In 1961 an international atomic standard of length was adopted. The wavelength in vacuum of the orange-red line from the spectrum of krypton 86 was chosen. One meter (m) is now defined to be 1,650,763.73 wavelengths of this light. The choice of an atomic standard offers advantages other than increased precision in length measurements. Krypton 86 can be obtained relatively easily and inexpensively everywhere, and all atoms of the material are identical and emit light of the same wavelength. The particular wavelength chosen is uniquely characteristic of krypton 86 and is very sharply defined. The definition of the yard, by international agreement, is 1 yard = 0.9144 m, exactly.[3] Thus, 1 inch = 25.4 mm, exactly; and 1 foot = 0.3048 m, exactly.

Similarly, time can be measured in a number of ways. Since the earliest times, the length of a day has been an accepted standard of time measurement.

---

[2]The United States has accepted the meter as a standard of length since 1893.

[3]"Guide for the Use of the International System of Units," National Institute of Standards and Technology (NIST) Special Publication 811, September 1991.

The internationally accepted standard unit of time—the *second* (s)—has been defined in the past at 1/86,400 of a mean solar day or 1/31,557,700 of a mean solar year. Time defined in terms of the rotation of the earth must be determined by astronomical observations. Since these observations require at least several weeks, a good secondary terrestrial measure, calibrated by astronomical observations, is used. Quartz crystal clocks, based on the electrically sustained natural periodic vibrations of a quartz wafer, have been used as secondary time standards. The best of these quartz clocks have kept time for a year with a maximum error of 0.02 second.

To meet the need for an even better time standard, an atomic clock has been developed that uses the periodic atomic vibrations of isotope cesium 133. The second based on this cesium clock is defined as the duration of 9,192,631,770 cycles of vibration. The cesium clock provides an improvement over the accuracy associated with astronomical methods by a factor of 200. Two cesium clocks will differ by no more than one second after running 3000 years.

The standard unit of mass, the *kilogram* (kg), is defined by a bar of platinum–iridium alloy that is kept at the International Bureau of Weights and Measures in Sèvres, France.

## The U.S. Customary System of Units

Most engineers in the United States use the U.S. customary system of units (USCS), sometimes called the British gravitational system, in which the base units are foot (ft) for length, pound (lb) for force, and second (s) for time. In this system, a foot is defined as 0.3048 m, exactly. The *pound* is defined as the weight at sea level and at a latitude of 45 degrees of a platinum standard that is kept at the Bureau of Standards in Washington, D.C. This platinum standard has a mass of 0.453,592,43 kg. The second is defined in the same manner as in the SI system.

In the U.S. customary system, the unit of mass is derived and is called a *slug*. One slug is the mass that is accelerated one foot per second squared by a force of one pound; that is, 1 slug equals 1 lb · $s^2$/ft. Since the weight of the platinum standard depends on the gravitational attraction of the earth, the U.S. customary system is a gravitational system of units rather than an absolute system of units.

## The International System of Units

The original metric system provided a set of units for the measurement of length, area, volume, capacity, and mass based on two fundamental units: the meter and the kilogram. With the addition of a unit of time, practical measurements began to be based on the meter–kilogram–second (MKS) system of units. In 1960, the Eleventh General Conference on Weights and Measures adopted as the international standard the Système International d'Unités (International System of Units), for which the abbreviation is SI in all languages.

The International System of Units adopted by the conference includes three classes of units: (1) base units, (2) supplementary units, and (3) derived units. The system is founded on the seven base units listed in Table 1-2.

Certain units of the International System have not been classified under either base units or derived units. These units, listed in Table 1-3, are called *supplementary units* and may be regarded as either base or derived units.

Derived units are expressed algebraically in terms of base units and/or

TABLE 1-2 Base Units and Their Symbols

| Quantity | Name of Unit | Symbol |
|---|---|---|
| Length | meter | m |
| Mass | kilogram | kg |
| Time | second | s |
| Electric current | ampere | A |
| Thermodynamic temperature | kelvin | K |
| Amount of substance | mole | mol |
| Luminous intensity | candela | cd |

TABLE 1-3 Supplementary Units and Their Symbols

| Quantity | Name of Unit | Symbol |
|---|---|---|
| Plane angle | radian | rad |
| Solid angle | steradian | sr |

supplementary units. Their symbols are obtained by means of the mathematical signs of multiplication and division. For example, the SI unit for velocity is meter per second (m/s) and the SI unit for angular velocity is radian per second (rad/s). In the SI system, the unit of force is derived and is called a *newton*. One newton is the force required to give one kilogram of mass an acceleration of one meter per second squared. Thus, $1 \text{ N} = 1 \text{ kg} \cdot \text{m/s}^2$. For some of the derived units, special names and symbols exist; those of interest in mechanics are listed in Table 1-4.

Prefixes are used to form names and symbols for decimal multiples and submultiples of SI names. The multiple usually should be chosen so that numerical values of the quantity will be between 0.1 and 1000. Only one prefix should be used in forming a multiple of a compound unit, and prefixes in the denominator should be avoided. Approved prefixes with their names and symbols are listed in Table 1-5.

TABLE 1-4 Derived Units and Their Symbols and Special Names

| Quantity | Derived SI Unit | Symbol | Special Name |
|---|---|---|---|
| Area | square meter | $\text{m}^2$ | |
| Volume | cubic meter | $\text{m}^3$ | |
| Linear velocity | meter per second | m/s | |
| Angular velocity | radian per second | rad/s | |
| Linear acceleration | meter per second squared | $\text{m/s}^2$ | |
| Frequency | (cycle) per second | Hz | hertz |
| Density | kilogram per cubic meter | $\text{kg/m}^3$ | |
| Force | kilogram · meter per second squared | N | newton |
| Moment of force | newton · meter | N · m | |
| Pressure | newton per meter squared | Pa | pascal |
| Stress | newton per meter squared | Pa | pascal |
| Work | newton · meter | J | joule |
| Energy | newton · meter | J | joule |
| Power | joule per second | W | watt |

TABLE 1-5 Multiples of SI Units

| Factor by Which Unit Is Multiplied | PREFIX | |
|---|---|---|
| | Name | Symbol |
| $10^{18}$ | exa | E |
| $10^{15}$ | peta | P |
| $10^{12}$ | tera | T |
| $10^{9}$ | giga | G |
| $10^{6}$ | mega | M |
| $10^{3}$ | kilo | k |
| $10^{2}$ | hecto* | h |
| 10 | deca* | da |
| $10^{-1}$ | deci* | d |
| $10^{-2}$ | centi* | c |
| $10^{-3}$ | milli | m |
| $10^{-6}$ | micro | $\mu$ |
| $10^{-9}$ | nano | n |
| $10^{-12}$ | pico | p |
| $10^{-15}$ | femto | f |
| $10^{-18}$ | atto | a |

*To be avoided when possible.

TABLE 1-6 Conversion Factors between the SI and U.S. Customary Systems

| Quantity | U.S. Customary to SI | SI to U.S. Customary |
|---|---|---|
| Length | 1 in. = 25.40 mm | 1 m = 39.37 in. |
|  | 1 ft = 0.3048 m | 1 m = 3.281 ft |
|  | 1 mi = 1.609 km | 1 km = 0.6214 mi |
| Area | $1 \text{ in}^2 = 645.2 \text{ mm}^2$ | $1 \text{ m}^2 = 1550 \text{ in}^2$ |
|  | $1 \text{ ft}^2 = 0.0929 \text{ m}^2$ | $1 \text{ m}^2 = 10.76 \text{ ft}^2$ |
| Volume | $1 \text{ in}^3 = 16.39(10^3) \text{ mm}^3$ | $1 \text{ mm}^3 = 61.02(10^{-6}) \text{ in}^3$ |
|  | $1 \text{ ft}^3 = 0.02832 \text{ m}^3$ | $1 \text{ m}^3 = 35.31 \text{ ft}^3$ |
|  | 1 gal = 3.785 L* | 1 L = 0.2642 gal |
| Velocity | 1 in./s = 0.0254 m/s | 1 m/s = 39.37 in./s |
|  | 1 ft/s = 0.3048 m/s | 1 m/s = 3.281 ft/s |
|  | 1 mi/h = 1.609 km/h | 1 km/h = 0.6214 mi/h |
| Acceleration | $1 \text{ in./s}^2 = 0.0254 \text{ m/s}^2$ | $1 \text{ m/s}^2 = 39.37 \text{ in./s}^2$ |
|  | $1 \text{ ft/s}^2 = 0.3048 \text{ m/s}^2$ | $1 \text{ m/s}^2 = 3.281 \text{ ft/s}^2$ |
| Mass | 1 slug = 14.59 kg | 1 kg = 0.06854 slug |
| Second moment of area | $1 \text{ in}^4 = 0.4162(10^6) \text{ mm}^4$ | $1 \text{ mm}^4 = 2.402(10^{-6}) \text{ in}^4$ |
| Force | 1 lb = 4.448 N | 1 N = 0.2248 lb |
| Distributed load | 1 lb/ft = 14.59 N/m | 1 kN/m = 68.54 lb/ft |
| Pressure or stress | 1 psi = 6.895 kPa | 1 kPa = 0.1450 psi |
|  | 1 ksi = 6.895 MPa | 1 MPa = 145.0 psi |
| Bending moment or torque | 1 lb · ft = 1.356 N · m | 1 N · m = 0.7376 ft · lb |
| Work or energy | 1 ft · lb = 1.356 J | 1 J = 0.7376 ft · lb |
| Power | 1 ft · lb/s = 1.356 W | 1 W = 0.7376 ft · lb/s |
|  | 1 hp = 745.7 W | 1 kW = 1.341 hp |

*Both L and l are accepted symbols for liter. Because the letter l can easily be confused with the numeral 1, the symbol L is recommended for United States use by the National Institute of Standards and Technology (see NITS Special Publication 811, September 1991).

As the use of the SI system becomes more commonplace in the United States, engineers will be required to be familiar with both the SI system and the U.S. customary system in common use today. As an aid to interpreting the physical significance of answers in SI units for those more accustomed to the U.S. customary system, some conversion factors for the quantities normally encountered in mechanics are provided in Table 1-6.

For the foreseeable future, engineers in the United States will be required to work with both the U.S. customary and SI systems of units; therefore, we have used both sets of units in examples and problems in this book.

## Example Problem 1-2

A manufacturer lists the fuel consumption for a new automobile as 15 kilometers per liter. Determine the fuel consumption in miles per gallon.

### SOLUTION

One accepted procedure for converting units is to write the associated units in abbreviated form with each of the numerical values used in the conversion. Like-unit symbols can then be canceled in the same manner as algebraic symbols. The conversion factors (see Table 1-6) needed for this example are

$$1 \text{ km} = 0.6214 \text{ mi}$$
$$1 \text{ gal} = 3.785 \text{ L}$$

Thus

$$15\frac{\text{km}}{\text{L}} \times 0.6214\frac{\text{mi}}{\text{km}} \times 3.785\frac{\text{L}}{\text{gal}} = 35.3 \text{ mi/gal} \quad \blacksquare \qquad \textbf{Ans.}$$

## Example Problem 1-3

The value of $G$ (universal gravitational constant) used for engineering computations in the U.S. customary system of units is $G = 3.439(10^{-8}) \text{ ft}^3/(\text{slug} \cdot \text{s}^2)$. Use the conversion factors listed in Table 1-6 to determine a value of $G$ with units of $\text{m}^3/(\text{kg} \cdot \text{s}^2)$ suitable for computations in the SI system of units.

### SOLUTION

The conversion factors (see Table 1-6) needed for this example are

$$1 \text{ ft}^3 = 0.02832 \text{ m}^3$$
$$1 \text{ kg} = 0.06854 \text{ slug}$$

Thus

$$G = 3.439(10^{-8})\frac{\text{ft}^3}{\text{slug} \cdot \text{s}^2} \times 0.02832\frac{\text{m}^3}{\text{ft}^3} \times 0.06854\frac{\text{slug}}{\text{kg}}$$

$$= 6.675(10^{-11})\frac{\text{m}^3}{\text{kg} \cdot \text{s}^2} \quad \blacksquare \qquad \textbf{Ans.}$$

## PROBLEMS

### Introductory Problems

**1-17\*** Determine the weight $W$, in U.S. customary units, of a 75-kg steel bar under standard conditions (sea level at a latitude of 45 degrees).

**1-18\*** Determine the mass $m$, in SI units, for a 500-lb steel beam under standard conditions (sea level at a latitude of 45 degrees).

**1-19\*** An automobile has a 440-cubic-inch engine displacement. Determine the engine displacement in liters.

**1-20** How many barrels of oil are contained in 100 kL of oil? One barrel (petroleum) equals 42.0 gal.

### Intermediate Problems

**1-21\*** Express the density, in SI units, of a specimen of material that has a specific weight of 0.025 lb/in³.

**1-22\*** The viscosity of crude oil under conditions of standard temperature and pressure is $7.13(10^{-3}) \text{ N} \cdot \text{s/m}^2$. Determine the viscosity of crude oil in U.S. Customary units.

**1-23** One acre equals 43,560 ft$^2$. One gallon equals 231 in$^3$. Determine the number of liters of water required to cover 2000 acres to a depth of 1 ft.

**1-24** The stress in a steel bar is 150 MPa. Express the stress in appropriate U.S. Customary units (ksi) by using the values listed in Table 1-6 for length and force as defined values.

**Challenging Problems**

**1-25** By definition, 1 hp = 33,000 ft · lb/min and 1 W = 1 N · m/s. Verify the conversion factors listed in Table 1-6 for converting power from U.S. Customary units to SI units by using the values listed for length and force as defined values.

**1-26*** The specific heat of air under standard atmospheric pressure, in SI units, is 1003 N · m/kg · °K. Determine the specific heat of air under standard atmospheric pressure in U.S. customary units (ft · lb/slug · °R).

# 1-4 DIMENSIONAL CONSIDERATIONS

All the physical quantities encountered in mechanics can be expressed dimensionally in terms of the three fundamental quantities: mass, length, and time, denoted respectively by $M$, $L$, and $T$. The dimensions of quantities other than the fundamental quantities follow from definitions or from physical laws. For example, the dimension of velocity, $L/T$, follows from the definition of velocity: rate of change of position with time. Similarly, acceleration is defined as the rate of change of velocity with time and has the dimension $L/T^2$. From Newton's second law [Eq. (1-1)], force is defined as the product of mass and acceleration; therefore, force has the dimension $ML/T^2$. The dimensions of a number of other physical quantities commonly encountered in mechanics are given in Table 1-7.

**Dimensional Homogeneity** An equation is said to be dimensionally homogeneous if the form of the equation does not depend on the units of measurement. For example, the equation describing the distance $h$ a body released from rest has fallen is $h = gt^2/2$, where $h$ is the distance traveled, $t$ is the time since release, and $g$ is the gravitational acceleration. This equation is valid whether length is measured in feet, meters, or inches and whether time is measured in hours, years, or seconds, provided $g$ is measured in the same units of length and time as $h$ and $t$. Therefore, by definition, this equation is dimensionally homogeneous.

If the value $g = 32.2$ ft/s$^2$ is substituted in the previous equation, the equation obtained is $h = 16.1t^2$ ft/s$^2$. This equation is not dimensionally homogeneous, since the equation applies only if length is measured in feet and time is measured in seconds. Dimensionally homogeneous equations are usually preferred in order to eliminate any uncertainty regarding units associated with constants appearing in dimensionally nonhomogeneous equations.

All like dimensions in a given equation should be measured with the same unit. For example, if the length dimension of a beam is measured in feet and the cross-sectional dimensions are measured in inches, all measurements should be converted to either feet or inches before they are used in a given equation. If this is done, the terms of the equation can be combined after the numerical values are substituted for the variables.

TABLE 1-7 Dimensions of the Physical Quantities of Mechanics

| Physical Quantity | Dimension | COMMON UNITS | |
| --- | --- | --- | --- |
| | | SI System | U.S. Customary System |
| Length | $L$ | m, mm | in., ft |
| Area | $L^2$ | $m^2$, $mm^2$ | $in^2$, $ft^2$ |
| Volume | $L^3$ | $m^3$, $mm^3$ | $in^3$, $ft^3$ |
| Angle | $1\ (L/L)$ | rad, degree | rad, degree |
| Time | $T$ | s | s |
| Linear velocity | $L/T$ | m/s | ft/s |
| Linear acceleration | $L/T^2$ | $m/s^2$ | $ft/s^2$ |
| Angular velocity | $1/T$ | rad/s | rad/s |
| Angular acceleration | $1/T^2$ | $rad/s^2$ | $rad/s^2$ |
| Mass | $M$ | kg | slug |
| Force | $ML/T^2$ | N | lb |
| Moment of a force | $ML^2/T^2$ | N · m | lb · ft |
| Pressure | $M/LT^2$ | Pa, kPa | psi, ksi |
| Stress | $M/LT^2$ | Pa, MPa | psi, ksi |
| Energy | $ML^2/T^2$ | J | ft · lb |
| Work | $ML^2/T^2$ | J | ft · lb |
| Power | $ML^2/T^3$ | W | hp |
| Linear impulse | $ML/T$ | N · s | lb · s |
| Momentum | $ML/T$ | N · s | lb · s |
| Specific weight | $M/L^2T^2$ | $N/m^3$ | $lb/ft^3$ |
| Density | $M/L^3$ | $kg/m^3$ | $slug/ft^3$ |
| Second moment of area | $L^4$ | $m^4$, $mm^4$ | $in^4$, $ft^4$ |
| Moment of inertia | $ML^2$ | kg · $m^2$ | slug · $ft^2$ |

## Example Problem 1-4

The angle of twist for a circular shaft subjected to a twisting moment is given by the equation $\theta = TL/GJ$. What are the dimensions of $J$ if $\theta$ is an angle in radians, $T$ is the moment of a force, $L$ is a length, and $G$ is a force per unit area?

### SOLUTION

The dimensions of the given quantities are

$$\theta = 1 \text{ (dimensionless)}$$

$$T = FL = \frac{ML}{T^2}(L) = \frac{ML^2}{T^2}$$

$$G = \frac{F}{A} = \frac{ML/T^2}{L^2} = \frac{M}{LT^2}$$

Therefore

$$J = \frac{TL}{G\theta} = \frac{(ML^2/T^2)(L)}{(M/LT^2)(1)} = L^4 \qquad \textbf{Ans.}$$

## Example Problem 1-5

Determine the dimensions of $I$, $R$, $w$, $M$, and $C$ in the dimensionally homogeneous equation

$$EIy = Rx^3 - P(x-a)^3 - wx^4 + Mx^2 + C$$

in which $x$ and $y$ are lengths, $P$ is a force, and $E$ is a force per unit area.

### SOLUTION

The equation can be written dimensionally as

$$\frac{F}{L^2}(I)(L) = R(L^3) - F(L-a)^3 - w(L^4) + M(L^2) + C$$

For this equation to be dimensionally homogeneous, $a$ must be a length; hence, all terms must have the dimensions $FL^3$. Thus,

$$(I)\frac{F}{L} = (R)L^3 = (w)L^4 = (M)L^2 = C = FL^3$$

The dimensions for each of the unknown quantities are obtained as follows:

$$I = \frac{L}{F}(FL^3) = L^4 \qquad \textbf{Ans.}$$

$$R = \frac{1}{L^3}(FL^3) = F \qquad \textbf{Ans.}$$

$$w = \frac{1}{L^4}(FL^3) = \frac{F}{L} \qquad \textbf{Ans.}$$

$$M = \frac{1}{L^2}(FL^3) = FL \qquad \textbf{Ans.}$$

$$C = FL^3 \ \blacksquare \qquad \textbf{Ans.}$$

## PROBLEMS

**Introductory Problems**

**1-27\*** Newton's law of gravitation can be expressed in equation form as

$$F = G\frac{m_1 m_2}{r^2}$$

If $F$ is a force, $m_1$ and $m_2$ are masses, and $r$ is a distance, determine the dimensions of $G$.

**1-28\*** The elongation of a bar of uniform cross section subjected to an axial force is given by the equation $\delta = PL/EA$. What are the dimensions of $E$ if $\delta$ and $L$ are lengths, $P$ is a force, and $A$ is an area?

**1-29** The period of oscillation of a simple pendulum is given by the equation $T = k\sqrt{L/g}$, where $T$ is in seconds, $L$ is in feet, $g$ is the acceleration due to gravity, and $k$ is a constant. What are the dimensions of $k$ for dimensional homogeneity?

**1-30** An important parameter in fluid flow problems involving thin films is the Weber number *(We)*, which can be expressed in equation form as

$$We = \frac{\rho v^2 L}{\sigma}$$

where $\rho$ is the density of the fluid, $v$ is a velocity, $L$ is a length, and $\sigma$ is the surface tension of the fluid. If the Weber number is dimensionless, what are the dimensions of the surface tension $\sigma$?

**Intermediate Problems**

**1-31\*** In the dimensionally homogeneous equation

$$\sigma = \frac{P}{A} + \frac{Mc}{I}$$

$\sigma$ is a stress, $A$ is an area, $M$ is a moment of a force, and $c$ is a length. Determine the dimensions of $P$ and $I$.

**1-32\*** In the dimensionally homogeneous equation

$$Pd = \tfrac{1}{2}mv^2 + \tfrac{1}{2}I\omega^2$$

$d$ is a length, $m$ is a mass, $v$ is a linear velocity, and $\omega$ is an angular velocity. Determine the dimensions of $P$ and $I$.

**1-33** In the dimensionally homogeneous equation

$$\tau = \frac{Tr}{J} + \frac{VQ}{Ib}$$

$\tau$ is a stress, $T$ is a torque (moment of a force), $V$ is a force, $r$ and $b$ are lengths, and $I$ is a second moment of an area. Determine the dimensions of $J$ and $Q$.

**1-34** In the dimensionally homogeneous equation

$$\tau = \frac{P}{A} + \frac{Tr}{J}$$

$\tau$ is a stress, $A$ is an area, $T$ is a torque (moment of a force), and $r$ is a length. Determine the dimensions of $P$ and $J$.

**Challenging Problems**

**1-35\*** The equation $x = Ae^{-t/b}\sin(at + \alpha)$ is dimensionally homogeneous. If $A$ is a length and $t$ is time, determine the dimensions of $x$, $a$, $b$, and $\alpha$.

**1-36\*** In the dimensionally homogeneous equation $w = x^3 + ax^2 + bx + a^2b/x$, if $x$ is a length, what are the dimensions of $a$, $b$, and $w$?

**1-37** Determine the dimensions of $a$, $b$, $c$, and $y$ in the dimensionally homogeneous equation

$$y = Ae^{-bt}\cos\left(\sqrt{1 - a^2}\,bt + c\right)$$

in which $A$ is a length and $t$ is time.

**1-38** Determine the dimensions of $c$, $\omega$, $k$ and $P$ in the differential equation

$$m\frac{d^2x}{dt^2} + c\frac{dx}{dt} + kx = P\cos\omega t$$

in which $m$ is a mass, $x$ is a length, and $t$ is time.

# 1-5 METHOD OF PROBLEM SOLVING

The principles of mechanics are few and relatively simple; however, the applications are infinite in number, variety, and complexity. Success in engineering mechanics depends to a large degree on a well-disciplined method of problem solving. Experience has shown that the development of good problem-solving methods and skills results from solving a large variety of problems. Professional problem solving consists of three phases: *problem definition and identification, model development and simplification,* and *mathematical solution and result in-*

*terpretation.* The problem-solving method outlined in this section will prove useful for the engineering mechanics courses that follow and for most situations encountered later in engineering practice.

Problems in engineering mechanics (statics, dynamics, and mechanics of deformable bodies) are concerned with the external and internal effects of a system of forces on a physical body. The approach usually used in solving a problem requires identification of all external forces acting on the body of interest. A carefully prepared drawing that shows the body of interest separated from all other interacting bodies and with all external forces applied is known as a *free-body diagram* (FBD).

---

*Most engineers consider an appropriate free-body diagram to be the single most important tool for the solution of mechanics problems.*

---

Given that the relationships between the external forces applied to a body and the motions or deformations that they produce are stated in mathematical form, the true physical situation must be represented by a mathematical model to obtain the required solution. Often, to simplify the solution, it is necessary to make assumptions or approximations in setting up this model. The most common approximation is to treat most of the bodies in statics and dynamics problems as rigid bodies. No real body is absolutely rigid; however, the changes in shape of a real body usually have a negligible effect upon the acceleration produced by a force system or upon the reactions required to maintain equilibrium of the body. Considerations of changes in shape under these circumstances would be an unnecessary complication of the problem. Similarly, the weights of many members can be neglected, since they are small with respect to the applied loads. A distributed force, which acts over a small area, can often be considered to be concentrated at a point.

Most actual physical problems cannot be solved exactly or completely. However, even in complicated problems, a simplified model can provide good qualitative results. Appropriate interpretation of such results can lead to approximate predictions of physical behavior or be used to verify the "reasonableness" of more sophisticated analytical, numerical, or experimental results. An engineer must always be aware of the actual physical problem under consideration and of any limitations associated with the mathematical model used. Assumptions must be continually evaluated to ensure that the mathematical problem solved provides an adequate representation of the physical process or device of interest.

As stated previously, the most effective way to learn the material contained in engineering mechanics courses is to solve a variety of problems. To become an effective engineer, the student must develop the ability to reduce complicated problems to simple parts that can be easily analyzed and to present the results of the work in a clear, logical, and neat manner. This can be accomplished by using the following sequence of steps.

### Problem definition and identification:

1. Read the problem carefully.
2. Identify the information given and the results requested.

**Model development and simplification:**

3. Identify the principles to be used to obtain the result.
4. Prepare a scaled sketch and tabulate the information provided.
5. Draw the appropriate free-body diagrams.

**Mathematical solution and result interpretation:**

6. Apply the appropriate principles and equations.
7. Report the answer with the appropriate number of significant figures and the appropriate units.
8. Study the answer and determine whether it is reasonable.

The development of an ability to apply an orderly approach to problem solving constitutes a significant part of an engineering education. Also, the problem-identification, model-simplification, and result-interpretation phases of engineering problem solving are often more important than the mathematical-solution phase.

## 1-6 SIGNIFICANCE OF NUMERICAL RESULTS

The accuracy of solutions to real engineering problems depends on three factors:

1. The accuracy of the known physical data.
2. The accuracy of the physical model.
3. The accuracy of the computations performed.

**The Accuracy of the Known Physical Data** Obviously, the solution can be no more accurate than the data that goes into it. Therefore, an engineer must first establish whether measurements were made to the nearest centimeter or the nearest millimeter, whether masses were determined to the nearest kilogram or the nearest gram, and so on. Since it is not possible to state error bounds on every number used in example and homework problems, it will be assumed that all data given in example and homework problems is accurate to at least four significant figures. That is, a weight given as 30 lb should be used as if it were written 30.00 lb.

Final answers should reflect the accuracy of the data. Reporting a final answer with too many significant figures is misleading, expresses too much confidence in the result, and could have dangerous consequences. Reporting a final answer with too few significant figures is also misleading and expresses too little confidence in the result.

Accuracy greater than 0.2% is seldom possible for practical engineering problems, since physical data are seldom known with any greater accuracy. A practical rule for rounding off the final numbers obtained in the computations involved in engineering analysis, which provides answers to approximately this degree of accuracy, is to retain four significant figures for numbers whose leading significant figure is small (for example, 12.34 and 234.5) and to retain three significant figures for numbers whose leading significant figure is large (for example, 6780 and 0.0876). If the digits being dropped are less than half of the last significant figure retained, the number is rounded down (7654 becomes 7650). If the digits being dropped are more than half of the last significant figure retained, the number is rounded up (123.456 becomes 123.5). If the digits being dropped are exactly half of the last figure retained, the last significant figure is rounded up if it is even and left unchanged if it is odd (12,345 and 12,355 will both become 12,350).

## The Accuracy of the Physical Model

Similarly, the solution can be no more accurate than the physical model on which it is based. For example, a common physical model is to treat a body as rigid. That is, the body is assumed to be so strong that it can withstand any forces applied without deforming (stretching, twisting, or bending) or breaking. Calculations of forces in members of a structure assume that the size and shape of the structure are independent of any forces applied to the structure and that the orientation and length of members can be determined from the initial geometry and are independent of any forces applied.

Of course, no real engineering material is perfectly rigid. Once the forces in the structure have been determined, the amount of deformation can be determined. If the amount of deformation is small, then the physical model was good and the solution should be accurate. If the amount of deformation is large, then the physical model was poor and the solution will not be very accurate. In this case, the deformation of the body must be included in the model used to calculate the forces in the members of the structure.

## The Accuracy of the Computations Performed

Pocket electronic calculators are widely used to perform the numerical computations required to solve engineering problems. The number of significant figures obtained when these calculators are used, however, should not be taken as an indication of the accuracy of the solution. As noted, previously, engineering data are seldom known to an accuracy greater than 0.2%; therefore, calculated results should always be rounded off to the number of significant figures that will yield the same degree of accuracy as the data on which they are based. Intermediate results, however, should be retained to a greater accuracy to prevent round-off error from accumulating and degrading the accuracy of the final results. These intermediate results are often retained in a calculator register or memory location and retain the full accuracy of the computational device. If written down on paper, intermediate results should be recorded with an accuracy of at least five or six significant figures.

For closed-form analytical solutions, the accuracy of the data and the adequacy of the model determine the accuracy of the results. For numerical solutions, the computational accuracy of the algorithms used also influences the accuracy of the results.

An error can be defined as the difference between two quantities. The difference, for example, might be between an experimentally measured value and a computed theoretical value. An error may also result from the rounding off of numbers during a calculation. One method for describing an error is to state a percent difference (%$D$). Thus, if $A$ represents the correct result and $B$ represents an approximation to $A$ (resulting from rounding errors, for example), then the percent difference between the two numbers is defined as

$$\%D = \frac{B - A}{A}(100)$$

In this equation, $A$ is the reference value to which $B$ is to be compared. A positive percent difference means that the approximation is too large; a negative percent difference, that the approximation is too small. The percent difference resulting from round-off error is illustrated in the following examples.

## Example Problem 1-6

Round off the number 12345 to two, three, and four significant figures. Find the percent difference between the rounded-off numbers and the original number by using the original number as the reference.

### SOLUTION

Rounding off the number 12345 to two, three, and four significant figures yields 12000, 12300, and 12350. The percent difference for each of these numbers is

$$\%D = \frac{A - B}{B}(100)$$

For 12,000,

$$\%D = \frac{12000 - 12345}{12345}(100) = -2.79\% \qquad \textbf{Ans.}$$

For 12,300,

$$\%D = \frac{12300 - 12345}{12345}(100) = -0.36\% \qquad \textbf{Ans.}$$

For 12,350,

$$\%D = \frac{12350 - 12345}{12345}(100) = +0.041\% \qquad \textbf{Ans.}$$

The minus signs associated with the percent differences indicate that the rounded-off numbers are smaller than the reference number. Similarly, a positive percent difference indicates that the rounded-off number is larger than the reference number. ■

## Example Problem 1-7

When engineers deal with angles, they are usually more interested in the sine or cosine of the angle than they are with the angle itself. Given that

$$\sin 5° = \cos 85° = \sin 175° = \sin 1085° = \cdots = 0.08716$$

the rounding of angles requires a different scheme than that described previously. Calculate and draw a graph of the percent error

**(a)** In the angle, caused by rounding the angle to four significant figures if the leading digit is 1 and to three significant figures if the leading digit is not 1.

**(b)** In the sine and cosine of the angle, caused by rounding the angle to four significant figures if the leading digit is 1 and to three significant figures if the leading digit is not 1.

**(c)** In the sine and cosine of the angle, caused by rounding the angle to two decimal places.

## SOLUTION

**(a)** Start by using a program such as MathCAD, MATLAB, or Excel to generate a series of angles between 1° and 89°. For angles between 1° and 1.999°, the maximum round-off error will be $\Delta\theta = 0.0005°$ (that is, 1.2345° and 1.2355° will both round to 1.235° while 1.23551° will round to 1.236°). For angles between 2° and 9.99°, the maximum round-off error will be $\Delta\theta = 0.005°$; for angles between 10° and 19.99°, the maximum round-off error will also be $\Delta\theta = 0.05°$; and for angles between 20° and 89.9° the maximum round-off error will be $\Delta\theta = 0.05°$. For each angle, divide the maximum round-off error by the original angle and multiply by 100 to get the percent error. For example, when $\theta = 35°$ the percent error is

$$\%D_\theta = \frac{\theta_{app} - \theta}{\theta} = \frac{\Delta\theta}{\theta} = \frac{0.05}{35}(100) = 0.1429\%$$

The result of computing the percent error for each angle between 1° and 89° and graphing the result is shown in Fig. 1-1. The maximum error is

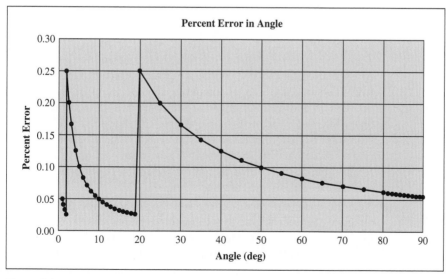

Figure 1.1

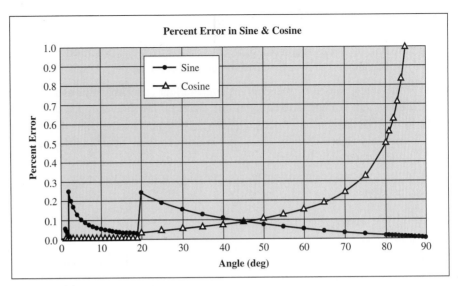

**Figure 1.2**

less than 0.2% for almost all of the numbers graphed. Keeping four significant figures for angles between 1° and 2.5° and between 10° and 25° would have kept the maximum error to less than 0.2% for the entire graph.

**(b)** For each of the angles in part (a), compute $\sin \theta$, $\sin (\theta \pm \Delta\theta)$, $\cos \theta$, and $\cos(\theta \pm \Delta\theta)$. Then compute the percent errors as above. Again, when $\theta = 35°$ the percent error is

$$\%D_{\sin} = \frac{\sin(\theta + \Delta\theta) - \sin\theta}{\sin\theta}(100) = \frac{\sin 35.05° - \sin 35°}{\sin 35°}(100)$$

$$= 0.1246\%$$

$$\%D_{\cos} = \frac{\cos(\theta - \Delta\theta) - \cos\theta}{\cos\theta}(100) = \frac{\cos 34.95° - \cos 35°}{\cos 35°}(100)$$

$$= 0.0611\%$$

The difference $(\theta - \Delta\theta)$ was used for the cosine and the difference $(\theta + \Delta\theta)$ was used for the sine so that the percent errors would both be positive numbers. The result of computing the percent errors for each angle between 1° and 89° and graphing the result is shown in Fig. 1-2. Note that while the error in the angle decreases from about 0.07% to about 0.06% for angles from 75° to 89°; the error in the cosine of the angle increases from about 0.3% to about 5% between the same angles. A different approach is needed to control the round-off error when the sine and cosine of the angles are important.

**(c)** Now the maximum round-off error in the angle will be $\Delta\theta = 0.005°$ for all angles between 1° and 89°. Computing the percent errors for each angle as in part (b) and graphing the result is shown in Fig. 1-3. Now the percent error is less than 0.2% for all but the very small angles and for angles very near 90°. Keeping three decimal places in the angles would keep the percent error less than 0.2% for nearly the entire graph.

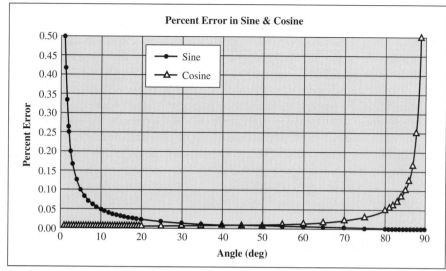

**Figure 1.3**

## PROBLEMS

### Intermediate Problems

Round off the numbers in the following problems to two significant figures. Find the percent difference between each rounded-off number and the original number by using the original number as the reference.

**1-39*** (a) 0.015362    (b) 55.33682    (c) 63,746.27

**1-40** (a) 0.837482    (b) 374.9371    (c) 937,284.9

Round off the numbers in the following problems to three significant figures. Find the percent difference between each rounded-off number and the original number by using the original number as the reference.

**1-41** (a) 0.034739    (b) 26.39473    (c) 55,129.92

**1-42*** (a) 0.472916    (b) 826.4836    (c) 339,872.8

Round off the numbers in the following problems to four significant figures. Find the percent difference between each rounded-off number and the original number by using the original number as the reference.

**1-43*** (a) 0.056623    (b) 74.82917    (c) 27,382.84

**1-44** (a) 0.664473    (b) 349.3378    (c) 274,918.2

## 1-7 SUMMARY

The foundations for studies in mechanics are the laws formulated by Sir Isaac Newton in 1687. The first law deals with conditions for equilibrium of a particle; therefore, it provides the foundation for the study of statics. The second law, which establishes a relationship between the force acting upon a particle and the motion of the particle, provides the foundation for the study of dynamics. The third law provides the foundation for understanding the concept of a force. In addition to the basic laws of motion, Newton also formulated the law of gravitation, which governs the mutual attraction between two bodies.

    Physical quantities used to express the laws of mechanics can be divided into fundamental quantities and derived quantities. The magnitude of each fundamental quantity is defined by an arbitrarily chosen unit or "standard." The units used in the SI system are the meter (m) for length, the kilogram (kg) for mass, and the second (s) for time. The unit of force is a derived unit called a newton (N). In the U.S. customary system of units, the units used are the foot (ft) for length, the pound (lb) for force, and the second (s) for time. The unit of mass is a derived unit called a slug.

The terms of an equation used to describe a physical process should not depend on the units of measurement (that is, they should be dimensionally homogeneous). If an equation is dimensionally homogeneous, the equation is valid for use with any system of units provided all quantities in the equation are measured in the same system. Use of dimensionally homogeneous equations eliminates the need for unit conversion factors.

Success in engineering depends to a large degree on a well-disciplined method of problem solving. Professional problem solving consists of three phases:

1. Problem definition and identification.
2. Model development and simplification.
3. Mathematical solution and result interpretation.

Problems in mechanics are concerned primarily with the effects of force systems on physical bodies. As a result, an extremely important part of the solution of any problem involves identification of the external forces acting upon the body. This is accomplished efficiently and accurately by using a free-body diagram. In obtaining a solution to most problems, the true physical situation must be represented by a mathematical model. A common approximation made in setting up this model is to treat the body as a rigid body. Even though no real body is absolutely rigid, changes in shape usually have a negligible effect upon accelerations produced by a force system or upon reactions required to maintain equilibrium of the body. Solution of some mechanics of materials problems, however, requires use of deformation characteristics of the body. Whenever a mathematical model is used in solving a problem, care must be exercised to ensure that the model and the associated mathematical problem being solved provide an adequate representation of the physical process or device that they represent.

The accuracy of solutions to real engineering problems depends on three factors:

1. The accuracy of the known physical data.
2. The accuracy of the physical model.
3. The accuracy of the computations performed.

Accuracy greater than 0.2% is seldom possible. Calculated results should always be "rounded off" to the number of significant figures that will yield the same degree of accuracy as the data on which they are based. In this book, final results beginning with a 1 will generally be rounded off to four significant figures and final results beginning with 2 through 9 will generally be rounded off to three significant figures. At least one additional figure will be retained in all intermediate results to maintain this degree of accuracy in the final results.

# REVIEW PROBLEMS

**1-45\*** The weight of the first Russian satellite, *Sputnik I,* was 184 lb on the surface of the earth. Determine the force exerted on the satellite by the earth at the low and high points of its orbit, which were 149 mi and 597 mi, respectively, above the surface of the earth.

**1-46*** The planet Jupiter has a mass of $1.90(10^{27})$ kg and a radius of $7.14(10^7)$ m. Determine the force of attraction between the earth and Jupiter when the minimum distance between the two planets is $6(10^{11})$ m.

**1-47*** Convert 640 acres (1 square mile) to hectares if 1 acre equals 4840 $yd^2$ and 1 hectare equals $10^4$ $m^2$.

**1-48** Determine the dimension of $c$ in the dimensionally homogeneous equation

$$v = \frac{mg}{c}\left(1 - e^{-ct/m}\right)$$

in which $v$ is a velocity, $m$ is a mass, $t$ is time, and $g$ is the gravitational acceleration.

**1-49** Develop an expression for the change in gravitational acceleration $\Delta g$ between the surface of the earth and a height $h$ when $h \ll R_e$.

# 2

# CONCURRENT FORCE SYSTEMS

## 2-1 INTRODUCTION

A force was defined in Section 1-2 as the action of one body on another. The action may be the result of direct physical contact between the bodies, or it may be the result of gravitational, electrical, or magnetic effects for bodies that are separated.

A force exerted on a body has two effects on the body: (1) the external effect, which is the tendency to change the motion of the body or to develop resisting forces (reactions) on the body, and (2) the internal effect, which is the tendency to deform the body. Both external and internal effects are discussed in this book. In many problems, the external effect is significant but the internal effect is not of interest. This is the case in many statics and dynamics problems when the body is assumed to be rigid. In other problems, when the body cannot be assumed to be rigid, the internal effects are important. Problems of this type are usually discussed in textbooks on Mechanics of Materials or Mechanics of Deformable Bodies.

When a number of forces are treated as a group, they are referred to as a *force system*. If a force system acting upon a body produces no external effect, the forces are said to be in balance and the body is said to be in equilibrium. If a body is acted upon by a force system that is not in balance, a change in motion of the body must occur. Such a force system is said to be unbalanced or to have a resultant.

Two force systems are said to be equivalent if they produce the same external effect when applied in turn to a given body. The resultant of a force system is the simplest equivalent system to which the original system will reduce. The process of reducing a force system to a simpler equivalent force system is called *reduction*. The process of expanding a force or a force system into a less simple equivalent system is called *resolution*. A component of a force is one of the two or more forces into which a given force may be resolved.

## 2-2 FORCES AND THEIR CHARACTERISTICS

The properties needed to describe a force are called the *characteristics of the force*. The characteristics of a force are

1. Its magnitude,
2. Its direction (line of action and sense), and
3. Its point of application.

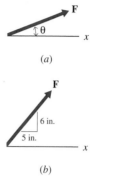

*(a)*

The magnitude (positive numerical value) of a force is the amount or size of the force. In this book, the magnitude of a force will be expressed in newtons (N) or kilonewtons (kN) when the SI system of units is used and in pounds (lb) or kilopounds (kip) when the U.S. customary system of units is used.

The direction of a force is the slope and sense of the line segment used to represent the force. In a two-dimensional problem, the slope can be specified by providing an angle, as shown in Fig. 2-1*a,* or by providing two dimensions, as shown in Fig. 2-1*b.* In a three-dimensional problem, the slope can be specified by providing three angles, as shown in Fig. 2-2*a,* or by providing three dimensions, as shown in Fig. 2-2*b.* The sense of the force can be specified by placing an arrowhead on the appropriate end of the line segment used to represent the force. Alternatively, a plus or minus sign can be used with the magnitude of a force to indicate the sense of the force.

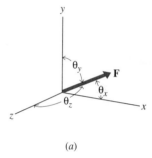

*(b)*

**Figure 2-1**

The point of application of a force is the point of contact between the two bodies. A straight line extending through the point of application in the direction of the force is called its *line of action.*

The three characteristics of a force are illustrated on the sketch of a block shown in Fig. 2-3. In this case, the force applied to the block can be described as a 100-lb (magnitude) force acting 30° upward and to the right (direction—slope and sense) through point *A* (point of application). A discussion of the manner in which these characteristics influence the reactions developed in holding a body at rest forms an important part of the study of statics. In a similar manner, a discussion of the manner in which these characteristics influence the change in motion of a body forms an important part of the study of kinetics.

## Scalar Quantities

Scalar quantities can be completely described by their magnitudes (numbers). Examples of scalar quantities in mechanics are mass, density, length, area, volume, speed, energy, time, and temperature. In mathematical operations, scalars follow the rules of elementary algebra. Symbols representing scalar quantities are printed in lightface italic form (*A*) in this book.

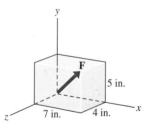

*(a)*

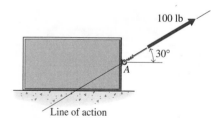

*(b)*

**Figure 2-2**

## Vector Quantities

A vector quantity has both a magnitude and a direction (line of action and sense) and obeys the parallelogram law of addition that will be described in Section 2-3. Examples of vector quantities in mechanics are force, displacement, velocity, and acceleration. A vector quantity can be represented graphically by using a directed line segment (arrow), as with the force vector shown in Figure 2-3. The magnitude of the vector is represented by the length of the arrow and by the label (100 lb) next to the arrow. The direction of the vector is specified by giving the angle $\theta$ between the arrow and some known reference direction. The sense is indicated by the arrowhead at the tip of the line segment.

Symbols representing vector quantities are printed in boldface type (**A**) to distinguish them from scalar quantities. The magnitude of a vector is a scalar quantity and the magnitude of vector quantities will be printed in lightface italic type ($A = |\mathbf{A}|$). In all handwritten work, it is important to distinguish between scalar and vector quantities, since vector quantities do not follow the rules of el-

**Figure 2-3**

ementary algebra in mathematical operations. A small arrow over the symbol for a vector quantity $(\vec{A})$ is often used in handwritten work to take the place of boldface type in printed material.

The use of scalars and vectors to represent physical quantities is a simple example of the modeling of physical quantities by mathematical methods. An engineer must be able to construct good mathematical models and be able to interpret correctly their physical meaning.

Vectors can be classified into three types—free, sliding, or fixed.

1. A free vector has a specific magnitude, slope, and sense, but its line of action does not pass through a unique point in space.

2. A sliding vector has a specific magnitude, slope, and sense, and its line of action passes through a unique point in space. The point of application of a sliding vector can be anywhere along its line of action.

3. A fixed vector has a specific magnitude, slope, and sense, and its line of action passes through a unique point in space. The point of application of a fixed vector is confined to a fixed point on its line of action.

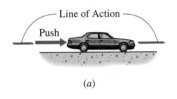

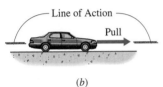

**Figure 2-4**

## Principle of Transmissibility

In most statics and dynamics problems, the assumption is made that the body is rigid. As a result, only the external effects of any force applied to the body are of interest. When this is the case, the force can be applied at any point along its line of action without changing the external effects of the force. For example, a stalled automobile can be moved by pushing on the back bumper (Fig. 2-4a) or pulling on the front bumper (Fig. 2-4b). If the magnitude, direction, and line of action of the two forces are identical, the external effect will be the same. Clearly in this case, the point of application of the force has no effect on the external effect (motion of the automobile).

This fact is formally expressed by the principle of transmissibility as "the external effect of a force on a rigid body is the same for all points of application of the force along its line of action." It should be noted that the external effect, only, remains unchanged. The internal effect of a force (stress and deformation) may be greatly influenced by a change in the point of application of the force along its line of action. For those cases where the principle of transmissibility applies (rigid-body mechanics), force can be treated as a sliding vector.

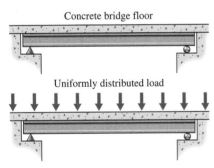

**Figure 2-5**

## Classification of Forces

A force has been defined as the action of one physical body on another. Since the interaction can occur when there is contact between the bodies or when the bodies are physically separated, forces can be classified under two general headings: (1) contacting or surface forces, such as a push or a pull produced by mechanical means, and (2) noncontacting or body forces, such as the gravitational pull of the earth on all physical bodies.

Forces may also be classified with respect to the area or volume over which they act. A force applied along a length, over an area, or over a volume is known as a *distributed force*. The distribution can be uniform or nonuniform. The weight (the sum of the gravity forces on each particle) of a concrete bridge floor of uni-

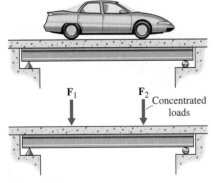

**Figure 2-6**

form thickness (Fig. 2-5) is an example of a uniformly distributed load. Any force applied to a relatively small area compared with the size of the loaded member can be assumed to be a concentrated force. For example, the force applied by a car wheel to the longitudinal members of a bridge (Fig. 2-6) can be considered to be a concentrated load.

  Any number of forces treated as a group constitute a force system. Force systems may be one-, two-, or three-dimensional. A force system is said to be *concurrent* if the action lines of all forces intersect at a common point (Fig. 2-7*a*) and *coplanar* when all the forces lie in the same plane (Fig. 2-7*b*). A *parallel* force system is one in which the action lines of the forces are parallel (Fig. 2-7*c*). In a parallel force system, the senses of the forces do not have to be the same. If the forces of a system have a common line of action, the system is said to be *collinear* (Fig. 2-7*d*).

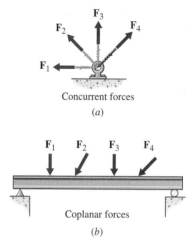

Concurrent forces

(*a*)

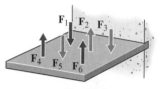

Coplanar forces

(*b*)

# 2-3 RESULTANT OF TWO OR MORE CONCURRENT FORCES

Any concurrent system of forces can be replaced by a single force **R,** called the resultant, which will produce the same effect on the body as the original system of forces. In the case of concurrent forces, the resultant is simply the vector sum of the forces that make up the system.

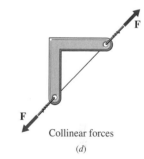

Parallel forces

(*c*)

**Addition of Vectors** Two free vectors **A** and **B** can be translated so that their tails meet at a common point, as shown in Fig. 2-8. The *parallelogram law of vector addition* states that the two vectors **A** and **B** are equivalent to the vector **R** that is the diagonal of a parallelogram constructed by using vectors **A** and **B** as the two adjacent sides (see Fig. 2-8). The vector **R** is the diagonal that passes through the point of intersection of the vectors **A** and **B**. The vector **R** is known as the resultant of the two vectors **A** and **B.** All vector quantities must obey this parallelogram law of addition. The resultant **R** can be represented mathematically by the vector equation

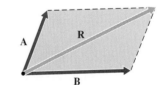

Collinear forces

(*d*)

**Figure 2-7**

$$\mathbf{R} = \mathbf{A} + \mathbf{B} \qquad (a)$$

The plus sign used in conjunction with vector quantities **A** and **B** (boldface type) indicates vector (parallelogram law) addition, not scalar (algebraic) addition.

  The resultant **R** of the two vectors can also be determined using half the parallelogram. Since half a parallelogram is a triangle, the method is called the *triangle law for addition of vectors.* When the triangle law is used to determine the resultant **R** of two vectors **A** and **B,** vector **A** is drawn to scale first; then vector **B** is drawn to scale from the tip of vector **A** in a head-to-tail fashion. The closing side of the triangle, drawn from the beginning of the first vector **A** to the tip of the second vector **B,** is the resultant of the two vectors. The process by which two vectors are added using the triangle law is illustrated in Fig. 2-9. In the case of force vectors, the triangle constructed using this method is called a *force triangle.*

**Figure 2-8**

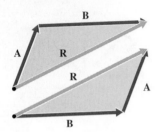

**Figure 2-9**

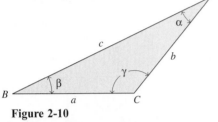

**Figure 2-10**

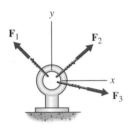

**Figure 2-11**

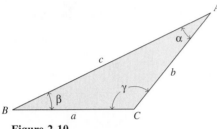

Alternatively, vector **B** can be drawn first; then vector **A** can be drawn from the tip of vector **B** in a head-to-tail fashion. Again, the resultant **R** of the two vectors is the closing side of the triangle. As indicated in Fig. 2-9

$$R = A + B = B + A \qquad (b)$$

The results shown in Fig. 2-9 indicate that the resultant **R** does not depend on the order in which the vectors **A** and **B** are added. Figure 2-9 is a graphical illustration of the commutative law of vector addition.

**Law of Sines and Law of Cosines** Numerical values for the magnitude and direction (length and angle) of a resultant force vector are obtained using trigonometric methods based on the law of sines and the law of cosines in conjunction with sketches of the force system. For example, consider the triangle shown in Fig. 2-10, which is similar to the vector triangles illustrated in Figs. 2-8 and 2-9. For this general triangle, the law of sines is

$$\frac{\sin \alpha}{a} = \frac{\sin \beta}{b} = \frac{\sin \gamma}{c} \qquad (c)$$

and the law of cosines is

$$c^2 = a^2 + b^2 - 2ab \cos \gamma \qquad (d)$$

**Resultant of Two Concurrent Forces** Application of the law of cosines to the determination of the magnitude of the resultant of the two force vectors **F₁** and **F₂** shown in Fig. 2-11 yields

$$R^2 = F_1^2 + F_2^2 - 2F_1F_2 \cos \gamma \qquad (e)$$

Since $\cos \gamma = \cos (180° - \phi) = -\cos \phi$, this equation can be written in alternate form as

$$R^2 = F_1^2 + F_2^2 + 2F_1F_2 \cos \phi \qquad (2\text{-}1)$$

where $\phi$ is the angle between the two force vectors. Note also from Fig. 2-11 and the law of sines that the orientation of the line of action of the resultant force **R** with respect to the force **F₁** is given by the expression

$$\sin \beta = \frac{F_2 \sin \phi}{R}$$

or

$$\beta = \sin^{-1}\left(\frac{F_2 \sin \phi}{R}\right) \qquad (2\text{-}2)$$

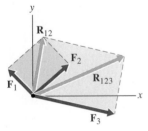

**Figure 2-12**

**Figure 2-13**

## Resultant of Three or More Concurrent Forces

The method can easily be extended to cover three or more forces. As an example, consider the case of three coplanar, concurrent forces acting on an eye bolt, as shown in Fig. 2-12. Application of the parallelogram law to forces $\mathbf{F}_1$ and $\mathbf{F}_2$, as shown graphically in Fig. 2-13, yields resultant $\mathbf{R}_{12}$. Then, combining resultant $\mathbf{R}_{12}$ with force $\mathbf{F}_3$ through a second graphical application of the parallelogram law yields resultant $\mathbf{R}_{123}$, which is the vector sum of the three forces.

In practice, numerical results for specific problems involving three or more forces are obtained algebraically by using the law of sines and the law of cosines in conjunction with sketches of the force system similar to those shown in Fig. 2-14. The sketches shown in Fig. 2-14 are known as *force polygons*. The order in which the forces are added can be arbitrary, as shown in Figs. 2-14*a* and *b*, where the forces are added in the order $\mathbf{F}_1$, $\mathbf{F}_2$, $\mathbf{F}_3$ in Fig. 2-14*a* and in the order $\mathbf{F}_3$, $\mathbf{F}_1$, $\mathbf{F}_2$ in Fig. 2-14*b*. Although the shape of the polygon changes, the resultant force remains the same. The fact that the sum of the three vectors is the same regardless of the order in which they are added illustrates the associative law of vector addition.

If there are more than three forces (as an example, Fig. 2-15 shows the eye bolt of Fig. 2-12 with four forces), the process of adding additional forces can be continued, as shown in Fig. 2-16, until all the forces are joined in head-to-tail fashion. The closing side of the polygon, drawn from the tail of the first vector to the head of the last vector, is the resultant of the force system.

Application of the parallelogram law to more than three forces requires extensive geometric and trigonometric calculation. Therefore, problems of this type are usually solved by using the rectangular-component method, which is developed in Section 2-6 of this book.

The procedure for determining the resultant $\mathbf{R}$ of a force system by using the law of sines and the law of cosines is demonstrated in the following examples.

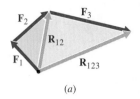

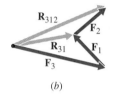

(a)

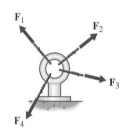

(b)

**Figure 2-14**

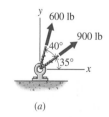

**Figure 2-15**

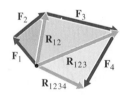

**Figure 2-16**

 **Example Problem 2-1**

Two forces are applied to a bracket as shown in Fig. 2-17*a*. Determine the magnitude of the resultant $\mathbf{R}$ of the two forces and the angle $\theta$ between the *x*-axis and the line of action of the resultant.

### SOLUTION

The two forces, the resultant $\mathbf{R}$, and the angle $\theta$ are shown in Fig. 2-17*b*. The triangle law for the addition of the two forces can be applied, as shown in Fig. 2-17*c*.

The magnitude $R$ of the resultant is obtained using the alternate form of the law of cosines as expressed by Eq. (2-1). Thus,

$$R^2 = F_1^2 + F_2^2 + 2F_1F_2 \cos \phi$$
$$= 900^2 + 600^2 + 2(900)(600) \cos 40°$$

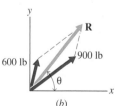

(a)

(b)

**Figure 2-17**

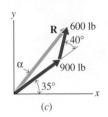

**Figure 2-17**

from which

$$R = 1413.3 \cong 1413 \text{ lb} \qquad \textbf{Ans.}$$

The line of action of the resultant **R** with respect to the 900-lb force is obtained using the law of sines

$$\frac{\sin \alpha}{600} = \frac{\sin(180° - 40°)}{1413.3}$$

from which

$$\alpha = 15.836°$$

Therefore, the angle $\theta$ between the $x$-axis and the line of action of the resultant is (Fig. 2-17$c$)

$$\theta = \alpha + 35° = 50.8° \qquad \textbf{Ans.}$$

## Example Problem 2-2

Determine the magnitude of the resultant **R** and the angle $\theta$ between the $x$-axis and the line of action of the resultant for the three forces shown in Fig. 2-18$a$.

### SOLUTION

**Resultant of $F_1$ and $F_2$.** First, the resultant $R_{12}$ of the two forces $F_1$ and $F_2$ is determined by adding the two forces vectorially using the parallelogram law as illustrated in Fig. 2-18$b$. The magnitude $R_{12}$ (using Eq. (2-1)) is

$$R_{12}^2 = F_1^2 + F_2^2 + 2F_1F_2 \cos \phi_1$$
$$= 500^2 + 600^2 + 2(500)(600) \cos 60°$$

from which

$$R_{12} = 953.9 \text{ lb}$$

The line of action of the resultant $R_{12}$ with respect to the force $F_2$ (see Fig. 2-18$b$) is obtained using the law of sines

$$\frac{\sin \beta_1}{500} = \frac{\sin(180° - 60°)}{953.9}$$

from which

$$\beta_1 = 27.00°$$

**Resultant of $R_{12}$ and $F_3$.** Next, the resultant **R** of the two forces $R_{12}$ and $F_3$ (see Fig. 2-18$c$) is determined by adding the two forces vectorially using the parallelogram law. The magnitude $R$ (using Eq. (2-1)) is

$$R^2 = R_{12}^2 + F_3^2 + 2R_{12}F_3 \cos \phi_2$$
$$= 953.9^2 + 700^2 + 2(953.9)(700) \cos 67°$$

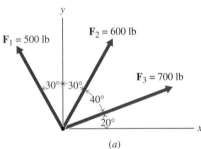

(a)

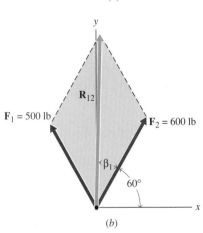

(b)

**Figure 2-18**

from which                                                    **Ans.**

$$R = 1386.3 \cong 1386 \text{ lb}$$

The line of action of the resultant **R** with respect to the force **F₃** (see Fig. 2-18c) is obtained using the law of sines

$$\frac{\sin \beta_2}{953.9} = \frac{\sin(180° - 67°)}{1386.3}$$

from which

$$\beta_2 = 39.30°$$

Therefore, the angle $\theta$ between the x-axis and the line of action of the resultant **R** is (Fig. 2-18c)

$$\theta = \beta_2 + 20° = 59.3° \qquad \text{**Ans.**}$$

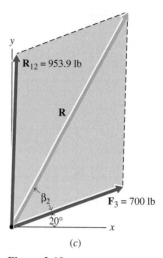

(c)

**Figure 2-18**

## Example Problem 2-3

Two forces **A** and **B** are applied to an eye bolt as shown in Fig. 2-19a. If the magnitudes of the two forces are $A = 100$ N and $B = 50$ N, calculate and plot the magnitude of the resultant **R** as a function of the angle $\theta_A$ ($0° \le \theta_A \le 180°$). Also calculate and plot the angle $\theta_R$ that the resultant makes with the force **B** as a function of the angle $\theta_A$.

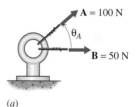

(a)

### SOLUTION

The resultant will be determined using the triangle law of addition. The force triangle is shown in Fig. 2-19b for an arbitrary angle $\theta_A$. The magnitude $R$ of the resultant is obtained using the alternate form of the law of cosines as expressed by Eq. (2-1),

$$R^2 = 50^2 + 100^2 + 2(50)(100) \cos \theta_A$$

from which

$$R = \sqrt{12{,}500 + 10{,}000 \cos \theta_A} \qquad (a)$$

(b)

**Figure 2-19**

The line of action of the resultant **R** (the angle $\theta_R$) is obtained using the law of sines

$$\frac{\sin \theta_R}{100} = \frac{\sin (180° - \theta_A)}{R}$$

from which

$$\theta_R = \sin^{-1}\left(\frac{100 \sin (180° - \theta_A)}{R}\right) \qquad (b)$$

For example, when $\theta_A = 60°$, Eq. (a) gives the magnitude of the resultant as

$$R = 132.288 \text{ N}$$

and Eq. (b) gives the angle as

$$\theta_R = 40.893°$$

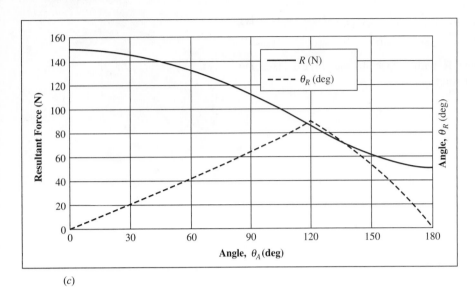

(c)

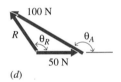

(d)

The required graph can be drawn using a program such as MathCAD, MATLAB, or Excel. Start by generating a series of angles $\theta_A$ between 0° and 180°. For each angle $\theta_A$, calculate the magnitude of the resultant using Eq. (a) and the angle $\theta_R$ using Eq. (b). The resulting graph is shown in Fig. 2-19c. The magnitude graph behaves as expected—decreasing from $R = A + B = 150$ N when $\theta_A = 0°$ (and both forces are in the $+x$-direction), to $R = 132.288$ N when $\theta_A = 60°$ (as calculated above), to $R = A - B = 50$ N when $\theta_A = 180°$ (and the forces are in opposite directions). The angle graph, however, has an unexpected "point" at $\theta_A = 120°$.

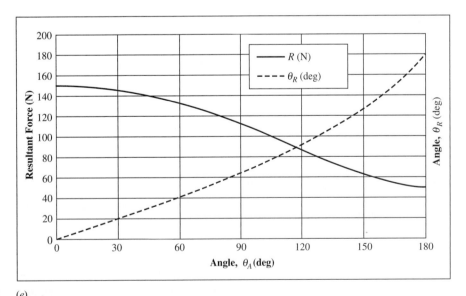

(e)

**Figure 2-19**

To see what is happening, we draw the force triangle for $\theta_A = 150°$ (Fig. 2-19$d$) and check the calculations. Using the law of cosines to get the magnitude

$$R^2 = 50^2 + 100^2 + 2(50)(100) \cos 150°$$

gives

$$R = 61.966 \text{ N}$$

which matches the graph of Fig. 2-19$c$. Using the law of sines to get the angle $\theta_R$

$$\frac{\sin \theta_R}{100} = \frac{\sin(180° - 150°)}{61.966}$$

gives

$$\theta_R = 53.794°$$

which also matches the graph of Fig. 2-19$c$ but doesn't look like a reasonable angle for the force triangle of Fig. 2-19$d$. The problem is with the inverse sine function, since $\sin(53.794°) = \sin(126.206°) = \sin(413.794°) = \cdots$. The angle we want is 126.206°, but our calculator and all computer programs always give an angle between $-90°$ and $90°$ for the inverse sine function.

Although it would be easy enough to adjust the angle $\theta_R$ for angles $\theta_A$ greater than 120°, we can avoid the problem completely by using the law of cosines to calculate $\theta_R$. Using the law of cosines and Fig. 2-19$b$

$$100^2 = R^2 + 50^2 - 2(R)(50) \cos \theta_R$$

gives

$$\theta_R = \cos^{-1}\left(\frac{R^2 - 7500}{100R}\right) \qquad (c)$$

When $\theta_A = 150°$ and $R = 61.966$ N, Eq. (c) gives the angle $\theta_R = 126.21°$, as expected. Graphing Eqs. (a) and (c) results in Fig. 2-19$e$. Now the angle $\theta_R$ increases smoothly from $\theta_R = 0°$ when $\theta_A = 0°$, to $\theta_R = 40.89°$ when $\theta_A = 60°$, to $\theta_R = 126.21°$ when $\theta_A = 150°$, to $\theta_R = 180°$ when $\theta_A = 180°$.

## ▌ PROBLEMS

Use the law of sines and the law of cosines, in conjunction with sketches of the force triangles, to solve the following problems.

**Introductory Problems**

Determine the magnitude of the resultant **R** and the angle $\theta$ between the $x$-axis and the line of action of the resultant for the following problems.

**2-1*** The two forces shown in Fig. P2-1.

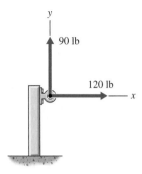

**Figure P2-1**

**2-2*** The two forces shown in Fig. P2-2.

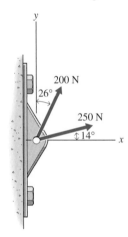

**Figure P2-2**

**2-3** The two forces shown in Fig. P2-3.

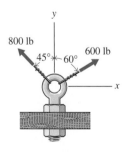

**Figure P2-3**

**2-4** The two forces shown in Fig. P2-4.

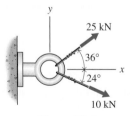

**Figure P2-4**

**Intermediate Problems**

**2-5*** The support ring of the traffic light shown in Fig. P2-5 is acted on by three forces–the weight of the traffic light (220 lb), a force in cable $A$ (280 lb), and a force in cable $B$ ($\mathbf{F}_B$). If the resultant of the three forces is zero, determine the magnitude and direction of $\mathbf{F}_B$.

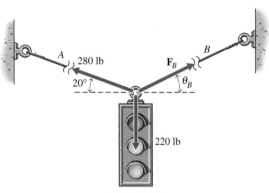

**Figure P2-5**

**2-6*** Determine the resultant of the three forces shown in Fig. P2-6. Locate the resultant with respect to the $x$-axis shown.

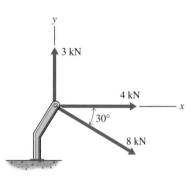

**Figure P2-6**

**2-7** A stalled automobile is being pulled by the two forces shown in Fig. P2-7. If the resultant pull is to be 120 lb in the $x$-direction, determine the magnitude and direction of the force **P.**

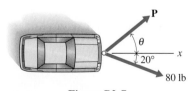

**Figure P2-7**

**2-8** The eye bolt shown in Fig. P2-8 is subjected to a 2700 N force and an unknown force **P**. If the resultant pull is 2000 N in the $x$-direction, determine the magnitude and direction of the force **P**.

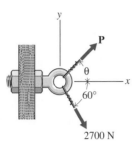

**Figure P2-8**

**2-9*** Two forces act on the bracket shown in Fig. P2-9. Determine the angle $\theta$ that will make the vertical component of the resultant of these two forces zero.

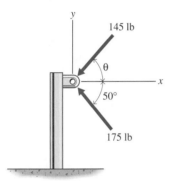

**Figure P2-9**

**2-10*** A 270-N force and a 400-N force act at point $B$ of the truss shown in Fig. P2-10. A third force **F** is to be applied at point $B$ so that the resultant of the three forces is zero. Determine the magnitude of **F** and its orientation with respect to the 400-N force.

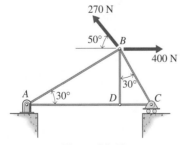

**Figure P2-10**

**Challenging Problems**

**2-11*** Three forces are applied to a bracket mounted on a post as shown in Fig. P2-11. Determine
(a) The magnitude and direction (angle $\theta_x$) of the resultant **R** of the three forces.

(b) The magnitudes of two other forces $\mathbf{F}_x$ and $\mathbf{F}_y$ that would have the same resultant.

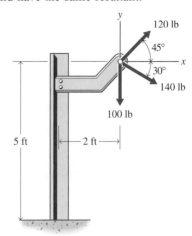

**Figure P2-11**

**2-12*** Four forces act on a small airplane in flight, as shown in Fig. P2-12; its weight (25 kN), the thrust provided by the engine (10 kN), the lift provided by the wings (24 kN), and the drag resulting from its motion through the air (3 kN). Determine the resultant of the four forces and its line of action with respect to the axis of the plane.

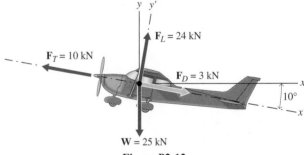

**Figure P2-12**

**2-13** Determine the resultant force **R** of the three forces applied to the gusset plate shown in Fig. P2-13.

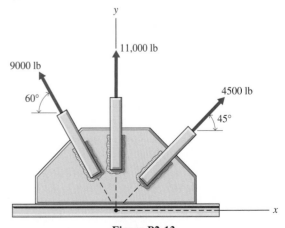

**Figure P2-13**

**2-14** Determine the magnitude and direction of the resultant of the three forces shown in Fig. P2-14.

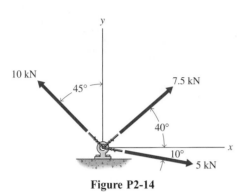

**Figure P2-14**

**Computer Problems**

**2-15** Two forces **A** and **B** are applied to an eye bolt as shown in Fig. P2-15. If the magnitudes of the two forces are $A = 50$ lb and $B = 100$ lb, calculate and plot the magnitude of the resultant **R** as a function of the angle $\theta_A$ ($0° \leq \theta_A \leq 180°$). Also calculate and plot the angle $\theta_R$ that the resultant makes with the force **B** as a function of the angle $\theta_A$. When is the resultant a maximum?

When is the resultant a minimum? When is the angle $\theta_R$ a maximum? Repeat for $A = 100$ lb and $B = 50$ lb.

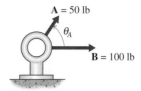

**Figure P2-15**

**2-16** Two forces **A** and **B** are applied to a bracket as shown in Fig. P2-16. If the magnitude of the force **B** is 325 N, calculate and plot the magnitude of the resultant **R** as a function of the magnitude of the force **A** ($0 \leq A \leq 900$ N).

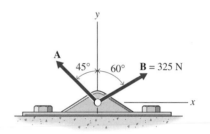

**Figure P2-16**

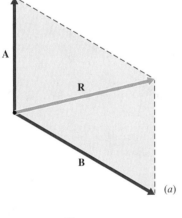

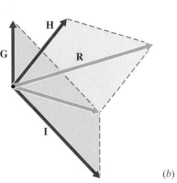

**Figure 2-20**

# 2-4 RESOLUTION OF A FORCE INTO COMPONENTS

In the previous section of this chapter, use of the parallelogram and triangle laws to determine the resultant **R** of two concurrent forces (vector quantities) $\mathbf{F}_1$ and $\mathbf{F}_2$ or three or more concurrent forces $\mathbf{F}_1$, $\mathbf{F}_2$, $\cdots$, $\mathbf{F}_n$ was discussed. In a similar manner, a single force **F** can be replaced by a system of two or more forces $\mathbf{F}_a$, $\mathbf{F}_b$, $\cdots$, $\mathbf{F}_n$. Forces $\mathbf{F}_a$, $\mathbf{F}_b$, $\cdots$, $\mathbf{F}_n$ are called *components of the original force*. In the most general case, the components of a force can be any system of forces that can be combined by the parallelogram law to produce the original force. Such components are not required to be concurrent or coplanar. Normally, however, the term *component* is used to specify either one of two coplanar concurrent forces or one of three noncoplanar concurrent forces that can be combined vectorially to produce the original force. The point of concurrency must be on the line of action of the original force. The process of replacing a force by two or more forces is called *resolution*.

The process of resolution does not produce a unique set of vector components. For example, consider the four coplanar sketches shown in Fig. 2-20. It is obvious from these sketches that

$$\mathbf{A} + \mathbf{B} = \mathbf{R} \qquad (\mathbf{G} + \mathbf{I}) + \mathbf{H} = \mathbf{R}$$
$$\mathbf{C} + \mathbf{D} = \mathbf{R} \qquad \mathbf{E} + \mathbf{F} = \mathbf{R}$$

where **R** is the same vector in each expression. Thus, an infinite number of sets of components exist for any vector.

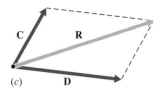

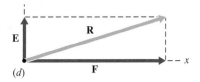

(c)  (d)

**Figure 2-20 (continued)**

The following examples illustrate use of the parallelogram and triangle laws to resolve a force (vector quantity) into components along any two oblique lines of action.

 **Example Problem 2-4**

Determine the magnitudes of the u- and v-components of the 900-N force shown in Fig. 2-21a.

**SOLUTION**

The magnitude and direction of the 900-N force are shown in Fig. 2-21b. The components $\mathbf{F}_u$ and $\mathbf{F}_v$ along the u- and v-axes can be determined by drawing lines parallel to the u- and v-axes through the head and tail of the vector used to represent the 900-N force. From the parallelogram thus produced, the law of sines can be applied to determine the forces $\mathbf{F}_u$ and $\mathbf{F}_v$, since all the angles for the two triangles which form the parallelogram are known. Thus,

$$\frac{F_u}{\sin 45°} = \frac{F_v}{\sin 25°} = \frac{900}{\sin 110°}$$

from which

$$F_u = |\mathbf{F}_u| = \frac{900 \sin 45°}{\sin 110°} = 677 \text{ N} \qquad \textbf{Ans.}$$

$$F_v = |\mathbf{F}_v| = \frac{900 \sin 25°}{\sin 110°} = 405 \text{ N} \qquad \textbf{Ans.}$$

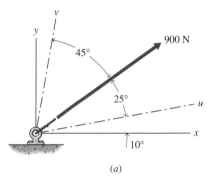

(a)

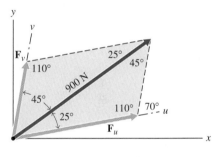

(b)

**Figure 2-21**

 **Example Problem 2-5**

A 100-kN force is resisted by an eye bar and a strut as shown in Fig. 2-22a. Determine the component $\mathbf{F}_u$ of the force along the axis of the eye bar AB and the component $\mathbf{F}_v$ of the force along the axis of the strut AC.

**SOLUTION**

The components $\mathbf{F}_u$ and $\mathbf{F}_v$ along the u- and v-axes can be determined by drawing lines parallel to the u- and v-axes through the head and tail of the vector

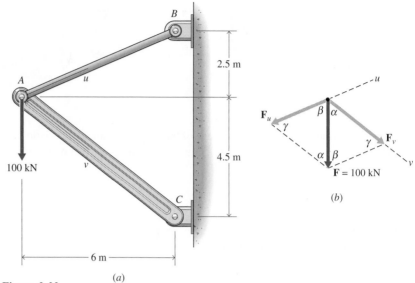

**Figure 2-22**

used to represent the 100-kN force. From the parallelogram thus produced (see Fig. 2-22b), the law of sines can be applied to determine the forces $\mathbf{F}_u$ and $\mathbf{F}_v$, since all of the angles for the two triangles that form the parallelogram can be determined. Thus,

$$\alpha = \tan^{-1} \frac{6}{4.5} = 53.13°$$

$$\beta = \tan^{-1} \frac{6}{2.5} = 67.38°$$

$$\gamma = 180 - 53.13 - 67.38 = 59.49°$$

From the law of sines

$$\frac{F_u}{\sin \alpha} = \frac{F_v}{\sin \beta} = \frac{100}{\sin \gamma}$$

$$F_u = \frac{100 \sin \alpha}{\sin \gamma} = \frac{100 \sin 53.13°}{\sin 59.49°} = 92.9 \text{ kN}$$

$$\mathbf{F_u} = 92.9 \text{ kN} \ \nearrow \ 22.62° \qquad \textbf{Ans.}$$

$$F_v = \frac{100 \sin \beta}{\sin \gamma} = \frac{100 \sin 67.38°}{\sin 59.49°} = 107.1 \text{ kN}$$

$$\mathbf{F_v} = 107.1 \text{ kN} \ \searrow \ 36.87° \ \blacksquare \qquad \textbf{Ans.}$$

# PROBLEMS

Use the law of sines and the law of cosines, in conjunction with sketches of the force triangles, to solve the following problems.

**Introductory Problems**

**2-17*** Determine the magnitudes of the $u$- and $v$-components of the 1000-lb force shown in Fig. P2-17.

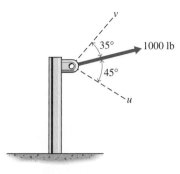

**Figure P2-17**

**2-18\*** Determine the components of the 3000-N force in the directions of members $AB$ and $BC$ of the truss shown in Fig. P2-18 when $\theta = 45°$.

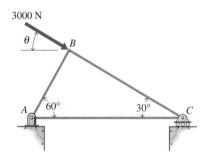

**Figure P2-18**

**2-19** Two cables are used to support a stop light as shown in Fig. P2-19. The resultant $\mathbf{R}$ of the cable forces $\mathbf{F}_u$ and $\mathbf{F}_v$ has a magnitude of 300 lb and its line of action is vertical. Determine the magnitudes of the forces $\mathbf{F}_u$ and $\mathbf{F}_v$.

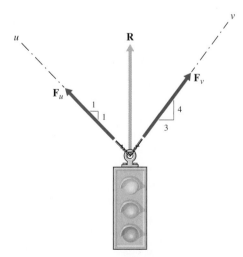

**Figure P2-19**

**2-20** Two ropes are used to tow a boat upstream as shown in Fig. P2-20. The resultant $\mathbf{R}$ of the rope forces $\mathbf{F}_u$ and $\mathbf{F}_v$ has a magnitude of 1500 N and its line of action is directed along the axis of the boat. Determine the magnitudes of the forces $\mathbf{F}_u$ and $\mathbf{F}_v$.

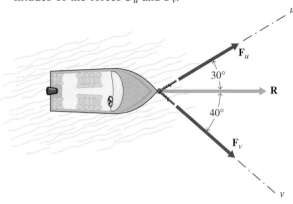

**Figure P2-20**

**2-21\*** Determine the $u$- and $v$-components of the 5200-lb force acting on the bracket shown in Fig. P2-21.

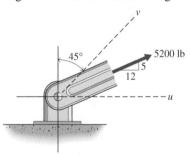

**Figure P2-21**

### Intermediate Problems

**2-22\*** A 5000-N force acts in the vertical direction on the block shown in Fig. P2-22. Determine the components of the force perpendicular to and parallel to the line $AB$.

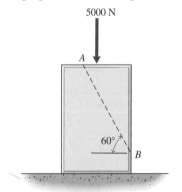

**Figure P2-22**

**2-23\*** Two forces $\mathbf{F}_u$ and $\mathbf{F}_v$ are applied to a bracket as shown in Fig. P2-23. If the resultant $\mathbf{R}$ of the two forces has a magnitude of 725 1b and a direction as shown on the figure, determine the magnitudes of the forces $\mathbf{F}_u$ and $\mathbf{F}_v$.

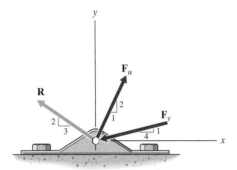

**Figure P2-23**

**2-24** A 20-kN force acts on the member shown in Fig. P2-24. Determine the components of the force in the horizontal and vertical directions ($\mathbf{F}_{AC}$ and $\mathbf{F}_{BC}$).

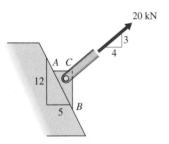

**Figure P2-24**

**2-25** A 500-1b force acts along the line $AB$ shown in Fig. P2-25. Determine the magnitude of the components in the direction of lines $AC$ and $AD$.

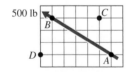

**Figure P2-25**

**2-26\*** Two forces $\mathbf{F}_u$ and $\mathbf{F}_v$ are applied to a bracket as shown in Fig. P2-26. If the resultant $\mathbf{R}$ of the two forces has a magnitude of 375 N and a direction as shown on the figure, determine the magnitudes of the forces $\mathbf{F}_u$ and $\mathbf{F}_v$.

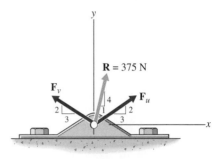

**Figure P2-26**

## Challenging Problems

**2-27\*** Three forces are applied to a bracket as shown in Fig. P2-27. The magnitude of the resultant $\mathbf{R}$ of the three forces is 50 kip. If the force $\mathbf{F}_1$ has a magnitude of 30 kip, determine the magnitudes of the forces $\mathbf{F}_2$ and $\mathbf{F}_3$.

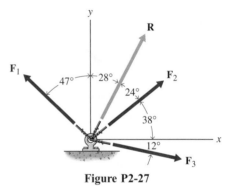

**Figure P2-27**

**2-28\*** A homogeneous cylinder is subjected to the three forces shown in Fig. P2-28. If the resultant of the three forces is zero, determine the magnitude of $\mathbf{N}_1$ and the angle $\theta$ of the surface to which $\mathbf{N}_1$ is normal.

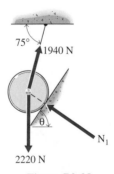

**Figure P2-28**

**2-29** Four forces act on the machine component shown in Fig. P2-29. If the resultant of $\mathbf{F}_1$, $\mathbf{F}_2$, and $\mathbf{F}_3$ is horizontal to the left and has a magnitude of 400 1b, determine the magnitudes of the forces $\mathbf{F}_2$ and $\mathbf{F}_3$.

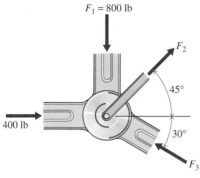

$F_1 = 800$ lb

$F_2$

45°

400 lb

30°

$F_3$

**Figure P2-29**

**2-30** A 140-N light is supported by two cables as shown in Fig. P2-30. If the resultant of the three forces is zero, determine the force $\mathbf{T}_1$ and the angle $\theta$.

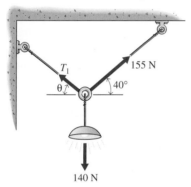

$T_1$

155 N

$\theta$

40°

140 N

**Figure P2-30**

**Computer Problems**

**2-31** Two forces $\mathbf{A}$ and $\mathbf{B}$ are applied to an eye bolt using ropes as shown in Fig. P2-31. The resultant $\mathbf{R}$ of the two forces has a magnitude $R = 4000$ lb and makes an angle of 30° with the force $\mathbf{A}$ as shown. If both of the forces pull on the eye bolt as shown (ropes cannot push on the eye bolt), what is the range of angles ($\theta_{min} \le \theta_B \le \theta_{max}$) for which this problem has a solution? Calculate and plot the required magnitudes $A$ and $B$ as functions of the angle angles $\theta_B$ ($\theta_{min} \le \theta_B \le \theta_{max}$). Why is the magnitude of $\mathbf{B}$ a minimum when $\theta_B = 90°$?

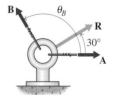

B

$\theta_B$

R

30°

A

**Figure P2-31**

**2-32** Three forces $\mathbf{A}$, $\mathbf{B}$, and $\mathbf{C}$ are applied to an eye bolt using ropes as shown in Fig. P2-32. Force $\mathbf{A}$ has a magnitude $A = 50$ N, and the resultant of the three forces is zero. If all of the forces pull on the eye bolt as shown (ropes cannot push on the eye bolt), what is the range of angles ($\theta_{min} \le \theta_C \le \theta_{max}$) for which this problem has a solution? Calculate and plot the required magnitudes $B$ and $C$ as functions of the angle $\theta_C$ ($\theta_{min} \le \theta_C \le \theta_{max}$). Why is the magnitude of $\mathbf{C}$ a minimum when $\theta_C = 90°$?

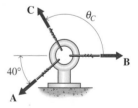

C

$\theta_C$

B

40°

A

**Figure P2-32**

# 2-5 RECTANGULAR COMPONENTS OF A FORCE

General oblique components of a force are not widely used for solving most practical engineering problems. Mutually perpendicular (rectangular) components, on the other hand, find wide usage. The process for obtaining rectangular components is greatly simplified, since the parallelogram used to represent the force and its components reduces to a rectangle and the law of cosines used to obtain numerical values of the components reduces to the Pythagorean theorem.

**Unit Vectors** Any vector quantity can be written as the product of its magnitude (a positive number) and its direction. Thus, for a vector A in the positive n-direction

$$\mathbf{A} = |\mathbf{A}|\mathbf{e}_n = A\mathbf{e}_n \qquad (2\text{-}3)$$

For a vector **B** in the negative *n*-direction

$$\mathbf{B} = |\mathbf{B}|(-\mathbf{e}_n) = -B\mathbf{e}_n \qquad (2\text{-}4)$$

It is obvious from Eq. (2-3) that the vector $\mathbf{e}_n$ can be expressed as

$$\mathbf{e}_n = \frac{\mathbf{A}}{|\mathbf{A}|} = \frac{\mathbf{A}}{A} \qquad (2\text{-}5)$$

It is also obvious that the vector $\mathbf{e}_n$ is dimensionless and has a magnitude of 1. It includes no information about the size or type of vector that **A** is, it is purely a directional entity.

A dimensionless vector of unit magnitude is called a *unit vector*. Unit vectors directed along the *x*-, *y*-, and *z*-axes of a Cartesian coordinate system are normally given the symbols **i, j,** and **k,** respectively. The sense of these unit vectors is indicated analytically by using a plus sign if the unit vector points in a positive *x*-, *y*-, or *z*-direction and a minus sign if the unit vector points in a negative *x*-, *y*-, or *z*-direction. The unit vectors shown in Fig. 2-23 are positive. In this book, the symbol **e** with a subscript is used to denote a unit vector in a direction other than a coordinate direction.

The Cartesian coordinate axes shown in Fig. 2-23 are arranged as a right-hand system. In a right-hand system, if the fingers of the right hand are curled about the *z*-axis in a direction from the positive *x*-axis toward the positive *y*-axis, then the thumb points in the positive *z*-direction, as shown in Fig. 2-24. All of the vector relationships developed in the remainder of this chapter will utilize a right-hand coordinate system.

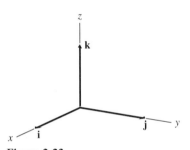

Figure 2-23

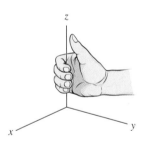

Figure 2-24

**Rectangular Components in Two Dimensions** A force vector **F** can be resolved into any number of components provided that the components sum to the given vector **F** by the parallelogram law. Mutually perpendicular components, called *rectangular components,* along the *x*- and *y*-coordinate axes are the most useful. The *x*- and *y*-axes are usually chosen horizontal and vertical, as shown in Fig. 2-25; however, they may be chosen in any two perpendicular directions. The choice is usually indicated by the geometry of the problem.[1]

Any vector **F** can be resolved into rectangular components $\mathbf{F}_x$ and $\mathbf{F}_y$ by constructing a rectangle (a right-angled parallelogram) with the vector **F** as the diagonal and rectangular components $\mathbf{F}_x = F_x \mathbf{i}$ along the *x*-axis and $\mathbf{F}_y = F_y \mathbf{j}$ along the *y*-axis as shown in Fig. 2-25. The vectors $\mathbf{F}_x$ and $\mathbf{F}_y$ are the *x* and *y* vector components of the vector **F,** and the scalars $F_x$ and $F_y$ are the *x* and *y* scalar components of the vector **F.** The scalar components $F_x$ and $F_y$ of the vector **F**

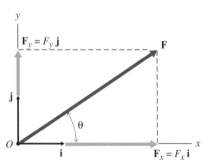

Figure 2-25

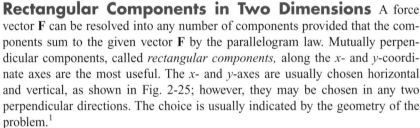

[1]Coordinate systems and axes are tools that may be used to advantage by the analyst. Machine components do not come inscribed with *x*- and *y*-axes; therefore, the analyst is free to select directions that are convenient for his or her work.

can be positive or negative, depending on the sense of the vector components $\mathbf{F}_x$ and $\mathbf{F}_y$. A scalar component is positive when the vector component has the same sense as the unit vector with which it is associated, and negative when the vector component has the opposite sense.

The vector $\mathbf{F} = \mathbf{F}_x + \mathbf{F}_y$ can be written in Cartesian vector form as

$$\mathbf{F} = F_x\,\mathbf{i} + F_y\,\mathbf{j} \qquad (2\text{-}6)$$

The scalar components $F_x$ and $F_y$ are related to the magnitude and direction of the vector by (Fig. 2-25)

$$F_x = F\cos\theta \qquad F_y = F\sin\theta \qquad (2\text{-}7)$$

The Pythagorean theorem can be used to express the magnitude of the vector $\mathbf{F}$ in terms of the scalar components by

$$F = \sqrt{F_x^2 + F_y^2} \qquad (2\text{-}8)$$

The slope of the vector $\mathbf{F}$ is obtained using the tangent function

$$\theta = \tan^{-1}\!\left(\frac{F_y}{F_x}\right) \qquad (2\text{-}9)$$

## Rectangular Components in Three Dimensions

Similarly, for problems requiring analysis in three dimensions, a force vector $\mathbf{F}$ in space can be resolved into three mutually perpendicular rectangular components $\mathbf{F}_x$, $\mathbf{F}_y$, and $\mathbf{F}_z$ by constructing a rectangular parallelepiped with the vector $\mathbf{F}$ as the diagonal and the rectangular components $\mathbf{F}_x = F_x\,\mathbf{i}$, $\mathbf{F}_y = F_y\,\mathbf{j}$, and $\mathbf{F}_z = F_z\,\mathbf{k}$ along $x$-, $y$-, and $z$-coordinate axes, as shown in Fig. 2-26. The vector $\mathbf{F}$ can be written in terms of its vector components, in terms of its scalar components, or in terms of its magnitude and direction

$$\mathbf{F} = \mathbf{F}_x + \mathbf{F}_y + \mathbf{F}_z$$
$$\mathbf{F} = F_x\,\mathbf{i} + F_y\,\mathbf{j} + F_z\,\mathbf{k} \qquad (2\text{-}10)$$
$$\mathbf{F} = F\,\mathbf{e}_n$$

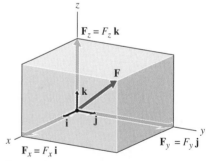

**Figure 2-26**

whichever is most convenient at the time.

The rectangular components are related to the magnitude of the vector $\mathbf{F}$ by

$$F_x = F\cos\theta_x \qquad F_y = F\cos\theta_y \qquad F_z = F\cos\theta_z \qquad (2\text{-}11)$$

where the direction angles $\theta_x$, $\theta_y$, and $\theta_z$ are the angles between the vector $\mathbf{F}$ and the coordinate axes (Fig. 2-27). The terms $\cos\theta_x$, $\cos\theta_y$, and $\cos\theta_z$ are called the *direction cosines* of the vector $\mathbf{F}$. These cosines are also the direction cosines of the unit vector $\mathbf{e}_n$, since its line of action coincides with the line of action of the vector $\mathbf{F}$.

The magnitude of the vector $\mathbf{F}$ can be expressed in terms of its rectangular components using the Pythagorean theorem

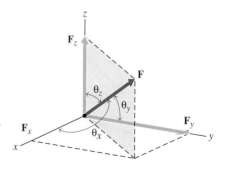

**Figure 2-27**

$$F = \sqrt{F_x^2 + F_y^2 + F_z^2} \tag{2-12}$$

Also, from Eq. (2-11), the direction cosines of a vector $\mathbf{F}$ can be expressed in terms of the magnitude of the vector $\mathbf{F}$ and its rectangular components as

$$\cos\theta_x = \frac{F_x}{F} \qquad \cos\theta_y = \frac{F_y}{F} \qquad \cos\theta_z = \frac{F_z}{F} \tag{2-13}$$

Finally, the unit vector $\mathbf{e}_n$ can be written in terms of the vector $\mathbf{F}$ and its rectangular components by using

$$\mathbf{e}_n = \frac{\mathbf{F}}{F} = \frac{F_x}{F}\mathbf{i} + \frac{F_y}{F}\mathbf{j} + \frac{F_z}{F}\mathbf{k}$$
$$= \cos\theta_x\,\mathbf{i} + \cos\theta_y\,\mathbf{j} + \cos\theta_z\,\mathbf{k} \tag{2-14}$$

Since the magnitude of a unit vector is 1, the direction cosines must satisfy

$$\cos^2\theta_x + \cos^2\theta_y + \cos^2\theta_z = 1 \tag{2-15}$$

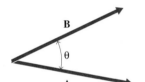

**Figure 2-28**

### The Scalar (Dot) Product and Rectangular Components
The dot or scalar product of two intersecting vectors is defined as the product of the magnitudes of the vectors and the cosine of the angle between them. Thus, by definition, for the vectors $\mathbf{A}$ and $\mathbf{B}$ shown in Fig. 2-28,

$$\mathbf{A} \cdot \mathbf{B} = \mathbf{B} \cdot \mathbf{A} = AB\cos\theta \tag{2-16}$$

where $0° \le \theta \le 180°$. This type of vector multiplication yields a scalar, not a vector. For $0° \le \theta \le 90°$, the scalar is positive. For $90° \le \theta \le 180°$, the scalar is negative. When $\theta = 90°$, the two vectors are perpendicular and the scalar is zero.

The dot product can be used to obtain the rectangular scalar component of a vector in a specified direction. For example, the rectangular scalar component of vector $\mathbf{A}$ along the $x$-axis is

$$A_x = \mathbf{A} \cdot \mathbf{i} = A(1)\cos\theta_x = A\cos\theta_x \tag{a}$$

Similarly, the rectangular scalar component of vector $\mathbf{A}$ in a direction $n$ (see Fig. 2-29a) is

$$A_n = \mathbf{A} \cdot \mathbf{e}_n = A\cos\theta_n \tag{2-17}$$

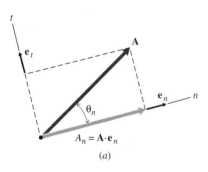

where $\mathbf{e}_n$ is the unit vector associated with the direction $n$. The vector component of $\mathbf{A}$ in the direction $n$ (see Fig. 2-29b) is then given by Eq. (2-3) as

$$\mathbf{A}_n = (\mathbf{A} \cdot \mathbf{e}_n)\mathbf{e}_n \tag{2-18}$$

The component of vector $\mathbf{A}$ perpendicular to the direction $n$ lies in the plane containing $\mathbf{A}$ and $n$, as shown in Fig. 2-29b, and can be obtained from the expression

$$\mathbf{A}_t = \mathbf{A} - \mathbf{A}_n \tag{b}$$

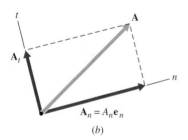

**Figure 2-29**

Once an expression for $\mathbf{A}_t$ has been obtained by using Eq. (b), the magnitude and direction of $\mathbf{A}_t$ can be determined using Eqs. (2-12) and (2-14).

If two vectors $\mathbf{A}$ and $\mathbf{B}$ are written in Cartesian vector form, the scalar product becomes

$$\begin{aligned}
\mathbf{A} \cdot \mathbf{B} &= (A_x\,\mathbf{i} + A_y\,\mathbf{j} + A_z\,\mathbf{k}) \cdot (B_x\,\mathbf{i} + B_y\,\mathbf{j} + B_z\,\mathbf{k}) \\
&= A_xB_x(\mathbf{i} \cdot \mathbf{i}) + A_xB_y(\mathbf{i} \cdot \mathbf{j}) + A_xB_z(\mathbf{i} \cdot \mathbf{k}) \\
&\quad + A_yB_x(\mathbf{j} \cdot \mathbf{i}) + A_yB_y(\mathbf{j} \cdot \mathbf{j}) + A_yB_z(\mathbf{j} \cdot \mathbf{k}) \\
&\quad + A_zB_x(\mathbf{k} \cdot \mathbf{i}) + A_zB_y(\mathbf{k} \cdot \mathbf{j}) + A_zB_z(\mathbf{k} \cdot \mathbf{k})
\end{aligned} \qquad (c)$$

Since the unit vectors $\mathbf{i}$, $\mathbf{j}$, and $\mathbf{k}$ are orthogonal,

$$\mathbf{i} \cdot \mathbf{j} = \mathbf{j} \cdot \mathbf{k} = \mathbf{k} \cdot \mathbf{i} = (1)(1)\cos 90° = 0$$
$$\mathbf{i} \cdot \mathbf{i} = \mathbf{j} \cdot \mathbf{j} = \mathbf{k} \cdot \mathbf{k} = (1)(1)\cos 0° = 1 \qquad (d)$$

Therefore

$$\mathbf{A} \cdot \mathbf{B} = A_xB_x + A_yB_y + A_zB_z \qquad (2\text{-}19)$$

Note also the special case

$$\mathbf{A} \cdot \mathbf{A} = A^2 \cos 0° = A^2 = A_x^2 + A_y^2 + A_z^2 \qquad (e)$$

which verifies Eq. (2-12) obtained by using the Pythagorean theorem.

Combining Eqs. (2-17), (2-14), and (2-19) gives the rectangular component of a force $\mathbf{F}$ in the direction $n$ as

$$F_n = \mathbf{F} \cdot \mathbf{e}_n = F_x \cos \theta_x + F_y \cos \theta_y + F_z \cos \theta_z \qquad (2\text{-}20)$$

An expression for the angle between two vectors can be obtained by combining Eqs. (2-16) and (2-19). Thus

$$\mathbf{A} \cdot \mathbf{B} = A_xB_x + A_yB_y + A_zB_z = AB \cos \theta \qquad (f)$$

or

$$\theta = \cos^{-1}\left(\frac{\mathbf{A} \cdot \mathbf{B}}{AB}\right) = \cos^{-1}\left(\frac{A_xB_x + A_yB_y + A_zB_z}{AB}\right) \qquad (2\text{-}21)$$

In mechanics, the scalar product is used to determine the rectangular component of a vector (force, moment, velocity, acceleration, etc.) along a line and to find the angle between two vectors (two forces, a force and a line, a force and an acceleration, etc.).

## Example Problem 2-6

A 450-N force $\mathbf{F}$ is applied at a point in a body as shown in Fig. 2-30.

(a) Determine the $x$ and $y$ scalar components of the force.
(b) Determine the $x'$ and $y'$ scalar components of the force.
(c) Express the force $\mathbf{F}$ in Cartesian vector form for the $xy$- and $x'y'$- axes.

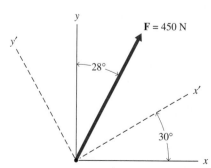

Figure 2-30

## SOLUTION

(a) The magnitude $F$ of the force is 450 N. The angle $\theta_x$ between the $x$-axis and the line of action of the force is

$$\theta_x = 90° - 28° = 62°$$

Thus,

$$F_x = F \cos \theta_x = 450 \cos 62° = +211 \text{ N} \qquad \textbf{Ans.}$$

$$F_y = F \sin \theta_x = 450 \sin 62° = +397 \text{ N} \qquad \textbf{Ans.}$$

(b) The magnitude $F$ of the force is 450 N. The angle $\theta_{x'}$ between the $x'$-axis and the line of action of the force is

$$\theta_{x'} = \theta_x - 30° = 62° - 30° = 32°$$

Thus,

$$F_{x'} = F \cos \theta_{x'} = 450 \cos 32° = +382 \text{ N} \qquad \textbf{Ans.}$$

$$F_{y'} = F \sin \theta_{x'} = 450 \sin 32° = +238 \text{ N} \qquad \textbf{Ans.}$$

As a check, note that

$$F = \sqrt{F_x^2 + F_y^2} = \sqrt{F_{x'}^2 + F_{y'}^2}$$
$$= \sqrt{211^2 + 397^2} = \sqrt{382^2 + 238^2} = 450 \text{ N}$$

(c) The force $\mathbf{F}$ expressed in Cartesian vector form for the $xy$- and $x'y'$-axes are

$$\mathbf{F} = F_x \mathbf{i} + F_y \mathbf{j} = 211 \, \mathbf{i} + 397 \, \mathbf{j} \text{ N} \qquad \textbf{Ans.}$$

$$\mathbf{F} = F_{x'} \mathbf{e}_{x'} + F_{y'} \mathbf{e}_{y'} = 382 \, \mathbf{e}_{x'} + 238 \, \mathbf{e}_{y'} \text{ N} \blacksquare \qquad \textbf{Ans.}$$

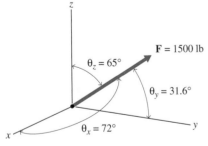

**Figure 2-31**

### ■ Example Problem 2-7

A 1500-lb force $\mathbf{F}$ is applied at a point in a body as shown in Fig. 2-31.

(a) Determine the $x$, $y$, and $z$ scalar components of the force.

(b) Express the force in Cartesian vector form.

## SOLUTION

(a) The magnitude $F$ of the force is 1500 lb. Thus,

$$F_x = F \cos \theta_x = 1500 \cos 72° = +464 \text{ lb} \qquad \textbf{Ans.}$$

$$F_y = F \cos \theta_y = 1500 \cos 31.6° = +1278 \text{ lb} \qquad \textbf{Ans.}$$

$$F_z = F \cos \theta_z = 1500 \cos 65° = +634 \text{ lb} \qquad \textbf{Ans.}$$

As a check, note that

$$F = \sqrt{F_x^2 + F_y^2 + F_z^2}$$
$$= \sqrt{464^2 + 1278^2 + 634^2} = 1500 \text{ lb}$$

(b) The force **F** expressed in Cartesian vector form is

$$\mathbf{F} = F_x \mathbf{i} + F_y \mathbf{j} + F_z \mathbf{k}$$
$$= 464 \mathbf{i} + 1278 \mathbf{j} + 634 \mathbf{k} \text{ lb} \blacksquare \qquad \textbf{Ans.}$$

---

## ▌Example Problem 2-8

A 25-kN force **F** is applied at a point in a body as shown in Fig. 2-32. Determine

(a) The angles $\theta_x$, $\theta_y$, and $\theta_z$.
(b) The $x$, $y$, and $z$ scalar components of the force.
(c) The rectangular component $F_n$ of the force along line $OA$.
(d) The rectangular component $F_t$ of the force perpendicular to line $OA$.

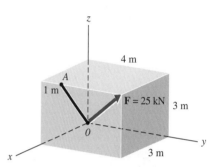

**Figure 2-32**

### SOLUTION

(a) The angles $\theta_x$, $\theta_y$, and $\theta_z$ can be determined from the geometry of the box shown in Fig. 2-32. The length of diagonal $d$ of the box is

$$d = \sqrt{x^2 + y^2 + z^2} = \sqrt{3^2 + 4^2 + 3^2} = 5.831 \text{ m}$$

Thus,

$$\theta_x = \cos^{-1} \frac{x}{d} = \cos^{-1} \frac{3}{5.831} = 59.0° \qquad \textbf{Ans.}$$

$$\theta_y = \cos^{-1} \frac{y}{d} = \cos^{-1} \frac{4}{5.831} = 46.7° \qquad \textbf{Ans.}$$

$$\theta_z = \cos^{-1} \frac{z}{d} = \cos^{-1} \frac{3}{5.831} = 59.0° \qquad \textbf{Ans.}$$

(b) The magnitude $F$ of the force is 25 kN. Thus,

$$F_x = F \cos \theta_x = 25 \left( \frac{3}{5.831} \right) = +12.862 \text{ kN} \qquad \textbf{Ans.}$$

$$F_y = F \cos \theta_y = 25 \left( \frac{4}{5.831} \right) = +17.150 \text{ kN} \qquad \textbf{Ans.}$$

$$F_z = F \cos \theta_z = 25 \left( \frac{3}{5.831} \right) = +12.862 \text{ kN} \qquad \textbf{Ans.}$$

(c) The angles $\theta_x'$, $\theta_y'$, and $\theta_z'$ between the $n$ direction (along $OA$) and the $x$-, $y$-, and $z$-axes can also be determined from the geometry of the box shown in Fig. 2-32. The length of diagonal $d'$ of the line from $O$ to $A$ is

$$d' = \sqrt{(x')^2 + (y')^2 + (z')^2} = \sqrt{3^2 + 1^2 + 3^2} = 4.359 \text{ m}$$

Thus,

$$\theta'_x = \cos^{-1}\frac{x'}{d'} = \cos^{-1}\frac{3}{4.359} = 46.5°$$

$$\theta'_y = \cos^{-1}\frac{y'}{d'} = \cos^{-1}\frac{1}{4.359} = 76.7°$$

$$\theta'_z = \cos^{-1}\frac{z'}{d'} = \cos^{-1}\frac{3}{4.359} = 46.5°$$

The unit vector $\mathbf{e}_n$ along line $OA$ is

$$\mathbf{e}_n = \cos\theta'_x\,\mathbf{i} + \cos\theta'_y\,\mathbf{j} + \cos\theta'_z\,\mathbf{k}$$
$$= 0.6882\,\mathbf{i} + 0.2294\,\mathbf{j} + 0.6882\,\mathbf{k}$$

The force $F$ expressed in Cartesian vector form is

$$\mathbf{F} = F_x\,\mathbf{i} + F_y\,\mathbf{j} + F_z\,\mathbf{k}$$
$$= 12.862\,\mathbf{i} + 17.150\,\mathbf{j} + 12.862\,\mathbf{k} \text{ kN}$$

Therefore,

$$F_n = \mathbf{F} \cdot \mathbf{e}_n$$
$$= (12.862\,\mathbf{i} + 17.150\,\mathbf{j} + 12.862\,\mathbf{k}) \cdot (0.6882\,\mathbf{i} + 0.2294\,\mathbf{j} + 0.6882\,\mathbf{k})$$
$$= 12.862(0.6882) + 17.150(0.2294) + 12.862(0.6882)$$
$$= 21.64 \cong 21.6 \text{ kN} \qquad \text{Ans.}$$

(d) The force $\mathbf{F}$ can be resolved into components $\mathbf{F}_n$ along $OA$ and $\mathbf{F}_t$ perpendicular to $OA$. Thus

$$\mathbf{F} = \mathbf{F}_n + \mathbf{F}_t \quad \text{and} \quad F = \sqrt{F_n^2 + F_t^2}$$

Therefore

$$F_t = \sqrt{F^2 - F_n^2} = \sqrt{25^2 - 21.64^2} = 12.52 \text{ kN} \ \blacksquare \qquad \text{Ans.}$$

## PROBLEMS

**Introductory Problems**

**2-33\*** Determine the $x$- and $y$-components of the 1000-lb force shown in Fig. P2-33.

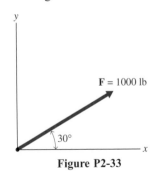

**Figure P2-33**

**2-34\*** Determine the $x$- and $y$-components of the 800-N force shown in Fig. P2-34.

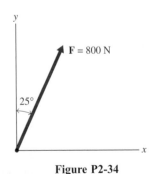

**Figure P2-34**

**2-35** Determine the $x$- and $y$-components of each force shown in Fig. P2-35.

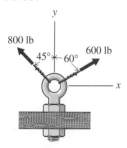

**Figure P2-35**

**2-36** Determine the $x$- and $y$-components of each force shown in Fig. P2-36.

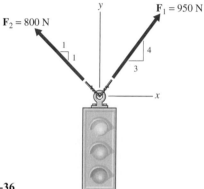

**Figure P2-36**

**Intermediate Problems**

**2-37\*** Two forces are applied to a post as shown in Fig. P2-37. Determine
(a) The $x$- and $y$-components of each force.
(b) The $x'$- and $y'$-components of each force.

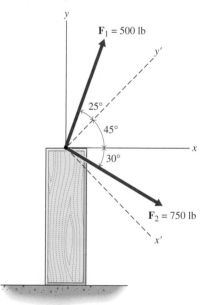

**Figure P2-37**

**2-38\*** For the 900-N force shown in Fig. P2-38,
(a) Determine the $x$, $y$, and $z$ scalar components of the force.
(b) Express the force in Cartesian vector form.

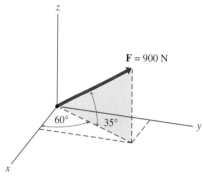

**Figure P2-38**

**2-39** As an automobile rounds a curve, a force **F** of magnitude 600 lb is exerted on one of the tires, as shown in Fig. P2-39. Determine the $x$, $y$, and $z$ scalar components of the force. The $xy$-plane is parallel to the roadway.

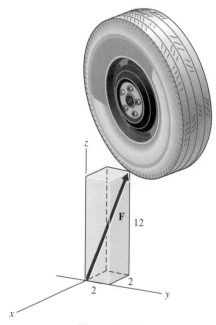

**Figure P2-39**

**2-40** A 50-kN force is applied to an eye bolt as shown in Fig. P2-40.
(a) Determine the direction angles $\theta_x$, $\theta_y$, and $\theta_z$.

(b) Determine the $x$, $y$, and $z$ scalar components of the force.

(c) Express the force in Cartesian vector form.

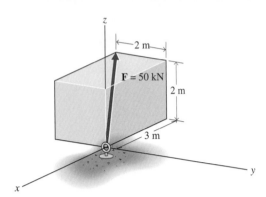

**Figure P2-40**

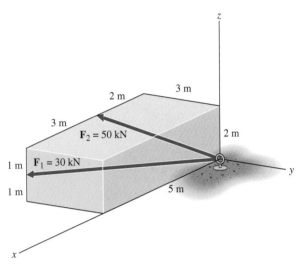

**Figure P2-42**

**2-41*** Two forces are applied at a point on a body as shown in Fig. P2-41. Determine

(a) The $x$- and $y$- components of each force.

(b) The $x'$- and $y'$- components of each force.

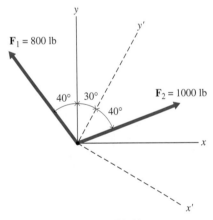

**Figure P2-41**

**Challenging Problems**

**2-42*** Two forces are applied to an eye bolt as shown in Fig. P2-42.

(a) Determine the $x$, $y$, and $z$ scalar components of the 30-kN force $\mathbf{F}_1$.

(b) Express the 30-kN force $\mathbf{F}_1$ in Cartesian vector form.

(c) Determine the magnitude of the rectangular component of the 30-kN force $\mathbf{F}_1$ along the line of action of the 50-kN force $\mathbf{F}_2$.

(d) Determine the angle $\alpha$ between the two forces.

**2-43*** A wire is stretched between two pylons, one of which is shown in Fig. P2-43. The 250-1b force in the wire is parallel to the $xy$-plane and makes an angle of $30°$ with the $y$-axis. Point $A$ lies in the $xz$-plane, and point $C$ lies in the $yz$-plane. Determine

(a) The magnitude of the rectangular component of the 250-1b force in the direction of member $AD$.

(b) The magnitude of the rectangular component of the 250-1b force in the direction of member $CD$.

(c) The angle $\alpha$ between members $AD$ and $CD$.

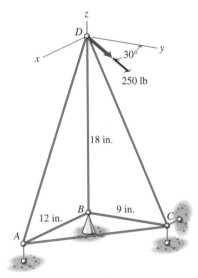

**Figure P2-43**

**2-44\*** The hot-air balloon shown in Fig. P2-44 is tethered with three mooring cables. The force $\mathbf{T}_A$ in cable $AD$ has a magnitude of 1860 N.
  (a) Express $\mathbf{T}_A$ in Cartesian vector form.
  (b) Determine the magnitude of the rectangular component of the force $\mathbf{T}_A$ along $BD$.
  (c) Determine the angle $\alpha$ between cables $AD$ and $BD$.

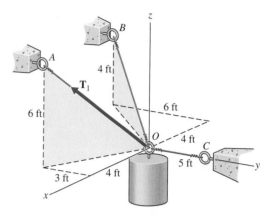

**Figure P2-45**

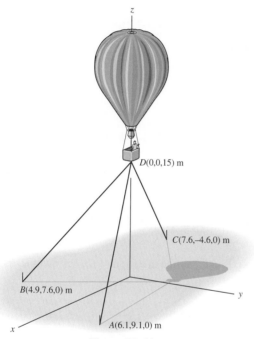

**Figure P2-44**

**2-45** A system of three cables supports the cylinder shown in Fig. P2-45. The magnitude of the force $\mathbf{T}_1$ in cable $AO$ is 2000 1b. Determine
  (a) The magnitude of the rectangular component of the force $\mathbf{T}_1$ along the line $OB$.
  (b) The angle $\alpha$ between the force $\mathbf{T}_1$ and the line $OB$.

**2-46** A 2000-N force $\mathbf{F}$ acts on a machine component as shown in Fig. P2-46.
  (a) Determine the $x$, $y$, and $z$ scalar components of the force.
  (b) Express the force in Cartesian vector form.
  (c) Determine the angle $\alpha$ between the force and the line $AB$.

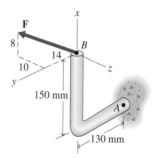

**Figure P2-46**

# 2-6 RESULTANTS BY RECTANGULAR COMPONENTS

Previous sections of this chapter discussed the use of parallelogram and triangle laws to determine the resultant $\mathbf{R}$ of two or more concurrent coplanar forces $\mathbf{F}_1$, $\mathbf{F}_2$, $\mathbf{F}_3$, $\cdots$, $\mathbf{F}_n$. Using the parallelogram law to add more than two forces is time-consuming and tedious, since the procedure requires extensive geometric and trigonometric calculation to determine the magnitude and locate the line of action of the resultant $\mathbf{R}$. Problems of this type, however, are

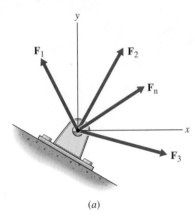

(a)

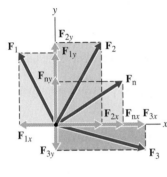

(b)

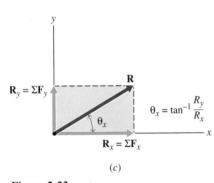

(c)

**Figure 2-33**

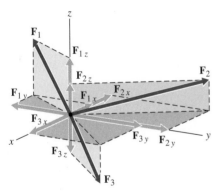

**Figure 2-34**

easily solved using the rectangular components of a force discussed in Section 2-5.

For any system of coplanar concurrent forces, such as the one shown in Fig. 2-33a, rectangular components $\mathbf{F}_{1x}$ and $\mathbf{F}_{1y}$, $\mathbf{F}_{2x}$ and $\mathbf{F}_{2y}$, $\mathbf{F}_{3x}$ and $\mathbf{F}_{3y}$, $\cdots$, and $\mathbf{F}_{nx}$ and $\mathbf{F}_{ny}$ can be determined as shown in Fig. 2-33b. Adding the respective x- and y-components yields

$$
\begin{aligned}
\mathbf{R}_x = \Sigma \mathbf{F}_x &= \mathbf{F}_{1x} + \mathbf{F}_{2x} + \mathbf{F}_{3x} + \cdots + \mathbf{F}_{nx} \\
&= (F_{1x} + F_{2x} + F_{3x} + \cdots + F_{nx})\,\mathbf{i} = R_x\,\mathbf{i} \\
\mathbf{R}_y = \Sigma \mathbf{F}_y &= \mathbf{F}_{1y} + \mathbf{F}_{2y} + \mathbf{F}_{3y} + \cdots + \mathbf{F}_{ny} \\
&= (F_{1y} + F_{2y} + F_{3y} + \cdots + F_{ny})\,\mathbf{j} = R_y\,\mathbf{j}
\end{aligned}
\tag{2-22}
$$

By the parallelogram law

$$
\mathbf{R} = \mathbf{R}_x + \mathbf{R}_y = R_x\,\mathbf{i} + R_y\,\mathbf{j}
\tag{2-23}
$$

The magnitude $R$ of the resultant can be determined from the Pythagorean theorem. Thus,

$$
R = \sqrt{R_x^2 + R_y^2} = \sqrt{(\Sigma F_x)^2 + (\Sigma F_y)^2}
\tag{2-24}
$$

The angle $\theta_x$ between the x-axis and the line of action of the resultant $\mathbf{R}$, as shown in Fig. 2-33c, is

$$
\theta_x = \tan^{-1}\frac{R_y}{R_x} = \tan^{-1}\frac{\Sigma F_y}{\Sigma F_x}
\tag{2-25}
$$

The angle $\theta_x$ can also be determined, if it is more convenient, from the equations

$$
\theta_x = \cos^{-1}\frac{\Sigma F_x}{R} \quad \text{or} \quad \theta_x = \sin^{-1}\frac{\Sigma F_y}{R}
\tag{2-26}
$$

The sense of each component must be designated in the summations by using a plus sign if the component acts in the positive x- or y-direction, and a minus sign if the component acts in the negative x- or y-direction.

In the general case of three or more concurrent forces in space, such as the three shown in Fig. 2-34, rectangular components $\mathbf{F}_{1x}$, $\mathbf{F}_{1y}$, and $\mathbf{F}_{1z}$; $\mathbf{F}_{2x}$, $\mathbf{F}_{2y}$, and $\mathbf{F}_{2z}$; and $\mathbf{F}_{3x}$, $\mathbf{F}_{3y}$, and $\mathbf{F}_{3z}$ can be determined. For $n$ forces, adding the respective x-, y-, and z-components yields

$$
\begin{aligned}
\mathbf{R}_x = \Sigma \mathbf{F}_x &= \mathbf{F}_{1x} + \mathbf{F}_{2x} + \mathbf{F}_{3x} + \cdots + \mathbf{F}_{nx} \\
&= (F_{1x} + F_{2x} + F_{3x} + \cdots + F_{nx})\,\mathbf{i} = R_x\,\mathbf{i} \\
\mathbf{R}_y = \Sigma \mathbf{F}_y &= \mathbf{F}_{1y} + \mathbf{F}_{2y} + \mathbf{F}_{3y} + \cdots + \mathbf{F}_{ny} \\
&= (F_{1y} + F_{2y} + F_{3y} + \cdots + F_{ny})\,\mathbf{j} = R_y\,\mathbf{j} \\
\mathbf{R}_z = \Sigma \mathbf{F}_z &= \mathbf{F}_{1z} + \mathbf{F}_{2z} + \mathbf{F}_{3z} + \cdots + \mathbf{F}_{nz} \\
&= (F_{1z} + F_{2z} + F_{3z} + \cdots + F_{nz})\,\mathbf{k} = R_z\,\mathbf{k}
\end{aligned}
\tag{2-27}
$$

The resultant **R** is then obtained from the expression

$$\mathbf{R} = \mathbf{R}_x + \mathbf{R}_y + \mathbf{R}_z = R_x\,\mathbf{i} + R_y\,\mathbf{j} + R_z\,\mathbf{k} \qquad (2\text{-}28)$$

Once the scalar components $R_x$, $R_y$, and $R_z$ are known, the magnitude $R$ of the resultant and the angles $\theta_x$, $\theta_y$, and $\theta_z$ between the line of action of the resultant and the positive coordinate axes can be obtained from the expressions

$$R = \sqrt{R_x^2 + R_y^2 + R_z^2} \qquad (2\text{-}29)$$

and

$$\theta_x = \cos^{-1}\frac{R_x}{R} \qquad \theta_y = \cos^{-1}\frac{R_y}{R} \qquad \theta_z = \cos^{-1}\frac{R_z}{R} \qquad (2\text{-}30)$$

## Example Problem 2-9

Determine the magnitude $R$ of the resultant of the four forces shown in Fig. 2-35$a$ and the angle $\theta_x$ between the $x$-axis and the line of action of the resultant.

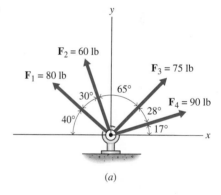

(a)

### SOLUTION

The magnitude $R$ of the resultant will be determined by using the rectangular components $F_x$ and $F_y$ of each of the forces. Thus

$$F_{1x} = 80 \cos 140° = -61.28 \text{ lb} \qquad F_{1y} = 80 \sin 140° = +51.42 \text{ lb}$$

$$F_{2x} = 60 \cos 110° = -20.52 \text{ lb} \qquad F_{2y} = 60 \sin 110° = +56.38 \text{ lb}$$

$$F_{3x} = 75 \cos 45° = +53.03 \text{ lb} \qquad F_{3y} = 75 \sin 45° = +59.03 \text{ lb}$$

$$F_{4x} = 90 \cos 17° = +86.07 \text{ lb} \qquad F_{4y} = 90 \sin 17° = +26.31 \text{ lb}$$

Once the rectangular components of the forces are known, the components $R_x$ and $R_y$ of the resultant are obtained from the expressions

$$R_x = \Sigma F_x = F_{1x} + F_{2x} + F_{3x} + F_{4x}$$
$$= -61.28 - 20.52 + 53.03 + 86.07 = +57.30 \text{ lb}$$

$$R_y = \Sigma F_y = F_{1y} + F_{2y} + F_{3y} + F_{4y}$$
$$= +51.42 + 56.38 + 53.03 + 26.31 = +187.14 \text{ lb}$$

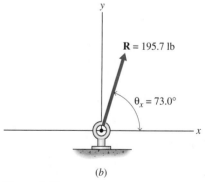

(b)

**Figure 2-35**

The magnitude $R$ of the resultant is

$$R = \sqrt{R_x^2 + R_y^2} = \sqrt{(57.30)^2 + (187.14)^2} = 195.7 \text{ lb} \qquad \textbf{Ans.}$$

The angle $\theta_x$ is obtained from the expression

$$\theta_x = \tan^{-1}\frac{R_y}{R_x} = \tan^{-1}\frac{+187.14}{+57.30} = 73.0° \qquad \textbf{Ans.}$$

The resultant **R** of the four forces of Fig. 2-35a is shown in Fig. 2-35b. ■

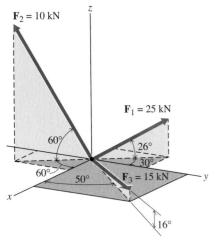

**Figure 2-36**

## Example Problem 2-10

Three forces are applied at a point on a body as shown in Fig. 2-36. Determine the magnitude $R$ of the resultant of the three forces and the angles $\theta_x$, $\theta_y$, and $\theta_z$ between the line of action of the resultant and the positive x-, y-, and z-coordinate axes.

### SOLUTION (USING SCALAR COMPONENTS)

The magnitude $R$ of the resultant will first be determined by using the rectangular components $F_x$, $F_y$, and $F_z$ of each of the forces. Thus

$$F_{1x} = 25 \cos 26° \cos 120° = -11.235 \text{ kN}$$

$$F_{1y} = 25 \cos 26° \sin 120° = +19.459 \text{ kN}$$

$$F_{1z} = 25 \sin 26° = +10.959 \text{ kN}$$

$$F_{2x} = 10 \cos 60° \cos (-60°) = +2.500 \text{ kN}$$

$$F_{2y} = 10 \cos 60° \sin (-60°) = -4.330 \text{ kN}$$

$$F_{2z} = 10 \sin 60° = +8.660 \text{ kN}$$

$$F_{3x} = 15 \cos 16° \cos 50° = +9.268 \text{ kN}$$

$$F_{3y} = 15 \cos 16° \sin 50° = +11.046 \text{ kN}$$

$$F_{3z} = 15 \sin 16° = +4.135 \text{ kN}$$

Once the rectangular components of the forces are known, the components $R_x$, $R_y$, and $R_z$ of the resultant are obtained from the expressions

$$R_x = \Sigma F_x = F_{1x} + F_{2x} + F_{3x}$$
$$= -11.235 + 2.500 + 9.268 = +0.533 \text{ kN}$$
$$R_y = \Sigma F_y = F_{1y} + F_{2y} + F_{3y}$$
$$= +19.459 - 4.330 + 11.046 = +26.175 \text{ kN}$$
$$R_z = \Sigma F_z = F_{1z} + F_{2z} + F_{3z}$$
$$= +10.959 + 8.660 + 4.135 = +23.754 \text{ kN}$$

The magnitude $R$ of the resultant is

$$R = \sqrt{R_x^2 + R_y^2 + R_z^2}$$
$$= \sqrt{(0.533)^2 + (26.175)^2 + (23.754)^2} = 35.35 \cong 35.4 \text{ kN} \qquad \textbf{Ans.}$$

The angles $\theta_x$, $\theta_y$, and $\theta_z$ are obtained from the expressions

$$\theta_x = \cos^{-1}\frac{R_x}{R} = \cos^{-1}\frac{+0.533}{35.35} = 89.1° \qquad \textbf{Ans.}$$

$$\theta_y = \cos^{-1}\frac{R_y}{R} = \cos^{-1}\frac{+26.175}{35.35} = 42.2° \qquad \textbf{Ans.}$$

$$\theta_z = \cos^{-1}\frac{R_z}{R} = \cos^{-1}\frac{+23.754}{35.35} = 47.8° \qquad \textbf{Ans.}$$

## SOLUTION (USING VECTOR NOTATION)

The solution can also be obtained using vector notation. Unit vectors $e_1$, $e_2$, and $e_3$, along the lines of action of forces $\mathbf{F}_1$, $\mathbf{F}_2$, and $\mathbf{F}_3$, respectively, are

$$\begin{aligned}
\mathbf{e}_1 &= (\cos 26° \cos 120°)\,\mathbf{i} + (\cos 26° \cos 30°)\,\mathbf{j} + (\cos 64°)\,\mathbf{k} \\
&= -0.4494\,\mathbf{i} + 0.7784\,\mathbf{j} + 0.4384\,\mathbf{k} \\
\mathbf{e}_2 &= (\cos 60° \cos 60°)\,\mathbf{i} + (\cos 60° \cos 150°)\,\mathbf{j} + (\cos 30°)\,\mathbf{k} \\
&= 0.2500\,\mathbf{i} - 0.4330\,\mathbf{j} + 0.8660\,\mathbf{k} \\
\mathbf{e}_3 &= (\cos 16° \cos 50°)\,\mathbf{i} + (\cos 16° \cos 40°)\,\mathbf{j} + (\cos 74°)\,\mathbf{k} \\
&= 0.6179\,\mathbf{i} + 0.7364\,\mathbf{j} + 0.2756\,\mathbf{k}
\end{aligned}$$

The forces are then written in Cartesian vector form as

$$\begin{aligned}
\mathbf{F}_1 = F_1\mathbf{e}_1 &= 25(-0.4494\,\mathbf{i} + 0.7784\,\mathbf{j} + 0.4384\,\mathbf{k}) \\
&= -11.235\,\mathbf{i} + 19.460\,\mathbf{j} + 10.960\,\mathbf{k} \text{ kN} \\
\mathbf{F}_2 = F_2\mathbf{e}_2 &= 10(0.2500\,\mathbf{i} - 0.4330\,\mathbf{j} + 0.8660\,\mathbf{k}) \\
&= 2.500\,\mathbf{i} - 4.330\,\mathbf{j} + 8.660\,\mathbf{k} \text{ kN} \\
\mathbf{F}_3 = F_3\mathbf{e}_3 &= 15(0.6179\,\mathbf{i} + 0.7364\,\mathbf{j} + 0.2756\,\mathbf{k}) \\
&= 9.269\,\mathbf{i} + 11.046\,\mathbf{j} + 4.134\,\mathbf{k} \text{ kN}
\end{aligned}$$

The resultant $\mathbf{R}$ of the three forces is

$$\begin{aligned}
\mathbf{R} = \mathbf{F}_1 + \mathbf{F}_2 + \mathbf{F}_3 &= \Sigma F_x\,\mathbf{i} + \Sigma F_y\,\mathbf{j} + \Sigma F_z\,\mathbf{k} \\
&= 0.533\,\mathbf{i} + 26.175\,\mathbf{j} + 23.754\,\mathbf{k} \text{ kN}
\end{aligned}$$

Once the scalar components of the resultant are determined, the magnitude of the resultant and the direction angles are determined as in the first part of the example. ■

## Example Problem 2-11

Determine the magnitude $R$ of the resultant of the three forces shown in Fig. 2-37 and the angles $\theta_x$, $\theta_y$, and $\theta_z$ between the line of action of the resultant and the positive x-, y-, and z-coordinate axes.

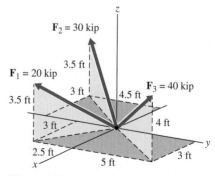

Figure 2-37

## SOLUTION

In addition to the origin of coordinates, the lines of action of forces $\mathbf{F}_1$, $\mathbf{F}_2$, and $\mathbf{F}_3$ pass through points $(3, -2.5, 3.5)$, $(-3, -4.5, 3.5)$, and $(3, 5, 4)$, respectively. Since the coordinates of these points in space are known, it is easy to determine the unit vectors associated with each of the forces. Thus,

$$\mathbf{e}_1 = \frac{3\,\mathbf{i} - 2.5\,\mathbf{j} + 3.5\,\mathbf{k}}{\sqrt{(3)^2 + (-2.5)^2 + (3.5)^2}} = +0.5721\,\mathbf{i} - 0.4767\,\mathbf{j} + 0.6674\,\mathbf{k}$$

$$\mathbf{e}_2 = \frac{-3\,\mathbf{i} - 4.5\,\mathbf{j} + 3.5\,\mathbf{k}}{\sqrt{(-3)^2 + (-4.5)^2 + (3.5)^2}} = -0.4657\,\mathbf{i} - 0.6985\,\mathbf{j} + 0.5433\,\mathbf{k}$$

$$\mathbf{e}_3 = \frac{3\,\mathbf{i} + 5\,\mathbf{j} + 4\,\mathbf{k}}{\sqrt{(3)^2 + (5)^2 + (4)^2}} = +0.4243\,\mathbf{i} + 0.7071\,\mathbf{j} + 0.5657\,\mathbf{k}$$

Once the unit vectors $\mathbf{e}_1$, $\mathbf{e}_2$, and $\mathbf{e}_3$ are known, the three forces can be expressed in Cartesian vector form as

$$\begin{aligned}
\mathbf{F}_1 = F_1\mathbf{e}_1 &= 20(+0.5721\,\mathbf{i} - 0.4767\,\mathbf{j} + 0.6674\,\mathbf{k}) \\
&= +11.442\,\mathbf{i} - 9.534\,\mathbf{j} + 13.348\,\mathbf{k} \text{ kip} \\
\mathbf{F}_2 = F_2\mathbf{e}_2 &= 30(-0.4657\,\mathbf{i} - 0.6985\,\mathbf{j} + 0.5433\,\mathbf{k}) \\
&= -13.971\,\mathbf{i} - 20.955\,\mathbf{j} + 16.299\,\mathbf{k} \text{ kip} \\
\mathbf{F}_3 = F_3\mathbf{e}_3 &= 40(+0.4243\,\mathbf{i} + 0.7071\,\mathbf{j} + 0.5657\,\mathbf{k}) \\
&= +16.972\,\mathbf{i} + 28.284\,\mathbf{j} + 22.628\,\mathbf{k} \text{ kip}
\end{aligned}$$

The resultant $\mathbf{R}$ of the three forces is

$$\mathbf{R} = \mathbf{F}_1 + \mathbf{F}_2 + \mathbf{F}_3 = R_x\,\mathbf{i} + R_y\,\mathbf{j} + R_z\,\mathbf{k} \text{ kip}$$

where

$$\begin{aligned}
R_x = \Sigma F_x = F_{1x} + F_{2x} + F_{3x} \\
= +11.442 - 13.971 + 16.972 = +14.443 \text{ kip}
\end{aligned}$$

$$\begin{aligned}
R_y = \Sigma F_y = F_{1y} + F_{2y} + F_{3y} \\
= -9.534 - 20.955 + 28.284 = -2.205 \text{ kip}
\end{aligned}$$

$$\begin{aligned}
R_z = \Sigma F_z = F_{1z} + F_{2z} + F_{3z} \\
= +13.348 + 16.299 + 22.628 = +52.28 \text{ kip}
\end{aligned}$$

Thus

$$\mathbf{R} = +14.443\,\mathbf{i} - 2.205\,\mathbf{j} + 52.28\,\mathbf{k} \text{ kip}$$

The magnitude $R$ of the resultant is

$$\begin{aligned}
R &= \sqrt{R_x^2 + R_y^2 + R_z^2} \\
&= \sqrt{(+14.443)^2 + (-2.205)^2 + (+52.28)^2} = 54.28 \cong 54.3 \text{ kip} \quad \textbf{Ans.}
\end{aligned}$$

The angles $\theta_x$, $\theta_y$, and $\theta_z$ are obtained from the expressions

$$\theta_x = \cos^{-1}\frac{R_x}{R} = \cos^{-1}\frac{+14.443}{54.28} = 74.6°$$ **Ans.**

$$\theta_y = \cos^{-1}\frac{R_y}{R} = \cos^{-1}\frac{-2.205}{54.28} = 92.3°$$ **Ans.**

$$\theta_z = \cos^{-1}\frac{R_z}{R} = \cos^{-1}\frac{+52.28}{54.28} = 15.60° \blacksquare$$ **Ans.**

# PROBLEMS

### Introductory Problems

**2-47\*** Two flower pots are supported with wires as shown in Fig. P2-47. If $\alpha = 6°$, the tension force $\mathbf{T}_1$ has a magnitude of 13 lb, and the tension force $\mathbf{T}_2$ has a magnitude of 9 lb, determine the magnitude and orientation of the resultant of the forces $\mathbf{T}_1$ and $\mathbf{T}_2$.

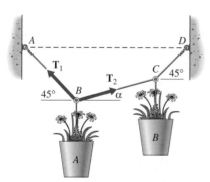

**Figure P2-47**

**2-48\*** Two forces are applied to an eye bolt as shown in Fig. P2-48. Determine the magnitude of the resultant $\mathbf{R}$ of the two forces and the angle $\theta_x$ between the line of action of the resultant and the $x$-axis.

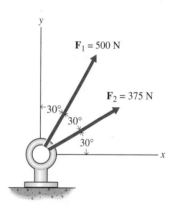

**Figure P2-48**

**2-49** An automobile stuck in a muddy field is being moved by using a cable fastened to a tree as shown in Fig. P2-49. The forces $\mathbf{T}_1$ and $\mathbf{T}_2$ in the two segments of the cable each have a magnitude of 650 lb. If the resultant of the three forces $\mathbf{T}_1$, $\mathbf{T}_2$, and $\mathbf{P}$ is to be zero, determine the magnitude of $\mathbf{P}$.

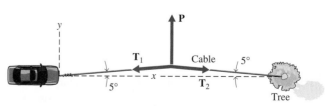

**Figure P2-49**

**2-50** Three forces act on the structural member shown in Fig. P2-50. Determine the resultant of the forces and express the result in Cartesian vector form.

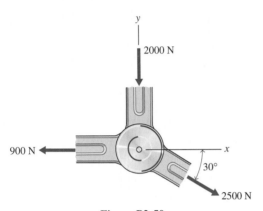

**Figure P2-50**

**Intermediate Problems**

**2-51*** Express the resultant of the two forces shown in Fig. P2-51 in Cartesian vector form, and determine the angles $\theta_x$, $\theta_y$, and $\theta_z$ between the line of action of the resultant and the positive coordinate axes.

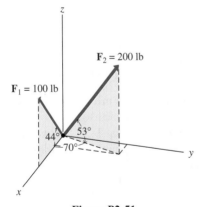

**Figure P2-51**

**2-52*** Determine the magnitude $R$ of the resultant of the two forces shown in Fig. P2-52. Also determine the angles $\theta_x$, $\theta_y$, and $\theta_z$ between the line of action of the resultant and the positive coordinate axes.

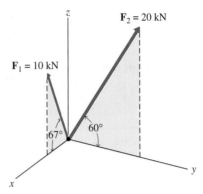

**Figure P2-52**

**2-53** Four forces are applied to the block shown in Fig. P2-53. Determine the magnitude of the resultant and the angle between the resultant and the x-axis.

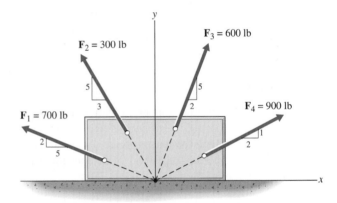

**Figure P2-53**

**2-54** Determine the magnitude and orientation of the resultant of the two forces applied to the eye bolt shown in Fig. P2-54.

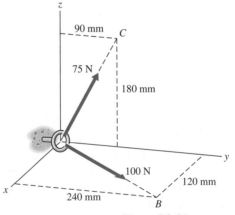

**Figure P2-54**

### Challenging Problems

**2-55\*** The three forces acting on the machine component shown in Fig. P2-55 have magnitudes $F_1 = 500$ lb, $F_2 = 300$ lb, and $F_3 = 200$ lb. Determine the resultant **R** of the three forces and express the resultant in Cartesian vector form. Also, determine the angles $\theta_x$, $\theta_y$, and $\theta_z$ between the line of action of the resultant and the positive coordinate axes.

**2-57** Three forces are applied at the corner of the box shown in Fig. P2-57. Determine the magnitude $R$ of the resultant and the angles $\theta_x$, $\theta_y$, and $\theta_z$ between the line of action of the resultant and the positive coordinate axes.

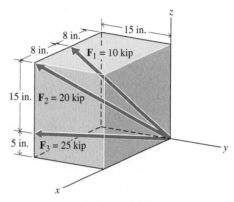

**Figure P2-57**

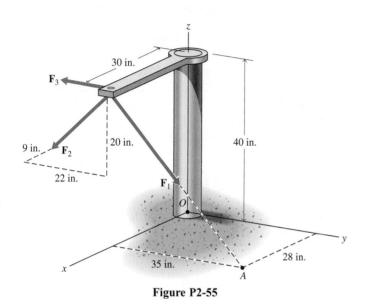

**Figure P2-55**

**2-56\*** Determine the magnitude $R$ of the resultant of the three forces shown in Fig. P2-56. Also determine the angles $\theta_x$, $\theta_y$, and $\theta_z$ between the line of action of the resultant and the positive coordinate axes.

**2-58** The pin $A$ shown in Fig. P2-58 supports a load **F** of magnitude 1250 N, and is held in place by a wire $AD$ and compression members $AB$ and $AC$. If the magnitudes of the forces $C_B$ and $C_C$ in the compression members are 995 N and 700 N, respectively, and the resultant of the four forces is zero, determine the magnitude of the force $T_D$.

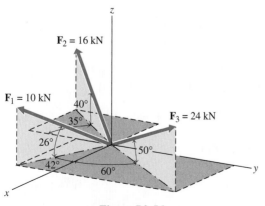

**Figure P2-56**

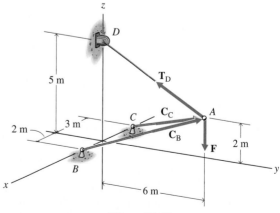

**Figure P2-58**

## 2-7 SUMMARY

A force is defined as the action of one physical body upon another. Since the interaction can occur when the bodies are in contact or when the bodies are physically separated, forces are classified as either surface forces (a push or a pull produced by mechanical means) or body forces (the gravitational pull of the earth). The characteristics of a force are its magnitude, its direction (line of action and sense), and its point of application. Since both a magnitude and a direction are needed to characterize a force and forces add according to the parallelogram law of addition, forces are vector quantities. A number of forces treated as a group constitute a force system.

The components of a force are any system of forces that can be combined by the parallelogram law to produce the original force. For most engineering problems, rectangular (mutually perpendicular) components of a force are more useful than general oblique components. The three rectangular components of a force $\mathbf{F}$ in space are $\mathbf{F}_x$, $\mathbf{F}_y$, and $\mathbf{F}_z$ along the $x$-, $y$-, and $z$-coordinate axes, respectively. The force $\mathbf{F}$ and its scalar components $F_x$, $F_y$, and $F_z$ can be written in Cartesian vector form by using unit vectors $\mathbf{i}$, $\mathbf{j}$, and $\mathbf{k}$ directed along the positive $x$-, $y$-, and $z$-coordinate axes as

$$\mathbf{F} = \mathbf{F}_x + \mathbf{F}_y + \mathbf{F}_z = F_x\,\mathbf{i} + F_y\,\mathbf{j} + F_z\,\mathbf{k} \tag{2-10}$$

where the scalar components $F_x$, $F_y$, and $F_z$ are related to the magnitude $F$ and the direction $\theta$ of the force $\mathbf{F}$ by the expressions

$$F_x = F\cos\theta_x \qquad F_y = F\cos\theta_y \qquad F_z = F\cos\theta_z \tag{2-11}$$

$$F = \sqrt{F_x^2 + F_y^2 + F_z^2} \tag{2-12}$$

$$\theta_x = \cos^{-1}\frac{F_x}{F} \qquad \theta_y = \cos^{-1}\frac{F_y}{F} \qquad \theta_z = \cos^{-1}\frac{F_z}{F} \tag{2-13}$$

and

$$\cos^2\theta_x + \cos^2\theta_y + \cos^2\theta_z = 1 \tag{2-15}$$

The rectangular component of a force $\mathbf{F}$ along an arbitrary direction $n$ can be obtained by using the vector dot product. Thus, if $\mathbf{e}_n$ is a unit vector in the specified direction $n$, the magnitude of the rectangular component $\mathbf{F}_n$ of the force $\mathbf{F}$ is

$$F_n = \mathbf{F} \cdot \mathbf{e}_n = (F_x\,\mathbf{i} + F_y\,\mathbf{j} + F_z\,\mathbf{k}) \cdot \mathbf{e}_n$$

If the angles between the direction $n$ and the $x$-, $y$-, and $z$-axes are $\theta_x$, $\theta_y$, and $\theta_z$, the unit vector $\mathbf{e}_n$ can be written in Cartesian vector form as

$$\mathbf{e}_n = \cos\theta_x\,\mathbf{i} + \cos\theta_y\,\mathbf{j} + \cos\theta_z\,\mathbf{k} \tag{2-14}$$

Thus, the magnitude of the force $\mathbf{F}_n$ is

$$F_n = \mathbf{F} \cdot \mathbf{e}_n = F_x\cos\theta_x + F_y\cos\theta_y + F_z\cos\theta_z \tag{2-20}$$

The rectangular component $\mathbf{F}_n$ of the force $\mathbf{F}$ is expressed in Cartesian vector form as

$$\mathbf{F}_n = (\mathbf{F} \cdot \mathbf{e}_n)\, \mathbf{e}_n = F_n\, \mathbf{e}_n = F_n(\cos\theta_x\, \mathbf{i} + \cos\theta_y\, \mathbf{j} + \cos\theta_z\, \mathbf{k})$$

The angle $\alpha$ between the line of action of the force $\mathbf{F}$ and the direction $n$ is determined by using the definition of a rectangular component of a force ($F_n = F\cos\alpha = \mathbf{F} \cdot \mathbf{e}_n$). Thus

$$\alpha = \cos^{-1}\frac{\mathbf{F} \cdot \mathbf{e}_n}{F} = \cos^{-1}\frac{F_n}{F} \tag{2-21}$$

A single force, called the resultant $\mathbf{R}$, will produce the same effect on a body as a system of concurrent forces. The resultant can be determined by adding the forces using the parallelogram law; however, this procedure is time-consuming and tedious when the system comprises more than two forces. Resultants are easily obtained, however, by using rectangular components of the forces. For the general case of two or more concurrent forces in space,

$$\mathbf{R}_x = \Sigma\mathbf{F}_x = R_x\, \mathbf{i} \qquad \mathbf{R}_y = \Sigma\mathbf{F}_y = R_y\, \mathbf{j} \qquad \mathbf{R}_z = \Sigma\mathbf{F}_z = R_z\, \mathbf{k} \tag{2-27}$$

The magnitude $R$ of the resultant and the angles $\theta_x$, $\theta_y$, and $\theta_z$ between the line of action of the resultant and the positive coordinate axes are

$$R = \sqrt{R_x^2 + R_y^2 + R_z^2} \tag{2-29}$$

$$\theta_x = \cos^{-1}\frac{R_x}{R} \qquad \theta_y = \cos^{-1}\frac{R_y}{R} \qquad \theta_z = \cos^{-1}\frac{R_z}{R} \tag{2-30}$$

## REVIEW PROBLEMS

**2-59\*** Three forces are applied to an eye bolt as shown in Fig. P2-59. Determine the magnitude $R$ of the resultant of the forces and the angle $\theta_x$ between the line of action of the resultant and the $x$-axis.

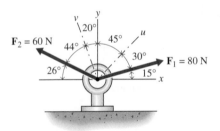

**Figure P2-60**

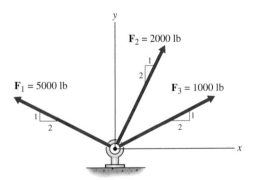

**Figure P2-59**

**2-60\*** Two forces, $\mathbf{F}_1$ and $\mathbf{F}_2$, are applied to an eye bolt as shown in Fig. P2-60. Determine
  (a) The magnitude and direction (angle $\theta_x$) of the resultant $\mathbf{R}$ of the two forces.
  (b) The magnitudes of two other forces $\mathbf{F}_u$ and $\mathbf{F}_v$ (along the axes $u$ and $v$) that would have the same resultant.

**2-61** Three forces are applied at a point on a body as shown in Fig. P2-61. Determine the resultant $\mathbf{R}$ of the three forces and the angles $\theta_x$, $\theta_y$, and $\theta_z$ between the line of action of the resultant and the positive $x$-, $y$-, and $z$-coordinate axes.

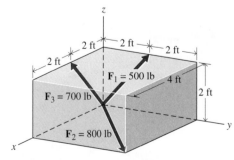

**Figure P2-61**

**2-62*** Three forces are applied at a point on a body as shown in Fig. P2-62. Determine the resultant **R** of the three forces and the angles $\theta_x$, $\theta_y$, and $\theta_z$ between the line of action of the resultant and the positive x-, y-, and z-coordinate axes.

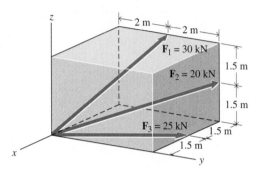

**Figure P2-62**

**2-63** Three forces are applied to a stalled automobile as shown in Fig. P2-63. Determine the magnitude of the force $\mathbf{F}_3$ and the magnitude of the resultant **R** if the line of action of the resultant is along the x-axis.

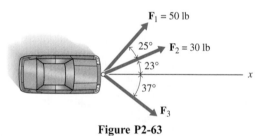

**Figure P2-63**

**2-64** Three cables are used to drag a heavy crate along a horizontal surface as shown in Fig. P2-64. The resultant **R** of the forces has a magnitude of 2800 N and its line of action is directed along the x-axis. Determine the magnitudes of the forces $\mathbf{F}_1$ and $\mathbf{F}_3$.

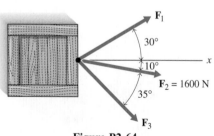

**Figure P2-64**

**2-65*** Two forces are applied at a point in a body as shown in Fig. P2-65. Determine
(a) The magnitude and direction (angles $\theta_x$, $\theta_y$, and $\theta_z$) of the resultant **R** of the two forces.
(b) The magnitude of the rectangular component of the force $\mathbf{F}_1$ along the line of action of the force $\mathbf{F}_2$.
(c) The angle $\alpha$ between forces $\mathbf{F}_1$ and $\mathbf{F}_2$.

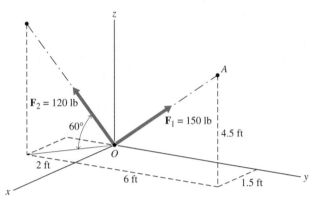

**Figure P2-65**

**2-66** Three forces are applied with cables to the anchor block shown in Fig. P2-66. Determine
(a) The magnitude and direction (angles $\theta_x$, $\theta_y$, and $\theta_z$) of the resultant **R** of the three forces.
(b) The magnitude of the rectangular component of the force $\mathbf{F}_1$ along the line of action of the force $\mathbf{F}_2$.
(c) The angle $\alpha$ between forces $\mathbf{F}_1$ and $\mathbf{F}_3$.

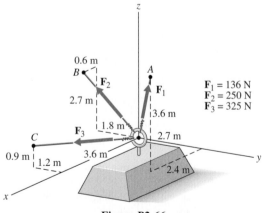

**Figure P2-66**

# EQUILIBRIUM: CONCURRENT FORCE SYSTEMS

# 3

## 3-1 INTRODUCTION

Statics was defined in Chapter 1 as the branch of rigid-body mechanics concerned with bodies that are acted upon by a balanced system of forces (the resultant of all forces acting on the body is zero) and hence are at rest or moving with a constant velocity in a straight line.

A body with negligible dimensions is commonly referred to as a *particle*. In mechanics, either large bodies or small bodies can be referred to as particles when the size and shape of the body have no effect on the response of the body to a system of forces. Under these conditions, the mass of the body can be assumed to be concentrated at a point. For example, the earth can be modeled as a particle for orbital motion studies because the size of the earth is insignificant when compared with the size of its orbit and the shape of the earth does not influence the description of its position or the action of forces applied to it.

Since it is assumed that the mass of a particle is concentrated at a point and that the size and shape of a particle can be neglected, a particle can be subjected only to a system of concurrent forces. Newton's first law of motion states that "in the absence of external forces ($\mathbf{R} = \mathbf{0}$), a particle originally at rest or moving with a constant velocity will remain at rest or continue to move with a constant velocity along a straight line." Thus, a necessary condition for equilibrium of a particle is

$$\mathbf{R} = \Sigma\mathbf{F} = \mathbf{0} \tag{3-1}$$

A particle in equilibrium must also satisfy Newton's second law of motion, which can be expressed in equation form [Eq. (1-1)] as

$$\mathbf{R} = \Sigma\mathbf{F} = m\mathbf{a} \tag{1-1}$$

In order to satisfy both Eqs. (1-1) and (3-1),

$$m\mathbf{a} = \mathbf{0}$$

Since the mass of the particle is not zero, the acceleration of a particle in equilibrium is zero ($\mathbf{a} = \mathbf{0}$). Thus, a particle initially at rest will remain at rest and a particle moving with a constant velocity will maintain that velocity. Therefore, Eq. (3-1) is both a necessary condition and a sufficient condition for equilibrium.

The particle assumption is valid for many practical applications and thus provides a means for introducing the student to some interesting engineering problems early in a statics course. For this reason, this short chapter on statics (equilibrium) of a particle has been introduced before consideration of the more difficult problems associated with equilibrium of a rigid body (which involves the concepts of moments and distributed loads).

The force system acting on a body in a typical statics problem consists of known forces and unknown forces. Both must be clearly identified before a solution to a specific problem is attempted. A method commonly used to identify all forces acting on a body in a given situation is described in the following section.

## 3-2 FREE-BODY DIAGRAMS

Problems in engineering mechanics (statics and dynamics) are concerned with the external effects produced when a system of forces acts on a physical body. The external effects typically include motion of the body in dynamics problems or development of support reactions to resist motion of the body in statics problems. Solving an engineering mechanics problem usually requires identification of all external forces acting on a "body of interest." A carefully prepared drawing or sketch that shows a "body of interest" separated from all interacting bodies is known as a *free-body diagram* (FBD). Once the body of interest is selected, the forces exerted by all other bodies on the one being considered must be determined and shown on the diagram. It is important that "all" forces acting "on" the body of interest be shown. Recall also that "a force cannot exist unless there is a body to exert the force." Frequently, the student will overlook and omit a force from the free-body diagram or show a force on the free-body diagram when there is no body present to exert the force.

The actual procedure for drawing a free-body diagram consists of two essential steps:

1. Make a decision regarding what body is to be isolated and analyzed. Prepare a sketch of the external boundary of the body selected.
2. Represent all forces, known and unknown, that are applied by other bodies to the isolated body with vectors in their correct positions.

The number of forces on a free-body diagram is determined by noting the number of bodies that exert forces on the body of interest. These forces may be either forces of contact or body forces. An important body force is the earth-pull on (or weight of) a body.

Each known force should be shown on a free-body diagram with its correct magnitude, slope, and sense. Letter symbols are used for the magnitudes of unknown forces. If a force has a known line of action but an unknown magnitude and sense, the sense of the force can be assumed. The correct sense will become apparent after solving for the unknown magnitude. By definition, the magnitude of a force is always positive; therefore, if the solution yields a negative magnitude, the minus sign indicates that the sense of the force is opposite to that assumed on the free-body diagram.

If both the magnitude and direction of a force acting on the body of interest are unknown (such as a pin reaction in a pin-connected structure), it is frequently convenient to show the rectangular components of the force on the free-body diagram instead of the actual force. In this way, one deals with two or three forces of unknown magnitude but known direction. After the rectangular components of the force are determined, the magnitude and direction of the actual force can easily be found. However, do not show both the force of unknown magnitude and its rectangular components on the same diagram.

The word *free* in the name *free-body diagram* emphasizes the idea that all bodies exerting forces on the body of interest are removed or withdrawn and are replaced by the forces they exert. Do not show both the bodies removed and the forces exerted by them on the free-body diagram. Sometimes it may be convenient to indicate, by lightweight dotted lines, the faint outlines of the bodies removed, in order to visualize the geometry and specify dimensions required for solution of the problem.

In drawing a free-body diagram of a given body, certain assumptions are made regarding the nature of the forces (reactions) exerted by other bodies on the body of interest. Two common assumptions are the following:

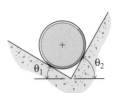

Cylinder supported by smooth surfaces

1. If a surface of contact at which a force is applied by one body to another body has only a small degree of roughness, it may be assumed to be smooth (frictionless), and hence, the action (or reaction) of the one body on the other is directed normal to the surface of contact as shown in Fig. 3-1.

2. A body that possesses only a small degree of bending stiffness (resistance to bending), such as a cord, rope, or chain, may be considered to be perfectly flexible, and hence the pull of such a body on any other body is directed along the axis of the flexible body as shown in Fig. 3-2.

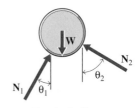

**Figure 3-1**   Free-body diagram

A large number of additional assumptions will be discussed in Section 6-2 in the chapter on equilibrium of rigid bodies, where the topic of idealization of supports and connections is considered.

The term *body of interest* used in the definition of a free-body diagram may mean any definite part of a structure or machine, such as an eye bar in a bridge truss or a connecting rod in an automobile engine. The body of interest may also be taken as a group of physical bodies joined together (considered as one body), such as an entire bridge or a complete engine.

The importance of drawing a free-body diagram before attempting to solve a mechanics problem cannot be overemphasized. A procedure that can be followed to construct a complete and correct free-body diagram contains the four steps shown at the top of the next page.

Application of these four steps to any mechanics problem should produce a complete and correct free-body diagram, which is an essential first step for the solution of any problem.

The free-body diagram is the "road map" for writing the equations of equilibrium. Every equation of equilibrium must be supported by a properly drawn, complete, free-body diagram. The symbols used in the equations of equilibrium must match the symbols used on the free-body diagram. For example, use $A \cos 30°$ rather than $A_x$ if $A$ is used on the free-body diagram to represent a force of known direction (30° with respect to the $x$ axis).

*Most engineers consider a free-body diagram to be the single most important tool for the solution of mechanics problems.* The free-body diagram clearly identifies the body to which the principles of mechanics are to be applied. It also clearly identifies all forces that act on the body and where they are applied. The use of a complete and proper free-body diagram can reduce many of the errors commonly made in solving mechanics problems.

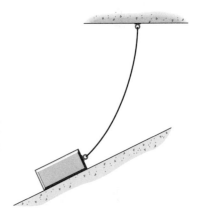

Block held on a smooth inclined surface with a flexible cable

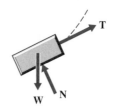

Free-body diagram

**Figure 3-2**

---

## CONSTRUCTING A FREE-BODY DIAGRAM

### Step 1

**Procedure:** Decide which body or combination of bodies is to be shown on the free-body diagram.

**Method:** The body must be separated from its surroundings at points where forces are to be determined. One half of the equal-opposite force pairs will be retained on the free-body diagram and will appear in the equations of equilibrium; the other half of the equal-opposite force pairs will be discarded with the surroundings.

### Step 2

**Procedure:** Prepare a drawing or sketch of the outline of this isolated or free body.

**Method:** Redraw the body including only those parts retained—do not show those parts that were discarded. Internal details are not important and can be omitted.

### Step 3

**Procedure:** Add forces to the sketch.

**Method:** Carefully trace around the boundary of the free body and identify all of the forces exerted by contacting or attracting bodies that were removed during the isolation process.

### Step 4

**Procedure:** Choose/show coordinate axes and dimensions on the free-body diagram.

**Method:** If coordinate axes are omitted, the x-axis will be assumed horizontal with positive to the right and the y-axis will be assumed vertical with positive up. Any other choice must be shown on the free-body diagram. It is often helpful to include on the diagram any dimensions required for solution of the problem.

---

# 3-3 EQUILIBRIUM OF A PARTICLE

In section 3-1 it was noted that the term *particle* is used in statics to describe a body when the size and shape of the body will not significantly affect the solution of the problem being considered and when the mass of the body can be assumed to be concentrated at a point. As a result, a particle can be subjected only to a system of concurrent forces, and the necessary and sufficient conditions for equilibrium can be expressed mathematically as

$$\mathbf{R} = \Sigma \mathbf{F} = \mathbf{0} \tag{3-1}$$

where $\Sigma \mathbf{F}$ is the vector sum of all forces acting on the particle.

## Two-Dimensional Problems

For a system of coplanar (say the $xy$-plane), concurrent forces, Eq. (3-1) can be written as

$$\mathbf{R} = \mathbf{R}_x + \mathbf{R}_y = \mathbf{R}_n + \mathbf{R}_t$$
$$= R_x\,\mathbf{i} + R_y\,\mathbf{j} = R_n\mathbf{e}_n + R_t\mathbf{e}_t$$
$$= \Sigma F_x\,\mathbf{i} + \Sigma F_y\,\mathbf{j} = \Sigma F_n\mathbf{e}_n + \Sigma F_t\mathbf{e}_t = \mathbf{0} \qquad (3\text{-}2)$$

Equation (3-2) is satisfied only if

$$\mathbf{R}_x = R_x\,\mathbf{i} = \Sigma F_x\,\mathbf{i} = \mathbf{0}$$
$$\mathbf{R}_y = R_y\,\mathbf{j} = \Sigma F_y\,\mathbf{j} = \mathbf{0}$$
or
$$\mathbf{R}_n = R_n\mathbf{e}_n = \Sigma F_n\mathbf{e}_n = \mathbf{0}$$
$$\mathbf{R}_t = R_t\mathbf{e}_t = \Sigma F_t\mathbf{e}_t = \mathbf{0}$$

In scalar form, these equations become

$$R_x = \Sigma F_x = 0$$
$$R_y = \Sigma F_y = 0$$
or
$$R_n = \Sigma F_n = 0$$
$$R_t = \Sigma F_t = 0$$
$$(3\text{-}3)$$

That is, the sum of the rectangular components of the forces in any direction must be zero. While this would appear to give an infinite number of equations, no more than two of the equations are independent. The remaining equations can be obtained from combinations of the two independent equations. It is sometimes convenient to use $\Sigma F_x = 0$ and $\Sigma F_n = 0$ as the two independent equations rather than $\Sigma F_x = 0$ and $\Sigma F_y = 0$ (see Examples 3-1 and 3-2). Equations (3-3) can be used to determine two unknown quantities (two magnitudes, two slopes, or a magnitude and a slope).

## Three-Dimensional Problems

For a three-dimensional system of concurrent forces, Eq. (3-1) can be written as

$$\mathbf{R} = \Sigma \mathbf{F} = \mathbf{R}_x + \mathbf{R}_y + \mathbf{R}_z$$
$$= R_x\,\mathbf{i} + R_y\,\mathbf{j} + R_z\,\mathbf{k}$$
$$= \Sigma F_x\,\mathbf{i} + \Sigma F_y\,\mathbf{j} + \Sigma F_z\,\mathbf{k} = \mathbf{0} \qquad (3\text{-}4)$$

Equation (3-4) is satisfied only if

$$\mathbf{R}_x = R_x\,\mathbf{i} = \Sigma F_x\,\mathbf{i} = \mathbf{0}$$
$$\mathbf{R}_y = R_y\,\mathbf{j} = \Sigma F_y\,\mathbf{j} = \mathbf{0} \qquad (3\text{-}5)$$
$$\mathbf{R}_z = R_z\,\mathbf{k} = \Sigma F_z\,\mathbf{k} = \mathbf{0}$$

In scalar form, these equations become

$$R_x = \Sigma F_x = 0$$
$$R_y = \Sigma F_y = 0 \qquad (3\text{-}6)$$
$$R_z = \Sigma F_z = 0$$

Equations (3-5) and (3-6) can be used to determine three unknown quantities (three magnitudes, three slopes, or any combination of three magnitudes and slopes). The procedure is illustrated in Examples 3-3 and 3-4. Example 3-3 illustrates the scalar method of solution for a three-dimensional problem. Example 3-4 illustrates the vector method of solution for a similar problem.

## Example Problem 3-1

A free-body diagram of a particle subjected to the action of four forces is shown in Fig. 3-3a. Determine the magnitudes of forces $\mathbf{F}_1$ and $\mathbf{F}_2$ so that the particle is in equilibrium.

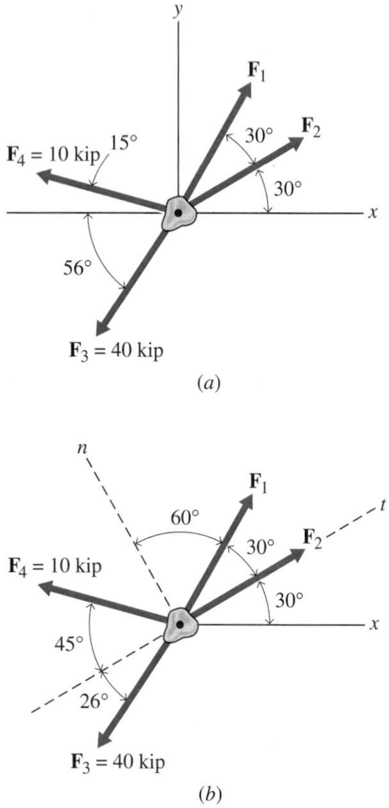

### SOLUTION (using x- and y-components of equilibrium)

In this example problem the particle has already been isolated from its surroundings. Figure 3-3a is the free-body diagram, and it shows all forces acting on the particle by the external world. The particle is subjected to a system of coplanar, concurrent forces, two of which are known ($\mathbf{F}_3$ and $\mathbf{F}_4$) and two of which are unknown ($\mathbf{F}_1$ and $\mathbf{F}_2$). For a concurrent system of forces in a plane, the vector equation of equilibrium [Eq. (3-1)] has only two independent scalar components. Writing the x-component of the equilibrium equation for the free-body diagram of Fig. 3-3a results in

$$+ \rightarrow \Sigma F_x = 0: \qquad F_{1x} + F_{2x} + F_{3x} + F_{4x} = 0$$

$$F_1 \cos 60° + F_2 \cos 30° - 40 \cos 56° - 10 \cos 15° = 0$$

$$0.5000 F_1 + 0.8660 F_2 - 22.368 - 9.659 = 0$$

from which

$$F_1 + 1.732 F_2 = 64.054 \qquad \text{(a)}$$

Writing the y-component of the equilibrium equation results in

$$+ \uparrow \Sigma F_y = 0: \qquad F_{1y} + F_{2y} + F_{3y} + F_{4y} = 0$$

$$F_1 \sin 60° + F_2 \sin 30° - 40 \sin 56° + 10 \sin 15° = 0$$

$$0.8660 F_1 + 0.5000 F_2 - 33.161 + 2.588 = 0$$

from which

$$F_1 + 0.5774 F_2 = 35.304 \qquad \text{(b)}$$

Solving Eqs. (a) and (b) simultaneously yields

$$F_1 = 20.9 \text{ kip} \qquad \textbf{Ans.}$$

$$F_2 = 24.9 \text{ kip} \qquad \textbf{Ans.}$$

### SOLUTION (using n- and x-components of equilibrium)

Using the x- and y-components of the equilibrium equation [Eq. (3-1)] resulted in a pair of equations that had to be solved simultaneously. An alternate choice of a coordinate system would be along and perpendicular to one of the unknown

(a)

(b)

**Figure 3-3**

forces (such as the *nt*-coordinate system of Fig. 3-3*b*). Since the force $\mathbf{F}_2$ has no *n*-component, it does not enter into the equation $\Sigma F_n = 0$, and the difficulty of solving simultaneous equations can be avoided. Writing the *n*-component of the equilibrium equation results in

$$+ \nwarrow \Sigma F_n = 0: \qquad F_{1n} + F_{2n} + F_{3n} + F_{4n} = 0$$

$$F_1 \sin 30° + 0 - 40 \sin 26° + 10 \sin 45° = 0$$

from which

$$F_1 = 20.93 \cong 20.9 \text{ kip} \qquad \textbf{Ans.}$$

Once force $\mathbf{F}_1$ is known, any other component of the equilibrium equation can be used to obtain the force $\mathbf{F}_2$. For example, writing the *x*-component of the equilibrium equation results in

$$+ \rightarrow \Sigma F_x = 0: \qquad F_{1x} + F_{2x} + F_{3x} + F_{4x} = 0$$

$$20.93 \cos 60° + F_2 \cos 30° - 40 \cos 56° - 10 \cos 15° = 0$$

from which

$$F_2 = 24.9 \text{ kip} \qquad \textbf{Ans.}$$

Summing forces in a direction perpendicular to one of the unknown forces eliminates the need to solve simultaneous equations in two-dimensional problems. Depending on the complexity of the problem and the complexity of the geometry, it may be more appropriate to solve the simultaneous equations. ■

## Example Problem 3-2

Two flexible and inextensible cables support a 220-lb traffic light as shown in Fig. 3-4*a*. Determine the tension in each of the cables.

### SOLUTION (using *x*- and *y*-components of equilibrium)

A flexible cable always exerts a tensile force with a line of action along the axis of the cable. Furthermore, if the weight of the cable may be neglected, the tensile force is the same at each and every location along the cable. Since we are trying to find the forces in the cables, we must "cut" the cables at some point along their length to break the "equal/opposite" force pairs in the cable. The stoplight is the one object that can be isolated which is acted on by both of the unknown tension forces and by the known weight force. The free-body diagram of the stoplight is drawn in Fig. 3-4*b*.

The tension forces in the cables and the weight of the traffic light are concurrent at the ring on the top of the traffic light. For a concurrent system of forces in a plane, the vector equation of equilibrium [Eq. (3-1)] has only two independent scalar components. Writing the *x*- and *y*-components of the equilibrium equation for the free-body diagram of Fig. 3-4*b* results in

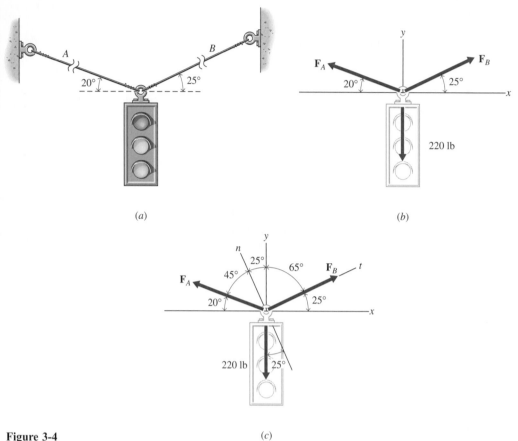

Figure 3-4

(a)

(b)

(c)

$$+ \rightarrow \Sigma F_x = 0: \qquad F_B \cos 25° - F_A \cos 20° = 0$$

$$F_B = 1.0368 F_A \qquad (a)$$

$$+ \uparrow \Sigma F_y = 0: \qquad F_B \sin 25° + F_A \sin 20° - 220 = 0 \qquad (b)$$

Combining Eqs. (a) and (b) gives

$$1.0368 F_A \sin 25° + F_A \sin 20° = 220 \qquad (c)$$

Solving Eq. (c) yields

$$F_A = 281.98 \cong 282 \text{ lb} \qquad \textbf{Ans.}$$

$$F_B = 292.36 \cong 292 \text{ lb} \qquad \textbf{Ans.}$$

## SOLUTION (using *n*- and *x*-components of equilibrium)

The solution of simultaneous equations [Eqs. (a) and (b)] can be avoided by using a coordinate system along and perpendicular to one of the unknown forces (such as the *nt*-coordinate system of Fig. 3-4c). Since the force $\mathbf{F}_B$ has no *n*-component, it does not enter into the equation $\Sigma F_n = 0$ and the difficulty of solving simultaneous equations can be avoided. Writing the $n$-component of the equilibrium equation results in

$$+ \nwarrow \Sigma F_n = 0: \qquad F_A \cos 45° - 220 \cos 25° = 0$$

from which

$$F_A = 281.98 \cong 282 \text{ lb} \qquad \qquad \textbf{Ans.}$$

Once force $\mathbf{F}_A$ is known, any other component of the equilibrium equation can be used to obtain the force $\mathbf{F}_B$. For example, writing the $x$-component of the equilibrium equation results in

$$+ \rightarrow \Sigma F_x = 0: \qquad F_B \cos 25° - 281.98 \cos 20° = 0 \qquad \qquad \text{(a)}$$

from which

$$F_B = 292.36 \cong 292 \text{ lb} \qquad \qquad \textbf{Ans.}$$

Again, summing forces in a direction perpendicular to one of the unknown forces eliminates the need to solve simultaneous equations in two-dimensional problems. Depending on the complexity of the problem and the complexity of the geometry, it may be more appropriate to solve the simultaneous equations. ∎

## Example Problem 3-3

A free-body diagram of a particle subjected to the action of four forces is shown in Fig 3-5. Determine the magnitude and the coordinate direction angles of the unknown force $\mathbf{F}_4$ so that the particle is in equilibrium.

### SOLUTION

In this example problem the particle has already been isolated from its surroundings. Figure 3-5 is the free-body diagram, and it shows all forces acting on the particle by the external world. The particle is subjected to a three-dimensional system of concurrent forces, three of which are known ($\mathbf{F}_1$, $\mathbf{F}_2$, and $\mathbf{F}_3$) and one of which is unknown ($\mathbf{F}_4$). For a concurrent system of forces in three dimensions, the vector equation of equilibrium [Eq. (3-1)] has three independent scalar components. Before we can write the $x$-, $y$-, and $z$-components of the equilibrium equation, we must first calculate the $x$-, $y$-, and $z$-components of each of the known forces on the free-body diagram. For $\mathbf{F}_1$:

$$F_{1x} = 0 \qquad F_{1y} = 0 \qquad F_{1z} = -200 \text{ lb}$$

For $\mathbf{F}_2$:

$$F_{2x} = \frac{1}{\sqrt{1^2+3^2}} (250) = 79.06 \text{ lb}$$

$$F_{2y} = \frac{3}{\sqrt{1^2+3^2}} (250) = 237.17 \text{ lb}$$

$$F_{2z} = 0$$

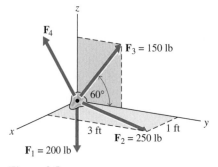

**Figure 3-5**

For $\mathbf{F}_3$:

$$F_{3x} = 0$$

$$F_{3y} = 150 \cos 60° = 75.00 \text{ lb}$$

$$F_{3z} = 150 \sin 60° = 129.90 \text{ lb}$$

Now, writing the $x$-, $y$-, and $z$-components of the equilibrium equation for the free-body diagram of Fig. 3-5 results in

$+ \swarrow \Sigma F_x = 0:$          $0 + 79.06 + 0 + F_{4x} = 0$             (a)

$+ \rightarrow \Sigma F_y = 0:$          $0 + 237.17 + 75.00 + F_{4y} = 0$      (b)

$+ \uparrow \Sigma F_z = 0:$          $-200 + 0 + 129.90 + F_{4z} = 0$      (c)

Equations (a), (b), and (c) can be solved sequentially to get

$$F_{4x} = -79.06 \text{ lb} \qquad F_{4y} = -312.17 \text{ lb} \qquad F_{4z} = 70.10 \text{ lb}$$

Once the rectangular components of the force $\mathbf{F}_4$ are known, Eqs. (2-12) and (2-13) can be used to determine its magnitude and coordinate direction angles. Thus

$$F_4 = \sqrt{F_{4x}^2 + F_{4y}^2 + F_{4z}^2} = \sqrt{(-79.06)^2 + (-312.17)^2 + (70.10)^2}$$

$$= 329.57 \cong 330 \text{ lb} \quad \textbf{Ans.}$$

$$\theta_x = \cos^{-1}\left(\frac{F_{4x}}{F_4}\right) = \cos^{-1}\left(\frac{-79.06}{329.57}\right) = 103.88° \quad \textbf{Ans.}$$

$$\theta_y = \cos^{-1}\left(\frac{F_{4y}}{F_4}\right) = \cos^{-1}\left(\frac{-312.17}{329.57}\right) = 161.30° \quad \textbf{Ans.}$$

$$\theta_z = \cos^{-1}\left(\frac{F_{4z}}{F_4}\right) = \cos^{-1}\left(\frac{70.10}{329.57}\right) = 77.72° \quad \textbf{Ans.}$$

## ■ Example Problem 3-4

A 500-N block is supported by a system of cables as shown in Fig. 3-6a. Determine the tensions in cables $A$, $B$, and $C$.

### SOLUTION

Since we are trying to find the forces in the cables, we must "cut" the cables at some point along their length to break the "equal/opposite" force pairs in the cable. The ring $D$ is acted on by all three of the unknown tension forces and by the known weight force. The free-body diagram of the ring is drawn in Fig. 3-6b. The coordinates of the support points for each of the cables are shown on the free-body diagram in $(x, y, z)$ format as an aid in writing vector equations for the cable tensions.

    The tension forces in the cables and the weight of the traffic light are concurrent at the ring $D$. For a concurrent system of forces in three dimensions,

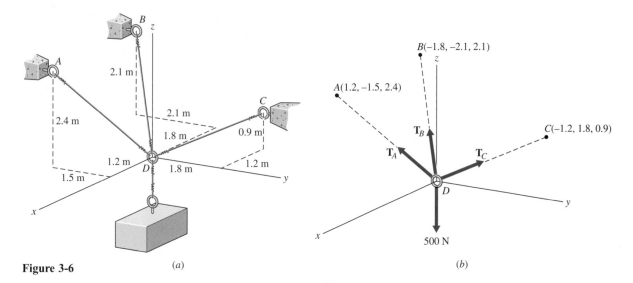

**Figure 3-6**

(a)

(b)

the vector equation of equilibrium [Eq. (3-1)] has three independent scalar components. Before writing the equilibrium equation, it is necessary to express the cable tensions and the weight of the block in Cartesian vector form. The tension $\mathbf{T}_A$ can be expressed in terms of its unknown magnitude $T_A$ and a unit vector $\mathbf{e}_{AD}$ in the direction of the line $AD$ as

$$\mathbf{T}_A = T_A \mathbf{e}_{AD} = T_A \left( \frac{1.2\,\mathbf{i} - 1.5\,\mathbf{j} + 2.4\,\mathbf{k}}{\sqrt{1.2^2 + 1.5^2 + 2.4^2}} \right)$$

$$= 0.3904 T_A\,\mathbf{i} - 0.4880 T_A\,\mathbf{j} + 0.7807 T_A\,\mathbf{k} \qquad (a)$$

Similarly for the tensions $\mathbf{T}_B$ and $\mathbf{T}_C$ and for the weight $\mathbf{W}$

$$\mathbf{T}_B = T_B \mathbf{e}_{BD} = T_B \left( \frac{-1.8\,\mathbf{i} - 2.1\,\mathbf{j} + 2.1\,\mathbf{k}}{\sqrt{1.8^2 + 2.1^2 + 2.1^2}} \right)$$

$$= -0.5183 T_B\,\mathbf{i} - 0.6047 T_B\,\mathbf{j} + 0.6047 T_B\,\mathbf{k} \qquad (b)$$

$$\mathbf{T}_C = T_C \mathbf{e}_{CD} = T_C \left( \frac{-1.2\,\mathbf{i} + 1.8\,\mathbf{j} + 0.9\,\mathbf{k}}{\sqrt{1.2^2 + 1.8^2 + 0.9^2}} \right)$$

$$= -0.5121 T_C\,\mathbf{i} + 0.7682 T_C\,\mathbf{j} + 0.3841 T_C\,\mathbf{k} \qquad (c)$$

$$\mathbf{W} = -500\,\mathbf{k}\ \text{N} \qquad (d)$$

Using Eqs. (a)–(d) and the free-body diagram (Fig. 3-6$b$) to write out the vector equilibrium equation results in

$$\Sigma \mathbf{F} = \mathbf{0}: \qquad 0.3904 T_A\,\mathbf{i} - 0.4880 T_A\,\mathbf{j} + 0.7807 T_A\,\mathbf{k}$$

$$-0.5183 T_B\,\mathbf{i} - 0.6047 T_B\,\mathbf{j} + 0.6047 T_B\,\mathbf{k}$$

$$-0.5121 T_C\,\mathbf{i} + 0.7682 T_C\,\mathbf{j} + 0.3841 T_C\,\mathbf{k} - 500\,\mathbf{k} = \mathbf{0} \qquad (e)$$

Since each of the components of Eq. (e) must be zero if the resultant is to be zero, the following three scalar equations must be satisfied.

$$0.3904T_A - 0.5183T_B - 0.5121T_C = 0$$
$$-0.4880T_A - 0.6047T_B + 0.7682T_C = 0$$
$$0.7807T_A + 0.6047T_B + 0.3841T_C = 500 \qquad \text{(f)}$$

The simultaneous solution of these three linear equations gives

$$T_A = 459 \text{ N} \qquad \qquad \textbf{Ans.}$$
$$T_B = 32.4 \text{ N} \qquad \qquad \textbf{Ans.}$$
$$T_C = 317 \text{ N} \blacksquare \qquad \qquad \textbf{Ans.}$$

## ▌ Example Problem 3-5

A 1000-lb load is securely fastened to a hoisting cable as shown in Fig. 3-7a. The tension in the flexible cable does not change as it passes around the small frictionless pulley at the right support. The weight of the cable may be neglected. Plot the tensions in the two cables ($T_{AB}$ and $P$) as a function of the sag distance $d$ ($0 \le d \le 10$ ft). Determine the minimum sag $d$ for which $P$ is less than

(a)  Twice the weight of the load.
(b)  Four times the weight of the load.
(c)  Eight times the weight of the load.

### SOLUTION

The ring $B$ holds the wires together, and it will be isolated to generate the free-body diagram shown in Fig. 3-7b. The tension forces in the cables and the weight of the load are concurrent at the ring $B$. For a concurrent system of forces in a plane, the vector equation of equilibrium [Eq. (3-1)] has only two independent scalar components. Writing the x- and y-components of the equilibrium equation for the free-body diagram of Fig. 3-7b results in

$$+ \rightarrow \Sigma F_x = 0: \qquad T_{BC} \cos \theta_C - T_{AB} \cos \theta_A = 0 \qquad \text{(a)}$$
$$+ \uparrow \Sigma F_y = 0: \qquad T_{AB} \sin \theta_A + T_{BC} \sin \theta_C - 1000 = 0 \qquad \text{(b)}$$

Solving Eq. (a) for $T_{AB}$ gives

$$T_{AB} = \frac{T_{BC} \cos \theta_C}{\cos \theta_A} \qquad \text{(c)}$$

and substituting Eq. (c) into Eq. (b) gives

$$T_{BC} \frac{\sin \theta_C \cos \theta_A + \sin \theta_A \cos \theta_C}{\cos \theta_A} = 1000 \qquad \text{(d)}$$

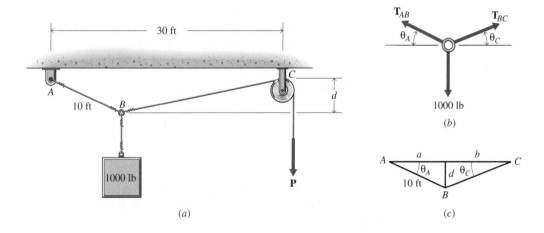

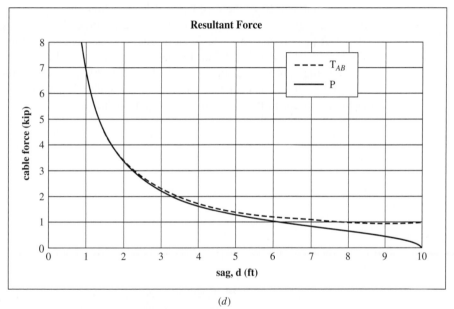

**Figure 3-7**

(d)

Before we can solve Eq. (d) for $T_{BC}$, we need to know how the angles $\theta_C$ and $\theta_A$ are related to the sag distance $d$. From the geometry of the triangles in Fig. 3-7c

$$\sin \theta_A = d/10$$
$$a = 10 \cos \theta_A$$
$$b = 30 - a \qquad (e)$$
$$\tan \theta_C = d/b$$

All that remains is to choose some values for $d$ and to solve Eqs. (a)–(e) for the tensions. For example, when $d = 6$ ft, Eqs. (e) give

$$\theta_A = \sin^{-1} \frac{6}{10} = 36.8699°$$
$$a = 10 \cos 36.8699° = 8 \text{ ft}$$

$$b = 30 - 8 = 22 \text{ ft}$$

$$\theta_C = \tan^{-1} \frac{6}{22} = 15.2551°$$

Then, Eqs. (d) and (c) give

$$T_{BC} = 1013.49 \text{ lb}$$

$$T_{AB} = 1222.22 \text{ lb}$$

where $T_{BC} = P$ because the tension in the hoisting cable does not change as the cable goes around the small pulley. Figure 3-7d shows the results of repeating this process for various values of the sag distance $d$ and graphing the results.

When $d = 10$ ft, the load hangs directly below the support $A$, cable $AB$ carries the entire load, and the hoisting cable is slack, $P = 0$ lb. As the load is raised ($d$ gets smaller), the force in both cables increases. At $d = 3.28$ ft, the force in the hoisting cable is twice the load; at $d = 1.66$ ft, the force in the hoisting cable is four times the load; and at $d = 0.833$ ft, the force in the hoisting cable is eight times the weight of the load being lifted. As $d$ goes to zero, the forces in the two cables both go to infinity. ∎

# PROBLEMS

### Introductory Problems

**3-1\*** Three forces act on a particle as shown in Fig. P3-1. Determine the magnitude of forces $\mathbf{F}_2$ and $\mathbf{F}_3$ so that the particle is in equilibrium.

**3-2\*** Four forces act on a particle as shown in Fig. P3-2. Determine the magnitudes of forces $\mathbf{F}_1$ and $\mathbf{F}_2$ so that the particle is in equilibrium.

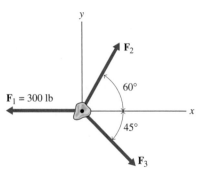

Figure P3-1

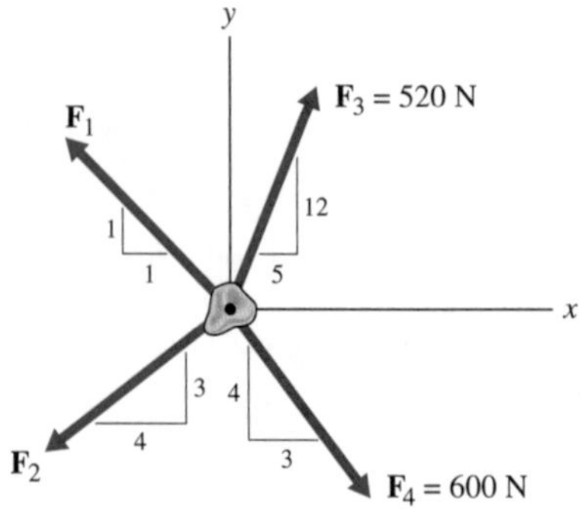

Figure P3-2

**3-3*** Block *A* of Fig. P3-3 rests on a smooth (frictionless) surface. If the block weighs 25 lb, determine the force exerted on the block by the surface and the force **P** parallel to the surface that is required to prevent motion of the block.

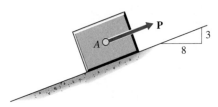

**Figure P3-3**

**3-4** A 750-kg body is supported by the flexible cable system shown in Fig. P3-4. Determine the tensions in cables *AC*, *BC*, and *CD*.

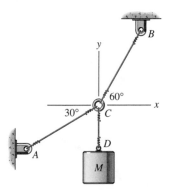

**Figure P3-4**

**3-5** An 800-lb homogeneous cylinder is supported by two rollers as shown in Fig. P3-5. Determine the forces exerted by the rollers on the cylinder. All surfaces are smooth (frictionless).

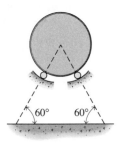

**Figure P3-5**

**3-6*** A worker is using a hoist and cable to lift a 175-kg engine from a car as shown in Fig. P3-6. Determine the forces in the three cables attached to the ring.

**Figure P3-6**

**3-7** The lightweight collar *A* shown in Fig. P3-7 is free to slide on the smooth rod *BC*. Determine the forces exerted on the collar by the cable and by the rod when the 900 lb downward force **F** is applied to the collar.

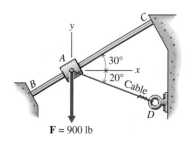

**Figure P3-7**

**3-8** An automobile stuck in a muddy field is being moved by using a cable fastened to a tree as shown in Fig. P3-8. When a 500-N sideways force **P** is applied to the cable, the cable is pulled 5° to the side as shown. For this position, determine the *x*- and *y*-components of the cable force being applied to the automobile.

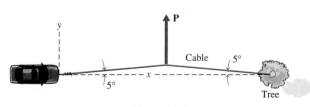

**Figure P3-8**

### Intermediate Problems

**3-9\*** Two flower pots are supported with cables as shown in Fig. P3-9. If pot $A$ weighs 10 lb and pot $B$ weighs 8 lb, determine the tension in each of the cables and the slope of cable $BC$.

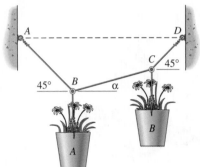

**Figure P3-9**

**3-10\*** Three smooth homogeneous cylinders $A$, $B$, and $C$ are stacked in a box as shown in Fig. P3-10. Each cylinder has a diameter of 250 mm and a mass of 245 kg. Determine
(a) The force exerted by cylinder $B$ on cylinder $A$.
(b) The forces exerted on cylinder $B$ by the vertical and horizontal surfaces at $D$ and $E$.

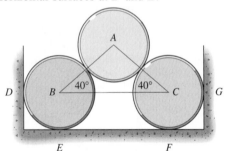

**Figure P3-10**

**3-11\*** Three smooth homogeneous cylinders $A$, $B$, and $C$ are stacked in a V-shaped trough as shown in Fig. P3-11. Each cylinder weighs 100 lb and has a diameter of 5 in. Determine the minimum angle $\theta$ for equilibrium.

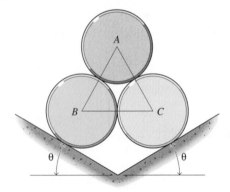

**Figure P3-11**

**3-12** A 250-kg body is supported by the flexible cable system shown in Fig. P3-12. Determine the tensions in cables $A$, $B$, $C$, and $D$.

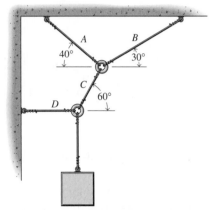

**Figure P3-12**

**3-13** A 500-lb lawn roller is to be pulled over a curb as shown in Fig. P3-13. Determine the minimum pulling force that must be applied by the man to just start the 3-ft-diameter roller over the curb. Also determine the angle $\theta$ that gives the minimum pulling force. Assume that the pulling force is along the handle, which makes an angle of $\theta$ with the horizontal.

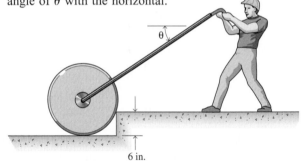

**Figure P3-13**

**3-14\*** In order to hold a 130-kg crate in a stationary position, a worker exerts a force $\mathbf{P}$ at an angle $\theta$ on a rope as shown in Fig. P3-14. Determine the force exerted by the worker when $\theta = 20°$.

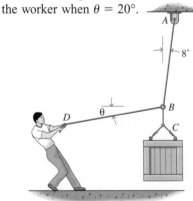

**Figure P3-14**

**3-15** A farmer is extracting a post from the ground using the structure shown in Fig. P3-15. What force must the farmer apply to the cable system if the force required to remove the post is 2000 lb?

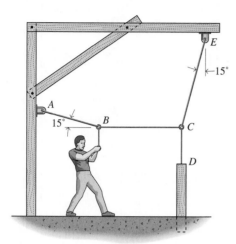

**Figure P3-15**

**3-16** A continuous cable is used to support two blocks as shown in Fig. P3-16. Block $A$ is supported by a small wheel that is free to roll on the cable. Determine the displacement $y$ of block $A$ for equilibrium if the masses of blocks $A$ and $B$ are 22 kg and 34 kg, respectively.

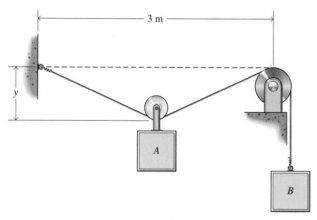

**Figure P3-16**

**Challenging Problems**

**3-17\*** Two bodies $W_1$ and $W_2$ weighing 200 lb and 150 lb, respectively, rest on a cylinder and are connected by a rope as shown in Fig. P3-17. If all surfaces are smooth, determine
(a) The reactions of the cylinder on the bodies.
(b) The tension in the rope.
(c) The angle $\theta$.

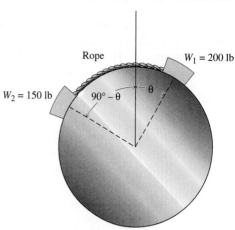

**Figure P3-17**

**3-18\*** Two bodies $A$ and $B$ weighing 800 N and 200 N, respectively, are held in equilibrium on perpendicular surfaces by a connecting flexible cable that makes an angle $\theta$ with the horizontal as shown in Fig. P3-18. If all surfaces are smooth, determine
(a) The reactions of the surfaces on the bodies.
(b) The tension in the cable $AB$.
(c) The angle $\theta$.

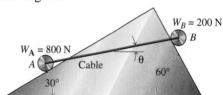

**Figure P3-18**

**3-19\*** Concrete is to be moved from a mixer to the second floor of a building under construction using a container as shown in Fig. P3-19. The container and its contents weigh 3000 lb, and it is supported by three cables equally spaced around the top of the 4-ft diameter container. Determine the force in each cable.

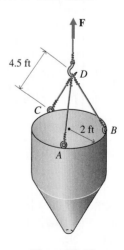

**Figure P3-19**

**3-20**  A mass $m$ is to be supported by two cables ($A$ and $B$) as shown Fig. P3-20. If the maximum force that the cables can withstand is 15 kN, determine the maximum mass $m$ that can be supported.

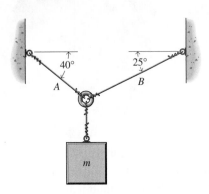

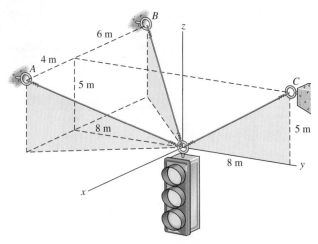

**Figure P3-22**

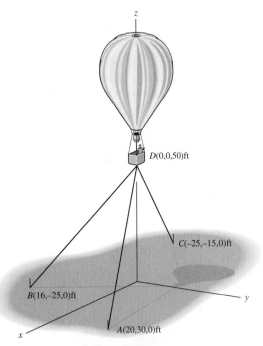

Figure P3-20

**3-21**  The hot-air balloon shown in Fig. P3-21 is tethered with three mooring cables. If the net lift of the balloon is 900 lb, determine the force exerted on the balloon by each of the three cables.

**3-23\***  A 250-lb force is applied to the joint at the top of the structure shown in Fig. P3-23. The joint is held in position by the slender members $AD$, $BD$, and $CD$, which can only exert forces that act along the members. If the 250-lb force is in the $xy$-plane, determine the forces in members $AD$, $BD$, and $CD$.

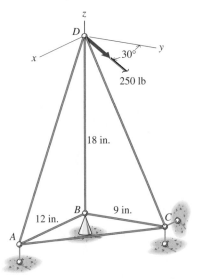

Figure P3-23

Figure P3-21

**3-22\***  A 100-kg traffic light is supported by a system of cables as shown in Fig. P3-22. Determine the tensions in each of the three cables.

**3-24**  Three cables are used to support a 250-kg homogeneous plate as shown in Fig. P3-24. Determine the force in each of the three cables.

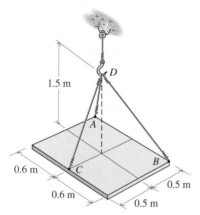

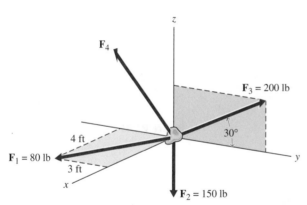

**Figure P3-24**

**3-25** A particle is in equilibrium under the action of four forces as shown on the free-body diagram of Fig. P3-25. Determine the magnitude and the coordinate direction angles of the unknown force $\mathbf{F}_4$.

**Figure P3-25**

**Computer Problems**

**3-26** A pair of steel pipes is stacked in a box as shown in Fig. P3-26. The masses and diameters of the smooth pipes are $m_A = 5$ kg, $m_B = 20$ kg, $d_A = 100$ mm, and $d_B = 200$ mm. Plot the two forces exerted on pipe $A$ (by pipe $B$ and by the side wall) as a function of the distance $b$ between the walls of the box (200 mm $\leq b$ $\leq$ 300 mm). Determine the range of $b$ for which

(a) The force at the side wall is less than $W_A$, the weight of pipe $A$.

(b) Neither of the two forces exceeds $2W_A$.

(c) Neither of the two forces exceeds $4W_A$.

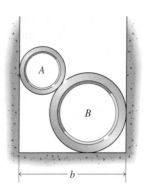

**Figure P3-26**

**3-27** A 75-lb stoplight is suspended between two poles as shown in Fig. P3-27. Neglect the weight of the flexible cables and plot the tension in both cables as a function of the sag distance $d$ ($0 \leq d \leq 8$ ft). Determine the minimum sag $d$ for which both tensions are less than

(a) 100 lb

(b) 250 lb

(c) 500 lb

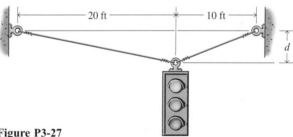

**Figure P3-27**

**3-28** A 50-kg load is suspended from a pulley as shown in Fig. P3-28. The tension in the flexible cable does not change as it passes around the small frictionless pulleys, and the weight of the cable may be neglected. Plot the force $P$ required for equilibrium as a function of the sag distance $d$ ($0 \leq d \leq 1$ m). Determine the minimum sag $d$ for which $P$ is less than

(a) Twice the weight of the load.

(b) Four times the weight of the load.

(c) Eight times the weight of the load.

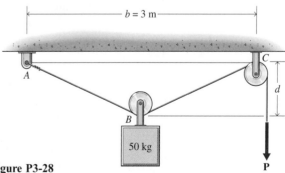

**Figure P3-28**

**3-29** A worker positions a 250-lb crate by pulling on the rope $BD$ as shown in Fig. P3-29. The 3-ft long rope $BD$ is horizontal ($\theta = 0$) when the 5-ft long rope $AB$ is vertical ($\phi = 0$).
(a) What is the maximum distance $b_{max}$ that the crate can be pulled to the side using this arrangement?
(b) Calculate and plot the forces in ropes $AB$ and $BD$ as a function of the distance $b$ for $0 \leq b \leq b_{max}$.
(c) How could the worker pull the crate to the side more than the $b_{max}$ calculated in part $a$?

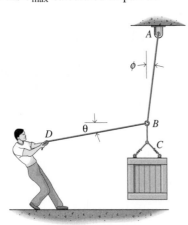

**Figure P3-29**

**3-30** Two small wheels are connected by a lightweight rigid rod as shown in Fig. P3-30. Plot the angle $\theta$ (between the rod and the horizontal) as a function of the weight $W_1$ ($0.25W_2 \leq W_1 \leq 10W_2$). Determine the weight $W_1$ for which

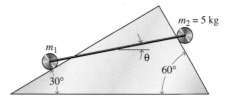

**Figure P3-30**

(a) $\theta = -50°$
(b) $\theta = 10°$
(c) $\theta = 25°$

**3-31** An automobile stuck in a muddy field is being moved by using a cable fastened to a tree as shown in Fig. P3-31. If the side force **P** has a magnitude of 150 lb,
(a) Calculate and plot $F_C$, the force applied to the car as a function of the angle $\theta$ ($0° \leq \theta \leq 45°$).
(b) What is the maximum angle $\theta$ for which this method is effective (that is, for which $P \leq F_C$)?.

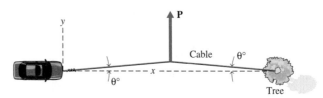

**Figure P3-31**

## 3-4 SUMMARY

The term *particle* is used in statics to describe a body when the size and shape of the body will not significantly affect the solution of the problem being considered and when the mass of the body can be assumed to be concentrated at a point. As a result, a particle can be subjected only to a system of concurrent forces and the necessary and sufficient conditions for equilibrium can be expressed mathematically as

$$\mathbf{R} = \Sigma \mathbf{F} = 0 \tag{3-1}$$

The force system acting on a body in a typical problem consists of known forces and unknown forces. Both must be clearly identified before a solution to a specific problem is attempted. A carefully prepared drawing that shows a "body of interest" separated from all other interacting bodies and with all external forces applied is known as a free-body diagram (FBD). The importance of drawing a free-body diagram before attempting to solve a mechanics problem cannot be overemphasized. A procedure that can be followed to construct a complete and correct free-body diagram contains the following four steps:

1. Decide which body or combination of bodies is to be shown on the free-body diagram.
2. Prepare a drawing or sketch of the outline of this isolated or free body.
3. Carefully trace around the boundary of the free body and identify all of the forces exerted by contacting or attracting bodies that were removed during the isolation process.
4. Choose the set of coordinate axes to be used in solving the problem and indicate their directions on the free-body diagram. Place any dimensions required for solution of the problem on the diagram.

Each known force should be shown on a free-body diagram with its correct magnitude, slope, and sense. Letter symbols can be used for the magnitudes of un-known forces. If a force has a known line of action but an unknown magnitude and sense, the sense of the force can be assumed. By definition, the magnitude of a force is always positive; therefore, if the solution yields a negative magnitude, the minus sign indicates that the sense of the force is opposite to that assumed on the free-body diagram. Most engineers consider an appropriate free-body diagram to be the single most important tool for the solution of mechanics problems.

For a three-dimensional system of concurrent forces, Eq. (3-1) can be written as

$$\mathbf{R} = \Sigma \mathbf{F} = \mathbf{R}_x + \mathbf{R}_y + \mathbf{R}_z$$
$$= R_x\,\mathbf{i} + R_y\,\mathbf{j} + R_z\,\mathbf{k}$$
$$= \Sigma F_x\,\mathbf{i} + \Sigma F_y\,\mathbf{j} + \Sigma F_z\,\mathbf{k} = \mathbf{0} \tag{3-4}$$

Equation (3-4) is satisfied only if

$$\Sigma F_x = 0 \qquad \Sigma F_y = 0 \qquad \Sigma F_z = 0 \tag{3-6}$$

Equations (3-6) can be used to determine three unknown quantities (three mag-nitudes, three slopes, or any combination of three magnitudes and slopes).

## REVIEW PROBLEMS

**3-32\*** A particle is in equilibrium under the action of four forces as shown on the free-body diagram of Fig. P3-32. Determine the magnitude and the direction angle $\theta$ of the unknown force $\mathbf{F}_4$.

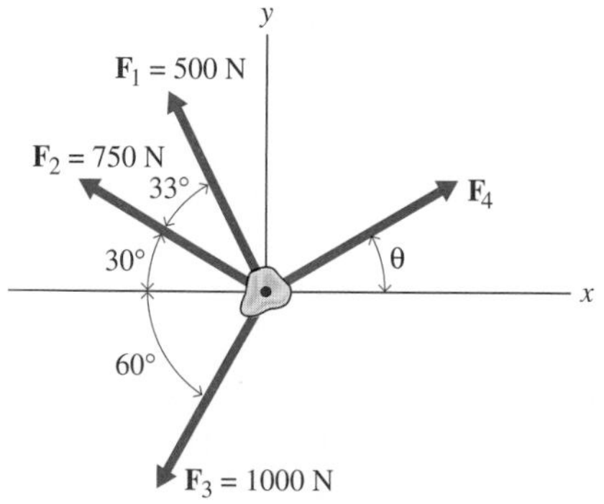

**Figure P3-32**

**3-33\*** Two 10-in.-diameter pipes and a 6-in.-diameter pipe are supported in a pipe rack as shown in Fig. P3-33. The

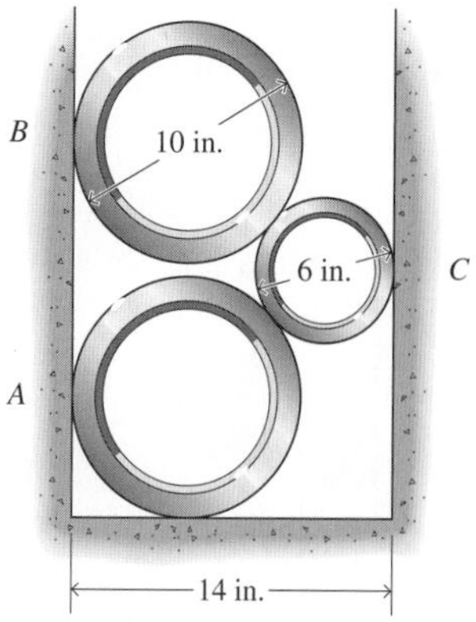

**Figure P3-33**

10-in.-diameter pipes weigh 300 lb each and the 6-in.-diameter pipe weights 175 lb. Determine the forces exerted on the pipes by the supports at the contact surfaces *A*, *B*, and *C*. Assume all surfaces to be smooth.

**3-34** The 250-kg block *A* of Fig. P3-34 is supported by a small wheel that is free to roll on the continuous cable between supports *B* and *C*. If the length of the cable is 42 m, determine the distance *x* and the tension *T* in the cable when the system is in equilibrium.

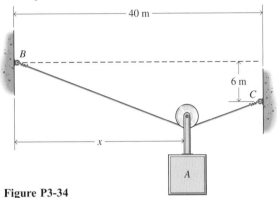

**Figure P3-34**

**3-35** A joint in a bridge truss is subjected to the forces shown in Fig. P3-35. Determine the forces **C** and **T** required for equilibrium.

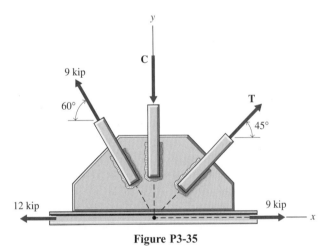

**Figure P3-35**

**3-36\*** Four forces act on a small airplane in flight as shown in Fig. P3-36; the weight **W** (25 kN), the thrust provided by the

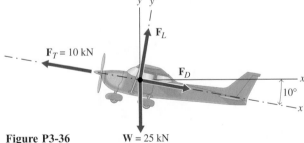

**Figure P3-36**

engine **F**$_T$ (10 kN), the lift provided by the wings **F**$_L$, and the drag resulting from motion through the air **F**$_D$. If the airplane is flying with a constant velocity in the negative *x'* direction, determine the magnitudes of the lift and drag forces.

**3-37\*** A 500-lb block is supported by three cables as shown in Fig. P3-37. Determine the tensions in cables *AB*, *AC*, and *AD*.

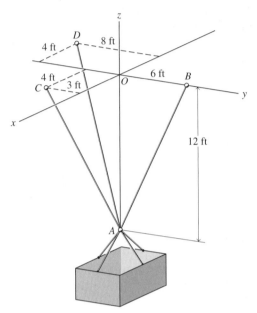

**Figure P3-37**

**3-38\*** A 1250-N force **F** is supported by a cable *AD* and by struts *AB* and *AC*, as shown in Fig. P3-38. If the struts can transmit only axial tensile or compressive forces, determine the forces in the struts and the tension in the cable.

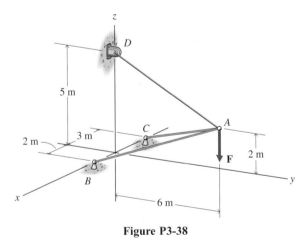

**Figure P3-38**

**3-39** A 75-lb force **F** is supported by a tripod as shown in Fig. P3-39. If the legs can transmit only axial tensile or compressive forces, determine the forces in the legs *AB*, *AC*, and *AD*.

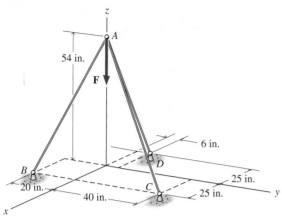

**Figure P3-39**

**3-40** A particle is in equilibrium under the action of four forces as shown on the free-body diagram of Fig. P3-40. Determine the magnitude of the unknown forces $F_1$, $F_2$, and $F_3$.

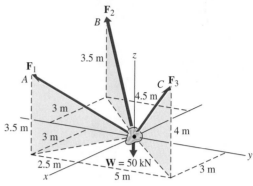

**Figure P3-40**

# STRESS, STRAIN, AND DEFORMATION: AXIAL LOADING

## 4-1 INTRODUCTION

Application of the equations of equilibrium is usually just the first step in solving engineering problems. Using these equations, an engineer can determine the forces exerted on a structure by its supports, the forces on bolts and rivets that connect parts of a machine together, the internal forces in cables or rods that either support the structure or are a part of the structure, and so on. A second and equally important step is determining the effect of the forces on the structure or machine. It is important, therefore, that all engineers understand the behavior of materials under the action of forces.

Safety and economy in a design are two considerations for which an engineer must accept responsibility. He or she must be able to calculate the intensity of the internal forces to which each part of a machine or structure is subjected and the deformation that each part experiences during the performance of its intended function. Then, by knowing the properties of the material from which the parts will be made, the engineer establishes the most effective size and shape of the individual parts, and the appropriate means of connecting them. Problems of this type are considered in courses commonly known as Mechanics of Materials, Strength of Materials, or Mechanics of Deformable Bodies.

In this chapter, the behavior of engineering materials subjected to uniaxial loading situations will be described. Several terms used to describe the material behavior are introduced and defined. An experimental setup used to determine the material properties is described. Finally, simple problems involving axially loaded members are solved.

## 4-2 AXIALLY LOADED MEMBERS—INTERNAL FORCES

Axial loading is produced by two or more collinear forces acting along the axis of a long slender member, such as the eyebar shown in Fig. 4-1. This type of loading occurs in many engineering elements including struts and connecting rods of engines, and in the individual members that make up bridge and building trusses. When a structural member or machine component is subjected to a system of external loads (applied loads and support reactions), a system of internal resisting forces develops within the member to balance the external forces.

For example, consider the eyebar shown in Fig. 4-1, which is subjected to a system of balanced, external, axial forces $\mathbf{F}_1$, $\mathbf{F}_2$, and $\mathbf{F}_3$. These forces tend to either crush the bar (*compression*) or pull it apart (*tension*). In either case, internal resisting forces develop within the bar to resist the crushing or pulling apart of the bar.

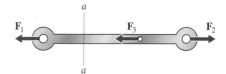

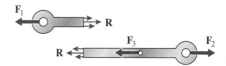

**Figure 4-1**

The internal forces, the resultant of which is $\mathbf{R}$, that develop on plane $a$–$a$ of Fig. 4-1 are shown in Fig. 4-2. These internal forces are the result of the mutual attraction (or repulsion) of the molecules on one side of the plane $a$–$a$ for the molecules on the other side of the plane $a$–$a$ and are distributed over the entire surface of the cutting plane. Since the bar is in equilibrium, both parts of the bar with plane $a$–$a$ as one of its bounding surfaces must also be in equilibrium under the action of the external collinear forces and the internal forces that develop on the plane. Thus, the resultant $\mathbf{R}$ of the internal forces on plane $a$–$a$ can be determined by using either the left or right part of the bar. The intensities of these internal forces (force per unit area) are called *stresses*. In general, forces acting over the small elements of area $dA$ that make up the total cross-sectional area $A$ of the bar at plane $a$–$a$ are not uniformly distributed; therefore, the stress distribution on plane $a$–$a$ does not have to be uniform.

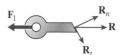

**Figure 4-2**

If the cutting plane $a$–$a$ is perpendicular to the axis of the bar, as shown in Fig. 4-2, then the internal stresses, the internal forces, and the resultant of the internal forces are all perpendicular to the cutting plane (along the axis of the bar). If the cutting plane $a$–$a$ cuts the bar at an arbitrary angle, as shown in Fig. 4-3, the resultant of the internal forces is still along the axis of the bar. However, experimental studies indicate that materials respond differently to forces that tend to pull surfaces apart than to forces that tend to slide surfaces relative to each other. Therefore, as shown in Fig. 4-3, the resultant $\mathbf{R}$ is usually resolved into a component $\mathbf{R}_n$ perpendicular to plane $a$–$a$ (a normal force from which a normal stress is determined) and a component $\mathbf{R}_t$ tangent to plane $a$–$a$ (a shear force from which a shear stress is determined).

**Figure 4-3**

## Normal Stress Under Axial Loading

In the simplest qualitative terms, stress is the intensity of internal force. A body must be able to withstand the intensity of internal force, or else the body may rupture or deform excessively. Force intensity (stress) is force divided by the area over which the force is distributed. Thus,

$$\text{Stress} = \frac{\text{Force}}{\text{Area}} \qquad (4\text{-}1)$$

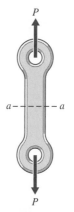

**Figure 4-4**

The forces shown in Fig. 4-4 are collinear with the axis of the eyebar and produce an axial tensile loading of the bar. When the eyebar is cut by a transverse plane, such as plane $a$–$a$ of Fig. 4-4, a free-body diagram of the bottom half of the bar can be drawn as shown in Fig. 4-5. Equilibrium of this portion of the bar is obtained with a distribution of internal force that develops on the exposed cross section. This distribution of internal force has a resultant $\mathbf{F}$ that is normal to the exposed surface, is equal in magnitude to $\mathbf{P}$, and has a line of action that is collinear with the line of action of $\mathbf{P}$. An average intensity of internal force, which is also known as the average normal stress $\sigma_{\text{avg}}$ on the cross section, can be computed as

$$\sigma_{\text{avg}} = \frac{F}{A} \qquad (4\text{-}2)$$

where $F$ is the magnitude of the force $\mathbf{F}$, and $A$ is the transverse cross-sectional area of the eyebar (the area over which the force $\mathbf{F}$ is distributed). Note that the force $\mathbf{F}$ and the area $A$ are perpendicular—thus the term *normal stress*.

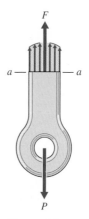

**Figure 4-5**

**Figure 4-6**

The Greek letter sigma ($\sigma$) will be used to denote a normal stress in this book. A positive sign will be used to indicate a tensile normal stress (member in tension), and a negative sign will be used to indicate a compressive normal stress (member in compression). This sign convention is independent of the selection of a coordinate system.

Consider now a small area $\Delta A$ on the exposed cross section of the bar and let $\Delta F$ represent the magnitude of the resultant of the internal forces transmitted by this small area as shown in Fig. 4-6. The average intensity of internal force being transmitted by area $\Delta A$ is obtained by dividing $\Delta F$ by $\Delta A$. If the internal forces transmitted across the section are assumed to be continuously distributed, the area $\Delta A$ can be made smaller and smaller and will approach a point on the exposed surface in the limit. The corresponding force $\Delta F$ also becomes smaller and smaller. The stress at the point on the cross section to which $\Delta A$ converges is defined as

$$\sigma = \lim_{\Delta A \to 0} \frac{\Delta F}{\Delta A} \tag{4-3}$$

In general, the stress $\sigma$ at a given point on a transverse cross section of an axially loaded bar will not be the same as the average stress computed by dividing the force $F$ by the cross-sectional area $A$. For long, slender, axially loaded members such as those found in trusses and similar structures, however, it is generally assumed that the normal stresses are uniformly distributed except in the vicinity of the points of application of the loads. The subject of nonuniform stress distributions under axial loading will be discussed in a later chapter of this book.

**Shearing Stress in Connections**  Loads applied to a structure or machine are generally transmitted to the individual members through connections that use rivets, bolts, pins, nails, or welds. In all of these connections, one of the most significant stresses induced is a shearing stress. The bolted and pinned connection shown in Fig. 4-7 will be used to introduce the concept of a shearing stress.

The method by which loads are transferred from one member of the connection to another is by means of a distribution of (internal) shearing force on a transverse cross section of the bolt or pin used to effect the connection. A free-body diagram of the left member of the connection of Fig. 4-7 is shown in Fig. 4-8. In this diagram, a transverse cut has been made through the bolt and the lower portion of the bolt remains in contact with the left member. The distribution of shearing force on the transverse cross section of the bolt has been replaced by a resultant shear force $V$. Since only one transverse cross section of the bolt is used to effect load transfer between the members, the bolt is said to be in *single shear*;

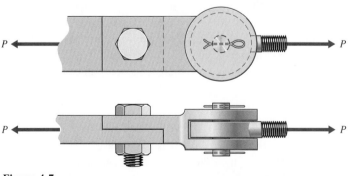

**Figure 4-7**

therefore, equilibrium requires that the resultant shear force $V$ equal the applied load $P$. A free-body diagram for the threaded eye bar at the right end of the connection of Fig. 4-7 is shown in Fig. 4-9. In this diagram, two transverse cuts have been made through the bolt and the middle portion of the bolt remains in contact with the eye bar. In this case, two transverse cross sections of the pin are used to effect load transfer between members of the connection and the pin is said to be in *double shear*. As a result, equilibrium requires that the resultant shear force $V$ on each cross section of the pin equals half the applied load $P$.

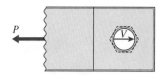

**Figure 4-8**

From the definition of stress given by Eq. (4-1), an average shearing stress on the transverse cross section of the bolt or pin can be computed as

$$\tau_{\text{avg}} = \frac{V}{A_s} \tag{4-4}$$

**Figure 4-9**

where $V$ is the magnitude of the shear force $\mathbf{V}$ and $A_s$ is the shear area (the transverse cross-sectional area of the bolt or pin over which the force $\mathbf{V}$ is distributed). Note that the force $\mathbf{V}$ is parallel to the area $A_s$—thus the term *shearing stress.*

The Greek letter tau ($\tau$) will be used to denote shearing stress in this book. A sign convention for shearing stress will be presented in a later section of the book.

The stress at a point on the transverse cross section of the bolt or pin can be obtained by using the same type of limit process that was used to obtain Eq. (4-3) for the normal stress at a point. Thus,

$$\tau = \lim_{\Delta A_s \to 0} \frac{\Delta V}{\Delta A_s} \tag{4-5}$$

Unlike the normal stress in long slender members, it can be shown that the shear stress $\tau$ cannot be uniformly distributed over the area. Therefore, the actual shear stress at any particular point and the maximum shear stress on a cross section will generally be different from the average shear stress calculated using Eq. (4-4). However, the design of simple connections is usually based on average stress considerations and this procedure will be followed in this book.

Another type of shear loading is termed *punching shear.* Examples of this type of loading include the action of a punch in forming rivet holes in a metal plate, the tendency of building columns to punch through footings, and the tendency of a tensile axial load on a bolt to pull the shank of the bolt through the head (Fig. 4-10). Under a punching shear load, the significant stress is the average shearing stress on the surface described by the periphery of the punching member and the thickness of the punched member; for example, the shaded cylindrical area $A_s = \pi d t$ shown extending through the head of the bolt in Fig. 4-10*b*.

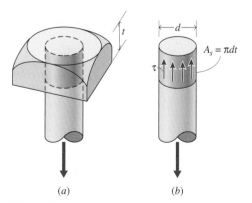

(a)                                        (b)

**Figure 4-10**

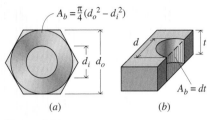

$A_b = \frac{\pi}{4}(d_o^2 - d_i^2)$

$d_i \ d_o$

$d$

$t$

$A_b = dt$

(a)

(b)

**Figure 4-11**

## Bearing Stress

Bearing stresses (compressive normal stresses) occur on the surface of contact between two interacting members. For the case of the connection shown in Fig. 4-7, bearing stresses occur on the surfaces of contact between the head of the bolt and the top plate and between the nut and the bottom plate. The force producing the stress is the axial tensile internal force **F** developed in the shank of the bolt as the nut is tightened. The area of interest for bearing stress calculations is the annular area $A_b = \frac{\pi}{4}(d_o^2 - d_i^2)$ of the bolt head or nut (see Fig. 4-11a) that is in contact with the plate. Thus, the average bearing stress $\sigma_b$ is expressed as

$$\sigma_b = \frac{F}{A_b} \tag{4-6}$$

Bearing stresses also develop on surfaces of contact where the shanks of bolts and pins are pressed against the sides of the hole through which they pass. Since the distribution of these forces is quite complicated, an average bearing stress $\sigma_b$ is often used for design purposes. This stress is computed by dividing the force $F$ transmitted across the surface of contact by the projected area $A_b = dt$ shown in Fig. 4-11b, instead of the actual contact area.

Stress, being the intensity of internal force, has the dimensions of force per unit area $(FL^{-2})$. Until recently, the commonly used unit for stress in the United States was the pound per square inch, abbreviated as psi. Since metals can sustain stresses of several thousand pounds per square inch, the unit ksi (kip per square inch) is also frequently used (1 ksi = 1000 psi). With the advent of the International System of Units (SI units), units of stress based on the international system are also being used in the United States and will undoubtedly come into wider use in the future. During the transition, both systems will be encountered by engineers; therefore, approximately half of the example problems and homework problems in this book are given using the U.S. customary system (pounds and inches) and the other half are given in SI units (newtons and meters). For problems with SI units, forces will be given in newtons (N) or kilonewtons (kN), dimensions in meters (m) or millimeters (mm), and masses in kilograms (kg). The SI unit for stress is a newton per square meter (N/m$^2$), also known as a pascal (Pa). Stress magnitudes normally encountered in engineering applications are expressed in meganewtons per square meter (MN/m$^2$), or megapascals (MPa).

## Example Problem 4-1

A flat steel bar has axial loads applied at points $A$, $B$, $C$, and $D$ as shown in Fig. 4-12a. If the bar has a cross-sectional area of 3 in$^2$, determine the axial stress in the bar

(a) On a cross section 20 in. to the right of point $A$.

(b) On a cross section 20 in. to the right of point $B$.

(c) On a cross section 20 in. to the right of point $C$.

### SOLUTION

The axial forces transmitted by cross-sections in intervals $AB$, $BC$, and $CD$ of the bar shown in Fig. 4-12a are obtained by using the free-body diagrams shown in Fig. 4-12b. The internal forces $F_{AB}$, $F_{BC}$, and $F_{CD}$ are all assumed to be in

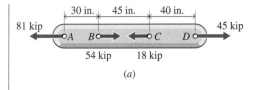

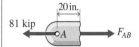

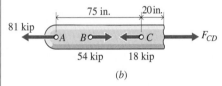

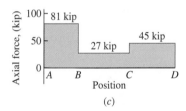

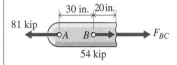

**Figure 4-12**

tension on the free-body diagrams. Summing forces along the axis of the bar on each free-body diagram yields

$$+ \rightarrow \Sigma F_x = 0: \qquad F_{AB} - 81 = 0 \qquad F_{AB} = 81 \text{ kip}$$

$$+ \rightarrow \Sigma F_x = 0: \qquad F_{BC} + 54 - 81 = 0 \qquad F_{BC} = 27 \text{ kip}$$

$$+ \rightarrow \Sigma F_x = 0: \qquad F_{CD} - 18 + 54 - 81 = 0 \qquad F_{CD} = 45 \text{ kip}$$

Since the internal forces are all positive, they are all tensile as was assumed on the free-body diagrams.

A pictorial representation of the distribution of axial force (internal) in bar *ABCD* is shown in Fig. 4-12*c*. This type of representation is known as an axial-force diagram and has been shown to be useful in solving problems involving axial-force distributions. The required stresses from Eq. (4-2) are

(a) $$\sigma_{AB} = \frac{F_{AB}}{3} = \frac{81}{3} = 27.0 \text{ ksi T} \qquad \textbf{Ans.}$$

(b) $$\sigma_{BC} = \frac{F_{BC}}{3} = \frac{27}{3} = 9.00 \text{ ksi T} \qquad \textbf{Ans.}$$

(c) $$\sigma_{CD} = \frac{F_{CD}}{3} = \frac{45}{3} = 15.00 \text{ ksi T} \qquad \textbf{Ans.}$$

where T indicates that the normal stress is tensile. ∎

## Example Problem 4-2

The round bar shown in Fig. 4-13a has steel (s), brass (b), and aluminum (a) sections. Axial loads are applied at cross sections A, B, C, and D. If the allowable axial normal stresses are 125 MPa in the steel, 70 MPa in the brass, and 85 MPa in the aluminum, determine the diameters required for each of the sections.

### SOLUTION

The axial forces transmitted by cross sections in intervals AB, BC, and CD of the bar shown in Fig. 4-13a are obtained by using the free-body diagrams shown in Fig. 4-13b. The internal forces $F_s$, $F_b$, and $F_a$ are all assumed to be in tension on the free-body diagrams. Summing forces along the axis of the bar on each free-body diagram yields

$$+ \rightarrow \Sigma F_x = 0: \qquad\qquad F_s + 270 = 0 \qquad F_s = -270 \text{ kN} = 270 \text{ kN C}$$
$$+ \rightarrow \Sigma F_x = 0: \qquad\qquad F_b + 270 - 245 = 0 \qquad F_b = -25 \text{ kN} = 25 \text{ kN C}$$
$$+ \rightarrow \Sigma F_x = 0: \qquad F_a + 270 - 245 + 200 = 0 \qquad F_a = -225 \text{ kN} = 225 \text{ kN C}$$

Since the internal forces are all negative, they are all opposite the direction assumed on the free-body diagrams. They are all compressive forces as designated by the C.

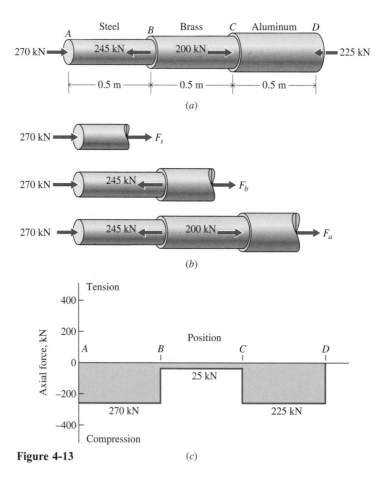

**Figure 4-13**

An axial-force diagram for bar *ABCD* is shown in Fig. 4-13c. The cross-sectional areas of the bar required to limit the stresses to the specified values are obtained from Eq. (4-2). Thus,

$$A_s = \frac{F_s}{\sigma_s} \qquad \frac{\pi}{4} d_s^2 = \frac{270(10^3)}{125(10^6)} \qquad d_s = 52.4(10^{-3}) \text{ m} = 52.4 \text{ mm} \qquad \textbf{Ans.}$$

$$A_b = \frac{F_b}{\sigma_b} \qquad \frac{\pi}{4} d_b^2 = \frac{25(10^3)}{70(10^6)} \qquad d_b = 21.3(10^{-3}) \text{ m} = 21.3 \text{ mm} \qquad \textbf{Ans.}$$

$$A_a = \frac{F_a}{\sigma_a} \qquad \frac{\pi}{4} d_a^2 = \frac{225(10^3)}{85(10^6)} \qquad d_a = 58.1(10^{-3}) \text{ m} = 58.1 \text{ mm} \quad ■ \quad \textbf{Ans.}$$

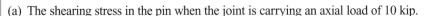

## Example Problem 4-3

A brass tube with an outside diameter of 2.00 in. and a wall thickness of 0.375 in. is connected to a steel tube with an inside diameter of 2.00 in. and a wall thickness of 0.250 in. by using a 0.750-in.-diameter pin as shown in Fig. 4-14a. Determine

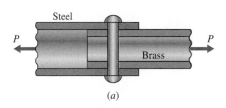

(a)

(a) The shearing stress in the pin when the joint is carrying an axial load of 10 kip.

(b) The length of joint required if the pin is replaced by a glued joint and the shearing stress in the glue must be limited to 250 psi.

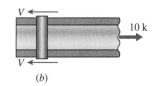

(b)

## SOLUTION

(a) A free-body diagram of the brass tube and pin is shown in Fig. 4-14b. The pin is in double shear; that is, in order to isolate the brass tube and pin, the pin had to be cut twice, exposing two shear forces *V*. The shear area $A_s$ over which each of the shear forces is distributed is the transverse cross-sectional area of the pin $A_s = (\pi/4)(0.750)^2 = 0.4418 \text{ in.}^2$. Summing forces along the axis of the tube on the free-body diagram yields

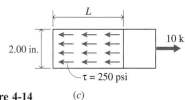

**Figure 4-14**     (c)

$$+ \rightarrow \Sigma F_x = 0: \qquad\qquad 10 - 2V = 0 \qquad V = 5 \text{ kip}$$

Therefore, from Eq. (4-4),

$$\tau = \frac{V}{A_s} \qquad \tau = \frac{5}{0.4418} = 11.32 \text{ ksi} \qquad\qquad \textbf{Ans.}$$

(b) A free-body diagram of the brass tube and glued joint is shown in Fig. 4-14c. For the glued joint, the shear area is $A_s = \pi dL = \pi(2)L = 2\pi L \text{ in.}^2$ and from Eq. (4-4) the shear force is $V = \tau A_s$. Summing forces along the axis of the tube on the free-body diagram yields

$$+ \rightarrow \Sigma F_x = 0: \qquad\qquad 10 - \tau A_s = 0 \qquad \tau A_s = 10 \text{ kip}$$

Therefore,

$$250(2\pi L) = 10(10^3)$$

from which

$$L = 6.37 \text{ in.} \quad ■ \qquad\qquad\qquad \textbf{Ans.}$$

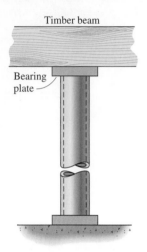

Timber beam

Bearing
plate

**Figure 4-15**

## Example Problem 4-4

The steel pipe column shown in Fig. 4-15 has an outside diameter of 150 mm and a wall thickness of 15 mm. The load imposed on the column by the timber beam is 150 kN. Determine

(a) The average bearing stress at the surface between the steel pipe column and the steel bearing plate.

(b) The diameter of a circular bearing plate if the average bearing stress between the steel bearing plate and the wood beam is not to exceed 3.25 MPa.

### SOLUTION

(a)   The annular area between the steel column and the bearing plate is

$$A_b = \frac{\pi}{4}(d_o^2 - d_i^2) = \frac{\pi}{4}[(150)^2 - (120)^2] = 6362 \text{ mm}^2$$

Thus, from Eq. (4-6),

$$\sigma_b = \frac{F}{A_b} = \frac{150(10^3)}{6362(10^{-6})} = 23.6(10^6) \text{ N/m}^2 = 23.6 \text{ MPa} \qquad \textbf{Ans.}$$

(b) The circular area between the bearing plate and the timber beam is

$$A_b = \frac{\pi}{4}d^2$$

Thus, from Eq. (4-6),

$$\sigma_b = \frac{F}{A_b} \qquad\qquad 3.25(10^6) = \frac{150(10^3)}{(\pi/4)d^2}$$

From which

$$d = 242(10^{-3}) \text{ m} = 242 \text{ mm} \quad \blacksquare \qquad \textbf{Ans.}$$

## Example Problem 4-5

A vertical shaft is supported by a thrust collar and bearing plate as shown in Fig. 4-16a. The force imposed on the bearing plate by the collar is 50 kip. If the bearing stress between the collar and the bearing plate must not exceed 10 ksi, determine the minimum diameter collar that must be used. Assume that the bearing stress is uniformly distributed over the surface of the collar.

   If the collar is not rigid, the stress between the collar and the bearing plate will not be uniform. If the stress varies as shown in Fig. 4-16b (decreasing linearly from $\sigma_{max}$ at the edge of the shaft to $\sigma_{max}/2$ at $r = 3$ in.), calculate and plot $\sigma_{max}$ versus the diameter $d_c$ of the collar ($2.5$ in. $\leq r_c \leq 5.0$ in.). Now what minimum diameter collar must be used if the bearing stress must not exceed 10 ksi? What

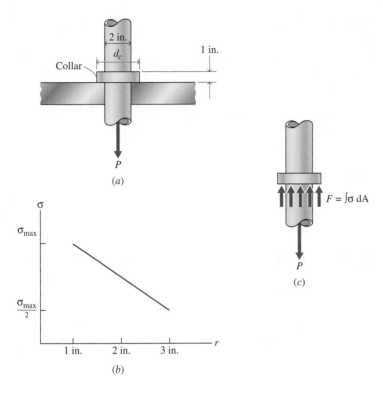

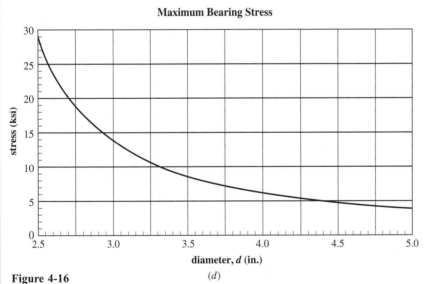

**Figure 4-16**

(d)

is the percent decrease in $\sigma_{max}$ for a 3-in.-diameter collar compared to a 2.5-in.-diameter collar? For a 4.0-in. diameter collar compared to a 3-in.-diameter collar?

## SOLUTION

A free-body diagram of the shaft is shown in Fig. 4-16c. Summing forces in the vertical direction gives

$$+\uparrow \Sigma F_y = 0: \qquad\qquad F - P = 0 \qquad\qquad F = P \qquad\quad (a)$$

where the force on the collar is the sum of the bearing stresses $F = \int dF = \int \sigma \, dA$.

**Rigid collar.** If the bearing stresses are uniformly distributed over the annular ring of the collar, then

$$P = \sigma \int dA = \sigma A = \sigma \frac{\pi}{4} (d_c^2 - 2^2)$$

Therefore, with $P = 50$ kip and $\sigma = 10$ ksi, the smallest diameter collar that can be used is

$$d_c = \sqrt{\frac{4(50)}{\pi(10)} + 4} = 3.22 \text{ in.} \qquad \textbf{Ans.}$$

**Flexible collar.** The equation for the linearly varying stress of Fig. 4-16b is

$$\sigma = \frac{\sigma_{max}}{4} (5 - r) \qquad (b)$$

Integrating the stresses of Eq. (b) over the annular ring of the collar gives

$$P = \int \sigma \, dA = \int_1^{r_c} \frac{\sigma_{max}}{4} (5 - r)(2\pi r \, dr)$$

$$= \frac{\pi \sigma_{max}}{4} \left[ 5(r_c^2 - 1) - \frac{2}{3}(r_c^3 - 1) \right]$$

Therefore, with $P = 50$ kip, the maximum stress on the collar is given by

$$\sigma_{max} = \frac{200}{\pi \left[ 5(r_c^2 - 1) - \frac{2}{3}(r_c^3 - 1) \right]} \qquad (c)$$

For example, when $r_c = 1.25$ in. (and $d_c = 2.5$ in.)

$$\sigma_{max} = \frac{200}{\pi \left[ 5(1.25^2 - 1) - \frac{2}{3}(1.25^3 - 1) \right]} = 29.242 \text{ ksi}$$

Calculating Eq. (c) for values of $d_c$ between 2.5 in. and 5.0 in. results in the graph of Fig. 4-16d.

From the graph of Fig. 4-16d, the smallest flexible collar for which $\sigma_{max} \leq 10$ ksi is

$$d_c = 3.32 \text{ in.} \qquad \textbf{Ans.}$$

which is only 3.1% larger than the diameter of the smallest rigid collar. From Eq. (c) or from the graph of Fig. 4-16d, when $d_c = 3$ in., the maximum stress is $\sigma_{max} = 13.642$ ksi and the percent decrease from when $d_c = 2.5$ in. is

$$\frac{29.242 - 13.642}{29.242}(100) = 53.4\% \qquad \textbf{Ans.}$$

The percent decrease in the maximum stress between $d_c = 3.0$ in. where the maximum stress is $\sigma_{max} = 13.642$ ksi and $d_c = 4.0$ in. where the maximum stress is $\sigma_{max} = 6.161$ ksi is

$$\frac{13.642 - 6.161}{13.642}(100) = 54.8\% \quad \blacksquare \qquad \textbf{Ans.}$$

# PROBLEMS

### Introductory Problems

**4-1\*** An aluminum tube with an outside diameter of 1.000 in. will be used to support a 10-kip load. If the axial stress in the member must be limited to 30 ksi T or C, determine the wall thickness required for the tube.

**4-2\*** Three steel bars with 25- × 15-mm cross sections are welded to a gusset plate as shown in Fig. P4-2. Determine the normal stresses in the bars when the forces shown are being applied to the plate.

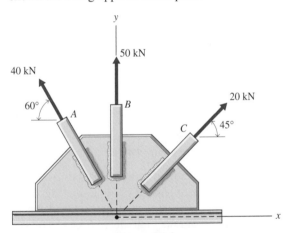

**Figure P4-2**

**4-3\*** Two $^1/_4$-in.-diameter steel cables $A$ and $B$ are used to support a 220-lb traffic light as shown in Fig. P4-3. Determine the normal stress in each of the cables.

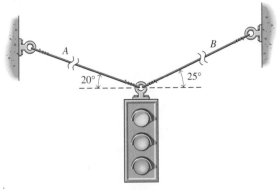

**Figure P4-3**

**4-4\*** A system of steel pipes is loaded and supported as shown in Fig. P4-4. If the normal stress in each pipe must not exceed 150 MPa, determine the cross-sectional areas required for each of the sections.

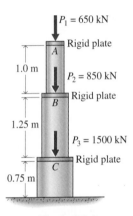

**Figure P4-4**

**4-5** Two 1-in.-diameter steel bars are welded to a gusset plate as shown in Fig. P4-5. Determine the normal stresses in the bars when forces $F_1 = 550$ lb and $F_2 = 750$ lb are applied to the plate.

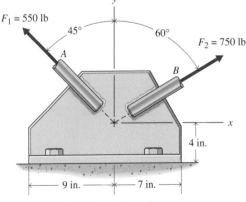

**Figure P4-5**

**4-6** Two strips of a plastic material are bonded together as shown in Fig. P4-6. The average shearing stress in the glue must be limited to 950 kPa. What length $L$ of splice plate is needed if the axial load carried by the joint is 50 kN?

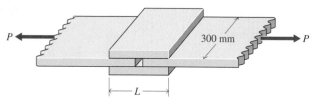

**Figure P4-6**

**4-7** A coupling is used to connect a 2-in.-diameter plastic rod to a 1.5-in.-diameter rod as shown in Fig. P4-7. If the average shearing stress in the adhesive must be limited to 500 psi, determine the minimum lengths $L_1$ and $L_2$ required for the joint if the applied axial load $P$ is 8000 lb.

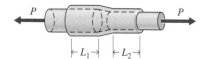

**Figure P4-7**

## Intermediate Problems

**4-8\*** Three plates are joined with a 12-mm-diameter pin as shown in Fig. P4-8. Determine the maximum load $P$ that can be transmitted by the joint if
(a) The maximum normal stress on a cross section at the pin must be limited to 350 MPa.
(b) The maximum bearing stress between a plate and the pin must be limited to 650 MPa.
(c) The maximum shearing stress on a cross section of the pin must be limited to 240 MPa.
(d) The punching shear resistance of the material in the top and bottom plates is 300 MPa.

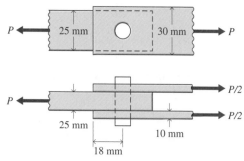

**Figure P4-8**

**4-9\*** A 100-ton hydraulic punch press is used to punch holes in a 0.50-in.-thick steel plate, as illustrated schematically in Fig. P4-9. If the average punching shear resistance of the steel plate is 40 ksi, determine the maximum diameter hole that can be punched.

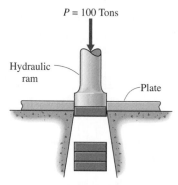

**Figure P4-9**

**4-10** A body with a mass of 250 kg is supported by five 15-mm-diameter cables, as shown in Fig. P4-10. Determine the normal stress in each of the cables.

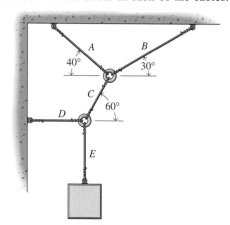

**Figure P4-10**

**4-11** The joint shown in Fig. P4-11 is used in a steel tension member which has a 2- × 1-in. rectangular cross section. If the allowable normal, bearing, and punching-shearing stresses in the joint are 13.5 ksi, 18.0 ksi, and 6.50 ksi, respectively, determine the maximum load $P$ that can be carried by the joint.

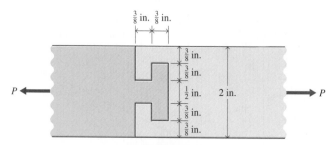

**Figure P4-11**

**4-12** A vertical shaft is supported by a thrust collar and bearing plate, as shown in Fig. P4-12. Determine the maximum axial load that can be applied to the shaft if the average punching shear stress in the collar and the

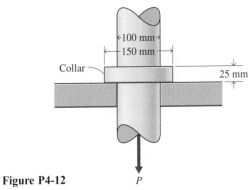

**Figure P4-12**

average bearing stress between the collar and the plate are limited to 75 and 100 MPa, respectively.

**4-13\*** A device for determining the shearing strength of wood is shown in Fig. P4-13. The dimensions of the wood specimen are 6 in. wide by 8 in. high by 2 in. thick. If a load of 16,800 lb is required to fail the specimen, determine the shearing strength of the wood.

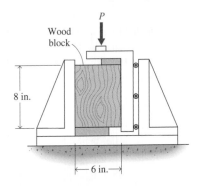

**Figure P4-13**

## Challenging Problems

**4-14\*** The 75-kg traffic light shown in Fig. P4-14 is supported by three cables of equal diameter. Determine the

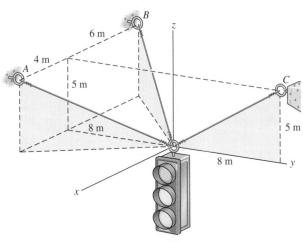

**Figure P4-14**

minimum diameter required if the normal stress in any cable must not exceed 160 MPa.

**4-15\*** A 1000-lb load is securely fastened to a hoisting rope as shown in Fig. P4-15. The force in the weightless flexible cable does not change as it passes around the small frictionless pulley at support $C$. The sag distance $d$ is 4 ft. Cables $AB$ and $BC$ have the same diameter. The normal stresses in these cables must not exceed 24 ksi, and the shearing stress in pin $A$ (double shear) must not exceed 12 ksi. Determine
(a) The minimum diameter of cables $AB$ and $BC$.
(b) The minimum diameter of the pin at $A$.

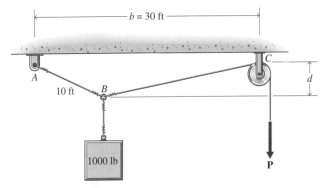

**Figure P4-15**

**4-16** A flat steel bar 100 mm wide by 25 mm thick has axial loads applied with 40-mm-diameter pins in double shear at points $A$, $B$, $C$, and $D$ as shown in Fig. P4-16. Determine
(a) The axial stress in the bar on a cross section at pin $B$.
(b) The average bearing stress on the bar at pin $B$.
(c) The shearing stress on the pin at $A$.

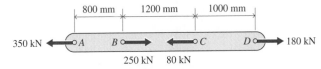

**Figure P4-16**

**4-17** Two flower pots, shown in Fig. P4-17, are supported with steel wires of equal diameter. Pot $A$ weighs 10 lb

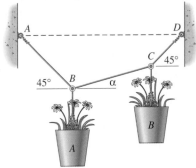

**Figure P4-17**

and pot *B* weighs 8 lb. Determine the minimum required diameter of the wires if the normal stress in the wires must not exceed 18 ksi.

## Computer Problems

**4-18** A steel pipe will be used to support a 40-kN load. If the wall thickness of the pipe is 10 percent of the pipe's outside diameter $d_o$, calculate and plot the normal stress in the pipe $\sigma$ as a function of the diameter $d_o$, (20 mm $\leq d_o \leq$ 75 mm). If the axial stress in the pipe must be limited to 250 MPa, what is the smallest size standard steel pipe (see Appendix A) that could be used?

**4-19** An aluminum tube with an outside diameter $d_o$ will be used to support a 10-kip load. If the axial stress in the tube must be limited to 30 ksi T or C, calculate and plot the required wall thickness $t$ as a function of $d_o$ (0.75 in. $\leq d_o \leq$ 4 in.). What diameter would be required for a solid aluminum shaft?

**4-20** The steel pipe column shown in Fig. P4-20a has an outside diameter of 150 mm and a wall thickness of 15 mm. The load imposed on the column by the

timber beam is 150 kN. If the bearing stress between the circular steel bearing plate and the timber beam is not to exceed 3.25 MPa, determine the minimum diameter bearing plate that must be used between the column and the beam. Assume that the bearing stress is uniformly distributed over the surface of the plate.

If the bearing plate is not rigid, the stress between the bearing plate and the timber beam will not be uniform. If the stress varies as shown in Fig. P4-20b (a uniform value of $\sigma_{max}$ above the column and decreasing linearly to $\sigma_{max}/5$ at the outside edge $r_p$ of the bearing plate), calculate and plot $\sigma_{max}$ versus the radius $r_p$ of the bearing plate (75 mm $\leq r_p \leq$ 500 mm). Now what minimum diameter bearing plate must be used if the bearing stress must not exceed 3.25 MPa? What is the percent decrease in $\sigma_{max}$ for a 400-mm-diameter bearing plate compared with a 150-mm-diameter bearing plate? For a 600-mm-diameter bearing plate compared with a 150-mm-diameter bearing plate?

**4-21** The tie rod shown in Fig. P4-21a has a diameter of 1.50 in and is used to resist the lateral pressure against the walls of a grain bin. The force imposed on the wall by the rod is 18,000 lb. If the bearing stress between the washer and the wall must not exceed 400 psi, determine the minimum diameter washer that must be used be-

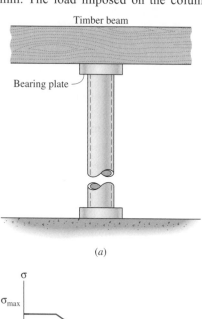

Timber beam

Bearing plate

(*a*)

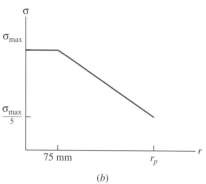

(*b*)

**Figure P4-20**

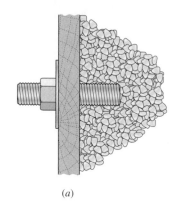

(*a*)

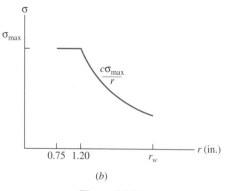

(*b*)

**Figure P4-21**

tween the head of the bolt and the grain bin wall. Assume that the bearing stress is uniformly distributed over the surface of the washer.

If the washer is not rigid, the stress between the washer and the wall will not be uniform. If the stress varies as shown in Fig. P4-21*b* (a uniform value of $\sigma_{max}$ under the 2.4-in.-diameter restraining nut and decreasing as $1/r$ to the outside edge $r_w$ of the

washer), calculate and plot $\sigma_{max}$ versus the radius $r_w$ of the washer (1 in. $\leq r_w \leq$ 8 in.). Now what minimum diameter washer must be used if the bearing stress must not exceed 400 psi? What is the percent decrease in $\sigma_{max}$ for an 8-in.-diameter washer compared with a 4-in.-diameter washer? For a 12-in.-diameter washer compared with an 8-in.-diameter washer?

# 4-3 STRESSES ON AN INCLINED PLANE IN AN AXIALLY LOADED MEMBER

In Section 4-2, normal, shear, and bearing stresses for axially loaded members were introduced. Stresses on planes inclined to the axis of axially loaded bars will now be considered. When the eyebar shown in Fig. 4-4 is cut by an inclined plane, a free-body diagram of the upper portion of the bar would appear as shown in Fig. 4-17. Equilibrium of this portion of the bar is established by placing a distribution of internal force on the cut section as shown in Fig. 4-18. The resultant $F$ of this distribution of internal force is equal in magnitude to the applied load $P$ and has a line of action that is coincident with the axis of the bar as shown in Fig. 4-18. An average total stress $S_{avg}$ on the inclined surface can be computed by using Eq. (4-2). This total stress conveys very little information that is useful for design purposes. The resultant $F$, however, can be replaced by normal and tangential components $N$ and $V$, as shown in Fig. 4-19. These components can be used to compute normal and shear stresses on the inclined surface by using Eqs. (4-2) and (4-4). The area $A_n$ of the inclined surface equals $A/\cos \theta$, where $A$ is the transverse cross-sectional area of the axially loaded member, the normal force is $N = P \cos \theta$, and the shear force is $V = P \sin \theta$. Therefore,

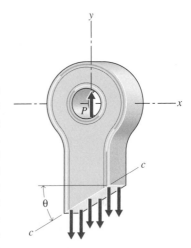

**Figure 4-17**

$$\sigma_n = \frac{N}{A_n} = \frac{P \cos \theta}{A/\cos \theta} = \frac{P}{A} \cos^2 \theta = \frac{P}{2A} (1 + \cos 2\theta) \qquad (4\text{-}7)$$

$$\tau_n = \frac{V}{A_n} = \frac{P \sin \theta}{A/\cos \theta} = \frac{P}{A} \sin \theta \cos \theta = \frac{P}{2A} \sin 2\theta \qquad (4\text{-}8)$$

In the preceding discussion, the assumption was made that the stress is uniformly distributed over the inclined surface. Nonuniform stress distribution under axial loading will be discussed later in this book.

Both the area of the inclined surface $A_n$ and the values for the normal and shear forces $N$ and $V$ on the surface depend on the angle $\theta$ of the inclined plane; therefore, the normal and shear stresses $\sigma_n$ and $\tau_n$ on the inclined plane also depend on the angle $\theta$. This dependence of stress on both force and area means that stress is not a vector quantity; therefore, the laws of vector addition do not apply to stresses that act on different planes. This need not be cause for concern if, in the application of the equations of equilibrium (or motion), one always replaces a stress with a total force (stress multiplied by the appropriate area), thus reducing the problem to one involving ordinary force vectors. However, stresses that act on a single particular plane can be treated as vectors because they all are associated with the same area.

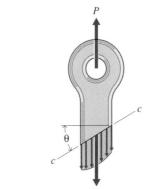

**Figure 4-18**

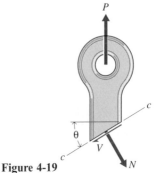

**Figure 4-19**

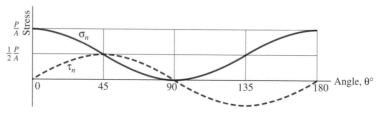

**Figure 4-20**

A graph showing the magnitudes of $\sigma_n$ and $\tau_n$ as a function of $\theta$ is shown in Fig. 4-20. These results indicate that $\sigma_n$ is maximum when $\theta$ is 0° or 180°, that $\tau_n$ is maximum when $\theta$ is 45° or 135°, and also that $\tau_{\max} = \sigma_{\max}/2$. Therefore, the maximum normal and shearing stresses for axial tensile or compressive loading are

$$\sigma_{\max} = P/A \qquad (4\text{-}9)$$

$$\tau_{\max} = P/2A \qquad (4\text{-}10)$$

Note that the normal stress is either maximum or minimum on planes for which the shearing stress is zero. It will be shown in Section 10-5 that the shearing stress is always zero on the planes of maximum or minimum normal stress. The concepts of maximum and minimum normal stress and maximum shearing stress for more general cases will be treated in later sections of this book.

Laboratory experiments indicate that both normal and shearing stresses under axial loading are important, since a brittle material loaded in tension will fail in tension on a transverse plane, whereas a ductile material loaded in tension will fail in shear on a 45° plane (Section 4-5).

The plot of normal and shear stresses for axial loading, shown in Fig. 4-20, indicates that the sign of the shearing stress changes when $\theta$ is greater than 90°. The magnitude of the shearing stress for any angle $\theta$, however, is the same as that for 90° + $\theta$. The sign change merely indicates that the shear force $V$ changes sense, being directed down to the right on plane 90° + $\theta$ instead of down to the left on plane $\theta$ as shown in Fig. 4-19. Later in the book it will be shown that if a shearing stress exists at a point on any plane, a shearing stress of the same magnitude must also exist at this point on an orthogonal plane in order to maintain equilibrium of the body.

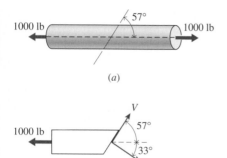

**Figure 4-21**

## ▌ Example Problem 4-6

A plastic bar with a circular cross section will be used to support an axial load of 1000 lb as shown in Fig. 4-21a. The allowable normal and shearing stresses in the adhesive joint used to connect the two parts of the bar are 675 psi and 350 psi, respectively. Determine the minimum diameter $d$ for the bar.

### SOLUTION

A free-body diagram for the portion of the bar to the left of the joint is shown in Fig. 4-21b. From a summation of forces normal and tangent to the inclined surface,

$+ \searrow F_n = 0:$        $N - 1000 \cos 33° = 0$        $N = 838.7$ lb

$+ \nearrow \Sigma F_t = 0:$        $V - 1000 \sin 33° = 0$        $V = 544.6$ lb

$$A = \frac{\pi d^2}{4} \qquad A_n = \frac{A}{\cos 33°} = \frac{\pi d^2/4}{\cos 33°} = 0.9365 d^2$$

From Eq. (4-1),

$$\sigma_n = \frac{N}{A_n} \qquad 675 = \frac{838.7}{0.9365 d^2} \qquad d = 1.1519 \text{ in.}$$

$$\tau_n = \frac{V}{A_n} \qquad 350 = \frac{544.6}{0.9365 d^2} \qquad d = 1.2890 \text{ in.}$$

Therefore

$$d_{min} = 1.289 \text{ in.} \qquad \textbf{Ans.}$$

Alternatively, from Eqs. (4-7) and (4-8),

$$\sigma_n = \frac{P}{2A}(1 + \cos 2\theta) = \frac{1000}{2(\pi d^2/4)}(1 + \cos 66°) = 675 \qquad d = 1.1518 \text{ in.}$$

$$\tau_n = \frac{P}{2A}(\sin 2\theta) = \frac{1000}{2(\pi d^2/4)}(\sin 66°) = 350 \qquad d = 1.2891 \text{ in.}$$

Therefore,

$$d_{min} = 1.289 \text{ in.} \qquad \blacksquare \qquad \textbf{Ans.}$$

## Example Problem 4-7

The block shown in Fig. 4-22a has a 200- $\times$ 100-mm rectangular cross section. The normal stress on plane $AB$ is 12.00 MPa C when the load $P$ is applied. If angle $\phi$ is 36°, determine

(a) The load $P$.
(b) The shearing stress on plane $AB$.
(c) The maximum normal and shearing stresses in the block.

### SOLUTION

$$A = 200(100) = 20,000 \text{ mm}^2 = 0.0200 \text{ m}^2$$
$$A_n = A/\cos \theta = 0.020/\cos 54° = 0.03403 \text{ m}^2$$

(a) A free-body diagram for the portion of the bar above plane $AB$ is shown in Fig. 4-22b. From Eq. (4-1),

$$N = \sigma_n A_n = 12(10^6)(0.03403) = 408.4(10^3) \text{ N} = 408.4 \text{ kN}$$

From a summation of forces normal to the plane,

$+ \nwarrow \Sigma F_n = 0:$        $408.4 - P \cos 54° = 0$

$$P = 694.8 \text{ kN} \cong 695 \text{ kN C} \qquad \textbf{Ans.}$$

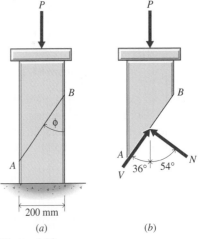

Figure 4-22

(b) From Eq. (4-8),

$$\tau_n = \frac{P}{2A}(\sin 2\theta) = \frac{694.8(10^3)}{2(0.0200)}(\sin 108°)$$

$$= 16.520(10^6) \text{ N/m}^2 = 16.52 \text{ MPa} \qquad \textbf{Ans.}$$

(c) From Eqs. (4-9) and (4-10),

$$\sigma_{max} = \frac{P}{A} = \frac{694.8(10^3)}{0.0200} = 34.74(10^6) \text{ N/m}^2 \cong 34.7 \text{ MPa C} \qquad \textbf{Ans.}$$

$$\tau_{max} = \frac{P}{2A} = \frac{694.8(10^3)}{2(0.0200)} = 17.37(10^6) \text{ N/m}^2 = 17.37 \text{ MPa} \blacksquare \qquad \textbf{Ans.}$$

---

# ■ PROBLEMS

## Introductory Problems

**4-22\*** The steel bar shown in Fig. P4-22 will be used to carry an axial tensile load of 400 kN. If the thickness of the bar is 45 mm, determine the normal and shearing stresses on plane $AB$.

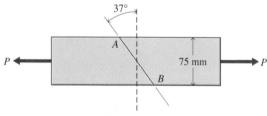

**Figure P4-22**

**4-23\*** An axial load $P$ is applied to a timber block with a 4- × 4-in. square cross section, as shown in Fig. P4-23. Determine the normal and shear stresses on the planes of the grain if $P = 5000$ lb.

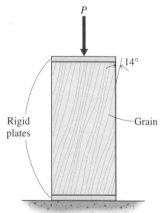

**Figure P4-23**

**4-24** A structural steel bar with a 20- × 25-mm rectangular cross section is subjected to an axial tensile load of 55 kN. Determine the maximum normal and shear stresses in the bar.

**4-25** A steel rod of circular cross section will be used to carry an axial tensile load of 50 kip. The maximum stresses in the rod must be limited to 25 ksi in tension and 15 ksi in shear. Determine the minimum diameter $d$ for the rod.

## Intermediate Problems

**4-26\*** Determine the maximum axial load $P$ that can be applied to the wood compression block shown in Fig. P4-26 if specifications require that the shear stress parallel to the grain not exceed 5.25 MPa, the compressive stress perpendicular to the grain not exceed 13.60 MPa, and the maximum shear stress in the block not exceed 8.75 MPa.

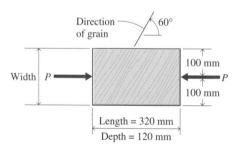

**Figure P4-26**

**4-27\*** A steel bar with a 4- × 1-in. rectangular cross section is being used to transmit an axial tensile load, as shown in Fig. P4-27. Normal and shear stresses on plane

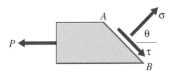

**Figure P4-27**

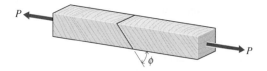

**Figure P4-30**

*AB* of the bar are 12 ksi tension and 9 ksi shear. Determine the angle $\theta$ and the applied load *P*.

**4-28** A steel bar with a butt-welded joint, as shown in Fig. P4-28 will be used to carry an axial tensile load of 400 kN. If the normal and shear stresses on the plane of the weld must be limited to 70 MPa and 45 MPa, respectively, determine the minimum thickness *t* required for the bar.

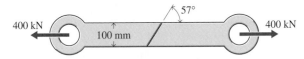

**Figure P4-28**

**4-29** The shearing stress on plane *AB* of the 4- × 8-in. rectangular block shown in Fig. P4-29 is 2 ksi when the axial load *P* is applied. If the angle $\phi$ is 35°, determine
(a) The load *P*.
(b) The normal stress on plane *AB*.
(c) The maximum normal and shearing stresses in the block.

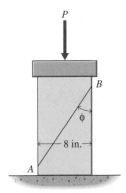

**Figure P4-29**

**Challenging Problems**

**4-30\*** A wood tension member with a 50- × 100-mm rectangular cross section will be fabricated with an inclined glued joint ($45° \le \phi \le 90°$) at its midsection, as shown in Fig. P4-30. If the allowable stresses for the glue are 5 MPa in tension and 3 MPa in shear, determine
(a) The optimum angle $\phi$ for the joint.
(b) The maximum safe load *P* for the member.

**4-31\*** The two parts of the eyebar shown in Fig. P4-31 are connected with two $^1/_2$-in.-diameter bolts (one on each side). Specifications for the bolts require that the axial tensile stress not exceed 12.0 ksi and that the shearing stress not exceed 8.0 ksi. Determine the maximum load *P* that can be applied to the eyebar without exceeding either specification.

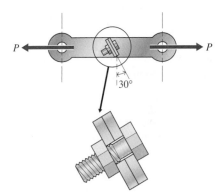

**Figure P4-31**

**4-32** The bar shown in Fig. P4-32 has a 200- × 100-mm rectangular cross section. Determine
(a) The normal and shearing stresses on plane *a–a*.
(b) The maximum normal and shearing stresses in the bar.

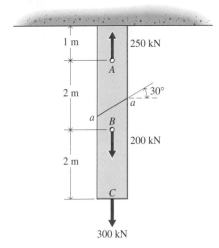

**Figure P4-32**

**4-33** A steel eyebar with a 4- × 1-in. rectangular cross section has been designed to transmit an axial tensile load. The length of the eyebar must be increased by welding a new center section in the bar ($45° \leq \phi \leq 90°$) as shown in Fig. P4-33. The stresses in the weld material must be limited to 12 ksi in tension and 9 ksi in shear. Determine
(a) The optimum angle $\phi$ for the joint.
(b) The maximum safe load $P$ for the redesigned member.

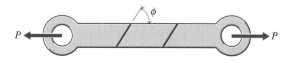

**Figure P4-33**

## Computer Problems

**4-34** A steel eyebar with a 100- × 25-mm rectangular cross section has been designed to transmit an axial tensile load $P$. The length of the eyebar must be increased by welding a new center section in the bar, as shown in Fig. P4-33. If $P = 250$ kN, calculate and plot the normal stress $\sigma_n$ and the shear stress $\tau_n$ in the weld material for weld angles $\phi$ ($30° \leq \phi \leq 90°$). If the stresses in the weld material must be limited to 80 MPa in tension and 60 MPa in shear, what ranges of $\phi$ would be acceptable for the joint? Repeat for $P = 305$ kN and for $P = 350$ kN. Are weld angles $\phi < 30°$ reasonable? Why or why not?

**4-35** Specifications for the rectangular (3- × 3- × 21-in.) block shown in Fig. P4-35 require that the normal and shearing stresses on plane $A$–$A$ not exceed 800 psi and 500 psi, respectively. If the plane $A$–$A$ makes an angle $\theta = 37°$ with the horizontal, calculate and plot the ratios $\sigma/\sigma_{max}$ and $\tau/\tau_{max}$ as a function of the load $P$ ($0 \leq P \leq 13$ kip). What is the maximum load $P_{max}$ that can be applied to the block? Which condition controls what the maximum load can be? Repeat for $\theta = 25°$. For what angle $\theta$ will the normal stress and the shear stress both reach their limiting values at the same time?

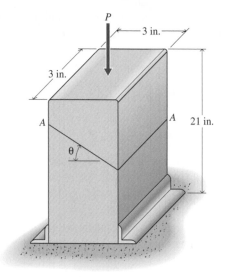

**Figure P4-35**

# 4-4 DISPLACEMENT, DEFORMATION, AND STRAIN

Relationships were developed in Sections 4-2 and 4-3 between forces and stresses and between stresses on planes having different orientations at a point using equilibrium considerations. No assumptions involving deformations or materials used in fabricating the body were made; therefore, the results are valid for an idealized rigid body or for a real deformable body. In the design of structural elements or machine components, the deformations experienced by the body as a result of the applied loads, often represent as important a design consideration as the stresses. For this reason, the nature of the deformations experienced by a real deformable body as a result of internal force or stress distributions will be studied and methods to measure or compute deformations will be established.

**Displacement.** When a system of loads is applied to a machine component or structural element, individual points of the body generally move. This movement of a point with respect to some convenient reference system of axes is a vector quantity known as a *displacement*. In some instances displacements are associated with a translation and/or rotation of the body as a whole, and neither the size nor the shape of the body is changed. The study of displacements in which neither the size nor the shape of the body is changed is the concern of courses in rigid-body mechanics. When displacements induced by applied loads cause the size and/or shape of a body to be altered, individual points of the body move relative to one another. The change in any dimension associated with these relative displacements is known as a *deformation* and will be designated by the Greek letter delta ($\delta$).

**Deformation.** Deformation is not uniquely related to force or stress, however. Two rods of identical material and identical cross-sectional area subjected to different loads (Fig. 4-23a), can have the same deformation if the second rod is half as long as the first. Similarly, if two rods of identical material and identical cross-sectional area are subjected to identical loads (Fig. 4-23b), the deformation in a 2-m-long rod will be twice as large as the deformation in a 1-m-long rod. Therefore, a quantitative measure of the intensity of the deformation is needed, just as stress is used to measure the intensity of an internal force (force per unit area).

**Strain.** *Strain* (deformation per unit length) is the quantity used to measure the intensity of a deformation just as stress (force per unit area) is used to measure the intensity of an internal force. In Section 4-2, two types of stresses were defined: normal stresses and shearing stresses. This same classification is used for strains. *Normal strain,* designated by the Greek letter epsilon ($\epsilon$), measures the change in size (elongation or contraction of an arbitrary line segment) of a body during deformation. *Shearing strain,* designated by the Greek letter gamma ($\gamma$), measures the change in shape (change in angle between two lines that are orthogonal in the undeformed state) of a body during deformation. The deformation or strain may be the result of a stress, of a change in temperature, or of other physical phenomena such as grain growth or shrinkage. In this book only strains resulting from changes in temperature or stress are considered.

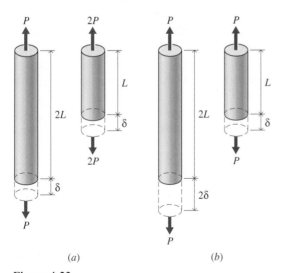

(a)                    (b)

**Figure 4-23**

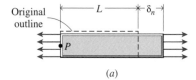

(a)

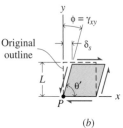

(b)

**Figure 4-24**

**Average Axial Strain.** The change in length (or width) of a simple bar under an axial load (Fig. 4-24a) can be used to illustrate the idea of a normal strain. The average axial strain (a normal strain, hereafter called *axial strain*) $\epsilon_{avg}$ over the length of the bar is obtained by dividing the axial deformation $\delta_n$ by the original length of the bar (the length $L$)

$$\epsilon_{avg} = \frac{\delta_n}{L} \qquad (4\text{-}11)$$

That is, the axial strain is the deformation $\delta_n$ in the direction of the length divided by the length $L$.

**True Axial Strain.** In cases in which the deformation is nonuniform along the length of the bar (a long bar hanging under its own weight, for example), the average axial strain given by Eq. (4-11) may be significantly different from the true axial strain at an arbitrary point $P$ along the bar. The true axial strain at a point can be determined by making the length over which the axial deformation is measured smaller and smaller. In the limit as $\Delta L \to 0$, a quantity defined as the axial strain at the point, $\epsilon(P)$, is obtained. This limit process is indicated by the expression

$$\epsilon(P) = \lim_{\Delta L \to 0} \frac{\Delta \delta_n}{\Delta L} = \frac{d\delta_n}{dL} \qquad (4\text{-}12)$$

**Shearing Strain.** In a similar manner a deformation involving a change in shape can be used to illustrate a shearing strain. The average shearing strain $\gamma_{avg}$ is obtained by dividing the deformation $\delta_s$ in a direction normal to the length $L$ by the length (Fig. 4-24b):

$$\gamma_{avg} = \frac{\delta_s}{L} = \tan \phi \qquad (4\text{-}13)$$

Since $\delta_s/L$ is usually very small (typically $\delta_s/L < 0.001$), $\sin \phi \cong \tan \phi \cong \phi$, where $\phi$ is measured in radians. Therefore $\gamma_{avg} = \phi = \delta_s/L$ is the decrease in the angle between two reference lines that are orthogonal in the undeformed state. Again, for those cases in which the deformation is nonuniform, the shearing strain at a point, $\gamma_{xy}(P)$, associated with two orthogonal reference lines $x$ and $y$, is obtained by measuring the shearing deformation as the size of the element is made smaller and smaller. In the limit as $\Delta L \to 0$,

$$\gamma_{xy}(P) = \lim_{\Delta L \to 0} \frac{\Delta \delta_s}{\Delta L} = \frac{d\delta_s}{dL} \qquad (4\text{-}14a)$$

The angle $\gamma_{xy}$ is difficult to observe and even more difficult to measure. An equivalent expression for shearing strain that is sometimes useful for calculations is

$$\gamma_{xy}(P) = \frac{\pi}{2} - \theta' \qquad (4\text{-}14b)$$

In this expression $\theta'$ is the angle in the deformed state between the two initially orthogonal reference lines.

**Units of Strain.** Equations (4-11) through (4-14) indicate that both normal and shearing strains are dimensionless quantities; however, normal strains are frequently expressed in units of in./in. or micro-in./in., while shearing strains are expressed in radians or micro-radians. The symbol $\mu$ is frequently used to indicate micro- $(10^{-6})$.

**Tensile/Compressive Strains.** From the definition of normal strain given by Eq. (4-11) or (4-12) it is evident that normal strain is positive when a line elongates and negative when the line contracts. In general, if the axial stress is tensile, the axial deformation will be an elongation. Therefore, positive normal strains are referred to as *tensile strains*. The reverse will be true for compressive axial stresses; therefore, negative normal strains are referred to as *compressive strains*. From Eq. (4-14b) it is evident that shearing strains will be positive if the angle between reference lines decreases. If the angle increases, the shearing strain is negative. Positive and negative shearing strains are not given special names. Normal and shearing strains for most engineering materials in the elastic range (see Section 4-5) seldom exceed values of 0.2 percent (0.002 in./in. or 0.002 rad).

## Example Problem 4-8

A 1.00-in.-diameter steel bar is 8 ft long. The diameter is reduced to $\frac{1}{2}$ in. in a 2-ft central portion of the bar. When an axial load is applied to the ends of the bar, the axial strain in the central portion of the bar is 960 $\mu$in./in., and the total elongation of the bar is 0.04032 in. Determine

(a) The elongation of the central portion of the bar.

(b) The axial strain in the end portions of the bar.

### SOLUTION

(a) The elongation of the central portion of the bar $\delta_C$ is obtained by using Eq. (4-11). Thus,

$$\delta_C = \epsilon_{\text{avg}} L = 960(10^{-6})(2)(12)$$

$$= 0.02304 \text{ in.} \cong 0.0230 \text{ in.} \qquad \textbf{Ans.}$$

(b) The elongation of the end portions of the bar $\delta_E$ is

$$\delta_E = \delta_{\text{total}} - \delta_C = 0.04032 - 0.02304 = 0.01728 \text{ in.}$$

The axial strain in the end portions of the bar is obtained by using Eq. (4-11):

$$\epsilon_E = \frac{\delta_E}{L} = \frac{0.01728}{6(12)} = 240(10^{-6}) = 240 \ \mu\text{in./in.} \ \blacksquare \qquad \textbf{Ans.}$$

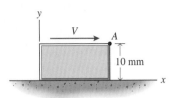

**Figure 4-25**

## Example Problem 4-9

The shear force $V$ shown in Fig. 4-25 produces an average shearing strain $\gamma_{avg}$ of 1000 $\mu$m/m in the block of material. Determine the horizontal movement of point $A$ resulting from application of the shear force $V$.

### SOLUTION

The horizontal movement of point $A$ is obtained by using Eq. (4-13). Thus,

$$\delta_A = \gamma_{avg}L = 1000(10^{-6})(10)$$
$$= 0.0100 \text{ mm} = 10.00 \ \mu\text{m} \quad \blacksquare \qquad \textbf{Ans.}$$

## PROBLEMS

### Introductory Problems

**4-36\*** Compression tests of concrete indicate that concrete fails when the axial compressive strain is 1200 $\mu$m/m. Determine the maximum change in length that a 200-mm-diameter $\times$ 400-mm-long concrete test specimen can tolerate before failure occurs.

**4-37\*** The 0.5- $\times$ 2- $\times$ 4-in. rubber mounts shown in Fig. P4-37 are used to isolate the vibrational motion of a machine from its supports. Determine the average shearing strain in the rubber mounts if the rigid frame displaces 0.01 in. vertically relative to the support.

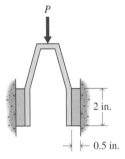

**Figure P4-37**

**4-38** A structural steel bar was loaded in tension to fracture. A 200-mm length of the bar was marked off in 25-mm lengths before loading. After the rod broke, the 25-mm segments were found to have lengthened to 30.0, 30.5, 31.5, 34.0, 44.5, 32.0, 31.0, and 30.0 mm, consecutively. Determine

(a) The average strain over the 200-mm length.
(b) The maximum average strain over any 50-mm length.

**4-39** Mutually perpendicular axes in an unstressed member were found to be oriented at 89.92° when the member was stressed. Determine the shearing strain associated with these axes in the stressed member.

### Intermediate Problems

**4-40\*** A thin triangular plate is uniformly deformed as shown in Fig. P4-40. Determine the shearing strain at $P$ associated with the two edges ($PQ$ and $PR$) that were orthogonal in the undeformed plate.

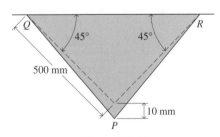

**Figure P4-40**

**4-41\*** The sanding-drum mandrel shown in Fig. P4-41 is made for use with a hand drill. The mandrel is made from a rubberlike material that expands when the nut is tightened to secure the sanding drum placed over the outside surface. If the diameter $D$ of the mandrel increases from 2.00 in. to 2.15 in. as the nut is tightened, determine
(a) The average normal strain along a diameter of the mandrel.
(b) The circumferential strain at the outside surface of the mandrel.

**Figure P4-41**

**4-42** A thin rectangular plate is uniformly deformed as shown by $PRSQ$ in Fig. P4-42. Determine the shearing stain $\gamma_{xy}$ at $P$.

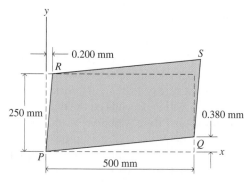

**Figure P4-42**

**4-43** A steel sleeve is connected to a steel shaft with a flexible rubber insert, as shown in Fig. P4-43. The in-

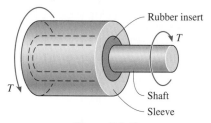

**Figure P4-43**

sert has an inside diameter of 3.3 in. and an outside diameter of 4.3 in. When the unit is subjected to a torque $T$, the shaft rotates 1.5° with respect to the sleeve. Assume that radial lines in the unstressed state remain straight as the rubber deforms. Determine the shearing strain $\gamma_{r\theta}$ in the rubber insert
(a) At the inside surface.
(b) At the outside surface.

**Challenging Problems**

**4-44\*** A steel rod is subjected to a nonuniform heating that produces an extensional (axial) strain that is proportional to the square of the distance from the unheated end ($\epsilon = kx^2$). If the strain is 1250 $\mu$m/m at the midpoint of a 3.00-m rod, determine
(a) The change in length of the rod.
(b) The average axial strain over the length $L$ of the rod.
(c) The maximum axial strain in the rod.

**4-45\*** The axial strain in a suspended bar of material of varying cross section due to its own weight, as shown in Fig. P4-45, is given by the expression $\gamma y/3E$, where $\gamma$ is the specific weight of the material, $y$ is the distance from the free (bottom) end of the bar, and $E$ is a material constant. Determine, in terms of $\gamma$, $L$, and $E$,
(a) The change in length of the bar due to its own weight.
(b) The average axial strain over the length $L$ of the bar.
(c) The maximum axial strain in the bar.

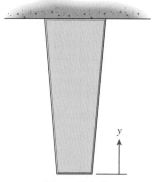

**Figure P4-45**

**4-46** A steel cable is used to tether an observation balloon. The force exerted on the cable by the balloon is sufficient to produce a uniform strain of 500 $\mu$m/m in the cable. In addition, at each point in the cable, the weight of the cable reduces the axial strain by an amount that is proportional to the length of the cable between the balloon and the point. When the balloon is directly overhead at an elevation of 300 m, the axial strain at the midlength of the cable is 350 $\mu$m/m. Determine
(a) The total elongation of the cable.
(b) The maximum height that the balloon could achieve

**4-47** A steel cable is used to support an elevator cage at the bottom of a 2000-ft-deep mine shaft. A uniform axial strain of 250 $\mu$in./in. is produced in the cable by the weight of the cage. At each point the weight of the cable produces an additional axial strain that is proportional to the length of the cable below the point. If the total axial strain in the cable at the cable drum (upper end of the cable) is 700 $\mu$in./in., determine
(a) The strain in the cable at a depth of 500 ft.
(b) The total elongation of the cable.

## 4-5 STRESS–STRAIN–TEMPERATURE RELATIONSHIPS

The satisfactory performance of a structure frequently is determined by the amount of deformation or distortion that can be permitted. A deformation of a few thousandths of an inch might make a boring machine useless, whereas the hook on a drag line might deform several inches without impairing its usefulness. It is often necessary to relate loads and temperature changes on a structure to the deformations produced by the loads and temperature changes. Experience has shown that the deformations caused by loads and by temperature effects are essentially independent of each other. The deformations due to the two effects may be computed separately and added together to get the total deformation.

**Stress–Strain Diagrams** The relationship between loads and deformation in a structure can be obtained by plotting diagrams showing loads and deformations for each member and each type of loading in a structure. However, the relationship between load and deformation depends on the dimensions of the members as well as on the type of material from which the members are made. For example, the graph of Fig. 4-26$a$ shows the relationship between the force required to stretch three bars of the same material but of different lengths and cross-sectional areas and the resulting deformations of the bars. It is not clear

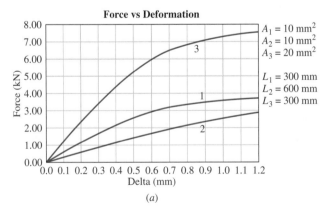

**Figure 4-26**

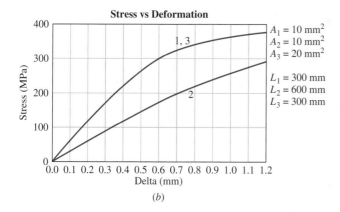

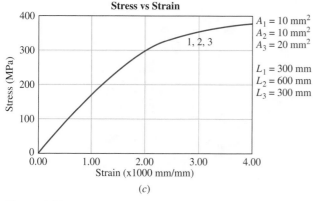

**Figure 4-26**  (*Continued*).

from this graph that these three curves all describe the same material behavior. However, if these curves are redrawn plotting stress versus deformation, as in Fig. 4-26*b*, the data for the first and third bars form a single line. If the curves are redrawn plotting stress versus strain as in Fig. 4-26*c*, the data for all three curves form a single line. That is, curves showing the relationship between stress and strain (such as Fig. 4-26*c*) are independent of the size and shape of the member and depend only on the type of material from which the members are made. Such diagrams are called *stress–strain diagrams.*

**The Tensile Test.** Data for stress–strain diagrams are obtained by applying an axial load to a test specimen and measuring the load and deformation simultaneously. A testing machine (Fig. 4-27) is used to deform the specimen and to measure the load required to produce the deformation. The stress is obtained by dividing the load by the initial cross-sectional area of the specimen, $\sigma = P/A$. The area will change somewhat during the loading, and the stress obtained using the initial area is obviously not the exact stress occurring at higher loads. However, it is the stress most commonly used in designing structures. Stress obtained by dividing the load by the actual area is frequently called the "true" stress and is useful in explaining the fundamental behavior of materials.

**Strain Measurement.** Strains are small in materials used in engineering structures, often less than 0.001, and their accurate determination requires special measuring equipment. Normal strain is obtained by measuring the deformation $\delta$ in a length $L$ and dividing $\delta$ by $L$. Instruments for measuring the defor-

**Figure 4-27** Hydraulic testing machine set up for a tension test (courtesy of MTS Systems Corporation).

mation $\delta$ are called *strain gages* or *extensometers,* and they obtain the desired accuracy by multiplying levers, dial indicators, beams of light, or other means. The electrical resistance strain gage is widely used for this type of measurement.

True strain, like true stress, is computed on the basis of the actual length of the test specimen during the test and is used primarily to study the fundamental properties of materials. The difference between nominal stress and strain, computed from initial dimensions of the specimen, and true stress and strain is negligible for stresses usually encountered in engineering structures, but sometimes the difference becomes important with larger stresses and strains. A more complete discussion of the experimental determination of stress and strain will be found in various books on experimental stress analysis.[1]

**Example Stress–Strain Diagrams.** Figures 4-28*a*, *b*, and *c* show tensile stress–strain diagrams for structural steel (a low-carbon steel), for a magnesium alloy, and for a gray cast iron, respectively. These diagrams will be used to explain a number of properties useful in the study of mechanics of materials. Although some of the relationships that follow have a basis in theory (for example, the linear relationship between stress and strain for small strain), others are purely empirical fits to experimental data. In either case, the values of specific constants for various materials must be experimentally determined.

---

[1]*Experimental Stress Analysis*, 3rd edition, J. W. Dally and W. F. Riley, McGraw-Hill, New York, 1991.

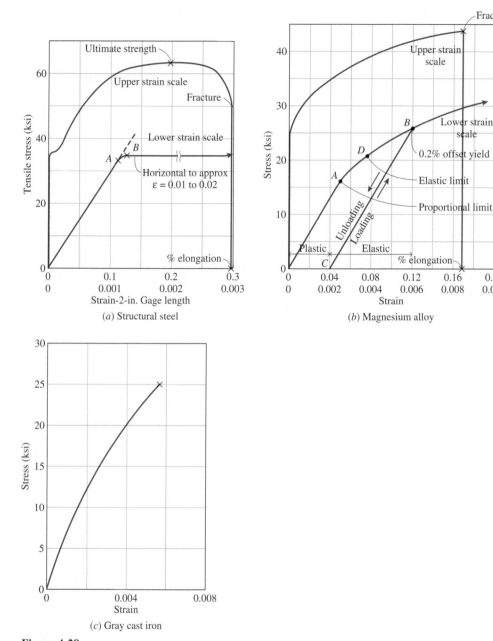

(a) Structural steel

(b) Magnesium alloy

(c) Gray cast iron

**Figure 4-28**

**Modulus of Elasticity.** The initial portion of the stress–strain diagram for most materials used in engineering structures (see Figs. 4-28a and b) is a straight line. The stress–strain diagrams for some materials, such as gray cast iron (see Fig. 4-28c) and concrete, show a slight curve even at very small stresses, but it is common practice to draw a straight line to average the data for the first part of the diagram and neglect the curvature. The proportionality of load to deflection was first recorded by Robert Hooke, who observed in 1678, "ut tensio sic vis" (as the stretch so the force); this is frequently referred to as Hooke's law

$$\sigma = E\epsilon \qquad (4\text{-}15a)$$

where $E$ is the slope of the straight-line portion of the stress–strain diagram. *It is important to realize that Hooke's law describes only the initial linear portion of the stress–strain diagram and is valid only for bars loaded in uniaxial extension, as in the testing machine of Fig. 4-27.* A modified version of Hooke's law valid for materials being stretched in two or three directions at the same time will be derived in Chapter 10.

Thomas Young, in 1807, suggested what amounts to using the ratio of stress to strain to measure the stiffness of a material. This ratio is called *Young's modulus* or the *modulus of elasticity,* and is the slope of the straight line portion of the stress–strain diagram. Young's modulus is written as

$$E = \sigma/\epsilon \qquad \text{(normal stress–strain)} \qquad (4\text{-}15b)$$

$$G = \tau/\gamma \qquad \text{(shear stress–strain)} \qquad (4\text{-}15c)$$

where $E$ is the modulus used for normal stress $\sigma$ and normal strain $\epsilon$ and $G$ (sometimes called the *shear modulus* or the *modulus of rigidity*) is the modulus used for shearing stress $\tau$ and shearing strain $\gamma$. The maximum stress for which stress and strain are proportional is called the *proportional limit* and is indicated by the ordinates at points $A$ on Fig. 4-28$a$ or $b$. The exact point of the proportional limit is difficult to determine from the stress–strain curve.

For points on the stress–strain curve beyond the proportional limit (such as point $C$ on Fig. 4-29), other quantities such as the tangent modulus and the secant modulus are used as measures of the stiffness of a material. The tangent modulus $E_t$ is defined as the slope of the stress–strain diagram at a particular stress level. Thus, the tangent modulus is a function of the stress (or strain) for stresses greater than the proportional limit. The secant modulus $E_s$ is the ratio of the stress to the strain at any point on the diagram. Young's modulus $E$, the tangent modulus $E_t$, and the secant modulus $E_s$ are all illustrated in Fig. 4-29. For stresses less than the proportional limit (point $A$), all three moduli are the same.

**Elastic Limit.** The action is said to be elastic if the strain resulting from loading disappears when the load is removed. The elastic limit (point $D$ in Fig. 4-28$b$) is the maximum stress for which the material acts elastically. For stresses above the elastic limit, some deformation (strain) remains when the load is removed (and the stress goes to zero). For most materials it is found that the stress–strain diagram for unloading (see line $BC$ in Fig. 4-28$b$) is approximately parallel to the linear loading portion (see line $OA$ in Fig. 4-28$b$). If the specimen is again loaded, the stress–strain diagram will usually follow the unloading curve until it reaches a stress a little less than the maximum stress attained during the initial loading, at which time it will start to curve in the direction of the initial loading curve. As indicated in Fig. 4-28$b$, the proportional limit for the second loading (point $B$) is greater than that for the initial loading. This phenomenon is called *strain hardening or work hardening.*

When the stress exceeds the elastic limit (or proportional limit for practical purposes), it is found that a portion of the deformation remains after the load is removed. The deformation remaining after an applied load is removed is called *plastic deformation* ($OC$ in Fig. 4-28$b$). Plastic deformation that depends only on the amount of load (stress) and is independent of the time duration of the applied load is known as *slip*. Plastic deformation that continues to increase under a constant stress is called *creep*. In many instances creep continues until fracture occurs; however, in other instances the rate of creep decreases and approaches zero

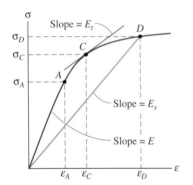

**Figure 4-29**

as a limit. Some materials are much more susceptible to creep than are others, but most materials used in engineering exhibit creep at elevated temperatures. The total strain is thus made up of elastic strain, possibly combined with plastic strain that results from slip, creep, or both. When the load is removed, the elastic portion of the strain is recovered, but the plastic part (slip and creep) remains as permanent set.[2]

**Yield Point.** A precise value for the proportional limit is difficult to obtain when the transition of the stress–strain diagram from a straight line to a curve is gradual. For this reason, other measures of stress that can be used as a practical elastic limit are required. The yield point and the yield strength for a specified offset are used for this purpose.

The *yield point* is the stress at which there is an appreciable increase in strain with no increase in stress, with the limitation that, if straining is continued, the stress will again increase. This latter specification indicates that there is a kink or "knee" in the stress–strain diagram, as indicated in Fig 4-28a. The yield point is easily determined without the aid of strain-measuring equipment because the load indicated by the testing machine ceases to rise (or may even drop) at the yield point. Unfortunately, few materials possess this property, the most common examples being low-carbon steels.

**Yield Strength.** The *yield strength* is defined as the stress that will induce a specified permanent set, usually 0.05 to 0.3 percent (which is equivalent to a strain of 0.0005 to 0.003), with 0.2 percent being the most commonly used value. The yield strength can be conveniently determined from a stress–strain diagram by laying off the specified offset (permanent set) on the strain axis (*OC* in Fig. 4-28b) and drawing a line *CB* parallel to *OA*. The stress indicated by the intersection of *CB* and the stress–strain diagram is the yield strength for the specified offset.

**Ultimate Strength.** The maximum stress (based on the original area) developed in a material before rupture is called the *ultimate strength* of the material (Fig. 4-28a), and the term may be modified as the ultimate tensile, compressive, or shearing strength of the material. Ductile materials undergo considerable plastic tensile or shearing deformation before rupture. When the ultimate strength of a ductile material is reached, the cross-sectional area of the test specimen starts to decrease or neck down (see Fig. 4-30), and the resultant load that can be carried by the specimen decreases. Thus, the stress based on the original area decreases beyond the ultimate strength of the material (Fig. 4-28a), although the true stress continues to increase until rupture.

Necking

**Figure 4-30**

**Elastoplastic Materials.** Most engineering structures are designed so that the stresses are less than the proportional limit; therefore, Young's modulus (the modulus of elasticity) provides a simple and convenient relationship between stress and strain. When the stress exceeds the proportional limit, no simple relation exists between stress and strain. Various empirical equations have been proposed relating the stress and strain beyond the proportional limit. A stress–strain dia-

---

[2]In some instances a portion of the strain that remains immediately after the stress is removed may disappear after a period of time. This reduction of strain is sometimes called *recovery*.

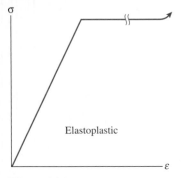

**Figure 4-31**

gram similar to the one shown in Fig. 4-31 (elastoplastic) is frequently assumed for mild steel or other materials with similar properties in order to simplify calculations.

**Ductility.** Strength and stiffness are not the only properties of interest to a design engineer. Another important property is *ductility*, defined as the capacity for plastic deformation in tension or shear. This property controls the amount of cold forming to which a material may be subjected. The forming of automobile bodies and the manufacture of fencing and other wire products all require ductile materials. Ductility is also an important property of materials used for fabricated structures. Under static loading, the presence of large stresses in the region of rivet holes or welds may be ignored, since ductility permits considerable plastic action to take place in the region of high stress, with a resulting redistribution of stress and the establishment of equilibrium. Two commonly used quantitative indices of ductility are the ultimate elongation (expressed as a percent elongation of length at rupture) and the reduction of cross-sectional area at the section where rupture occurs (expressed as a percentage of the original area).

**Creep Limit.** The property indicating the resistance of a material to failure by creep is known as the *creep limit* and is defined as the maximum stress for which the plastic strain will not exceed a specified amount during a specified time interval at a specified temperature. The creep limit is important when designing parts to be fabricated with polymeric materials (commonly known as plastics) and when designing metal parts that will be subjected to high temperatures and sustained loads (for example, the turbine blades in a turbojet engine).

**Poisson's Ratio.** A material loaded in one direction will undergo strains perpendicular to the direction of the load in addition to those parallel to the load. The ratio of the lateral or transverse strain ($\epsilon_{lat}$ or $\epsilon_t$) to the longitudinal or axial strain ($\epsilon_{long}$ or $\epsilon_a$) is called *Poisson's ratio* after Siméon D. Poisson, who identified the constant in 1811. Poisson's ratio is a constant for stresses below the proportional limit and has a value between $\frac{1}{4}$ and $\frac{1}{3}$ for most metals. The Greek letter nu ($\nu$) is used for Poisson's ratio, which is given by the equation

$$\nu = -\frac{\epsilon_{lat}}{\epsilon_{long}} = -\frac{\epsilon_t}{\epsilon_a} \tag{4-16}$$

*The ratio $\nu = -\epsilon_t/\epsilon_a$ is valid only for a uniaxial state of stress.* It will be shown later in this book that Poisson's ratio is related to $E$ and $G$ by the formula

$$E = 2(1 + \nu)G \tag{4-17}$$

The properties discussed in this section are primarily concerned with static or continuous loading or with slowly varying loading at room temperature.

**Effect of Composition.** The alloy content of a material affects the stress–strain behavior of the material. For example, Fig. 4-32 shows the effects of various alloy content on the stress–strain curves for steels of different strength levels ranging from a very hard, strong, brittle steel (A) to a relatively soft, ductile steel (E). The alloy content does not affect the modulus of elasticity but does affect the elastic limit, the ultimate strength, the fracture strength, and the ductility of the steel.

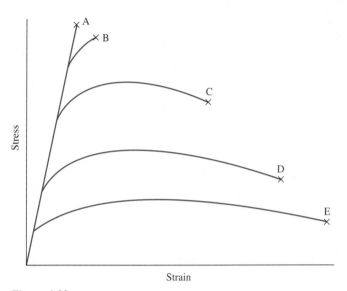

**Figure 4-32**

**Effect of Temperature.**  Temperature also affects the stress–strain behavior of a material. The stress-strain diagrams shown in Fig. 4-28 were for room temperature. Figure 4-33 shows the effect of temperature on the tensile stress–strain diagram for a class 40 gray iron. The ductility of the material increases as the temperature increases, whereas the ultimate strength decreases as the temperature increases.

**Effect of Tension or Compression.**  The stress–strain behavior of some materials depends upon whether the axial load is tension or compression. For duc-

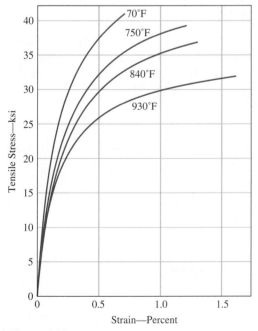

**Figure 4-33**

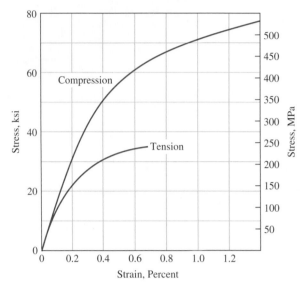

**Figure 4-34**

tile materials the tension and compression behavior are usually assumed to be the same. For brittle materials the stress–strain curve obtained in a tension test differs from the curve obtained in a compression test. For example, Fig. 4-34 shows the tension and compression stress–strain curves for a class 35 gray iron. The linear range of stress–strain behavior for the compression curve is larger than the linear range for the tension curve. The ultimate strength in compression is also greater than the ultimate strength in tension.

## Example Problem 4-10

A 100-kip axial load is applied to a 1- × 4- × 90-in. rectangular bar. When loaded, the 4-in. side measures 3.9986 in. and the length has increased 0.09 in. Determine Poisson's ratio, the modulus of elasticity, and the modulus of rigidity of the material.

### SOLUTION

The lateral and longitudinal deformations and strains and the axial stress for the bar are

$$\delta_{\text{lat}} = 3.9986 - 4 = -0.0014 \text{ in.}$$

$$\epsilon_{\text{lat}} = \frac{\delta_{\text{lat}}}{L} = \frac{-0.0014}{4} = -0.00035$$

$$\epsilon_{\text{long}} = \frac{\delta_{\text{long}}}{L} = \frac{0.09}{90} = 0.00100$$

$$\sigma = \frac{P}{A} = \frac{100}{4(1)} = 25 \text{ ksi}$$

Poisson's ratio is obtained by using Eq. (4-16):

$$\nu = -\frac{\epsilon_{lat}}{\epsilon_{long}} = -\frac{-0.00035}{0.00100} = 0.35$$

**Ans.**

The modulus of elasticity is obtained by using Eq. (4-15b):

$$E = \frac{\sigma}{\epsilon} = \frac{25}{0.00100} = 25,000 \text{ ksi}$$

**Ans.**

The modulus of rigidity is obtained by using Eq. (4-17):

$$G = \frac{E}{2(1 + \nu)} = \frac{25,000}{2(1 + 0.35)} = 9260 \text{ ksi} \blacksquare$$

**Ans.**

# PROBLEMS

### Introductory Problems

**4-48*** At the proportional limit, a 200-mm-gage length of a 15-mm-diameter alloy bar has elongated 0.90 mm and the diameter has been reduced 0.022 mm. The total axial load carried was 62.6 kN. Determine the modulus of elasticity, Poisson's ratio, and the proportional limit for the material.

**4-49*** A 1.50-in.-diameter rod 20 ft long elongates 0.48 in. under a load of 53 kip. The diameter of the rod decreases 0.001 in. during the loading. Determine the modulus of elasticity, Poisson's ratio, and the modulus of rigidity for the material.

### Intermediate Problems

**4-50*** A tensile test specimen having a diameter of 5.64 mm and a gage length of 50 mm was tested to fracture. Stress and strain values, which were calculated from load and deformation data obtained during the test, are shown in Fig. P4-50. Determine
(a) The modulus of elasticity.
(b) The proportional limit.
(c) The ultimate strength.
(d) The yield strength (0.05% offset).
(e) The yield strength (0.2% offset).
(f) The fracture stress.
(g) The true fracture stress if Poisson's ratio $\nu = 0.30$ remains constant.
(h) The tangent modulus at a stress level of 400 MPa.
(i) The secant modulus at a stress level of 400 MPa.

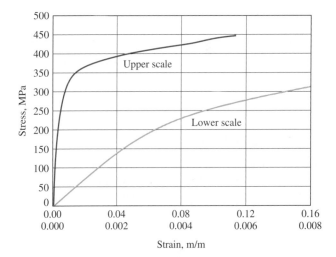

**Figure P4-50**

**4-51** A tensile test specimen having a diameter of 0.250 in. and a gage length of 2.000 in. was tested to fracture. Stress and strain values, which were calculated from load and deformation data obtained during the test, are shown in Fig. P4-51. Determine
(a) The modulus of elasticity.
(b) The proportional limit.
(c) The ultimate strength.

(d) The yield strength (0.05% offset).
(e) The yield strength (0.2% offset).
(f) The fracture stress.
(g) The true fracture stress if the final diameter of the specimen at the location of the fracture was 0.212 in.
(h) The tangent modulus at a stress level of 56 ksi.
(i) The secant modulus at a stress level of 56 ksi.

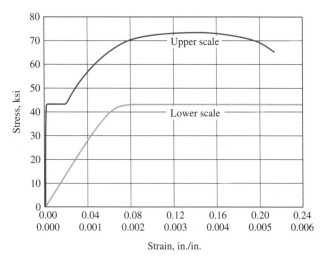

**Figure P4-51**

### Challenging Problems

**4-52** A tensile test specimen having a diameter of 11.28 mm and a gage length of 50 mm was tested to fracture. Load and deformation data obtained during the test were as follows:

| Load (kN) | Change in Length (mm) | Load (kN) | Change in Length (mm) |
|---|---|---|---|
| 0 | 0 | 43.8 | 1.50 |
| 7.6 | 0.02 | 45.8 | 2.00 |
| 14.9 | 0.04 | 48.3 | 3.00 |
| 22.2 | 0.06 | 49.7 | 4.00 |
| 28.5 | 0.08 | 50.4 | 5.00 |
| 29.9 | 0.10 | 50.7 | 6.00 |
| 30.6 | 0.12 | 50.4 | 7.00 |
| 32.0 | 0.16 | 50.0 | 8.00 |
| 33.0 | 0.20 | 49.7 | 9.00 |
| 33.3 | 0.24 | 47.9 | 10.00 |
| 36.8 | 0.50 | 45.1 | Fracture |
| 41.0 | 1.00 | | |

Determine
(a) The modulus of elasticity.
(b) The proportional limit.
(c) The ultimate strength.
(d) The yield strength (0.05% offset).
(e) The yield strength (0.2% offset).
(f) The fracture stress.
(g) The true fracture stress if the final diameter of the specimen at the location of the fracture was 9.50 mm.
(h) The tangent modulus at a stress level of 315 MPa.
(i) The secant modulus at a stress level of 315 MPa.

**4-53*** A tensile test specimen having a diameter of 0.505 in. and a gage length of 2.00 in. was tested to fracture. Load and deformation data obtained during the test were as follows:

| Load (lb) | Change in Length (in.) | Load (lb) | Change in Length (in.) |
|---|---|---|---|
| 0 | 0 | 12,600 | 0.0600 |
| 2,200 | 0.0008 | 13,200 | 0.0800 |
| 4,300 | 0.0016 | 13,900 | 0.1200 |
| 6,400 | 0.0024 | 14,300 | 0.1600 |
| 8,200 | 0.0032 | 14,500 | 0.2000 |
| 8,600 | 0.0040 | 14,600 | 0.2400 |
| 8,800 | 0.0048 | 14,500 | 0.2800 |
| 9,200 | 0.0064 | 14,400 | 0.3200 |
| 9,500 | 0.0080 | 14,300 | 0.3600 |
| 9,600 | 0.0096 | 13,800 | 0.4000 |
| 10,600 | 0.0200 | 13,000 | Fracture |
| 11,800 | 0.0400 | | |

Determine
(a) The modulus of elasticity.
(b) The proportional limit.
(c) The ultimate strength.
(d) The yield strength (0.05% offset).
(e) The yield strength (0.2% offset).
(f) The fracture stress.
(g) The true fracture stress if the final diameter of the specimen at the location of the fracture was 0.425 in.
(h) The tangent modulus at a stress level of 46,000 psi.
(i) The secant modulus at a stress level of 46,000 psi.

# 4-6 THERMAL STRAIN

Most engineering materials when unrestrained expand when heated and contract when cooled. The thermal strain due to a one degree (1°) change in temperature is designated by $\alpha$ and is known as the *coefficient of thermal expansion*. The thermal strain due to a temperature change of $\Delta T$ degrees is

$$\epsilon_T = \alpha \, \Delta T \qquad (4\text{-}18)$$

Like the constants described in the last section, the value of $\alpha$ for various materials must be determined experimentally. The coefficient of thermal expansion is approximately constant for a large range of temperatures (in general, the coefficient increases with an increase of temperature). For a homogeneous, isotropic material,[3] the coefficient applies to all dimensions (all directions). Values of the coefficient of thermal expansion for several materials are included in Appendix A.

**Total Strains.** Strains caused by temperature changes and strains caused by applied loads are essentially independent. The total normal strain in a body acted on by both temperature changes and applied loads[4] is given by

$$\epsilon_{\text{total}} = \epsilon_\sigma + \epsilon_T = \frac{\sigma}{E} + \alpha \, \Delta T \qquad (4\text{-}19)$$

Since homogeneous, isotropic materials, when unrestrained, expand uniformly in all directions when heated (and contract uniformly when cooled), neither the shape of the body nor the shearing stresses and shearing strains are affected by temperature changes.

## Example Problem 4-11

A $\frac{1}{2}$-in.-diameter steel [$E = 30{,}000$ ksi, $\alpha = 6.5(10^{-6})/°F$] rod has an initial length of 6 ft. Determine the change in length of the rod after a tensile load of 5000 lb is applied to the rod and the temperature of the rod decreases 50°F.

## SOLUTION

Strain $\epsilon$ is the ratio of change in length $\delta$ and initial length $L$, and stress $\sigma$ is the ratio of force $P$ and area $A$. Therefore, Eq. (4-19) can be written

$$\epsilon = \frac{\delta}{L} = \frac{\sigma}{E} + \alpha \, \Delta T = \frac{P}{EA} + \alpha \, \Delta T$$

---

[3]In a homogeneous material, material properties such as modulus of elasticity and Poisson's ratio do not vary from point to point. Examples of nonhomogeneous materials are concrete (which consists of sand and rocks held together by cement) and particle board (which consists of sawdust and wood chips held together by glue). In an isotropic material, material properties such as modulus of elasticity and Poisson's ratio are independent of direction within the material. Examples of nonisotropic materials are fiber-reinforced composites and many crystalline materials.

[4]Assuming the deformation remains in the linearly elastic range so that Hooke's law [Eq. (4-15)] applies.

and the rod will stretch

$$\delta = \left( \frac{P}{EA} + \alpha \, \Delta T \right) L$$

$$= \left[ \frac{5000}{30(10^6)(\pi/4)(1/2)^2} + 6.5(10^{-6})(-50) \right](6)$$

$$= 0.003143 \text{ ft} = 0.03772 \text{ in.} \cong 0.0377 \text{ in.} \quad \blacksquare \qquad \text{Ans.}$$

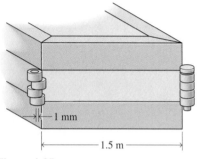

**Figure 4-35**

## Example Problem 4-12

A 1.5-m-long brass [$E = 100$ GPa, $\alpha = 17.6(10^{-6})/°$C] strap is intended to hold one side of a box together, as shown in Fig. 4-35. However, the box is slightly too large and the eyelets miss lining up by 1 mm.

(a) Determine how much the strap would have to be heated to lengthen the strap enough to insert the pin through the eyelets.

(b) Determine the stress that would exist in the strap after the strap cooled back to room temperature.

### SOLUTION

(a) Since strain $\epsilon$ is the ratio of change in length $\delta$ and initial length $L$, Eq. (4-19) can be written

$$\epsilon = \frac{\delta}{L} = \frac{\sigma}{E} + \alpha \, \Delta T$$

If no stress is applied to the strap ($\sigma = 0$) and the temperature is raised to stretch the strap by 1 mm = 0.001 m, then

$$\delta = \left( \frac{\sigma}{E} + \alpha \, \Delta T \right) L$$

$$= [0 + (17.6)(10^{-6})\Delta T](1.5) = 0.001 \text{ m}$$

Solving for $\Delta T$ yields

$$\Delta T = 37.9°C \qquad \text{Ans.}$$

(b) If the strap stays stretched by 1 mm after it cools down ($\Delta T = 0$), then the stress must increase to maintain the stretch

$$\delta = \left( \frac{\sigma}{E} + \alpha \, \Delta T \right) L$$

$$= \left[ \frac{\sigma}{100(10^9)} + 0 \right](1.5) = 0.001 \text{ m}$$

Solving for $\sigma$ yields

$$\sigma = 66.67(10^6) \text{ N/m}^2 \cong 66.7 \text{ MPa T} \quad \blacksquare \qquad \text{Ans.}$$

# PROBLEMS

## Introductory Problems

**4-54*** A cast iron pipe has an inside diameter of 70 mm and an outside diameter of 105 mm. The length of the pipe is 2.5 m. The coefficient of thermal expansion for cast iron is $\alpha = 12.1(10^{-6})/°C$. Determine the dimension changes caused by
(a) An increase in temperature of 70°C.
(b) A decrease in temperature of 85°C.

**4-55*** A large cement kiln has a length of 225 ft and a diameter of 12 ft. Determine the change in length and diameter of the structural steel shell [$\alpha = 6.5(10^{-6})/°F$] caused by an increase in temperature of 250°F.

**4-56** An airplane has a wing span of 40 m. Determine the change in length of the aluminum alloy [$\alpha = 22.5(10^{-6})/°C$] wing spar if the plane leaves the ground at a temperature of 40°C and climbs to an altitude where the temperature is −40°C.

## Intermediate Problems

**4-57*** Determine the movement of the pointer of Fig. P4-57 with respect to the scale zero when the temperature increases 80°F. The coefficients of thermal expansion are $6.6(10^{-6})/°F$ for the steel and $12.5(10^{-6})/°F$ for the aluminum.

**4-58** A bronze [$\alpha_B = 16.9(10^{-6})/°C$] sleeve with an inside diameter of 99.8 mm is to be placed over a solid steel [$\alpha_s = 11.9(10^{-6})/°C$] cylinder, which has an outside diameter of 100 mm. If the temperatures of the cylinder and sleeve remain equal, how much must the temperature be increased in order for the bronze sleeve to slip over the steel cylinder?

## Challenging Problems

**4-59** A steel [$E = 30,000$ ksi and $\alpha = 6.5(10^{-6})/°F$] surveyor's tape $\frac{1}{2}$ in. wide $\times$ $\frac{1}{32}$ in. thick is exactly 100 ft long at 72°F and under a pull of 10 lb. What correction should be introduced if the tape is used to make a 100-ft measurement at a temperature of 100°F and under a pull of 25 lb.

**4-60*** A 25-mm-diameter aluminum [$\alpha = 22.5(10^{-6})/°C$, $E = 73$ GPa, and $\nu = 0.33$] rod hangs vertically while suspended from one end. A 2500-kg mass is attached at the other end. After the load is applied, the temperature decreases 50°C. Determine
(a) The axial stress in the rod.
(b) The axial strain in the rod.
(c) The change in diameter of the rod.

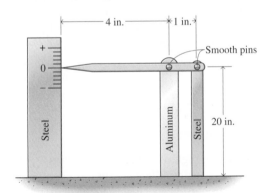

**Figure P4-57**

# 4-7 DEFORMATION OF AXIALLY LOADED MEMBERS

**Uniform Member.** When a straight bar of uniform cross section is axially loaded by forces applied at the ends, the axial strain along the length of the bar is assumed to have a constant value,[5] and the elongation (or contraction) of the

---

[5]In Section 6-3 it will be shown that the forces at the ends of such members must be equal in magnitude, opposite in direction, and directed along the axis of the member. Furthermore, the internal forces at any position along the member must be the same as the forces at the ends of the member and also must act along the axis of the member.

bar resulting from the axial load **P** may be expressed as $\delta = \epsilon L$ (by the definition of average axial strain). If Hooke's law [Eq. (4-15a)] applies, the axial deformation may be expressed in terms of either stress or load as

$$\delta = \epsilon L = \frac{\sigma L}{E} \tag{4-20a}$$

or

$$\delta = \frac{PL}{EA} \tag{4-20b}$$

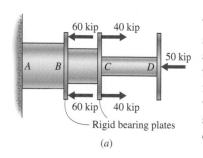

(a)

The first form will be convenient in elastic problems in which the limiting axial stress and axial deformation are both specified and either the maximum allowable load or the required size (cross-sectional area) of the member is to be determined. The stress corresponding to the specified deformation can be obtained from Eq. (4-20a) and compared to the allowable stress. The smaller of the two values can then be used to compute the allowable load or the required cross-sectional area. In general, Eq. (4-20a) is preferred when the problem involves the determination or comparison of stresses.

**Multiple Loads/Sizes.** Equation (4-20b), which gives the elongation (or contraction) $\delta$ occurring over some length $L$, applies only to uniform members for which $P$, $A$, and $E$ are constant over the entire length $L$. If a bar is subjected to a number of axial loads at different points along the bar, or if the bar consists of parts having different cross-sectional areas or of parts composed of different materials (Fig. 4-36a), then the change in length of each part can be computed by using Eq. (4-20b). The changes in length of the various parts of the bar can then be added algebraically to give the total change in length of the complete bar:

$$\delta = \sum_{i=1}^{n} \delta_i = \sum_{i=1}^{n} \frac{P_i L_i}{E_i A_i} \tag{4-21}$$

where $A_i$, $P_i$, and $E_i$ are all constant on segment $i$ and the force $P_i$ is the internal force in segment $i$ of the bar and is usually different than the forces applied at the ends of the segment. These forces must be calculated from equilibrium of the segment and are often shown on an axial force diagram such as Fig. 4-36b.

**Nonuniform Deformation.** For cases in which the axial force or the cross-sectional area varies continuously along the length of the bar (Fig. 4-37), Eq. (4-20b) is not valid. The axial strain at a point for the case of nonuniform deformation was defined in Section 4-4 as $\epsilon = d\delta/dL$. Thus, the increment of deformation associated with a differential element of length $dL = dx$ may be expressed as $d\delta = \epsilon \, dx$. If Hooke's law applies, the strain may be again expressed as $\epsilon = \sigma/E$, where $\sigma = P_x/A_x$. The subscripts indicate that both the applied load $P_x$ and the cross-sectional area $A_x$ may be functions of position $x$ along the bar. Thus,

$$d\delta = \frac{P_x}{EA_x} dx \tag{a}$$

**Figure 4-36**

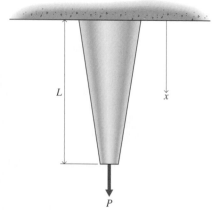

**Figure 4-37**

(The axial force diagram (b) shows tension and compression axial force in kip, with $P_{BC}$ positive and $P_{AB}$, $P_{CD}$ negative.)

Integrating Eq. (a) yields the following expression for the total elongation (or contraction) of the bar:

$$\delta = \int_0^L d\delta = \int_0^L \frac{P_x}{EA_x} \, dx \qquad (4\text{-}22)$$

Equation (4-22) gives acceptable results for tapered bars, provided the angle between the sides of the bar does not exceed 20°.

## Example Problem 4-13

The compression member shown in Fig. 4-38a consists of a solid aluminum bar $A$, which has an outside diameter of 100 mm; a brass tube $B$, which has an outside diameter of 150 mm and an inside diameter of 100 mm; and a steel pipe $C$, which has an outside diameter of 200 mm and an inside diameter of 125 mm. The moduli of elasticity of the aluminum, brass, and steel are 73, 100, and 210 GPa, respectively. Determine the overall shortening of the member under the action of the indicated loads.

### SOLUTION

The forces transmitted by cross sections in parts $A$, $B$, and $C$ of the member shown in Fig. 4-38a are obtained by using the free-body diagrams shown in Fig. 4-38b. Summing forces along the axis of the bar yields

$$+ \uparrow \Sigma F = 0: \qquad -P_A - 650 = 0 \qquad P_A = -650 \text{ kN} = 650 \text{ kN C}$$
$$+ \uparrow \Sigma F = 0: \qquad -P_B - 650 - 850 = 0 \qquad P_B = -1500 \text{ kN} = 1500 \text{ kN C}$$
$$+ \uparrow \Sigma F = 0: \qquad -P_C - 650 - 850 - 1500 = 0 \qquad P_C = -3000 \text{ kN} = 3000 \text{ kN C}$$

A pictorial representation of the distribution of axial, or internal force in the member is shown in Fig. 4-38c. The cross-sectional areas of the aluminum, brass, and steel are

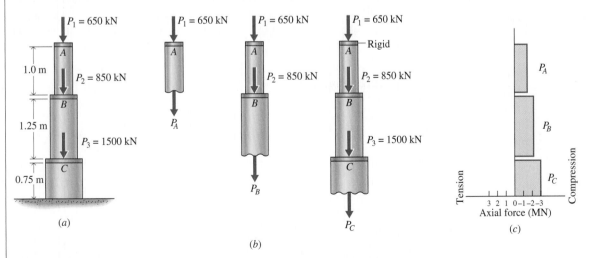

(a)

(b)

(c)

**Figure 4-38**

$$A_A = \frac{\pi}{4}d^2 = \frac{\pi}{4}(100)^2 = 7854 \text{ mm}^2 = 0.007854 \text{ m}^2$$

$$A_B = \frac{\pi}{4}(d_o^2 - d_i^2) = \frac{\pi}{4}(150^2 - 100^2) = 9817 \text{ mm}^2 = 0.009817 \text{ m}^2$$

$$A_C = \frac{\pi}{4}(d_o^2 - d_i^2) = \frac{\pi}{4}(200^2 - 125^2) = 19{,}144 \text{ mm}^2 = 0.019144 \text{ m}^2$$

The changes in length of the different parts are obtained by using Eq. (4-20b). Thus,

$$\delta_A = \frac{P_A L_A}{E_A A_A} = \frac{-650(10^3)(1.0)}{73(10^9)(0.007854)} = -1.1337(10^{-3}) \text{ m} = -1.1337 \text{ mm}$$

$$\delta_B = \frac{P_B L_B}{E_B A_B} = \frac{-1500(10^3)(1.25)}{100(10^9)(0.009817)} = -1.9100(10^{-3}) \text{ m} = -1.9100 \text{ mm}$$

$$\delta_C = \frac{P_C L_C}{E_C A_C} = \frac{-3000(10^3)(0.75)}{210(10^9)(0.019144)} = -0.5597(10^{-3}) \text{ m} = -0.5597 \text{ mm}$$

The total change in length of the complete bar is given by Eq. (4-21) as

$$\delta_{\text{total}} = \delta_A + \delta_B + \delta_C$$

$$= -1.1337 - 1.9100 - 0.5597 = -3.6034 \text{ mm} \cong -3.60 \text{ mm} \blacksquare \quad \textbf{Ans.}$$

## Example Problem 4-14

A homogeneous bar of uniform cross section $A$ hangs vertically while suspended from one end as shown in Fig. 4-39$a$. Determine the elongation of the bar due to its own weight $W$ in terms of $W$, $L$, $A$, and $E$.

### SOLUTION

A free-body diagram of a segment of the bar, Fig. 4-39$b$, shows that the axial force is a function of $x$, the distance from the free end of the bar. Thus, Eq. (4-22) is applicable. The weight of the segment of the bar shown in Fig. 4-39$b$ is

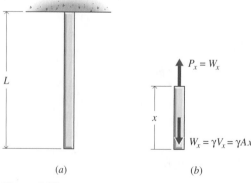

(a)                          (b)

**Figure 4-39**

$W_x = \gamma V_x = \gamma A x$, where $\gamma$ is the specific weight of the material of which the bar is made, and $V_x$ is the volume of the bar segment. From Eq. (4-22),

$$\delta = \int_0^L \frac{P_x}{EA_x}\,dx = \frac{1}{EA}\int_0^L \gamma A x\,dx = \frac{\gamma}{E}\int_0^L x\,dx$$

where $\gamma$, $A$, and $E$ are each constant. The elongation of the bar is

$$\delta = \frac{\gamma}{E}\int_0^L x\,dx = \frac{\gamma x^2}{2E}\Big]_0^L = \frac{\gamma L^2}{2E}$$

The weight of the bar is $W = \gamma A L$, from which $\gamma = \dfrac{W}{AL}$. Thus, the elongation of the bar is

$$\delta = \frac{\gamma L^2}{2E} = \frac{W}{AL}\left[\frac{L^2}{2E}\right] = \frac{WL}{2AE}\ \blacksquare \qquad\qquad \textbf{Ans.}$$

## PROBLEMS

### Introductory Problems

**4-61\*** The rigid yokes $B$ and $C$ of Fig. P4-61 are securely fastened to the 2-in.-square steel ($E = 30,000$ ksi) bar $AD$. Determine
(a) The maximum normal stress in the bar.
(b) The change in length of segment $AB$.
(c) The change in length of segment $BC$.
(d) The change in length of the complete bar.

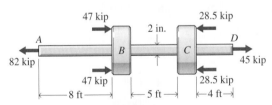

**Figure P4-61**

**4-62\*** A structural tension member of aluminum alloy ($E = 70$ GPa) has a rectangular cross section of $25 \times 75$ mm and is 2 m long. Determine the maximum axial load that may be applied if the axial stress is not to exceed 100 MPa and the total elongation is not to exceed 4 mm.

**4-63** The tension member of Fig. P4-63 consists of a steel ($E = 30,000$ ksi) pipe $A$, which has an outside diameter of 6 in. and an inside diameter of 4.5 in., and a solid aluminum alloy ($E = 10,600$ ksi) bar $B$, which has a diameter of 4 in. Determine the overall elongation of the member.

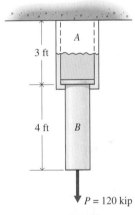

**Figure P4-63**

**4-64** A steel ($E = 200$ GPa) rod, which has a diameter of 30 mm and a length of 1.0 m, is attached to the end of a Monel ($E = 180$ GPa) tube, which has an internal diameter of 40 mm, a wall thickness of 10 mm, and a length of 2.0 m, as shown in Fig. P4-64. Determine the load required to stretch the assembly 3.00 mm.

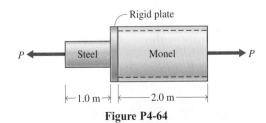

**Figure P4-64**

**Intermediate Problems**

**4-65\*** The floor of a warehouse is supported by an air-dried, red oak timber column (see Appendix A for properties) as shown in Fig. P4-65. Contents of the warehouse subject the 12- × 12-in. column to an axial load of 200 kip. If the column is 30 in. long, determine
(a) The deformation of the column.
(b) The normal stress in the column.
(c) The bearing stress between the column and the lower bearing plate.
(d) The maximum shearing stress in the column.

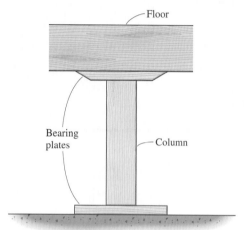

**Figure P4-65**

**4-66\*** The roof and second floor of a building are supported by the column shown in Fig. P4-66. The column is a structural steel (see Appendix A for properties) section having a cross-sectional area of 5700 mm². The roof and floor subject the column to the axial forces shown. Determine
(a) The amount that the first floor will settle.
(b) The amount that the roof will settle.

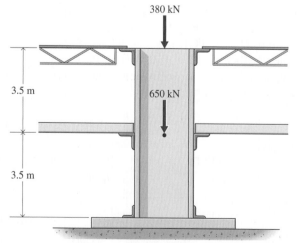

**Figure P4-66**

**4-67** The tension member of Fig. P4-67 consists of a structural steel pipe *A*, which has an outside diameter of 6 in. and an inside diameter of 4.5 in., and a solid 2014-T4 aluminum alloy bar *B*, which has a diameter of 4 in. (see Appendix A for properties). Determine
(a) The change in length of the steel pipe.
(b) The overall deflection of the member.
(c) The maximum normal and shearing stresses in the aluminum bar.

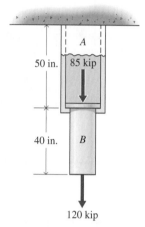

**Figure P4-67**

**4-68** An aluminum alloy ($E = 73$ GPa) tube *A* with an outside diameter of 75 mm is used to support a 25-mm-diameter steel ($E = 200$ GPa) rod *B*, as shown in Fig. P4-68. Determine the minimum thickness *t* required for the tube if the maximum deflection of the loaded end of the rod must be limited to 0.40 mm.

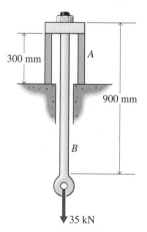

**Figure P4-68**

**Challenging Problems**

**4-69\*** A structural steel (see Appendix A for properties) bar of rectangular cross section consists of uniform and tapered sections as shown in Fig. P4-69. The width of the tapered section varies linearly from 2 in. at the bottom to 5 in. at the top. The bar has a constant thickness of ½ in. Determine the elongation of the bar resulting from application of the 30-kip load $P$. Neglect the weight of the bar.

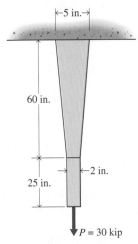

**Figure P4-69**

**4-70** Determine the change in length of the homogeneous conical bar of Fig. P4-70 due to its own weight. Express the results in terms of $L$, $E$, and the specific weight $\gamma$ of the material. The taper of the bar is slight enough for the assumption of a uniform axial stress distribution over a cross section to be valid.

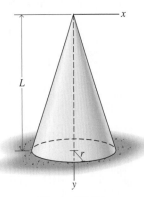

**Figure P4-70**

**4-71** The bar shown in Fig. P4-71 is made of annealed bronze (see Appendix A for properties). In addition to its own weight, the bar is subjected to an axial tensile load $P$ of 5000 lb at its lower end. Determine the elongation of the bar due to the combined effects of its weight and the load $P$. Let $r = 4$ in. and $L = 60$ in.

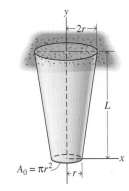

**Figure P4-71**

# 4-8 STATICALLY INDETERMINATE AXIALLY LOADED MEMBERS

If it is possible to find the internal forces in any member of a structure or mechanical system by means of the equations of equilibrium alone, then the structure is called *statically determinate*. If the number of unknown forces in any structure or mechanical system exceeds the number of independent equations of equilibrium, the structure is said to be *statically indeterminate*. Problems of this type can be analyzed by supplementing the equilibrium equations with additional equations involving the geometry of the deformations in the members of the structure or system. The following outline of procedure will be helpful in the analysis of problems involving statically indeterminate situations.

1. Draw a free-body diagram or diagrams.
2. Write the equations of equilibrium relating the forces on the free-body diagram.

3. If the number of unknowns exceeds the number of equilibrium equations, draw a displacement diagram.

4. Write equations (*compatibility equations*) relating the deformations of various parts of the structure or mechanical system using the displacement diagram.

5. Write deformation equations relating the forces in the equations of equilibrium and the displacements in the compatibility equations.

6. When the number of independent equilibrium equations and deformation equations equals the number of unknowns, the equations can be solved simultaneously.

Hooke's law [Eq. (4-15)] and the definitions of stress and strain can be used to relate deformations and forces when all stresses are less than the corresponding proportional limits of the materials used in the fabrication of the members. If some of the stresses exceed the proportional limits of the materials, stress–strain diagrams can be used to relate the loads and deformations. In this text, problems will be limited to the region of elastic action of the materials.

It is recommended that a displacement diagram be drawn showing deformations to assist in obtaining the correct deformation equation. The displacement diagram should be as simple as possible (a line diagram), with the deformations indicated with exaggerated magnitudes and clearly dimensioned. Note that an equilibrium equation and the corresponding deformation equation must be compatible; that is, when a tensile force is assumed for a member in the free-body diagram, a tensile deformation must be indicated for the same member in the deformation diagram. If the diagrams are compatible, a negative result will indicate that the assumption was wrong; however, the magnitude of the result will be correct.

## Example Problem 4-15

A rigid plate $C$ is used to transfer a 20-kip load $P$ to a steel ($E = 30,000$ ksi) rod $A$ and to an aluminum alloy ($E = 10,000$ ksi) pipe $B$, as shown in Fig. 4-40a. The supports at the top of the rod and at the bottom of the pipe are rigid,

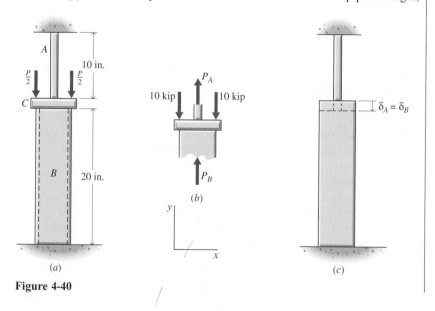

*(a)*

*(b)*

*(c)*

**Figure 4-40**

and there are no stresses in the rod or pipe before the load $P$ is applied. The cross-sectional areas of rod $A$ and pipe $B$ are 0.800 in$^2$ and 3.00 in$^2$, respectively. Determine

(a) The normal stresses in rod $A$ and in pipe $B$.

(b) The displacement of plate $C$.

## SOLUTION

(a) A free-body diagram of plate $C$ and portions of the rod $A$ and the pipe $B$ is shown in Fig. 4-40b. The free-body diagram contains two unknown forces; $P_A = \sigma_A A_A$ (which is drawn as a pulling or tensile force), and $P_B = \sigma_B A_B$ (which is drawn as a pushing or compressive force). Since only one equation of equilibrium is available,

$$+ \uparrow \Sigma F_y = 0: \qquad P_A + P_B - 20 = 0$$

$$0.800\sigma_A + 3.00\sigma_B = 20 \qquad\qquad (a)$$

the problem is statically indeterminate. The additional equation needed to obtain a solution to the problem is obtained from the deformation diagram shown in Fig. 4-40c.

When the 20-kip load is applied to the rigid plate $C$, the plate will move down a distance $\delta_C$; the steel rod $A$ will stretch an amount $\delta_A = \delta_C = (PL/EA)_A = (\sigma L/E)_A$; and the aluminum pipe $B$ will shrink an equal amount $\delta_B = \delta_C = (PL/EA)_B = (\sigma L/E)_B$. (Note that the forces and deformations are compatible; that is, $P_A$ is a pulling force on the free-body diagram and in the equilibrium equation and $\delta_A$ is a stretch on the deformation diagram and in the deformation equations, $P_B$ is a pushing force on the free-body diagram and in the equilibrium equation and $\delta_B$ is a shrink on the deformation diagram and in the deformation equations.) Therefore, compatibility of the deformations ($\delta_A = \delta_B = \delta_C$) requires that

$$\left(\frac{\sigma L}{E}\right)_A = \left(\frac{\sigma L}{E}\right)_B$$

$$\frac{\sigma_A(10)}{30,000} = \frac{\sigma_B(20)}{10,000}$$

from which

$$\sigma_A = 6\sigma_B \qquad\qquad (b)$$

Solving Eqs. (a) and (b) simultaneously yields

$$\sigma_A = 15.384 \text{ ksi} \cong 15.38 \text{ ksi T} \qquad\qquad \textbf{Ans.}$$

$$\sigma_B = 2.564 \text{ ksi} \cong 2.56 \text{ ksi C} \qquad\qquad \textbf{Ans.}$$

Note that the equilibrium and deformation equations could just as easily have been written and solved in terms of the forces $P_A$ and $P_B$ rather than in terms of the stresses $\sigma_A$ and $\sigma_B$. Stresses were selected because the problem asked for the stresses but did not ask for the forces.

(b) The displacement of the plate $C$ is the same as the deformation of rod $A$ and the deformation of pipe $B$ ($\delta_A = \delta_B = \delta_C$). Therefore

$$\delta_C = \delta_A = \frac{\sigma_A L_A}{E_A} = \frac{(15.384)(10)}{30{,}000} = 0.00513 \text{ in. } \downarrow \ \blacksquare \qquad \textbf{Ans.}$$

---

## Example Problem 4-16

Nine 25-mm-diameter steel ($E = 200$ GPa) reinforcing bars are used in the short concrete ($E = 30$ GPa) pier shown in Fig. 4-41a. An axial load $P$ of 650 kN is applied to the pier through a rigid capping plate. Determine

(a) The normal stresses in the concrete and in the steel bars.
(b) The shortening of the pier.

### SOLUTION

(a) A free-body diagram of the rigid capping plate is shown in Fig. 4-41b. The free-body diagram contains two unknown forces, the resultant force $P_C$ exerted by the concrete and the resultant force $P_R$ exerted by the rods. Both forces are drawn as *pushing* on the rigid capping plate. The capping plate will exert equal pushing forces back on the concrete pier and the steel reinforcing bars. Thus, the forces $P_C$ and $P_R$ are both compressive forces and they will lead to shortening of both the pier and the rods. Since only one equation of equilibrium is available,

$$+ \uparrow \ \Sigma F_y = 0: \qquad P_R + P_C - P = 0 \qquad P_R + P_C = 650 \text{ kN} \qquad \text{(a)}$$

the problem is statically indeterminate. The additional equation needed to obtain a solution to the problem is obtained from the deformation diagram shown in Fig. 4-41c.

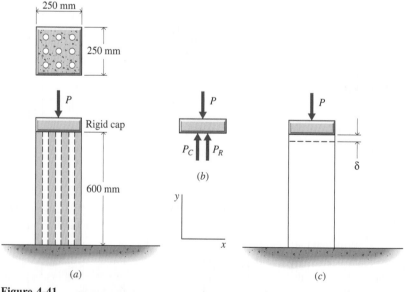

(a)   (b)   (c)

**Figure 4-41**

As the load $P$ is applied to the rigid capping plate, it moves downward an amount $\delta$, which is the same as the shrink experienced by the steel rods $\delta_R$ and by the concrete $\delta_C$. The relationship between the loads and the deformations for axial loading is given by Eq. (4-20b), $\delta_R = (PL/EA)_R$ and $\delta_C = (PL/EA)_C$. (Note that the forces and deformations are compatible; that is, $P_R$ and $P_C$ are both pushing forces on the free-body diagram and in the equilibrium equation, and $\delta_R$ and $\delta_C$ are both shrinks on the deformation diagram and in the deformation equations.)

The total cross-sectional area $A_R$ for the nine steel rods is

$$A_R = 9\left[\frac{\pi}{4}(25^2)\right] = 4418 \text{ mm}^2 \qquad (b)$$

The cross-sectional area $A_C$ for the concrete is

$$A_C = (250)^2 - 4418 = 58,080 \text{ mm}^2 \qquad (c)$$

Therefore, compatibility of the deformations $\delta_R = \delta_C = \delta$ requires that

$$\frac{P_R (0.600)}{(200)(10^9)(4418)(10^{-6})} = \frac{P_C (0.600)}{(30)(10^9)(58,080)(10^{-6})} \qquad (d)$$

from which

$$P_R = 0.5071 P_C \qquad (e)$$

Solving Eqs. (a) and (e) simultaneously yields

$$P_R = 218.7 \text{ kN C}$$

$$P_C = 431.3 \text{ kN C}$$

The normal stresses in the rods and in the concrete are obtained by using the definition of stress, Eq. (4-2). Thus

$$\sigma_R = \frac{P_R}{A_R} = \frac{218.7(10^3)}{4418(10^{-6})} = 49.50(10^6) \text{ N/m}^2 \cong 49.5 \text{ MPa C} \qquad \textbf{Ans.}$$

$$\sigma_C = \frac{P_C}{A_C} = \frac{431.3(10^3)}{58,080(10^{-6})} = 7.426(10^6) \text{ N/m}^2 \cong 7.43 \text{ MPa C} \qquad \textbf{Ans.}$$

(b) The shortening of the pier is obtained from either the deformation of the rods or the deformation of the concrete since they are equal. Thus, from the deformation of the rods,

$$\delta = \delta_R = \frac{218.7(10^3)(0.600)}{(200)(10^9)(4418)(10^{-6})}$$

$$= 0.1485(10^{-3}) \text{ m} = 0.1485 \text{ mm} \quad \blacksquare \qquad \textbf{Ans.}$$

# PROBLEMS

### Introductory Problems

**4-72\*** A hollow brass ($E = 100$ GPa) tube $A$ with an outside diameter of 100 mm and an inside diameter of 50 mm is fastened to a 50-mm-diameter steel ($E = 200$ GPa) rod $B$, as shown in Fig. P4-72. The supports at the top and bottom of the assembly and the collar $C$ used to apply the 500-kN load $P$ are rigid. Determine
(a) The normal stresses in each of the members.
(b) The deflection of the collar $C$.

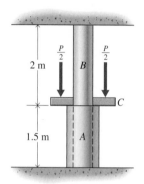

**Figure P4-72**

**4-73\*** The 7.5- $\times$ 7.5- $\times$ 20-in. oak ($E = 1800$ ksi) block shown in Fig. P4-73 was reinforced by bolting two 2- $\times$ 7.5- $\times$ 20-in. steel ($E = 29,000$ ksi) plates to opposite sides of the block. If the stresses in the wood and the steel are to be limited to 4.6 ksi and 22 ksi, respectively, determine
(a) The maximum axial compressive load $P$ that can be applied to the reinforced block.
(b) The shortening of the block when the load of part (a) is applied.

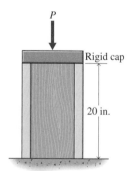

**Figure P4-73**

**4-74** Five 25-mm-diameter steel ($E = 200$ GPa) reinforcing bars will be used in a 1-m-long concrete ($E = 31$ GPa) pier with a square cross section, as shown in

Fig. P4-74. The allowable strengths in compression for steel and concrete are 130 MPa and 9.5 MPa, respectively. Determine the minimum size of pier required to support a 900-kN axial load.

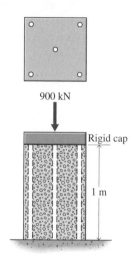

**Figure P4-74**

### Intermediate Problems

**4-75\*** A hollow steel ($E = 30,000$ ksi) tube $A$ with an outside diameter of 2.5 in. and an inside diameter of 2 in. is fastened to an aluminum ($E = 10,000$ ksi) bar $B$ that has a 2-in. diameter over one-half of its length and a 1-in. diameter over the other half. The assembly is attached to unyielding supports at the left and right ends and is loaded as shown in Fig. P4-75. Determine
(a) The normal stresses in all parts of the bar.
(b) The deflection of cross-section $a - a$.

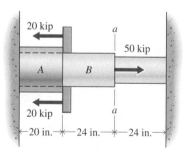

**Figure P4-75**

**4-76\*** A 150-mm-diameter $\times$ 200-mm-long polymer ($E = 2.10$ GPa) cylinder will be attached to a 45-mm-diameter $\times$ 400-mm-long brass ($E = 100$ GPa) rod by using the flange type of connection shown in Fig. P4-76. A 0.15-mm clearance exists between the parts as a result of a machining error. If the bolts are inserted and tightened, determine

(a) The normal stresses produced in each of the members.

(b) The final position of the flange – polymer interface after assembly with respect to the left support.

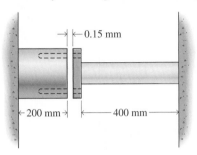

**Figure P4-76**

**4-77** The assembly shown in Fig. P4-77 consists of a steel bar $A(E_s = 30{,}000$ ksi and $A_s = 1.25$ in$^2$), a rigid bearing plate $C$ that is securely fastened to bar $A$, and a bronze bar $B$ ($E_B = 15{,}000$ ksi and $A_B = 3.75$ in$^2$). A clearance of 0.015 in. exists between the bearing plate $C$ and bar $B$ before the assembly is loaded. After a load $P$ of 95 kip is applied to the bearing plate, determine

(a) The normal stresses in bars $A$ and $B$.

(b) The vertical displacement of the bearing plate $C$.

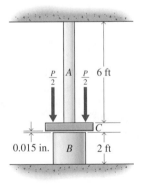

**Figure P4-77**

**4-78** A column similar to Fig. P4-74 is being designed to carry a load of 4450 kN. The column, which will have a 500- × 500-mm square cross section, will be made of concrete ($E = 20$ GPa) and will be reinforced with 50-mm-diameter steel ($E = 200$ GPa) bars. If the allowable stresses are 120 MPa in the steel and 8 MPa in the concrete, determine

(a) The number of steel bars required.

(b) The stresses in the steel and concrete when the bars of part $a$ are used.

(c) The change in length of a 3-m-long column when the bars of part $a$ are used.

## Challenging Problems

**4-79*** A $\frac{1}{2}$-in-diameter alloy-steel bolt ($E = 30{,}000$ ksi) passes through a cold-rolled brass sleeve ($E = 15{,}000$ ksi) as shown in Fig. P4-79. The cross-sectional area of the sleeve is 0.375 in$^2$. Determine the normal stresses produced in the bolt and sleeves by tightening the nut $\frac{1}{4}$ turn (0.020 in.).

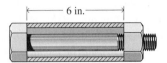

**Figure P4-79**

**4-80** The two faces of the clamp shown in Fig. P4-80 are 250 mm apart when the two stainless-steel ($E_s = 190$ GPa) bolts connecting them are unstretched. A force $P$ is applied to separate the faces of the clamp so that an aluminum alloy ($E_a = 73$ GPa) bar with a length of 251 mm can be inserted as shown. Each of the bolts has a cross-sectional area of 120 mm$^2$, and the bar has a cross-sectional area of 625 mm$^2$. After the load $P$ is removed, determine

(a) The axial stresses in the bolts and in the bar.

(b) The change in length of the aluminum alloy bar.

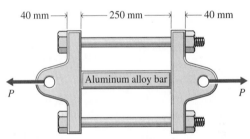

**Figure P4-80**

## Computer Problems

**4-81** A high-strength steel bolt ($E_s = 30{,}000$ ksi and $A_s = 0.785$ in$^2$) passes through a brass sleeve ($E_b = 15{,}000$ ksi and $A_b = 1.767$ in$^2$), as shown in Fig. P4-81. As the nut is tightened, it advances a distance of 0.125 in along the bolt for each complete turn of the nut. Compute and plot

(a) The axial stresses $\sigma_s$ (in the steel bolt) and $\sigma_b$ (in the brass sleeve) as functions of the angle of twist $\theta$ of the nut ($0° \leq \theta \leq 180°$).

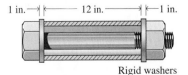

**Figure P4-81**

(b) The elongations $\delta_s$ (of the steel bolt) and $\delta_b$ (of the brass sleeve) as function of $\theta$ ($0° \leq \theta \leq 180°$).

(c) The distance $L$ between the two washers as a function of $\theta$ ($0° \leq \theta \leq 180°$).

**4-82** The short pier shown in Fig. P4-82 is reinforced with nine steel ($E = 210$ GPa) reinforcing bars. An axial compressive load $P$ is applied to the pier through the rigid capping plate. The axial load carried by the matrix material is a function of $R$, the percentage of the cross section taken up by the steel reinforcement bars. The load is also a function of the modulus ratio $E_R/E_M$, where $E_R$ and $E_M$ are the modulus of elasticity for the reinforcement material and the matrix material, respectively. For the three matrix-reinforcement combinations listed, compute and plot the percentage of the load carried by the matrix as a function of $R$ ($0 \leq R \leq$ 100%).

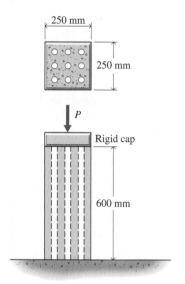

**Figure P4-82**

| | Matrix | Reinforcement | $E_R/E_M$ |
|---|---|---|---|
| (a) | rubber | steel | 50,000 |
| (b) | wood | steel | 20 |
| (c) | concrete | steel | 7.5 |

## 4-9 THERMAL EFFECTS

When a temperature change takes place while a member is restrained (free movement restricted or prevented), stresses (referred to as "thermal stresses") are induced in the member. For example, the bar $AB$ of Fig. 4-42a is securely fastened to rigid supports at the ends and is subjected to a temperature change. Since the ends of the bar are fixed, the total deformation of the bar must be zero.

$$\delta_{\text{total}} = \delta_T + \delta_\sigma = \epsilon_T L + \epsilon_\sigma L$$

$$0 = \alpha \Delta T L + \frac{\sigma}{E} L$$

in which the term $\delta_T$ is the deformation due to a temperature change and $\delta_\sigma$ is the deformation due to an axial load. If the temperature of the bar increases ($\Delta T$ positive), then the induced stress must be negative and the wall must push on the ends of the rod. If the temperature of the bar decreases ($\Delta T$ negative), then the induced stress must be positive and the wall must pull on the ends of the rod.

That is, if end $B$ were not attached to the wall and the temperature drops, end $B$ would move to $B'$, a distance $|\delta_T| = |\epsilon_T L| = |\alpha \Delta T L|$, as indicated in Fig. 4-42b. Therefore, for the total deformation of the bar to be zero, the wall at $B$ must apply a force $P = \sigma A$ (Fig. 4-42c) of sufficient magnitude to move end $B$ through a dis-

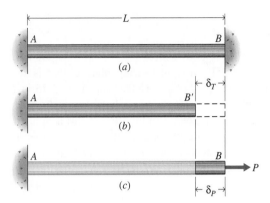

**Figure 4-42**

tance $\delta_P = \epsilon_\sigma L = (\sigma/E)L$ so that the length of the bar is again $L$, the distance between the walls. Since the walls do not move, $|\delta_T| = \delta_P$, or

$$\delta_P - |\delta_T| = \delta_P + \delta_T = 0$$

and thus the total deformation of the bar is zero.

## Example Problem 4-17

A 10-m section of steel [$E = 200$ GPa and $\alpha = 11.9(10^{-6})/°C$] rail has a cross-sectional area of 7500 mm². Both ends of the rail are tight against adjacent rails that, for this problem, can be assumed to be rigid. The rail is supported against lateral movement. For an increase in temperature of 50°C, determine

(a) The normal stress in the rail.
(b) The internal force on a cross section of the rail.

### SOLUTION

(a) The change in length of the rail resulting from the temperature change is given by modifying Eq. (4-18) as

$$\delta = \epsilon_T L = \alpha L \Delta T = 11.9(10^{-6})(10)(50) = 5.95(10^{-3}) \text{ m} = 5.95 \text{ mm}$$

The stress required to resist a change in length of 5.95 mm is given by Eq. (4-20a) as

$$\sigma = \frac{E\delta}{L} = \frac{200(10^9)(5.95)(10^{-3})}{10} = 119.0(10^6) \text{ N/m}^2 = 119.0 \text{ MPa C} \quad \textbf{Ans.}$$

(b) The internal force on a cross section of the rail is

$$F = \sigma A = 119.0(10^6)(7500)(10^{-6}) = 892.5(10^3) \text{ N} \cong 893 \text{ kN C} \quad \blacksquare \ \textbf{Ans.}$$

## █ Example Problem 4-18

The assembly shown in Fig. 4-43$a$ consists of a steel rod $A$ ($E_A$ = 30,000 ksi, $A_A$ = 2.50 in$^2$, and $\alpha_A$ = 6.6 × 10$^{-6}$/°F), a rigid bearing plate $C$ that is securely fastened to bar $A$, and a bronze bar $B$ ($E_B$ = 15,000 ksi, $A_B$ = 3.75 in$^2$, and $\alpha_B$ = 9.4 × 10$^{-6}$/°F). A clearance of 0.015 in. exists between the bearing plate $C$ and bar $B$ before the assembly is loaded. If a load $P$ = 5 kip is applied to the bearing plate and then the temperature of the assembly is slowly raised, calculate and plot the stresses $\sigma_A$ in the steel rod and $\sigma_B$ in the bronze bar as a function of the temperature increase $\Delta T$ for 0°F < $\Delta T$ < 50°F.

### SOLUTION

The first step is to draw a free-body diagram. When the force $P$ is applied to the bearing plate $C$, we expect the plate to be pushed down (causing a tensile force in $A$) and press against the bar $B$ (causing a compressive force in $B$). The free-body diagram is drawn accordingly (Fig. 4-43$b$). The only equation of equilibrium that gives any useful information is the sum of forces in the vertical direction

$$+\uparrow \Sigma F_y = 0: \qquad T_A + P_B - 5 = 0 \qquad T_A + P_B = 5 \text{ kip}$$

$$2.5\sigma_A + 3.75\sigma_B = 5000 \text{ lb} \qquad\qquad (a)$$

Therefore, the problem is statically indeterminate and we have to write a compatibility equation relating the deformations to solve for the stresses.

Since $T_A$ is a tensile force, the stress deformation $(\sigma L/E)_A$ will represent a stretch of the rod $A$. When the temperature of $A$ increases, the stretch of $A$ will be even greater. Therefore, the total stretch of $A$ is the sum of the stress deformation (a stretch) and the temperature deformation (also a stretch)

$$\delta_A = \left(\frac{\sigma L}{E} + \alpha \Delta T L\right)_A \qquad\qquad (b)$$

Likewise, since $P_B$ is a compressive force, the stress deformation $(\sigma L/E)_B$ will represent a shrink of the bar $B$. When the temperature increases, bar $B$ will stretch, thus reducing the shrink caused by the force $P_B$. Therefore, the total

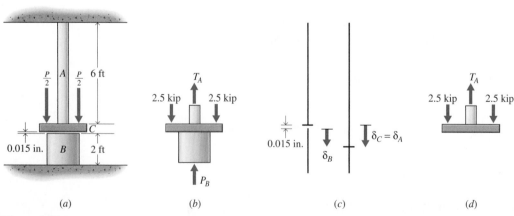

*(a)*      *(b)*      *(c)*      *(d)*

**Figure 4-43**

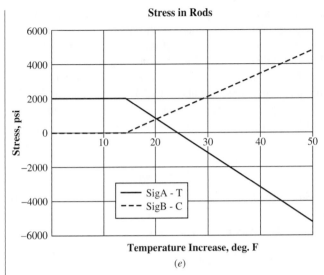

Stress in Rods

Temperature Increase, deg. F

(e)

**Figure 4-43** (*Continued*).

shrink of $B$ is the difference of the stress deformation (a shrink) and the temperature deformation (a stretch)

$$\delta_B = \left(\frac{\sigma L}{E} - \alpha \Delta T L\right)_B \tag{c}$$

The deformation diagram (Fig. 4-43c) relates the stretch $\delta_A$, the shrink $\delta_B$, and the initial gap of 0.015 in.

$$\delta_A = \delta_B + 0.015 \tag{d}$$

Combining equations (b), (c), and (d) gives

$$\left(\frac{\sigma L}{E} + \alpha \Delta T L\right)_A = \left(\frac{\sigma L}{E} - \alpha \Delta T L\right)_B + 0.015 \tag{e}$$

or, putting in numbers,

$$\left[\frac{\sigma_A(72)}{(30)(10^6)} + (6.6)(10^{-6})\Delta T(72)\right] =$$

$$\left[\frac{\sigma_B(24)}{(15)(10^6)} - (9.4)(10^{-6})\Delta T(24)\right] + 0.015 \tag{f}$$

Multiplying through by $30 \times 10^6$ and rearranging gives

$$72\sigma_A - 48\sigma_B = 450{,}000 - 21{,}024\Delta T \tag{g}$$

Finally, solving equations (a) and (g) simultaneously gives the stresses

$$\sigma_A = 4942.308 - 202.154\Delta T$$

$$\sigma_B = 134.769\Delta T - 1961.539 \tag{h}$$

which shows that both $\sigma_A$ and $\sigma_B$ are linear functions of the temperature increase $\Delta T$. However, this says that for a temperature increase of less than about 14.5°F, the stress in bar $B$ is negative or tension (opposite what was initially assumed). But there is nothing pulling on bar $B$ that could put it in tension, and this solution cannot be valid. (Actually, this solution would apply if the plate $C$ were pulled down to the bar $B$, welded to it, and then released. The tension stress above would be the tension in the weld until the temperature increase exceeded about 14.5°F.) Therefore, for a temperature increase of less than about 14.5°F, the assumption that the applied force closes the gap and the plate pushes on the bar $B$ is not correct.

Starting over with a new free-body diagram (Fig. 4-43$d$) gives the equilibrium equation

$$+ \uparrow \; \Sigma F_y = 0: \qquad T_A - 5 = 0 \qquad T_A = 5 \text{ kip}$$

$$2.5\sigma_A = 5000 \text{ lb} \tag{i}$$

So now the stresses in the two bars are (for $\Delta T < 14.5°$)

$$\sigma_A = 2000 \text{ psi}$$

$$\sigma_B = 0 \text{ psi} \tag{j}$$

For temperature increases greater than $\Delta T > 14.5°$, the stresses are given by Eq. (h). The stresses are shown in the graph of Fig. 4-43$e$. Positive stresses represent tension in bar $A$ and compression in bar $B$ (as assumed on the free-body diagrams), and negative stresses represent compression in bar $A$ and tension in bar $B$ (opposite what was assumed on the free-body diagrams). ■

# PROBLEMS

## Introductory Problems

**4-83\*** A 3-in.-diameter × 80-in.-long aluminum alloy bar is stress free after being attached to rigid supports, as shown in Fig. P4-83. Determine the normal stress in the bar after the temperature drops 100°F. Use $E = 10,600$ ksi and $\alpha = 12.5(10^{-6})/°F$.

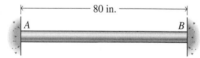

**Figure P4-83**

**4-84\*** A 6-m-long × 50-mm-diameter rod of aluminum alloy $[E = 70 \text{ GPa}, \nu = 0.346, \text{ and } \alpha = 22.5(10^{-6})/°C]$ is attached at the ends to supports that yield to permit a change in length of 1.00 mm in the rod when stressed. When the temperature is 35°C, there is no stress in the rod. After the temperature of the rod drops to −20°C, determine

(a) The normal stress in the rod.
(b) The change in diameter of the rod.

**4-85** A bar consists of 3-in.-diameter aluminum alloy $[E = 10,600 \text{ ksi}, \nu = 0.33, \text{ and } \alpha = 12.5(10^{-6})/°F]$ and 4-in.-

diameter steel $[E = 30,000 \text{ ksi}, \nu = 0.30, \text{ and } \alpha = 6.6(10^{-6})/°F]$ parts, as shown in Fig. P4-85. If end supports are rigid and the bar is stress free at 0°F, determine

(a) The normal stress in both parts of the bar at 80°F.
(b) The change in diameter of the steel part of the bar.

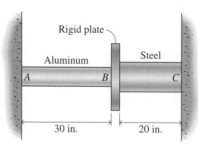

**Figure P4-85**

## Intermediate Problems

**4-86\*** A steel tie rod containing a rigid turnbuckle (see Fig. P4-86) has its ends attached to rigid walls. During the summer, when the temperature is 30°C, the turnbuckle is tightened to produce a stress in the rod of 15 MPa. Determine the normal stress in the rod in the winter when the temperature is $-10°C$. Use $E = 200$ GPa and $\alpha = 11.9(10^{-6})/°C$.

**Figure P4-86**

**4-87** Nine $\frac{3}{4}$-in.-diameter steel $(E = 30,000$ ksi$)$ reinforcing bars were used when the short concrete $(E = 4500$ ksi$)$ pier shown in Fig. P4-87 was constructed. After a load $P$ of 150 kip was applied to the pier, the temperature increased 100°F. The coefficients of thermal expansion for steel and concrete are $6.6(10^{-6})/°F$ and $6.0(10^{-6})/°F$, respectively. Determine

(a) The normal stresses in the concrete and in the steel bars after the temperature increases.

(b) The change in length of the pier resulting from the combined effects of the temperature change and the load.

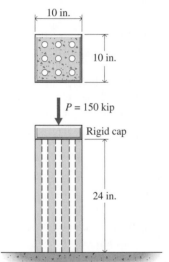

**Figure P4-87**

**4-88\*** The assembly shown in Fig. P4-88 consists of a steel $(E = 210$ GPa$)$ cylinder $A$, a rigid bearing plate $C$, and an aluminum alloy $(E = 71$ GPa$)$ bar $B$. Cylinder $A$ has a cross-sectional area of 1850 mm², and bar $B$ has a cross-sectional area of 2500 mm². After an axial load of 600 kN is applied, the temperature of cylinder $A$ decreases 50°C and the temperature of bar $B$ increases 25°C. The coefficients of thermal expansion are $11.9(10^{-6})/°C$ for the steel and $22.5(10^{-6})/°C$ for the aluminum. Determine

(a) The normal stresses in the cylinder and in the bar after the load is applied and the temperatures change.

(b) The displacement of plate $C$ after the load is applied and the temperatures change.

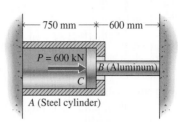

**Figure P4-88**

**Challenging Problems**

**4-89\*** A high-strength steel $[E_s = 30,000$ ksi, $A_s = 0.785$ in², and $\alpha_s = 6.6(10^{-6})/°F]$ bolt passes through a brass $[E_b = 15,000$ ksi, $A_b = 1.767$ in², and $\alpha_b = 9.8(10^{-6})/°F]$ sleeve, as shown in Fig. P4-89. After the unit is assembled at 40°F, the temperature is increased to 100°F. If the unit is free of stress at 40°F, determine the normal stresses in the bolt and in the sleeve at 100°F.

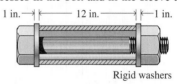

**Figure P4-89**

**4-90** The two faces of the clamp of Fig. P4-90 are 250 mm apart when the two stainless-steel $[E_s = 190$ GPa, $A_s = 115$ mm² (each), and $\alpha_s = 17.3(10^{-6})/°C]$ bolts connecting them are unstretched. A force $P$ is applied to separate the faces of the clamp so that an aluminum alloy $[E_a = 73$GPa, $A_a = 625$mm², and $\alpha_a = 22.5(10^{-6})/°C]$ bar with a length of 250.50 mm can be inserted as shown. After the load $P$ is removed, the temperature is raised 100°C. Determine the normal stresses in the bolts and in the bar, and the distance between the faces of the clamps.

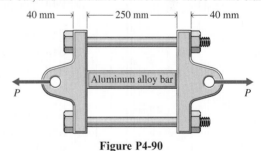

**Figure P4-90**

**4-91** A prismatic bar $[E = 10,000$ ksi and $\alpha = 12.5(10^{-6})/°F]$, free of stress at room temperature, is fastened to rigid walls at its ends. One end of the bar is heated 200°F above room temperature while the other end is maintained at room temperature. The change in temperature $\Delta T$ along the bar is proportional to the square of the distance from the unheated end. Determine the normal stress in the bar after the change in temperature.

**Computer Problems**

**4-92**  The two faces of the clamp shown in Fig. P4-90 are 250 mm apart when the two steel [$E_s$ = 190 GPa, $A_s$ = 115 mm$^2$ (each), and $\alpha_s$ = 17.3(10$^{-6}$)/°C] bolts connecting them are unstretched. A force $P$ is applied to separate the faces of the clamp so that a brass [$E_b$ = 100 GPa, $A_b$ = 625 mm$^2$, and $\alpha_b$ = 17.6(10$^{-6}$)/°C] bar with a length of 250.50 mm can be inserted as shown. After the load $P$ is removed, the temperature of the system is slowly raised. Compute and plot

(a) The axial stress $\sigma_s$ in the steel bolts and the axial stress $\sigma_b$ in the brass bar as functions of the temperature rise $\Delta T$ (0° $\leq \Delta T \leq$ 100°C).

(b) The elongation $\delta_s$ of the steel bolts and the elongation $\delta_b$ of the brass bar as functions of the temperature rise $\Delta T$ (0° $\leq \Delta T \leq$ 100°C).

(c) The distance $L$ between the faces of the clamp as a function of the temperature rise $\Delta T$ (0° $\leq \Delta T \leq$ 100°C).

**4-93**  An aluminum ($E_{al}$ = 10,000 ksi, $\alpha_{al}$ = 12.5 × 10$^{-6}$/°F, $A_{al}$ = 1.4 in$^2$) bolt passes through a steel ($E_{st}$ = 30,000 ksi, $\alpha_{st}$ = 6.6 × 10$^{-6}$/°F, $A_{st}$ = 0.4 in$^2$) sleeve as shown in Fig. P4-89. Initially, the nut is tightened against the washer at room temperature until the bolt has a tensile force of 3500 lb. Then the temperature of the assembly is slowly raised. Calculate and plot.

(a) The stress $\sigma_{al}$ in the aluminum bolt and the stress $\sigma_{st}$ in the steel sleeve as functions of the temperature increase $\Delta T$ (0°F $\leq \Delta T \leq$ 100°F).

(b) The change in length of the aluminum bolt $\delta_{al}$ and the change in length of the steel sleeve $\delta_{st}$ as functions of the temperature increase $\Delta T$ (0°F $\leq \Delta T \leq$ 100°F).

## 4-10 DESIGN

A designer must select a material and properly proportion a member to perform a specified function without failure. *Failure* is defined as the state or condition in which a member or structure no longer functions as intended. To accomplish the design task, one must anticipate the type of failure (*failure mode*) that may occur. Once the failure mode has been determined, the significant material property that controls failure is established. Design computations are performed using mathematical relationships between load and stress or load and deformation.

**Modes of Failure.**  The *mode of failure* of a member depends upon many factors: the type of material, the manner of loading, the rate of loading, and environmental conditions. Discussion in this book will be limited to members subjected to static or slowly applied loads at room temperature. Furthermore, failure modes are limited to *elastic failure*, which occurs as a result of excessive elastic deformation; *yielding* (sometimes referred to as slip failure), characterized by excessive plastic deformation; and *failure by fracture* (complete separation of the material).

**Significant Material Property.**  Associated with each mode of failure is a significant material property. When a structure is designed to avoid elastic failure, the modulus of elasticity is the significant material property. Since yielding is characterized by excessive plastic deformation, the significant material property is the yield strength. Failure by fracture may be due to sudden fracture of a brittle material, fracture of a material with cracks or flaws, or fracture due to repeated loading. In this book, fracture failure will be limited to sudden fracture, where the significant material property is the ultimate strength.

**Mathematical Analysis.**  A failure criterion is needed to perform design computations. The failure criterion may be based upon a probabilistic model or an allowable stress model. Only the allowable stress model (called *allowable stress* *design*, ASD) will be discussed in this book. Once the mode of failure is established, the ASD model states that the design is satisfactory so long as

$$\text{Strength} \geq \text{Stress} \qquad (4\text{-}23)$$

in which "strength" is the significant material property, and "stress" refers to the computed stress in the member. For example, stress $= \sigma = P/A$ for an axially loaded member.

**Factor of Safety.** Most design problems involve many unknown variables. The load that the structure or machine must carry is usually estimated. The actual load may vary considerably from the estimate, especially when loads at some future time must be considered. Since testing usually damages a material, the properties of a material used in a structure cannot be evaluated directly but are normally determined by testing specimens of a similar material. Furthermore, the actual stresses that will exist in a structure are unknown because the calculations are based on assumptions about the distribution of stresses in the material. Because of these and other unknown variables it is customary to write Eq. (4-23) as

$$\text{Strength} \geq (\text{Factor of safety})(\text{Stress}) \qquad (4\text{-}24)$$

where the *factor of safety* (FS) takes into account the imponderables.

As an aid to understanding the use of Eq. (4-24), consider an axially loaded rod that is to be designed (find the required diameter) so that the material does not yield. Then, since the mode of failure is yielding, the significant material property is the yield strength $\sigma_y$, that is, strength equals $\sigma_y$. Then, Eq. (4-24) becomes

$$\sigma_y \geq (\text{FS})(F/A)$$

or

$$\sigma_y \geq (\text{FS})\left(\frac{F}{\pi d^2/4}\right)$$

Solving for the diameter gives

$$d \geq \sqrt{4(\text{FS})(F)/(\pi \sigma_y)}$$

For a given material ($\sigma_y$ known), a given factor of safety, and a given load, the minimum required diameter would be

$$d_{\min} = \sqrt{4(\text{FS})(F)/(\pi \sigma_y)}$$

The following example problems illustrate the use of the design principles previously discussed.

## Example Problem 4-19

An axially loaded circular bar is subjected to a load of 6500 lb. The bar is made of structural steel, and the factor of safety is to be 1.5. Determine the minimum diameter bar required if yielding is to be avoided.

## SOLUTION

Since the mode of failure is yielding, the significant material property (strength) is the yield strength. Using Table A-17, the yield strength of structural steel is $36(10^3)$ psi. Equation (4-24) then gives

$$\sigma_y \geq (FS)(F/A)$$

$$\sigma_y \geq (FS)\left(\frac{F}{\pi d^2/4}\right)$$

$$d^2 \geq \frac{4F(FS)}{\pi \sigma_y}$$

$$d^2 \geq \frac{4(6500)(1.5)}{\pi(36)(10^3)}$$

Therefore

$$d \geq 0.587 \text{ in.} \qquad \textbf{Ans.}$$

The minimum required diameter is 0.587 in. If rods are commercially available in increments of $\frac{1}{8}$ in., a rod of diameter $\frac{5}{8}$ in. would be selected. ∎

## Example Problem 4-20

An axially loaded circular bar made of structural steel has a constant cross-sectional area and is subjected to the forces shown in Fig. 4-44a. The factor of safety, based on failure by yielding, is to be 1.8. Determine the minimum permissible diameter of the bar required to support the loads.

## SOLUTION

The forces transmitted by sections $AB$ and $BC$ are obtained from free-body diagrams of portions of the bar isolated by using cutting planes to the right of pin $A$ and to the left of pin $C$ and drawing the axial force diagram shown in Fig. 4-44b. Thus, the maximum load transmitted by any cross section is $F_{AB} = 36$ kN. Since the criterion for failure is yielding, the significant material property is the yield strength. From Table A-18, $\sigma_y = 250$ MPa. Proceeding as in the previous example

$$d^2 \geq \frac{4F(FS)}{\pi \sigma_y}$$

where $F = F_{AB}$ is the largest internal force in the constant diameter bar. Substituting the numerical values

$$d^2 \geq \frac{4(36)(10^3)(1.8)}{\pi(250)(10^6)}$$

$$d \geq 0.01817 \text{ m}$$

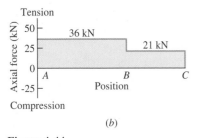

**Figure 4-44**

Therefore

$$d_{\min} = 18.17 \text{ mm} \blacksquare$$ **Ans.**

## Example Problem 4-21

A 40-lb light is supported at the midpoint of a 10-ft length of wire that is made of 0.2% C hardened steel, as shown in Fig. 4-45a. For reasons of safety, a factor of safety of 3 based on the yield strength of the wire is specified. Spools of wire are available with diameters of 10, 20, 30, 40, and 50 mil (1 mil = 0.001 in.). What spool size would you select for suspending the light?

### SOLUTION

The forces transmitted by cables $AB$ and $AC$ are obtained from the free-body diagram of joint $A$ of the cable system shown in Fig. 4-45b. The wire angle is

$$\theta = \cos^{-1}\frac{4}{5} = 36.87°$$

From the horizontal component of the equilibrium equation

$$+ \rightarrow \Sigma F_x = 0: \qquad T_{AC} \cos\theta - T_{AB} \cos\theta = 0 \qquad (a)$$

we get that the two tension forces must be equal, $T_{AC} = T_{AB}$. Then, from the vertical component of the equilibrium equation

$$+ \uparrow \Sigma F_y = 0: \qquad T_{AC} \sin 36.87° + T_{AB} \sin 36.87° - 40 = 0 \qquad (b)$$

Substituting $T_{AC} = T_{AB}$ into Eq. (b) gives

$$T_{AC} = T_{AB} = 33.33 \text{ lb}$$

For 0.2% C hardened steel (see Table A-17), $\sigma_y = 62$ ksi. Therefore, proceeding as in the previous examples

$$d^2 \geq \frac{4F(\text{FS})}{\pi\sigma_y}$$

$$d^2 \geq \frac{4(33.33)(3)}{\pi(62)(10^3)}$$

$$d \geq 0.0453 \text{ in.}$$

The required spool size is 50 (use 50-mil wire). $\blacksquare$ **Ans.**

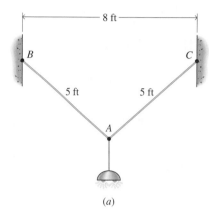
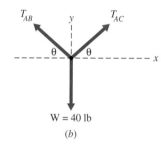

(a)

(b)

**Figure 4-45**

## Example Problem 4-22

The short post shown in Fig. 4-46a is subjected to an axial compressive load $P = 150$ kip. The load is applied to the post through a rigid steel plate. The core of the post is annealed bronze, and the outer segment of the post is composed of

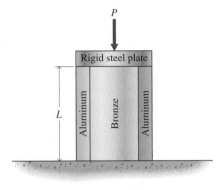

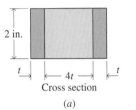

$t$  $4t$  $t$
Cross section

*(a)*

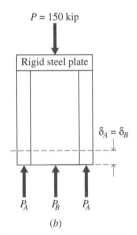

$P = 150$ kip

Rigid steel plate

$\delta_A = \delta_B$

$P_A$  $P_B$  $P_A$

*(b)*

**Figure 4-46**

two symmetrically placed plates of 2024-T4 aluminum. Each of the aluminum plates is one fourth as thick as the bronze core. Determine the minimum thickness $t$ required if the factor of safety based on failure by yielding is 1.5.

## SOLUTION

Since the criterion for failure is yielding, the significant property for each material is the yield strength. From Table A-17, the modulus of elasticity and the yield strength are $E_B = 15,000$ ksi and $\sigma_y = 20$ ksi for the bronze and $E_A = 10,600$ ksi and $\sigma_y = 48$ ksi for the aluminum. The free-body diagram (Fig. 4-46b) has two unknown forces; $P_A$, the force in the aluminum plates, and $P_B$, the force in the bronze core. Since only one equation of equilibrium is available,

$$+ \uparrow \Sigma F_y = 0: \qquad 2P_A + P_B - 150 = 0$$
$$2P_A + P_B = 150 \text{ kip} \qquad (a)$$

the problem is statically indeterminate. As the rigid steel plate pushes down on the top of the post, the bronze core and the two aluminum plates will all shorten the same amount. Therefore, the deformation equation is $\delta_A = \delta_B$ which gives

$$\left(\frac{PL}{EA}\right)_A = \left(\frac{PL}{EA}\right)_B$$

$$\frac{P_A L}{(10,600)2t} = \frac{P_B L}{(15,000)8t}$$

or

$$P_A = 0.17667 \, P_B \qquad (b)$$

Solving Eqs. (a) and (b) yields

$$P_A = 19.582 \text{ kip}$$
$$P_B = 110.84 \text{ kip}$$

The failure criterion is

$$\sigma_y \geq (FS)(P/A)$$

$$A \geq \frac{P(FS)}{\sigma_y}$$

Applying the failure criterion to each member of the structure yields

$$2t \geq \frac{19.582(1.5)}{48} \qquad t \geq 0.306 \text{ in.}$$

for the aluminum and

$$8t \geq \frac{110.84(1.5)}{20} \qquad t \geq 1.039 \text{ in.}$$

for the bronze. Therefore, the minimum thickness is

$$t_{min} = 1.039 \text{ in.} \quad \blacksquare \qquad \textbf{Ans.}$$

# PROBLEMS

### Introductory Problems

**4-94*** A short standard-weight steel pipe (see Appendix A) is used to support an axial compressive load of 100 kN. If yielding ($\sigma_y = 250$ MPa) should not occur and the factor of safety is to be 1.6, determine the smallest nominal diameter pipe that may be used to support the load.

**4-95*** A short column made of structural steel is used to support the floor beams of a building, as shown in Fig. P4-95. Each floor beam (*A* and *B*) transmits a force of 40 kip to the column. The column has the shape of a wide-flange (W) section (see Appendix A). The factor of safety based on failure by yielding is 3.0. Select the lightest wide-flange section that will support the given loads.

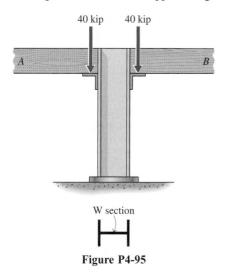

**Figure P4-95**

**4-96** The two structural steel (see Appendix A) rods *A* and *B* shown in Fig. P4-96 are used to support a mass *m* = 2000 kg. If failure is by yielding and a factor of safety of 1.75 is specified, determine the diameters of the rods (to the nearest 1 mm) that must be used to support the mass. Both rods are to have the same diameter.

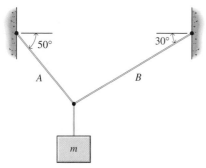

**Figure P4-96**

### Intermediate Problems

**4-97*** The machine component shown in Fig. P4-97 is made of hot-rolled Monel. The forces at *B* are applied to the component with a rigid collar that is firmly attached to the component. If the mode of failure is yielding and the factor of safety is 1.5, determine the minimum permissible diameter of each segment of the machine component.

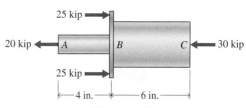

**Figure P4-97**

**4-98*** An axial load *P* = 1000 kN is applied to the rigid steel bearing plate on the top of the short column shown in Fig. P4-98. The outside segment of the column is made of structural steel. The inside core is made of fairly high-strength concrete. Both segments are square. The failure modes are yielding for the steel and fracture for the concrete. The factor of safety is to be 1.4. If the area of the concrete is to be 10 times the area of the steel, determine the required dimensions.

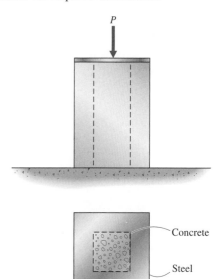

**Figure P4-98**

**4-99** Four axial forces are applied to the 1-in.-thick, 0.4% C hot-rolled steel bar as shown in Fig. P4-99. The factor of safety for failure by yielding is 1.75. Determine the minimum width *w* of the constant cross-sectional area bar.

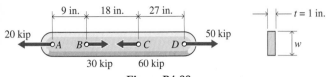

**Figure P4-99**

## Challenging Problems

**4-100\*** The two parts of the eyebar shown in Fig. P4-100 are connected by two bolts (one on each side of the eye bar). The bolts are made of a grade of steel with a tensile yield strength of 1035 MPa and a shear yield strength of 620 MPa. The eyebar is subjected to the

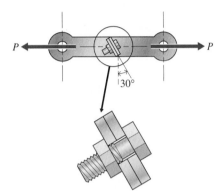

**Figure P4-100**

forces $P = 85$ kN. Determine the minimum bolt diameter required to safely support the forces if the mode of failure is yielding and the factor of safety is 1.5.

**4-101** The two solid rods shown in Fig. P4-101 are pin-connected at the ends and support a weight of 10 kip. The rods are made of SAE 4340 heat-treated steel. The factor of safety for failure by yielding is to be 1.5. For a minimum weight of rod design, determine

(a) The optimum angle $\theta$.
(b) The required diameter for the rods.
(c) The weight of each rod. Is it reasonable to neglect the weight of the rods in the design?

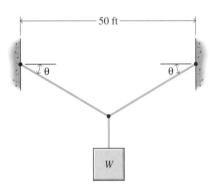

**Figure P4-101**

## 4-11 SUMMARY

When a structural member or machine component is subjected to a system of external loads (applied loads and support reactions), a system of internal resisting forces develops within the member to balance the external forces. These forces tend to either crush the member (compression) or pull it apart (tension). The intensities of these internal forces (forces per unit area) are called stresses.

For long, slender, axially loaded members, it is generally assumed that the normal stresses on a transverse plane are uniformly distributed except near the points of load application. In these members an average normal stress on the cross section can be computed as

$$\sigma_{\text{avg}} = \frac{F}{A} \tag{4-2}$$

where $F$ is the magnitude of the force transmitted by the cross section and $A$ is the cross-sectional area of the member. A positive sign indicates a tensile normal stress, and a negative sign indicates a compressive normal stress. This sign convention is independent of the selection of a coordinate system.

Experimental studies indicate that materials respond differently to forces that tend to pull surfaces apart than they do to forces that tend to slide surfaces relative to each other. Stresses on planes inclined to the axes of axially loaded bars are obtained by resolving the internal forces on the inclined plane into a component perpendicular to the plane (a normal force $N$ from which a normal stress $\sigma$ is determined) and a component tangent to the plane (a shear force $V$ from which a shear stress $\tau$ is determined). Both normal and shearing stresses under axial loading are important since a brittle material loaded in tension will fail in tension on a transverse plane (where the normal stress is a maximum):

$$\sigma_{max} = \frac{P}{A} \qquad (4\text{-}9)$$

whereas a ductile material loaded in tension will fail in shear on a 45° plane (where the shear stress is a maximum):

$$\tau_{max} = \frac{P}{2A} \qquad (4\text{-}10)$$

Loads applied to a structure or machine are generally transmitted to the individual members through connections that use rivets, bolts, or pins. In all of these connections, one of the most significant stresses induced is a shearing stress. From the definition of stress, an average shearing stress on the transverse cross section of the rivet, bolt, or pin can be computed as

$$\tau_{avg} = \frac{V}{A_s} \qquad (4\text{-}4)$$

where $V$ is the magnitude of the shear force transmitted by the rivet, bolt, or pin and $A_s$ is the cross-sectional area of the rivet, bolt, or pin.

Deformation is a measure of the change in size or shape of a body. Strain (deformation per unit length) is the quantity used to measure the intensity of a deformation, just as stress (force per unit area) is used to measure the intensity of an internal force. Normal strain,

$$\epsilon_{avg} = \frac{\delta_n}{L} \qquad (4\text{-}11)$$

measures the change in size (elongation or contraction of an arbitrary line segment) of a body during deformation. Normal strain is positive when the line elongates and negative when the line contracts. In general, if the axial stress is tensile, the axial deformation will be an elongation. Therefore, positive normal strains are referred to as tensile strains and negative normal strains are referred to as compressive strains.

Shearing strain,

$$\gamma_{avg} = \frac{\delta_s}{L} = \tan\phi \qquad (4\text{-}13)$$

$$\gamma_{xy}(P) = \frac{\pi}{2} - \theta' \qquad (4\text{-}14b)$$

measures the change in shape (change in angle between two lines that are orthogonal in the undeformed state) of a body during a deformation. Shearing strains will be positive if the angle between reference lines decreases and negative if the angle increases. Normal and shearing strains for most engineering materials in the elastic range seldom exceed values of 0.2 percent (0.002 in./in. or 0.002 rad).

Curves showing the relationship between stress and strain (called stress–strain diagrams) are independent of the size and shape of the member and depend only on the type of material from which the member is made. Data for stress–strain diagrams are obtained by applying an axial load to a test specimen and measuring the load and deformation simultaneously. The initial portion of the stress–strain diagram for most materials used in engineering structures is a straight line and is represented by Hooke's law

$$\sigma = E\epsilon \tag{4-15a}$$

where the constant of proportionality (modulus of elasticity, $E$) must be determined from the experimental data. Although the stress–strain diagram for some materials such as gray cast iron and concrete show a slight curve even at very small stresses, it is common practice to draw a straight line to average the data for the first part of the diagram and neglect the curvature. It is important to realize that Hooke's Law only describes the initial linear portion of the stress–strain diagram and is valid only for uniaxally loaded bars.

A body loaded in one direction will undergo strains perpendicular to the direction of the load in addition to those parallel to the load. The ratio of the lateral or transverse strain to the longitudinal or axial strain is called Poisson's ratio:

$$\nu = -\frac{\epsilon_{\text{lat}}}{\epsilon_{\text{long}}} = -\frac{\epsilon_t}{\epsilon_a} \tag{4-16}$$

and is a constant for stresses below the proportional limit. Poisson's ratio is related to $E$ and $G$ by

$$E = 2(1 + \nu)G \tag{4-17}$$

and has a value between $\frac{1}{4}$ and $\frac{1}{3}$ for most metals. The ratio $\nu = -\epsilon_t/\epsilon_a$ is valid only for a uniaxial state of stress.

When unrestrained, most engineering materials expand when heated and contract when cooled. The thermal strain of an unrestrained body due to a temperature change of $\Delta T$ degrees is

$$\epsilon_T = \alpha \Delta T \tag{4-18}$$

where $\alpha$ is known as the coefficient of thermal expansion.

Strains caused by temperature changes and strains caused by applied loads are essentially independent. The total normal strain in a body acted upon by both temperature changes and axially applied loads is given by

$$\epsilon_{\text{total}} = \epsilon_\sigma + \epsilon_T = \frac{\sigma}{E} + \alpha \Delta T \tag{4-19}$$

Since homogeneous, isotropic materials expand uniformly in all directions when heated (and contract uniformly when cooled), neither the shape of the body nor the shearing stresses and shearing strains are affected by temperature changes, if the body is unrestrained.

When a straight bar of uniform cross section is axially loaded by forces applied at the ends, the axial strain along the length of the bar is assumed to have a constant value, and the elongation (or contraction) of the bar resulting from the axial load $P$ may be expressed as $\delta = \epsilon L$ (by the definition of average axial strain). If Hooke's law [Eq. (4-15a)] applies, the axial deformation may be expressed in terms of either stress or load as

$$\delta = \epsilon L = \frac{\sigma L}{E} \tag{4-20a}$$

or

$$\delta = \frac{PL}{EA} \tag{4-20b}$$

where the force in the member $P$, the cross-sectional area of the member $A$, and modulus of elasticity $E$ are all constant over the length $L$. If a bar is subjected to a number of axial loads at different points along the bar, or if the bar consists of parts having different cross-sectional areas or of parts composed of different materials, then the change in length of each part can be computed using Eq. (4-20b). The changes in length of the various parts of the bar can then be added algebraically to give the total change in length of the complete bar

$$\delta = \sum_{i=1}^{n} \delta_i = \sum_{i=1}^{n} \frac{P_i L_i}{E_i A_i} \tag{4-21}$$

where $A_i$ and $E_i$ are constant on segment $i$ of length $L_i$ and the force $P_i$ is the internal force in segment $i$ of the bar. Force $P_i$ is usually different than the forces applied at the ends of the segment. These forces must be calculated from equilibrium of the segment and are often shown on an axial force diagram.

A designer is required to select a material and properly proportion a member to perform a specified function without failure. Failure is defined as the state or condition in which a member or structure no longer functions as intended. Elastic failure occurs as a result of excessive elastic deformation. When a structure is designed to avoid elastic failure, the stiffness of the material, indicated by modulus of elasticity, is the significant property. Failure by yielding is characterized by excessive plastic deformation. Yield strength, yield point, and proportional limit are used as indices of strength with respect to failure by yielding for members subjected to static loads. Failure by fracture is a complete separation of the material. The ultimate strength of a material is the index of resistance to failure by fracture under static loads in which creep is not involved.

Design computations are performed using the allowable stress design model, which states that the design is satisfactory so long as

$$\text{Strength} \geq (\text{Factor of safety}) \, (\text{Stress}) \tag{4-24}$$

in which stress refers to the computed stress in the member being designed and strength is the significant material property that depends upon the mode of failure. The factor of safety takes into account the imponderables.

# REVIEW PROBLEMS

**4-102*** A tension member consists of a 50-mm-diameter brass ($E = 100$ GPa) bar connected to a 32-mm-diameter stainless steel ($E = 190$ GPa) bar, as shown in Fig. P4-102. For an applied load $P = 50$ kN, determine
(a) The normal stresses in each segment of the member.
(b) The elongation of the member.

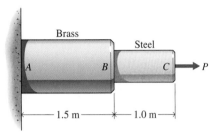

**Figure P4-102**

**4-103*** An alloy steel ($E = 30,000$ ksi) bar is loaded and supported as shown in Fig. P4-103. The loading collar at $B$ is free to slide on section $BC$. The diameters of sections $AB$, $BC$, and $CD$ are 2.50 in., 1.50 in., and 1.00 in., respectively. The lengths of all three segments are 15 in. Determine the normal stresses in each section and the overall change in length of the bar.

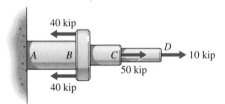

**Figure P4-103**

**4-104*** The floor beams of a storage shed are supported as shown in Fig. P4-104. Each of the floor beams $B$ and $C$ transmits a 50 kN load to post $A$. Post $A$, the baseplate, and the footing have cross-sectional areas of 15,000 mm², 30,000 mm², and 260,000 mm², respectively. Determine

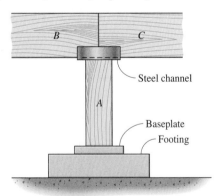

**Figure P4-104**

(a) The normal stress in post $A$.
(b) The bearing stress between the post and the baseplate.
(c) The bearing stress between the baseplate and the footing.
(d) The bearing stress between the footing and the ground.

**4-105** A 1-in.-diameter steel [$\alpha = 6.5(10^{-6})/°F$, $E = 30,000$ ksi, and $\nu = 0.30$] bar is subjected to a temperature decrease of 150°F. The ends of the bar are supported by two walls that displace a small amount during the temperature change. If the measured strain in the bar is $-600$ $\mu$in./in. after the temperature change, determine the load being transmitted to the walls.

**4-106** A 90-mm-diameter brass ($E = 100$ GPa) bar is securely fastened to a 50-mm-diameter steel ($E = 200$ GPa) bar. The ends of the composite bar are then attached to rigid supports, as shown in Fig. P4-106. Determine the stresses in the brass and the steel after a temperature drop of 70°C occurs. The thermal coefficients of expansion for the brass and the steel are $17.6(10^{-6})/°C$ and $11.9(10^{-6})/°C$, respectively.

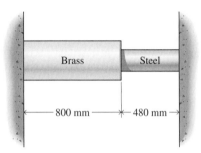

**Figure P4-106**

**4-107*** A steel ($E = 30,000$ ksi) pipe column with an outside diameter of 3 in. and an inside diameter of 2.5 in. is attached to unyielding supports at the top and bottom as shown in Fig. P4-107. A rigid collar $C$ is used to apply a 50-kip load $P$. Determine
(a) The normal stresses in the top and bottom portions of the pipe.
(b) The deflection of the collar $C$.

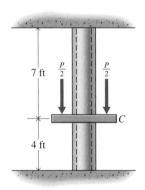

**Figure P4-107**

**4-108\*** A 3-mm-diameter cord ($E = 7$ GPa) that is covered with a 0.5-mm-thick plastic sheath ($E = 14$ GPa) is subjected to an axial tensile load $P$, as shown in Fig. P4-108. The load is transferred to the cord and sheath by rigid blocks attached to the ends of the assembly. The yield strengths for the cord and sheath are 15 MPa and 56 MPa, respectively. Determine the maximum allowable load if a factor of safety of 3 with respect to failure by yielding is specified.

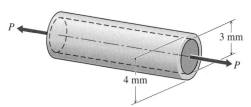

$P$    3 mm    $P$    4 mm

**Figure P4-108**

**4-109** The assembly shown in Fig. P4-109 consists of a steel ($E_s = 30,000$ ksi, $A_s = 1.25$ in$^2$) bar $A$, a rigid bearing plate $C$ that is securely fastened to bar $A$, and a bronze ($E_b = 15,000$ ksi, $A_b = 3.75$ in$^2$) bar $B$. A clearance of 0.025 in. exists before the assembly is loaded by a force $P = 15$ kip. Determine, for each segment of the assembly

(a) The normal stress.
(b) The change in length.

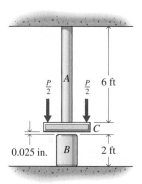

**Figure P4-109**

**4-110** An 80-kN force $P$ is applied to the 150- $\times$ 180-mm wood block shown in Fig. P4-110. Determine the normal stress perpendicular to the grain of the wood and the shearing stress parallel to the grain of the wood.

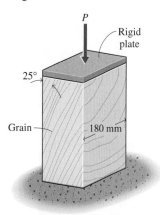

**Figure P4-110**

**4-111** A 4000-lb force $P$ is applied to the square structural steel block shown in Fig. P4-111. Determine

(a) The change in length of the block.
(b) The maximum normal stress in the block.
(c) The maximum shearing stress in the block.
(d) The normal stress perpendicular to the plane $A - A$.
(e) The shearing stress parallel to the plane $A - A$.

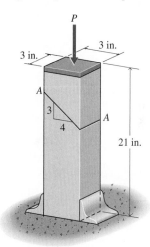

**Figure P4-111**

# EQUIVALENT FORCE/ MOMENT SYSTEMS

## 5-1 INTRODUCTION

The resultant force **R** of a system of two or more concurrent forces $\mathbf{F}_1$, $\mathbf{F}_2$, . . . , $\mathbf{F}_n$ was defined in Chapter 2 as the single force that will produce the same effect on a body as the original system of forces. When the resultant force **R** of a concurrent force system is zero, the body on which the system of forces acts is in equilibrium and the force system is said to be balanced. Methods to determine resultants of concurrent force systems were discussed in Chapter 2 and applied to equilibrium of a particle in Chapter 3.

For the case of a three-dimensional body that has a definite size and shape, the particle idealization discussed in Chapter 3 is no longer valid, in general, since the forces acting on the body are usually not concurrent. For these more general force systems, the condition $\mathbf{R} = \mathbf{0}$ is a necessary but not a sufficient condition for equilibrium of the body. A second restriction related to the tendency of a force to produce rotation of a body must also be satisfied and gives rise to the concept of a moment. In this chapter, the moment of a force about a point and the moment of a force about a line (axis) will be defined and methods will be developed for finding the resultant forces and the resultant moments for force systems that are not concurrent.

## 5-2 MOMENTS AND THEIR CHARACTERISTICS

The moment of a force about a point or axis is a measure of the tendency of the force to rotate a body about that point or axis. For example, the moment of force **F** about point $O$ in Fig. 5-1$a$ is a measure of the tendency of the force to rotate the body about line $A$–$A$. Line $A$–$A$ is perpendicular to the plane containing force **F** and point $O$.

A moment has both a magnitude and a direction, and adds according to the parallelogram law of addition; therefore, it is a vector quantity. The magnitude of a moment $|\mathbf{M}|$ is defined as the product of the magnitude of a force $|\mathbf{F}|$ and the perpendicular distance $d$ from the line of action of the force to the axis. Thus, in Fig. 5-1$b$, the magnitude of the moment of the force **F** about point $O$

(actually about axis $A$–$A$, which is perpendicular to the page and passes through point $O$) is

$$M_O = |\mathbf{M}_O| = |\mathbf{F}|d \qquad (5\text{-}1)$$

Point $O$ is called the *moment center*, distance $d$ is called the *moment arm*, and line $A$–$A$ is called the *axis of the moment*.

The direction (sense) of a moment in a two-dimensional problem can be specified by using a small curved arrow about the point, as shown in Fig. 5-1*b*. If the force tends to produce a counterclockwise rotation, the moment is assumed to be positive. In a similar manner, if the force tends to produce a clockwise rotation, the moment is negative.

Since the magnitude of a moment of a force is the product of a force and a length, the dimensional expression for a moment is $FL$. In the U.S. customary system, the units commonly used for moments are lb · ft and lb · in. or in. · lb, ft · lb, and ft · kip. In the SI system, the units commonly used for moments are N · m, kN · m, and so on. It is immaterial whether the unit of force or the unit of length is stated first. However, in this book the unit of force will be stated first, followed by the unit of length.

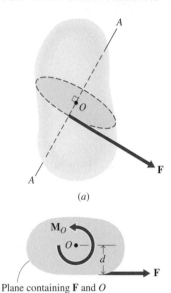

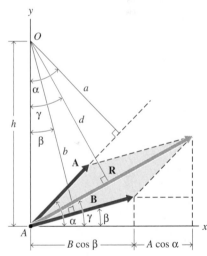

Plane containing **F** and $O$

(*b*)

**Figure 5-1**

## Principle of Moments—Varignon's Theorem

A concept often used in solving mechanics (statics, dynamics, mechanics of materials) problems is the principle of moments. This principle, when applied to a system of forces, states that the moment **M** of the resultant **R** of a system of forces with respect to any axis or point is equal to the vector sum of the moments of the individual forces of the system with respect to the same axis or point. Application of this principle to a pair of concurrent forces is known as Varignon's theorem. Varignon's theorem can be illustrated by using the concurrent force system, shown in Fig. 5-2, where **R** is the resultant of forces **A** and **B** that lie in the $xy$-plane. The point of concurrency $A$ and the moment center $O$ have been arbitrarily selected to lie on the $y$-axis. The distances $d$, $a$, and $b$ are the perpendicular distances from the moment center $O$ to the lines of action of forces **R**, **A**, and **B**, respectively. The angles $\gamma$, $\alpha$, and $\beta$ (measured from the $x$-axis) locate the forces **R**, **A**, and **B**, respectively.

The magnitudes of the moments produced by the resultant **R** and by the two forces **A** and **B** with respect to point $O$ are

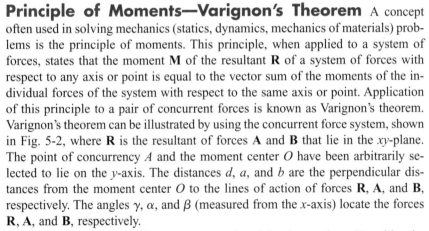

**Figure 5-2**

$$M_R = Rd = R(h \cos \gamma)$$
$$M_A = Aa = A(h \cos \alpha) \qquad (a)$$
$$M_B = Bb = B(h \cos \beta)$$

From Fig. 5-2 note also that

$$R \cos \gamma = A \cos \alpha + B \cos \beta \qquad (b)$$

Substituting Eqs. (a) into Eq. (b) and multiplying both sides of the equation by $h$ yields

$$M_R = M_A + M_B \qquad (5\text{-}2)$$

Equation (5-2) indicates that the moment of the resultant $\mathbf{R}$ with respect to a point $O$ is equal to the sum of the moments of the forces $\mathbf{A}$ and $\mathbf{B}$ with respect to the same point $O$.

## Example Problem 5-1

Three forces are applied to a triangular plate as shown in Fig. 5-3. Determine

(a) The moment of force $\mathbf{F}_3$ about point $C$.
(b) The moment of force $\mathbf{F}_2$ about point $B$.
(c) The moment of force $\mathbf{F}_1$ about point $B$.
(d) The moment of force $\mathbf{F}_3$ about point $E$.

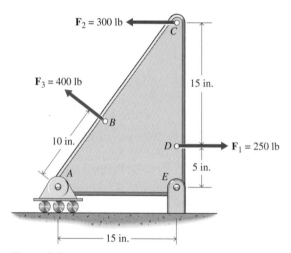

**Figure 5-3**

## SOLUTION

The angle of inclination $\phi$ of the inclined edge of the plate with respect to a horizontal line is

$$\phi = \tan^{-1}\frac{20}{15} = 53.13°$$

$$L_{AC} = \sqrt{20^2 + 15^2} = 25 \text{ in.}$$

From Eq. (5-1):

$$M = |\mathbf{F}|\, d$$

(a) $M_C = F_3 L_{BC} = 400(25 - 10) = 6000$ lb · in.

$$\mathbf{M}_C = 6000 \text{ lb · in. } \circlearrowleft \qquad \text{Ans.}$$

(b) $M_B = F_2 d_2 = 300(15 \sin 53.13°) = 3600$ lb · in.

$$\mathbf{M}_B = 3600 \text{ lb} \cdot \text{in. } \circlearrowleft \qquad \textbf{Ans.}$$

(c) $M_B = F_1 d_1 = 250(15 - 15 \sin 53.13°) = 750$ lb · in.

$$\mathbf{M}_B = 750 \text{ lb} \cdot \text{in. } \circlearrowleft \qquad \textbf{Ans.}$$

(d) $M_E = F_3 d_3 = 400(10 - 15 \cos 53.13°) = 400$ lb · in.

$$\mathbf{M}_E = 400 \text{ lb} \cdot \text{in. } \circlearrowleft \ \blacksquare \qquad \textbf{Ans.}$$

## Example Problem 5-2

Use the principle of moments to determine the moment about point $B$ of the 300-N force shown in Fig. 5-4a.

### SOLUTION

The magnitudes of the rectangular components of the 300-N force are

$$F_x = F \cos 30° = 300 \cos 30° = 259.8 \text{ N}$$
$$F_y = F \sin 30° = 300 \sin 30° = 150.0 \text{ N}$$

Once the forces $F_x$ and $F_y$ are known, the moment $M_B$ is

$$+\!\!\downarrow M_B = -F_x(0.250) - F_y(0.200)$$
$$= -259.8(0.250) - 150.0(0.200) = -95.0 \text{ N} \cdot \text{m}$$

$$\mathbf{M}_B = 95.0 \text{ N} \cdot \text{m } \circlearrowleft \qquad \textbf{Ans.}$$

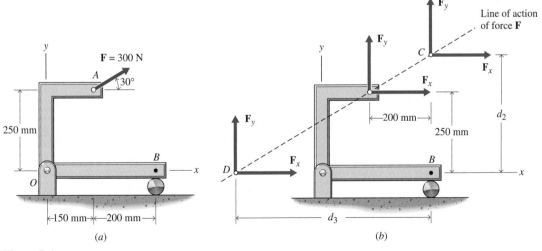

(a)                                    (b)

**Figure 5-4**

The moment about point $B$ can also be determined by moving the force $\mathbf{F}$ along its line of action(principle of transmissibility) to points $C$ or $D$, as shown in Fig. 5-4b. For both of these points, one component of the force produces no moment about point $B$. Thus,

For point $C$:

$$d_2 = 250 + 200 \tan 30° = 365.5 \text{ mm}$$

$$+\lcurvearrowdown M_B = -F_x d_2 = -259.8(0.3655) = -95.0 \text{ N} \cdot \text{m}$$

$$\mathbf{M}_B = 95.0 \text{ N} \cdot \text{m} \downdownarrows \qquad \textbf{Ans.}$$

For point $D$:

$$d_3 = 200 + 250 \cot 30° = 633.0 \text{ mm}$$

$$+\lcurvearrowdown M_B = -F_y d_3 = -150.0(0.633) = -95.0 \text{ N} \cdot \text{m}$$

$$\mathbf{M}_B = 95.0 \text{ N} \cdot \text{m} \downdownarrows \blacksquare \qquad \textbf{Ans.}$$

# PROBLEMS

### Introductory Problems

**5-1\*** Two forces are applied to a bracket as shown in Fig. P5-1. Determine
(a) The moment of force $\mathbf{F}_1$ about point $O$.
(b) The moment of force $\mathbf{F}_2$ about point $O$.

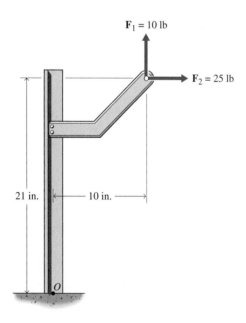

**Figure P5-1**

**5-2\*** Determine the moments of the 225-N force shown in Fig. P5-2 about points $A$, $B$, and $C$.

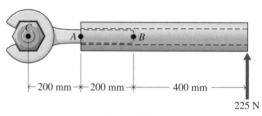

**Figure P5-2**

**5-3** Two forces are applied at a point in the plane of a rigid steel plate as shown in Fig. P5-3. Determine the moments of
(a) The 500-lb force about points $A$ and $B$.
(b) The 300-lb force about points $B$ and $C$.

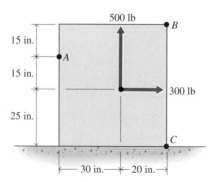

**Figure P5-3**

**5-4** Two forces are applied to the bridge truss shown in Fig. P5-4. Determine the moments of
(a) The 3.6-kN force about points $A$ and $D$.
(b) The 2.7-kN force about point $A$.

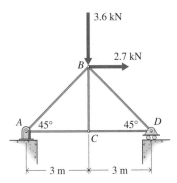

**Figure P5-4**

## Intermediate Problems

**5-5\*** A 50-lb force is applied to the handle of a lug wrench, which is being used to tighten the nuts on the rim of an automobile tire as shown in Fig. P5-5. The diameter of the bolt circle is $5\frac{1}{2}$ in. Determine the moments of the force about the axle of the wheel (point $O$) and about the point of contact of the wheel with the pavement (point $A$).

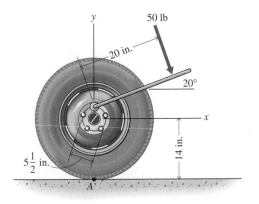

**Figure P5-5**

**5-6\*** A 160-N force is applied to the handle of a door as shown in Fig. P5-6. Determine the moments of the force about the hinges $A$ and $B$.

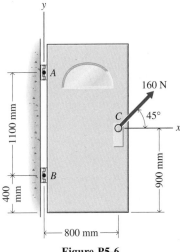

**Figure P5-6**

**5-7** Determine the moment of the 425-lb force shown in Fig. P5-7 about point $B$.

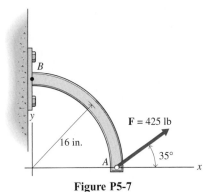

**Figure P5-7**

**5-8** A pry bar is used to extract a nail from a board as shown in Fig. P5-8. Determine the moment of the 120-N force
(a) About point $A$.
(b) About point $B$.

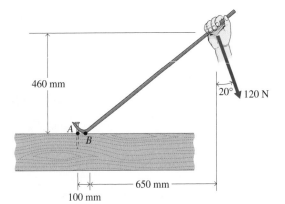

**Figure P5-8**

**Challenging Problems**

**5-9\*** A man exerts a force **P** to hold a 60-ft pole in the position shown in Fig. P5-9. If the moment of the force **P** about point $A$ is 4000 lb · ft, determine the magnitude of the force **P**.

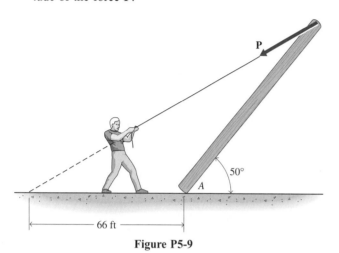

**Figure P5-9**

**5-10\*** Two forces act on a truss as shown in Fig. P5-10. Member AC of the truss is perpendicular to members BC and CD. Determine the moment of
(a) The 4-kN force about point $A$.
(b) The 3-kN force about point $D$.

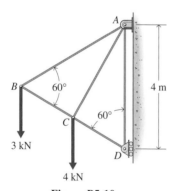

**Figure P5-10**

**5-11** The crane and boom shown in Fig. P5-11 is lifting a 4000-lb load. Determine the moment of the load about point $C$.

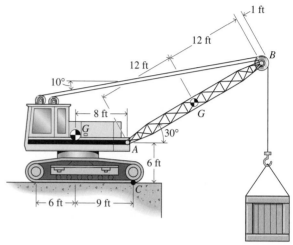

**Figure P5-11**

**5-12** Due to combustion, a compressive force **P** is exerted on the connecting rod of an automobile engine as shown in Fig. P5-12. The lengths of the crank throw $AB$ and connecting rod $BC$ are 75 mm and 225 mm, respectively. Determine the moment of the force **P** about the bearing at $A$ in terms of the crank angle $\theta$.

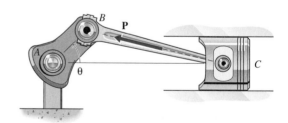

**Figure P5-12**

## 5-3 VECTOR REPRESENTATION OF A MOMENT

For some two-dimensional problems and for most three-dimensional problems, use of Eq. (5-1) for moment determinations is not convenient owing to difficulties in determining the perpendicular distance $d$ between the line of action of the force and the moment center $O$. For these types of problems, a vector approach simplifies moment calculations.

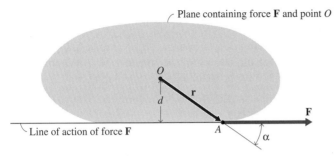

**Figure 5-5**

In Fig. 5-1, the moment of the force **F** about point $O$ can be represented by the expression

$$\mathbf{M}_O = \mathbf{r} \times \mathbf{F} \qquad (5\text{-}3)$$

where **r** is a position vector from the moment center $O$ to a point $A$ on the line of action of the force **F**, as shown in Fig. 5-5. By definition, the cross product (also called the vector product) of the two intersecting vectors **r** and **F** yields a vector that has a magnitude that is the product of the magnitudes of the vectors **r** and **F** and the sine of the angle $\alpha$ between them. The direction of the vector product is perpendicular to the plane containing the vectors **r** and **F**. Therefore

$$\mathbf{M}_O = \mathbf{r} \times \mathbf{F}$$
$$= |\mathbf{r}| \, |\mathbf{F}| \sin \alpha \; \mathbf{e} \qquad (5\text{-}4)$$

where $\alpha$ is the angle ($0 \le \alpha \le 180°$) between the two intersecting vectors **r** and **F** and **e** is a unit vector perpendicular to the plane containing vectors **r** and **F**. It is obvious from Fig. 5-5 that the term $|\mathbf{r}| \sin \alpha$ equals the perpendicular distance $d$ from the line of action of the force to the moment center $O$. Note also from Fig. 5-6 that the distance $d$ is independent of the position $A$ along the line of action of the force, since

$$|\mathbf{r}_1| \sin \alpha_1 = |\mathbf{r}_2| \sin \alpha_2 = |\mathbf{r}_3| \sin \alpha_3 = d$$

Thus, Eq. (5-4) can be written

$$\mathbf{M}_O = |\mathbf{F}|d \; \mathbf{e} = Fd \; \mathbf{e} = M_O \, \mathbf{e} \qquad (5\text{-}5)$$

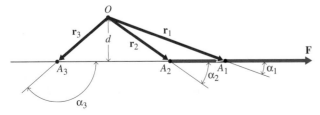

**Figure 5-6**

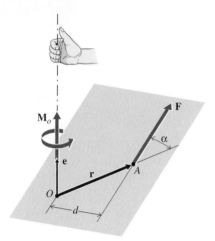

**Figure 5-7**

In Eq. (5-5), the direction of the unit vector **e** is determined (see Fig. 5-7) by using the right-hand rule (fingers of the right hand curl from positive **r** to positive **F** and the thumb points in the direction of positive $\mathbf{M}_O$). Thus, Eq. (5-3) yields both the magnitude $M_O$ and the direction **e** of the moment $\mathbf{M}_O$. It is important to note that the sequence $\mathbf{r} \times \mathbf{F}$ must be maintained in calculating moments since the sequence $\mathbf{F} \times \mathbf{r}$ will produce a moment with the opposite sense. The vector product is *not* a commutative operation.

**Cartesian Representation of the Vector Product.** If the two vectors **r** and **F** are written in Cartesian vector form, the cross product becomes

$$\mathbf{r} \times \mathbf{F} = (r_x \mathbf{i} + r_y \mathbf{j} + r_z \mathbf{k}) \times (F_x \mathbf{i} + F_y \mathbf{j} + F_z \mathbf{k})$$
$$= r_x F_x (\mathbf{i} \times \mathbf{i}) + r_x F_y (\mathbf{i} \times \mathbf{j}) + r_x F_z (\mathbf{i} \times \mathbf{k})$$
$$+ r_y F_x (\mathbf{j} \times \mathbf{i}) + r_y F_y (\mathbf{j} \times \mathbf{j}) + r_y F_z (\mathbf{j} \times \mathbf{k})$$
$$+ r_z F_x (\mathbf{k} \times \mathbf{i}) + r_z F_y (\mathbf{k} \times \mathbf{j}) + r_z F_z (\mathbf{k} \times \mathbf{k}) \qquad \text{(a)}$$

Since **i**, **j**, and **k** are orthogonal,

$$\mathbf{i} \times \mathbf{i} = [(1)(1)\sin 0°] \mathbf{k} = \mathbf{0}$$
$$\mathbf{i} \times \mathbf{j} = [(1)(1)\sin 90°] \mathbf{k} = \mathbf{k}$$

Similarly

$$\begin{array}{lll}
\mathbf{i} \times \mathbf{i} = \mathbf{0} & \mathbf{j} \times \mathbf{i} = -\mathbf{k} & \mathbf{k} \times \mathbf{i} = \mathbf{j} \\
\mathbf{i} \times \mathbf{j} = \mathbf{k} & \mathbf{j} \times \mathbf{j} = \mathbf{0} & \mathbf{k} \times \mathbf{j} = -\mathbf{i} \\
\mathbf{i} \times \mathbf{k} = -\mathbf{j} & \mathbf{j} \times \mathbf{k} = \mathbf{i} & \mathbf{k} \times \mathbf{k} = \mathbf{0}
\end{array} \qquad \text{(b)}$$

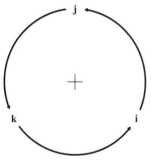

**Figure 5-8**

Equations (b) can be represented graphically by arranging the unit vectors **i**, **j**, and **k** in a circle in a counterclockwise order, as shown in Fig. 5-8. The product of two unit vectors will be positive if they follow each other in counterclockwise order, and negative if they follow each other in clockwise order. Substituting Eqs. (b) into Eq. (a) yields

$$\mathbf{r} \times \mathbf{F} = r_x F_y (\mathbf{k}) + r_x F_z (-\mathbf{j}) + r_y F_x (-\mathbf{k}) + r_y F_z (\mathbf{i}) + r_z F_x (\mathbf{j}) + r_z F_y (-\mathbf{i})$$
$$= (r_y F_z - r_z F_y) \mathbf{i} + (r_z F_x - r_x F_z) \mathbf{j} + (r_x F_y - r_y F_x) \mathbf{k} \qquad \text{(5-6)}$$

which is the expanded form of the determinant

$$\mathbf{r} \times \mathbf{F} = \begin{vmatrix} \mathbf{i} & \mathbf{j} & \mathbf{k} \\ r_x & r_y & r_z \\ F_x & F_y & F_z \end{vmatrix} \qquad \text{(5-7)}$$

Note carefully the arrangement of terms in the determinant, which places the unit vectors **i**, **j**, and **k** in the first row, the components $r_x$, $r_y$, and $r_z$ of **r** in the second row, and the components $F_x, F_y, F_z$ of **F** in the third row. If the two bottom rows of a determinant are interchanged, the sign of the determinant will change. Therefore

$$\mathbf{r} \times \mathbf{F} = -\mathbf{F} \times \mathbf{r}$$

## Moment of a Force About a Point
The position vector **r** from the point about which the moment is to be determined (say, point $B$ in Fig. 5-9) to any point on the line of action of the force **F** (say, point $A$) can be expressed in terms of the unit vectors **i**, **j**, and **k**, and the coordinates $(x_A, y_A, z_A)$ and $(x_B, y_B, z_B)$ of points $A$ and $B$, respectively. Thus,

$$\mathbf{r} = \mathbf{r}_{A/B} = \mathbf{r}_A - \mathbf{r}_B = (x_A - x_B)\,\mathbf{i} + (y_A - y_B)\,\mathbf{j} + (z_A - z_B)\,\mathbf{k} \quad (5\text{-}8)$$

where the subscript $A/B$ indicates $A$ with respect to $B$.

Equation (5-3) is applicable for both the two-dimensional case (forces in, say, the $xy$-plane) and the three-dimensional case (forces with arbitrary space orientations).

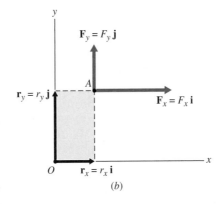

**Figure 5-9**

### The Two-Dimensional Case.
Consider first the moment $\mathbf{M}_O$ about the origin of coordinates (see Fig. 5-10a) produced by a force **F** in the $xy$-plane. The line of action of the force passes through point $A$. For this special case (see Fig. 5-10b).

$$\mathbf{F} = F_x\,\mathbf{i} + F_y\,\mathbf{j}$$

and the position vector **r** from the origin $O$ to point $A$ is

$$\mathbf{r} = r_x\,\mathbf{i} + r_y\,\mathbf{j}$$

The vector product $\mathbf{r} \times \mathbf{F}$ for this two-dimensional case can be written in determinant form as

$$\mathbf{M}_O = \mathbf{r} \times \mathbf{F} = \begin{vmatrix} \mathbf{i} & \mathbf{j} & \mathbf{k} \\ r_x & r_y & 0 \\ F_x & F_y & 0 \end{vmatrix} = (r_x F_y - r_y F_x)\mathbf{k} = M_z\,\mathbf{k} \quad (5\text{-}9)$$

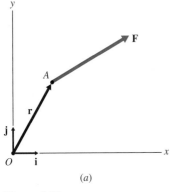

(a)

(b)

**Figure 5-10**

Thus, for the two-dimensional case, the moment $\mathbf{M}_O$ about point $O$ due to a force $\mathbf{F}$ in the $xy$-plane is perpendicular to the plane (directed along the $z$-axis). The moment is completely defined by the scalar quantity

$$M_O = M_{Oz} = r_x F_y - r_y F_x \tag{5-10}$$

since a positive value for $M_O$ indicates a tendency to rotate the body in a counterclockwise direction, which, by the right-hand rule, is along the positive $z$-axis. Similarly, a negative value indicates a tendency to rotate the body in a clockwise direction, which requires a moment in the negative $z$-direction.

**The Three-Dimensional Case.** The moment $\mathbf{M}_O$ about the origin of coordinates $O$ produced by a force $\mathbf{F}$ with a space (three-dimensional) orientation can also be determined by using Eq. (5-3). For this general case (see Fig. 5-11), the force $\mathbf{F}$ can be expressed in Cartesian vector form as

$$\mathbf{F} = F_x \mathbf{i} + F_y \mathbf{j} + F_z \mathbf{k}$$

and the position vector $\mathbf{r}$ from the origin $O$ to an arbitrary point $A$ on the line of action of the force as

$$\mathbf{r} = r_x \mathbf{i} + r_y \mathbf{j} + r_z \mathbf{k}$$

The vector product $\mathbf{r} \times \mathbf{F}$ for this three-dimensional case can be written in determinant form as

$$\mathbf{M}_O = \mathbf{r} \times \mathbf{F} = \begin{vmatrix} \mathbf{i} & \mathbf{j} & \mathbf{k} \\ r_x & r_y & r_z \\ F_x & F_y & F_z \end{vmatrix}$$

$$= (r_y F_z - r_z F_y)\,\mathbf{i} + (r_z F_x - r_x F_z)\,\mathbf{j} + (r_x F_y - r_y F_x)\,\mathbf{k}$$

$$= M_{Ox}\,\mathbf{i} + M_{Oy}\,\mathbf{j} + M_{Oz}\,\mathbf{k} \tag{5-11}$$

where

$$M_{Ox} = r_y F_z - r_z F_y$$
$$M_{Oy} = r_z F_x - r_x F_z \tag{5-12}$$
$$M_{Oz} = r_x F_y - r_y F_x$$

are the three scalar components of the moment of force $\mathbf{F}$ about point $O$. The magnitude of the moment $|\mathbf{M}_O|$ (see Fig. 5-12) is

$$|\mathbf{M}_O| = \sqrt{M_{Ox}^2 + M_{Oy}^2 + M_{Oz}^2} \tag{5-13}$$

Alternatively, the moment $\mathbf{M}_O$ can be written as

$$\mathbf{M}_O = M_O\,\mathbf{e} \tag{5-14}$$

where

$$\mathbf{e} = \cos\theta_x\,\mathbf{i} + \cos\theta_y\,\mathbf{j} + \cos\theta_z\,\mathbf{k} \tag{5-15}$$

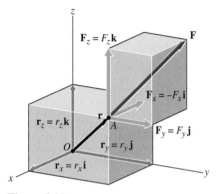

**Figure 5-11**

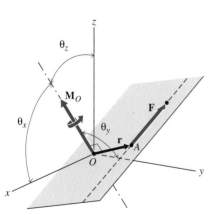

**Figure 5-12**

The direction cosines associated with the unit vector **e** are

$$\cos \theta_x = \frac{M_{Ox}}{|\mathbf{M}_O|} \qquad \cos \theta_y = \frac{M_{Oy}}{|\mathbf{M}_O|} \qquad \cos \theta_z = \frac{M_{Oz}}{|\mathbf{M}_O|} \qquad (5\text{-}16)$$

A moment obeys all the rules of vector combination and can be considered a sliding vector with a line of action coinciding with the moment axis.

The principle of moments discussed in Section 5-2 is not restricted to two concurrent forces but may be extended to any force system. The proof for an arbitrary number of concurrent forces follows from the distributive property of the vector product. Thus,

$$\mathbf{M}_O = \mathbf{r} \times \mathbf{R}$$

where

$$\mathbf{R} = \mathbf{F}_1 + \mathbf{F}_2 + \cdots + \mathbf{F}_n$$

therefore

$$\mathbf{M}_O = \mathbf{r} \times (\mathbf{F}_1 + \mathbf{F}_2 + \cdots + \mathbf{F}_n)$$
$$= (\mathbf{r} \times \mathbf{F}_1) + (\mathbf{r} \times \mathbf{F}_2) + \cdots + (\mathbf{r} \times \mathbf{F}_n)$$

Thus

$$\mathbf{M}_O = \mathbf{M}_{OR} = \mathbf{M}_{O1} + \mathbf{M}_{O2} + \cdots + \mathbf{M}_{On} \qquad (5\text{-}17)$$

Equation (5-17) indicates that the moment of the resultant of any number of forces, $\mathbf{M}_{OR}$, is equal to the sum of the moments of the individual forces.

## Example Problem 5-3

A 1000-lb force is applied to a beam cross section as shown in Fig. 5-13. Determine

(a) The moment of the force about point $O$.
(b) The perpendicular distance $d$ from point $B$ to the line of action of the force.

### SOLUTION

### Scalar Analysis

(a) The scalar components of the force **F** are

$$F_x = \frac{4}{5}F = \frac{4}{5}(1000) = 800 \text{ lb}$$

$$F_y = \frac{3}{5}F = \frac{3}{5}(1000) = 600 \text{ lb}$$

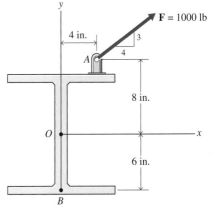

**Figure 5-13**

The moment of the force **F** about point $O$ is equal to the sum of the moments of its components about $O$;

$$+\circlearrowleft M_O = \Sigma M_O = (600)(4) - (800)(8)$$
$$= -4000 \text{ lb} \cdot \text{in.} = 4000 \text{ lb} \cdot \text{in.} \circlearrowright \qquad \textbf{Ans.}$$

(b) The moment of the force **F** about point $B$ is equal to the sum of the moments of its components about $B$;

$$+ \circlearrowleft M_B = \Sigma M_B = (600)(4) - (800)(14)$$

$$= -8800 \text{ lb} \cdot \text{in.} = 8800 \text{ lb} \cdot \text{in.} \circlearrowright$$

$$d = \frac{M_B}{F} = \frac{8800}{1000} = 8.80 \text{ in.} \qquad \textbf{Ans.}$$

### Vector Analysis

(a) The force **F** and the position vector **r** from point $O$ to point $A$ can be expressed in Cartesian vector form as

$$\mathbf{F} = 1000(0.80\,\mathbf{i} + 0.60\,\mathbf{j}) = (800\,\mathbf{i} + 600\,\mathbf{j}) \text{ lb}$$

$$\mathbf{r} = \mathbf{r}_{A/O} = (4\,\mathbf{i} + 8\,\mathbf{j}) \text{ in.}$$

From Eq. (5-9)

$$\mathbf{M}_O = \mathbf{r} \times \mathbf{F} = \begin{vmatrix} \mathbf{i} & \mathbf{j} & \mathbf{k} \\ r_x & r_y & 0 \\ F_x & F_y & 0 \end{vmatrix} = (r_x F_y - r_y F_x)\mathbf{k} = M_{Oz}\,\mathbf{k}$$

$$\mathbf{M}_O = (r_x F_y - r_y F_x)\,\mathbf{k} = [(4)(600) - (8)(800)]\,\mathbf{k}$$

$$= -4000\,\mathbf{k} \text{ lb} \cdot \text{in.} = 4000 \text{ lb} \cdot \text{in.} \downarrow \qquad \textbf{Ans.}$$

(b) The position vector **r** from point $B$ to point $A$ is

$$\mathbf{r} = \mathbf{r}_{A/B} = (4\,\mathbf{i} + 14\,\mathbf{j}) \text{ in.}$$

$$\mathbf{M}_B = (r_x F_y - r_y F_x)\,\mathbf{k} = [(4)(600) - (14)(800)]\,\mathbf{k}$$

$$= -8800\,\mathbf{k} \text{ lb} \cdot \text{in.} = 8800 \text{ lb} \cdot \text{in.} \downarrow$$

$$d = \frac{|\mathbf{M}_B|}{|\mathbf{F}|} = \frac{8800}{1000} = 8.80 \text{ in.} \quad \blacksquare \qquad \textbf{Ans.}$$

## ▌ Example Problem 5-4

A force **F** of magnitude 840 N acts at a point as shown in Fig. 5-14. Determine

(a) The moment of the force about point $B$.

(b) The direction angles associated with the unit vector **e** along the axis of the moment.

(c) The perpendicular distance $d$ from point $B$ to the line of action of the force.

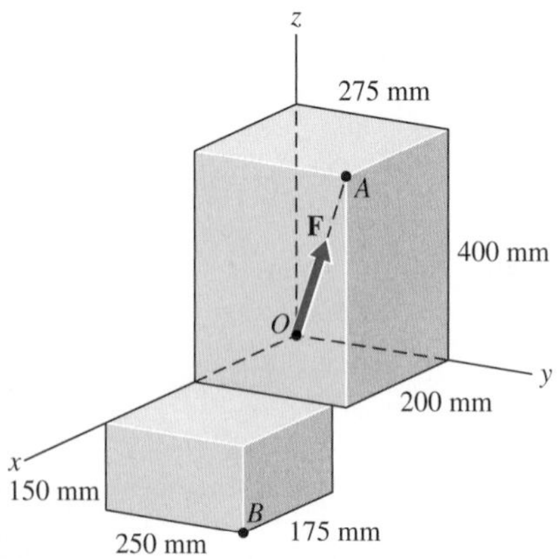

### SOLUTION

### Scalar Analysis

(a)

$$d_{OA} = \sqrt{(x_A)^2 + (y_A)^2 + (z_A)^2}$$

$$= \sqrt{(200)^2 + (275)^2 + (400)^2} = 525 \text{ mm}$$

**Figure 5-14**

$$F_x = \frac{200}{525}(F) = \frac{200}{525}(840) = 320 \text{ N}$$

$$F_y = \frac{275}{525}(F) = \frac{275}{525}(840) = 440 \text{ N}$$

$$F_z = \frac{400}{525}(F) = \frac{400}{525}(840) = 640 \text{ N}$$

At point $B$ (positive moments counterclockwise when viewing in the negative coordinate direction):

$$M_{Bx} = -F_y z_B - F_z y_B = -440(0.150) - 640(0.250) = -226 \text{ N} \cdot \text{m}$$

$$M_{By} = +F_z x_B + F_x z_B = +640(0.375) + 320(0.150) = +288 \text{ N} \cdot \text{m}$$

$$M_{Bz} = +F_x y_B - F_y x_B = +320(0.250) - 440(0.375) = -85.0 \text{ N} \cdot \text{m}$$

$$\mathbf{M}_B = (-226 \, \mathbf{i} + 288 \, \mathbf{j} - 85.0 \, \mathbf{k}) \text{ N} \cdot \text{m} \qquad \textbf{Ans.}$$

## Vector Analysis

The force $\mathbf{F}$ and the position vector $\mathbf{r}$ from point $B$ to point $A$ can be written in Cartesian vector form as

$$\mathbf{F} = 840[(200/525) \, \mathbf{i} + (275/525) \, \mathbf{j} + (400/525) \, \mathbf{k}]$$

$$= (320 \, \mathbf{i} + 440 \, \mathbf{j} + 640 \, \mathbf{k}) \text{ N}$$

$$\mathbf{r}_{A/B} = (-0.175 \, \mathbf{i} + 0.025 \, \mathbf{j} + 0.550 \, \mathbf{k}) \text{ m}$$

The moment $\mathbf{M}_B$ is given by Eq. (5-11) as

$$\mathbf{M}_B = \mathbf{r}_{A/B} \times \mathbf{F} = \begin{vmatrix} \mathbf{i} & \mathbf{j} & \mathbf{k} \\ r_x & r_y & r_z \\ F_x & F_y & F_z \end{vmatrix} = \begin{vmatrix} \mathbf{i} & \mathbf{j} & \mathbf{k} \\ -0.175 & 0.025 & 0.550 \\ 320 & 440 & 640 \end{vmatrix}$$

$$= (-226 \, \mathbf{i} + 288 \, \mathbf{j} - 85.0 \, \mathbf{k}) \text{ N} \cdot \text{m} \qquad \textbf{Ans.}$$

Alternatively, the position vector $\mathbf{r}$ can be written from point $B$ to point $O$ as

$$\mathbf{r}_{O/B} = (-0.375 \, \mathbf{i} - 0.250 \, \mathbf{j} + 0.150 \, \mathbf{k}) \text{ m}$$

The moment $\mathbf{M}_B$ is then given by Eq. (5-11) as

$$\mathbf{M}_B = \mathbf{r}_{O/B} \times \mathbf{F} = \begin{vmatrix} \mathbf{i} & \mathbf{j} & \mathbf{k} \\ r_x & r_y & r_z \\ F_x & F_y & F_z \end{vmatrix} = \begin{vmatrix} \mathbf{i} & \mathbf{j} & \mathbf{k} \\ -0.375 & -0.250 & 0.150 \\ 320 & 440 & 640 \end{vmatrix}$$

$$= (-226 \, \mathbf{i} + 288 \, \mathbf{j} - 85.0 \, \mathbf{k}) \text{ N} \cdot \text{m} \qquad \textbf{Ans.}$$

(b) The magnitude of moment $\mathbf{M}_B$ is obtained by using Eq. (5-13). Thus

$$|\mathbf{M}_B| = \sqrt{(-226)^2 + (288)^2 + (-85.0)^2} = 375.8 \text{ N} \cdot \text{m}$$

The direction angles are obtained by using Eq. (5-16). Thus

$$\theta_x = \cos^{-1} \frac{M_x}{|M_B|} = \cos^{-1} \frac{-226}{375.8} = 127.0° \qquad \text{Ans.}$$

$$\theta_y = \cos^{-1} \frac{M_y}{|M_B|} = \cos^{-1} \frac{288}{375.8} = 40.0° \qquad \text{Ans.}$$

$$\theta_z = \cos^{-1} \frac{M_z}{|M_B|} = \cos^{-1} \frac{-85.0}{375.8} = 103.1° \qquad \text{Ans.}$$

(c) The distance $d$ is obtained by using the definition of a moment. Thus

$$d = \frac{M}{F} = \frac{375.8}{840} = 0.447 \text{ m} = 447 \text{ mm} \qquad \blacksquare \qquad \text{Ans.}$$

## PROBLEMS

**Introductory Problems**

**5-13\*** A 760-lb force acts on a bracket as shown in Fig. P5-13. Determine the moment of the force about point $A$
(a) Using the vector approach.
(b) Using the scalar approach.

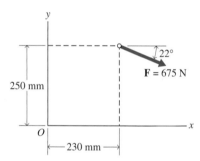

**Figure P5-14**

**5-15** Two forces $F_1$ and $F_2$ are applied to a gusset plate as shown in Fig. P5-15. Determine the moment
(a) Of force $F_1$ about point $A$.
(b) Of force $F_2$ about point $B$.

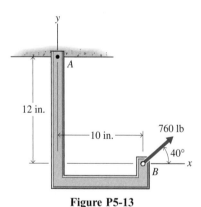

**Figure P5-13**

**5-14\*** Determine the moment of the 675-N force shown in Fig. P5-14 about point $O$
(a) Using the vector approach.
(b) Using the scalar approach.

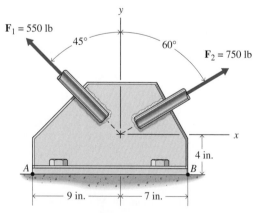

**Figure P5-15**

**5-16** Two forces $\mathbf{F}_1$ and $\mathbf{F}_2$ are applied to a bracket as shown in Fig. P5-16. Determine the moment
(a) Of force $\mathbf{F}_1$ about point $O$.
(b) Of force $\mathbf{F}_2$ about point $A$.

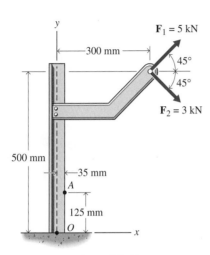

$F_1 = 5$ kN
$300$ mm
$45°$
$45°$
$F_2 = 3$ kN
$y$
$500$ mm
$35$ mm
$A$
$125$ mm
$O$
$x$

**Figure P5-16**

**Intermediate Problems**

**5-17\*** Determine the moment of the contact force $\mathbf{F}$ of the crutch shown in Fig. P5-17 about point $A$. The length of the crutch is 5 ft.

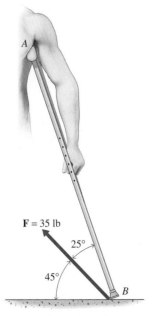

$A$
$F = 35$ lb
$25°$
$45°$
$B$

**Figure P5-17**

**5-18\*** The moment of the force $\mathbf{F}$ shown in Fig. P5-18 about point $A$ is $-2700$ **k** N · mm, and its moment about point $C$ is $-7500$ **k** N · mm. Determine the magnitude and orientation (angle $\theta$) of the force $\mathbf{F}$.

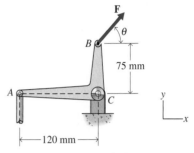

$F$
$\theta$
$B$
$75$ mm
$A$
$C$
$y$
$x$
$120$ mm

**Figure P5-18**

**5-19** A 970-lb force acts at a point in a body as shown in Fig. P5-19. Determine the moment of the force about point $C$.

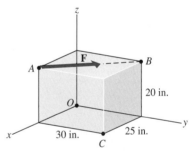

$z$
$A$
$F$
$B$
$20$ in.
$O$
$y$
$x$
$30$ in.
$25$ in.
$C$

**Figure P5-19**

**5-20** A 200-N force is applied to a pipe wrench as shown in Fig. P5-20. Determine the moment of the force about point $A$. Express the result in Cartesian vector form.

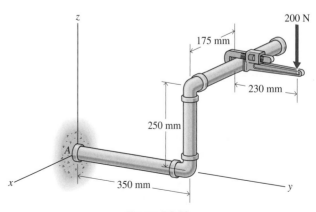

$z$
$200$ N
$175$ mm
$230$ mm
$250$ mm
$A$
$x$
$350$ mm
$y$

**Figure P5-20**

**5-21\*** A 650-lb force acts at a point in a body as shown in Fig. P5-21. Determine
(a) The moment of the force about point $A$.
(b) The direction angles associated with the moment vector.

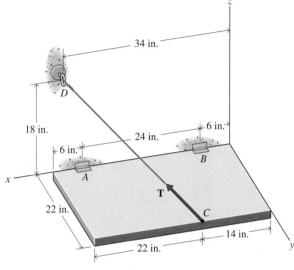

Figure P5-23

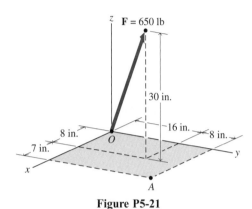

Figure P5-21

**5-22** A pipe bracket is loaded as shown in Fig. P5-22. Determine the moment of the force $\mathbf{F}$ about point $O$.

**5-24\*** A 450-N force $\mathbf{F}$ acts on a machine component as shown in Fig. P5-24. The direction angles of $\mathbf{F}$ are $\theta_x = 70°$, $\theta_y = 30°$, and $\theta_z = 69°$. Determine the moment of the force about point $A$. Express your answer in Cartesian vector form.

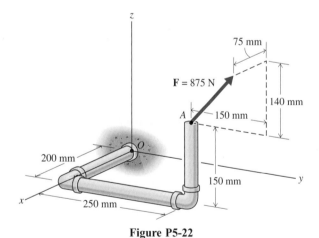

Figure P5-22

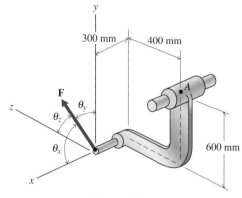

Figure P5-24

**Challenging Problems**

**5-23\*** The magnitude of the tension $\mathbf{T}$ in cable $CD$ of Fig. P5-23 is 150 lb. Determine the moment of $\mathbf{T}$ about point $B$
(a) Using a position vector from $B$ to $D$.
(b) Using a position vector from $B$ to $C$.

**5-25\*** The magnitude of the force $\mathbf{F}$ shown in Fig. P5-25 is 100 lb. Determine the moment of $\mathbf{F}$ about the bearing at $C$.

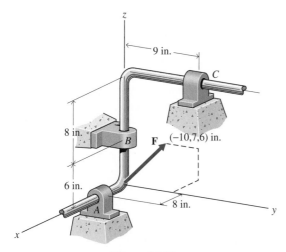

**Figure P5-25**

**5-26** Determine the moment of the 800-N force shown in Fig. P5-26 about point $D$
(a) Using a position vector from $D$ to $B$.
(b) Using a position vector from $D$ to $E$.

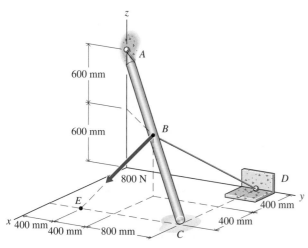

**Figure P5-26**

**5-27** The magnitudes of the forces $\mathbf{F}_1$, $\mathbf{F}_2$, and $\mathbf{F}_3$ shown in Fig. P5-27 are 550 lb, 300 lb, and 600 lb, respectively. Determine the sum of the moments of the three forces about point $B$.

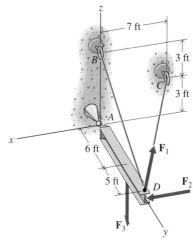

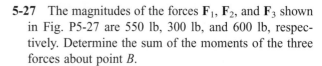

**Figure P5-27**

**5-28** If the magnitude of the moment of the cable force $\mathbf{T}$ shown in Fig. P5-28 about the hinge at $B$ is $M_B = 1150$ N · m, determine the magnitude of the force $\mathbf{T}$.

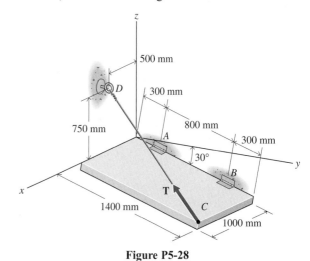

**Figure P5-28**

# 5-4 MOMENT OF A FORCE ABOUT A LINE (AXIS)

The moment $\mathbf{M}_O$ of a force $\mathbf{F}$ about a point $O$ was defined as the vector product

$$\mathbf{M}_O = \mathbf{r} \times \mathbf{F} \qquad\qquad (5\text{-}3)$$

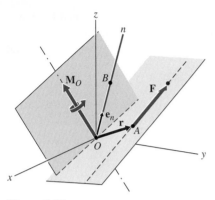

**Figure 5-15**

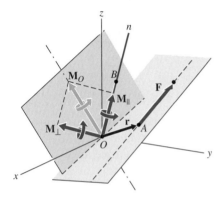

**Figure 5-16**

While it is possible, mathematically, to define the moment of a force about a point, the quantity has no physical significance in mechanics, since bodies rotate about axes (as illustrated in Fig. 5-1) and not points. The vector definition of a moment about a point [Eq. (5-3)] is only an intermediate step in a process that allows us to find the moment about an axis that passes through the point.

The moment $\mathbf{M}_{OB}$ of a force with respect to a line (say line $OB$ in Fig. 5-15) can be determined by first calculating the moment $\mathbf{M}_O$ about point $O$ on the line (or about any other point on the line). Then, the moment vector $\mathbf{M}_O$ can be resolved into components that are parallel $\mathbf{M}_{\parallel}$ and perpendicular $\mathbf{M}_{\perp}$ to the line $OB$, as shown in Fig. 5-16. If $\mathbf{e}_n$ is a unit vector in the $n$-direction along line $OB$, as shown in Fig. 5-15, then

$$\mathbf{M}_{OB} = \mathbf{M}_{\parallel} = (\mathbf{M}_O \cdot \mathbf{e}_n) \, \mathbf{e}_n$$

$$= [(\mathbf{r} \times \mathbf{F}) \cdot \mathbf{e}_n] \, \mathbf{e}_n = M_{OB} \, \mathbf{e}_n \qquad (5\text{-}18)$$

These two operations, the vector product $\mathbf{r} \times \mathbf{F}$ of the position vector $\mathbf{r}$ and the force $\mathbf{F}$ to obtain the moment $\mathbf{M}_O$ about point $O$, followed by the scalar product $\mathbf{M}_O \cdot \mathbf{e}_n$ of the moment $\mathbf{M}_O$ about point $O$ and the unit vector $\mathbf{e}_n$ along the desired moment axis, to obtain the moment $M_{OB}$ can be performed in sequence or combined into one operation. The quantity inside the brackets of Eq. (5-18) is called the *triple scalar product*. The triple scalar product can be written in determinant form as

$$M_{OB} = \mathbf{M}_O \cdot \mathbf{e}_n = (\mathbf{r} \times \mathbf{F}) \cdot \mathbf{e}_n = \begin{vmatrix} \mathbf{i} & \mathbf{j} & \mathbf{k} \\ r_x & r_y & r_z \\ F_x & F_y & F_z \end{vmatrix} \cdot \mathbf{e}_n \qquad (5\text{-}19)$$

or alternatively as

$$M_{OB} = \mathbf{M}_O \cdot \mathbf{e}_n = (\mathbf{r} \times \mathbf{F}) \cdot \mathbf{e}_n = \begin{vmatrix} e_{nx} & e_{ny} & e_{nz} \\ r_x & r_y & r_z \\ F_x & F_y & F_z \end{vmatrix} \qquad (5\text{-}20)$$

where $e_{nx}$, $e_{ny}$, and $e_{nz}$ are the Cartesian components (direction cosines) of the unit vector $\mathbf{e}_n$. The unit vector $\mathbf{e}_n$ is usually selected in the direction from $O$ toward $B$. A positive coefficient of $\mathbf{e}_n$ in the expression $\mathbf{M}_{OB} = M_{OB} \, \mathbf{e}_n$ means that the moment vector has the same sense as that selected for $\mathbf{e}_n$, whereas a negative sign indicates that $\mathbf{M}_{OB}$ is opposite to the sense of $\mathbf{e}_n$.

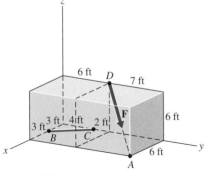

**Figure 5-17**

## Example Problem 5-5

The force $\mathbf{F}$ in Fig. 5-17 has a magnitude of 440 lb. Determine

(a) The moment $\mathbf{M}_B$ of the force about point $B$.

(b) The component of moment $\mathbf{M}_B$ parallel to line $BC$.

(c) The component of moment $\mathbf{M}_B$ perpendicular to line $BC$.

(d) The unit vector associated with the component of moment $\mathbf{M}_B$ perpendicular to line $BC$.

## SOLUTION

(a) The force $\mathbf{F}$ and the position vector $\mathbf{r}$ from point $B$ to point $A$ can be written in Cartesian vector form as

$$\mathbf{F} = 440[(6/11)\,\mathbf{i} + (7/11)\,\mathbf{j} - (6/11)\,\mathbf{k}] = (240\,\mathbf{i} + 280\,\mathbf{j} - 240\,\mathbf{k})\ \text{lb}$$

$$\mathbf{r}_{A/B} = (3\,\mathbf{i} + 13\,\mathbf{j})\ \text{ft}$$

The moment $\mathbf{M}_B$ is given by Eq. (5-11) as

$$\mathbf{M}_B = \mathbf{r}_{A/B} \times \mathbf{F} = \begin{vmatrix} \mathbf{i} & \mathbf{j} & \mathbf{k} \\ r_x & r_y & r_z \\ F_x & F_y & F_z \end{vmatrix} = \begin{vmatrix} \mathbf{i} & \mathbf{j} & \mathbf{k} \\ 3 & 13 & 0 \\ 240 & 280 & -240 \end{vmatrix}$$

$$= (-3120\,\mathbf{i} + 720\,\mathbf{j} - 2280\,\mathbf{k})\ \text{lb} \cdot \text{ft} \qquad \textbf{Ans.}$$

(b) The unit vector $\mathbf{e}_{BC}$ associated with line $BC$ is

$$\mathbf{e}_{BC} = -0.60\,\mathbf{i} + 0.80\,\mathbf{j}$$

The component of moment $\mathbf{M}_B$ parallel to line $BC$ is given by Eq. (5-18) as

$$\mathbf{M}_{BC} = \mathbf{M}_\parallel = (\mathbf{M}_B \cdot \mathbf{e}_{BC})\,\mathbf{e}_{BC} = M_{BC}\,\mathbf{e}_{BC}$$

$$M_{BC} = \mathbf{M}_B \cdot \mathbf{e}_{BC}$$
$$= (-3120\,\mathbf{i} + 720\,\mathbf{j} - 2280\,\mathbf{k}) \cdot (-0.60\,\mathbf{i} + 0.80\,\mathbf{j})$$
$$= (-3120)(-0.60) + (720)(0.80) = 2448\ \text{lb} \cdot \text{ft}$$

$$\mathbf{M}_{BC} = M_{BC}\,\mathbf{e}_{BC}$$
$$= 2448(-0.60\,\mathbf{i} + 0.80\,\mathbf{j})$$
$$= (-1469\,\mathbf{i} + 1958\,\mathbf{j})\ \text{lb} \cdot \text{ft} \qquad \textbf{Ans.}$$

Alternatively, the moment $M_{BC}$ can be determined in a single operation by using Eq. (5-20). A different point on the line can also be used. For example, consider point $C$ and the position vector $\mathbf{r}_{D/C}$.

$$\mathbf{r}_{D/C} = (2\,\mathbf{j} + 6\,\mathbf{k})\ \text{ft}$$

From Eq. (5-20):

$$M_{BC} = \begin{vmatrix} e_{nx} & e_{ny} & e_{nz} \\ r_x & r_y & r_z \\ F_x & F_y & F_z \end{vmatrix} = \begin{vmatrix} -0.60 & 0.80 & 0 \\ 0 & 2 & 6 \\ 240 & 280 & -240 \end{vmatrix}$$

$$= (-0.60)(-2160) - (0.80)(-1440) = 2448\ \text{lb} \cdot \text{ft}$$

(c) The moment $\mathbf{M}_\perp$ is obtained as the difference between $\mathbf{M}_B$ and $\mathbf{M}_\parallel$ since $\mathbf{M}_\parallel$ and $\mathbf{M}_\perp$ are the two rectangular components of $\mathbf{M}_B$. Thus,

$$\mathbf{M}_\perp = \mathbf{M}_B - \mathbf{M}_\parallel = \mathbf{M}_B - \mathbf{M}_{BC}$$
$$= (-3120\,\mathbf{i} + 720\,\mathbf{j} - 2280\,\mathbf{k}) - (-1469\,\mathbf{i} + 1958\,\mathbf{j})$$
$$= (-1651\,\mathbf{i} - 1238\,\mathbf{j} - 2280\,\mathbf{k})\ \text{lb} \cdot \text{ft} \qquad \textbf{Ans.}$$

(d) The magnitude of moment $\mathbf{M}_\perp$ is

$$|\mathbf{M}_\perp| = \sqrt{(-1651)^2 + (-1238)^2 + (-2280)^2} = 3075 \text{ lb} \cdot \text{ft}$$

Therefore

$$\mathbf{e}_\perp = (-1651/3075)\,\mathbf{i} + (-1238/3075)\,\mathbf{j} + (-2280/3075)\,\mathbf{k}$$
$$= -0.537\,\mathbf{i} - 0.403\,\mathbf{j} - 0.741\,\mathbf{k} \qquad\qquad \textbf{Ans.}$$

As a check:

$$\mathbf{e}_\parallel \cdot \mathbf{e}_\perp = (-0.60\,\mathbf{i} + 0.80\,\mathbf{j}) \cdot (-0.5369\,\mathbf{i} - 0.4026\,\mathbf{j} - 0.7415\,\mathbf{k})$$
$$= 0.00006 \cong 0$$

which verifies, except for round-off in the expression for $\mathbf{e}_\perp$, that the moment components $\mathbf{M}_\parallel$ and $\mathbf{M}_\perp$ are perpendicular. ■

## Example Problem 5-6

The magnitude of force $\mathbf{F}$ in Fig. 5-18 is 500 N. Determine the scalar component of the moment at point $O$ about line $OC$.

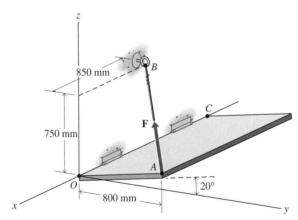

**Figure 5-18**

### SOLUTION

The moment of force $\mathbf{F}$ about line $OC$ can be determined by using a two-step process. The $z$-coordinate of point $A$ that will be needed to write a vector equation for force $\mathbf{F}$ is $z_A = 800 \tan 20° = 291.2$ mm. The force $\mathbf{F}$ can be written in Cartesian vector form as

$$\mathbf{F} = 500\left[\frac{-850\,\mathbf{i} - 800\,\mathbf{j} + 458.8\,\mathbf{k}}{\sqrt{(-850)^2 + (-800)^2 + (458.8)^2}}\right]$$
$$= -338.86\,\mathbf{i} - 318.93\,\mathbf{j} + 182.91\,\mathbf{k} \text{ N}$$

The position vector $\mathbf{r}_{B/O}$ can be written in Cartesian vector form as

$$\mathbf{r}_{B/O} = -0.850\,\mathbf{i} + 0.750\,\mathbf{k} \text{ m}$$

The moment of force $\mathbf{F}$ about point $O$ is then given by the expression

$$\mathbf{M}_O = \mathbf{r}_{B/O} \times \mathbf{F} = \begin{vmatrix} \mathbf{i} & \mathbf{j} & \mathbf{k} \\ -0.850 & 0 & 0.750 \\ -338.86 & -318.93 & 182.91 \end{vmatrix}$$

$$= 239.20 \, \mathbf{i} - 98.67 \, \mathbf{j} + 271.09 \, \mathbf{k} \; \text{N} \cdot \text{m}$$

A unit vector $\mathbf{e}_{OC}$ along line $OC$ with a sense from $O$ to $C$ is

$$\mathbf{e}_{OC} = -1.000 \, \mathbf{i}$$

The scalar component of moment $\mathbf{M}_O$ along line $OC$ is then determined by using the dot product

$$M_{OC} = \mathbf{M}_O \cdot \mathbf{e}_{OC}$$

$$= (239.20 \, \mathbf{i} - 98.67 \, \mathbf{j} + 271.09 \, \mathbf{k}) \cdot (-1.000 \, \mathbf{i})$$

$$= -239.20 \, \text{N} \cdot \text{m} \cong -239 \, \text{N} \cdot \text{m} \qquad \textbf{Ans.}$$

Alternatively, the scalar moment component $M_{OC}$ can be determined in a single step by using Eq. (5-20)

$$M_{OC} = \begin{vmatrix} -1.000 & 0 & 0 \\ -0.850 & 0 & 0.750 \\ -338.86 & -318.93 & 182.91 \end{vmatrix}$$

$$= -239.20 \, \text{N} \cdot \text{m} \cong -239 \, \text{N} \cdot \text{m} \; \blacksquare \qquad \textbf{Ans.}$$

## PROBLEMS

### Introductory Problems

**5-29\*** The force $\mathbf{F}$ in Fig. P5-29 can be expressed in Cartesian vector form as $\mathbf{F} = (60 \, \mathbf{i} + 100 \, \mathbf{j} + 120 \, \mathbf{k})$ lb. Determine the scalar component of the moment at point $B$ about line $BC$.

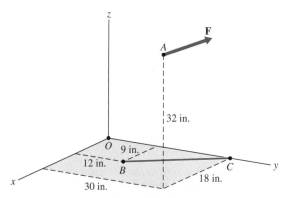

**Figure P5-29**

**5-30\*** A 3000-N force is applied to the pipe shown in Fig. P5-30. Determine the scalar component of the moment of the force about the $x$-axis.

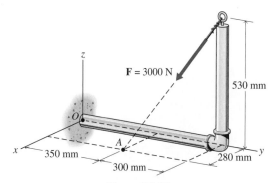

**Figure P5-30**

**5-31** A force $\mathbf{F} = (-30 \, \mathbf{i} + 50 \, \mathbf{j} - 40 \, \mathbf{k})$ lb is applied to the machine component shown in Fig. P5-31. Determine the scalar component of the moment of the force about the $z$-axis.

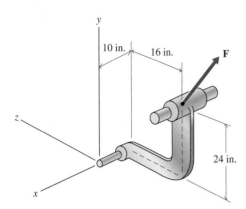

**Figure P5-31**

**5-32** The magnitude of the force **F** in Fig. P5-32 is 595 N. Determine the scalar component of the moment at point $O$ about line $OC$.

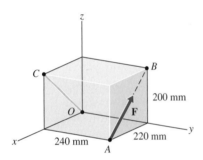

**Figure P5-32**

**Intermediate Problems**

**5-33*** A 40-lb vertical force **F** is applied to a lug wrench as shown in Fig. P5-33. Determine the magnitude of the component of the moment that would tighten the lug nut.

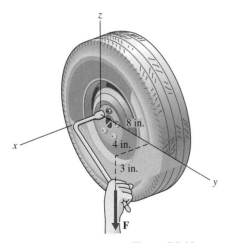

**Figure P5-33**

**5-34*** If the magnitude of the force **T** shown in Fig. P5-34 is 1000 N, determine the scalar component of the moment of the force about the line $CD$.

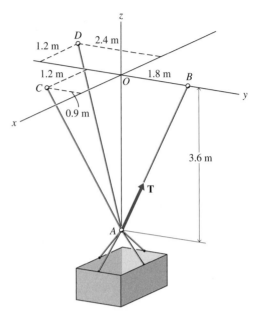

**Figure P5-34**

**5-35** A 120-lb force **F** is applied to a lever-shaft assembly as shown in Fig. P5-35. Determine the moment of the force about each coordinate axis.

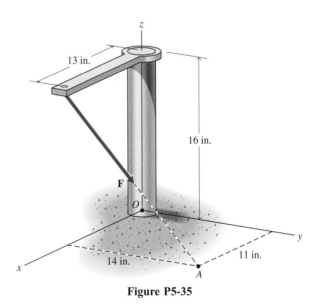

**Figure P5-35**

**5-36** A 650-N force acts on the awning structure shown in Fig. P5-36. Determine the moment of the force about line $BC$. Express the result in Cartesian vector form.

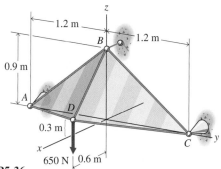

**Figure P5-36**

## Challenging Problems

**5-37\*** The magnitude of the force **F** in Fig. P5-37 is 107 lb. Determine the component of the moment of the force that rotates the door about the axis of the hinges.

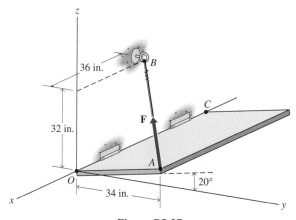

**Figure P5-37**

**5-38\*** A bracket is rigidly attached to a wall at $O$ and is subjected to a 384-N force **F** as shown in Fig. P5-38. Determine the component of the moment of the force
(a) That twists the bracket about the $y$-axis.
(b) That bends the bracket about the $x$-axis.

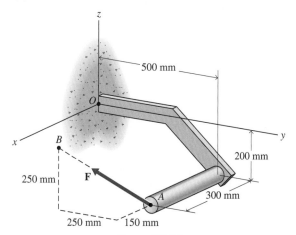

**Figure P5-38**

**5-39** A bar is bent in a circular arc of radius $R = 6$ ft and is subjected to a 660-lb force **F** as shown in Fig. P5-39. The force tends to twist and bend the member about the coordinate axes. Determine the twisting and bending moments and state the axes about which each occurs.

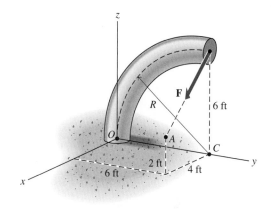

**Figure P5-39**

**5-40\*** The magnitude of the force **F** in Fig. P5-40 is 976 N. Determine
(a) The component of the moment at point $C$ parallel to line $CE$.
(b) The component of the moment at point $C$ perpendicular to line $CE$ and the direction angles associated with this moment vector.

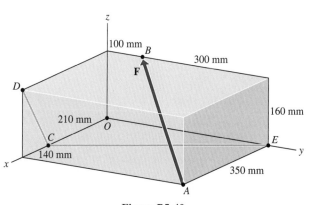

**Figure P5-40**

**5-41** The magnitude of the force $\mathbf{F}$ in Fig. P5-41 is 781 lb. Determine
  (a) The component of the moment at point $C$ parallel to line $CD$.
  (b) The component of the moment at point $C$ perpendicular to line $CD$ and the direction angles associated with this moment vector.

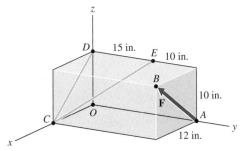

**Figure P5-41**

## 5-5 COUPLES

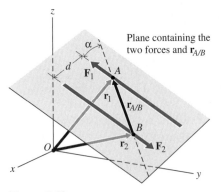

**Figure 5-19**

Two equal, noncollinear, parallel forces of opposite sense (see Fig. 5-19) are called a *couple*. Since the two forces are equal, parallel, and of opposite sense, the sum of the forces in any direction is zero. Therefore, a couple tends only to rotate the body on which it acts. The moment of a couple is defined as the sum of the moments of the pair of forces that comprise the couple. For the two forces $\mathbf{F}_1$ and $\mathbf{F}_2$ shown in Fig. 5-19, the moments of the forces about points $A$ and $B$ in the plane of the couple are

$$M_A = |\mathbf{F}_2|d \qquad M_B = |\mathbf{F}_1|d$$

However,

$$|\mathbf{F}_1| = |\mathbf{F}_2| = F$$

therefore,

$$M_A = M_B = Fd$$

which indicates that the magnitude of the moment of a couple about a point in the plane of the couple is equal to the magnitude of one of the forces multiplied by the perpendicular distance between the forces. The point $A$ or $B$ does not have to lie on one of the forces but must lie in the plane of the forces.

Other characteristics of a couple can be determined by considering two parallel forces in space such as those shown in Fig. 5-19. The sum of the moments of the two forces about any point $O$ is

$$\mathbf{M}_O = \mathbf{r}_1 \times \mathbf{F}_1 + \mathbf{r}_2 \times \mathbf{F}_2$$

or, since $\mathbf{F}_2$ equals $-\mathbf{F}_1$,

$$\mathbf{M}_O = \mathbf{r}_1 \times \mathbf{F}_1 + \mathbf{r}_2 \times (-\mathbf{F}_1)$$
$$= (\mathbf{r}_1 - \mathbf{r}_2) \times \mathbf{F}_1 = \mathbf{r}_{A/B} \times \mathbf{F}_1$$

where $\mathbf{r}_{A/B}$ is the position vector from any point $B$ on the line of action of force $\mathbf{F}_2$ to any point $A$ on the line of action of force $\mathbf{F}_1$. Therefore, from the definition of the vector cross product,

$$\mathbf{M}_O = \mathbf{r}_{A/B} \times \mathbf{F}_1$$
$$= |\mathbf{r}_{A/B}|\,|\mathbf{F}_1|\sin\alpha\,\mathbf{e} = F_1 d\,\mathbf{e} \qquad (5\text{-}21)$$

where $d$ is the perpendicular distance between the forces of the couple and $\mathbf{e}$ is a unit vector perpendicular to the plane of the couple with its sense in the direction specified for moments by the right-hand rule. It is obvious from Eq. (5-21) that the moment of a couple does not depend upon the location of the moment center $O$. Thus, the moment of a couple is a free vector.

The characteristics of a couple, which control its "external effect" (overall tendency to translate and to rotate) on a rigid body, are

1. The magnitude of the moment of the couple.
2. The sense (direction of rotation) of the couple.
3. The aspect of the plane of the couple; that is, the direction or slope of the plane (not its location) as defined by a normal $n$ to the plane.

Equation (5-21) indicates that several transformations of a couple can be made without changing any of the external effects of the couple on the body. For example,

1. A couple can be translated to a parallel position in its plane or to any parallel plane [since position vectors $\mathbf{r}_1$ and $\mathbf{r}_2$ do not appear in Eq. (5-21)].
2. A couple can be rotated in its plane.
3. The magnitude of the two forces of a couple and the distance between them can be changed provided the product $Fd$ remains constant.

For two-dimensional problems, a couple is frequently represented by a curved arrow on a sketch of the body. The magnitude of the moment of the couple $|\mathbf{M}|$ ($M = Fd$) is provided and the curved arrow indicates the sense of the couple.

Any number of couples $\mathbf{C}_1, \mathbf{C}_2, \ldots, \mathbf{C}_n$ in a plane can be combined to yield a resultant couple $\mathbf{C}$ equal to the algebraic sum of the individual couples. A system of couples in space (see Fig. 5-20$a$) can be combined into a single resultant couple $\mathbf{C}$, by representing each couple of the system (since a couple is a free vector) by a vector, drawn for convenience, from the origin of a set of rectangular axes. Each couple can then be resolved into components $\mathbf{C}_x$, $\mathbf{C}_y$, and $\mathbf{C}_z$ along the coordinate axes. These vector components represent couples lying in the $yz$-, $zx$-, and $xy$-planes, respectively. The $x$-, $y$-, and $z$-components of the resultant couple $\mathbf{C}$ are obtained as $\Sigma\mathbf{C}_x$, $\Sigma\mathbf{C}_y$, and $\Sigma\mathbf{C}_z$, as shown in Fig. 5-20$b$.

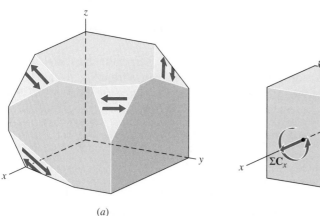

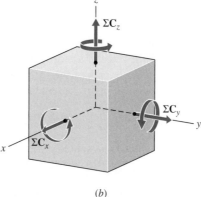

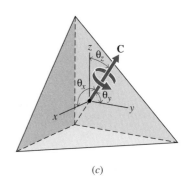

(a)                    (b)                    (c)

**Figure 5-20**

The original system of couples is thus reduced to three couples lying in the co-ordinate planes. The resultant couple $\mathbf{C}$ for the system (see Fig. 5-20c) can be written in vector form as

$$\mathbf{C} = \Sigma\mathbf{C}_x + \Sigma\mathbf{C}_y + \Sigma\mathbf{C}_z = \Sigma C_x\,\mathbf{i} + \Sigma C_y\,\mathbf{j} + \Sigma C_z\,\mathbf{k} \qquad (5\text{-}22)$$

The magnitude of the couple $\mathbf{C}$ is

$$|\mathbf{C}| = \sqrt{(\Sigma C_x)^2 + (\Sigma C_y)^2 + (\Sigma C_z)^2} \qquad (5\text{-}23)$$

Alternatively, the couple $\mathbf{C}$ can be written as

$$\mathbf{C} = |\mathbf{C}|\,\mathbf{e} = C\,\mathbf{e} \qquad (5\text{-}24)$$

where

$$\mathbf{e} = \cos\theta_x\,\mathbf{i} + \cos\theta_y\,\mathbf{j} + \cos\theta_z\,\mathbf{k}$$

The direction cosines associated with the unit vector $\mathbf{e}$ are

$$\theta_x = \cos^{-1}\frac{\Sigma C_x}{|\mathbf{C}|} \qquad \theta_y = \cos^{-1}\frac{\Sigma C_y}{|\mathbf{C}|} \qquad \theta_z = \cos^{-1}\frac{\Sigma C_z}{|\mathbf{C}|} \qquad (5\text{-}25)$$

## Example Problem 5-7

A beam is loaded with a system of forces as shown in Fig. 5-21. Express the resultant of the force system in Cartesian vector form.

### SOLUTION

An examination of Fig. 5-21 indicates that the force system consists of a system of three couples in the same plane.

### Scalar Analysis

The forces at $A$ and $D$ form a couple. The perpendicular distance between their lines of action is $d_1 = 300$ mm $= 0.3$ m. Therefore, the moment of this couple is

$$M_1 = F_A d_1 = 500(0.300) = 150.0 \text{ N} \cdot \text{m} \downarrow = -150.0 \text{ N} \cdot \text{m}$$

The forces at $B$ and $C$ also form a couple. The perpendicular distance between the lines of action of these forces is not obvious from Fig. 5-21. Therefore, components of the forces will be used to determine the moments of this couple.

$$F_{Bx} = F_B \cos 60° = 750 \cos 60° = 375 \text{ N}$$

$$F_{By} = F_B \sin 60° = 750 \sin 60° = 649.5 \text{ N}$$

$$M_2 = F_{Bx} d_2 = 375(0.250) = 93.75 \text{ N} \cdot \text{m} \downarrow = -93.75 \text{ N} \cdot \text{m}$$

$$M_3 = F_{By} d_3 = 649.5(0.300) = 194.85 \text{ N} \cdot \text{m} \uparrow = +194.85 \text{ N} \cdot \text{m}$$

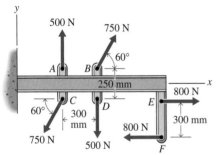

**Figure 5-21**

The forces at $E$ and $F$ also form a couple. The perpendicular distance between their lines of action is $d_4 = 300$ mm $= 0.3$ m. Therefore, the moment of this couple is

$$M_4 = F_E d_4 = 800(0.300) = 240 \text{ N} \cdot \text{m} \, \downarrow = -240 \text{ N} \cdot \text{m}$$

For any number of couples in a plane, the resultant couple $\mathbf{C}$ is equal to the algebraic sum of the individual couples. Thus,

$$C = \Sigma M = -150.0 - 93.75 + 194.85 - 240 = -288.9 \text{ N} \cdot \text{m} \cong -289 \text{ N} \cdot \text{m}$$

$$\mathbf{C} = -289 \text{ k N} \cdot \text{m} \qquad \text{Ans.}$$

## Vector Analysis

$$\mathbf{C} = (\mathbf{r}_{D/A} \times \mathbf{F}_D) + (\mathbf{r}_{B/C} \times \mathbf{F}_B) + (\mathbf{r}_{E/F} \times \mathbf{F}_E)$$

$$= [(0.300 \, \mathbf{i} - 0.250 \, \mathbf{j}) \times (-500 \, \mathbf{j})]$$

$$+ [(0.300 \, \mathbf{i} + 0.250 \, \mathbf{j}) \times (375 \, \mathbf{i} + 649.5 \, \mathbf{j})] + [(0.300 \, \mathbf{j}) \times (800 \, \mathbf{i})]$$

$$= -288.9 \text{ k N} \cdot \text{m} \cong -289 \text{ k N} \cdot \text{m} \quad \blacksquare \qquad \text{Ans.}$$

## Example Problem 5-8

The magnitudes of the four couples applied to the block shown in Fig. 5-22 are $|\mathbf{C}_1| = 75$ lb $\cdot$ ft, $|\mathbf{C}_2| = 50$ lb $\cdot$ ft, $|\mathbf{C}_3| = 60$ lb $\cdot$ ft, and $|\mathbf{C}_4| = 90$ lb $\cdot$ ft. Determine the magnitude of the resultant couple $\mathbf{C}$ and the direction angles associated with the unit vector $\mathbf{e}$ used to describe the normal to the plane of the resultant couple $\mathbf{C}$.

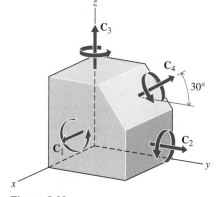

**Figure 5-22**

## SOLUTION

The $x$-, $y$-, and $z$-components of the resultant couple $\mathbf{C}$ are

$$\Sigma \mathbf{C}_x = \mathbf{C}_1 = 75 \, \mathbf{i} \text{ lb} \cdot \text{ft}$$

$$\Sigma \mathbf{C}_y = \mathbf{C}_2 + \mathbf{C}_4 \cos 30° = 50 \, \mathbf{j} + 90 \cos 30° \, \mathbf{j} = 127.9 \, \mathbf{j} \text{ lb} \cdot \text{ft}$$

$$\Sigma \mathbf{C}_z = \mathbf{C}_3 + \mathbf{C}_4 \sin 30° = 60 \, \mathbf{k} + 90 \sin 30° \, \mathbf{k} = 105.0 \, \mathbf{k} \text{ lb} \cdot \text{ft}$$

The resultant couple $\mathbf{C}$ for the system can be written in vector form as

$$\mathbf{C} = \Sigma \mathbf{C}_x + \Sigma \mathbf{C}_y + \Sigma \mathbf{C}_z = (75 \, \mathbf{i} + 127.9 \, \mathbf{j} + 105.0 \, \mathbf{k}) \text{ lb} \cdot \text{ft}$$

The magnitude of the couple $\mathbf{C}$ is

$$|\mathbf{C}| = \sqrt{(\Sigma C_x)^2 + (\Sigma C_y)^2 + (\Sigma C_z)^2}$$

$$= \sqrt{(75)^2 + (127.9)^2 + (105.0)^2} = 181.7 \text{ lb} \cdot \text{ft} \qquad \text{Ans.}$$

The direction angles are

$$\theta_x = \cos^{-1} \frac{\Sigma C_x}{|\mathbf{C}|} = \cos^{-1} \frac{75}{181.7} = 65.6° \qquad \text{Ans.}$$

$$\theta_y = \cos^{-1}\frac{\Sigma C_y}{|\mathbf{C}|} = \cos^{-1}\frac{127.9}{181.7} = 45.3° \qquad \textbf{Ans.}$$

$$\theta_z = \cos^{-1}\frac{\Sigma C_z}{|\mathbf{C}|} = \cos^{-1}\frac{105.0}{181.7} = 54.7° \;\blacksquare \qquad \textbf{Ans.}$$

## ▪ PROBLEMS

### Introductory Problems

**5-42\*** Determine the moment of the couple shown in Fig. P5-42 and the perpendicular distance between the two forces.

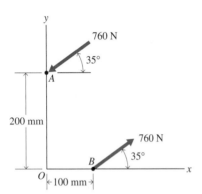

**Figure P5-42**

**5-43\*** A lug wrench is being used to tighten a lug nut on an automobile wheel as shown in Fig. P5-43. Two equal magnitude, parallel forces of opposite sense are applied to the wrench. If the magnitude of each force is 25 lb, determine the couple applied to a lug nut and express the result in Cartesian vector form.

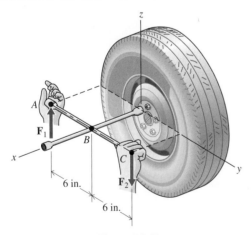

**Figure P5-43**

**5-44** To open a valve to a steam line in a power plant requires a couple of magnitude 54 N · m. Determine the

magnitude of each force **F** shown in Fig. P5-44 required to open the valve.

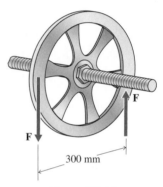

**Figure P5-44**

### Intermediate Problems

**5-45\*** A beam is loaded with a system of three couples as shown in Fig. P5-45. Express the resultant of the couple system in Cartesian vector form.

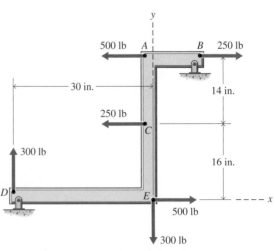

**Figure P5-45**

**5-46\*** The input and output torques (couples) from a gear box are shown in Fig. P5-46. Determine the magnitude and direction of the resultant torque **T**.

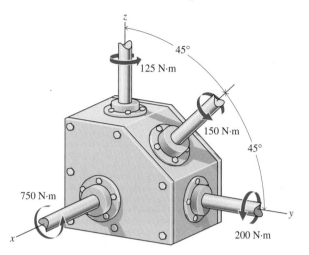

**Figure P5-46**

**5-47** Three couples are applied to a rectangular block as shown in Fig. P5-47. Determine the magnitude of the resultant couple **C** and the direction angles associated with the resultant couple vector.

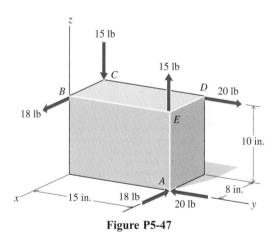

**Figure P5-47**

**Challenging Problems**

**5-48\*** Three couples are applied to a bent bar as shown in Fig. P5-48. Determine
(a) The magnitude of the resultant couple **C** and the direction angles associated with the resultant couple vector.
(b) The scalar component of the resultant couple **C** about line *OA*.

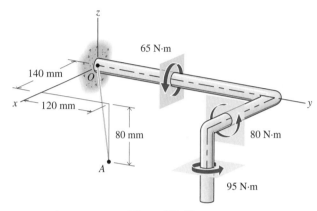

**Figure P5-48**

**5-49** Three couples are applied to a rectangular block as shown in Fig. P5-49. Determine
(a) The magnitude of the resultant couple **C** and the direction angles associated with the resultant couple vector.
(b) The scalar component of the resultant couple **C** about line *OA*.

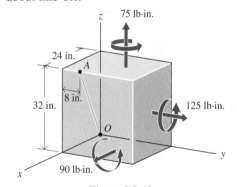

**Figure P5-49**

# 5-6 EQUIVALENT FORCE-COUPLE SYSTEMS

In many problems in mechanics it is convenient to resolve a force **F** into a parallel force **F** and a couple **C** (called an *equivalent force-couple*). In Fig. 5-23*a*, let **F** represent a force acting on a body at point *A*. An arbitrary point *O* in the body and the plane containing both force **F** and point *O* are shown shaded in Fig.

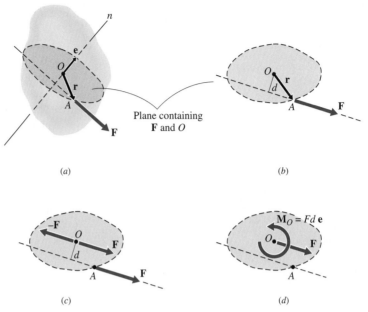

**Figure 5-23**

5-23*a*. The aspect of the shaded plane in the body can be described by using its outer normal *n* and a unit vector **e** along the outer normal. A two-dimensional representation of the shaded plane is shown in Fig. 5-23*b*. If two equal and opposite collinear forces **F**, parallel to the original force, are introduced at point *O*, as shown in Fig. 5-23*c*, the three forces have the same external effect on the body as the original force **F**, since the effects of the two equal, opposite, and collinear forces cancel. The new three-force system can be considered to be a force **F** acting at point *O* (parallel to the original force and of the same magnitude and sense), and a couple **C**, which has the same moment as the moment of the original force about point *O*, as shown in Fig. 5-23*d*. The magnitudes and action lines of the forces of the couple can be changed in accordance with the transformations of a couple discussed in Section 5-5.

Since a force can be resolved into a force and a couple lying in the same plane, it follows conversely that a force and a couple lying in the same plane can be combined into a single force in the plane by reversing the procedure. Thus, the sole external effect of combining a couple with a force is to move the action line of the force into a parallel position. The magnitude and sense of the force remain unchanged.

Two different force systems are *equivalent* if they produce the same external effect when applied to a rigid body. The *resultant* of any force system is the simplest equivalent system to which the given system will reduce. For some systems, the resultant is a single force. For other systems, the simplest equivalent system is a couple. Still other force systems reduce to a force and a couple as the simplest equivalent system. As will be shown in Chapter 6, for a rigid body to be in equilibrium, the resultant must vanish.

## Coplanar Force Systems
The resultant of a system of coplanar forces $\mathbf{F}_1$, $\mathbf{F}_2$, $\mathbf{F}_3$, ..., $\mathbf{F}_n$ can be determined by using the rectangular components of the forces in any two convenient perpendicular directions. Thus, for a

system of forces in the $xy$-plane, such as the one shown in Fig. 5-24$a$, the resultant force **R** can be expressed as

$$\mathbf{R} = \mathbf{R}_x + \mathbf{R}_y = R_x\,\mathbf{i} + R_y\,\mathbf{j} = R\,\mathbf{e}$$

where

$$R_x = \Sigma F_x \qquad R_y = \Sigma F_y$$

$$R = |\mathbf{R}| = \sqrt{(\Sigma F_x)^2 + (\Sigma F_y)^2}$$

$$\mathbf{e} = \cos\theta_x\,\mathbf{i} + \cos\theta_y\,\mathbf{j} \qquad (5\text{-}26)$$

$$\cos\theta_x = \frac{\Sigma F_x}{R} \qquad \cos\theta_y = \frac{\Sigma F_y}{R}$$

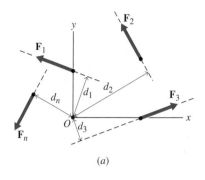

(a)

The location of the line of action of **R** with respect to an arbitrary point $O$ (say the origin of the $xy$-coordinate system) can be computed by using the principle of moments. The moment of **R** about point $O$ must equal the sum of the moments of the forces $\mathbf{F}_1$, $\mathbf{F}_2$, $\mathbf{F}_3$, $\ldots$, $\mathbf{F}_n$ of the original system about the same point $O$, as shown in Fig. 5-24$b$. Thus,

$$Rd_R = F_1 d_1 + F_2 d_2 + F_3 d_3 + \cdots + F_n d_n = \Sigma M_O$$

Therefore,

$$d_R = \frac{\Sigma M_O}{R} \qquad (5\text{-}27)$$

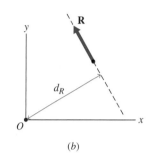

(b)

where $\Sigma M_O$ stands for the algebraic sum of the moments of the forces of the original system about point $O$. The direction of $d_R$ is selected so that the product $Rd_R$ produces a moment about point $O$ with the same sense (clockwise or counterclockwise) as the algebraic sum of the moments of the forces of the original system about point $O$.

The location of the line of action of **R** with respect to point $O$ can also be specified by determining the intercept of the line of action of the force with one of the coordinate axes. For example, in Fig. 5-24$c$, the intercept $x_R$ is determined from the equation

$$x_R = \frac{\Sigma M_O}{R_y} \qquad (5\text{-}28)$$

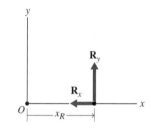

**Figure 5-24**  (c)

since the component $R_x$ of the resultant force **R** does not produce a moment about point $O$. The special case for a system of coplanar parallel forces is illustrated in Fig. 5-25$a$ and $b$.

In the event that the resultant force **R** of a system of coplanar forces $\mathbf{F}_1$, $\mathbf{F}_2$, $\ldots$, $\mathbf{F}_n$ is zero but the moment $\Sigma M_O$ is not zero, the resultant is a couple **C** whose vector is perpendicular to the plane containing the forces (the $xy$-plane for the case being discussed). Thus, the resultant of a coplanar system of forces may be either a force **R** or a couple **C**.

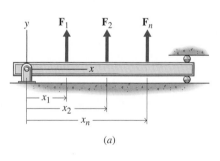

(a)

## Noncoplanar Parallel Force Systems

If all forces of a three-dimensional system are parallel, the resultant force **R** is the algebraic sum of the forces of the system. The line of action of the resultant is determined by using

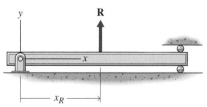

**Figure 5-25**  (b)

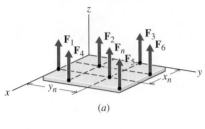

(a)

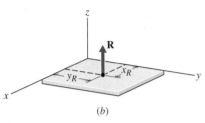

(b)

**Figure 5-26**

the principle of moments. Thus, as shown in Fig. 5-26a, for a system of forces $\mathbf{F}_1, \mathbf{F}_2, \ldots, \mathbf{F}_n$, perpendicular to the $xy$-plane,

$$\mathbf{R} = \mathbf{F}_1 + \mathbf{F}_2 + \cdots + \mathbf{F}_n = R\,\mathbf{k} = \Sigma F\,\mathbf{k} \tag{5-29}$$

$$\mathbf{M}_O = \mathbf{r} \times \mathbf{R} = \mathbf{r}_1 \times \mathbf{F}_1 + \mathbf{r}_2 \times \mathbf{F}_2 + \mathbf{r}_3 \times \mathbf{F}_3 + \cdots + \mathbf{r}_n \times \mathbf{F}_n \tag{5-30}$$

The intersection of the line of action of the resultant force $\mathbf{R}$ with the $xy$-plane (see Fig. 5-26b) is located by equating the moments of $\mathbf{R}$ about the $x$- and $y$-axes to the sums of the moments of the forces of the system $\mathbf{F}_1, \mathbf{F}_2, \ldots, \mathbf{F}_n$ about the $x$- and $y$-axes. It is important to note from Eqs. (5-12) that $M_x = F_z r_y$, while $M_y = -F_z r_x$ (since $F_x = F_y = r_z = 0$). Therefore,

$$Rx_R = F_1 x_1 + F_2 x_2 + \cdots + F_n x_n = -\Sigma M_y$$

$$Ry_R = F_1 y_1 + F_2 y_2 + \cdots + F_n y_n = \Sigma M_x$$

which gives

$$x_R = -\frac{\Sigma M_y}{R} \qquad y_R = \frac{\Sigma M_x}{R} \tag{5-31}$$

In the event that the resultant force $\mathbf{R}$ of the system of parallel forces is zero but the moments $\Sigma M_x$ and $\Sigma M_y$ are not zero, the resultant is a couple $\mathbf{C}$ whose vector lies in a plane perpendicular to the forces (in the $xy$-plane for the system of forces illustrated in Fig. 5-26). Thus, the resultant of a noncoplanar system of parallel forces may be either a force $\mathbf{R}$ or a couple $\mathbf{C}$.

**General Force Systems**  The resultant $\mathbf{R}$ of a general, three-dimensional system of forces $\mathbf{F}_1, \mathbf{F}_2, \mathbf{F}_3, \ldots, \mathbf{F}_n$, such as the one shown in Fig. 5-27a, can be determined by resolving each force of the system into an equal parallel force through any point (taken for convenience as the origin $O$ of a system of coordinate axes) and a couple. Figure 5-27b shows the resolution for force $\mathbf{F}_1$. Thus, as shown in Fig. 5-27c, the given system is replaced by two systems:

1. A system of noncoplanar, concurrent forces through the origin $O$ that have the same magnitudes and directions as the forces of the original system.
2. A system of noncoplanar couples.

Each force and couple of the two systems can be resolved into components along the coordinate axes as shown in Figs. 5-27d and e. The resultant of the concurrent force system is a force through the origin $O$, which can be expressed as

$$\mathbf{R} = \mathbf{R}_x + \mathbf{R}_y + \mathbf{R}_z = R_x\,\mathbf{i} + R_y\,\mathbf{j} + R_z\,\mathbf{k} = R\,\mathbf{e} \tag{5-32}$$

where

$$R_x = \Sigma F_x \qquad R_y = \Sigma F_y \qquad R_z = \Sigma F_z$$

$$R = |\mathbf{R}| = \sqrt{(\Sigma F_x)^2 + (\Sigma F_y)^2 + (\Sigma F_z)^2}$$

$$\mathbf{e} = \cos \theta_x\,\mathbf{i} + \cos \theta_y\,\mathbf{j} + \cos \theta_z\,\mathbf{k}$$

$$\cos \theta_x = \frac{\Sigma F_x}{|\mathbf{R}|} \qquad \cos \theta_y = \frac{\Sigma F_y}{|\mathbf{R}|} \qquad \cos \theta_z = \frac{\Sigma F_z}{|\mathbf{R}|}$$

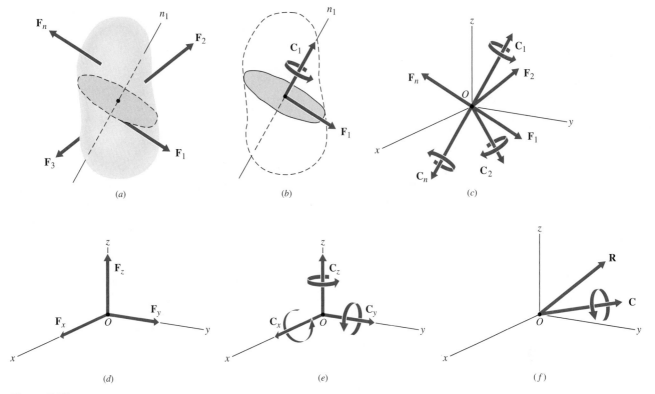

**Figure 5-27**

The resultant of the system of noncoplanar couples is a couple $\mathbf{C}$ that can be expressed as

$$\mathbf{C} = \Sigma\mathbf{C}_x + \Sigma\mathbf{C}_y + \Sigma\mathbf{C}_z = \Sigma C_x\,\mathbf{i} + \Sigma C_y\,\mathbf{j} + \Sigma C_z\,\mathbf{k} = C\,\mathbf{e} \quad (5\text{-}33)$$

where

$$C = |\mathbf{C}| = \sqrt{(\Sigma C_x)^2 + (\Sigma C_y)^2 + (\Sigma C_z)^2}$$

$$\mathbf{e} = \cos\theta_x\,\mathbf{i} + \cos\theta_y\,\mathbf{j} + \cos\theta_z\,\mathbf{k}$$

$$\cos\theta_x = \frac{\Sigma C_x}{|\mathbf{C}|} \qquad \cos\theta_y = \frac{\Sigma C_y}{|\mathbf{C}|} \qquad \cos\theta_z = \frac{\Sigma C_z}{|\mathbf{C}|}$$

The resultant force $\mathbf{R}$ and the resultant couple $\mathbf{C}$ shown in Fig. 5-27$f$ together constitute the resultant of the system with respect to point $O$. In special cases, the resultant couple $\mathbf{C}$ may vanish, leaving the force $\mathbf{R}$ as the resultant of the system. Again in special cases, the resultant force $\mathbf{R}$ may vanish, leaving the couple $\mathbf{C}$ as the resultant of the system. If the resultant force $\mathbf{R}$ and the resultant couple $\mathbf{C}$ both vanish, the resultant of the system is zero and the system is in equilibrium. Thus, the resultant of a general force system may be a force $\mathbf{R}$, a couple $\mathbf{C}$, or both a force $\mathbf{R}$ and a couple $\mathbf{C}$.

When the couple $\mathbf{C}$ is perpendicular to the resultant force $\mathbf{R}$, as shown in Fig. 5-28, the two can be combined to form a single force $\mathbf{R}$ whose line of action is a distance $d = C/R$ from point $O$ in a direction that makes the direction of the moment of $R$ about $O$ the same as that of $\mathbf{C}$.

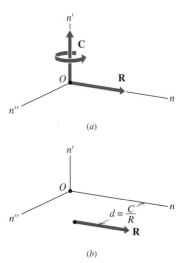

**Figure 5-28**

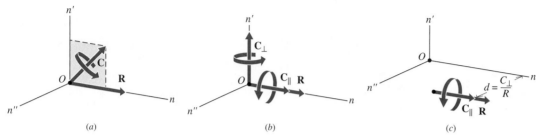

**Figure 5-29**

Another transformation of the resultant (**R** and **C**) of a general force system is illustrated in Figs. 5-29 and 5-30. In this case, the couple **C** is resolved into components parallel and perpendicular to the resultant force **R**, as shown in Fig. 5-29*b*. The resultant force **R** and the perpendicular component of the couple **C**$_\perp$ can be combined as illustrated in Fig. 5-28. In addition, the parallel component of the couple **C**$_\parallel$ can be translated to coincide with the line of action of the resultant force **R**, as shown in Fig. 5-29*c*. The combination of couple **C**$_\parallel$ and resultant force **R** is known as a *wrench*. The action may be described as a push (or pull) and a twist about an axis parallel to the push (or pull). When the force and moment vectors have the same sense, as shown in Fig. 5-29*c*, the wrench is positive. When the vectors have the opposite sense, as shown in Fig. 5-30*c*, the wrench is negative. A screwdriver is an example of a wrench.

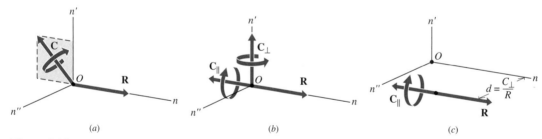

**Figure 5-30**

## Example Problem 5-9

An 800-N force is applied to a bracket as shown in Fig. 5-31*a*. Replace the force by a force at point *B* and a couple. Express your answer in Cartesian vector form.

### SOLUTION
### Scalar Analysis

The force **F** of magnitude 800 N can be resolved into the components

$$F_x = \frac{4}{5}F = \frac{4}{5}(800) = 640 \text{ N}$$

$$F_y = -\frac{3}{5}F = -\frac{3}{5}(800) = -480 \text{ N}$$

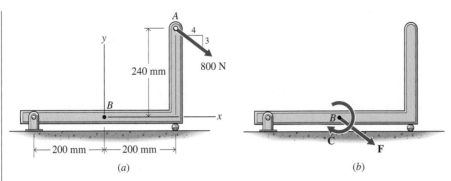

**Figure 5-31**

The moment of the force about point $B$ is

$$M_B = -F_x(0.240) + F_y(0.200)$$
$$= -640(0.240) + (-480)(0.200)$$
$$= -249.6 \text{ N} \cdot \text{m} \cong 250 \text{ N} \cdot \text{m} \downarrow$$

The force and moment in Cartesian vector form are

$$\mathbf{F} = 640 \, \mathbf{i} - 480 \, \mathbf{j} \text{ N} \qquad \textbf{Ans.}$$
$$\mathbf{C} = 250 \text{ N} \cdot \text{m} \downarrow = -250 \, \mathbf{k} \text{ N} \cdot \text{m} \qquad \textbf{Ans.}$$

The result is shown in Fig. 5-31$b$.

## Vector Analysis

The force $\mathbf{F}$ can be expressed in Cartesian vector form as

$$\mathbf{F} = 800\left[\frac{4}{5}\,\mathbf{i} - \frac{3}{5}\,\mathbf{j}\right] = 640\,\mathbf{i} - 480\,\mathbf{j} \text{ N} \qquad \textbf{Ans.}$$

The position vector $\mathbf{r}_{A/B}$ is expressed as

$$\mathbf{r}_{A/B} = 0.200\,\mathbf{i} + 0.240\,\mathbf{j} \text{ m}$$

From Eq. (5-21):

$$\mathbf{C} = \mathbf{r}_{A/B} \times \mathbf{F} = [0.200\,\mathbf{i} + 0.240\,\mathbf{j}] \times [640\,\mathbf{i} - 480\,\mathbf{j}]$$
$$= -249.6\,\mathbf{k} \text{ N} \cdot \text{m} \cong -250\,\mathbf{k} \text{ N} \cdot \text{m} \quad \blacksquare \qquad \textbf{Ans.}$$

## Example Problem 5-10

The force $\mathbf{F}$ shown in Fig. 5-32 has a magnitude of 780 lb. Replace the force $\mathbf{F}$ by a force $\mathbf{F}_O$ at point $O$ and a couple $\mathbf{C}$.

(a) Express the force $\mathbf{F}_O$ and the couple $\mathbf{C}$ in Cartesian vector form.

(b) Determine the direction angles associated with the unit vector $\mathbf{e}$ that describes the aspect of the plane of the couple.

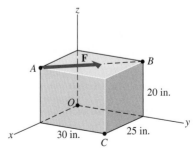

**Figure 5-32**

## SOLUTION

(a)
$$\mathbf{F}_O = F\mathbf{e} = 780\left[\frac{-25\,\mathbf{i} + 30\,\mathbf{j}}{\sqrt{1525}}\right]$$
$$= -499.3\,\mathbf{i} + 599.2\,\mathbf{j}\ \text{lb} \cong -499\,\mathbf{i} + 599\,\mathbf{j}\ \text{lb} \qquad \textbf{Ans.}$$

$$\mathbf{r}_{A/O} = 25\,\mathbf{i} + 20\,\mathbf{k}\ \text{in.}$$

From Eq. (5-21):

$$\mathbf{C} = \mathbf{M}_O = \mathbf{r}_{A/O} \times \mathbf{F}$$
$$= [25\,\mathbf{i} + 20\,\mathbf{k}] \times [-499.3\,\mathbf{i} + 599.2\,\mathbf{j}]$$
$$= -11{,}984\,\mathbf{i} - 9986\,\mathbf{j} + 14{,}980\,\mathbf{k}\ \text{lb}\cdot\text{in.}$$
$$\cong -11.98\,\mathbf{i} - 9.99\,\mathbf{j} + 14.98\,\mathbf{k}\ \text{kip}\cdot\text{in.} \qquad \textbf{Ans.}$$

(b) From Eq. (5-23):

$$C = \sqrt{(-11{,}984)^2 + (-9986)^2 + (14{,}980)^2} = 21{,}627\ \text{lb}\cdot\text{in.}$$

From Eq. (5-25):

$$\theta_x = \cos^{-1}\frac{C_x}{C} = \cos^{-1}\frac{-11{,}984}{21{,}627} = 123.7° \qquad \textbf{Ans.}$$

$$\theta_y = \cos^{-1}\frac{C_y}{C} = \cos^{-1}\frac{-9986}{21{,}627} = 117.5° \qquad \textbf{Ans.}$$

$$\theta_z = \cos^{-1}\frac{C_z}{C} = \cos^{-1}\frac{14{,}980}{21{,}627} = 46.2° \ \blacksquare \qquad \textbf{Ans.}$$

---

## Example Problem 5-11

Three forces and a couple are applied to a bracket as shown in Fig. 5-33a. Determine

(a) The magnitude and direction of the resultant.
(b) The perpendicular distance $d_R$ from point $O$ to the line of action of the resultant.
(c) The distance $x_R$ from point $O$ to the intercept of the line of action of the resultant with the x-axis.

## SOLUTION

(a) The resultant of a system of coplanar forces is either a force $\mathbf{R}$ or a couple $\mathbf{C}$. The resultant force is obtained by using Eqs. (5-26). Thus,

$$R_x = \Sigma F_x = 500\cos 60° + 300 = 550\ \text{N}$$
$$R_y = \Sigma F_y = 500\sin 60° + 200 = 633\ \text{N}$$

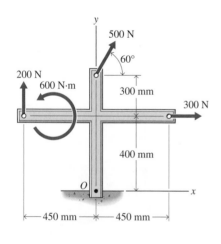

**Figure 5-33**

(a)

$$R = |\mathbf{R}| = \sqrt{(\Sigma F_x)^2 + (\Sigma F_y)^2} = \sqrt{(550)^2 + (633)^2} = 838.6 \text{ N} \cong 839 \text{ N} \textbf{ Ans.}$$

$$\theta_x = \cos^{-1} \frac{\Sigma F_x}{R} = \cos^{-1} \frac{550}{838.6} = 49.0° \qquad \textbf{Ans.}$$

(b) From Eq. (5-27):

$$+\!\downarrow \Sigma M_O = -300(0.400) - 500 \cos 60° \, (0.700)$$

$$-200(0.450) + 600 = 215 \text{ N} \cdot \text{m}$$

$$d_R = \frac{\Sigma M_O}{R} = \frac{215}{838.6} = 0.2564 \text{ m} \cong 256 \text{ mm} \qquad \textbf{Ans.}$$

(c) From Eq. (5-28):

$$x_R = \frac{\Sigma M_O}{R_y} = \frac{215}{633} = 0.3397 \text{ m} \cong 340 \text{ mm} \qquad \textbf{Ans.}$$

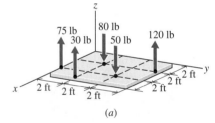

(b)

**Figure 5-33**

The results are illustrated in Fig. 5-33*b*. ∎

---

## Example Problem 5-12

Determine the resultant of the parallel force system shown in Fig. 5-34*a* and locate the intersection of the line of action of the resultant with the *xy*-plane.

### SOLUTION

The resultant **R** of the force system is determined by using Eq. (5-29). Thus,

$$\mathbf{R} = \Sigma F \, \mathbf{k}$$

$$= (75 + 30 - 80 - 50 + 120) \, \mathbf{k} = 95 \, \mathbf{k} \text{ lb} \qquad \textbf{Ans.}$$

The intersection $(x_R, y_R)$ of the line of action of the resultant with the *xy*-plane is determined by using Eqs. (5-31). With positive moments defined by using the right-hand rule,

$$\Sigma M_y = F_1 x_1 + F_2 x_2 + \cdots + F_n x_n$$

$$= -75(4) - 30(6) + 80(2) + 50(4) - 120(2) = -360 \text{ lb} \cdot \text{ft}$$

$$\Sigma M_x = F_1 y_1 + F_2 y_2 + \cdots + F_n y_n$$

$$= 75(0) + 30(2) - 80(2) - 50(4) + 120(6) = 420 \text{ lb} \cdot \text{ft}$$

Therefore,

$$x_R = -\frac{\Sigma M_y}{R} = -\frac{-360}{95} = 3.79 \text{ ft} \qquad \textbf{Ans.}$$

$$y_R = \frac{\Sigma M_x}{R} = \frac{420}{95} = 4.42 \text{ ft} \qquad \textbf{Ans.}$$

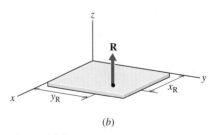

(a)

(b)

**Figure 5-34**

The coordinates $(x_R, y_R)$ can also be determined by using a vector analysis. Thus,

$$\mathbf{M}_O = \mathbf{r} \times \mathbf{R} = (\mathbf{r}_1 \times \mathbf{F}_1) + (\mathbf{r}_2 \times \mathbf{F}_2) + (\mathbf{r}_3 \times \mathbf{F}_3) + (\mathbf{r}_4 \times \mathbf{F}_4) + (\mathbf{r}_5 \times \mathbf{F}_5)$$
$$= [(x_R\,\mathbf{i} + y_R\,\mathbf{j}) \times (95\,\mathbf{k})]$$
$$= [(4\,\mathbf{i}) \times (75\,\mathbf{k})] + [(6\,\mathbf{i} + 2\,\mathbf{j}) \times (30\,\mathbf{k})] + [(2\,\mathbf{i} + 2\,\mathbf{j}) \times (-80\,\mathbf{k})]$$
$$+ [(4\,\mathbf{i} + 4\,\mathbf{j}) \times (-50\,\mathbf{k})] + [(2\,\mathbf{i} + 6\,\mathbf{j}) \times (120\,\mathbf{k})]$$
$$= -95x_R\,\mathbf{j} + 95y_R\,\mathbf{i} = 420\,\mathbf{i} - 360\,\mathbf{j}$$

Solving for $x_R$ and $y_R$ yields

$$x_R = \frac{360}{95} = 3.79 \text{ ft} \qquad \textbf{Ans.}$$

$$y_R = \frac{420}{95} = 4.42 \text{ ft} \qquad \textbf{Ans.}$$

The resultant is shown in Fig. 5-34*b*. ∎

## Example Problem 5-13

Replace the force system shown in Fig. 5-35*a* with a force **R** through point $O$ and a couple **C**.

### SOLUTION

The three forces and their positions with respect to point $O$ can be written in Cartesian vector form as

$$\mathbf{F}_A = 250\left[-\frac{1.5}{2.5}\,\mathbf{i} + \frac{2}{2.5}\,\mathbf{k}\right] = (-150\,\mathbf{i} + 200\,\mathbf{k}) \text{ N} \quad \mathbf{r}_{A/O} = (1.5\,\mathbf{i} + 3\,\mathbf{j}) \text{ m}$$

$$\mathbf{F}_B = 335\left[\frac{1.5}{3.354}\,\mathbf{i} - \frac{3}{3.354}\,\mathbf{j}\right] = (150\,\mathbf{i} - 300\,\mathbf{j}) \text{ N} \quad \mathbf{r}_{B/O} = (3\,\mathbf{j} + 2\,\mathbf{k}) \text{ m}$$

$$\mathbf{F}_C = 360\left[\frac{3}{3.606}\,\mathbf{j} - \frac{2}{3.606}\,\mathbf{k}\right] = (300\,\mathbf{j} - 200\,\mathbf{k}) \text{ N} \quad \mathbf{r}_{C/O} = (1.5\,\mathbf{i} + 2\,\mathbf{k}) \text{ m}$$

Each of the three forces can be replaced by an equal force through point $O$ and a couple. The vector sum of the concurrent forces is

$$\mathbf{R} = \mathbf{F}_A + \mathbf{F}_B + \mathbf{F}_C$$
$$= -150\,\mathbf{i} + 200\,\mathbf{k} + 150\,\mathbf{i} - 300\,\mathbf{j} + 300\,\mathbf{j} - 200\,\mathbf{k} = 0 \qquad \textbf{Ans.}$$

The moment $\mathbf{M}_O$ of the resultant couple is

$$\mathbf{C} = \mathbf{M}_O = (\mathbf{r}_{A/O} \times \mathbf{F}_A) + (\mathbf{r}_{B/O} \times \mathbf{F}_B) + (\mathbf{r}_{C/O} \times \mathbf{F}_C)$$

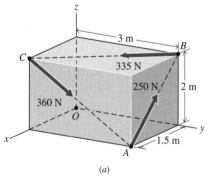

(a)

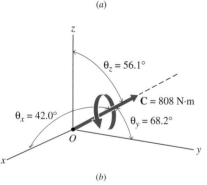

(b)

**Figure 5-35**

$$= \begin{vmatrix} \mathbf{i} & \mathbf{j} & \mathbf{k} \\ 1.5 & 3 & 0 \\ -150 & 0 & 200 \end{vmatrix} + \begin{vmatrix} \mathbf{i} & \mathbf{j} & \mathbf{k} \\ 0 & 3 & 2 \\ 150 & -300 & 0 \end{vmatrix} + \begin{vmatrix} \mathbf{i} & \mathbf{j} & \mathbf{k} \\ 1.5 & 0 & 2 \\ 0 & 300 & -200 \end{vmatrix}$$

$$= (600 \, \mathbf{i} + 300 \, \mathbf{j} + 450 \, \mathbf{k}) \, \text{N} \cdot \text{m} \qquad \textbf{Ans.}$$

The magnitude of the resultant couple is

$$|\mathbf{C}| = \sqrt{C_x^2 + C_y^2 + C_z^2}$$
$$= \sqrt{(600)^2 + (300)^2 + (450)^2} = 807.8 \cong 808 \, \text{N} \cdot \text{m} \qquad \textbf{Ans.}$$

Finally, the direction angles that locate the axis of the couple are

$$\theta_x = \cos^{-1} \frac{C_x}{|\mathbf{C}|} = \cos^{-1} \frac{600}{807.8} = 42.0° \qquad \textbf{Ans.}$$

$$\theta_y = \cos^{-1} \frac{C_y}{|\mathbf{C}|} = \cos^{-1} \frac{300}{807.8} = 68.2° \qquad \textbf{Ans.}$$

$$\theta_z = \cos^{-1} \frac{C_z}{|\mathbf{C}|} = \cos^{-1} \frac{450}{807.8} = 56.1° \qquad \textbf{Ans.}$$

The resultant of the force system is the couple shown in Fig. 5-35*b*. ∎

## PROBLEMS

**Introductory Problems**

**5-50\*** Replace the 3-kN force shown in Fig. P5-50 by a force at point $A$ and a couple.

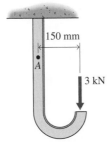

150 mm

$A$

3 kN

**Figure P5-50**

**5-51\*** Replace the 50-lb force shown in Fig. P5-51 by a force at point $A$ and a couple. Express your answer in Cartesian vector form.

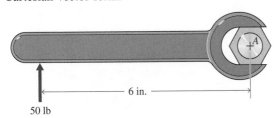

6 in.

50 lb

**Figure P5-51**

**5-52** Replace the 130-N vertical force shown in Fig. P5-52 by the mechanic's hand to the wrench by an equivalent force-couple system at the lug nut.

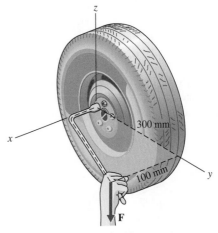

$z$

300 mm

$x$

100 mm

$y$

**F**

**Figure P5-52**

**5-53** A gusset plate is riveted to a beam by three rivets as shown in Fig. P5-53. Replace the 2500-lb force by a force-couple system at the top rivet.

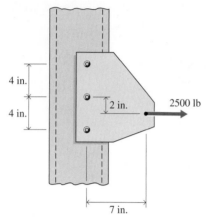

**Figure P5-53**

**5-54\*** Four forces are applied to a truss as shown in Fig. P5-54. Determine the magnitude and direction of the resultant of the four forces and the perpendicular distance $d_R$ from point $A$ to the line of action of the resultant.

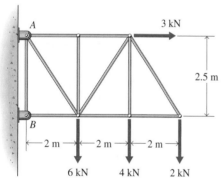

**Figure P5-54**

**5-55\*** Three forces are applied to the locked pulley shown in Fig. P5-55. Determine the magnitude and direction of the resultant of the three forces and the perpendicular distance from the axle of the pulley to the line of action of the resultant.

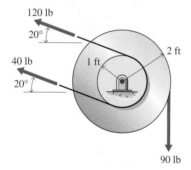

**Figure P5-55**

**5-56\*** Determine the resultant of the four forces acting on the bell crank shown in Fig. P5-56, and determine where the resultant intersects the $x$-axis.

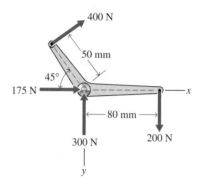

**Figure P5-56**

**5-57** Three 75-lb traffic lights are suspended over a roadway as shown in Fig. P5-57. Determine the resultant of the weights of the traffic lights and locate the resultant with respect to point $A$.

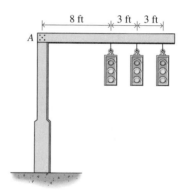

**Figure P5-57**

**5-58** Four parallel forces act on a concrete slab as shown in Fig. P5-58. Determine the resultant of the forces and locate the intersection of the line of action of the resultant with the $xy$-plane.

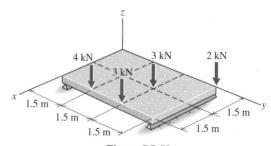

**Figure P5-58**

**5-59** The forces exerted on the wheels of an airplane by the runway are shown in Fig. P5-59. Determine the resultant of the three forces and locate the intersection of the line of action of the resultant with the *xy*-plane (the runway).

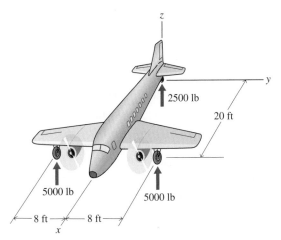

Figure P5-59

**Intermediate Problems**

**5-60\*** The magnitude of the force **F** acting on the casting shown in Fig. P5-60 is 2 kN. Replace the force with a force **R** through point *A* and a couple **C**.

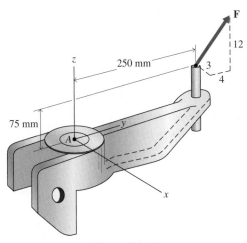

Figure P5-60

**5-61\*** A force $\mathbf{F} = (50\,\mathbf{i} + 50\,\mathbf{j} - 200\,\mathbf{k})$ lb acts on the wall bracket shown in Fig. P5-61. Replace the force by a force-couple system at rivet *B*. The eye bolt is small, and the force **F** may be considered as acting at point *C*.

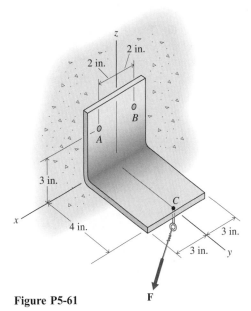

Figure P5-61

**5-62** The homogeneous plate shown in Fig. P5-62 has a mass of 90 kg. The magnitude of the force **T** in cable *BC* is 800 N. Replace the weight and cable forces by an equivalent force-couple system at hinge *A*.

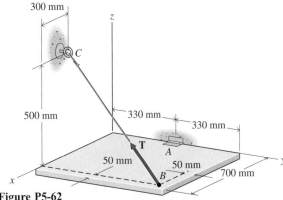

Figure P5-62

**5-63** Forces act at *A*, *D*, and *E* of the member shown in Fig. P5-63. If the equivalent force-couple system at *B* is $\mathbf{R} = (-100\,\mathbf{i} - 100\,\mathbf{j})$ lb and a couple **C**, determine the magnitude of $\mathbf{F}_E$, the angle $\theta$, and the required couple **C**.

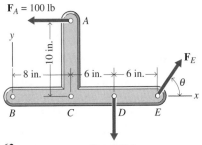

Figure P5-63

**5-64\***  Four forces and a couple are applied to a rectangular plate as shown in Fig. P5-64. Determine the magnitude and direction of the resultant of the force-couple system and the distance $x_R$ from point $O$ to the intercept of the line of action of the resultant with the $x$-axis.

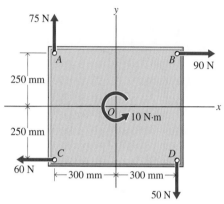

**Figure P5-64**

**5-65**  The magnitude of the resultant $\mathbf{R}$ of the four forces acting on the legs of the table shown in Fig. P5-65 is 80 lb. Determine the magnitude of $\mathbf{F}$ and the location, $x_R$ and $y_R$, of the line of action of $\mathbf{R}$.

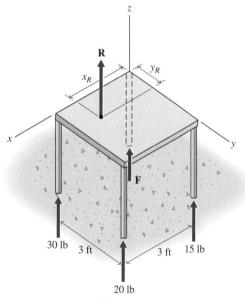

**Figure P5-65**

**Challenging Problems**

**5-66\***  Figure P5-66 shows a crankshaft-flywheel arrangement of a one-cylinder engine. A 1000-N force $\mathbf{P}$ is supplied by the connecting rod, and a couple $\mathbf{C}$ of magnitude 250 N · m is delivered to the crankshaft by the flywheel. Replace the force and couple by an equivalent force-couple system at the bearing $A$.

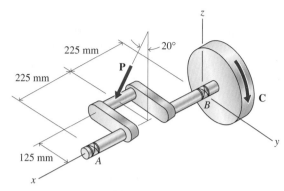

**Figure P5-66**

**5-67\***  A farmer is using the hand winch shown in Fig. P5-67 to raise a 40-lb bucket of water from a well. When the force $\mathbf{P}$ and the weight of the bucket are replaced by an equivalent force-couple system at bearing $D$, the result is $\mathbf{R} = -10\,\mathbf{j} - 60\,\mathbf{k}$ lb and a couple $\mathbf{C}_D$. Determine the force $\mathbf{P}$ applied to the handle of the winch and the couple $\mathbf{C}_D$.

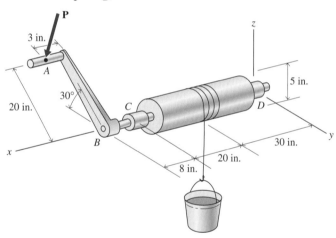

**Figure P5-67**

**5-68**  In order to remove a rusty screw from a steel plate, a worker attaches a screwdriver to the bent bar shown in Fig. P5-68. To hold the screwdriver in place, the

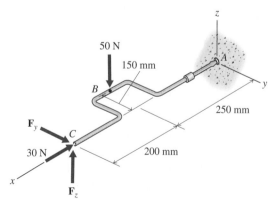

**Figure P5-68**

worker applies a 30-N force at $C$; a 50-N force is applied to the handle of the bar in an attempt to remove the screw. Forces $\mathbf{F}_y$ and $\mathbf{F}_z$ are also applied at $C$. It is desirable that the screwdriver not bend about the $y$- and $z$-axes.

(a) Determine the forces $\mathbf{F}_y$ and $\mathbf{F}_z$.

(b) Replace the forces at $B$ and $C$ by an equivalent force-couple system at $A$.

**5-69** Reduce the forces shown in Fig. P5-69 to a wrench and locate the intersection of the wrench with the $xy$-plane.

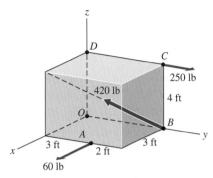

**Figure P5-69**

# 5-7 CENTER OF GRAVITY AND CENTER OF MASS

**Center of Gravity** An immediate application of equivalent force-couples is the replacement of distributed forces such as gravitational forces with a single equivalent force. For example, consider the set of $n$ particles shown in Fig. 5-36$a$. Approximating the weights of the $n$ particles as a parallel force system (Fig. 5-36$b$),[1] the external effect of the $n$ weights is the same as that of a single equivalent force $\mathbf{W}$ acting at point $G$ as shown in Fig. 5-36$c$. The equivalent force $\mathbf{W}$ is called the *weight of the body*, and the point $G$ through which the force acts is called the *center of gravity*. As in Section 5-6, the direction of the weight force $\mathbf{W}$ is parallel to all of the individual weights $\mathbf{W}_i$, and its magnitude is the sum of the magnitudes of the individual weights (so that the sum of forces is the same on Figs. 5-36$b$ and $c$)

$$W = \sum_{i=1}^{n} W_i \tag{5-34a}$$

and the location of $G$ (so that the sum of moments is the same on Figs. 5-36$b$ and $c$)[2] is

$$M_{yz} = Wx_G = \sum_{i=1}^{n} W_i x_i \quad \text{or} \quad x_G = \frac{1}{W} \sum_{i=1}^{n} W_i x_i \tag{5-34b}$$

$$M_{zx} = Wy_G = \sum_{i=1}^{n} W_i y_i \quad \text{or} \quad y_G = \frac{1}{W} \sum_{i=1}^{n} W_i y_i \tag{5-34c}$$

$$M_{xy} = Wz_G = \sum_{i=1}^{n} W_i z_i \quad \text{or} \quad z_G = \frac{1}{W} \sum_{i=1}^{n} W_i z_i \tag{5-34d}$$

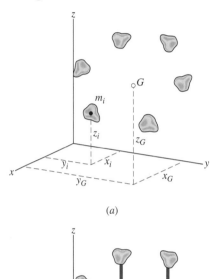

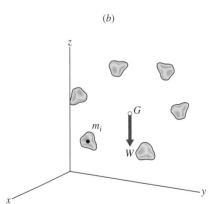

---

[1] The weight forces are actually concurrent at the center of the earth. However, most engineering structures are small compared to the radius of the earth, and the error introduced by assuming the forces are parallel is very small.

[2] Equation (5-34d) is not directly obtainable from Figs. 5-36$b$ and $c$. By the Principle of Transmissibility, the force $\mathbf{W}$ can act anywhere along the vertical line through $G$ without affecting its moment. Equation (5-34d) is obtained by rotating the figures, both the masses and the coordinate system, so that gravity acts along the $y$-axis instead of the $z$-axis. Then the sum of moments gives Eq. (5-34d).

**Figure 5-36**   (c)

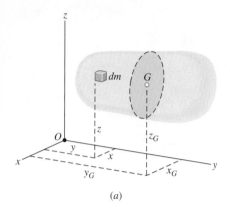

(a)

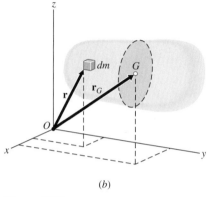

(b)

**Figure 5-37**

where $x_i$, $y_i$, $z_i$, and $x_G$, $y_G$, $z_G$ are shown on Fig. 5-36a. The moments $M_{yz}$, $M_{zx}$, and $M_{xy}$ are called *first moments of the weight forces* relative to the $yz$-, $zx$-, and $xy$-planes, respectively.

If the particles form a continuous body, as shown in Fig. 5-37a, the summations must be replaced by integrals over the mass of the body, giving

$$W = \int dW \tag{5-35a}$$

$$M_{yz} = Wx_G = \int x \, dW \quad \text{or} \quad x_G = \frac{1}{W} \int x \, dW \tag{5-35b}$$

$$M_{zx} = Wy_G = \int y \, dW \quad \text{or} \quad y_G = \frac{1}{W} \int y \, dW \tag{5-35c}$$

$$M_{xy} = Wz_G = \int z \, dW \quad \text{or} \quad z_G = \frac{1}{W} \int z \, dW \tag{5-35d}$$

Equations (5-34b), (5-34c), and (5-34d) can be combined into a single vector equation by multiplying the first, second, and third equations by **i**, **j**, and **k**, respectively, and adding. Thus,

$$Wx_G \, \mathbf{i} + Wy_G \, \mathbf{j} + Wz_G \, \mathbf{k} = \sum_{i=1}^{n} W_i x_i \, \mathbf{i} + \sum_{i=1}^{n} W_i y_i \, \mathbf{j} + \sum_{i=1}^{n} W_i z_i \, \mathbf{k}$$

from which

$$W(x_G \, \mathbf{i} + y_G \, \mathbf{j} + z_G \, \mathbf{k}) = \sum_{i=1}^{n} W_i (x_i \, \mathbf{i} + y_i \, \mathbf{j} + z_i \, \mathbf{k})$$

which reduces to

$$\mathbf{M}_O = W\mathbf{r}_G = \sum_{i=1}^{n} W_i \mathbf{r}_i \quad \text{or} \quad \mathbf{r}_G = \frac{1}{W} \sum_{i=1}^{n} W_i \mathbf{r}_i \tag{5-36}$$

where $\mathbf{r}_i = x_i \, \mathbf{i} + y_i \, \mathbf{j} + z_i \, \mathbf{k}$ is the position vector from the origin $O$ to the $i$th particle (Fig. 5-37b) and $\mathbf{r}_G = x_G \, \mathbf{i} + y_G \, \mathbf{j} + z_G \, \mathbf{k}$ is the position vector from the origin to the center of gravity. Similarly for Eqs. (5-35) if the particles form a continuous body:

$$W\mathbf{r}_G = \int \mathbf{r} \, dW \quad \text{or} \quad \mathbf{r}_G = \frac{1}{W} \int \mathbf{r} \, dW \tag{5-37}$$

The center of gravity $G$ located by Eqs. (5-34) through (5-37) represents the point at which all of the weight of the body could be concentrated without changing the external effects on the body.

**Center of Mass**   For practical engineering work in which the size of a body is small in comparison to the size of the earth, all of the particles that make up the body can be assumed to be at the same distance from the center of the earth; therefore, they experience the same gravitational acceleration $g$. Dividing Eqs. (5-35) by the gravitational acceleration gives

$$m = \int dm \tag{5-38a}$$

$$mx_G = \int x \, dm \quad \text{or} \quad x_G = \frac{1}{m} \int x \, dm \tag{5-38b}$$

$$my_G = \int y \, dm \quad \text{or} \quad y_G = \frac{1}{m} \int y \, dm \qquad \text{(5-38c)}$$

$$mz_G = \int z \, dm \quad \text{or} \quad z_G = \frac{1}{m} \int z \, dm \qquad \text{(5-38d)}$$

The location of the center of mass $(x_G, y_G, z_G)$ is the same as the center of gravity, and both points will be labeled $G$. However, the center of mass is defined equally well in the weightlessness of space as it is on the surface of the earth.

The moments $M_{yz}$, $M_{zx}$, and $M_{xy}$ in Eqs. (5-34) and (5-35) are called *first moments of the weight forces* relative to the $yz$-, $zx$-, and $xy$-planes, respectively. The integrand in each case is the first power of the distance to the respective plane. Similarly, the integrals $\int x \, dm$, $\int y \, dm$, and $\int z \, dm$ are called the *first moments of the mass* (and integrals of the form $\int x \, dA$ are called *first moments of area*) although they are not strictly moments, since $dm$ (and $dA$) are not actually forces. Later, integrals of the form $\int x^2 \, dm$ (and $\int x^2 \, dA$) will be introduced. Such integrals are called *second moments of mass* (and *second moments of area*) since the second power of the distance appears in these expressions.

The following example illustrates the procedure used to locate the "center of gravity" or the "center of mass" of a system of particles.

## Example Problem 5-14

Four bodies $A$, $B$, $C$, and $D$ (which can be treated as particles) are attached to a shaft as shown in Fig. 5-38. The masses of the bodies are 0.2, 0.4, 0.6, and 0.8 slug, respectively, and the distances from the $z$-axis of the shaft to their mass centers (end view) are 1.50, 2.50, 2.00, and 1.25 ft, respectively. Find the mass center for the four bodies.

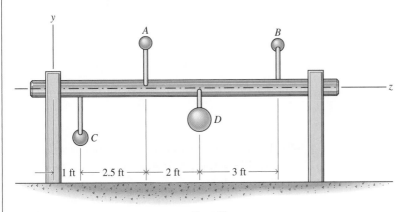

Front View

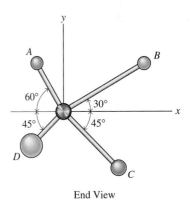

End View

**Figure 5-38**

## SOLUTION

A typical equation from Eqs. (5-34) that is used to locate the mass center of a system of particles is

$$x_G = \frac{1}{m} \sum_{i=1}^{n} m_i x_i \quad \text{where} \quad m = \sum_{i=1}^{n} m_i$$

Thus, for the system of four bodies shown in Fig. 5-38:

$$\Sigma m_i = 0.2 + 0.4 + 0.6 + 0.8 = 2.0 \text{ slug}$$

$$\Sigma m_i x_i = m_A x_A + m_B x_B + m_C x_C + m_D x_D$$

$$= 0.2(-1.50 \cos 60°) + 0.4(2.50 \cos 30°) + 0.6(2.00 \cos 45°)$$

$$+ 0.8(-1.25 \cos 45°) = 0.8574 \text{ slug} \cdot \text{ft}$$

$$\Sigma m_i y_i = m_A y_A + m_B y_B + m_C y_C + m_D y_D$$

$$= 0.2(1.50 \sin 60°) + 0.4(2.50 \sin 30°) + 0.6(-2.00 \sin 45°)$$

$$+ 0.8(-1.25 \sin 45°) = -0.7958 \text{ slug} \cdot \text{ft}$$

$$\Sigma m_i z_i = m_A z_A + m_B z_B + m_C z_C + m_D z_D$$

$$= 0.2(3.5) + 0.4(8.5) + 0.6(1.0) + 0.8(5.5) = 9.10 \text{ slug} \cdot \text{ft}$$

$$x_G = \frac{\Sigma m_i x_i}{m} = \frac{0.8574}{2.00} = 0.429 \text{ ft} \qquad \textbf{Ans.}$$

$$y_G = \frac{\Sigma m_i y_i}{m} = \frac{-0.7958}{2.00} = -0.398 \text{ ft} \qquad \textbf{Ans.}$$

$$z_G = \frac{\Sigma m_i z_i}{m} = \frac{9.10}{2.00} = 4.55 \text{ ft} \blacksquare \qquad \textbf{Ans.}$$

# PROBLEMS

## Introductory Problems

**5-70*** Locate the center of mass for the three particles shown in Fig. P5-70 if $m_A = 26$ kg, $m_B = 21$ kg, and $m_C = 36$ kg.

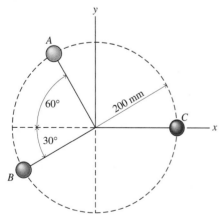

**Figure P5-70**

**5-71*** Locate the center of gravity for the four particles shown in Fig. P5-71 if $W_A = 20$ lb, $W_B = 25$ lb, $W_C = 30$ lb, and $W_D = 40$ lb.

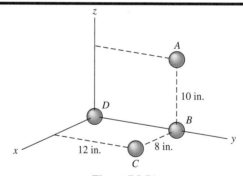

**Figure P5-71**

**5-72** Locate the center of mass for the four particles shown in Fig. P5-72 if $m_A = 16$ kg, $m_B = 24$ kg, $m_C = 14$ kg, and $m_D = 36$ kg.

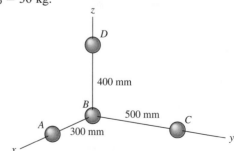

**Figure P5-72**

**Intermediate Problems**

**5-73*** Locate the center of gravity for the five particles shown in Fig. P5-73 if $W_A = 25$ lb, $W_B = 35$ lb, $W_C = 15$ lb, $W_D = 28$ lb, and $W_E = 16$ lb.

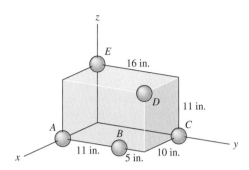

**Figure P5-73**

**5-74*** Locate the center of mass for the five particles shown in Fig. P5-74 if $m_A = 2$ kg, $m_B = 3$ kg, $m_C = 4$ kg, $m_D = 3$ kg, and $m_E = 2$ kg.

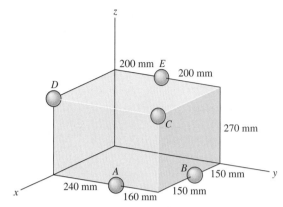

**Figure P5-74**

**5-75** Three bodies with masses of 2, 4, and 6 slugs are located at points $(2, 3, 4)$, $(3, -4, 5)$, and $(-3, 4, 6)$, respectively. Locate the mass center of the system if the distances are measured in feet.

---

# 5-8 CENTROIDS OF VOLUMES, AREAS, AND LINES

**Centroids of Volumes**  If the density $\rho$ (mass per unit volume) of the material that makes up a body is constant (the same for every particle that makes up the body), then $m = \rho V$, $dm = \rho\, dV$, and Eqs. (5-38) can be divided by the density $\rho$ to get

$$x_C = \frac{1}{V}\int x\, dV \qquad y_C = \frac{1}{V}\int y\, dV \qquad z_C = \frac{1}{V}\int z\, dV \qquad (5\text{-}39)$$

where $V$ is the volume of the body. The coordinates $x_C$, $y_C$, and $z_C$, defined by Eq. (5-39), depend only on the geometry of the body and are independent of the physical properties. The point located by such a set of coordinates is known as the *centroid C* of the volume of the body. The term *centroid* is usually used in connection with geometrical figures (volumes, areas, and lines), whereas the terms *center of mass* and *center of gravity* are used in connection with physical bodies.

Note that the centroid $C$ of a body can be defined whether the body is homogeneous (made of a uniform material; constant density) or not. If the body is homogeneous, the centroid will have the same position as the center of mass and the center of gravity. If the density of the material varies from point to point within the body, the center of gravity of the body and the centroid of the volume occupied by the body will usually be at different points, as indicated in Fig. 5-39. Since the density of the lower portion of the cone in Fig. 5-39 is greater than the density of the upper portion, the center of gravity $G$ (which depends on the weights of the two parts) will be below the centroid $C$ (which depends only on the volume of the two parts).

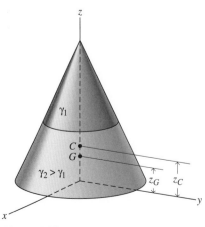

**Figure 5-39**

## Centroids of Areas

**Centroids of Areas**  If the body consists of a homogeneous, thin plate of uniform thickness $t$ and surface area $A$, then $m = \rho t A$, $dm = \rho t \, dA$, and Eq. (5-38) can be divided by $\rho t$ to get

$$x_C = \frac{1}{A} \int x \, dA \qquad y_C = \frac{1}{A} \int y \, dA \qquad z_C = \frac{1}{A} \int z \, dA \qquad (5\text{-}40)$$

For a thin three-dimensional shell (such as the dome covering a football stadium), three coordinates $x_C$, $y_C$, and $z_C$ are required to specify the location of the centroid $C$ of the shell. The location of the centroid in this case often does not lie on the shell. For example, the centroid of a basketball is not located on the rubber shell of the ball but is located at the center of the hollow cavity inside the shell. For a flat plate in the $xy$-plane, only two coordinates in the plane of the plate $x_C$ and $y_C$ are required to specify the location of the centroid $C$ of the plate. Even in this case, the centroid may lie inside or outside the plane area.

## Centroids of Lines

**Centroids of Lines**  If the body consists of a homogeneous curved wire with a small uniform cross-sectional area $A$ and length $L$, then $m = \rho A L$, $dm = \rho A \, dL$, and Eqs. (5-38) can be divided by $\rho A$ to get

$$x_C = \frac{1}{L} \int x \, dL \qquad y_C = \frac{1}{L} \int y \, dL \qquad z_C = \frac{1}{L} \int z \, dL \qquad (5\text{-}41)$$

Two or three coordinates, depending on the shape, are required to specify the location of the centroid of the line defining the shape of the wire. Unless the line is straight, the centroid will usually not lie on the line.

## Centroid, Center of Mass, or Center of Gravity by Integration

**Centroid, Center of Mass, or Center of Gravity by Integration**  The procedure involved in the determination, by integration, of the coordinates of the centroid, center of mass, or center of gravity of a body can be summarized as follows:

1. Prepare a sketch of the body approximately to scale.
2. Establish a coordinate system. Rectangular coordinates are used with most shapes that have flat planes for boundaries. Polar coordinates are usually used for shapes with circular boundaries. Whenever a line or plane of symmetry exists in a body, a coordinate axis or plane should be chosen to coincide with this line or plane. The centroid, center of mass, or center of gravity will always lie on such a line or plane since the moments of symmetrically located pairs of elements (one with a positive coordinate and the other with an equal negative coordinate) will always cancel.
3. Select an element of volume, area, or length. For center of mass or center of gravity determinations, determine the mass or weight of the element by using the appropriate expression (constant or variable) for the density or specific weight. The element can frequently be selected so that only single integration is required for the complete body or for the several parts into which the body can be divided. Sometimes, however, it may be necessary to use double integration or perhaps triple integration for some shapes. If possible, the element should be chosen so that all parts are the same distance from the reference axis or plane. This distance will be the moment arm for first-

moment determinations. When the parts of the element are at different distances from the reference axis or plane, the location of the centroid, center of mass, or center of gravity of the element must be known in order to establish the moment arm for moment calculations. Integrate the expression to determine the volume, area, length, mass, or weight of the body.

4. Write an expression for the first moment of the element with respect to one of the reference axes or planes. Integrate the expression to determine the first moment with respect to the reference axis or plane.

5. Use the appropriate equation [Eqs. (5-34), (5-35), etc.] to obtain the coordinate of the centroid, center of mass, or center of gravity with respect to the reference axis or plane.

6. Repeat steps 3 to 5, using different reference axes or planes for the other coordinates of the centroid, center of mass, or center of gravity.

7. Locate the centroid, center of mass, or center of gravity on the sketch. Gross errors are often detected by using this last step.

The following examples illustrate the procedures for locating centroids (of areas, lines, and volumes) and centers of mass or centers of gravity of bodies by integration.

## Example Problem 5-15

Locate the centroid of the rectangular area shown in Fig. 5-40a.

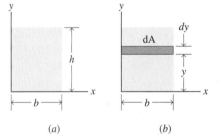

(a)                     (b)

**Figure 5-40**

### SOLUTION

Symmetry considerations require that the centroid of a rectangular area be located at the center of the rectangle. Thus, for the rectangular area shown in Fig. 5-40a, $x_C = b/2$ and $y_C = h/2$. These results are established by integration in the following fashion. For the differential element of area shown in Fig. 5-40b, $dA = b\,dy$. The element $dA$ is located a distance $y$ from the $x$-axis; therefore, the moment of the area about the $x$-axis is

$$M_x = \int_A y\,dA = \int_0^h y\,(b\,dy) = b\left[\frac{y^2}{2}\right]_0^h = \frac{bh^2}{2}$$

From Eq. (5-40):

$$y_C = \frac{M_x}{A} = \frac{bh^2/2}{bh} = \frac{h}{2} \qquad \textbf{Ans.}$$

In a similar manner by using an element of area $dA = h\,dx$, the moment of the area about the $y$-axis is

$$M_y = \int_A x\,dA = \int_0^b x\,(h\,dx) = h\left[\frac{x^2}{2}\right]_0^b = \frac{hb^2}{2}$$

From Eq. (5-40):

$$x_C = \frac{M_y}{A} = \frac{hb^2/2}{bh} = \frac{b}{2} \qquad \textbf{Ans.}$$

The element of area $dA = b \, dy$, used to calculate $M_x$, was not used to calculate $M_y$ since all parts of the horizontal strip are located at different distances $x$ from the $y$-axis. As a result of this example, it is now known that $x_C = b/2$ for the element of area $dA = b \, dy$ shown in Fig. 5-40$b$. This result will be used frequently in later examples to simplify the integrations. ∎

## Example Problem 5-16

Locate the $y$-coordinate of the centroid of the area of the quarter circle shown in Fig. 5-41$a$.

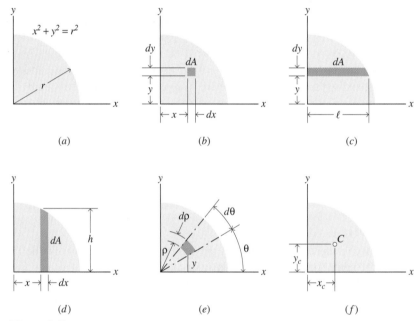

Figure 5-41

## SOLUTION

Four different elements will be used to solve this problem.

### Element 1: Double Integral using Rectangular Coordinates

For the differential element shown in Fig. 5-41$b$, $dA = dy \, dx$. The element $dA$ is at a distance $y$ from the $x$-axis; therefore, the moment of the area about the $x$-axis is

$$M_x = \int_A y \, dA = \int_0^r \int_0^{\sqrt{r^2 - x^2}} y \, dy \, dx$$

$$= \int_0^r \left[ \frac{y^2}{2} \right]_0^{\sqrt{r^2 - x^2}} dx = \int_0^r \frac{r^2 - x^2}{2} \, dx = \left[ \frac{r^2 x}{2} - \frac{x^3}{6} \right]_0^r = \frac{r^3}{3}$$

From Eq. (5-40):

$$y_C = \frac{M_x}{A} = \frac{r^3/3}{\pi r^2/4} = \frac{4r}{3\pi}$$     **Ans.**

### Element 2: Single Integral Using a Horizontal Strip

Alternatively, the element of area can be selected as shown in Fig. 5-41c. For this element, which is located a distance $y$ from the $x$-axis, $dA = \ell \, dy = \sqrt{r^2 - y^2} \, dy$. Therefore, the moment of the area about the $x$-axis is

$$M_x = \int_A y \, dA = \int_0^r y \sqrt{r^2 - y^2} \, dy = \left[ -\frac{(r^2 - y^2)^{3/2}}{3} \right]_0^r = \frac{r^3}{3}$$

From Eq. (5-40):

$$y_C = \frac{M_x}{A} = \frac{r^3/3}{\pi r^2/4} = \frac{4r}{3\pi} \qquad\qquad \textbf{Ans.}$$

### Element 3: Single Integral Using a Vertical Strip

The element of area could also be selected as shown in Fig. 5-41d. For this element, $dA = h \, dx = \sqrt{r^2 - x^2} \, dx$; however, all parts of the element are at different distances $y$ from the $x$-axis. For this type of element, the results of Example Problem 5-15 can be used to compute a moment $dM_x$ that can be integrated to yield moment $M_x$. Thus

$$dM_x = \frac{h}{2} \, dA = \frac{h}{2} h \, dx = \frac{h^2}{2} \, dx = \frac{r^2 - x^2}{2} \, dx$$

$$M_x = \int_A dM_x = \int_0^r \frac{r^2 - x^2}{2} \, dx = \left[ \frac{r^2 x}{2} - \frac{x^3}{6} \right]_0^r = \frac{r^3}{3}$$

From Eq. (5-40):

$$y_C = \frac{M_x}{A} = \frac{r^3/3}{\pi r^2/4} = \frac{4r}{3\pi} \qquad\qquad \textbf{Ans.}$$

### Element 4: Double Integral Using Polar Coordinates

Finally, polar coordinates can be used to locate the centroid of the quarter circle. With polar coordinates, the element of area is $dA = \rho \, d\theta \, d\rho$ and the distance from the $x$-axis to the element is $y = \rho \sin \theta$, as shown in Fig. 5-41e. Thus

$$M_x = \int_A y \, dA = \int_0^r \int_0^{\pi/2} \rho^2 \sin \theta \, d\theta \, d\rho$$

$$= \int_0^r \rho^2 \left[ -\cos \theta \right]_0^{\pi/2} d\rho = \int_0^r \rho^2 \, d\rho = \left[ \frac{\rho^3}{3} \right]_0^r = \frac{r^3}{3}$$

From Eq. (5-40):

$$y_C = \frac{M_x}{A} = \frac{r^3/3}{\pi r^2/4} = \frac{4r}{3\pi} \qquad\qquad \textbf{Ans.}$$

In a completely similar manner, the $x$-coordinate of the centroid is obtained as

$$x_C = \frac{M_y}{A} = \frac{r^3/3}{\pi r^2/4} = \frac{4r}{3\pi}$$

The results are illustrated in Fig. 5-41f. ∎

## Example Problem 5-17

A circular arc of thin homogeneous wire is shown in Fig. 5-42a.

(a) Locate the x- and y-coordinates of the mass center.
(b) Use the results of part a to determine the coordinates of the mass center for a half circle.

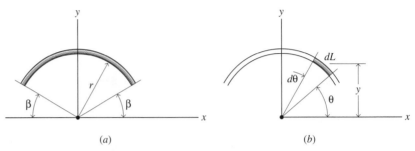

(a)                              (b)

**Figure 5-42**

### SOLUTION

(a) The wire can be assumed to consist of a large number of differential elements of length $dL$, as shown in Fig. 5-42b. The mass of each of these elements is

$$dm = \rho \, dV = \rho A \, dL = \rho A(r \, d\theta)$$

Therefore, the total mass of the wire is

$$m = \int dm = \int_{\beta}^{\pi - \beta} \rho A r \, d\theta = \rho A r \int_{\beta}^{\pi - \beta} d\theta = \rho A r(\pi - 2\beta)$$

The distance $y$ from the x-axis to the element $dm$ of mass is $y = r \sin \theta$. Thus,

$$my_G = \int y \, dm = \int_{\beta}^{\pi - \beta} (r \sin \theta)(\rho A r \, d\theta)$$

$$= \rho A r^2 \int_{\beta}^{\pi - \beta} \sin \theta \, d\theta = \rho A r^2 (2 \cos \beta)$$

where $A$ is the cross-sectional area of the wire. Therefore,

$$y_G = \frac{2\rho A r^2 \cos \beta}{\rho A r(\pi - 2\beta)} = \frac{2r \cos \beta}{\pi - 2\beta} \qquad \text{Ans.}$$

Since the length of wire is symmetric about the y-axis,

$$x_G = 0 \qquad \text{Ans.}$$

(b) For the half circle, $\beta = 0$

$$y_G = \frac{2r}{\pi} \qquad \text{Ans.}$$

$$x_G = 0 \quad \blacksquare \qquad \text{Ans.}$$

### Example Problem 5-18

Locate the center of gravity $G$ of the homogeneous right circular cone shown in Fig. 5-43a, which has an altitude $h$, radius $r$, and is made of a material with a specific weight $\gamma$.

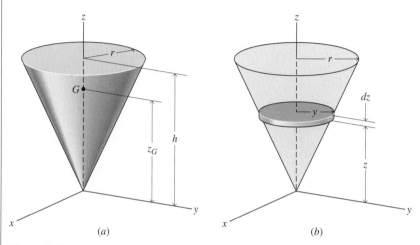

(a)                    (b)

**Figure 5-43**

### SOLUTION

From symmetry, it is obvious that $x_G = y_G = 0$. The coordinate $z_G$ of the center of gravity $G$ of the cone can be located by using the differential element of volume shown in Fig. 5-43b. The weight $dW$ of the differential element is

$$dW = \gamma\, dV = \gamma\, (\pi y^2)\, dz = \gamma\pi \left[\frac{rz}{h}\right]^2 dz = \frac{\gamma\pi r^2}{h^2}\, z^2 dz$$

From Eq. (5-35):

$$z_G = \frac{1}{W} \int z\, dW$$

Thus,

$$W z_G = \int z\, dW = \int_0^h \frac{\gamma\pi r^2}{h^2}\, z^3 dz = \frac{1}{4}\gamma\pi r^2 h^2$$

The weight of the cone is

$$W = \int_V \gamma\, dV = \int_0^h \frac{\gamma\pi r^2}{h^2}\, z^2 dz = \frac{\gamma\pi r^2}{h^2} \left[\frac{z^3}{3}\right]_0^h = \frac{1}{3}\gamma\pi r^2 h$$

Therefore

$$z_G = \frac{\gamma\pi r^2 h^2/4}{\gamma\pi r^2 h/3} = \frac{3h}{4} \qquad \textbf{Ans.}$$

Since the $xz$-plane and the $yz$-plane are planes of symmetry,

$$x_G = y_G = 0 \quad\blacksquare \qquad \textbf{Ans.}$$

## ◼ PROBLEMS

### Introductory Problems

**5-76\***  Locate the centroid of the shaded area shown in Fig. P5-76 if $b = 200$ mm and $h = 300$ mm.

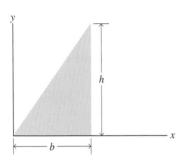

**Figure P5-76**

**5-77\***  Determine the $y$-coordinate of the centroid of the shaded area shown in Fig. P5-77.

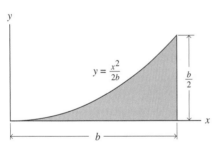

**Figure P5-77**

**5-78**  Determine the $x$-coordinate of the centroid of the shaded area shown in Fig. P5-78.

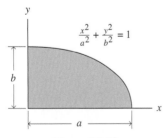

**Figure P5-78**

**5-79**  Locate the centroid of the shaded area shown in Fig. P5-79.

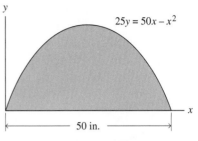

**Figure P5-79**

**5-80\***  Determine the $y$-coordinate of the centroid of the shaded area shown in Fig. P5-80.

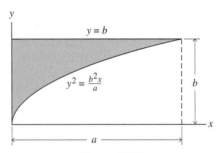

**Figure P5-80**

### Intermediate Problems

**5-81\***  Locate the centroid of the shaded area shown in Fig. P5-81.

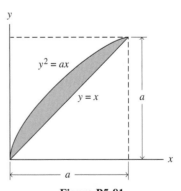

**Figure P5-81**

**5-82\***  Determine the $y$-coordinate of the centroid of the shaded area shown in Fig. P5-82.

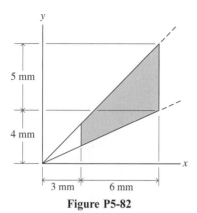

Figure P5-82

**5-83** Determine the $x$-coordinate of the shaded area shown in Fig. P5-83.

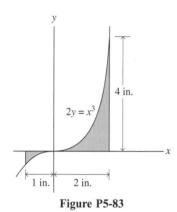

Figure P5-83

**5-84** Locate the centroid of the shaded area shown in Fig. P5-84.

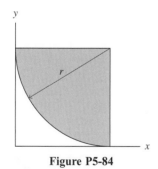

Figure P5-84

**5-85\*** Locate the centroid of the volume obtained by revolving the shaded area shown in Fig. P5-85 about the $x$-axis.

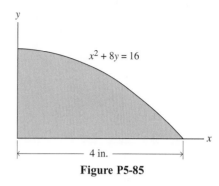

Figure P5-85

**5-86** Locate the centroid of the volume obtained by revolving the shaded area shown in Fig. P5-86 about the $x$-axis.

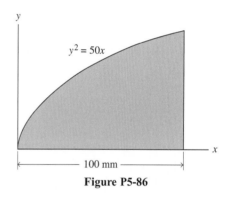

Figure P5-86

**Challenging Problems**

**5-87** Locate the centroid of the curved homogeneous slender rod shown in Fig. P5-87.

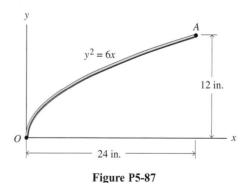

Figure P5-87

**5-88\*** Locate the centroid of the curved homogeneous slender rod shown in Fig. P5-88 if $b = 50$ mm.

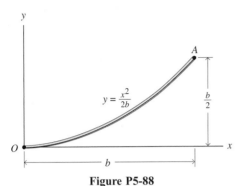

$$y = \frac{x^2}{2b}$$

**Figure P5-88**

**5-89\*** Locate the centroid of the volume of the portion of a right circular cone shown in Fig. P5-89.

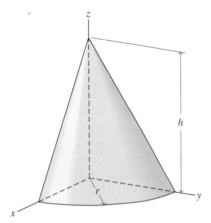

**Figure P5-89**

**5-90\*** Locate the mass center of the hemisphere shown in Fig. P5-90 if the density $\rho$ at any point $P$ is proportional to the distance from the $xy$-plane to the point $P$.

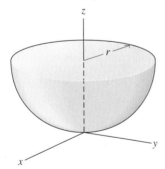

**Figure P5-90**

**5-91** Locate the mass center of the right circular cone shown in Fig. P5-91 if the density $\rho$ at any point $P$ is proportional to the distance from the $xy$-plane to the point $P$.

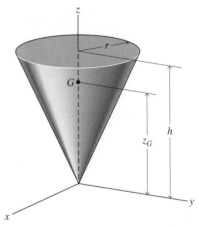

**Figure P5-91**

**5-92** Locate the centroid of the volume of the tetrahedron shown in Fig. P5-92.

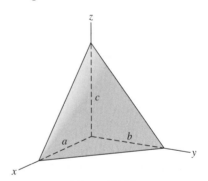

**Figure P5-92**

## 5-9 CENTROIDS OF COMPOSITE BODIES

The centroid of an area, line, or volume is a property of the shape of the area, line, or volume and is independent of the coordinate system used to compute it. For example, no matter what coordinate system is used to compute the location of the centroid of a rectangle, it will always be found to be halfway between the opposite sides, as shown in Fig. 5-44a. Similarly, the centroid of a quarter circle

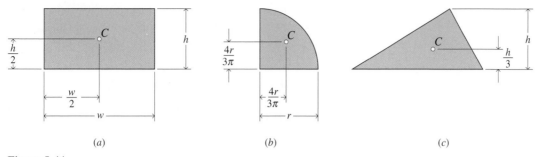

(a)  (b)  (c)

**Figure 5-44**

will always be $4r/3\pi$ from the radial lines bounding the quarter circle (Fig. 5-44b) and the centroid of a triangle will always be one third of the way (measured perpendicularly to the side) from any side of the triangle to the opposite vertex (Fig. 5-44c). Once the centroids of these common shapes are known, they need not be continually recalculated. In fact, the known location of the centroids of common simple shapes can be used to locate the centroids of more complicated shapes.

For example, the area shown in Fig. 5-45 consists of two parts—a triangle and a rectangle. Integrating to find the area and the centroid gives

$$A = \int_{x=0}^{x=a} dA + \int_{x=a}^{x=b} dA = A_{\text{tri}} + A_{\text{rect}}$$

$$M_y = x_C A = \int_{x=0}^{x=a} x\, dA + \int_{x=a}^{x=b} x\, dA$$

$$= (x_C A)_{\text{tri}} + (x_C A)_{\text{rect}}$$

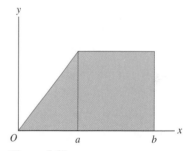

**Figure 5-45**

The location of the centroid of the composite area, $x_C$, consisting of the triangle and rectangle is the sum of the first moments of the two component parts divided by the total area of the two component parts. In general, if an area is composed of $n$ simple shapes whose centroid locations, $x_{Ci}$, are known (or can be looked up easily), the centroid of the larger area can be found from

$$x_C A = \sum_{i=1}^{n} x_{Ci} A_i \qquad A = \sum_{i=1}^{n} A_i \qquad (5\text{-}42)$$

Composite areas with holes are also handled easily by using Eq. (5-42). Consider the composite area of Fig. 5-46b, which consists of a circular hole cut out of a rectangle and triangle. Denoting the triangle by $t$, the rectangle by $r$, the circular hole by $h$, and the composite body by $c$, the centroid location of the solid triangle–rectangle area (Fig. 5-46a) is found from

$$(x_C A)_t + (x_C A)_r = (x_C A)_c + (x_C A)_h$$

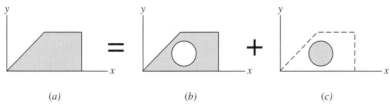

(a)  (b)  (c)

**Figure 5-46**

The first, second, and last terms are easily computed using known properties of simple shapes. Therefore, the centroid of the composite body can be found from

$$(x_C A)_c = (x_C A)_t + (x_C A)_r - (x_C A)_h$$

$$A_c = A_t + A_r - A_h$$

These equations are identical to Eq. (5-42) if the area of the hole is considered to be a negative quantity.

Similar equations can be developed for composite lines, volumes, masses, and weights. The final results would show the $A$'s of Eqs. (5-42) replaced with $L$'s, $V$'s, $m$'s, and $W$'s, respectively. Tables 5-1 and 5-2 contain a listing of centroid locations for some common shapes.

TABLE 5-1 Centroid Locations For A Few Common Line Segments And Areas

Circular arc

$$L = 2r\alpha$$

$$x_C = \frac{r \sin \alpha}{\alpha}$$

$$y_C = 0$$

Circular sector

$$A = r^2\alpha$$

$$x_C = \frac{2r \sin \alpha}{3\alpha}$$

$$y_C = 0$$

Quarter circular arc

$$L = \frac{\pi r}{2}$$

$$x_C = \frac{2r}{\pi}$$

$$y_C = \frac{2r}{\pi}$$

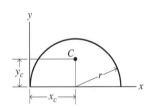

Quadrant of a circle

$$A = \frac{\pi r^2}{4}$$

$$x_C = \frac{4r}{3\pi}$$

$$y_C = \frac{4r}{3\pi}$$

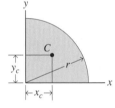

Semicircular arc

$$L = \pi r$$

$$x_C = r$$

$$y_C = \frac{2r}{\pi}$$

Semicircular area

$$A = \frac{\pi r^2}{2}$$

$$x_C = r$$

$$y_C = \frac{4r}{3\pi}$$

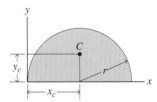

Rectangular area

$$A = bh$$

$$x_C = \frac{b}{2}$$

$$y_C = \frac{h}{2}$$

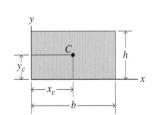

Quadrant of an ellipse

$$A = \frac{\pi ab}{4}$$

$$x_C = \frac{4a}{3\pi}$$

$$y_C = \frac{4b}{3\pi}$$

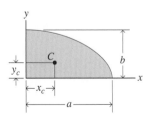

TABLE 5-1 (Continued)

Triangular area

$$A = \frac{bh}{2}$$

$$x_C = \frac{2b}{3}$$

$$y_C = \frac{h}{3}$$

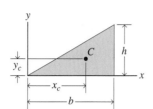

Parabolic spandrel

$$A = \frac{bh}{3}$$

$$x_C = \frac{3b}{4}$$

$$y_C = \frac{3h}{10}$$

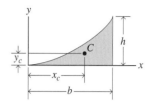

Triangular area

$$A = \frac{bh}{2}$$

$$x_C = \frac{a + b}{3}$$

$$y_C = \frac{h}{3}$$

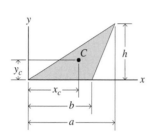

Quadrant of a parabola

$$A = \frac{2bh}{3}$$

$$x_C = \frac{5b}{8}$$

$$y_C = \frac{2h}{5}$$

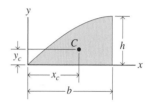

TABLE 5-2 Centroid Locations For A Few Common Volumes

Rectangular parallelepiped

$$V = abc$$

$$x_C = \frac{a}{2}$$

$$y_C = \frac{b}{2}$$

$$z_C = \frac{c}{2}$$

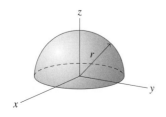

Rectangular tetrahedron

$$V = \frac{abc}{6}$$

$$x_C = \frac{a}{4}$$

$$y_C = \frac{b}{4}$$

$$z_C = \frac{c}{4}$$

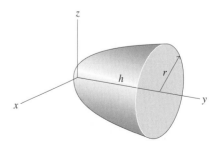

Circular cylinder

$$V = \pi r^2 L$$

$$x_C = 0$$

$$y_C = \frac{L}{2}$$

$$z_C = 0$$

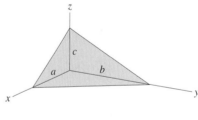

Semicylinder

$$V = \frac{\pi r^2 L}{2}$$

$$x_C = 0$$

$$y_C = \frac{L}{2}$$

$$z_C = \frac{4r}{3\pi}$$

Hemisphere

$$V = \frac{2\pi r^3}{3}$$

$$x_C = 0$$

$$y_C = 0$$

$$z_C = \frac{3r}{8}$$

Paraboloid

$$V = \frac{\pi r^2 h}{2}$$

$$x_C = 0$$

$$y_C = \frac{2h}{3}$$

$$z_C = 0$$

TABLE 5-2 (Continued)

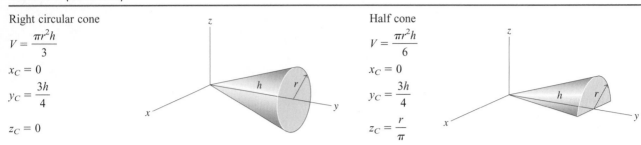

Right circular cone

$$V = \frac{\pi r^2 h}{3}$$

$x_C = 0$

$$y_C = \frac{3h}{4}$$

$z_C = 0$

Half cone

$$V = \frac{\pi r^2 h}{6}$$

$x_C = 0$

$$y_C = \frac{3h}{4}$$

$$z_C = \frac{r}{\pi}$$

The following examples illustrate the procedure for determining the locations of centroids of composite lines, areas, and volumes and centers of mass and centers of gravity for composite bodies.

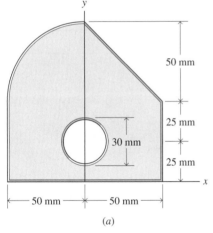

(a)

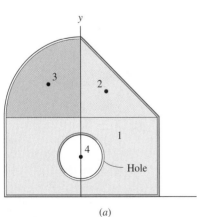

(a)

**Figure 5-47**

## Example Problem 5-19

Locate the centroid of the composite area shown in Fig. 5-47a.

### SOLUTION

The composite area can be divided into four parts: a rectangle, a triangle, a quarter circle, and a circle, as shown in Fig. 5-47b. Recall that the area of the hole is negative since it must be subtracted from the area of the rectangle. The centroid locations for each of these parts can be determined from the relationships listed in Table 5-1:

For the triangle:

$$x_C = \frac{b}{3} = \frac{50}{3} = 16.67 \text{ mm}$$

$$y_C = 50 + \frac{h}{3} = 50 + \frac{50}{3} = 66.67 \text{ mm}$$

For the quarter circle:

$$x_C = -\frac{4r}{3\pi} = -\frac{4(50)}{3\pi} = -21.22 \text{ mm}$$

$$y_C = 50 + \frac{4r}{3\pi} = 50 + \frac{4(50)}{3\pi} = 71.22 \text{ mm}$$

The centroid for the composite area is determined by listing the area, centroid location, and first moments for the individual parts in a table and applying Eqs. (5-42). Thus,

| Part | $A_i$ (mm²) | $x_{Ci}$ (mm) | $M_y$ (mm³) | $y_{Ci}$ (mm) | $M_x$ (mm³) |
|------|------------|---------------|-------------|---------------|-------------|
| 1 | 5000 | 0 | 0 | 25 | 125,000 |
| 2 | 1250 | 16.67 | 20,838 | 66.67 | 83,338 |
| 3 | 1963 | −21.22 | −41,655 | 71.22 | 139,805 |
| 4 | −707 | 0 | 0 | 25 | −17,675 |
|   | 7506 |   | −20,817 |   | 330,468 |

From Eqs. (5-42),

$$x_C = \frac{M_y}{A} = \frac{-20,817}{7506} = -2.77 \text{ mm}$$  **Ans.**

$$y_C = \frac{M_x}{A} = \frac{330,468}{7506} = 44.0 \text{ mm}$$  ■  **Ans.**

## ▌ Example Problem 5-20

A slender steel rod is bent into the shape shown in Fig. 5-48a. Locate the centroid of the rod.

### SOLUTION

The rod can be divided into three parts as shown in Fig. 5-48b. The centroid locations for each of these parts are known or can be determined from the relationships listed in Table 5-1. For the semicircular arc,

$$L_3 = \pi r = \pi(9.9) = 31.1 \text{ in.}$$

$$y_{C3} = 7 + \frac{r \sin \alpha}{\alpha} \cos 45° = 7 + \frac{9.9 \sin (\pi/2)}{\pi/2} \cos 45° = 11.457 \text{ in.}$$

$$z_{C3} = 7 + \frac{r \sin \alpha}{\alpha} \sin 45° = 7 + \frac{9.9 \sin (\pi/2)}{\pi/2} \sin 45° = 11.457 \text{ in.}$$

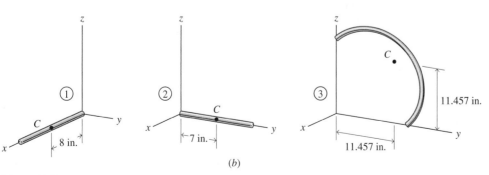

(b)

**Figure 5-48**

The centroid for the composite rod can be determined by listing the length, centroid location, and first moments for the individual parts in a table and applying Eqs. (5-42). Thus,

| Part | $L_i$ (in.) | $x_{Ci}$ (in.) | $M_{yz}$ (in²) | $y_{Ci}$ (in.) | $M_{zx}$ (in²) | $z_{Ci}$ (in.) | $M_{xy}$ (in²) |
|------|-------------|----------------|----------------|----------------|----------------|----------------|----------------|
| 1 | 16 | 8 | 128 | 0 | 0 | 0 | 0 |
| 2 | 14 | 0 | 0 | 7 | 98 | 0 | 0 |
| 3 | 31.1 | 0 | 0 | 11.457 | 356.3 | 11.457 | 356.3 |
|   | 61.1 |   | 128 |   | 454.3 |   | 356.3 |

From Eqs. (5-42),

$$x_C = \frac{M_{yz}}{L} = \frac{128}{61.1} = 2.09 \text{ in.} \qquad \textbf{Ans.}$$

$$y_C = \frac{M_{zx}}{L} = \frac{454.3}{61.1} = 7.44 \text{ in.} \qquad \textbf{Ans.}$$

$$z_C = \frac{M_{xy}}{L} = \frac{356.3}{61.1} = 5.83 \text{ in.} \ \blacksquare \qquad \textbf{Ans.}$$

## Example Problem 5-21

A cylinder with a hemispherical cavity and a conical cap is shown in Fig. 5-49.

(a) Locate the centroid of the composite volume if $R = 140$ mm, $L = 250$ mm, and $h = 300$ mm.

(b) Locate the center of mass of the composite volume if the cylinder is made of steel ($\rho = 7870$ kg/m³) and the cap is made of aluminum ($\rho = 2770$ kg/m³).

**SOLUTION**

The composite body can be divided into three parts: a cone, a cylinder, and a hemisphere. The volume of the hemisphere is negative, since it must be subtracted from the volume of the cylinder. The $xz$- and $yz$-planes are planes of symmetry for both the volume of the composite body and for the mass of the composite body. Therefore, $x_C$, $y_C$, $x_G$, and $y_G$ are all zero.

(a) The $z$-coordinate of the centroid is given by Eq. (5-42), with $A$ replaced with $V$:

$$z_C V = \sum_{i=1}^{3} z_{Ci} V_i \qquad V = \sum_{i=1}^{3} V_i \qquad \text{(a)}$$

Centroid locations and volumes of the three pieces are determined using Table 5-2. For the cone,

$$z_{C1} = L + \left(h - \frac{3h}{4}\right) = 250 + \frac{300}{4} = 325 \text{ mm}$$

$$V_1 = \frac{\pi R^2 h}{3} = \frac{\pi (140)^2 (300)}{3} = 6.158(10^6) \text{ mm}^3$$

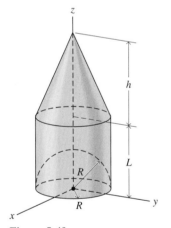

**Figure 5-49**

For the cylinder,

$$z_{C2} = \frac{L}{2} = \frac{250}{2} = 125 \text{ mm}$$

$$V_2 = \pi R^2 L = \pi(140)^2 (250) = 15.394(10^6) \text{ mm}^3$$

For the hemisphere,

$$z_{C3} = \frac{3R}{8} = \frac{3(140)}{8} = 52.5 \text{ mm}$$

$$V_3 = -\frac{2\pi R^3}{3} = -\frac{2\pi(140)^3}{3} = -5.747(10^6) \text{ mm}^3$$

Using Eq. (a) for the composite body gives

$$V = 15.805(10^6) \text{ mm}^3$$

$$z_C(15.805)(10^6) = 325(6.158)(10^6) + 125(15.394)(10^6)$$
$$+52.5(-5.747)(10^6)$$

Therefore,

$$z_C = 229 \text{ mm} \qquad\qquad \textbf{Ans.}$$

and from symmetry

$$x_C = y_C = 0 \text{ mm} \qquad\qquad \textbf{Ans.}$$

(b) The center of mass is found using

$$z_G m = \sum_{i=1}^{3} z_{Gi} m_i \qquad m = \sum_{i=1}^{3} m_i \qquad\qquad \text{(b)}$$

Even though the entire composite body is not homogeneous, each part of the body is homogeneous. Therefore, the $z_{Gi}$ for each part of the body are identical to the $z_{Ci}$ found in part $a$. For the cone,

$$m_1 = \rho V = 2770(6.158)(10^{-3}) = 17.058 \text{ kg}$$

For the cylinder,

$$m_2 = \rho V = 7870(15.394)(10^{-3}) = 121.15 \text{ kg}$$

For the hemisphere,

$$m_3 = \rho V = 7870(-5.747)(10^{-3}) = -45.23 \text{ kg}$$

Using Eq. (b) for the composite body gives

$$m = 92.98 \text{ kg}$$

$$z_G(92.98) = 325(17.058) + 125(121.15) + 52.5(-45.23)$$

Therefore,

$$z_G = 197.0 \text{ mm}$$   **Ans.**

and from symmetry

$$x_G = y_G = 0 \text{ mm} \blacksquare$$   **Ans.**

## PROBLEMS

### Introductory Problems

**5-93\***   Locate the centroid of the shaded area shown in Fig. P5-93.

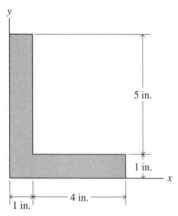

**Figure P5-93**

**5-94\***   Locate the centroid of the shaded area shown in Fig. P5-94.

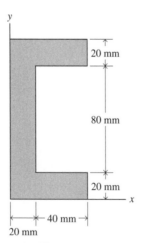

**Figure P5-94**

**5-95**   Locate the centroid of the shaded area shown in Fig. P5-95.

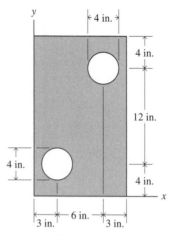

**Figure P5-95**

**5-96\***   Locate the centroid of the slender rod shown in Fig. P5-96.

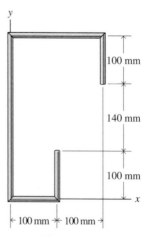

**Figure P5-96**

**5-97**   Locate the centroid of the slender rod shown in Fig. P5-97.

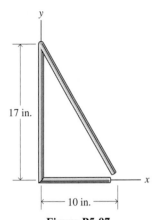

**Figure P5-97**

### Intermediate Problems

**5-98\*** Locate the centroid of the slender rod shown in Fig. P5-98.

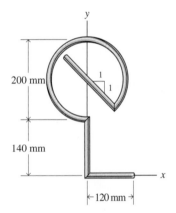

**Figure P5-98**

**5-99\*** Locate the centroid of the shaded area shown in Fig. P5-99.

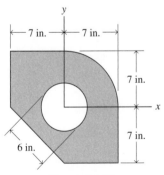

**Figure P5-99**

**5-100** Locate the centroid of the shaded area shown in Fig. P5-100.

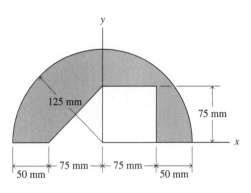

**Figure P5-100**

**5-101\*** Locate the centroid of the slender rod shown in Fig. P5-101.

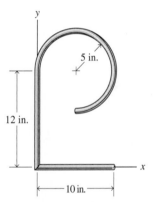

**Figure P5-101**

**5-102** Locate the centroid of the shaded area shown in Fig. P5-102.

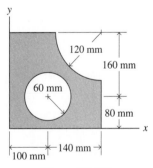

**Figure P5-102**

### Challenging Problems

**5-103\*** Locate the center of gravity of the bracket shown in Fig. P5-103 if all three holes have 6-in. diameters.

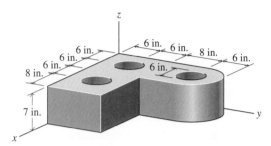

**Figure P5-103**

**5-104\***  Locate the center of mass of the machine component shown in Fig. P5-104. The brass ($\rho = 8750$ kg/m³)
disk $C$ is mounted on the steel ($\rho = 7870$ kg/m³) shafts
$A$ and $B$.

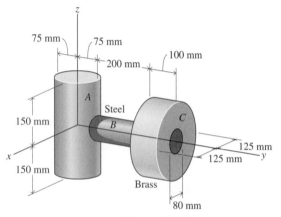

**Figure P5-104**

**5-105**  A bracket is made of brass ($\gamma = 0.316$ lb/in³) and
aluminum ($\gamma = 0.100$ lb/in³) plates as shown in Fig.
P5-105. Locate
(a)  The centroid of the bracket.
(b)  The center of gravity of the bracket.

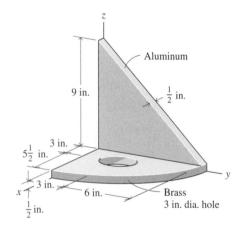

**Figure P5-105**

**5-106\***  A cylinder with a conical cavity and a hemispherical cap is shown in Fig. P5-106. Locate
(a)  The centroid of the composite volume if $R = 200$
mm, and $h = 250$ mm.
(b)  The center of gravity of the composite volume if
the cylinder is made of brass ($\rho = 8750$ kg/m³) and
the cap is made of aluminum ($\rho = 2770$ kg/m³).

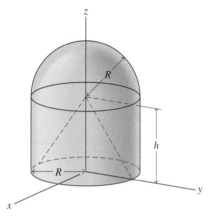

**Figure P5-106**

**5-107**  A slender rod is made of two materials; the segment
along the $x$-axis is aluminum ($\gamma = 0.100$ lb/in³) and the
remainder is brass ($\gamma = 0.316$ lb/in³). Locate the center
of gravity of the slender rod shown in Fig. P5-107.

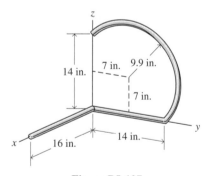

**Figure P5-107**

# 5-10 DISTRIBUTED LOADS ON STRUCTURAL MEMBERS

When a load is applied to a rigid body such as a structural member, it is often distributed along a line or over an area $A$. In many instances, it is convenient to replace this distributed load with a single equivalent force that is externally equivalent to the distributed load. The methods used to replace the distributed weight force with an equivalent single force work for any distributed force as long as the elements of the distributed force are all parallel to each other. The quantities to be determined are the magnitude of the equivalent force (the direction of the force is the same as that of all of the components) and the location of its line of action.

Consider the beam shown in Fig. 5-50$a$. A beam is a structural member whose length is large compared to its cross-sectional dimensions and which is loaded and supported in the direction transverse to the axis of the member. Since distributed loads on beams do not vary across the width of the cross section of the beam, the actual load intensity on the beam can be multiplied by the width of the beam to yield a distributed line load $w$ (N/m or lb/ft) whose magnitude

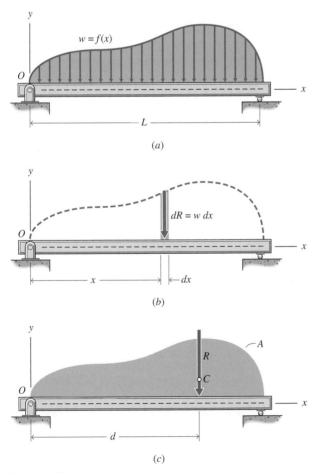

Figure 5-50

varies only with position $x$ along the beam. The distributed load $w$ versus position $x$ graph (see Fig. 5-50a) is known as a *load diagram*. The magnitude of the differential force $d\mathbf{R}$ exerted on the beam by the distributed load $w$ in an increment of length $dx$ (see Fig. 5-50b) is

$$dR = w \, dx$$

Therefore, the magnitude of the single concentrated resultant force $\mathbf{R}$ that is equivalent to the distributed load $w$ is

$$R = \int_0^L w \, dx \tag{5-43}$$

That is, the magnitude of the resultant force $\mathbf{R}$ is equal to the area $A$ under the load diagram, as shown in Fig. 5-50c.

The location of the line of action of the equivalent force $\mathbf{R}$ can be determined (using the principle of moments) by equating the moment of the equivalent force $\mathbf{R}$ about an arbitrary point $O$ to the moment of the distributed load about the same point $O$. The moment produced by the force $dR = w \, dx$ about point $O$ of Fig. 5-50b is

$$dM_O = x \, dR$$

and the total moment produced by the distributed load $w$ about point $O$ is

$$M_O = \int dM_O = \int x \, dR = x_C R \tag{a}$$

where $x_C$ is the distance along the beam from point $O$ to the centroid $C$ of the area $A$ under the load diagram. The moment produced by the equivalent force $\mathbf{R}$ about point $O$ is

$$M_O = Rd \tag{b}$$

where $d$ is the distance along the beam from point $O$ to the line of action of the equivalent force $\mathbf{R}$. Setting Eqs. (a) and (b) equal gives

$$M_O = Rd = \int x \, dR = x_C R$$

or

$$d = x_C = \frac{M_O}{R} \tag{5-44}$$

That is, for the purpose of calculating the sum of forces and the sum of moments, the distributed force is equivalent to a single force whose magnitude is equal to the area under the load diagram and whose the line of action passes through the centroid of the load diagram.

The following examples illustrate the procedure for determining the single concentrated resultant force $\mathbf{R}$ that is equivalent to the distributed load $w$ and the location of its line of action.

## Example Problem 5-22

A beam is subjected to a system of loads that can be represented by the load diagram shown in Fig. 5-51a. Determine the resultant of this system of distributed loads and locate its line of action with respect to the left support of the beam.

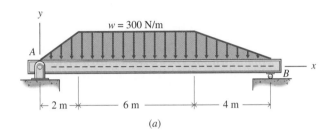

(a)

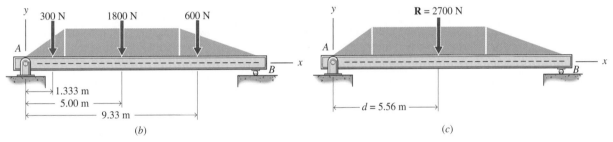

(b)                                                    (c)

**Figure 5-51**

## SOLUTION

The magnitude of the resultant **R** of the distributed load shown in Fig. 5-51a is equal to the area under the load diagram. The load diagram can be divided into two triangles and a rectangle. Thus, from the relationships listed in Table 5-1:

$$F_1 = A_1 = \frac{1}{2}b_1h_1 = \frac{1}{2}(2)(300) = 300 \text{ N}$$

$$x_{C1} = \frac{2}{3}b_1 = \frac{2}{3}(2) = 1.333 \text{ m}$$

$$F_2 = A_2 = b_2h_2 = (6)(300) = 1800 \text{ N}$$

$$x_{C2} = 2 + \frac{1}{2}b_2 = 2 + \frac{1}{2}(6) = 5.00 \text{ m}$$

$$F_3 = A_3 = \frac{1}{2}b_3h_3 = \frac{1}{2}(4)(300) = 600 \text{ N}$$

$$x_{C3} = 8 + \frac{1}{3}b_3 = 8 + \frac{1}{3}(4) = 9.33 \text{ m}$$

The equivalent forces for the three different areas and the locations of their lines of action are shown in Fig. 5-51b. Thus,

$$R = F_1 + F_2 + F_3 = 300 + 1800 + 600 = 2700 \text{ N} \qquad \textbf{Ans.}$$

The line of action of the resultant with respect to the left support is located by summing moments about point $A$. Thus,

$$M_A = Rd = F_1 x_{C1} + F_2 x_{C2} + F_3 x_{C3}$$

$$= 300(1.333) + 1800(5.00) + 600(9.33) = 15,000 \text{ N} \cdot \text{m}$$

Finally,

$$d = x_C = \frac{M_A}{R} = \frac{15,000}{2700} = 5.56 \text{ m} \qquad \text{**Ans.**}$$

The resultant force $\mathbf{R}$ and its line of action are shown in Fig. 5-51$c$. ∎

## Example Problem 5-23

A beam is subjected to a system of loads that can be represented by the load diagram shown in Fig. 5-52$a$. Determine the resultant of this system of loads and locate its line of action with respect to the left support of the beam.

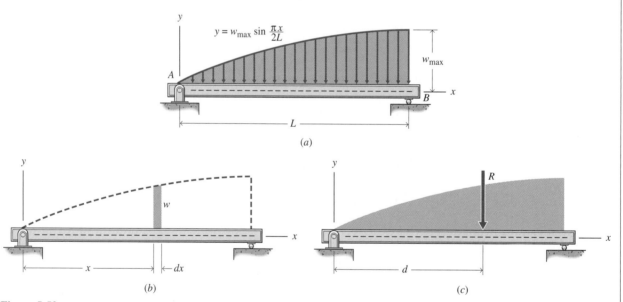

**Figure 5-52**

## SOLUTION

The magnitude of the resultant $\mathbf{R}$ of the distributed load shown in Fig. 5-52$a$ is equal to the area under the load diagram, and the line of action of the resultant passes through the centroid of the area. Since the area under this load diagram and the location of its centroid do not normally appear in tables of areas and centroid locations, it is necessary to use integration methods to determine the magnitude of the resultant and the location of its line of action.

The area under the load diagram is determined by using the element of area shown in Fig. 5-52*b*. Thus,

$$A = \int_A w \, dx = \int_0^L w_{max} \sin \frac{\pi x}{2L} \, dx = \frac{2Lw_{max}}{\pi}\left[-\cos \frac{\pi x}{2L}\right]_0^L = \frac{2Lw_{max}}{\pi}$$

Thus,

$$R = A = \frac{2Lw_{max}}{\pi} = 0.637 w_{max}L \qquad \text{Ans.}$$

The moment of the area about support $A$ is

$$M_A = \int_A x \, (w \, dx) = \int_0^L w_{max} \, x \sin \frac{\pi x}{2L} \, dx$$

$$= w_{max}\left[\frac{4L^2}{\pi^2} \sin \frac{\pi x}{2L} - \frac{2L}{\pi} x \cos \frac{\pi x}{2L}\right]_0^L = \frac{4L^2 w_{max}}{\pi^2}$$

From Eq. (5-44):

$$d = x_C = \frac{M_A}{A} = \frac{4L^2 w_{max}/\pi^2}{2Lw_{max}/\pi} = \frac{2L}{\pi} = 0.637 \, L \qquad \text{Ans.}$$

The results are shown in Fig. 5-52*c*. ∎

## PROBLEMS

### Introductory Problems

**5-108\*** The loads acting on a beam are distributed in a triangular manner as shown in Fig. P5-108. Determine and locate the resultant with respect to the left end of the beam.

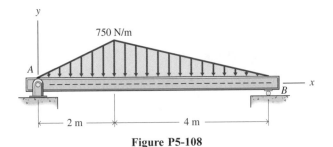

**Figure P5-108**

**5-109\*** Determine the resultant of the distributed loads acting on the beam shown in Fig. P5-109, and locate its line of action with respect to the support at $A$.

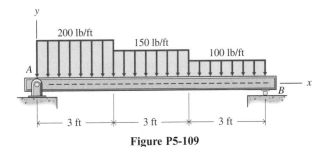

**Figure P5-109**

**5-110** A distributed load acts on the beam shown in Fig. P5-110. Determine the resultant of the distributed load and locate its line of action with respect to the support at $A$.

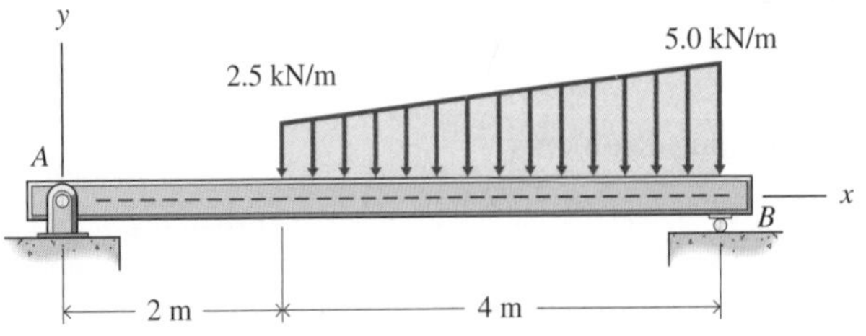

**Figure P5-110**

## Intermediate Problems

**5-111\***  A distributed load acts on a beam as shown in Fig. P5-111. Determine and locate the resultant of the distributed load with respect to the support at $A$.

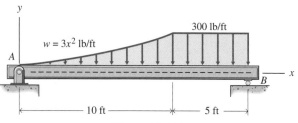

**Figure P5-111**

**5-112\***  Determine the resultant of the distributed load acting on the beam shown in Fig. P5-112 and locate its line of action with respect to the support.

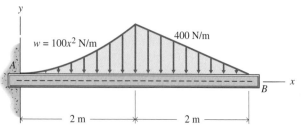

**Figure P5-112**

**5-113**  Determine the resultant of the distributed load acting on the beam shown in Fig. P5-113 and locate its line of action with respect to the support.

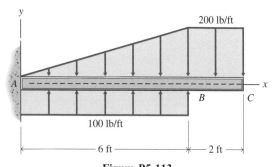

**Figure P5-113**

## Challenging Problems

**5-114\***  A gate is used to hold water as shown in Fig. P5-114. If the width of the gate is 2 m, determine the resultant of the water pressure acting on the gate and locate its line of action with respect to the bottom of the gate.

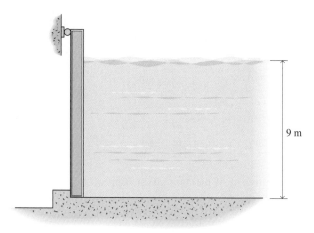

**Figure P5-114**

**5-115\***  A flexible cable is used to tether the balloon shown in Fig. P5-115. The cable weighs 1.2 lb/ft along its length. Determine the magnitude of the resultant force and its location with respect to $A$.

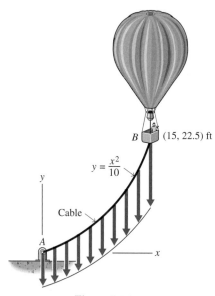

**Figure P5-115**

**5-116**  A loaded rivet causes a force distribution or pressure $p$ on a plate, as shown in Fig. P5-116. The rivet has a diameter of 25 mm, and the plate is 15 mm thick. Determine the magnitude of the resultant force acting on the plate. Let $p_o = 400$ N/m$^2$.

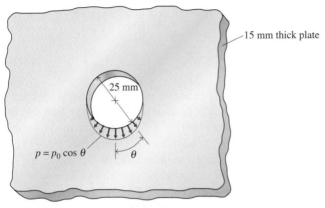

Figure P5-116

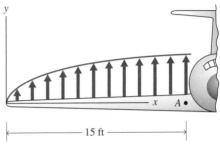

**5-117** The lift on the wing of an airplane due to aerodynamic forces is shown in Fig. P5-117. The lift $L$ may be described by the function $L = 25 \sqrt{x}$ lb/ft. Determine the moment of the resultant lift force about point $A$.

Figure P5-117

## 5-11 SUMMARY

For a three-dimensional body, the particle idealization discussed in Chapter 3 is not valid, in general, because the forces acting on the body are usually not concurrent. For a general force system, $\mathbf{R} = 0$ is a necessary but not a sufficient condition for equilibrium of the body. A second restriction related to the tendency of a force to produce rotation of a body must also be satisfied and gives rise to the concept of a moment. In this chapter, the moment of a force about a point and the moment of a force about a line (axis) were defined and methods were developed for finding the resultant forces and the resultant moments (couples) for any general force system that may be applied to a body.

A moment is a vector quantity because it has both a magnitude and a direction, and adds according to the parallelogram law of addition. The moment of a force $\mathbf{F}$ about a point $O$ can be represented by the vector cross product

$$\mathbf{M}_O = \mathbf{r} \times \mathbf{F} \tag{5-3}$$

where $\mathbf{r}$ is a position vector from point $O$ to any point on the line of action of the force $\mathbf{F}$. The cross product of the two intersecting vectors $\mathbf{r}$ and $\mathbf{F}$ is

$$\mathbf{M}_O = \mathbf{r} \times \mathbf{F} = M_x \mathbf{i} + M_y \mathbf{j} + M_z \mathbf{k} = M_O \mathbf{e} \tag{5-11}$$

where

$$M_x = r_y F_z - r_z F_y \qquad M_y = r_z F_x - r_x F_z \qquad M_z = r_x F_y - r_y F_x$$

are the three scalar components of the moment. The magnitude of the moment $\mathbf{M}_O$ is

$$M_O = |\mathbf{M}_O| = \sqrt{M_x^2 + M_y^2 + M_z^2} \tag{5-13}$$

The direction cosines associated with the unit vector $\mathbf{e}$ are

$$\cos \theta_x = \frac{M_x}{|\mathbf{M}_O|} \qquad \cos \theta_y = \frac{M_y}{|\mathbf{M}_O|} \qquad \cos \theta_z = \frac{M_z}{|\mathbf{M}_O|} \tag{5-16}$$

The moment $\mathbf{M}_{OB}$ of a force $\mathbf{F}$ about a line $OB$ in a direction $n$ specified by the unit vector $\mathbf{e}_n$ is

$$\mathbf{M}_{OB} = (\mathbf{M}_O \cdot \mathbf{e}_n) \, \mathbf{e}_n = [(\mathbf{r} \times \mathbf{F}) \cdot \mathbf{e}_n] \, \mathbf{e}_n = M_{OB} \, \mathbf{e}_n \qquad (5\text{-}18)$$

Two equal parallel forces of opposite sense are called a couple. A couple has no tendency to translate a body in any direction but tends only to rotate the body on which it acts. Several transformations of a couple can be made without changing any of the external effects of the couple on the body. A couple (1) can be translated to a parallel position in its plane or to any parallel plane and (2) can be rotated in its plane. Also, (3) the magnitude $F$ of the two forces of a couple and the distance $d$ between them can be changed provided the product $Fd$ remains constant. Any system of couples in a plane or in space can be combined into a single resultant couple $\mathbf{C}$.

Any force $\mathbf{F}$ can be resolved into a parallel force $\mathbf{F}$ and a couple $\mathbf{C}$. Alternatively, a force and a couple in the same plane can be combined into a single force. The sole effect of combining a couple with a force is to move the action line of the force into a parallel position. The magnitude and sense of the force remain unchanged.

The resultant of a force system acting on a rigid body is the simplest force system that can replace the original system without altering the external effect of the system on the body. For a coplanar system of forces, the resultant is either a force $\mathbf{R}$ or a couple $\mathbf{C}$. For a three-dimensional system of forces, the resultant may be a force $\mathbf{R}$, a couple $\mathbf{C}$, or both a force $\mathbf{R}$ and a couple $\mathbf{C}$.

In many instances, surface loads on a body are not concentrated at a point but are distributed along a length or over an area. Other forces, known as body forces, are distributed over the volume of the body. A distributed force at any point is characterized by its intensity and its direction.

Previously, moments of forces about points or axes were considered. In engineering analysis, equations are also encountered that represent moments of masses, forces, volumes, areas, or lines with respect to axes or planes. Such moments are called first moments of the quantity being considered, since the first power of a distance is used in the expression.

The term *center of mass* is used to denote the point in a physical body where the mass can be conceived to be concentrated so that the moment of the concentrated mass with respect to an axis or plane equals the moment of the distributed mass with respect to the same axis or plane. The term *center of gravity* is used to denote the point in the body through which the weight of the body acts, regardless of the position (or orientation) of the body. The location of the center of gravity in a body is determined by using equations of the form

$$M_{yz} = Wx_G = \int x \, dW \qquad \text{or} \qquad x_G = \frac{1}{W} \int x \, dW \qquad (5\text{-}35b)$$

If the density $\rho$ of a body is constant, Eqs. (5-35) reduce to

$$x_C = \frac{1}{V} \int x \, dV \qquad y_C = \frac{1}{V} \int y \, dV \qquad z_C = \frac{1}{V} \int z \, dV \qquad (5\text{-}39)$$

Equations (5-39) indicate that the coordinates $x_C$, $y_C$, and $z_C$ depend only on the geometry of the body and are independent of the physical properties. The point located by such a set of coordinates is known as the centroid of the volume of the body. The term *centroid* is usually used in connection with geometrical figures (volumes, areas, and lines), whereas the terms *center of mass* and *center of gravity* are used in connection with physical bodies. The centroid of a volume has the same position as the center of gravity of the body if the body is homogeneous. If the density is variable, the center of gravity of the body and the centroid of the volume will be at different points.

When a load, applied to a rigid body such as a structural member, is distributed along a line or over an area $A$, it is often convenient for purposes of static analysis to replace this distributed load with a resultant force $\mathbf{R}$ that is equivalent to the distributed load $w$. For a beam with a distributed load along its length, the magnitude of the resultant force is determined from the expression

$$R = \int_0^L w \, dx \tag{5-43}$$

which indicates that the magnitude of the resultant force is equal to the area under the load diagram used to represent the distributed load. The line of action of the resultant force passes through the centroid of the area under the load diagram.

## REVIEW PROBLEMS

**5-118*** Determine the moment of the 1650-N force shown in Fig. P5-118 about point $O$.

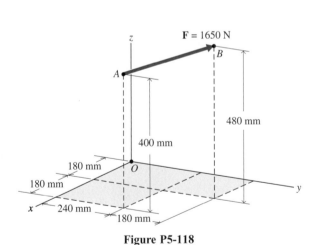

**Figure P5-118**

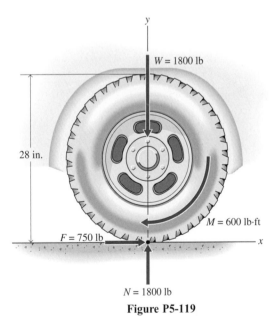

**Figure P5-119**

**5-119*** The driving wheel of a truck is subjected to the force-couple system shown in Fig. P5-119. Replace this system by an equivalent single force and determine the point of application of the force along the vertical diameter of the wheel.

**5-120** A 200-N force is applied at corner $B$ of a rectangular plate as shown in Fig. P5-120. Determine the moment of the force
(a) About point $O$.
(b) About line $OD$.

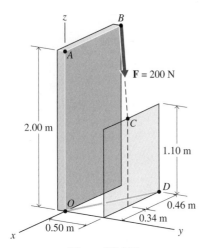

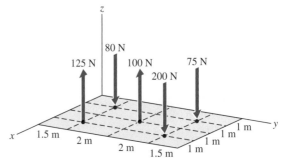

Figure P5-120

**5-121** A 2500-lb jet engine is suspended from the wing of an airplane as shown in Fig. P5-121. Determine the moment produced by the engine at point $A$ in the wing when the plane is

(a) On the ground with the engine not operating.
(b) In flight with the engine developing a thrust $T$ of 15,000 lb.

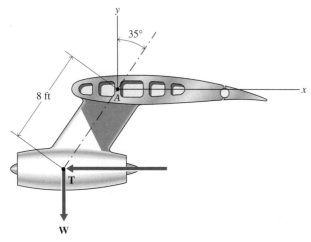

Figure P5-121

**5-122** Two forces and a couple act on a beam as shown in Fig. P5-122. Determine the resultant $R$ and its location $x$.

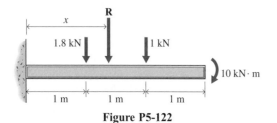

Figure P5-122

**5-123\*** Replace the force and couple acting on the wall bracket shown in Fig. P5-123 by a force-couple system at point $A$.

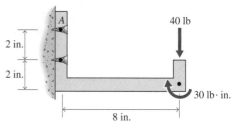

Figure P5-123

**5-124\*** Determine the resultant of the parallel force system shown in Fig. P5-124 and locate the intersection of its line of action with the $xy$-plane.

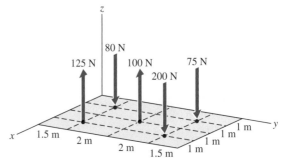

Figure P5-124

**5-125** The concrete floor of a building supports four wood columns as shown in Fig. P5-125. The resultant of the forces transmitted through the columns to the floor is $R = 75$ kip at the location shown in the figure. Determine the magnitude of the forces $F_1$ and $F_2$.

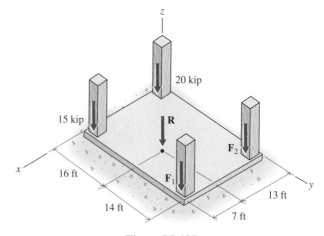

Figure P5-125

**5-126** A bent rod supports a 450-kN force **F** as shown in Fig. P5-126.
(a) Replace the 450-N force with a force **R** through the coordinate origin $O$ and a couple **C**.
(b) Determine the twisting moments produced by force **F** in the three different segments of the rod.

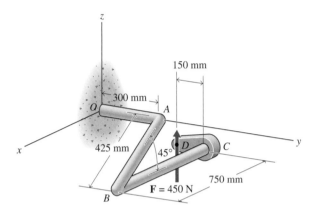

**Figure P5-126**

**5-127** Locate the centroid of the shaded area shown in Fig. P5-127.

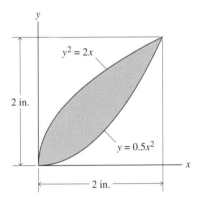

**Figure P5-127**

**5-128\*** Two channel sections and a plate are used to form the cross section shown in Fig. P5-128. Each of the channels has a cross-sectional area of 2605 mm². Locate the $y$-coordinate of the centroid of the composite section with respect to the top surface of the plate.

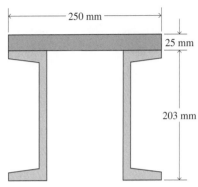

**Figure P5-128**

**5-129** Locate the centroid of the volume shown in Fig. P5-129 if $R = 10$ in. and $h = 32$ in.

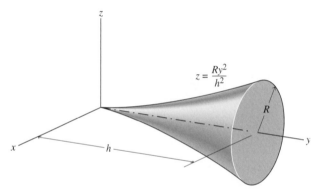

**Figure P5-129**

**5-130\*** Locate the centroid and the mass center of the volume shown in Fig. P5-130, which consists of an aluminum cylinder ($\rho = 2770$ kg/m³) and a steel ($\rho = 7870$ kg/m³) cylinder and sphere.

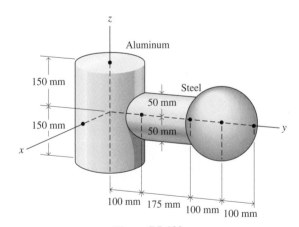

**Figure P5-130**

**5-131\*** Determine the resultant **R** of the system of distributed loads on the beam of Fig. P5-131, and locate its line of action with respect to the left support of the beam.

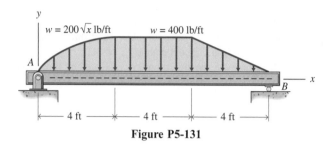

**Figure P5-131**

# EQUILIBRIUM: RIGID AND DEFORMABLE BODIES

**6**

## 6-1 INTRODUCTION

The concept of equilibrium was introduced in Chapter 3 and applied to a system of forces acting on a particle. Since any system of forces acting on a particle is a concurrent force system, a particle is in equilibrium if the resultant force **R** of the force system acting on the particle is zero. For the case of a rigid body, it was shown in Chapter 5 that the most general force system can be expressed in terms of a resultant force **R** and a resultant couple **C**. Therefore, for a rigid body to be in equilibrium, both the resultant force **R** and the resultant couple **C** must vanish. These two conditions are expressed by the two vector equations

$$\mathbf{R} = \Sigma F_x \mathbf{i} + \Sigma F_y \mathbf{j} + \Sigma F_z \mathbf{k} = \mathbf{0}$$
$$\mathbf{C} = \Sigma M_x \mathbf{i} + \Sigma M_y \mathbf{j} + \Sigma M_z \mathbf{k} = \mathbf{0}$$

(6-1)

Equations (6-1) can be expressed in scalar form as

$$\Sigma F_x = 0 \qquad \Sigma F_y = 0 \qquad \Sigma F_z = 0$$
$$\Sigma M_x = 0 \qquad \Sigma M_y = 0 \qquad \Sigma M_z = 0$$

(6-2)

Equations (6-2) are the necessary conditions for equilibrium of a rigid body. If all of the forces acting on a body can be determined from these equations, then they are also the sufficient conditions for equilibrium.

    The forces and moments that act on a rigid body are either external or internal. Forces applied to a rigid body by another body or by the earth are external forces. Fluid pressure on the wall of a tank or the force applied by a truck wheel to a road surface are examples of external forces. The weight of a body is another example of an external force. If the body of interest is composed of several parts, the forces holding the parts together are defined as internal forces. In addition, internal forces hold the particles forming the body together. In Chapter 4, the intensity of these internal forces was called *stress*. If the intensity of the internal force is too large, the body may rupture or deform excessively.

    External forces can be divided into applied forces and reaction forces. External forces that tend to cause a body to move in some direction or rotate about some axis are applied forces. External forces exerted on a body by supports or connections are reaction forces. Reaction forces oppose the tendency of applied forces to cause motion. They hold the body in equilibrium. Our concern in this chapter is with external forces and the moments these external forces produce. Since internal forces occur as equal and opposite pairs, they have no effect on equilibrium of the overall rigid body.

The best way to identify all forces acting on a body of interest is to use the free-body diagram approach introduced in Chapter 3, where equilibrium of a particle was considered. This free-body diagram of the body of interest must show all of the applied forces and all of the reaction forces exerted on the body by the supports. In Section 6-2, free-body diagrams for a number of different rigid bodies are considered and the reaction forces exerted on them by a number of different types of supports are specified.

Once the free-body diagram has been drawn and the reaction forces have been found using the equations of equilibrium, the body of interest can be sectioned and the resultant of the internal forces can be determined. The ability of the body to withstand the intensity of the internal forces can be investigated, and the deformation characteristics of the body can be studied.

## 6-2 FREE-BODY DIAGRAMS

The concept of a free-body diagram was introduced in Chapter 3 and used to solve equilibrium problems involving bodies that could be idealized as a particle. In all these problems, the external forces (applied forces and reaction forces) could be represented as a concurrent force system. In the more general case of a rigid body, systems of forces other than concurrent force systems are encountered and the free-body diagrams become much more complicated. The basic procedure for drawing the diagram, however, remains the same and consists essentially of the following four steps (slightly modified from Section 3-2 to account for a rigid body):

**Step 1.** Decide which body or combination of bodies is to be shown on the free-body diagram.

**Step 2.** Prepare a drawing or sketch of the outline of this isolated or free body.

**Step 3.** Carefully trace around the boundary of the free body and identify all the forces and moments exerted by contacting or interacting bodies that were removed during the isolation process.

**Step 4.** Choose the set of coordinate axes to be used in solving the problem and indicate these directions on the free-body diagram.

Application of these four steps to a statics problem will produce a complete and correct free-body diagram, which is *an essential first step for the solution of any problem.*

Forces that are known should be added to the diagram and labeled with their proper magnitudes and directions. Letter symbols can be used to represent magnitudes of forces that are unknown. If the correct sense of an unknown force is not obvious, the sense can be arbitrarily assumed. The algebraic sign of the calculated value of the unknown force will indicate the sense of the force. A plus sign indicates that the force is in the direction assumed, and a minus sign indicates that the force is in a direction opposite from that assumed.

When connections or supports are removed from the isolated body, the actions of these connections or supports must be represented by forces and/or moments on the free-body diagram. The forces and moments used to represent the actions of common connections and supports used with bodies subjected to two-dimensional force systems are identified and discussed in Table 6-1. A similar discussion for connections and supports used with bodies subjected to three-dimensional force systems is presented in Table 6-2.

A common type of force, regardless of the type of support or connection, is the weight of the body, which is the gravitational attraction of the earth on the body. As shown in Fig. 6-1, the line of action of the weight, **W,** passes through the center of gravity of the body and is directed toward the center of the earth.

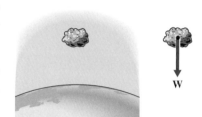

**Figure 6-1**

## Idealization of Two-Dimensional Supports and Connections

Common types of supports and connections used with rigid bodies subjected to two-dimensional force systems, together with the forces and moments used to represent the actions of these supports and connections on a free body, are listed in Table 6-1.

TABLE 6-1 Two-Dimensional Reactions at Supports and Connections

---

**(1) FLEXIBLE CORD, ROPE, CHAIN, OR CABLE**

**Figure 6-2**

A flexible cord, rope, chain, or cable (Fig. 6-2) always exerts a tensile force **R** on the body. The line of action of the force **R** is known; it is tangent to the cord, rope, chain, or cable at the point of attachment.

**(2) RIGID LINK**

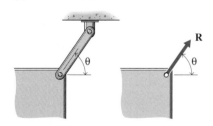

**Figure 6-3**

A rigid link (Fig. 6-3) can exert either a tensile or a compressive force **R** on the body. The line of action of the force **R** is known; it must be directed along the axis of the link (see Section 6-3 for proof).

**(3) BALL, ROLLER, OR ROCKER**

**Figure 6-4**

A ball, roller, or rocker (Fig. 6-4) can exert a compressive force **R** on the body. The line of action of the force **R** is perpendicular to the surface supporting the ball, roller, or rocker.

TABLE 6-1 (*Continued*)

### (4) SMOOTH SURFACE

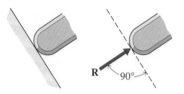

Figure 6-5

A smooth surface, either flat or curved (Fig. 6-5), can exert a compressive force **R** on the body. The line of action of the force **R** is perpendicular to the smooth surface at the point of contact between the body and the smooth surface.

### (5) SMOOTH PIN

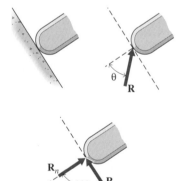

Figure 6-6

A smooth pin (Fig. 6-6) can exert a force **R** of unknown magnitude $R$ and direction $\theta$ on the body. As a result, the force **R** is usually represented on a free-body diagram by its rectangular components $\mathbf{R}_x$ and $\mathbf{R}_y$.

### (6) ROUGH SURFACE

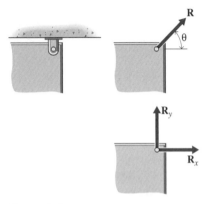

Figure 6-7

Rough surfaces (Fig. 6-7) are capable of supporting a tangential frictional force $\mathbf{R}_t$ as well as a compressive normal force $\mathbf{R}_n$. As a result, the force **R** exerted on a body by a rough surface is a compressive force **R** at an unknown angle $\theta$. The force **R** is usually represented on a free-body diagram by its rectangular components $\mathbf{R}_n$ and $\mathbf{R}_t$.

### (7) PIN IN A SMOOTH GUIDE

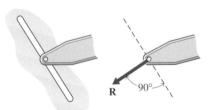

Figure 6-8

A pin in a smooth guide (Fig. 6-8) can transmit only a force **R** perpendicular to the surfaces of the guide. The sense of **R** is assumed on the figure and may be either downward to the left or upward to the right.

TABLE 6-1 (*Continued*)

### (8) COLLAR ON A SMOOTH SHAFT

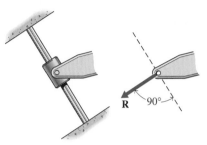

**Figure 6-9**

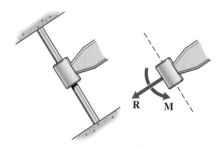

**Figure 6-10**

A collar on a smooth shaft (Fig. 6-9) that is pin-connected to a body can transmit only a force **R** perpendicular to the axis of the shaft. When the connection between the collar and the body is fixed (Fig. 6-10), the collar can transmit both a force **R** and a moment **M** perpendicular to the axis of the shaft. If the shaft is not smooth, a tangential frictional force $\mathbf{R}_t$ as well as a normal force $\mathbf{R}_n$ can be transmitted.

### (9) FIXED SUPPORT

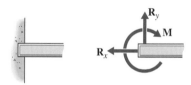

**Figure 6-11**

A fixed support (Fig. 6-11) can exert both a force **R** and a couple **C** on the body. The magnitude $R$ and the direction $\theta$ of the force **R** are not known. Therefore, the force **R** is usually represented on a free-body diagram by its rectangular components $\mathbf{R}_x$ and $\mathbf{R}_y$ and the couple **C** by its moment **M**.

### (10) LINEAR ELASTIC SPRING

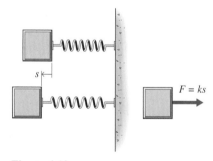

**Figure 6-12**

The force **F** exerted on a body by a linear elastic spring (see Fig. 6-12) is proportional to the change in length of the spring. The spring will exert a tensile force if lengthened and a compressive force if shortened. The line of action of the force is along the axis of the spring.

### (11) IDEAL PULLEY

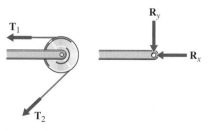

**Figure 6-13**

Pulleys (Fig. 6-13) are used to change the direction of a rope or cable. The pin connecting an ideal pulley to a member can exert a force **R** of unknown magnitude $R$ and direction $\theta$ on the body. The force **R** is usually represented on a free-body diagram by its rectangular components $\mathbf{R}_x$ and $\mathbf{R}_y$. Also, since the pin is smooth (frictionless), the tension $T$ in the cable must remain constant to satisfy moment equilibrium about the axis of the pulley.

## Idealization of Three-Dimensional Supports and Connections

Common types of supports and connections used with rigid bodies subjected to three-dimensional force systems, together with the forces and moments used to represent the actions of these supports and connections on a free-body diagram, are listed in Table 6-2.

---

TABLE 6-2  Three-Dimensional Reactions at Supports and Connections

### (1) BALL AND SOCKET

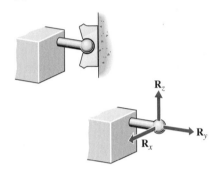

A ball and socket joint (Fig. 6-14) can transmit a force **R** but no moment. The force **R** is usually represented on a free-body diagram by its three rectangular components $\mathbf{R}_x$, $\mathbf{R}_y$, and $\mathbf{R}_z$.

**Figure 6-14**

### (2) HINGE

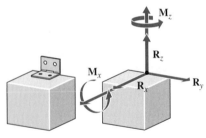

A hinge (Fig. 6-15) is normally designed to transmit a force **R** in a direction perpendicular to the axis of the hinge pin. The design may also permit a force component to be transmitted along the axis of the pin. Individual hinges also have the ability to transmit small moments about axes perpendicular to the axis of the pin. However, properly aligned pairs of hinges transmit only forces under normal conditions of use. Thus, the action of a hinge is represented on a free-body diagram by the force components $\mathbf{R}_x$, $\mathbf{R}_y$, and $\mathbf{R}_z$ and the moments $\mathbf{M}_x$, and $\mathbf{M}_z$ when the axis of the pin is in the $y$-direction.

**Figure 6-15**

### (3) BALL BEARING

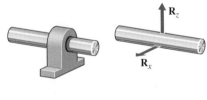

Ideal (smooth) ball bearings (Fig. 6-16) are designed to transmit a force **R** in a direction perpendicular to the axis of the bearing. The action of the bearing is represented on a free-body diagram by the force components $\mathbf{R}_x$ and $\mathbf{R}_z$ when the axis of the bearing is in the $y$-direction.

**Figure 6-16**

TABLE 6-2 (*Continued*)

### (4) JOURNAL BEARING

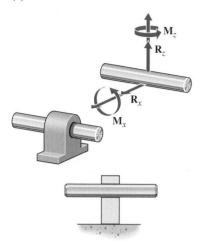

Figure 6-17

Journal bearings (Fig. 6-17) are designed to transmit a force **R** in a direction perpendicular to the axis of the bearing. Individual journal bearings also have the ability to transmit small moments about axes perpendicular to the axis of the shaft. However, properly aligned pairs of bearings transmit only forces perpendicular to the axis of the shaft under normal conditions of use. Therefore, the action of a journal bearing is represented on a free-body diagram by the force components $R_x$ and $R_z$ and the couple moments $M_x$ and $M_z$ when the axis of the bearing is in the $y$-direction.

### (5) THRUST BEARING

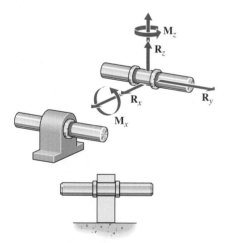

Figure 6-18

A thrust bearing (Fig. 6-18), as the name implies, is designed to transmit force components both perpendicular and parallel (thrust) to the axis of the bearing. Individual thrust bearings also have the ability to transmit small moments about axes perpendicular to the axis of the shaft. However, properly aligned pairs of bearing transmit only forces under normal conditions of use. Therefore, the action of a thrust bearing is represented on a free-body diagram by the force components $R_x$, $R_y$, and $R_z$ and the couple moments $M_x$ and $M_z$ when the axis of the bearing is in the $y$-direction.

### (6) SMOOTH PIN AND BRACKET

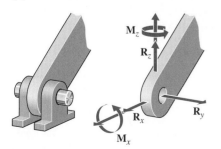

Figure 6-19

A pin and bracket (Fig. 6-19) is designed to transmit a force **R** in a direction perpendicular to the axis of the pin but may also transmit a force component along the axis of the pin. The unit also has the ability to transmit small moments about axes perpendicular to the axis of the pin. Therefore, the action of a smooth pin and bracket is represented on a free-body diagram by the force components $R_x$, $R_y$, and $R_z$ and the couple moments $M_x$ and $M_z$ when the axis of the pin is in the $y$-direction.

TABLE 6-2 (*Continued*)

**(7) FIXED SUPPORT**

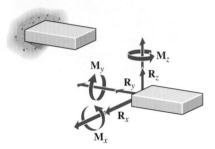

A fixed support (Fig. 6-20) can resist both a force **R** and a couple **C**. The magnitudes and directions of the force and couple are not known. Thus, the action of a fixed support is represented on a free-body diagram by the force components $R_x$, $R_y$, and $R_z$ and the moment components $M_x$, $M_y$, and $M_z$.

**Figure 6-20**

---

## Example Problem 6-1

Draw a free-body diagram for the beam shown in Fig. 6-21*a*.

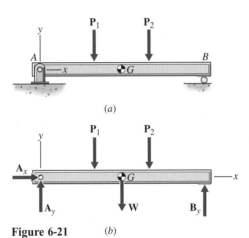

**Figure 6-21**

### SOLUTION

Two concentrated forces $P_1$ and $P_2$ are applied to the beam. The weight of the beam is represented by the force **W**, which has a line of action that passes through the center of gravity *G* of the beam. The beam is supported at the left end with a smooth pin and bracket and at the right end with a roller. The action of the left support is represented by the forces $A_x$ and $A_y$ (see Fig. 6-6). The action of the roller is represented by the force $B_y$, which acts normal to the surface of the beam (see Fig. 6-5). A complete free-body diagram for the beam is shown in Fig. 6-21*b*. ∎

---

## Example Problem 6-2

A cylinder is supported on a smooth inclined surface by a two-bar frame as shown in Fig. 6-22*a*. Assume that the cylinder has weight *W* and that the two bars have negligible weight. Draw a free-body diagram for

(a) The cylinder.   (b) The two-bar frame.   (c) The pin at *C*.

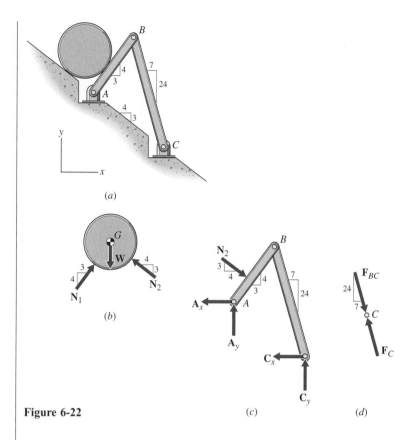

**Figure 6-22**

## SOLUTION

(a) The free-body diagram for the cylinder is shown in Fig. 6-22*b*. The weight **W** of the cylinder acts through the center of gravity *G*. The forces **N**$_1$ and **N**$_2$ act normal to the smooth surfaces at the points of contact.

(b) The free-body diagram for the two-bar frame is shown in Fig. 6-22*c*. The actions of the smooth pin and bracket supports at points *A* and *C* are represented by forces **A**$_x$ and **A**$_y$ and **C**$_x$ and **C**$_y$, respectively. Note that the pin forces at *B* are internal and do not appear on the free-body diagram shown in Fig. 6-22*c*.

(c) Since bar *BC* is a link (see Fig. 6-3), the resultant **F**$_C$ of forces **C**$_x$ and **C**$_y$ must have a line of action along the axis of the link. As a result, the free-body diagram for pin *C* can be drawn as shown in Fig. 6-22*d*. ∎

---

## Example Problem 6-3

Draw the free-body diagram for the curved bar *AC* shown in Fig. 6-23*a*, which is supported by a ball and socket joint at *A*, a flexible cable at *B*, and a pin and bracket at *C*. Neglect the weight of the bar.

### SOLUTION

The action of the ball and socket joint (see Fig. 6-14) at support *A* is represented by three rectangular force components **A**$_x$, **A**$_y$, and **A**$_z$. The action of the pin and bracket (see Fig. 6-19) at support *C* can be represented by force com-

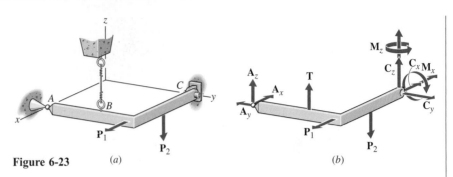

**Figure 6-23**   (a)   (b)

ponents $\mathbf{C}_x$, $\mathbf{C}_y$, and $\mathbf{C}_z$ and moment components $\mathbf{M}_x$ and $\mathbf{M}_z$. The action of the cable (see Fig. 6-2) is represented by the cable tension $\mathbf{T}$. A complete free-body diagram for bar $AC$ is shown in Fig. 6-23b. ∎

## PROBLEMS

### Introductory Problems

**6-1**  Draw a free-body diagram of the angle bracket shown in Fig. P6-1.

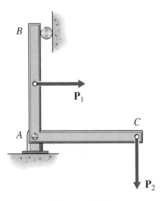

**Figure P6-1**

**6-2**  Draw a free-body diagram of the diving board shown in Fig. P6-2. The surface at $B$ is smooth. Neglect the weight of the diving board.

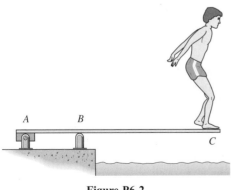

**Figure P6-2**

**6-3**  Draw a free-body diagram of the lawn mower shown in Fig. P6-3. The lawn mower has a weight $\mathbf{W}$ and is resting on a rough surface.

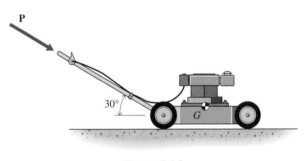

**Figure P6-3**

**6-4**  Draw a free-body diagram of the sled shown in Fig. P6-4.

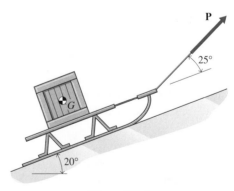

**Figure P6-4**

## Intermediate Problems

**6-5** Draw a free-body diagram of the bracket shown in Fig. P6-5. The contact surfaces between the cylinders and bracket are smooth.

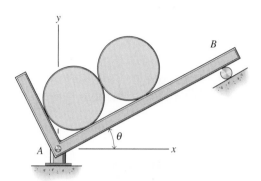

**Figure P6-5**

**6-6** Forces **P** are applied to the handles of the pipe pliers shown in Fig. P6-6. Draw free-body diagrams of each handle.

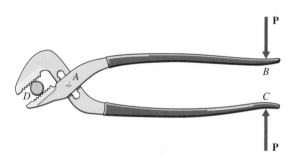

**Figure P6-6**

**6-7** The man shown in Fig. P6-7 has a weight $W_1$; the beam has a weight $W_2$ and a center of gravity $G$. Draw a free-body diagram of the beam.

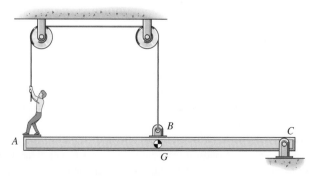

**Figure P6-7**

**6-8** Forces **P** are applied to the handles of the bolt cutter shown in Fig. P6-8. Draw a free-body diagram of
(a) The lower handle.
(b) The lower cutter jaw.

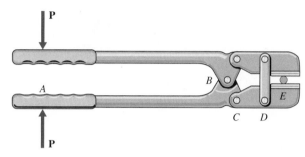

**Figure P6-8**

## Challenging Problems

**6-9** Draw a free-body diagram of the door shown in Fig. P6-9. The homogeneous door has weight **W**.

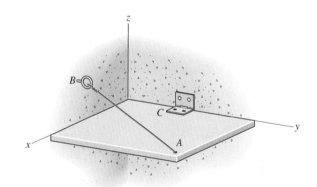

**Figure P6-9**

**6-10** Draw a free-body diagram of the bent bar shown in Fig. P6-10. The support at $A$ is fixed, and the bar has negligible mass.

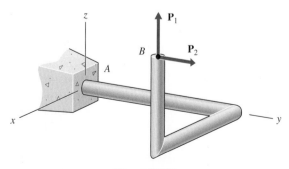

**Figure P6-10**

**6-11** Draw a free-body diagram of the bar shown in Fig. P6-11. The support at $A$ is a journal bearing and the supports at $B$ and $C$ are ball bearings.

**6-12** Draw a free-body diagram of the shaft shown in Fig. P6-12. The bearing at $A$ is a thrust bearing, and the bearing at $D$ is a ball bearing. Neglect the weights of the shaft and the levers.

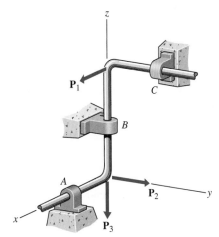

**Figure P6-11**

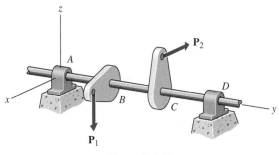

**Figure P6-12**

## 6-3 EQUILIBRIUM IN TWO DIMENSIONS

The term *two-dimensional* is used to describe problems in which the forces involved are contained in a plane (say the $xy$-plane) and the axes of all couples are perpendicular to the plane containing the forces. For two-dimensional problems, since a force in the $xy$-plane has no $z$-component and produces no moments about the $x$- or $y$-axes, Eqs. (6-1) reduce to

$$\mathbf{R} = \Sigma F_x \mathbf{i} + \Sigma F_y \mathbf{j} = \mathbf{0}$$
$$\mathbf{C} = \Sigma M_z \mathbf{k} = \mathbf{0} \tag{6-3}$$

Thus, three of the six scalar equations of equilibrium [Eqs. (6-2)] are automatically satisfied; namely,

$$\Sigma F_z = 0 \qquad \Sigma M_x = 0 \qquad \Sigma M_y = 0$$

Therefore, there are only three *independent* scalar equations of equilibrium for a rigid body subjected to a two-dimensional system of forces. The three equations can be expressed as

$$\Sigma F_x = 0 \qquad \Sigma F_y = 0 \qquad \Sigma M_A = 0 \tag{6-4}$$

The third equation represents the sum of the moments of all forces about a $z$-axis through any point $A$ on or off the body. Equations (6-4) are both the necessary and sufficient conditions for equilibrium of a rigid body subjected to a two-dimensional system of forces.

There are two additional ways in which the equations of equilibrium can be expressed for a body subjected to a two-dimensional system of forces. The resultant force $\mathbf{R}$ and the resultant couple $\mathbf{C}$ at point $A$ of a rigid body subjected to a general two-dimensional force system are shown in Fig. 6-24a. The resultant can be represented in terms of its components as shown in Fig. 6-24b. If the condition $\Sigma M_A = 0$ is satisfied, $\mathbf{C} = \mathbf{0}$. If, in addition, the condition $\Sigma F_x = 0$ is satisfied, $\mathbf{R} = \Sigma F_y \mathbf{j}$. For any point $B$ on or off the body that does not lie on the $y$-axis, the equation $\Sigma M_B = 0$ can be satisfied only if $\Sigma F_y = 0$. Thus, an alternative set of *independent* scalar equilibrium equations for two-dimensional problems is

$$\Sigma F_x = 0 \qquad \Sigma M_A = 0 \qquad \Sigma M_B = 0 \qquad (6\text{-}5)$$

where points $A$ and $B$ must have different $x$-coordinates.

The conditions of equilibrium for a two-dimensional force system can also be expressed by using three moment equations. Again, if the condition $\Sigma M_A = 0$ is satisfied, $\mathbf{C} = \mathbf{0}$. In addition, for a point $B$ (see Fig. 6-24c) on the $x$-axis on or off the body (except at point $A$), the equation $\Sigma M_B = 0$ can be satisfied, once $\Sigma M_A = 0$, only if $\Sigma F_y = 0$. Thus, the resultant can only have an $x$-component, $\mathbf{R} = \Sigma F_x \mathbf{i}$. Finally, for any point $C$ (see Fig. 6-24) on or off the body that does not lie on the $x$-axis, the equation $\Sigma M_C = 0$ can be satisfied only if $\Sigma F_x = 0$. Thus, there is a second set of alternative *independent* scalar equilibrium equations for two-dimensional problems:

$$\Sigma M_A = 0 \qquad \Sigma M_B = 0 \qquad \Sigma M_C = 0 \qquad (6\text{-}6)$$

where $A$, $B$, and $C$ are any three points not on the same straight line.

## Two-Force Members
Equilibrium of a body under the action of two forces occurs with sufficient frequency to warrant special attention. For example, consider a link $AB$ with negligible weight, as shown in Fig. 6-25a. Any forces exerted on the link by the frictionless pins at $A$ and $B$ can be resolved into components along and perpendicular to the axis of the link, as shown in Fig. 6-25b. From the equilibrium equations

$$+\nearrow \Sigma F_x = 0: \qquad A_x - B_x = 0 \qquad A_x = B_x$$
$$+\nwarrow \Sigma F_y = 0: \qquad A_y - B_y = 0 \qquad A_y = B_y$$

Forces $A_y$ and $B_y$, however, form a couple that must be zero when the link is in equilibrium; therefore, $A_y = B_y = 0$. Thus, for two-force members, equilibrium requires that the forces be equal, opposite, and collinear, as shown in Fig. 6-25c.

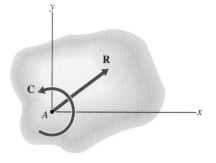

(a)

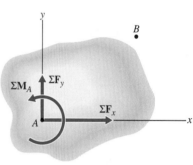

(b)

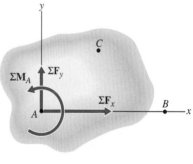

(c)

**Figure 6-24**

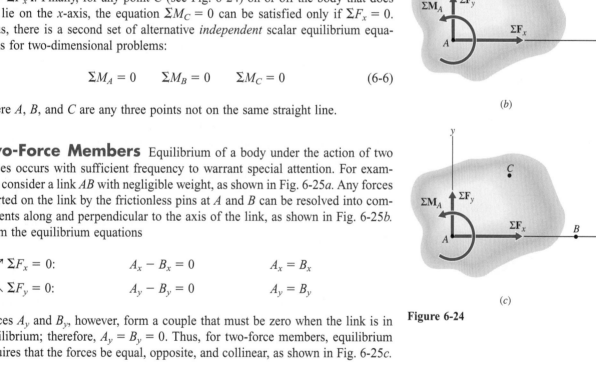

(a)      (b)      (c)      (d)

**Figure 6-25**

**Figure 6-26**

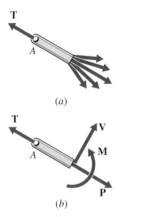

(a)

(b)

**Figure 6-27**

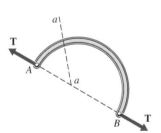

**Figure 6-28**

The shape of the member, as shown in Fig. 6-25d, has no effect on this simple requirement; however, the weights of the members must be negligible.

When forces are directed away from the ends of a straight two-force member as shown in Fig. 6-26, the member is said to be in tension and the force **T** is referred to as the tension force in the member and not just the tension force on the end of the member. That this description of the force is correct is easily shown by considering equilibrium of a portion of the member to the left of section *a–a*. In general, there will be some complex force distribution acting on the cut surface as shown on the free-body diagram of Fig. 6-27a. However, this system of forces can be replaced with an equivalent force-couple as shown on the free-body diagram of Fig. 6-27b. Then equilibrium of forces in the direction perpendicular to the axis of the member requires that **V**, the shear component of the equivalent force-couple, must be zero. Similarly, equilibrium of forces in the direction along the axis of the member requires that **P**, the axial component of the equivalent force-couple, must be equal in magnitude and opposite in direction to **T**, the force applied to the end of the member. Finally, moment equilibrium requires that **M**, the couple component of the equivalent force-couple, must be zero. That is, if the forces at the ends of a straight two-force member are pulling on the member, then the resultant of the forces on any cut section of the member must also be an axial force pulling on the cut section of the member, regardless of where the member is cut. Therefore, it is proper to talk about the force **T** as the *tension* in the member. If the forces are directed toward the ends of a straight two-force member, the member is said to be in *compression*.

Axially loaded structural members such as the straight two-force member were discussed in Chapter 4. Once **T** (Fig. 6-26) has been found using equilibrium principles, the internal normal force **P** on a transverse cross section is known since **P** = **T**. The internal resistance and deformation of the member are then found using the equations of Chapter 4—that is, $\sigma = P/A$ (Eq. 4-2) and $\delta = PL/EA$ (Eq. 4-20b)—where $\sigma$ is the normal stress on a transverse cross section and $\delta$ is the elongation (or contraction) of the member having cross-sectional area $A$, length $L$, and modulus of elasticity $E$. Thus, for straight two-force members the principles developed in Chapter 4 are sufficient to determine stress and deformation.

When a two-force member is curved, however, the forces at the ends of the member do not act along the axis of the member. Instead, the forces act along the line joining the points where the forces are applied, as shown in Fig. 6-28. If the member is cut normal to its axis at section *a–a*, there will be a complex force distribution acting on the cut surface, as shown in Fig. 6-29a. This system of

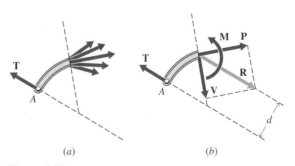

(a)                          (b)

**Figure 6-29**

forces can be replaced with an equivalent force-couple, as shown on the free-body diagram of Fig. 6-29b. However, force equilibrium now requires that the resultant **R** of the axial and shear components **P** and **V** of the equivalent force-couple be equal in magnitude and opposite in direction to the force **T** at the end of the member, as shown in Fig. 6-29b. Since the forces **R** and **T** are not collinear, moment equilibrium now requires that $M = Td \neq 0$.

As a result of the previous discussion it is obvious that the design of straight two-force members need only consider axial forces, but curved two-force members must be designed to withstand shearing forces $V$ and bending moments $M$ as well as axial forces $P$. To further complicate the problem, the magnitudes of the shear forces, bending moments, and axial forces in curved members depend on where the member is cut. Analyzing the internal resistance of a curved two-force member requires concepts more advanced than those discussed in Chapter 4. The stress and deformation characteristics of these members will be developed throughout the remainder of this book.

## Statically Indeterminate Reactions and Partial Constraints

Consider a body subjected to a system of coplanar forces $\mathbf{F}_1$, $\mathbf{F}_2$, $\mathbf{F}_3$, $\mathbf{F}_4$, . . . , $\mathbf{F}_n$, as shown in Fig. 6-30a. This system of forces can be replaced by an equivalent force-couple system at an arbitrary point $A$ as shown in Fig. 6-30b. The resultant force has been represented by its rectangular components $\mathbf{R}_x$ and $\mathbf{R}_y$ and the couple by $\mathbf{M}_A$. For the body to be in equilibrium, the supports must be capable of exerting an equal and opposite force-couple system on the body. As an example, consider the supports shown in Fig. 6-31a. The pin support at $A$, as shown in Fig. 6-31b, can exert forces in the $x$- and $y$-directions to prevent a translation of the body but no moment to prevent a rotation about an axis through $A$. The link at $B$, which exerts a force in the $y$-direction, produces the moment about point $A$ required to prevent rotation. Thus, all motion of the body is prevented, and the free-body diagram shown in Fig. 6-31b is in equilibrium under the action of the forces shown. The reaction forces exerted on the body by the supports are called constraint forces. When the equations of equilibrium are sufficient to determine the unknown forces at the supports, as in Fig. 6-31, the body is said to be statically determinate with adequate (proper) constraints.

Three support reactions for a body subjected to a coplanar system of forces does not always guarantee that the body is statically determinate with adequate

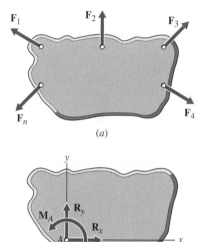

(a)

(b)

**Figure 6-30**

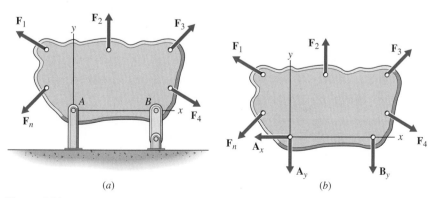

(a)                    (b)

**Figure 6-31**

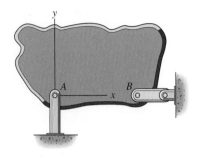

(a)

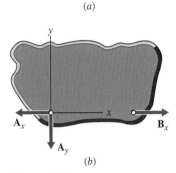

(b)

**Figure 6-32**

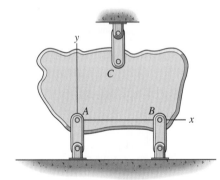

(a)

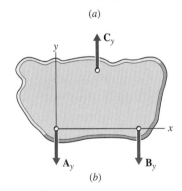

(b)

**Figure 6-33**

constraints. For example, consider the body and supports shown in Fig. 6-32a. The pin support at $A$ (see Fig. 6-32b) can exert forces in the $x$- and $y$-directions to prevent a translation of the body, but since the line of action of force $\mathbf{B}_x$ passes through point $A$, it does not exert the moment required to prevent rotation of the body about point $A$. Similarly, the three links shown in Fig. 6-33a can prevent rotation of the body about any point (see Fig. 6-33b) and translation of the body in the $y$-direction but not translation of the body in the $x$-direction. The bodies in Figs. 6-32 and 6-33 are partially (improperly) constrained and the equations of equilibrium are not sufficient to determine all of the unknown reactions. A body with an adequate number of reactions is improperly constrained when the constraints are arranged in such a way that support forces are either concurrent or parallel. Partially constrained bodies can be in equilibrium for specific systems of forces. For example, the reactions $R_A$ and $R_B$ for the beam shown in Fig. 6-34a can be determined by using the equilibrium equations $\Sigma F_y = 0$ and $\Sigma M_A = 0$. The beam is improperly constrained, however, since motion in the $x$-direction would occur if any of the applied loads had a small $x$-component.

Finally, if the link at $B$ in Fig. 6-31a is replaced with a pin support, as shown in Fig. 6-35a, an additional reaction $\mathbf{B}_x$ (see Fig. 6-35b) is obtained that is not required to prevent movement of the body. Obviously, the three independent equations of equilibrium will not provide sufficient information to determine the four unknowns. Constrained bodies with extra supports are statically indeterminate because the equations of equilibrium are not sufficient to solve for all of the support reactions. Relations involving physical properties of the body, in addition to the equations of equilibrium, are required to determine some of the unknown reactions. This was done in Chapter 4 for axially loaded bodies. The supports not required to maintain equilibrium of the body are called *redundant supports*. Typical examples of redundant supports for beams are shown in Fig. 6-36. The roller at support $B$ of the cantilever beam shown in Fig. 6-36a can be removed and the beam will remain in equilibrium. Similarly, the roller support at either $B$ or $C$ (but not both) could be removed from the beam of Fig. 6-36b and the beam would remain in equilibrium.

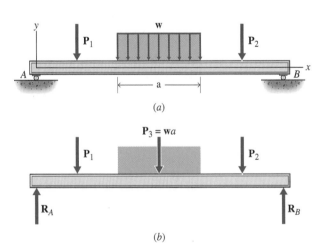

(a)

(b)

**Figure 6-34**

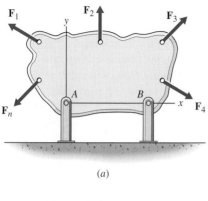

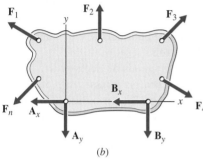

Figure 6-35

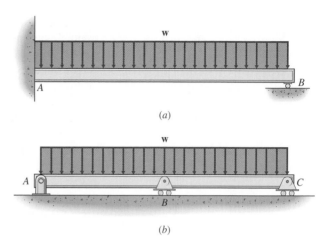

Figure 6-36

**Problem Solving.** In Section 1-5, a procedure was outlined for solving engineering problems. The procedure consisted of three phases:

**Problem definition and identification:**

1. Read the problem carefully. *Many student difficulties arise from failure to observe this preliminary step.*
2. Identify the information given and the results requested.

**Model development and simplification:**

3. Identify the principles to be used to obtain the result.
4. Prepare a scaled sketch and tabulate the information provided.
5. Draw the appropriate free-body diagrams. *Carefully label all applied forces and support reactions. Establish a convenient set of coordinate axes. Use a right-handed system in case vector cross products must be employed. Compare the number of unknowns on the free-body diagram with the number of independent equations of equilibrium. Draw additional diagram(s) if needed.*

**Mathematical solution and result interpretation:**

6. Apply the appropriate principles and equations.
7. Report the answer with the appropriate number of significant figures and the appropriate units.
8. Study the answer and determine whether it is reasonable. *As a check, write some other equilibrium equations and see if they are satisfied by the solution.*

Bodies subjected to coplanar force systems are not very complex, so a scalar solution is usually suitable for analysis. This results from the fact that moments can be expressed as scalars instead of vectors. For the more general case of rigid bodies subjected to three-dimensional force systems (discussed in a later section of this chapter), vector analysis methods are usually more appropriate.

In certain situations, normal stress, bearing stress, shearing stress, and deformation can be determined. For example, if a member is axially loaded (a straight two-force member), the methods developed in Sections 4-2 and 4-3 can be used to compute normal stress and shearing stress on a section of the member. If a two-force member is curved, the methods developed in Sections 4-2 and 4-3 are not sufficient to calculate the stresses.

In addition, the deformation of a straight two-force member can be found using the methods developed in Section 4-7. The deformation of members with a combination of axial forces, shear forces, and bending moments cannot be found using the results of Section 4-7; methods developed throughout the remainder of this book are needed to solve this class of problems.

Regardless of how a body is loaded, the methods of Sections 4-2 can be used to determine shearing stresses in pins and bearing stresses at pinned connections. For example, the reaction at $A$ of Fig. 6-37$a$ can be found using the principles of rigid-body equilibrium applied to the free-body diagram shown in Fig. 6-37$b$. The shearing stress in the pin at $A$ can then be found using the results of Section 4-2. The bearing stress between the pin at $A$ and beam $AB$, or between the pin at $A$ and the support bracket, can also be found using the results of Section 4-2.

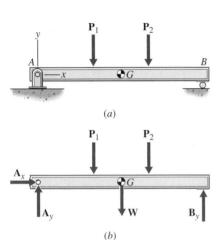

(a)

(b)

**Figure 6-37**

### Equations of Equilibrium Applied to a Deformable Body.

In most engineering applications, bodies are assumed to be rigid when the equations of equilibrium are used to determine support reactions even though it is a fact that bodies deform when the loads are applied. If the amount of deformation is small, the geometry of the deformed body will differ from that of the undeformed body by a negligible amount and the equations of equilibrium for the undeformed body will give good results.

If the amount of deformation is not small, then the geometry of the deformed body must be used to write the equations of equilibrium. An iterative procedure is often used in this case.

1. Assume that the body is rigid and draw the free-body diagram.
2. Write the equations of equilibrium from the free-body diagram, and solve for the forces of interest.
3. Calculate the deformation that the body would experience as a result of the forces calculated in step 2.
4. Draw a new free-body diagram, taking into account the change in geometry caused by the deformation calculated in step 3.
5. Repeat steps 2 through 4 until the answer no longer changes.

Equilibrium requirements should always be satisfied when a body is in the deformed configuration. However, it will be shown in Example Problem 6-11 that for most engineering problems, forces may be determined, within engineering accuracy, using the equilibrium equations and a free-body diagram in the undeformed configuration. These forces may then be used to determine stresses and deformations with sufficient accuracy for most engineering applications.

## Example Problem 6-4

A pin-connected truss is loaded and supported as shown in Fig. 6-38a. The body W has a mass of 100 kg. Determine the components of the reactions at supports A and B. Neglect the masses of the members of the truss.

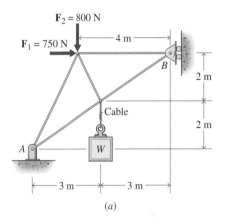

(a)

### SOLUTION

The tension T in the cable between the body W and the truss is

$$T = mg = 100(9.81) = 981 \text{ N}$$

A free-body diagram of the truss is shown in Fig. 6-38b. The action of the pin support at A is represented by force components $\mathbf{A}_x$ and $\mathbf{A}_y$. The action of the roller support at B is represented by the force $\mathbf{B}_x$, which acts perpendicular to the vertical surface at B. Since the truss is subjected to a general coplanar force system, three independent equilibrium equations are available to solve for the unknown magnitudes of the three forces $\mathbf{A}_x$, $\mathbf{A}_y$, and $\mathbf{B}_x$.

### Solution by Scalar Analysis

Determination of $\mathbf{B}_x$:

$$+\circlearrowleft \Sigma M_A = 0: \qquad B_x(4) - 800(2) - 750(4) - 981(3) = 0$$

$$B_x = +1885.8 \text{ N} \qquad \mathbf{B}_x = 1886 \text{ N} \leftarrow \qquad \textbf{Ans.}$$

Determination of $\mathbf{A}_x$:

$$+\rightarrow \Sigma F_x = 0: \qquad A_x + 750 - B_x = A_x + 750 - 1885.8 = 0$$

$$A_x = +1135.8 \text{ N} \qquad \mathbf{A}_x = 1136 \text{ N} \rightarrow \qquad \textbf{Ans.}$$

Determination of $\mathbf{A}_y$:

$$+\uparrow \Sigma F_y = 0: \qquad A_y - 800 - 981 = 0$$

$$A_y = +1781 \text{ N} \qquad \mathbf{A}_y = 1781 \text{ N} \uparrow \qquad \textbf{Ans.}$$

The component $\mathbf{A}_x$ can also be determined directly by summing moments about point C.

$$+\downdownarrows \Sigma M_C = 0: \qquad A_x(4) - 800(2) - 981(3) = 0$$

$$A_x = +1135.8 \text{ N} \qquad \mathbf{A}_x = 1136 \text{ N} \rightarrow \qquad \textbf{Ans.}$$

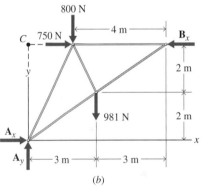

(b)

**Figure 6-38**

## Solution by Vector Analysis

From Eqs. (6-3):

$$\mathbf{R} = \Sigma F_x \mathbf{i} + \Sigma F_y \mathbf{j} = 0 \qquad \mathbf{C} = \Sigma M_z \mathbf{k} = 0$$

The applied loads can be written in Cartesian vector form as

$$\mathbf{F}_1 = 750 \; \mathbf{i} \; \text{N} \qquad \mathbf{F}_2 = -800 \; \mathbf{j} \; \text{N} \qquad \mathbf{F}_3 = -981 \; \mathbf{j} \; \text{N}$$

The reactions can be written as

$$\mathbf{A} = A_x \mathbf{i} + A_y \mathbf{j} \qquad \mathbf{B} = -B_x \mathbf{i}$$

Therefore, from Eqs. (6-3),

$$
\begin{aligned}
\mathbf{R} &= \Sigma F_x \mathbf{i} + \Sigma F_y \mathbf{j} \\
&= (A_x - B_x + 750) \; \mathbf{i} + (A_y - 800 - 981) \; \mathbf{j} = 0
\end{aligned}
$$

Summing moments about point $A$ to minimize the number of unknowns:

$$
\begin{aligned}
\mathbf{C} &= \Sigma M_A \mathbf{k} \\
&= \mathbf{r}_1 \times \mathbf{F}_1 + \mathbf{r}_2 \times \mathbf{F}_2 + \mathbf{r}_3 \times \mathbf{F}_3 + \mathbf{r}_B \times \mathbf{B} \\
&= [(2 \; \mathbf{i} + 4 \; \mathbf{j}) \times (750 \; \mathbf{i})] + [(2 \; \mathbf{i} + 4 \; \mathbf{j}) \times (-800 \; \mathbf{j})] \\
&\quad + [(3 \; \mathbf{i} + 2 \; \mathbf{j}) \times (-981 \; \mathbf{j})] + [(6 \; \mathbf{i} + 4 \; \mathbf{j}) \times (-B_x \mathbf{i}] \\
&= -3000 \; \mathbf{k} - 1600 \; \mathbf{k} - 2943 \; \mathbf{k} + 4B_x \; \mathbf{k} = (-7543 + 4B_x) \; \mathbf{k} = 0
\end{aligned}
$$

Equating the coefficients of $\mathbf{i}$, $\mathbf{j}$, and $\mathbf{k}$ to zero gives

$$A_x - B_x + 750 = 0$$
$$A_y - 800 - 981 = 0$$
$$-7543 + 4B_x = 0$$

Solving simultaneously yields:

$$A_x = 1136 \; \text{N} \qquad A_y = 1781 \; \text{N} \qquad B_x = 1886 \; \text{N}$$

Therefore:

$$\mathbf{A} = 1136 \; \mathbf{i} + 1781 \; \mathbf{j} \; \text{N} \qquad \mathbf{B} = -1886 \; \mathbf{i} \; \text{N} \qquad \text{Ans.}$$

The results are obviously identical to those obtained using a scalar analysis. For two-dimensional problems, scalar methods are usually preferred. ∎

## Example Problem 6-5

A ladder weighing 250 lb is supported by a post and held in place by a cable as shown in Fig. 6-39a. Assume that all surfaces are smooth. Determine the tension in the cable and the forces on the ladder at the contacting surfaces.

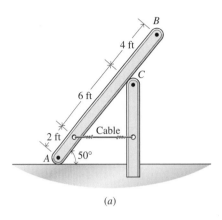

(a)

### SOLUTION

A free-body diagram of the ladder is shown in Fig. 6-39b. All surfaces are smooth; therefore, the reaction at $A$ is a vertical force $\mathbf{A}_y$ and the reaction at $C$ is a force $\mathbf{C}$ perpendicular to the ladder. The cable exerts a tension $\mathbf{T}$ on the ladder in the direction of the cable. Since the ladder is subjected to a general coplanar force system, three independent equilibrium equations are available to solve for the unknown magnitudes of forces $\mathbf{A}_y$, $\mathbf{C}$, and $\mathbf{T}$.

### Solution Using Eqs. (6-4):

$$+\rightarrow \Sigma F_x = 0: \qquad T - C \sin 50° = 0$$

$$T - 0.7660C = 0 \qquad \text{(a)}$$

$$+\uparrow \Sigma F_y = 0: \qquad A_y + C \cos 50° - 250 = 0$$

$$A_y + 0.6428C = 250 \qquad \text{(b)}$$

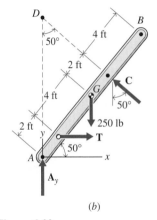

(b)   **Figure 6-39**

$$+\downarrow \Sigma M_A = 0: \qquad C(8) - T(2 \sin 50°) - 250(6 \cos 50°) = 0$$

$$8C - 1.5321T = 964.2 \qquad \text{(c)}$$

Solving Eqs. (a) and (c) simultaneously yields

$$T = +108.2 \text{ lb} \qquad \mathbf{T} = 108.2 \text{ lb} \rightarrow \qquad \textbf{Ans.}$$

$$C = +141.2 \text{ lb} \qquad \mathbf{C} = 141.2 \text{ lb} \searrow 40° \qquad \textbf{Ans.}$$

Once $C$ is known, $A_y$ is determined from Eq. (b) as

$$A_y = 250 - 0.6428C = 250 - 0.6428(141.2)$$

$$A_y = +159.2 \text{ lb} \qquad \mathbf{A}_y = 159.2 \text{ lb} \uparrow \qquad \textbf{Ans.}$$

### Solution Using Eqs. (6-6):

Determination of **T**:
The determination of the cable tension **T** can be simplified by taking moments about the point of concurrence (off the bar) of forces $\mathbf{A}_y$ and $\mathbf{C}$. Thus,

$$+\downarrow \Sigma M_D = 0: \qquad T[(8/\sin 50°) - 2 \sin 50°] - 250(6 \cos 50°) = 0$$

$$T = +108.2 \text{ lb} \qquad \mathbf{T} = 108.2 \text{ lb} \rightarrow \qquad \textbf{Ans.}$$

Determination of $\mathbf{A}_y$:

$+\!\!\downarrow \Sigma M_C = 0$:     $-A_y(8 \cos 50°) + T(6 \sin 50°) + 250(2 \cos 50°) = 0$

$-A_y(8 \cos 50°) + 108.2(6 \sin 50°) + 250(2 \cos 50°) = 0$

$A_y = +159.2$ lb     $\mathbf{A}_y = 159.2$ lb $\uparrow$     **Ans.**

Determination of $\mathbf{C}$:

$+\!\!\downarrow \Sigma M_A = 0$:     $-T(2 \sin 50°) - 250(6 \cos 50°) + C(8) = 0$

$-108.2(2 \sin 50°) - 250(6 \cos 50°) + C(8) = 0$

$C = +141.2$ lb     $\mathbf{C} = 141.2$ lb $\searrow 40°$     **Ans.**

In many problems, use of Eqs. (6-6) eliminates the need to solve simultaneous equations, as illustrated in this example problem. Once $T$ is known, the equilibrium equations $\Sigma F_x = 0$ and $\Sigma F_y = 0$ could be used to determine $\mathbf{C}$ and $\mathbf{A}_y$ instead of using the equations $\Sigma M_A = 0$ and $\Sigma M_C = 0$. Thus,

$+\!\!\rightarrow \Sigma F_x = 0$:     $T - C \sin 50° = 0$

$$C = \frac{T}{\sin 50°} = \frac{108.2}{\sin 50°}$$

$C = +141.2$ lb     $\mathbf{C} = 141.2$ lb $\searrow 40°$     **Ans.**

$+\!\!\uparrow \Sigma F_y = 0$:     $A_y + C \cos 50° - 250 = 0$

$A_y = 250 - C \cos 50° = 250 - 141.2 \cos 50°$

$A_y = +159.2$ lb     $\mathbf{A}_y = 159.2$ lb $\uparrow$     **Ans.**

Regardless of the set of equilibrium equations used, one must keep in mind that there are no more than three independent equations of equilibrium in two-dimensional problems. ∎

## Example Problem 6-6

A pin-connected two-bar frame is loaded and supported as shown in Fig. 6-40a. Determine the reactions at supports $A$ and $B$. The masses of the two bars are negligible.

### SOLUTION

At first glance it may appear that the two-bar frame shown in Fig. 6-40a is statically indeterminate, with support reactions $A_x$, $A_y$, $B_x$, and $B_y$. Once it is observed, however, that loads are applied to member $AC$ only at pins $A$ and $C$, member $AC$ is identified as a two-force member with the support force at $A$ having a line of action along the line joining the two pins. Note that member $BC$ is not a two-force member.

A free-body diagram of the frame is shown in Fig. 6-40b. No force is shown at C because members AC and BC are not disconnected at pin C. The action of the pin support at A is represented by the single force **A** with a slope

$$\theta_A = \tan^{-1}\frac{200}{300} = 33.69°$$

The action of the pin support at B is represented by force components $B_x$ and $B_y$. Since the frame is subjected to a general coplanar force system, three independent equilibrium equations are available to solve for the unknown magnitudes of forces **A**, $B_x$, and $B_y$.

Determination of **A**:

$$+\!\!\downarrow \Sigma M_B = 0: \qquad -A\sin 33.69°(0.6) + 400(0.16) + 600(0.1) = 0$$

$$A = +372.6 \text{ N} \qquad \mathbf{A} = 373 \text{ N} \measuredangle 33.7° \qquad \textbf{Ans.}$$

Determination of $B_x$:

$$+\!\!\rightarrow \Sigma F_x = 0: \qquad A\cos 33.69° + B_x - 600 = 0$$

$$372.6\cos 33.69° + B_x - 600 = 0$$

$$B_x = +290 \text{ N} \qquad \mathbf{B}_x = 290 \text{ N} \rightarrow$$

Determination of $B_y$:

$$+\!\!\downarrow \Sigma M_A = 0: \qquad B_y(0.6) - 400(0.44) + 600(0.1) = 0$$

$$B_y = +193.3 \text{ N} \qquad \mathbf{B}_y = 193.3 \text{ N} \uparrow$$

The reaction at support B is

$$B = \sqrt{(B_x)^2 + (B_y)^2} = \sqrt{(290)^2 + (193.3)^2} = 348.5 \text{ N} \cong 349 \text{ N}$$

$$\theta_B = \tan^{-1}\frac{B_y}{B_x} = \tan^{-1}\frac{193.3}{290} = 33.69°$$

$$\mathbf{B} = 349 \text{ N} \measuredangle 33.7° \qquad \textbf{Ans.}$$

The results are shown in Fig. 6-40c. ■

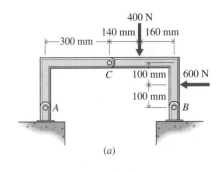

(a)

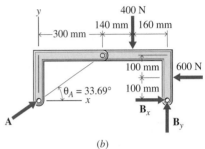

(b)

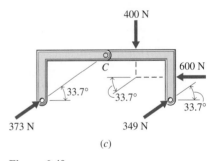

(c)

**Figure 6-40**

---

## Example Problem 6-7

A pin-connected three-bar frame is loaded and supported as shown in Fig. 6-41a. Determine

(a) The reactions at supports A and B.

(b) The shearing stress in the pin at B, which is double shear and has a diameter of 0.5 in.

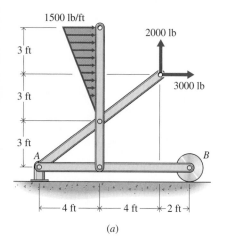

1500 lb/ft

2000 lb

3 ft

3000 lb

3 ft

3 ft

A

B

├── 4 ft ──┼── 4 ft ──┼─2 ft─┤

(a)

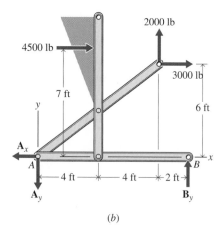

2000 lb

4500 lb

3000 lb

7 ft

6 ft

$y$

$A_x$

$x$

A

B

├── 4 ft ──┼── 4 ft ──┼─2 ft─┤

$A_y$

$B_y$

(b)

**Figure 6-41**

## SOLUTION

(a) A free-body diagram of the frame is shown in Fig. 6-41$b$. The support bracket at $A$ is removed from the pin, but the pin is not removed from the two members of the frame. Thus, the force components $A_x$ and $A_y$ represent the reaction on the pin due to the support bracket. The action of the roller support at $B$ is represented by the force $\mathbf{B}_y$, which acts perpendicular to the horizontal surface at $B$. The distributed load can be represented temporarily on the free-body diagram by a resultant $\mathbf{R}$ (area under load diagram), with a line of action at a distance $y_C$ (centroid of area under load diagram) above support $A$. Thus,

$$R = \text{Area} = \frac{1}{2}(1500)(6) = 4500 \text{ lb} \qquad y_C = 3 + \frac{2}{3}(6) = 7 \text{ ft}$$

Since the frame is subjected to a general coplanar force system, three independent equilibrium equations are available to solve for $A_x$, $A_y$, and $B_y$. Determination of $\mathbf{B}_y$:

$+\!\!\downarrow \Sigma M_A = 0:$   $B_y(10) - 4500(7) + 2000(8) - 3000(6) = 0$

$B_y = +3350 \text{ lb}$   $\mathbf{B} = \mathbf{B}_y = 3350 \text{ lb} \uparrow$   **Ans.**

Determination of $\mathbf{A}_x$:

$+\!\!\rightarrow \Sigma F_x = 0:$   $-A_x + 4500 + 3000 = 0$

$A_x = +7500 \text{ lb}$   $\mathbf{A}_x = 7500 \text{ lb} \leftarrow$

Determination of $\mathbf{A}_y$:

$+\!\!\downarrow \Sigma M_B = 0:$   $A_y(10) - 4500(7) - 2000(2) - 3000(6) = 0$

$A_y = +5350 \text{ lb}$   $\mathbf{A}_y = 5350 \text{ lb} \downarrow$

The reaction at support $A$ is

$$A = \sqrt{(A_x)^2 + (A_y)^2} = \sqrt{(7500)^2 + (5350)^2} = 9213 \text{ lb} \cong 9210 \text{ lb}$$

$$\theta = \tan^{-1}\frac{A_y}{A_x} = \tan^{-1}\frac{5350}{7500} = 35.5°$$

$\mathbf{A} = 9210 \text{ lb} \nearrow 35.5°$   **Ans.**

(b) The shearing stress on the pin at $B$ (double shear) is given by Eq. (4-4) as

$$\tau = \frac{V}{A_s} = \frac{B}{2(\pi d^2/4)} = \frac{3350}{2\pi(0.5)^2/4} = 8530 \text{ psi} \blacksquare$$   **Ans.**

## Example Problem 6-8

A beam is loaded and supported as shown in Fig. 6-42$a$. Determine

(a) The components of the reactions at supports $A$ and $B$.

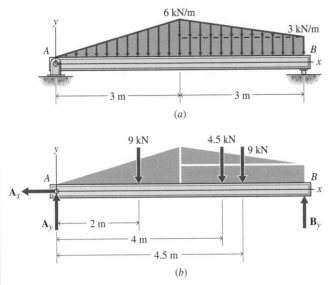

**Figure 6-42**

(b) The shearing stress on a cross section of the 40-mm-diameter pin at support $A$ if the pin is in double shear.

(c) The bearing stress between the beam and the 40- $\times$ 200-mm bearing plate at support $B$.

## SOLUTION

(a) A free-body diagram of the beam is shown in Fig. 6-42$b$. The action of the pin support at $A$ is represented by force components $\mathbf{A}_x$ and $\mathbf{A}_y$. The action of the roller support at $B$ is represented by the force $\mathbf{B}_y$, which acts perpendicular to the horizontal surface at $B$. The distributed loads can be represented temporarily on the free-body diagram by resultants $\mathbf{R}_1$, $\mathbf{R}_2$, and $\mathbf{R}_3$ with lines of action at distances $x_{C1}$, $x_{C2}$, and $x_{C3}$, respectively, from the left support. Thus,

$$R_1 = A_1 = \frac{1}{2}(6)(3) = 9 \text{ kN} \qquad x_{C1} = \frac{2}{3}(3) = 2 \text{ m}$$

$$R_2 = A_2 = \frac{1}{2}(3)(3) = 4.5 \text{ kN} \qquad x_{C2} = 3 + \frac{1}{3}(3) = 4 \text{ m}$$

$$R_3 = A_3 = (3)(3) = 9 \text{ kN} \qquad x_{C3} = 3 + \frac{1}{2}(3) = 4.5 \text{ m}$$

The beam is subjected to a coplanar system of parallel applied forces in the $y$-direction; therefore, $\mathbf{A}_x = \mathbf{0}$. The two remaining equilibrium equations are available to solve for $\mathbf{A}_y$ and $\mathbf{B}_y$.
Determination of $\mathbf{B}_y$:

$$+\zeta \; \Sigma M_A = 0: \qquad B_y(6) - 9(2) - 4.5(4) - 9(4.5) = 0$$

$$B_y = +12.75 \text{ kN} \qquad \mathbf{B}_y = 12.75 \text{ kN} \uparrow \qquad \textbf{Ans.}$$

Determination of $A_y$:

$$+\downarrow \Sigma M_B = 0: \qquad -A_y(6) + 9(4) + 4.5(2) + 9(1.5) = 0$$

$$A_y = +9.75 \text{ kN} \qquad \mathbf{A}_y = 9.75 \text{ kN} \uparrow \qquad \text{Ans.}$$

Alternatively (or as a check):

$$+\uparrow \Sigma F_y = 0: \qquad A_y - 9 - 4.5 - 9 + 12.75 = 0$$

$$A_y = +9.75 \text{ kN} \qquad \mathbf{A}_y = 9.75 \text{ kN} \uparrow$$

(b) The shearing stress on the pin at $A$ is given by Eq. (4-4) as

$$\tau = \frac{V}{A_s} = \frac{A}{2(\pi/4)d^2} = \frac{9.75(10^3)}{2(\pi/4)(0.040)^2}$$

$$= 3.88(10^6) \text{ N/m}^2 = 3.88 \text{ MPa} \qquad \text{Ans.}$$

(c) The bearing stress on the beam at $B$ is given by Eq. (4-6) as

$$\sigma_b = \frac{F}{A_b} = \frac{B}{dt} = \frac{12.75(10^3)}{0.040(0.200)}$$

$$= 1.594(10^6) \text{ N/m}^2 = 1.594 \text{ MPa C} \blacksquare \qquad \text{Ans.}$$

## Example Problem 6-9

The structure shown in Fig. 6-43$a$ is used to support a 9100-lb load. If the diameter of member $AB$ is 1.25 in. and the diameter of the pin at $C$ is 1.75 in., determine

(a) The normal stress on a cross section of member $AB$.
(b) The shearing stress on a cross section of pin $C$.
(c) The deformation of member $AB$ if it is made of steel having a modulus of elasticity of 29,000 ksi.

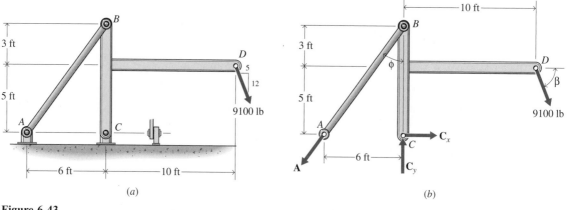

(a)                    (b)

**Figure 6-43**

## SOLUTION

A free-body diagram of the complete structure is shown in Fig. 6-43b. From the geometry of the structure

$$\beta = \tan^{-1}\frac{12}{5} = 67.38° \qquad \phi = \tan^{-1}\frac{6}{8} = 36.87°$$

The action of the pin at $A$ is represented by a force $\mathbf{A}$ with a line of action along the axis of member $AB$, since member $AB$ is a two-force member. The pin at $C$ is represented by force components $\mathbf{C}_x$ and $\mathbf{C}_y$. Since the structure is subjected to a general coplanar force system, three independent equilibrium equations are available to solve for $\mathbf{A}$, $\mathbf{C}_x$, and $\mathbf{C}_y$.

Determination of $\mathbf{A}$:

$+\circlearrowleft \Sigma M_C = 0$:     $A \cos \phi \, (6) - 9100 \sin \beta \, (10) - 9100 \cos \beta \, (5) = 0$

$A \cos 36.87° \, (6) - 9100 \sin 67.38° \, (10) - 9100 \cos 67.38° \, (5) = 0$

$$A = +21{,}150 \text{ lb}$$

Determination of $\mathbf{C}_x$:

$+\!\lfloor \Sigma M_B = 0$:     $C_x(8) - 9100 \sin \beta \, (10) + 9100 \cos \beta \, (3) = 0$

$C_x(8) - 9100 \sin 67.38° \, (10) + 9100 \cos 67.38° \, (3) = 0$

$$C_x = +9187 \text{ lb}$$

Determination of $\mathbf{C}_y$:

$+\!\uparrow \Sigma F_y = 0$:     $C_y - 9100 \sin \beta - A \cos \phi = 0$

$C_y - 9100 \sin 67.38° - 21{,}150 \cos 36.87° = 0$

$$C_y = +25{,}320 \text{ lb}$$

The reaction at support $C$ is

$$C = \sqrt{(C_x)^2 + (C_y)^2} = \sqrt{(9187)^2 + (25{,}320)^2}$$

$$= 26{,}940 \text{ lb}$$

(a) The normal stress on a cross section of member $AB$ is given by Eq. (4-2) as

$$\sigma_{AB} = \frac{F_{AB}}{A_{AB}} = \frac{21{,}150}{(\pi/4)(1.25)^2} = 17.23(10^3) \text{ psi} = 17.23 \text{ ksi T} \qquad \textbf{Ans.}$$

(b) The shearing stress on a cross section of pin $C$ is given by Eq. (4-4). Since the pin is in single shear (see Fig. 6-43a),

$$\tau = \frac{C}{A_s} = \frac{26{,}940}{(\pi/4)(1.75)^2} = 11.20(10^3) \text{ psi} = 11.20 \text{ ksi} \qquad \textbf{Ans.}$$

(c) The deformation of member $AB$ is given by Eq. (4-20b) as

$$\delta = \frac{F_{AB}L_{AB}}{E_{AB}A_{AB}} = \frac{21{,}150(120)}{29{,}000{,}000(\pi/4)(1.25)^2} = 0.0713 \text{ in.} \quad \blacksquare \qquad \textbf{Ans.}$$

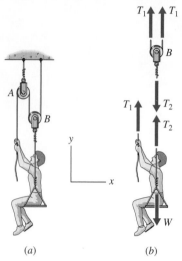

(a)

(b)

**Figure 6-44**

## Example Problem 6-10

The mass of the man shown in Fig. 6-44a is 80 kg. How much force must he exert on the rope of the chair lift in order to support himself?

### SOLUTION

Free-body diagrams for the man and for pulley $B$ of the lift system are shown in Fig. 6-44b. Summing moments about the axis of the pulley verifies that the tension force in a continuous rope that passes over a frictionless pulley has a constant magnitude. Summing forces in a vertical direction then yields

$$+\uparrow \Sigma F_y = 0: \qquad T_1 + T_1 - T_2 = 0$$

from which $T_2 = 2T_1$.

Turning now to the free-body diagram for the man and summing forces in a vertical direction yields

$$+\uparrow \Sigma F_y = 0: \qquad T_1 + T_2 - W = 0$$
$$T_1 + 2T_1 - mg = 0$$
$$3T_1 - 80(9.81) = 0$$

from which

$$T_1 = 262 \text{ N} \quad \blacksquare$$

**Ans.**

## Example Problem 6-11

Lever $ABC$ supports a 100-lb load $W$ as shown in Fig. 6-45a. The rigid lever is held in a horizontal position by the cable $AD$. Using $a = 30$ in., $R = 15$ in., and $L = 45$ in., calculate

(a) The tension in the cable $AD$ if the cable is rigid.
(b) The tension in the cable $AD$ if the cable is a $\frac{3}{32}$-in.-diameter steel ($E = 29,000$ ksi) wire.
(c) The tension in the cable $AD$ if the cable is a $\frac{3}{32}$-in.-diameter aluminum ($E = 10,600$ ksi) wire.
(d) The percent difference in the tension for the steel and aluminum wires compared to the rigid cable.

### SOLUTION

(a) When both the cable and the lever are assumed to be rigid, the free-body diagram for the lever is as shown in Fig. 6-45b. The moment equilibrium equation for lever $ABC$ is

$$+ \circlearrowleft \Sigma M_B = 0: \qquad T(15) - 100(30) = 0$$

Thus the tension in the rigid cable is

$$T = 200 \text{ lb}$$

**Ans.**

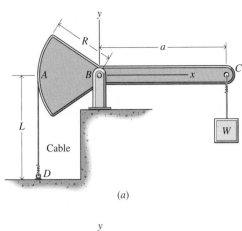

(a)

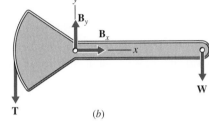

(b)

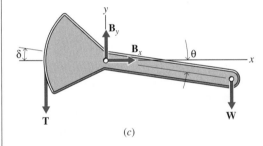

(c)

**Figure 6-45**

(b) When the cable is assumed to be deformable (a $\frac{3}{32}$-in.-diameter steel wire), the free-body diagram for the lever after the cable deforms (deformed state of equilibrium) is as shown in Fig. 6-45c. The moment equilibrium equation for lever $ABC$ is

$$+ \circlearrowleft \Sigma M_B = 0: \qquad T(15) - 100(30 \cos \theta) = 0$$

and the tension in the deformed cable is

$$T = 200 \cos \theta \qquad\qquad\qquad\qquad (a)$$

Equation (a) cannot be solved for $T$ since $\theta$ is unknown. Since the remaining equations of equilibrium do not provide the additional information needed to solve for $T$, the problem is statically indeterminate. A review of Section 4-8 reveals that statically indeterminate problems are solved by using the equilibrium equations [in this case Eq. (a)] together with equations obtained from the deformation of the member.

The deformation (stretch) of the cable is given by Eq. (4-20b) as

$$\delta = \frac{TL}{EA} = \frac{T(45)}{(29)(10^6)\pi(3/32)^2/4} = 224.79(10^{-6})T \tag{b}$$

Combining Eqs. (a) and (b) gives

$$\delta = 44.959(10^{-3}) \cos \theta \tag{c}$$

However, the cable wraps around a circular sector on the lever, so

$$\delta = 15\theta \tag{d}$$

in which $\theta$ is in radians. Combining Eqs. (c) and (d) gives

$$15\theta = 44.959(10^{-3}) \cos \theta \tag{e}$$

which can be solved by using trial and error, by using numerical methods, or by graphing both sides of Eq. (e) and locating the intersection of the two curves. The solution to Eq. (e) is

$$\theta = 0.00300 \text{ rad} \cong 0.1717°$$

Once $\theta$ is known, Eq. (a) gives the tension in the cable

$$T = 199.999 \text{ lb} \qquad \textbf{Ans.}$$

A check on the normal stress in the steel wire ($\sigma = P/A$) yields $\sigma = 29.0$ ksi. This stress is within the linear range of the stress-strain behavior of the steels used to produce wire products. The normal stress must be in the linear range of the stress-strain diagram for Eq. (b) to be valid.

(c) Changing the modulus of elasticity in Eq. (b) to 10,600 ksi causes Eq. (e) to become

$$15\theta = 122.999(10^{-3}) \cos \theta$$

which gives

$$\theta = 0.008266 \text{ rad} \cong 0.474°$$

Once $\theta$ is known, Eq. (a) gives the tension in the cable

$$T = 199.993 \text{ lb} \qquad \textbf{Ans.}$$

A check on the normal stress in the aluminum wire yields $\sigma = 29.0$ ksi, which is again within the linear range of the stress–strain behavior of the aluminum so that Eq. (b) is valid.

(d) The percent difference between the tension in the steel wire and the rigid cable is

$$\%D = \frac{200 - 199.999}{199.999}(100) = 0.0005\% \qquad \textbf{Ans.}$$

The percent difference between the tension in the aluminum wire and the rigid cable is

$$\%D = \frac{200 - 199.993}{199.993}(100) = 0.0035\% \qquad \textbf{Ans.}$$

Equilibrium requirements should always be satisfied when a body is in the deformed configuration. However, this example shows that *for most engineering problems, forces may be determined, within engineering accuracy, using the equilibrium equations and a free-body diagram in the undeformed configuration. These forces may then be used to determine stresses and deformations with sufficient accuracy for most engineering applications.* ■

## Example Problem 6-12

The tower crane shown in Fig. 6-46a is used to lift construction materials. The counterweight A weighs 31,000 N, the motor M weighs 4500 N, the weight of the boom AB is 36,000 N and can be considered acting at point G, and the weight of the tower is 23,000 N and can be considered as acting at its midpoint.

(a) If the tower is lifting a 9000-N load, calculate and plot the reaction forces on the feet at D and E as a function of the distance x at which the weight is being lifted (4 ≤ x ≤ 36 m).

(b) It is desired that the reaction forces at the feet D and E always be greater than 4500 N to ensure that the tower is never in danger of tipping over. Plot $W_{max}$, the maximum load that can be lifted as a function of the distance x (4 ≤ x ≤ 36 m).

## SOLUTION

The free-body diagram of the crane is shown in Fig. 6-46b. There is nothing trying to push the crane in the horizontal direction. Therefore, no horizontal reaction is needed on the footpads to prevent horizontal motion, and the reaction forces on the footpads are shown as a pair of vertical forces. This leaves only two equations of equilibrium relating the four variables D, E, W, and x. Writing force equilibrium in the y-direction and moment equilibrium about point H, which is midway between the two footpads (where x = 0), gives

$$+\uparrow \ \Sigma F_y = 0: \qquad D + E - W - 94{,}500 = 0 \qquad (a)$$

$$+ \circlearrowright \ \Sigma M_H = 0: \qquad 3E - 3D - Wx + 31{,}000(5)$$

$$+ 4500(3) - 36{,}000(6) = 0 \qquad (b)$$

Solving Eqs. (a) and (b) yields

$$D = 39{,}333.3 + \frac{W(3 - x)}{6} \ N \qquad (c)$$

$$E = 55{,}166.5 + \frac{W(x + 3)}{6} \ N \qquad (d)$$

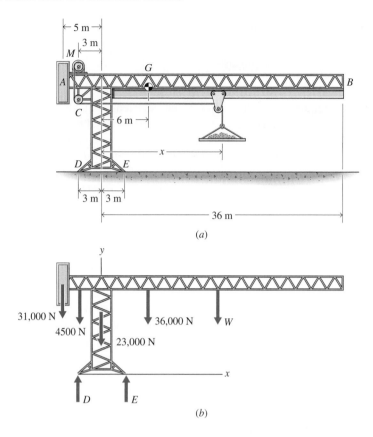

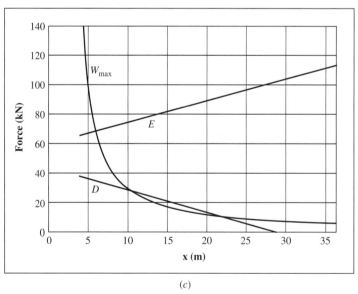

**Figure 6-46**

(a) If $W = 9000$ N, Eqs. (c) and (d) give

$$D = 43,833.3 - 1500x \text{ N} \tag{e}$$

$$E = 59,666.6 + 1500x \text{ N} \tag{f}$$

The force $E$ increases linearly with $x$ and the force $D$ decreases linearly with $x$. This result is shown on the graph of Fig. 6-46c. Note that the force $D = 0$ and the crane will tip if

$$x \geq \frac{43{,}833.3}{1500} = 29.22 \text{ m}$$

(b) If $D = 4500$, Eq. (c) gives the maximum weight that can be lifted

$$W_{\max} = \frac{209{,}000}{x - 3} \text{ N} \tag{g}$$

The maximum weight decreases rapidly from about 105 kN when $x = 5$ m to about 6 kN when $x = 36$ m. This result is also shown on the graph of Fig. 6-46c. ■

# PROBLEMS

### Introductory Problems

**6-13\*** A beam is loaded and supported as shown in Fig. P6-13. Determine the reactions at supports $A$ and $B$.

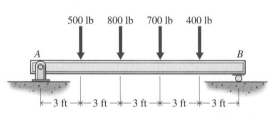

**Figure P6-13**

**6-14\*** A beam is loaded and supported as shown in Fig. P6-14. Determine the reaction at support $A$.

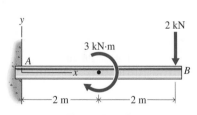

**Figure P6-14**

**6-15\*** A 30-lb force **P** is applied to the brake pedal of an automobile as shown in Fig. P6-15. Determine the force **Q** applied to the brake cylinder and the reaction at support $A$.

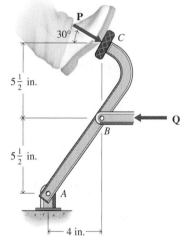

**Figure P6-15**

**6-16** A beam is loaded and supported as shown in Fig. P6-16. Determine the reactions at supports $A$ and $B$.

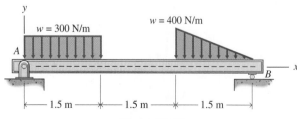

**Figure P6-16**

**6-17** A rope and pulley system is used to support a body $W$ as shown in Fig. P6-17. Each pulley is free to rotate, and the rope is continuous over the pulleys. Determine the tension $T$ in the rope required to hold body $W$ in equilibrium if the weight of body $W$ is 400 lb. Assume that all rope segments are vertical.

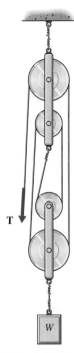

**Figure P6-17**

**6-18\*** Pulleys 1 and 2 of the rope and pulley system shown in Fig. P6-18 are connected and rotate as a unit. The radii of pulleys 1 and 2 are 100 mm and 300 mm, respectively. Rope $A$ is wrapped around pulley 1 and is fastened to pulley 1 at point $A'$. Rope $B$ is wrapped around pulley 2 and is fastened to pulley 2 at point $B'$. Rope $C$ is continuous over pulleys 3 and 4. Determine the tension $T$ in rope $C$ required to hold body $W$ in equilibrium if the mass of body $W$ is 225 kg.

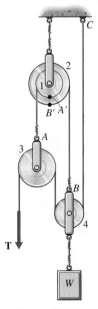

**Figure P6-18**

**6-19** Three pipes are supported in a pipe rack as shown in Fig. P6-19. Each pipe weighs 100 lb. Determine the reactions at supports $A$ and $B$.

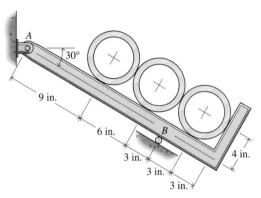

**Figure P6-19**

**6-20** The man shown in Fig. P6-20 has a mass of 75 kg; the beam has a mass of 40 kg. The beam is in equilibrium with the man standing at the end and pulling on the cable. Determine the force exerted on the cable by the man and the reaction at support $C$.

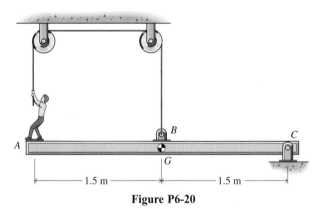

**Figure P6-20**

**Intermediate Problems**

**6-21\*** A 75-lb load is supported by an angle bracket, pulley, and cable as shown in Fig. P6-21. Determine
(a) The force exerted on the bracket by the pin at $C$.
(b) The reactions at supports $A$ and $B$ of the bracket.
(c) The shearing stress on a cross section of the $\frac{1}{4}$-in.-diameter pin at $C$, which is in double shear.

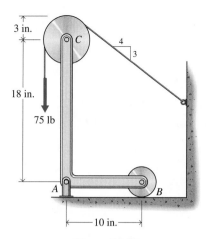

**Figure P6-21**

**6-22\*** A pipe strut *BC* is loaded and supported as shown in Fig. P6-22. Determine
(a) The reactions at supports *A* and *C*.
(b) The shearing stress on a cross section of the 10-mm-diameter pin at *C*, which is in double shear.
(c) The elongation of cable *AB* if it is made of aluminum alloy ($E = 73$ GPa) and has a diameter of 6 mm.

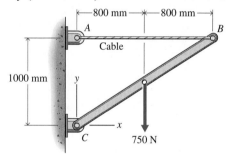

**Figure P6-22**

**6-23** The lawn mower shown in Fig. P6-23 weighs 35 lb and has a center of gravity at *G*. Determine
(a) The magnitude of the force **P** required to push the mower at a constant velocity.
(b) The forces exerted on the front and rear wheels by the inclined surface.
(c) The shearing stresses in the $\frac{1}{2}$-in.-diameter shoulder bolts of the front and rear wheels.

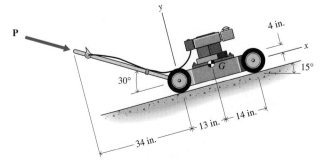

**Figure P6-23**

**6-24** The coal wagon shown in Fig. P6-24 is used to haul coal from a mine. If the mass of the coal and wagon is 2000 kg (with its center of mass at *G*), determine
(a) The magnitude of the force **P** required to move the wagon at a constant velocity.
(b) The force exerted on each of the front wheels by the inclined surface.
(c) The shearing stress in each of the 25-mm-diameter front axles.

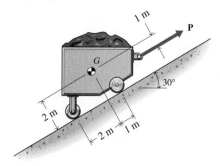

**Figure P6-24**

**6-25\*** A lever is loaded and supported as shown in Fig. P6-25. Determine
(a) The reactions at *A* and *C*.
(b) The normal stress in the $\frac{1}{2}$-in.-diameter rod *CD*.
(c) The shearing stress in the $\frac{1}{2}$-in.-diameter pin at *A*, which is in double shear.
(d) The change in length of rod *CD*, which is made of a material with a modulus of elasticity of $30(10^6)$ psi.

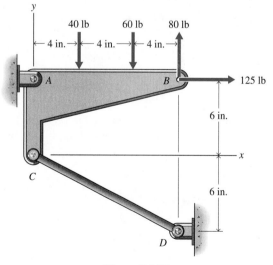

**Figure P6-25**

**6-26** The wood plane shown in Fig. P6-26 moves with a constant velocity when subjected to the forces shown. Determine
(a) The shearing force of the wood on the plane.
(b) The normal force, and its location, of the wood on the plane.

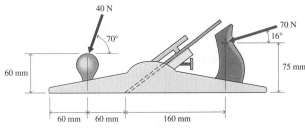

**Figure P6-26**

**6-27** A bracket of negligible weight is used to support the distributed load shown in Fig. P6-27. Determine

(a) The reactions at the supports $A$ and $B$.

(b) The shearing stress in the 0.25-in.-diameter pin at $A$, which is in single shear.

(c) The bearing stress between the bracket and the 1- × 1-in. bearing plate at $B$.

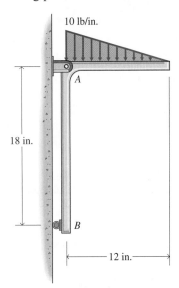

**Figure P6-27**

**6-28*** Determine the force **P** required to push the 135-kg cylinder over the small block shown in Fig. P6-28.

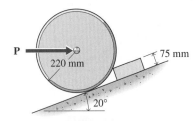

**Figure P6-28**

**6-29** The electric motor shown in Fig. P6-29 weighs 25 lb. Due to friction between the belt and pulley, the belt forces have magnitudes of $T_1 = 21$ lb and $T_2 = 1$ lb. Determine

(a) The support reactions at $A$ and $B$.

(b) The shearing stress in the $\frac{1}{4}$-in.-diameter pin at $A$, which is in single shear.

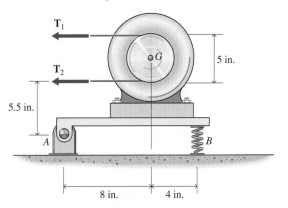

**Figure P6-29**

**Challenging Problems**

**6-30*** Bar $AB$ of Fig. P6-30 has a uniform cross section, a mass of 25 kg, and a length of 1 m. Determine the angle $\theta$ for equilibrium.

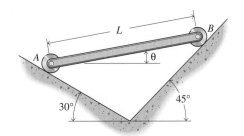

**Figure P6-30**

**6-31*** The wrecker truck of Fig. P6-31 has a weight of 15,000 lb and a center of gravity at $G$. The force exerted on the rear (drive) wheels by the ground consists of both a normal component $B_y$ and a tangential component $B_x$, while the force exerted on the front wheels consists of a normal force $A_y$ only. Determine the maximum pull $P$ that the wrecker can exert when $\theta = 30°$ if $B_x$ cannot exceed $0.8B_y$ (because of friction considerations) and the wrecker does not tip over backward (the front wheels remain in contact with the ground).

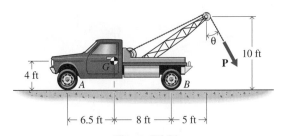

**Figure P6-31**

**6-32\*** Pulleys *A* and *B* of the chain hoist shown in Fig. P6-32 are connected and rotate as a unit. The chain is continuous, and each of the pulleys contains slots that prevent the chain from slipping. Determine the force **F** required to hold a 450-kg block *W* in equilibrium if the radii of pulleys *A* and *B* are 90 mm and 100 mm, respectively.

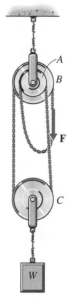

**Figure P6-32**

**6-33** The crane and boom shown in Fig. P6-33 weigh 12,000 lb and 600 lb, respectively. When the boom is in the position shown, determine
(a) The maximum load that can be lifted by the crane.
(b) The tension in the cable used to raise and lower the boom when the load being lifted is 3600 lb.
(c) The pin reaction at boom support *A* when the load being lifted is 3600 lb.
(d) The required size of pin *A* (which is in double shear) if the shearing stress must not exceed 12 ksi.

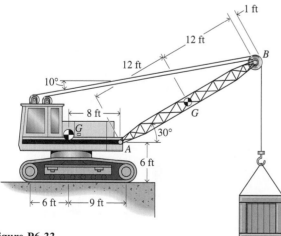

**Figure P6-33**

**6-34** A bar *AB* of negligible mass is supported in a horizontal position by two cables as shown in Fig. P6-34. Determine
(a) The magnitude of force **P** and the force in each cable.
(b) The change in length of the 15-mm-diameter cable *BD*, if it is initially 1 m long and is made of steel with a modulus of elasticity of 200 GPa.

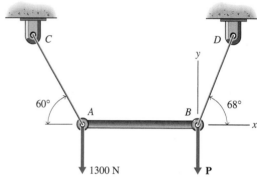

**Figure P6-34**

**6-35\*** The garage door *ABCD* shown in Fig. P6-35 is being raised by a cable *DE*. The one-piece door is a homogeneous rectangular slab weighing 225 lb. Frictionless rollers *B* and *C* run in tracks at each side of the door as shown. Determine
(a) The force in the cable and the reactions at the rollers when *d* = 75 in.
(b) The shearing stress (single shear) in the $\frac{3}{8}$-in.-diameter rods that connect the rollers to the door.

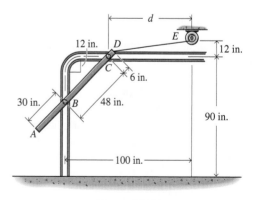

**Figure P6-35**

**6-36** The lever shown in Fig. P6-36 is formed in a quarter circular arc of radius 450 mm. Determine the angle $\theta$ if neither of the reactions at *A* or *B* can exceed 200 N.

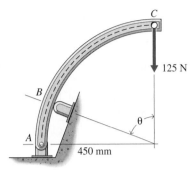

**Figure P6-36**

**6-37** A man is slowly raising a 20-ft-long homogeneous pole weighing 150 lb as shown in Fig. P6-37. The lower end of the pole is kept in place by smooth surfaces. Determine the force exerted by the man to hold the pole in the position shown.

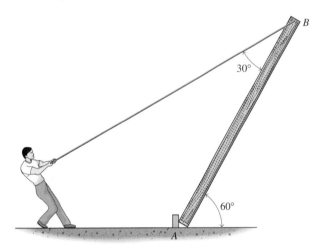

**Figure P6-37**

## Computer Problems

**6-38** The lever shown in Fig. P6-36 is a quarter circular arc of radius 450 mm. The 25-mm-diameter pin at $A$ is smooth and frictionless and is in single shear.
(a) Plot $A$, the magnitude of the pin force at $A$, and $B$, the force on the smooth support $B$, as functions of $\theta$ ($10° \leq \theta \leq 80°$), the angle at which the support is located.
(b) Plot $\tau$, the average shear stress in the pin at $A$, as a function of $\theta$ ($10° \leq \theta \leq 80°$).

**6-39** The crane and boom shown in Fig. P6-39 weigh 12,000 lb and 600 lb, respectively. The pulleys at $D$ and $E$ are small, and the cables attached to them remain essentially parallel. The 3-in.-diameter pin at $A$ is smooth and frictionless, and is in double shear. The distributed force exerted on the treads by the ground is equivalent to a single resultant force $N$ acting at some distance $d$ behind point $C$.

(a) Plot $d$, the location of the equivalent force $N$ relative to point $C$, as a function of the boom angle $\theta$ ($0° \leq \theta \leq 80°$) when the crane is lifting a 3600-lb load.
(b) Plot $\tau$, the average shear stress in the pin at $A$, as a function of $\theta$ ($0° \leq \theta \leq 80°$) when the crane is lifting a 3600-lb load.
(c) It is desired that the resultant force on the tread always be at least 1 ft behind $C$ to ensure that the crane is never in danger of tipping over. Plot $W_{max}$, the maximum load that may be lifted, as a function of $\theta$ ($0° \leq \theta \leq 80°$).

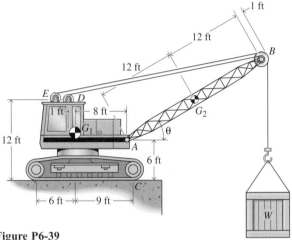

**Figure P6-39**

**6-40** An extension ladder arrangement is being raised into position by a hydraulic cylinder as shown in Fig. P6-40. The center of gravity of the 500-kg ladder is at $G$. The 25-mm-diameter pin at $A$ is smooth, and in double shear.
(a) Plot $\sigma_n$, the normal stress in the 40-mm-diameter piston rod of the hydraulic cylinder, as a function of the angle $\theta$ ($10° \leq \theta \leq 90°$).
(b) Plot $\tau$, the shear stress in the pin at $A$, as a function of $\theta$ ($10° \leq \theta \leq 90°$).

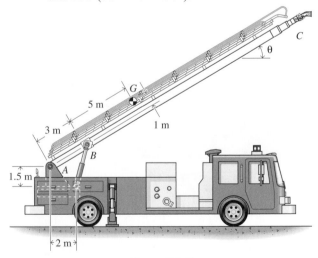

**Figure P6-40**

**6-41** The hydraulic cylinder $BC$ is used to tip the box of the dump truck shown in Fig. P6-41. The pins at $A$ (one on each side of the box) each have a diameter of 1.5 in. and are in double shear. If the combined weight of the box and the load is 22,000 lb and acts through the center of gravity $G$,
  (a) Plot $C$, the force in the hydraulic cylinder, as a function of the angle $\theta$ ($0° \le \theta \le 80°$).
  (b) Plot $\tau$, the shear stress in the pin at $A$, as a function of $\theta$ ($0° \le \theta \le 80°$).

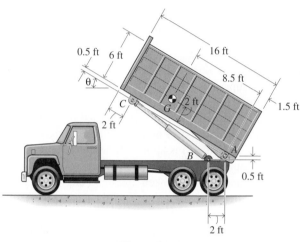

**Figure P6-41**

**6-42** The wrecker truck of Fig. P6-42 has a mass of 6800 kg and a center of gravity at **G**. The force exerted on the rear (drive) wheels by the ground consists of both a normal component $B_y$ and a tangential component $B_x$, while the force exerted on the front wheels consists of a normal force $A_y$ only. Plot $P$, the maximum pull that the wrecker can exert, as a function of $\theta$ ($0° \le \theta \le 90°$) if $B_x$ cannot exceed $0.8B_y$ (because of friction consid-

erations) and the wrecker does not tip over backward (the front wheels remain in contact with the ground).

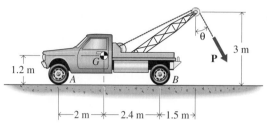

**Figure P6-42**

**6-43** The garage door $ABCD$ shown in Fig. P6-43 is being raised by a cable $DE$. The one-piece door is a homogeneous rectangular slab weighing 225 lb. Frictionless rollers $B$ and $C$ run in tracks at each side of the door as shown.
  (a) Plot $T$, the tension in the cable, as a function of $d$ ($0 \le d \le 100$ in.).
  (b) Plot $B$ and $C$, the forces on the frictionless rollers, as a function of $d$ ($0 \le d \le 100$ in.).

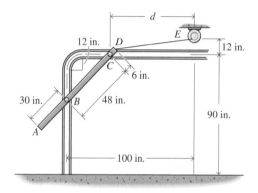

**Figure P6-43**

# 6-4 FRAMES AND MACHINES

Structures that contain members other than two-force members are called *frames* or *machines*. While frames and machines may also contain one or more two-force members, they always contain at least one member that is acted upon by forces at more than two points or is acted upon by both forces and moments.

The main distinction between frames and machines is that frames are rigid structures while machines are not. For example, the structure shown in Fig. 6-47a is a frame. Since it is a rigid body, three support reactions (Fig. 6-47b) are sufficient to fix it in place and overall equilibrium is sufficient to determine the three support reactions.

The structure of Fig. 6-48a is a machine although it is also referred to as a nonrigid frame or a linkage. It is a nonrigid in the sense that it depends on its supports to maintain its shape. The lack of internal rigidity is compensated for

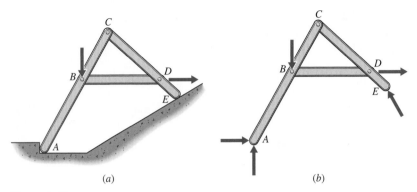

**Figure 6-47**

by an extra support reaction (Fig. 6-48b). In this case, overall equilibrium is not sufficient to determine all four support reactions. The structure must be taken apart and analyzed even if the only information desired is the support reactions.

In a more specific sense, the term *machine* is usually used to describe devices such as pliers, clamps, nutcrackers, and so on that are used to magnify the effect of forces. In each case, a force (input) is applied to the handle of the device and a much larger force (output) is applied by the device somewhere else. Like nonrigid frames, these machines must be taken apart and analyzed even if the only information desired is the relationship between the input and output forces.

The method of solution for frames and machines consists of taking the structures apart, drawing free-body diagrams of each of the components, and writing the equations of equilibrium for each of the free-body diagrams. Since some of the members of frames and machines are not two-force members, the directions of the forces on these members are not known. The analysis of frames and machines will consist of solving for the equilibrium of a system of rigid bodies rather than a system of particles.

**Frames** The method of analysis for frames can be demonstrated by using the table shown in Fig. 6-49a. None of the members that make up the table are two-force members. Although the table can be folded up by unhooking the top from the leg, in normal use the table is a stable, rigid structure. Therefore, the table is a frame.

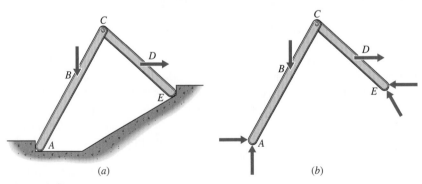

**Figure 6-48**

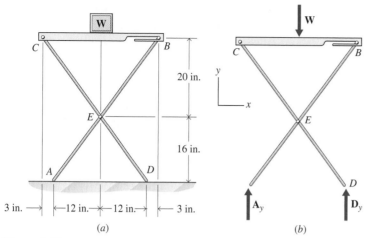

**Figure 6-49**

The analysis will be started by first drawing the free-body diagram of the entire table (Fig. 6-49*b*) for which the equations of equilibrium

$$+\uparrow \Sigma F_y = 0: \qquad A_y + D_y - W = 0$$

$$+\curvearrowleft \Sigma M_A = 0: \qquad 24D_y - 12W = 0$$

yield the support reactions

$$A_y = W/2 \qquad \text{and} \qquad D_y = W/2$$

Next, the table is taken apart and free-body diagrams of each of its parts are drawn (Fig. 6-50). Since none of the members are two-force members, the directions of the forces at joints *B*, *C*, and *E* are not known—they are not directed along the members! Although the forces may be represented in terms of any convenient components, the free-body diagrams must take into account Newton's third law of action and reaction. That is, when drawing free-body diagrams, the forces exerted by one member on a second must be equal in magnitude and opposite in direction to the forces exerted by the second member on the first. For Fig. 6-50 this is effected by showing the components of the force exerted by member *AB* on member *CD* at joint *E* (Fig. 6-50*c*) to have equal magnitude and opposite direction to the components of the force exerted by member *CD* on member *AB* at joint *E* (Fig. 6-50*b*) and similarly for the other joints.

The smooth floor at *D* can only exert an upward force on the leg, and the force should be shown as such on the free-body diagrams. Similarly, the horizontal component of force exerted by the slot on the leg *AB* can only act to the left and should be shown as such. If the values of these forces turn out to be negative, either the solution is in error or the table is not in equilibrium.

The proper direction of the other force components are not as clear. While it is easy to guess that the vertical force components $B_y$ and $C_y$ act upward on the table top *BC* and downward on the legs, it is not as easy to decide whether to draw $E_y$ acting upward or downward on leg *AB*. At this point it doesn't matter, since the frictionless pin connections can support a force in either direction. The direction in which a force is shown on one member is unimportant as long

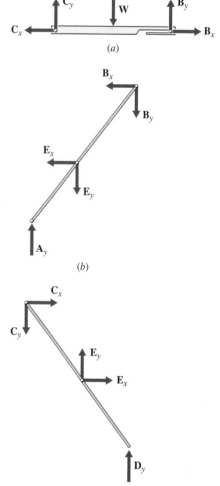

**Figure 6-50**

as the force is represented consistent with Newton's third law on each part of the structure. If a force component is assumed in the wrong direction, its symbol will simply end up having a negative value. This can be accounted for in the report of final answers as shown in the Example Problems.

Although not all of the members of a frame can be two-force members, it is possible and quite likely that one or more of the members will be two-force members. Take advantage of any such members and show that force as acting in its known direction. But be sure that all forces are not directed along the members. Perhaps one of the most common mistakes that students make is to treat the members of frames as two-force members; that is, to draw all forces as acting along the members.

In most cases, it doesn't matter to which member a pin is attached when the structure is taken apart. There are, however, a few special situations in which it does matter:

> When a pin connects a support and two or more members, the pin must be assigned to (left attached to) one of the members. The support reactions are applied to the pin on this member.

> When a pin connects two or more members and a load is applied to the pin, the pin must be assigned to one of the members. The load is applied to the pin on this member.

Following these simple rules will avoid confusion as to where the loads and support reactions should be applied.

Finally, the equations of equilibrium are written for each part of the frame and are solved for the joint forces. There are three independent equations of equilibrium for each part; hence, for the table parts of Fig. 6-50 there will be nine equations to solve for the six remaining unknown forces ($B_x$, $B_y$, $C_x$, $C_y$, $E_x$, and $E_y$). Prior solution of the overall equilibrium of the frame for the support reactions will have reduced three of these equations to a check of the consistency of the answers.

## Machines
The method described for frames is also used to analyze machines and other nonrigid structures. In each case, the structure is taken apart, free-body diagrams are drawn for each part, and the equations of equilibrium are applied to each free-body diagram. For machines and nonrigid structures, however, the structure must be taken apart and analyzed even if the only information desired is the support reactions or the relationship between the external forces acting upon it.

The method of analysis for machines can be demonstrated by using the simple garlic press shown in Fig. 6-51a. Forces $H_1$ and $H_2$ applied to the handles (the input forces) are converted into forces $G_1$ and $G_2$ applied to the garlic clove (the output forces). Equilibrium of the entire press only gives that $H_1 = H_2$; it gives no information about the relationship between the input forces and the output forces.

To determine the relationship between the input forces and the output forces, the machine must be taken apart and free-body diagrams drawn for each of its parts, as shown in Fig. 6-51b. Then the sum of moments about $B$ of the top handle gives

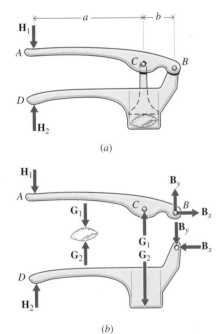

(a)

(b)

**Figure 6-51**

$$+\!\downarrow \Sigma M_B = 0: \qquad (a + b)H_1 - b\,G_1 = 0$$

or

$$G_1 = \frac{a+b}{b} H_1$$

The ratio of the output and input forces is called the mechanical advantage (M.A.) of the machine:

$$\text{Mechanical advantage} = \frac{\text{Output force}}{\text{Input force}}$$

For the garlic press, the mechanical advantage is just

$$\text{M.A.} = \frac{a+b}{b}$$

## Stress and Deformation: Frames and Machines

As stated previously, frames and machines contain members other than two-force members. However, frames and machines may also contain one or more two-force members. In Chapter 4, methods were developed to determine stress and deformation for two-force members. For members more complex than two-force members—for example, any of the members of the frame shown in Fig. 6-49, or members $AB$ or $DB$ of the machine of Fig. 6-51—the calculations of stress and deformation are more complicated than the methods presented in Chapter 4. Methods to solve these problems will be developed later in this book. However, certain stress calculations for frames and machines can be made. For example, shearing stress can be found for a pin, and the bearing stress between a pin and a member can be determined. Example Problem 6-14 will illustrate stress and deformation calculations for a frame.

## Example Problem 6-13

A bag of potatoes is sitting on the chair of Fig. 6-52a. The force exerted by the potatoes on the frame at one side of the chair is equivalent to horizontal and vertical forces of 24 N and 84 N, respectively, at $E$ and a force of 28 N perpendicular to member $BH$ at $G$ (as shown in the free-body diagram of Fig. 6-52b). Find the forces acting on member $BH$.

### SOLUTION

The equations of equilibrium for the entire chair (Fig. 6-52b) are

$$+\rightarrow \Sigma F_x = 0: \qquad\qquad\qquad 24 - 28 \cos\theta = 0$$

$$+\uparrow \Sigma F_y = 0: \qquad\qquad A + B - 84 - 28 \sin\theta = 0$$

$$+\downarrow \Sigma M_B = 0: \qquad 0.2(84) - 0.5(24) - 0.4A + \left(0.3 + \frac{0.5}{\cos\theta}\right)(28) = 0$$

where $\theta = \tan^{-1}(\tfrac{3}{5}) = 30.96°$. The first equation is satisfied identically. The remaining two equations give

$$A = 73.82 \text{ N} \qquad B = 24.58 \text{ N}$$

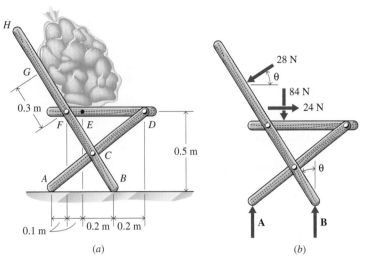

(a)                                                    (b)

**Figure 6-52**

Next the chair is disassembled and free-body diagrams are drawn for each part (Fig. 6-53). For member $DF$, the equilibrium equations can be written

$$+\rightarrow \Sigma F_x=0: \qquad D_x - F_x + 24 = 0$$

$$+\uparrow \Sigma F_y = 0: \qquad F_y + D_y - 84 = 0$$

$$+\downarrow \Sigma M_D = 0: \qquad 0.4(84) - 0.5\,F_y = 0$$

which gives

$$F_y = 67.2 \text{ N} \qquad D_y = 16.80 \text{ N} \qquad D_x = F_x - 24 \text{ N}$$

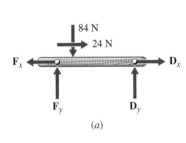

(a)

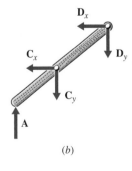

(b)

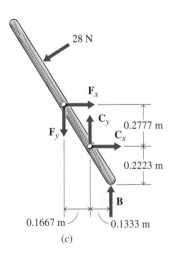

(c)

**Figure 6-53**

Now the equations of equilibrium for member $BH$ are

$$+\rightarrow \Sigma F_x = 0: \qquad F_x + C_x - 28 \cos \theta = 0$$

$$+\uparrow \Sigma F_y = 0: \qquad 24.58 + C_y - 67.2 - 28 \sin \theta = 0$$

$$+\downarrow \Sigma M_C = 0: \qquad \left(0.3 + \frac{0.1667}{\sin \theta}\right)(28) + 0.1333(24.58)$$

$$+ 0.1667(67.2) - 0.2777F_x = 0$$

which have only three unknowns remaining and can be solved to get

$$F_x = 115.1 \text{ N} \qquad C_x = -91.0 \text{ N} \qquad C_y = 57.0 \text{ N}$$

Then the forces acting on member $BH$ are

$$\mathbf{B} = 24.6\ \mathbf{j} \text{ N} \qquad\qquad \textbf{Ans.}$$

$$\mathbf{C} = -91.0\ \mathbf{i} + 57.0\ \mathbf{j} \text{ N} \qquad\qquad \textbf{Ans.}$$

$$\mathbf{F} = 115.1\ \mathbf{i} - 67.2\ \mathbf{j} \text{ N} \qquad\qquad \textbf{Ans.}$$

plus the applied force of 28 N perpendicular to the bar at $G$. These forces are shown on the "report diagram" of Fig. 6-54. ∎

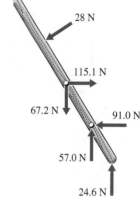

**Figure 6-54**

## Example Problem 6-14

The weight of books on a shelf bracket is equivalent to a vertical force of 75 lb as shown on Fig. 6-55$a$. All members are made of 195-T6 cast aluminum and all pins have $\frac{1}{4}$-in. diameters. Determine

(a) All forces acting on all three members of this frame.

(b) The shearing stress on a cross section of pin $B$, which is in single shear.

(c) The change in length of member $AC$ as a result of the loads if the member has a $\frac{1}{8}$- $\times$ $\frac{1}{2}$-in. rectangular cross section.

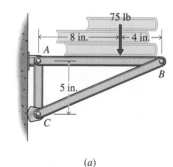

(a)

### SOLUTION

(a) First draw the free-body diagram of the entire shelf bracket as in Fig. 6-55$b$. The equations of overall equilibrium are

$$+ \circlearrowright \Sigma M_A = 0: \qquad 5C - 8(75) = 0$$

$$+\rightarrow \Sigma F_x = 0: \qquad A_x + C = 0$$

$$+\uparrow \Sigma F_y = 0: \qquad A_y - 75 = 0$$

which are solved to get the support reactions

$$A_x = -120.0 \text{ lb} \qquad A_y = 75.0 \text{ lb} \qquad C = 120.0 \text{ lb} \qquad \textbf{Ans.}$$

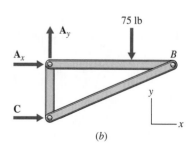

(b)

**Figure 6-55**

Next, dismember the bracket and draw separate free-body diagrams of each member (Fig. 6-56). Members $AC$ and $BC$ are straight two-force members, and

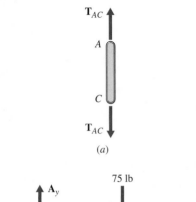

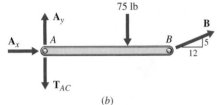

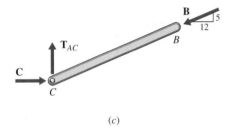

**Figure 6-56**

thus the forces in these members must act along the members. Pin $A$ connects a support and two members. Since member $AC$ is a two-force member, pin $A$ will be assigned to (left attached to) member $AB$. The forces that act on pin $A$ are the support reactions $A_x$ and $A_y$ and a vertical force due to the two-force member $AC$. Similarly, pin $B$ connects two members, one of which is a two-force member. Therefore, pin $B$ is assigned to member $AB$, and the only force on pin $B$ is along the two-force member $BC$. Pin $C$ connects a support and two-members. Since both of the members are two-force members, pin $C$ is arbitrarily assigned to member $BC$, and thus the force on pin $C$ due to member $AC$ is vertical. Then the equations of equilibrium can be written for member $AB$ (Fig. 6-56$b$)

$$+ \circlearrowleft \ \Sigma M_B = 0: \qquad 4(75) + 12T_{AC} - 12(75.0) = 0$$

$$+ \circlearrowleft \ \Sigma M_A = 0: \qquad 12[(^5/_{13})B] - 8(75) = 0$$

from which

$$T_{AC} = 50.0 \text{ lb} \qquad B = 130.0 \text{ lb} \qquad \textbf{Ans.}$$

It is easily verified that these values also satisfy the equations of equilibrium for the other free-body diagrams. These forces are all shown on the "report diagrams" of Fig. 6-57.

(b) The force transmitted by a cross section of pin $B$ is $B = 130.0$ lb. The shearing stress on a cross section of pin $B$ is determined by using Eq. (4-4):

$$\tau = \frac{B}{A_{\text{pin}}} = \frac{130.0}{\pi(^1/_4)^2/4} = 2650 \text{ psi} \qquad \textbf{Ans.}$$

(c) The change in length of member $AC$ is determined by using Eq. (4-20b):

$$\delta = \left(\frac{TL}{EA}\right)_{AC} = \frac{50.0(5)}{(10.3)(10^6)(1/8)(1/2)} = 0.000388 \text{ in.} \quad \blacksquare \qquad \textbf{Ans.}$$

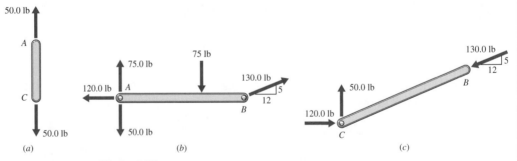

**Figure 6-57**

## Example Problem 6-15

The cord of Fig. 6-58$a$ is wrapped around a frictionless pulley and supports a 100-kg mass $m$. Determine

(a) The reaction components at supports $A$ and $E$, and the force exerted on bar $ABC$ by the pin at $B$.

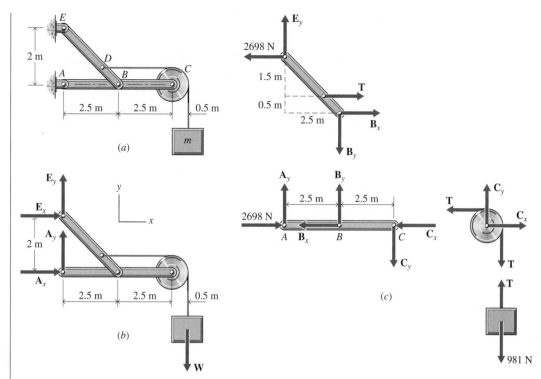

Figure 6-58

(b) The shearing stress in the 25-mm-diameter pin at $B$, which is in single shear.

## SOLUTION

(a) An examination of the free-body diagram for the complete system shown in Fig. 6-58b shows four reaction components; therefore, all of the components cannot be determined by applying the three independent equilibrium equations available for this coplanar system of forces. The force $W$ shown on the diagram is

$$W = mg = 100(9.81) = 981 \text{ N}$$

The reaction component $E_x$ can be determined by summing moments about point $A$. Thus,

$$+ \circlearrowleft \Sigma M_A = 0: \qquad -E_x(2) - (981)(5.5) = 0$$
$$E_x = -2698 \text{ N} \cong 2700 \text{ N} \leftarrow$$

The reaction component $A_x$ can then be determined by summing forces in the $x$-direction. Thus,

$$+ \rightarrow \Sigma F_x = 0: \qquad A_x + E_x = 0$$
$$A_x = -(-2698) = +2698 \text{ N} \cong 2700 \text{ N} \rightarrow$$

Finally, summing forces in the $y$-direction yields

$$+\uparrow \Sigma F_y = 0: \qquad A_y + E_y - 981 = 0$$

$$A_y + E_y = 981 \text{ N}$$

In order to determine forces $A_y$ and $E_y$, the system must be taken apart and analyzed. A complete set of free-body diagrams for all members of the system are shown in Fig. 6-58c. Note that neither $ABC$ nor $BDE$ is a two-force member, and none of the forces that act at pins $A$, $B$, $C$, and $E$ act along either member. Newton's third law (of action and reaction) must be carefully observed when placing pin forces on mating members. From the free-body diagram of mass $m$ whose weight is 981 N,

$$+\uparrow \Sigma F_y = 0: \qquad T - 981 = 0 \qquad T = 981 \text{ N}$$

From a free-body diagram of the pulley,

$$+\rightarrow \Sigma F_x = 0: \qquad C_x - T = 0 \qquad C_x = 981 \text{ N}$$

$$+\uparrow \Sigma F_y = 0: \qquad C_y - T = 0 \qquad C_y = 981 \text{ N}$$

From a free-body diagram for bar $ABC$,

$$+\circlearrowleft \Sigma M_A = 0: \qquad B_y(2.5) - C_y(5) = 0 \qquad B_y = 1962 = 1962 \text{ N} \uparrow$$

$$+\circlearrowleft \Sigma M_B = 0: \qquad -A_y(2.5) - C_y(2.5) = 0 \qquad A_y = -981 \text{ N} = 981 \text{ N} \downarrow$$

$$+\rightarrow \Sigma F_x = 0: \qquad -B_x - C_x + 2698 = 0 \qquad B_x = 1717 \text{ N} = 1717 \text{ N} \leftarrow$$

$$B = \sqrt{(B_x)^2 + (B_y)^2} = \sqrt{(1717)^2 + (1962)^2} = 2607 \text{ N} \cong 2610 \text{ N}$$

$$\theta_x = \tan^{-1} \frac{B_y}{B_x} = \tan^{-1} \frac{1962}{1717} = 48.81°$$

Finally, from the free-body diagram for bar $BDE$,

$$+\uparrow \Sigma F_y = 0: \qquad E_y - B_y = 0 \qquad E_y = 1962 \text{ N} = 1962 \text{ N} \uparrow$$

Therefore, the desired answers are

$$\mathbf{A}_x = 2700 \text{ N} \rightarrow \qquad \mathbf{A}_y = 981 \text{ N} \downarrow \qquad \qquad \textbf{Ans.}$$

$$\mathbf{E}_x = 2700 \text{ N} \leftarrow \qquad \mathbf{E}_y = 1962 \text{ N} \uparrow \qquad \qquad \textbf{Ans.}$$

$$\mathbf{B} = 2610 \text{ N} \searrow 48.8° \qquad \qquad \textbf{Ans.}$$

(b) The shearing stress on a cross section of the pin at $B$ is given by Eq. (4-4),

$$\tau = \frac{B}{A_{\text{pin}}} = \frac{2607}{\pi(0.025)^2/4} = 5.31(10^6) \text{ N/m}^2 = 5.31 \text{ MPa} \blacksquare \qquad \textbf{Ans.}$$

## PROBLEMS

### Introductory Problems

**6-44\*** Determine all forces acting on member *ABE* of the frame of Fig. P6-44.

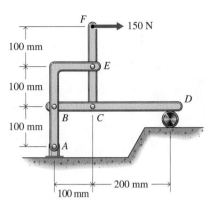

Figure P6-44

**6-45\*** In the linkage of Fig. P6-45, $a = 2.0$ ft, $b = 1.5$ ft, $\theta = 30°$, and $P = 40$ lb. Determine all forces acting on member *BCD*.

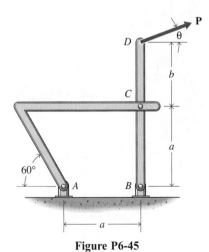

Figure P6-45

**6-46\*** Forces of 5 N are applied to the handles of the paper punch of Fig. P6-46. Determine the force exerted on the paper at *D* and the force exerted on the pin at *B* by handle *ABC*.

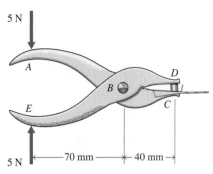

Figure P6-46

**6-47** Forces of 25 lb are applied to the handles of the pipe pliers shown in Fig. P6-47. Determine the force exerted on the pipe at *D* and the force exerted on handle *DAB* by the pin at *A*.

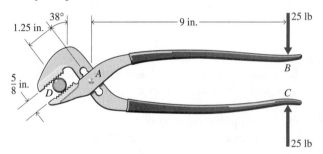

Figure P6-47

**6-48** The jaws and bolts of the wood clamp in Fig. P6-48 are parallel. The bolts pass through swivel mounts so that no moments act on them. The clamp exerts forces of 300 N on each side of the board. Treat the forces on the boards as uniformly distributed over the contact areas and determine the forces in each of the bolts. Show on a sketch all forces acting on the upper jaw of the clamp.

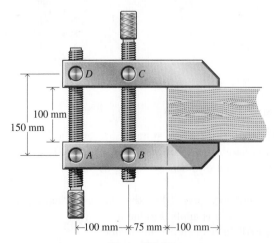

Figure P6-48

**6-49\*** Determine all forces acting on member *ABCD* of the frame of Fig. P6-49.

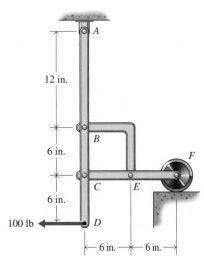

**Figure P6-49**

**6-50** In Fig. P6-50, a cable is attached to the structure at *E*, passes around the 0.8-m-diameter, frictionless pulley at *A*, and then is attached to a 1000-N weight *W*. Determine

(a) The support reaction at *G*.

(b) All forces acting on member *ABCD*.

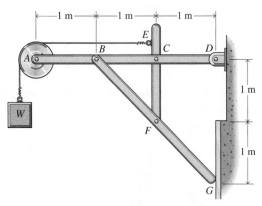

**Figure P6-50**

**Intermediate Problems**

**6-51\*** A pin-connected system of levers and bars is used as a toggle for a press as shown in Fig. P6-51. Three members are joined by pin *D*, as shown in the insert. Determine

(a) The force exerted on the can at *A* when a force of 1000 lb is applied to the lever at *G*.

(b) All forces that act on member *CD*.

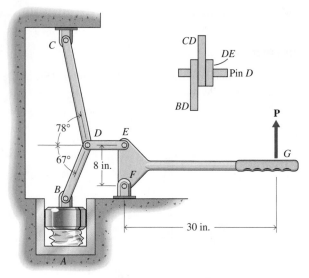

**Figure P6-51**

**6-52\*** The front-wheel suspension of an automobile is shown in Fig. P6-52. The pavement exerts a vertical force of 2700 N on the tire. Determine the force in the spring and the forces at *A*, *B*, and *D*.

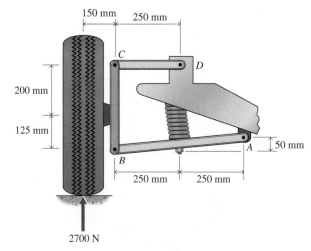

**Figure P6-52**

**6-53** The fold-down chair of Fig. P6-53 weighs 30 lb and has its center of gravity at *G*. Determine

(a) All forces acting on member *ABC*.

(b) The shearing stress on a cross section of the $^3/_8$-in.-diameter pin at *B*, which is in single shear.

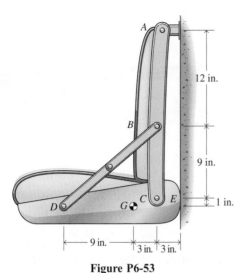

**Figure P6-53**

**6-54** A scissors jack for an automobile is shown in Fig. P6-54. The screw threads exert a force **F** on the blocks at joints $A$ and $B$. Determine
(a) The force **P** exerted on the automobile if $F = 800$ N and $\theta = 15°$, $\theta = 30°$, and $\theta = 45°$.
(b) The shearing stress on a cross section of the 10-mm-diameter pin at $C$, which is in single shear. Solve for each angle in part $a$.

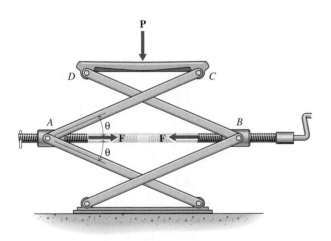

**Figure P6-54**

**6-55\*** A force of 20 lb is required to pull the stopper $DE$ in Fig. P6-55. Determine
(a) All forces acting on member $BCD$.
(b) The shearing stress on the cross section of the $\frac{1}{8}$-in.-diameter pin at $B$, which is in single shear.
(c) The deformation of the $\frac{1}{8}$- $\times$ $\frac{3}{8}$-in. member $AB$, which is made of steel with a modulus of elasticity of $30(10^6)$ psi.

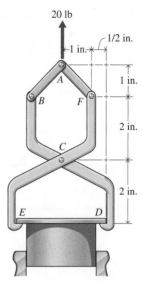

**Figure P6-55**

**6-56** Member $BD$ of the frame shown in Fig. P6-56 is made of structural steel ($E = 200$ GPa) and has a rectangular cross section 50 mm wide by 15 mm thick. All pins have 15-mm diameters. Determine
(a) The axial stress in member $BD$.
(b) The shearing stress on a cross section of pin $C$ if it is loaded in double shear.
(c) The change in length of member $BD$.

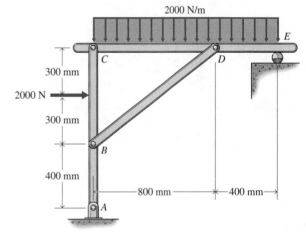

**Figure P6-56**

**6-57** The hoist pulley structure of Fig. P6-57 is rigidly attached to the wall at $C$. A load of sand hangs from the cable that passes around the 1-ft-diameter, frictionless pulley at $D$. The weight of the sand can be treated as a triangular distributed load with a maximum intensity of 70 lb/ft. Determine
(a) All forces acting on member $ABC$.

(b) The shearing stress on the cross section of the $\frac{1}{2}$-in.-diameter pin $D$, which is in double shear.

(c) The change in length of the $\frac{1}{4}$- × 1-in. member $BE$ [$E = 29(10^6)$ psi].

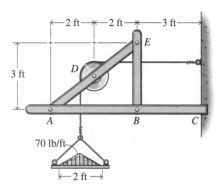

**Figure P6-57**

## Challenging Problems

**6-58\*** A pair of vice grip pliers is shown in Fig. P6-58. Determine the force exerted on the bolt by the jaws of the pliers when a force of magnitude 100 N is applied to the handle. Let $d = 30$ mm.

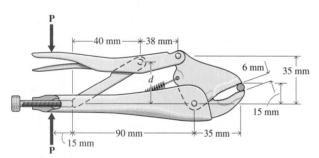

**Figure P6-58**

**6-59\*** Forces of 50 lb are applied to the handles of the bolt cutter of Fig. P6-59. Determine

(a) All forces acting on the handle $ABC$.

(b) The force exerted on the bolt at $E$.

(c) The axial stress in the links at $D$ (one on each side) if each has a $\frac{1}{8}$- × $\frac{3}{4}$-in. rectangular cross section.

(d) The change in length of the links at $D$ if they are 4 in. long and made of SAE 4340 heat-treated steel.

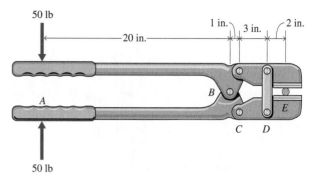

**Figure P6-59**

**6-60\*** The tower crane of Fig. P6-60 is rigidly attached to the building at $F$. A cable is attached at $D$ and passes over small frictionless pulleys at $A$ and $E$. The object suspended from $C$ has a mass of 1530 kg. Determine

(a) All forces acting on member $ABCD$.

(b) The support reactions at $F$.

(c) The shearing stress on a cross section of pin $A$ if it has a diameter of 15 mm and is loaded in double shear.

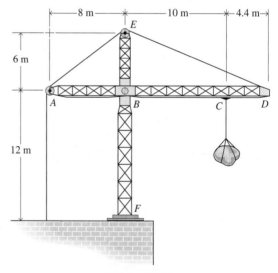

**Figure P6-60**

**6-61** Three bars are connected with smooth pins to form the frame shown in Fig. P6-61. The weights of the bars are negligible. Determine

(a) The reactions at supports $A$ and $E$.

(b) The resultant forces at pins $B$, $C$, and $D$.

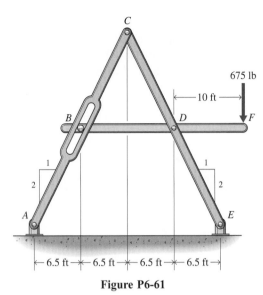

**Figure P6-61**

**6-62** Figure P6-62 is a simplified sketch of the mechanism used to raise the bucket of a bulldozer. The bucket and its contents weigh 10 kN and have a center of gravity at *H*. Arm *ABCD* has a weight of 2 kN and a center of gravity at *B*; arm *DEFG* has a weight of 1 kN and a center of gravity at *E*. The weight of the hydraulic cylinders can be ignored.

(a) Calculate the force in the horizontal cylinders *CJ* and *EI* and all forces acting on arm *DEFG* for the position shown.

(b) Determine the required diameter of the pin at *E* if the shearing stress cannot exceed 120 MPa. The pin is in double shear.

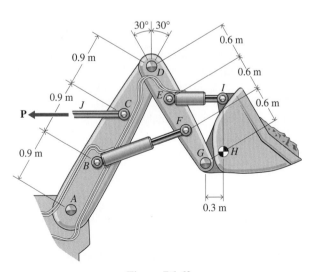

**Figure P6-62**

**6-63** The mechanism of Fig. P6-63 is designed to keep its load level while raising it. A pin on the rim of the 4-ft-diameter pulley fits in a slot on arm *ABC*. Arms *ABC* and *DE* are each 4 feet long, and the package being lifted weighs 80 lb. The mechanism is raised by pulling on the rope that is wrapped around the pulley. Determine the force *P* applied to the rope and all forces acting on the arm *ABC* when the package has been lifted 4 feet, as shown.

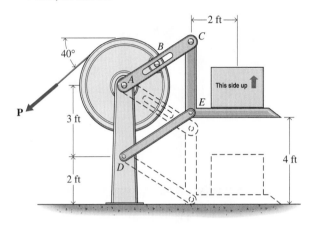

**Figure P6-63**

**6-64** A drum of oil with a mass of 200 kg is supported by a frame (of which there are two) as shown in Fig. P6-64. Determine

(a) All forces acting on member *ACE*.

(b) The elongation of the 20-mm-diameter wire if it is made of a material with a modulus of elasticity of 200 GPa.

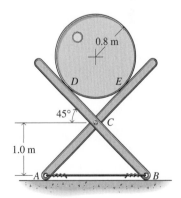

**Figure P6-64**

**Computer Problems**

**6-65** The mechanism shown in Fig. P6-65 is designed to keep its load level while raising it. A pin on the rim of the 4-ft-diameter pulley fits in a slot on arm *ABC*. Arms

*ABC* and *DE* are each 4 ft long, and the package being lifted weighs 80 lb. The mechanism is raised by pulling on the rope that is wrapped around the pulley.

(a) Plot *P*, the force required to hold the platform as a function of the platform height *h* ($0 \le h \le 5.75$ ft).

(b) Plot *A*, *C*, and *E*, the magnitudes of the pin reaction forces at *A*, *C*, and *E* as a function of *h* ($0 \le h \le 5.75$ ft).

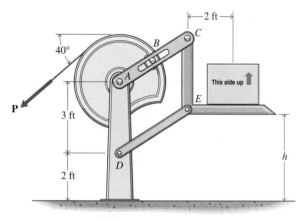

**Figure P6-65**

**6-66** Forces of $P = 100$ N are being applied to the handles of the vise grip pliers shown in Fig. P6-66. Plot the force applied on the bolt by the jaws as a function of the distance *d* ($20$ mm $\le d \le 30$ mm).

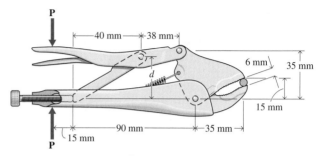

**Figure P6-66**

## 6-5 STATICALLY INDETERMINATE PROBLEMS

In many simple structures (and mechanical systems) constructed with axially loaded members, it is possible to find the reactions at the supports and the forces in the individual members by drawing free-body diagrams and solving equilibrium equations. Such structures (and systems) are referred to as being statically determinate.

For many other structures (and mechanical systems), the equations of equilibrium are not sufficient for the determination of axial forces in the members and reactions at the supports; these structures (and systems) are referred to as being statically indeterminate. Problems of this type can be analyzed by supplementing the equilibrium equations with additional equations involving the geometry of the deformations in the members of the structure or system. The following outline of procedure will be helpful in the analysis of problems involving statically indeterminate situations.

1. Draw a free-body diagram.
2. Note the number of unknowns involved (magnitudes and positions).
3. Recognize the type of force system on the free-body diagram and note the number of independent equations of equilibrium available for this system.
4. If the number of unknowns exceeds the number of equilibrium equations, a deformation equation must be written for each extra unknown.
5. When the number of independent equilibrium equations and deformation equations equals the number of unknowns, the equations can be solved si-

multaneously. Deformations and forces must be related in order to solve the equations simultaneously.

Equation (4-20b) can be used to relate deformations and forces when all stresses are less than the corresponding proportional limits of the materials used in the fabrication of the members. It is recommended that a displacement diagram be drawn showing deformations to assist in obtaining the correct deformation equation. The displacement diagram should be as simple as possible (a line diagram), with the deformations indicated with exaggerated magnitudes and clearly dimensioned. Loading members (members assumed to be rigid), especially, should be indicated by single lines. Note that an equilibrium equation and the corresponding deformation equation must be compatible; that is, when a tensile force is assumed for a member in the free-body diagram, a tensile deformation (stretch) must be indicated for the same member in the deformation diagram. If the diagrams are compatible, a negative result will indicate that the assumption was wrong; however, the magnitude of the result will be correct.

## Example Problem 6-16

A pin-connected structure is loaded and supported as shown in Fig. 6-59a. Member $CD$ is rigid and is horizontal before the load $P$ is applied. Member $A$ is an aluminum alloy bar with a modulus of elasticity of 75 GPa and a cross-sectional area of 1000 mm$^2$. Member $B$ is a structural steel bar with a modulus of elasticity of 200 GPa and a cross-sectional area of 500 mm$^2$. Determine

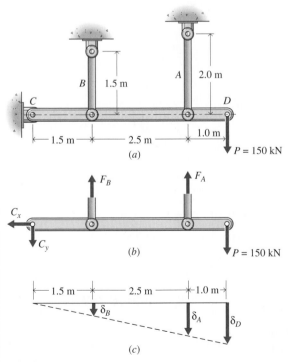

**Figure 6-59**

(a) The normal stresses in bars $A$ and $B$.

(b) The deflection of pin $D$.

## SOLUTION

(a) A free body diagram of member $CD$ and portions of members $A$ and $B$ is shown in Fig. 6-59b. The free-body diagram contains four unknown forces: $C_x$, $C_y$, $F_A$, and $F_B$. Since only three independent equilibrium equations are available, the problem is statically indeterminate. As the load $P$ is applied to member $CD$, it will tend to rotate clockwise about pin $C$ and produce deformations in members $A$ and $B$, as shown in Fig. 6-59c. The extensions shown in Fig. 6-59c are compatible with the tensile forces shown in members $A$ and $B$ in Fig. 6-59b. The unknown reaction at $C$ is not needed to complete the solution of the problem and can be eliminated from further consideration by summing moments about pin $C$

$$+ \circlearrowleft \Sigma M_C = 0: \qquad -P(5) + F_A(4) + F_B(1.5) = 0$$

from which

$$4F_A + 1.5F_B = 150(10^3)(5) = 750(10^3) \text{ N} \qquad (a)$$

As the rigid bar $CD$ rotates about pin $C$, the pins $B$, $A$, and $D$ move on circular arcs about pin $C$. If the angle of rotation is small, these motions can be approximated as vertical (perpendicular to the bar $CD$) displacements, as shown in Fig. 6-59c. Then, the deformation equation is obtained using similar triangles

$$\frac{\delta_A}{4} = \frac{\delta_B}{1.5}$$

$$\frac{F_A L_A}{4 E_A A_A} = \frac{F_B L_B}{1.5 E_B A_B}$$

$$\frac{F_A (2)}{4(75)(10^9)(1000)(10^{-6})} = \frac{F_B (1.5)}{1.5(200)(10^9)(500)(10^{-6})}$$

or

$$F_A = 1.5 F_B \qquad (b)$$

Solving Eqs. (a) and (b) simultaneously yields

$$F_A = 150.0(10^3) \text{ N} = 150.0 \text{ kN}$$

$$F_B = 100.0(10^3) \text{ N} = 100.0 \text{ kN}$$

The normal stresses in the two bars are

$$\sigma_A = \frac{F_A}{A_A} = \frac{150.0(10^3)}{1000(10^{-6})} = 150.0(10^6) \text{ N/m}^2 = 150.0 \text{ MPa T} \qquad \textbf{Ans.}$$

$$\sigma_B = \frac{F_B}{A_B} = \frac{100.0(10^3)}{500(10^{-6})} = 200(10^6) \text{ N/m}^2 = 200 \text{ MPa T} \qquad \textbf{Ans.}$$

(b) Since bar $CD$ rotates as a rigid body, the deflection of pin $D$ is (again using similar triangles),

$$\delta_D = \frac{5}{4}\delta_A = \frac{5(150.0)(10^3)(2)}{4(75)(10^9)(1000)(10^{-6})}$$
$$= 5.00(10^{-3}) \text{ m} = 5.00 \text{ mm} \downarrow \qquad \textbf{Ans.}$$

Note that when pin $D$ moves down 5.00 mm, the rigid bar rotates clockwise 0.001 rad $\cong 0.0573°$. This slight angle will have a negligible effect on the free-body diagram and the equilibrium equation. ■

## Example Problem 6-17

A pin-connected structure is loaded and supported as shown in Fig. 6-60$a$. Member $CD$ is rigid and is horizontal before the load $P$ is applied. After the 150-kN load is applied, the temperature increases by 100°C. Member $A$ is an aluminum alloy bar with a modulus of elasticity of 75 GPa, a cross-sectional area of 1000 mm$^2$, and a thermal coefficient of expansion of 22(10$^{-6}$)/°C. Member $B$ is a structural steel bar with a modulus of elasticity of 200 GPa, a cross-sectional area of 500 mm$^2$, and a thermal coefficient of expansion of 12(10$^{-6}$)/°C. Determine

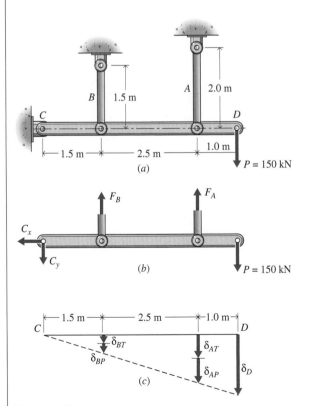

Figure 6-60

(a) The normal stresses in bars $A$ and $B$.

(b) The deflection of pin $D$.

## SOLUTION

(a) A free-body diagram of member $CD$ and portions of members $A$ and $B$ is shown in Fig. 6-60b. The free-body diagram contains four unknown forces: $C_x$, $C_y$, $F_A$, and $F_B$. Since only three independent equilibrium equations are available, the problem is statically indeterminate. As the load $P$ is applied to member $CD$, it will tend to rotate clockwise about pin $C$ and produce deformations in members $A$ and $B$, as shown in Fig. 6-60c. The deformations in bars $A$ and $B$ are functions of both load and temperature change

$$\delta_A = \delta_{AP} + \delta_{AT}$$

$$\delta_B = \delta_{BP} + \delta_{BT} \tag{a}$$

where the subscripts $P$ and $T$ refer to load and temperature respectively. Summing moments about pin $C$

$$+ \circlearrowleft \Sigma M_c = 0: \qquad -P(5) + F_A(4) + F_B(1.5) = 0$$

gives

$$4F_A + 1.5F_B = 150(10^3)(5) = 750(10^3) \tag{b}$$

The deformation equation is obtained using similar triangles

$$\frac{\delta_A}{4} = \frac{\delta_B}{1.5}$$

$$\frac{F_A L_A}{4E_A A_A} + \frac{\alpha_A L_A (\Delta T)}{4} = \frac{F_B L_B}{1.5E_B A_B} + \frac{\alpha_B L_B (\Delta T)}{1.5}$$

$$\frac{F_A (2)}{4(75)(10^9)(1000)(10^{-6})} + \frac{22(10^{-6})(2.0)(100)}{4}$$

$$= \frac{F_B (1.5)}{1.5(200)(10^9)(500)(10^{-6})} + \frac{12(10^{-6})(1.5)(100)}{1.5}$$

from which

$$F_A = 1.5F_B + 15(10^3) \tag{c}$$

Solving Eqs. (b) and (c) simultaneously yields

$$F_A = 153.00(10^3) \text{ N} = 153.00 \text{ kN}$$

$$F_B = 92.000(10^3) \text{ N} = 92.00 \text{ kN}$$

The normal stresses in the two bars are

$$\sigma_A = \frac{F_A}{A_A} = \frac{153.00(10^3)}{1000(10^{-6})} = 153.0(10^6) \text{ N/m}^2 = 153.0 \text{ MPa T} \qquad \textbf{Ans.}$$

$$\sigma_B = \frac{F_B}{A_B} = \frac{92.00(10^3)}{500(10^{-6})} = 184.0(10^6) \text{ N/m}^2 = 184.0 \text{ MPa T} \qquad \textbf{Ans.}$$

(b) Since bar $CD$ rotates as a rigid body, the deflection of pin $D$ is (again using similar triangles),

$$\delta_D = \frac{5}{4} \delta_A = \frac{5 F_A L_A}{4 E_A A_A} + \frac{5}{4} \alpha_A L_A (\Delta T)$$

$$= \frac{5(153.00)(10^3)(2)}{4(75)(10^9)(1000)(10^{-6})} + \frac{5}{4} (22)(10^{-6})(2.0)(100)$$

$$= 10.60(10^{-3}) \text{ m} = 10.60 \text{ mm} \downarrow$$          **Ans.**

# PROBLEMS

## Introductory Problems

**6-67\*** A load $P$ will be supported by a structure consisting of a rigid bar $A$, two aluminum alloy ($E = 10,600$ ksi) bars $B$, and a stainless steel ($E = 28,000$ ksi) bar $C$, as shown in Fig. P6-67. Each bar has a cross-sectional area of 2.00 in². If the bars are unstressed before the load $P$ is applied, determine the normal stresses in the bars after a 40-kip load is applied.

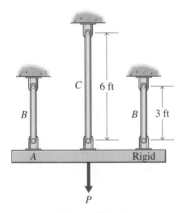

**Figure P6-67**

**6-68\*** The rigid bar $CDE$, shown in Fig. P6-68, is horizontal before the load $P$ is applied. Tie rod $A$ is a hot-rolled steel ($E = 210$ GPa) bar with a length of 450 mm and a cross-sectional area of 300 mm². Post $B$ is an oak timber ($E = 12$ GPa) with a length of 375 mm and a cross-sectional area of 4500 mm². After the 225-kN load $P$ is applied, determine
(a) The normal stresses in bar $A$ and post $B$.
(b) The vertical displacement of point $D$

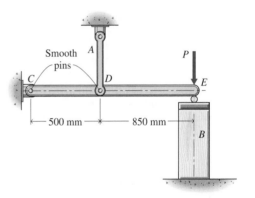

**Figure P6-68**

**6-69** A pin-connected structure is loaded and supported as shown in Fig. P6-69. Member $CD$ is rigid and is horizontal before the load $P$ is applied. Member $A$ is an aluminum alloy bar with a modulus of elasticity of 10,600 ksi and a cross-sectional area of 2.25 in². Member $B$ is a stainless steel bar with a modulus of elasticity of 28,000 ksi and a cross-sectional area of 1.75 in². After the load is applied to the structure, determine
(a) The normal stresses in bars $A$ and $B$.
(b) The vertical displacement of point $D$.

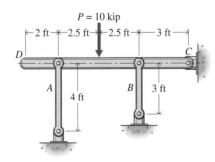

**Figure P6-69**

**6-70** Bar *B* of the pin-connected system of Fig. P6-70 is made of an aluminum alloy [$E_a = 70$ GPa, $A_a = 300$ mm$^2$, and $\alpha_a = 22.5(10^{-6})/°C$], and bar *A* is made of a hardened carbon steel [$E_s = 210$ GPa, $A_s = 1200$ mm$^2$, and $\alpha_s = 11.9(10^{-6})/°C$]. Bar *CDE* is to be considered rigid. When the system is unloaded at 40°C, bars *A* and *B* are unstressed. After load *P* is applied, the temperature of both bars decreases to 15°C. Determine

(a) The normal stresses in bars *A* and *B*.

(b) The vertical displacement (deflection) of pin *E*.

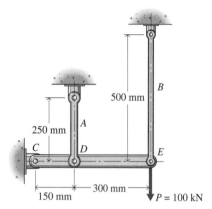

**Figure P6-70**

**Intermediate Problems**

**6-71*** A rigid bar *CD* is loaded and supported as shown in Fig. P6-71. Bars *A* and *B* are unstressed before the 30-kip load *P* is applied. Bar *A* is made of steel ($E = 30,000$ ksi) and has a cross-sectional area of 2 in$^2$. Bar *B* is made of brass ($E = 15,000$ ksi) and has a cross-sectional area of 1.5 in$^2$. Determine

(a) The stresses in bars *A* and *B*.

(b) The vertical displacement (deflection) of pin *C*.

(c) The shearing stress on a cross section of the $^3/_4$-in.-diameter pin at *D*, which is in double shear.

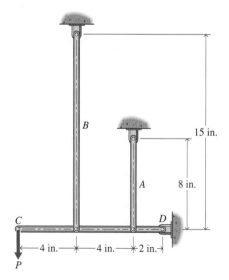

**Figure P6-71**

**6-72*** A rigid bar *CD* is loaded and supported as shown in Fig. P6-72. Bars *A* and *B* are unstressed before the 150-kN load *P* is applied. Bar *A* is made of stainless steel ($E = 190$ GPa) and has a cross-sectional area of 750 mm$^2$. Bar *B* is made of an aluminum alloy ($E = 73$ GPa) and has a cross-sectional area of 1250 mm$^2$. Determine

(a) The stresses in bars *A* and *B*.

(b) The vertical displacement (deflection) of pin *D*.

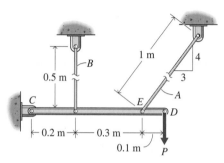

**Figure P6-72**

**6-73** A 40-kip load *P* will be supported by a structure consisting of a rigid bar *A*, two aluminum alloy [$E_a = 10,600$ ksi and $\alpha_a = 12.5(10^{-6})/°F$] bars *B*, and a stainless steel [$E_s = 28,000$ ksi and $\alpha_s = 9.6(10^{-6})/°F$] bar *C*, as shown in Fig. P6-73. The bars are unstressed when the structure is assembled at 72°F. Each bar has a cross-sectional area of 2.00 in$^2$. Determine the normal stresses in the bars after the 40-kip load is applied and the temperature is increased to 250°F.

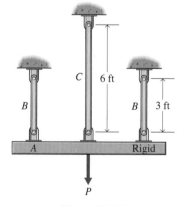

**Figure P6-73**

**Challenging Problems**

**6-74*** A pin-connected structure is loaded and supported as shown in Fig. P6-74. Member *CD* is rigid and is horizontal before the 75-kN load *P* is applied. Bar *A* is made of structural steel ($E = 200$ GPa), and bar *B* is made of an aluminum alloy ($E = 73$ GPa). The cross-sectional areas of members $A$ and $B$ are 625 mm$^2$ and 2570 mm$^2$, respectively. Determine

(a) The vertical displacement of the pin used to apply the load.

(b) The shearing stress on a cross section of the 25-mm-diameter pin at $C$, which is in double shear.

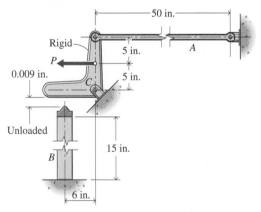

**Figure P6-74**

**6-75\*** Before the 20 kip load $P$ is applied, the arms of the crank $C$ shown in Fig. P6-75 are horizontal and vertical; there is a 0.009-in. gap between the horizontal arm and the brass ($E_b = 15,000$ ksi and $A_b = 12$ in$^2$) post $B$; and the aluminum ($E_a = 10,000$ ksi and $A_a = 2$ in$^2$) rod $A$ is horizontal. If the crank $C$ is rigid, determine

(a) The stresses in members $A$ and $B$.

(b) The change in length of members $A$ and $B$.

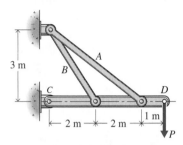

**Figure P6-75**

**6-76** The mechanism of Fig. P6-76 consists of a structural steel ($E = 200$ GPa) rod $A$ with a cross-sectional area of 350 mm$^2$, a cold-rolled brass ($E = 100$ GPa) rod $B$ with a cross-sectional area of 750 mm$^2$, and a rigid bar $C$. The nuts at the top ends of rods $A$ and $B$ are initially tightened to the point where all slack is removed from the mechanism but the bars remain free of stress. If a nut advances 2.5 mm with each full turn (360°), determine

(a) The stresses in the rods when the nut at the top end of rod $B$ rotates 180°.

(b) The vertical displacement at the top end of rod $A$ for part (a).

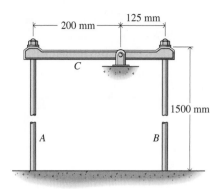

**Figure P6-76**

**6-77** The pin connected structure shown in Fig. P6-77 consists of a cold-rolled bronze [$E_b = 15,000$ ksi and $\alpha_b = 9.4(10^{-6})/°F$] bar $A$, which has a cross-sectional area of 3.00 in$^2$, and two 0.2% C hardened steel [$E_s = 30,000$ ksi and $\alpha_s = 6.6(10^{-6})/°F$] bars $B$, which have cross-sectional areas of 2.50 in$^2$. If the temperature of bar $A$ decreases by 50°F and the temperature of bars $B$ increases by 30°F after the 200-kip load is applied, determine

(a) The normal stresses in the bars.

(b) The displacement of pin $C$.

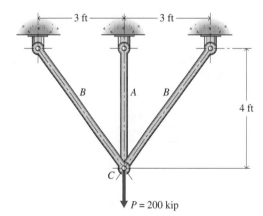

**Figure P6-77**

## Computer Problems

**6-78** Three bars are connected by smooth frictionless pins as shown in Fig. P6-78. Bar $BCD$ is rigid, bar $AB$ is aluminum ($E = 73$ GPa; $L = 750$ mm; $d = 40$ mm; $\sigma_{max} = 240$ MPa), and bar $DE$ is steel ($E = 200$ GPa; $L = 500$ mm; $d = 30$ mm; $\sigma_{max} = 400$ MPa). The 50-mm-diameter pivot pin $C$ is aluminum ($\tau_{max} = 180$ MPa) and is in double shear. After the unit is assembled, the nut $D$ is slowly tightened.

(a) Plot $\sigma_{AB}$, $\sigma_{DE}$, and $\tau_C$ as functions of the distance that the nut advances ($0 \leq \delta_{nut} \leq 2$ mm).

(b) Plot $\delta_{AB}$ and $\delta_{DE}$ as function of $\delta_{nut}$ ($0 \leq \delta_{nut} \leq 2$ mm).

(c) What is the maximum amount $\delta_{nut}$ that the nut can be tightened?

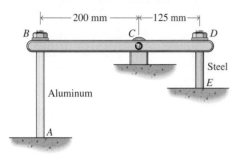

**Figure P6-78**

**6-79** Three bars are connected by smooth frictionless pins as shown in Fig. P6-79. Bar $ABCD$ is rigid, bar $AE$ is aluminum ($E = 10,600$ ksi; $L = 24$ in.; $d = \frac{1}{2}$ in.; $\sigma_{max} = 40$ ksi), and bar $CF$ is steel ($E = 30,000$ ksi; $L = 18$ in.; $d = \frac{3}{4}$ in.; $\sigma_{max} = 50$ ksi). The $\frac{3}{4}$-in.-diameter pivot pin $B$ is steel ($\tau_{max} = 25$ ksi) and is in single shear. The holes in bar $CF$ are slightly overdrilled so that pin $C$ moves down 0.06 in. before contact is made with bar $CF$.

(a) Plot $\sigma_{AE}$, $\sigma_{CF}$, and $\tau_B$ as functions of $P$ ($0 \leq P \leq 10$ kip).
(b) Plot $\delta_{AE}$ and $\delta_{CF}$ as functions of $P$ ($0 \leq P \leq 10$ kip).
(c) What is the maximum force $P$ that the system can withstand?

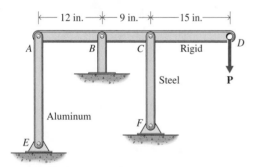

**Figure P6-79**

**6-80** Member $ABCD$ of the pin-connected structure shown in Fig. P6-80 is rigid, bar $BF$ is made of steel [$E_{BF} = 210$ GPa; $A_{BF} = 1200$ mm²; $\alpha_{BF} = 11.9(10^{-6})/°C$], and bar $CE$ is made of an aluminum alloy [$E_{CE} = 73$ GPa; $A_{CE} = 900$ mm²; $\alpha_{CE} = 22.5(10^{-6})/°C$]. As a result of a misalignment of the pin holes at $A$, $B$, and $C$, bar $CE$ must be heated by 80°C (after pins $A$ and $B$ are in place) to permit insertion of pin $C$. Compute and plot

(a) The axial stresses $\sigma_{BF}$ (in the steel bar) and $\sigma_{CE}$ (in the aluminum bar) as functions of the temperature decrease (as bar $CE$ cools back down to room temperature) $\Delta T$ ($0°C \geq \Delta T \geq -80°C$).
(b) The elongations $\delta_{BF}$ (of the steel bar) and $\delta_{CE}$ (of the aluminum bar) as functions of the temperature decrease $\Delta T$ ($0°C \geq \Delta T \geq -80°C$).

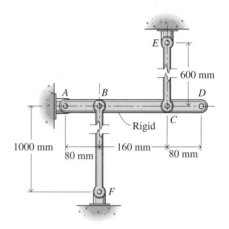

**Figure P6-80**

**6-81** Three bars are connected by smooth frictionless pins as shown in Fig. P6-81. Bar $BCD$ is rigid, bar $AC$ is aluminum [$E = 12,000$ ksi; $L = 36$ in.; $d = 1$ in.; $\sigma_{max} = 20$ ksi; $\alpha = 12.5(10^{-6})/°F$], and bar $DE$ is steel [$E = 30,000$ ksi; $L = 30$ in.; $d = \frac{1}{2}$ in.; $\sigma_{max} = 10$ ksi; $\alpha = 6.6(10^{-6})/°F$]. The $\frac{5}{8}$-in.-diameter pivot pin $B$ is steel ($\tau_{max} = 5$ ksi) and is in single shear. If the temperature drops after the unit is assembled,

(a) Plot $\sigma_{AC}$, $\sigma_{DE}$, and $\tau_B$ as functions of the temperature drop $\Delta T$ ($0 \geq \Delta T \geq -60°F$).
(b) Plot $\delta_{AC}$ and $\delta_{DE}$ as functions of $\Delta T$ ($0 \geq \Delta T \geq -60°F$).
(c) What is the maximum temperature drop that the system can withstand?

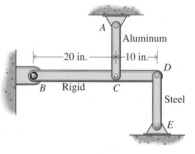

**Figure P6-81**

# 6-6 PLANE TRUSSES

A truss is a structure composed of straight members joined together at their end points and loaded only at the joints (see Fig. 6-61). The airy structure of a truss provides greater strength over large spans than would more solid types of structures. Trusses are commonly seen supporting the roofs of buildings as well as television towers, antennas, aircraft frames and highway bridges. Although not commonly seen, trusses also form the skeletal structure of many large buildings.

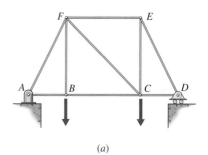

(a)

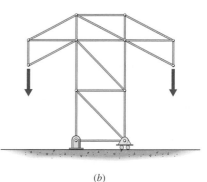

(b)

**Figure 6-61**

> Planar trusses lie in a single plane, and all applied loads must lie in the same plane. Planar trusses are often used in pairs to support bridges, as shown in Fig. 6-62. All members of the truss *ABCDEF* lie in the same vertical plane. Loads on the floor of the bridge are carried by means of the floor construction to the joints *A*, *B*, *C*, and *D*. The loads thus transmitted to the joints lie in the same vertical plane as the truss.
>
>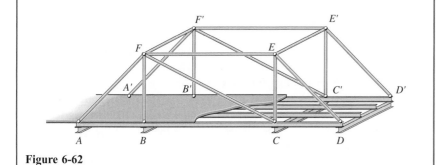
>
> **Figure 6-62**

Four main assumptions are made in the analysis of trusses. One result of the assumptions is that all members of the idealized truss are two-force members. Although the assumptions are idealizations of actual structures, real trusses behave according to these idealizations to a high degree of approximation. The resulting error is usually small enough to justify the assumptions.

Truss members are connected at their ends only.

The first assumption means that the truss of Figs. 6-61*a* and 6-62 should be drawn as shown in Fig. 6-63. In actual practice, the main top and bottom chords frequently consist of members that span several joints, such as member *ABCD* of Fig. 6-62, rather than a series of shorter members between joints. The members of a truss are usually long and slender, however, and can support little lateral load or bending moment. Hence, the noncontinuous member assumption is usually acceptable.

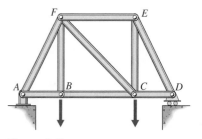

**Figure 6-63**

Truss members are connected by frictionless pins.

In real trusses, the members are usually bolted, welded, or riveted to a gusset plate, as shown in Fig. 6-64*a*, rather than connected by an idealized frictionless pin, as shown in Fig. 6-64*b*. However, experience has shown the frictionless

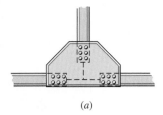

*(a)*

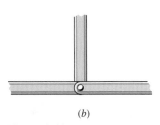

*(b)*

**Figure 6-64**

**Figure 6-65**

pin to be an acceptable idealization as long as the axes of the members all intersect at a single point.

Trusses are loaded only at the joints.

As stated earlier, the members that make up a truss are usually long and slender. Like cables, such members can withstand large tensile (axial) load but cannot withstand moments or large lateral loads. Loads must either be applied directly to the joints as indicated in the diagrams of Fig. 6-61 or must be carried to the joints by a floor structure as shown in Fig. 6-62.

The weights of the members may be neglected.

Frequently in the analysis of trusses the weights of the members are neglected. While this may be acceptable for small trusses, it may not be acceptable for a large bridge truss. Again because the members can withstand little bending moment or lateral load, experience has shown that little error results from assuming the load acts at the joints of the truss. Common practice is to assume that half of the weight of each member acts at each joint.

The result of these four assumptions is that forces act only at the ends of the members. Also, because the pins are assumed to be frictionless, there is no moment applied to the ends of the members. Therefore, by the analysis of Section 6-3, each member is a two-force member supporting only an axial force, as shown in Fig. 6-65. In its simplest form, a truss (such as that shown in Fig. 6-66a) consists of a collection of two-force members held together by frictionless pins as shown in Fig. 6-66b. The forces acting on the pins of the truss shown in Fig. 6-66a are also shown in Fig. 6-66b.

For general two-force members, the forces act along the line joining the points where the forces are applied. Since truss members are usually straight, however, the forces will act along the axis of the member as shown in Figs. 6-65 and

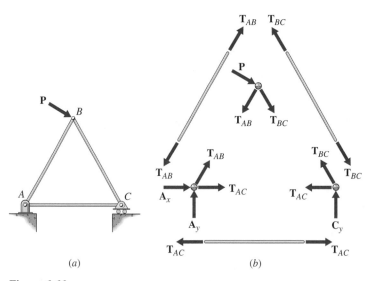

*(a)*          *(b)*

**Figure 6-66**

6-66.[1] Forces that pull on the ends of a member (as in Figs. 6-65 and 6-66) are called *tensile* and tend to elongate the member. Such forces are called *tensile forces*, and the member is said to be in *tension*. Forces that push on the ends of the member tend to shorten the member. Such forces are called *compressive forces*, and the member is said to be in *compression*. It may be noted from Fig. 6-66 that when a joint exerts a force that pulls on the end of a member, the member exerts a force that pulls on the joint.

It is important to distinguish between truss members that are in tension and those that are in compression. The long slender members that make up a truss are very strong in tension but tend to bend or buckle under large compressive loads. Truss members in compression either must be made thicker than the other truss members or must be braced to prevent buckling. Buckling will be discussed in Chapter 11.

One end of a bridge truss is usually allowed to "float" on a rocker or roller support, as shown in Fig. 6-61. Aside from the mathematical requirement (in a planar equilibrium problem, only three support reactions can be determined), such a support is needed to allow for the expansion or contraction of the structure due to temperature variations.

To retain their shape and support the large loads applied to them, trusses must be rigid structures. The simplest structure that is rigid (independent of how it is supported) is a triangle. Of course, the word "rigid" does not mean that a truss will not deform under loading. It will undergo very small deformations, but will very nearly retain its original shape.

"Rigid" is often interpreted also to mean that the truss will retain its shape when removed from its supports or when one of the supports is free to slide. In this sense, the truss of Fig. 6-67 is rigid, whereas the truss of Fig. 6-68 is not. The truss of Fig. 6-68 is called a *compound truss*, and the lack of internal rigidity is made up for by an extra external support reaction.

The basic building block of all trusses is a triangle. Large trusses are constructed by attaching several triangles together. One method of construction starts with a basic triangular element, such as triangle *ABC* of Fig. 6-69. Additional triangular elements are added one at a time by attaching one new joint (for example, *D*) to the truss, using two new members (for example, *BD* and *CD*). A truss that can be constructed in this fashion is called a *simple truss*. While it might appear that all trusses composed of triangles are simple trusses, such is not the case. For example, neither of the trusses of Figs. 6-67 or 6-68 is a simple truss.

The importance of a simple truss is that it allows a simple way to check the rigidity and solvability of a truss. Clearly, since a simple truss is constructed solely of triangular elements, it is always rigid. Also, since each new joint brings two new members with it, a simple relationship exists between the number of joints *j* and the number of members *m* in a simple plane truss:

$$m = 2j - 3 \qquad (6\text{-}7)$$

In the discussion of the method of joints that follows, this will be seen to be exactly the condition necessary to guarantee that the number of equations to be

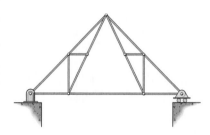

**Figure 6-67**

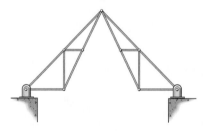

**Figure 6-68**

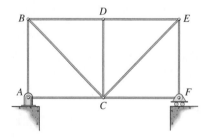

**Figure 6-69**

---

[1]For curved two-force members, however, the line joining the ends is not the axis of the member. All of the trusses considered in this chapter will contain only straight two-force members.

solved (2*j*) is the same as the number of unknowns to be solved for (*m* member forces and 3 support reactions).

Although Eq. (6-7) ensures that a simple plane truss is rigid and solvable, it is neither sufficient nor necessary to ensure that a nonsimple plane truss is rigid and solvable. For example, the nonsimple plane trusses of Figs. 6-67 and 6-68 are both rigid (at least while attached to their supports) and solvable, although one (Fig. 6-67) satisfies Eq. (6-7) and the other (Fig. 6-68) does not. A tempting generalization of Eq. (6-7) is:

$$m = 2j - r \qquad (6\text{-}8)$$

where *r* is the number of support reactions. The trusses in Figs. 6-67 and 6-68 satisfy Eq. (6-8), as do all simple trusses that have the customary three support reactions. However, constructions can be envisioned for which even Eq. (6-8) is not a proper test of the solvability of a truss.

## Method of Joints

Consider the truss of Fig. 6-70*a*, whose free-body diagram is shown in Fig. 6-70*b*. Since the entire truss is a rigid body in equilibrium, each part must also be in equilibrium. The method of joints consists of taking the truss apart, drawing separate free-body diagrams of each part—each member and each pin, as in Fig. 6-71—and applying the equations of equilibrium to each part of the truss in turn.

The free-body diagrams of the members in Fig. 6-71 have only axial forces applied to their ends because of assumptions about how a truss is constructed and loaded. The symbol $T_{BC}$ is used to represent the unknown force in member *BC*. (No significance is attached to the order of the subscripts; that is, $T_{BC} = T_{CB}$). Since the lines of action of the member forces are all known, the force in each member is completely specified by giving its magnitude and sense; that is, whether the force points away from the member as in Fig. 6-71 or toward the member. Thus the force (a vector) is represented by the scalar symbol $T_{BC}$. The sense of the force will be taken from the sign of $T_{BC}$; positive indicates the direction drawn on the free-body diagram, negative indicates the opposite direction.

Forces that point away from a member, as in Fig. 6-71, tend to stretch the member and are called *tensile*. Forces that point toward a member tend to compress the member and are called *compressive*. Whether a member is in compression or tension is usually not known ahead of time. Although some people try to guess and draw some of the forces in tension and others in compression, it is not necessary to do so. In this book, all free-body diagrams will be drawn as though all members are in tension. A negative value for a force in the solution will indicate that the member was really in compression. This can be reported either by saying that $T_{BC} = -2500$ lb or by saying that $T_{BC} = 2500$ lb (C). The latter is preferable, since it does not depend on whether member *BC* was assumed to be in tension or in compression in the free-body diagram.

According to Newton's third law (of action and reaction), the force exerted on a member by a pin and the force exerted on a pin by a member are equal and opposite. Therefore, the same symbol $T_{AB}$ is used for the force exerted by the member *AB* on pin *B* and for the force exerted by pin *B* on member *AB*. Having drawn the free-body diagrams of the members as two-force members ensure that the members are in equilibrium. No further information is obtained from the free-body diagrams of the two-force members; therefore, they may be discarded for

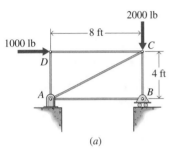

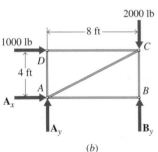

**Figure 6-70**

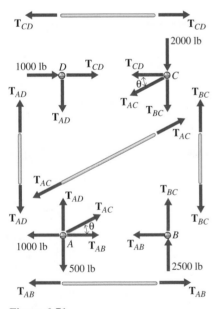

**Figure 6-71**

the remainder of the analysis. The analysis of the truss reduces to considering the equilibrium of the joints that make up the truss—hence the name Method of Joints.

Equilibrium of the joints that make up the truss is expressed by drawing a separate free-body diagram for each joint and writing the equilibrium equation

$$\Sigma \mathbf{F} = \mathbf{0} \qquad (6\text{-}9)$$

for each joint. Since each joint consists of concurrent forces in a plane, moment equilibrium gives no useful information and Eq. (6-9) has only two independent components. Therefore, for a plane truss containing $j$ pins, there will be a total of $2j$ independent scalar equations available. But according to Eq. (6-7), this is precisely the number of independent equations needed to solve for the $m$ member forces and 3 support reactions of a simple truss.

Solution of the $2j$ equations is significantly simplified if a joint can be found on which only two unknown forces and one or more known forces act (for example, joint $D$ of Fig. 6-71). In this case the two equations for this joint can be solved independently of the rest of the equations. If such a joint is not readily available, one can usually be created by solving the equilibrium equations for the entire truss first, that is, solve for the support reactions (see joint $B$ of Fig. 6-71). Once two of the unknown forces have been determined, they can be treated as known forces on the free-body diagrams of the other joints. The joints are solved sequentially in this fashion until all forces are known.

As mentioned earlier, a negative value for a force indicates that the member is in compression rather than in tension. It is unnecessary to go back to the free-body diagram and change the direction of the arrow. In fact, doing so is likely only to cause confusion. The free-body diagrams should all be drawn consistently. A negative value for a symbol on one free-body diagram translates to the same negative value for the same symbol on another free-body diagram.

Once all of the forces have been determined, a summary should be made listing the magnitude of the force in each member and whether the member is in tension or compression (see the Example Problems).

Finally, it must be noted that the equations of equilibrium for the entire truss are contained in the equations of equilibrium of the joints (see Problem 6-94). That is, if all the joints are in equilibrium and all the members are in equilibrium, then the entire truss is also in equilibrium. A consequence of this is that the three support reactions can be determined along with the $m$ member forces from the $2j$ equations of equilibrium of the joints. Overall equilibrium in this case may be used as a check of the solution. However, if overall equilibrium is used first to determine the support reactions and help start the method of joints, then three of the $2j$ joint equations of equilibrium will be redundant and may be used as a check of the solution.

## Zero-Force Members 
Frequently, certain members of a given truss carry no load. Zero-force members in a truss usually arise in one of two general ways. The first is

> When only two members form a noncollinear truss joint and no external load or support reaction is applied to the joint, then both members must be zero-force members.

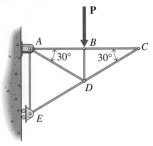

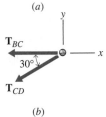

**Figure 6-72**

The truss of Fig. 6-72*a* is an example of this condition. The free-body diagram of pin *C* is drawn in Fig. 6-72*b*. The equations of equilibrium for this joint,

$$+\rightarrow \Sigma F_x = 0: \qquad -T_{BC} - T_{CD}\cos 30° = 0$$

$$+\uparrow \Sigma F_y = 0: \qquad -T_{CD}\sin 30° = 0$$

are trivially solved to get

$$T_{CD} = 0 \quad \text{and} \quad T_{BC} = 0$$

That is, for this particular truss and for this particular loading, the two members *BC* and *CD* could be removed without affecting the solution or even (in this particular case) the stability of the truss.

The second way in which zero-force members normally arise in a truss is as follows:

> When three members form a truss joint for which two of the members are collinear and the third forms an angle with the first two, then the noncollinear member is a zero-force member, provided no external force or support reaction is applied to that joint. The two collinear members carry equal loads (either both tension or both compression).

Such a condition arises, for example, when the load of Fig. 6-72*a* is moved from pin *B* to pin *C*, as in Fig. 6-73*a*. The free-body diagram of pin *B* is drawn in Fig. 6-73*b*. The equations of equilibrium for this joint are

$$+\rightarrow \Sigma F_x = 0: \qquad -T_{AB} + T_{BC} = 0$$

$$+\uparrow \Sigma F_y = 0: \qquad -T_{BD} = 0$$

Thus, since joint *B* is now unloaded, the force in member *BD* vanishes and the forces in members *AB* and *BC* are equal in magnitude—either both tension (both positive) or both compression (both negative).

Once it is known that *BD* is a zero-force member, the same reasoning can then be used to show that member *AD* carries no load. The free-body diagram of pin *D* is drawn in Fig. 6-73*c*. To simplify the calculations, coordinate axes are chosen along and normal to the collinear members *CD* and *DE*. The equations of equilibrium are then

$$+\nearrow \Sigma F_{x'} = 0: \qquad -T_{DE} - T_{AD}\cos 60° + T_{BD}\cos 60° + T_{CD} = 0$$

$$+\nwarrow \Sigma F_{y'} = 0: \qquad T_{AD}\sin 60° + T_{BD}\sin 60° = 0$$

But since $T_{BD} = 0$ (*BD* is already known to be a zero-force member), then

$$T_{AD} = 0 \quad \text{and} \quad T_{DE} = T_{CD}$$

Thus, for the loading of Fig. 6-73*a*, both members *AD* and *BD* are zero-force members.

These *zero-force members cannot simply be removed from the truss and discarded; they are needed to guarantee the stability of the truss.* If members *AD* and *BD* were removed, there would be nothing to prevent some small disturbance from moving pin *D* slightly out of alignment, as in Fig. 6-74*a*. Then the free-

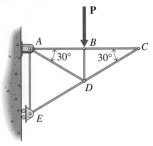

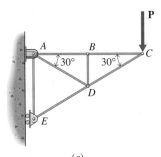

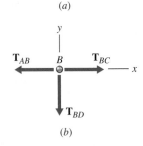

**Figure 6-73**

body diagram of pin $D$ would look like Fig. 6-74$b$. Again choosing axes along and normal to the line $CE$ gives the equilibrium equations

$$+\nearrow \Sigma F_{x'} = 0: \qquad -T_{DE} \cos \phi + T_{CD} \cos \phi = 0$$

$$+\nwarrow \Sigma F_{y'} = 0: \qquad T_{DE} \sin \phi + T_{CD} \sin \phi = 0$$

The first of these equations requires that $T_{CD} = T_{DE}$, while the second requires that $T_{CD} = -T_{DE}$. The only way both of these equations can be satisfied is if both forces equal zero. But equilibrium of pin $C$ requires that $T_{CD}$ not be zero. What has happened, of course, is that the truss is no longer in static equilibrium. Pin $D$ will continue to buckle outward and the truss will collapse.

A seemingly trivial solution to the stability problem would be to replace the two members $CD$ and $DE$ with a single member $CE$ and to replace the two members $AB$ and $BC$ with a single member $AC$. While this solution would satisfy the statics part of the problem, it would not take care of the tendency for long slender members to buckle when subjected to large compressive loads. Therefore, long members such as member $CE$ of Fig. 6-73 are usually replaced by a pair of shorter members and the mid-joint braced if analysis of the truss indicates the member is likely to be in compression for some expected loading. Long members such as member $AC$ of Fig. 6-72 must also be replaced by a pair of shorter members and the mid-joint braced if it is ever desired to load the truss at some point along the long member.

Thus one must not be too quick to discard truss members just because they carry no load for a given configuration. These members are often needed to carry part of the load when the applied loading changes and they are almost always needed to guarantee the stability of the truss.

While recognizing these and other special joint-loading conditions can simplify the analysis of a truss, such recognition is not required to solve the truss. If one does not recognize that a member is a zero-force member, drawing the free-body diagram and writing the equilibrium equations will immediately show that it is a zero-force member. Also, these shortcuts should be applied with care. If there is any doubt about whether or not a member is a zero-force member, the prudent choice is to draw the free-body diagram and solve for the member force.

## Method of Sections

As stated in the section on the Method of Joints, if an entire truss is in equilibrium, then each and every part of the truss is also in equilibrium. That does not mean, however, that the truss must be broken up into its most elemental parts—individual members and pins. In the method of sections, the truss will be divided up into just two pieces. Each of these pieces is also a rigid body in equilibrium.

For example, the truss of Fig. 6-75$a$ can be divided into two parts by passing an imaginary section $a$–$a$ through some of its members. Of course, the section must pass entirely through the truss so that complete free-body diagrams can be drawn for each of the two pieces. Since the whole truss is in equilibrium, the part of the truss to the left of section $a$–$a$ and the part of the truss to the right of section $a$–$a$ are both in equilibrium also.

Free-body diagrams of the two parts are drawn in Figs. 6-75$b$ and 6-75$c$, and include the forces on each cut member that was exerted by the other part of the member, which was cut away. Since the members are all straight two-force members, the forces in these members must act along the members as shown.

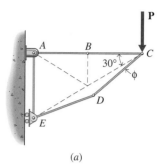

(a)

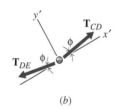

(b)

**Figure 6-74**

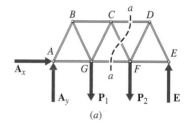

(a)

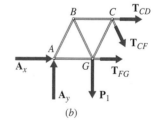

(b)

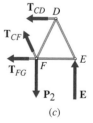

(c)

**Figure 6-75**

Forces in members that have not been cut are internal to the rigid bodies and are not shown on the free-body diagrams. Thus, in order to determine the force in a member, the section must cut through that member.

As is the case with any rigid body in plane equilibrium, three independent equations of equilibrium can be written for each rigid body. The resulting six equations are sufficient to solve for the six unknowns—the forces in the three cut members and the three support reactions. As with the method of joints, the solution of the equations can be simplified if the support reactions are determined from overall equilibrium before the truss is sectioned. Then the equilibrium equations for either free-body diagram will yield the remaining three unknown forces. In this case, the equilibrium equations for the remaining portion of the truss give no new information; they merely repeat the other equilibrium equations (see Problem 6-95).

If a section cuts through four or more members whose forces are unknown, then the method of sections will not generate enough equations of equilibrium to solve for all of the unknown forces. While it still might be possible to obtain values for one or two of the forces (see Example Problem 6-21), it is usually best to use a section that cuts through no more than three members whose forces are unknown.

It will often happen that a section that cuts no more than three members and that passes through a given member of interest cannot be found. In such a case it may be necessary to draw a section through a nearby member and solve for the forces in it first. Then the method of joints can be used to find the forces in the members next to the cut section, or the truss can be further sectioned to find the force in the member of interest.

One of the principal advantages of using the method of sections is that the force in a member near the center of a large truss usually can be determined without first obtaining the forces in the rest of the truss. As a result, the calculation of the force is independent of any errors in other internal forces previously calculated. To find the same force using the method of joints, however, would require that the force in a large number of other members be determined first. Any errors made in the determination of one member force will cause all subsequent forces to be in error as well.

Finally, the method of sections may be used as a spot check when the method of joints or a computer program is used to solve a truss problem with a large number of members. Although it is unlikely that a computer will make an error in its computation, it is quite possible that the input data may be in error. Most often these errors occur when an operator incorrectly enters the coordinates of a joint, incorrectly specifies how the joints are connected, or incorrectly applies a load to the truss. In such cases, the method of sections can be used to check independently the forces in one or two interior members.

Since the members of a truss are straight two-force members, the methods of Sections 4-2 and 4-7 may be used to determine stresses (in the pins and members) and deformations (elongations or contractions) of the members, respectively. The procedure will be illustrated in Example Problem 6-22.

## Example Problem 6-18

Use the method of joints to find the force in each member of the truss of Fig. 6-76a.

## SOLUTION

Truss $ABCD$ is a simple truss with $m = 5$ members and $j = 4$ joints. Therefore, the eight equations obtained from equilibrium of the four joints can be solved for the three support reactions as well as the forces in all five members.

The first step is to draw a free-body diagram of the entire truss (Fig. 6-76b) and write the equilibrium equations

$+\rightarrow \Sigma F_x = 0$: $\qquad\qquad 1000 + A_x = 0$

$+\uparrow \Sigma F_y = 0$: $\qquad\qquad -2000 + A_y + B_y = 0$

$+\downarrow \Sigma M_A = 0$: $\qquad -4(1000) - 8(2000) + 8B_y = 0$

These equations can be solved to get

$$A_x = -1000 \text{ lb} \qquad B_y = 2500 \text{ lb} \qquad A_y = -500 \text{ lb}$$

Next, draw a free-body diagram of pin $D$ (Fig. 6-77d) and solve the equilibrium equations

$+\rightarrow \Sigma F_x = 0$: $\qquad 1000 + T_{CD} = 0$

$+\uparrow \Sigma F_y = 0$: $\qquad -T_{AD} = 0$

to get

$$T_{CD} = -1000 \text{ lb} \qquad T_{AD} = 0 \text{ lb}$$

Next draw a free-body diagram of pin $C$ (Fig. 6-77c) and solve the equilibrium equations

$+\rightarrow \Sigma F_x = 0$: $\qquad -T_{CD} - T_{AC} \cos \theta = 0$

$+\uparrow \Sigma F_y = 0$: $\qquad -2000 - T_{BC} - T_{AC} \sin \theta = 0$

where

$$\sin \theta = \frac{AD}{AC} = \frac{4}{\sqrt{4^2 + 8^2}} = 0.4472$$

and

$$\cos \theta = \frac{CD}{AC} = \frac{8}{\sqrt{4^2 + 8^2}} = 0.8944$$

But $T_{CD} = -1000$ lb, so

$$T_{AC} = -\frac{T_{CD}}{\cos \theta} = -\frac{-1000}{0.8944} = 1118 \text{ lb}$$

and

$$T_{BC} = -2000 - 1118(0.4472) = -2500 \text{ lb}$$

Next, draw a free-body diagram of pin $B$ (Fig. 6-77b) and write the equilibrium equations

$+\rightarrow \Sigma F_x = 0$: $\qquad -T_{AB} = 0$

$+\uparrow \Sigma F_y = 0$: $\qquad T_{BC} + 2500 = 0$

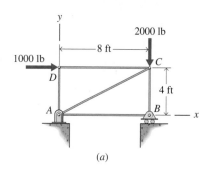

(a)

(b)

**Figure 6-76**

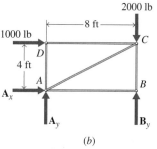

(d)

(c)

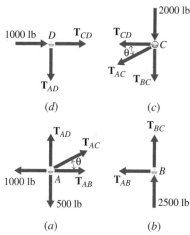

(a)

(b)

**Figure 6-77**

The first of these equations can be solved to get

$$T_{AB} = 0 \text{ lb}$$

The second equation contains no unknowns because the value of $T_{BC}$ has already been found. The second equation can be used to check the consistency of the answers

$$-2500 + 2500 = 0 \text{ (check)}$$

Finally, draw a free-body diagram of pin $A$ (Fig. 6-77$a$) and write the equilibrium equations

$+\rightarrow \Sigma F_x = 0$:          $T_{AB} + T_{AC} \cos \theta - 1000 = 0$

$+\uparrow \Sigma F_y = 0$:          $T_{AD} + T_{AC} \sin \theta - 500 = 0$

Again, there are no unknowns in these equations because the values of $T_{AB}$, $T_{AC}$, and $T_{AD}$ have already been found. These two equations again reduce to a check of the consistency of the solution:

$$0 + 1118(0.8944) - 1000 = -0.0608$$

$$0 + 1118(0.4472) - 500 = -0.0304$$

The difference is less than the rounding performed on $T_{AB}$, $T_{AC}$, and $T_{AD}$ and so the solution checks. The desired answers then are

| | | |
|---|---|---|
| $AB, AD$: | 0 lb | **Ans.** |
| $AC$: | 1118 lb T | **Ans.** |
| $BC$: | 2500 lb C | **Ans.** |
| $CD$: | 1000 lb C | **Ans.** |

The fact that $T_{AB}$ and $T_{AD}$ both came out zero is a peculiarity of the loading and does not mean that members $AB$ and $AD$ should be eliminated from the truss. For a slightly different loading situation, the forces in these members will not be zero. Even for the given loading condition, members $AB$ and $AD$ are necessary to ensure the rigidity of the truss. Without member $AB$, for example, the truss would collapse if the roller support at $B$ is disturbed slightly to the right or left. ■

## Example Problem 6-19

The truss shown in Fig. 6-78$a$ supports one side of a bridge; an identical truss supports the other side. Floor beams carry vehicle loads to the truss joints. A 2000-kg car is stopped on the bridge. Calculate the force in each member of the truss using the method of joints.

### SOLUTION

The truss of Fig. 6-78$a$ is a simple truss with $m = 7$ members and $j = 5$ joints. Therefore, the ten equations obtained from equilibrium of the five joints can be

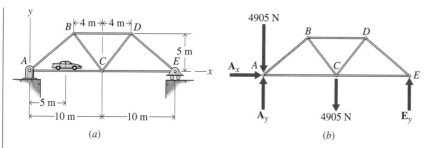

Figure 6-78

solved for the three support reactions as well as the forces in all seven members.

The first step is to divide the weight of the car between the joints of the truss. Half of the car's weight—$\frac{1}{2}(2000)(9.81) = 9810$ N—is carried by the truss shown and the other half is carried by the truss on the other side of the bridge. Since the car is midway between joints $A$ and $C$, $\frac{1}{2}(9810) = 4905$ N will be applied to joint $A$ and 4905 N will be applied to joint $C$.

The next step is to draw a free-body diagram of the entire truss (Fig. 6-78$b$) and write the equilibrium equations

$$+\rightarrow \Sigma F_x = 0: \qquad\qquad A_x = 0$$

$$+\uparrow \Sigma F_y = 0: \qquad A_y - 4905 - 4905 + E_y = 0$$

$$+\downarrow \Sigma M_A = 0: \qquad 20E_y - 10(4905) = 0$$

These equations can be solved immediately to get

$$A_x = 0 \text{ N} \qquad E_y = 2453 \text{ N} \qquad A_y = 7357 \text{ N}$$

Now the free-body diagram of pin $A$ (Fig. 6-79$a$) is drawn and the equilibrium equations are written

$$+\rightarrow \Sigma F_x = 0: \qquad A_x + T_{AC} + T_{AB} \cos \theta = 0$$

$$+\uparrow \Sigma F_y = 0: \qquad A_y - 4905 + T_{AB} \sin \theta = 0$$

where $A_x = 0$ N, $A_y = 7357$ N, and $\theta = \tan^{-1}(\frac{5}{6}) = 39.81°$. This gives

$$T_{AB} = -3,830 \text{ N} \qquad T_{AC} = 2,942 \text{ N}$$

Since one of the three forces applied at pin $B$ is now known, the free-body diagram of pin $B$ (Fig. 6-79$b$) is drawn next. The equilibrium equations for this pin are

$$+\rightarrow \Sigma F_x = 0: \qquad -T_{AB} \cos \theta + T_{BD} + T_{BC} \cos \phi = 0$$

$$+\uparrow \Sigma F_y = 0: \qquad -T_{AB} \sin \theta - T_{BC} \sin \phi = 0$$

where $T_{AB} = -3,830$ N and $\phi = \tan^{-1}(\frac{5}{4}) = 51.34°$. The equations are solved to get

$$T_{BC} = 3140 \text{ N} \qquad T_{BD} = -4904 \text{ N}$$

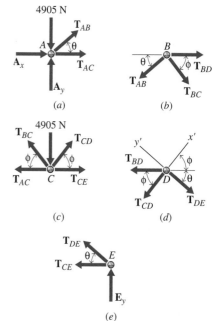

Figure 6-79

At this point, either pin $C$, pin $D$, or pin $E$ could be used, since each of these pins has only two forces whose values have not been determined. For this example the free-body diagram of pin $D$ (Fig. 6-79$d$) will be considered next. Writing the standard horizontal and vertical components of the equilibrium equations gives

$$+\rightarrow \Sigma F_x = 0: \qquad -(-4904) - T_{CD} \cos \phi + T_{DE} \cos \theta = 0$$

$$+\uparrow \Sigma F_y = 0: \qquad -T_{CD} \sin \phi - T_{DE} \sin \theta = 0$$

Both of these equations contain the unknown forces $T_{CD}$ and $T_{DE}$. While the solution of this pair of equations is not particularly difficult, the calculations can be simplified if the equilibrium equations are written in terms of components that are along and perpendicular to member $CD$. This gives

$$+\nearrow \Sigma F_{x'} = 0: \qquad -(-4904) \cos \phi - T_{CD} + T_{DE} \cos (\theta + \phi) = 0$$

$$+\nwarrow \Sigma F_{y'} = 0: \qquad -4904 \sin \phi - T_{DE} \sin (\theta + \phi) = 0$$

The second of these equations can be solved immediately to get

$$T_{DE} = -3830 \text{ N}$$

Then

$$T_{CD} = 3140 \text{ N}$$

Moving to pin $C$, the free-body diagram (Fig. 6-79$c$) is drawn and the equilibrium equations are written

$$+\rightarrow \Sigma F_x = 0: \quad -2942 - 3140 \cos \phi + 3140 \cos \phi + T_{CE} = 0$$

$$+\uparrow \Sigma F_y = 0: \quad 3140 \sin \phi - 4905 + 3140 \sin \phi = 0$$

The first of these equations gives

$$T_{CE} = 2942 \text{ N}$$

Since the values of all of the forces in the second equation have already been found, this equation reduces to a check of the consistency of the results:

$$3140 \sin 51.34° - 4905 + 3140 \sin 51.34° = -1.1570$$

The small number $-1.1570$ is due to rounding all of the intermediate answers to four significant figures. Keeping more accuracy in the intermediate values would reduce the residual and so the solution checks.

Finally, draw the free-body diagram for pin $E$ (Fig. 6-79$e$) and write the equilibrium equations

$$+\rightarrow \Sigma F_x = 0: \qquad -T_{CE} - T_{DE} \cos \theta = 0$$

$$+\uparrow \Sigma F_y = 0: \qquad T_{DE} \sin \theta + E_y = 0$$

Again, there are no unknowns left to be solved for. These equations are used simply as a check:

$$-2942 - (-3830) \cos 39.81° = 0.09796$$

$$-3830 \sin 39.81° + 2453 = 0.8663$$

and again the solution checks. The required answers (to three significant figures) are

| | | |
|---|---|---|
| $AB, DE$: | 3830 N C | **Ans.** |
| $AC, CE$: | 2940 N T | **Ans.** |
| $BC, CD$: | 3140 N T | **Ans.** |
| $BD$: | 4900 N C ∎ | **Ans.** |

## Example Problem 6-20

The roof truss of Fig. 6-80a is composed of 30°–60°–90° right triangles and is loaded as shown. Determine the forces in members $CD$, $CE$, and $EF$ using the method of sections.

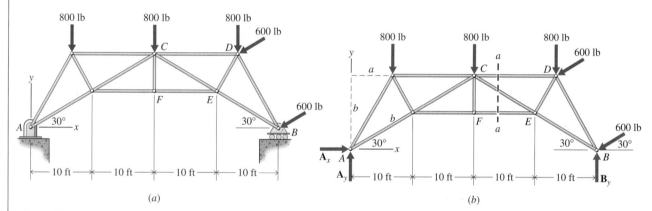

**Figure 6-80**

## SOLUTION

A free-body diagram of the entire truss (Fig. 6-80b) is used to solve for the support reactions at $A$ and $B$. Dimensions $a$ and $b$ needed for these calculations are

$$b = 10/\cos 30° = 11.547 \text{ ft}$$

$$a = 11.547 \tan 30° = 6.667 \text{ ft}$$

Summing moments about $A$ gives

$$+\circlearrowleft \Sigma M_A = 0:$$

$$B_y(40) - 600 \sin 30° (40) + 600 \cos 30° (11.547)$$

$$- 600 \sin 30° (33.333) - 800(33.333) - 800(20) - 800(6.667) = 0$$

which can be solved for $B_y$ to get

$$B_y = 1600 \text{ lb} \uparrow$$

Summing forces in the $x$- and $y$-directions gives

$$+\rightarrow\Sigma F_x = 0: \qquad A_x - 2(600 \cos 30°) = 0$$
$$A_x = 1039.2 \text{ lb} \cong 1039 \text{ lb} \rightarrow$$

$$+\uparrow \Sigma F_y = 0: \qquad A_y + 1600 - 3(800) - 2(600 \sin 30°) = 0$$
$$A_y = 1400 \text{ lb} \uparrow$$

Section $a$–$a$ of Fig. 6-80$b$ passes through members $CD$, $CE$, and $EF$. A free-body diagram for a part of the truss to the right of this section is shown in Fig. 6-80$c$. The force in member $CE$ can be found by summing forces in the $y$-direction. Thus,

$$+ \uparrow \Sigma F_y = 0: \qquad T_{CE} \sin 30° + 1600 - 800 - 2(600) \sin 30° = 0$$
$$T_{CE} = -400.0 \text{ lb}$$

The force in member $EF$ can be found by summing moments about point $C$ in Fig. 6-80$c$. Thus,

$$+\Big\lfloor \Sigma M_C = 0: \qquad -T_{EF}(10 \tan 30°) - 800(13.333) - 600 \sin 30° \, (13.333)$$
$$- 600 \cos 30° \, (20 \tan 30°) - 600 \sin 30°(20) + 1600(20) = 0$$
$$T_{EF} = 923.8 \text{ lb}$$

Finally, the force in member $CD$ can be found by summing forces in the $x$-direction or by summing moments about $E$. Summing forces in the $x$-direction and using Fig. 6-80$c$ gives

$$+\rightarrow\Sigma F_x = 0: \qquad -T_{CD} - T_{EF} - T_{CE} \cos 30° - 2(600 \cos 30°) = 0$$
$$-T_{CD} - 923.8 + 400 \cos 30° - 2(600 \cos 30°) = 0$$
$$T_{CD} = -1616.6 \text{ lb}$$

Alternatively, summing moments about $E$ gives

$$+\Big\lfloor \Sigma M_E = 0: \qquad T_{CD}(5.774) - 800(3.333) - 600 \sin 30°(3.333)$$
$$+ 600 \cos 30°(5.774) - 600 \sin 30°(10)$$
$$- 600 \cos 30°(5.774) + 1600(10) = 0$$
$$T_{CD} = -1616.5 \text{ lb}$$

The desired answers are

| | | |
|---|---|---|
| $CD$: | 1617 lb C | **Ans.** |
| $CE$: | 400 lb C | **Ans.** |
| $EF$: | 924 lb T | **Ans.** |

Since overall equilibrium was first used to find the support reactions, equilibrium of either the part of the truss to the left of section $a$–$a$ or the part of the truss to the right of section $a$–$a$ can be used to solve for the member forces. Usually the part with the fewest forces acting on it will result in the simplest equations of equilibrium. ■

## Example Problem 6-21

Use the method of sections to find the forces in members $CD$ and $FG$ of the truss in Fig. 6-81a.

### SOLUTION

Cut a section through members $CD$, $DE$, $EF$, and $FG$ as shown in Fig. 6-81a and draw a free-body diagram for the upper part of the truss (Fig. 6-81b). Summing moments about $D$

$+\curvearrowleft \Sigma M_D = 0$: $\quad 4(500 \cos 30°) - 6(500 \sin 30°) - 8T_{FG}$

$\quad\quad\quad\quad\quad -12(500 \cos 30°) - 6(500 \sin 30°) = 0$

gives

$$T_{FG} = -808.0 \text{ lb}$$

Then summing moments about $F$

$+\curvearrowleft \Sigma M_F = 0$: $\quad 12(500 \cos 30°) - 6(500 \sin 30°) + 8T_{CD}$

$\quad\quad\quad\quad\quad - 4(500 \cos 30°) - 6(500 \sin 30°) = 0$

gives

$$T_{CD} = -58.01 \text{ lb}$$

The consistency of these answers can be checked by summing forces in the $y$-direction. Thus,

$+\uparrow \Sigma F_y = 0$: $\quad -2(500 \cos 30°) - (-808.0) - (-58.01) = -0.01540 \cong 0$

which is within the accuracy of the calculations above. The desired answers are

$$CD: \quad 58.0 \text{ lb C} \quad\quad\quad \textbf{Ans.}$$
$$FG: \quad 808 \text{ lb C} \quad\quad\quad \textbf{Ans.}$$

Note that it was not necessary in this problem to first find the support reactions using overall equilibrium. Also note that neither $T_{DE}$ nor $T_{EF}$ can be found from this section. Either additional sections or the method of joints would be needed to find these forces if they needed to be found. ■

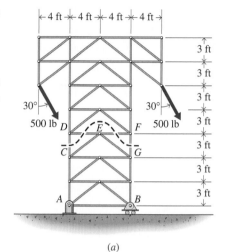

(a)

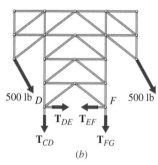

(b)

**Figure 6-81**

## Example Problem 6-22

All members of the inverted Mansard truss of Fig. 6-82a are made of structural steel. Determine

(a) The axial stress in member $CH$ if the cross-sectional area of this member is 2.50 in.$^2$.

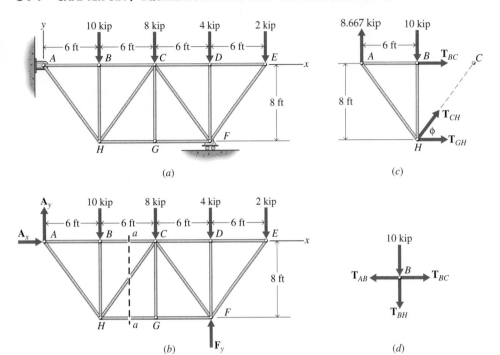

**Figure 6-82**

(b) The deformation of member *BH* if the cross-sectional area of this member is 2.50 in².

## SOLUTION

First find the support reactions by drawing a free-body diagram of the entire truss (Fig. 6-82b) and writing the equilibrium equations:

$+\!\!\downarrow \Sigma M_A = 0:$   $\qquad F_y(18) - 10(6) - 8(12) - 4(18) - 2(24) = 0$

$\qquad\qquad\qquad\qquad F_y = 15.333 \text{ kip} = 15.333 \text{ kip} \uparrow$

$+\!\!\rightarrow \Sigma F_x = 0:$   $\qquad A_x = 0$

$\qquad\qquad\qquad\qquad\qquad A_x = 0 \text{ kip}$

$+\!\!\uparrow \Sigma F_y = 0:$   $\qquad A_y - 10 - 8 - 4 - 2 + F_y = 0$

$\qquad\qquad\qquad\qquad A_y = 8.667 \text{ kip} = 8.667 \text{ kip} \uparrow$

(a) Section *a–a* of Fig. 6-82b passes through members *BC*, *CH*, and *GH*. A free-body diagram for the part of the truss to the left of this section is shown in Fig. 6-82c. The force in member *CH* can be found by summing forces in the y-direction. Thus,

$$\phi = \tan^{-1}\frac{8}{6} = 53.13°$$

$+\!\!\uparrow \Sigma F_y = 0:$   $\qquad T_{CH} \sin 53.13° + 8.667 - 10 = 0$

$$T_{CH} = 1.6663 \text{ kip} = 1666.3 \text{ lb T}$$

The axial stress in member $CH$ is given by Eq. (4-2) as

$$\sigma = \frac{T_{CH}}{A_{CH}} = \frac{1666.3}{2.5} = 666.5 \text{ psi} \cong 667 \text{ psi T} \qquad \textbf{Ans.}$$

(b) The force in member $BH$ can be determined from a free-body diagram of the pin at $B$ (Fig. 6-82d) and the equilibrium equation $\Sigma F_y = 0$. Thus,

$$+\uparrow \Sigma F_y = 0: \qquad -T_{BH} - 10 = 0$$

$$T_{BH} = -10.00 \text{ kip} = 10,000 \text{ lb C}$$

The deformation of member $BH$ is given by Eq. (4-20b) as

$$\delta = \frac{T_{BH}L_{BH}}{E_{BH}A_{BH}} = \frac{-10,000(8)(12)}{29,000,000(2.5)} = -0.01324 \text{ in. } \blacksquare \qquad \textbf{Ans.}$$

# PROBLEMS

## Introductory Problems

**6-82\*** Use the method of joints and determine the force in each member of the truss shown in Fig. P6-82, if $a = 2$ m and $P = 5$ kN.

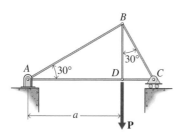

**Figure P6-82**

**6-83\*** A 4000-lb crate is attached by light, inextensible cables to the truss of Fig. P6-83. Determine the force in each member of the truss using the method of joints.

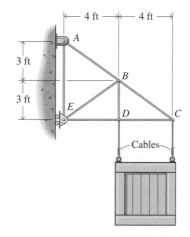

**Figure P6-83**

**6-84** Use the method of joints and determine the force in each member of the truss shown in Fig. P6-84. All members are 3 m long.

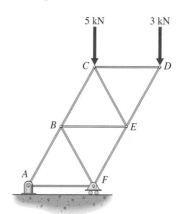

**Figure P6-84**

**6-85** Each truss member in Fig. P6-85 is 5 ft long. Find the forces in members $CD$ and $EF$ using the method of sections.

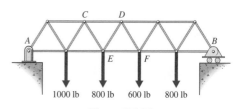

**Figure P6-85**

**6-86\*** Find the forces in members $CJ$ and $KJ$ of the roof truss shown in Fig. P6-86 using the method of sections.

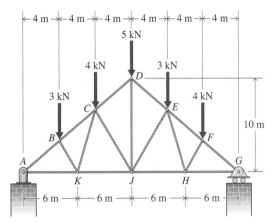

**Figure P6-86**

**6-87** Find the forces in members $BC$, $BF$, and $AF$, of the stairs truss of Fig. P6-87 using the method of sections.

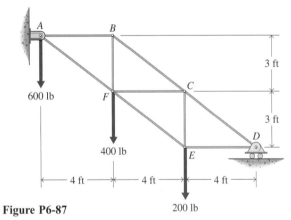

**Figure P6-87**

### Intermediate Problems

**6-88\*** The Gambrel truss shown in Fig. P6-88 supports one side of a bridge; an identical truss supports the other side. Floor beams carry vehicle loads to the truss joints. Calculate the forces in members $BC$, $BG$, and $DE$ when a truck having a mass of 3500 kg is stopped in the middle of the bridge as shown. The center of gravity of the truck is 1 m in front of the rear wheels.

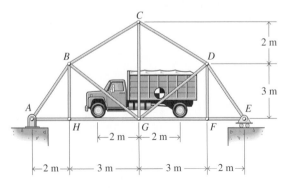

**Figure P6-88**

**6-89\*** A truss is loaded and supported as shown in Fig. P6-89. Determine

(a) The normal stress in member $CD$ if it has a diameter of $\frac{1}{2}$ in.

(b) The change in length of member $CF$ if it has a diameter of $\frac{1}{2}$ in. and a modulus of elasticity of $29(10^6)$ psi.

(c) The change in length of member $EF$ if it has the same diameter and modulus of elasticity as member $CF$.

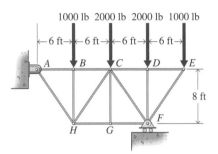

**Figure P6-89**

**6-90** A transmission line truss supports a 5-kN load, as shown in Fig. P6-90. All members of the truss are made of structural steel (see Appendix A for properties). Determine the change in length of members $FG$ and $CD$ if they have cross-sectional areas of 300 mm$^2$ each.

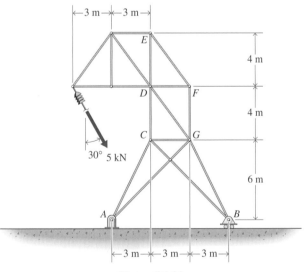

**Figure P6-90**

**6-91\*** The truss shown in Fig. P6-91 supports a sign that weighs 3000 lb. The sign is connected to the truss at joints $E$, $G$, and $H$, and the connecting links are adjusted so that each joint carries one third of the load. All members of the truss are made of structural steel (see Appendix A for properties), and each has a cross-sectional area of 0.564 in². Determine the axial stresses and strains in members $CD$, $CF$, $CG$, and $FG$ of the truss.

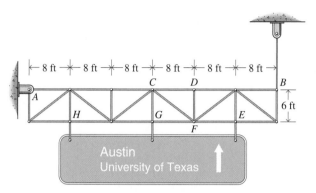

**Figure P6-91**

**6-92** All members of the truss shown in Fig. P6-92 are made of structural steel ($E = 200$ GPa) and are 25 mm in diameter. Determine the normal stresses in and the change in length of members $CD$, $DI$, and $HI$.

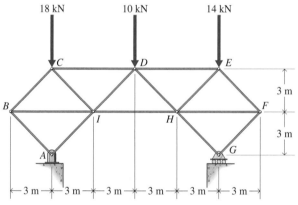

**Figure P6-92**

**6-93** Determine the force in member $BD$ of the truss shown in Fig. P6-93.

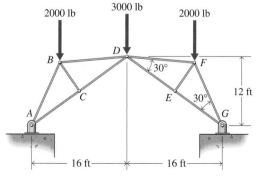

**Figure P6-93**

**Challenging Problems**

**6-94** For the simple truss of Fig. P6-94 show that overall equilibrium of the truss is a consequence of equilibrium of all of the pins; hence the equations of overall equilibrium give no new information. (*Hint*: Write equations of equilibrium for each of the pins and eliminate the unknown member forces from these equations.)

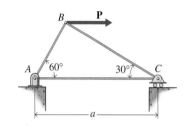

**Figure P6-94**

**6-95** Show that the overall equilibrium of a truss is a consequence of the equilibrium of the two separate parts generated by the method of sections. That is, section the bridge truss of Fig. P6-95 as indicated and write the equilbrium equations for each piece. Eliminate the member forces from the resulting six equations and show that the result is equivalent to the equilibrium of the whole truss.

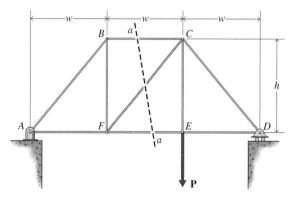

**Figure P6-95**

**6-96\*** The flat roof of a building is supported by a series of parallel plane trusses spaced 2 m apart (only one such truss is shown in Fig. P6-96). Calculate the forces in all the members of a typical truss when water collects to a depth of 0.2 m as shown. The density of water is 1000 kg/m³.

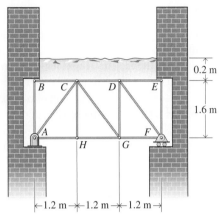

**Figure P6-96**

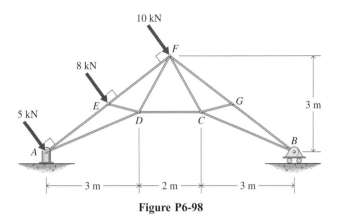

**Figure P6-98**

**6-99\*** Find the forces in members $DE$, $DJ$, and $JK$ of the truss of Fig. P6-99.

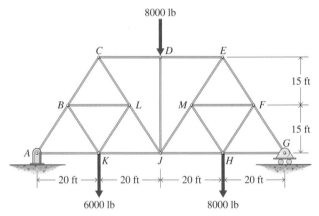

**Figure P6-99**

**6-97** Snow on a roof supported by the Howe truss of Fig. P6-97 can be approximated as a distributed load of 20 lb/ft (measured along the roof). Treat the distributed load as you would the weight of the members; that is, replace the total load on each of the upper members as a vertical force, half applied to the joint at each end of the member. Determine the forces in members $BC$, $BG$, $CG$, and $GH$.

**Computer Problems**

**6-100** The Gambrel truss shown in Fig. P6-100 supports one side of a bridge; an identical truss supports the other side. A 3400-kg truck is stopped on the bridge at the location shown and floor beams carry the vehicle load to the truss joints. If the center of gravity of the truck is located 1.5 m in front of the rear wheels, plot the force in members $BC$, $BG$, and $GH$ as a function of the truck's location $d$ ($0 \leq d \leq 20$ m).

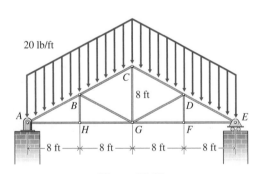

**Figure P6-97**

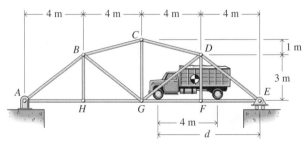

**6-98\*** Determine the force in members $CD$, $DE$, and $DF$ of the roof truss shown in Fig. P6-98. Triangle $CDF$ is an equilateral triangle, and joints $E$ and $G$ are at the midpoints of their respective sides.

**Figure P6-100**

**6-101** An overhead crane consists of an I-beam supported by a simple truss as shown in Fig. P6-101. If the uniform I-beam weighs 400 lb and is supporting a load of 1000 lb at a distance $d$ from its left end, plot the force in member $AB$, $BC$, $EF$, and $FG$ as a function of the position $d$ ($0 \leq d \leq 8$ ft).

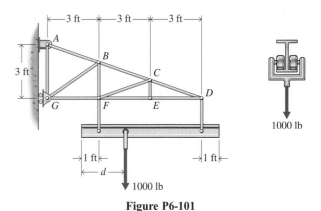

Figure P6-101

---

# 6-7 EQUILIBRIUM IN THREE DIMENSIONS

Any three-dimensional system of forces $\mathbf{F}_1, \mathbf{F}_2, \ldots, \mathbf{F}_n$ and couples $\mathbf{C}_1, \mathbf{C}_2, \ldots, \mathbf{C}_n$ can be replaced by an equivalent system that consists of three mutually perpendicular concurrent forces and three mutually perpendicular couples. The resultant $\mathbf{R}$ of the concurrent force system can be expressed as

$$\mathbf{R} = \Sigma F_x \mathbf{i} + \Sigma F_y \mathbf{j} + \Sigma F_z \mathbf{k} \qquad (6\text{-}10)$$

The resultant $\mathbf{C}$ of the system of couples can be expressed as

$$\mathbf{C} = \Sigma M_x \mathbf{i} + \Sigma M_y \mathbf{j} + \Sigma M_z \mathbf{k} \qquad (6\text{-}11)$$

The resultant force $\mathbf{R}$ and the resultant couple $\mathbf{C}$, together, constitute the resultant of the general three-dimensional force system. Equations (6-10) and (6-11) indicate that the resultant of the force system may be a force $\mathbf{R}$, a couple $\mathbf{C}$, or both a force $\mathbf{R}$ and a couple $\mathbf{C}$. Thus, a rigid body subjected to a general three-dimensional system of forces will be in equilibrium if $\mathbf{R} = \mathbf{C} = \mathbf{0}$, which requires that

$$\Sigma F_x = 0 \qquad \Sigma F_y = 0 \qquad \Sigma F_z = 0$$

and $\hspace{11cm}$ (6-2)

$$\Sigma M_x = 0 \qquad \Sigma M_y = 0 \qquad \Sigma M_z = 0$$

Thus, there are six independent scalar equations of equilibrium for a rigid body subjected to a general three-dimensional system of forces. The first three equations express the requirement that the $x$-, $y$-, and $z$-components of the resultant force $\mathbf{R}$ must be zero for a body to be in equilibrium. The second three equations express the further equilibrium requirement that there be no couple components acting on the body about any of the coordinate axes or about axes parallel to the coordinate axes. These six equations are both the necessary and sufficient conditions for equilibrium of the body. The six equations are independent since each can be satisfied independently of the others.

The following examples illustrate the solution of three-dimensional problems. A vector analysis is used for Example Problem 6-23, while a scalar analysis is used for Example Problem 6-24. Simple stress and deformation calculations are illustrated in Example Problem 6-23.

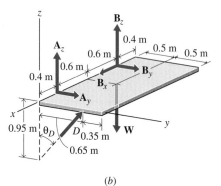

*(a)*

**Figure 6-83**

## Example Problem 6-23

The homogeneous door shown in Fig. 6-83a has a mass of 25 kg and is supported in a horizontal position by two hinges and a bar. The hinges have been properly aligned; therefore, they exert only force reactions on the door. Assume that the hinge at $B$ resists any force along the axis of the hinge pins. Determine

(a) The reactions at supports $A$, $B$, and $D$.

(b) The axial stress and deformation in bar $CD$ if it is made of structural steel and has a 25- $\times$ 8-mm rectangular cross section.

(c) The shearing stress on a cross section of the pin at $D$ if the diameter of the pin is 6 mm.

### SOLUTION

(a) A free-body diagram of the door is shown in Fig. 6-83b. The weight, the hinge reactions at $A$ and $B$, and the bar reaction at $D$ can be written as

$$\mathbf{W} = -mg\,\mathbf{k} = -25(9.81)\,\mathbf{k} = -245.3\,\mathbf{k}\ \text{N}$$

$$\mathbf{A} = A_y\,\mathbf{j} + A_z\,\mathbf{k}\ \text{N}$$

$$\mathbf{B} = B_x\,\mathbf{i} + B_y\,\mathbf{j} + B_z\,\mathbf{k}\ \text{N}$$

$$\mathbf{D} = 0.5647D\,\mathbf{j} + 0.8253D\,\mathbf{k}\ \text{N}$$

For the door to be in equilibrium, Eq. (6-10) must be satisfied. Thus,

$$\mathbf{R} = \Sigma F_x\,\mathbf{i} + \Sigma F_y\,\mathbf{j} + \Sigma F_z\,\mathbf{k} = 0$$

$$(B_x)\,\mathbf{i} + (A_y + B_y + 0.5647D)\,\mathbf{j}$$

$$+(A_z + B_z + 0.8253D - 245.3)\,\mathbf{k} = 0 \quad\quad (a)$$

Summing moments about $B$ to eliminate the maximum number of unknowns

$$\mathbf{C} = \Sigma C_x\,\mathbf{i} + \Sigma C_y\,\mathbf{j} + \Sigma C_z\,\mathbf{k} = 0$$

$$(\mathbf{r}_{A/B} \times \mathbf{A}) + (\mathbf{r}_{W/B} \times \mathbf{W}) + (\mathbf{r}_{D/B} \times \mathbf{D}) = 0$$

$$\begin{vmatrix} \mathbf{i} & \mathbf{j} & \mathbf{k} \\ 1.2 & 0 & 0 \\ 0 & A_y & A_z \end{vmatrix} + \begin{vmatrix} \mathbf{i} & \mathbf{j} & \mathbf{k} \\ 0.6 & 0.5 & 0 \\ 0 & 0 & -245.3 \end{vmatrix} + \begin{vmatrix} \mathbf{i} & \mathbf{j} & \mathbf{k} \\ 1.6 & 0.65 & 0 \\ 0 & 0.5647D & 0.8253D \end{vmatrix}$$

$$= (-122.65 + 0.5364D)\,\mathbf{i} + (-1.2A_z + 147.18 - 1.3205D)\,\mathbf{j}$$

$$+ (1.2A_y + 0.9035D)\,\mathbf{k} = 0 \quad\quad (b)$$

Equating the coefficients of $\mathbf{i}$, $\mathbf{j}$, and $\mathbf{k}$ to zero in Eqs. (a) and (b) and solving yields

$$B_x = 0 \quad\quad B_x = 0$$

$$A_y + B_y + 0.5647D = 0 \quad\quad B_y = 43.03\ \text{N}$$

$$A_z + B_z + 0.8253D - 245.3 = 0 \qquad B_z = 185.56 \text{ N}$$

$$-122.65 + 0.5364D = 0 \qquad D = 228.65 \text{ N}$$

$$-1.2A_z + 147.18 - 1.3205D = 0 \qquad A_z = -128.96 \text{ N}$$

$$1.2A_y + 0.9035D = 0 \qquad A_y = -172.15 \text{ N}$$

The reactions at hinges $A$ and $B$ and the force exerted by the bar at $D$ are

$$\mathbf{A} = -172.2\,\mathbf{j} - 129.0\,\mathbf{k} \text{ N} \qquad |\mathbf{A}| = 215 \text{ N} \qquad \textbf{Ans.}$$

$$\mathbf{B} = 43.0\,\mathbf{j} + 185.6\,\mathbf{k} \text{ N} \qquad |\mathbf{B}| = 190.5 \text{ N} \qquad \textbf{Ans.}$$

$$\mathbf{D} = 129.1\,\mathbf{j} + 188.7\,\mathbf{k} \text{ N} \qquad |\mathbf{D}| = 229 \text{ N} \qquad \textbf{Ans.}$$

A scalar analysis can frequently be used in three-dimensional problems when a single unknown is the only quantity required. For example, the force exerted by the bar at support $D$ can be determined by summing moments about the axis of the hinges. Thus,

$$+\!\!\downarrow\ \Sigma M_x = 0: \qquad\qquad D(0.95 \sin \theta_D) - W(0.5) = 0$$

$$D(0.95 \sin 34.38°) - 245.3(0.5) = 0$$

$$D = 228.6 \cong 229 \text{ N}$$

(b) The axial stress and deformation in bar $CD$ is given by Eqs. (4-2) and (4-20b) as

$$\sigma = \frac{F_{CD}}{A_{CD}} = \frac{228.6}{0.025(0.008)} = 1.143(10^6) \text{ N/m}^2 = 1.143 \text{ MPa} \qquad \textbf{Ans.}$$

$$L_{CD} = \sqrt{(0.65)^2 + (0.95)^2} = 1.151 \text{ m}$$

$$E_{CD} = 200 \text{ GPa (see Table A-18)}$$

$$\delta = \frac{F_{CD}L_{CD}}{E_{CD}A_{CD}} = \frac{228.6(1.151)}{200(10^9)(0.025)(0.008)}$$

$$= 0.00658(10^{-3}) \text{ m} = 0.00658 \text{ mm} \qquad \textbf{Ans.}$$

(c) The shearing stress on a cross section of the pin at $D$, which is in single shear, is given by Eq. (4-4) as

$$\tau = \frac{F_{CD}}{A_s} = \frac{228.6}{(\pi/4)(0.006)^2} = 8.085(10^6) \text{ N/m}^2 \cong 8.09 \text{ MPa} \quad \blacksquare \qquad \textbf{Ans.}$$

## Example Problem 6-24

A homogeneous flat plate that weighs 500 lb is supported by a shaft $AB$ and a cable $C$ as shown in Fig. 6-84a. The bearing at $A$ is a ball bearing and the bearing at $B$ is a thrust bearing. The bearings are properly aligned; therefore, they

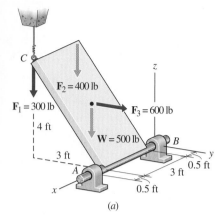

(a)

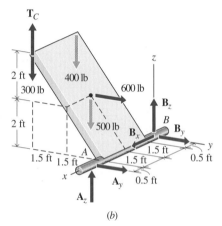

(b)

**Figure 6-84**

transmit only force components. When the three forces shown in Fig. 6-84a are applied to the plate, determine the reactions at bearings $A$ and $B$ and the tension in cable $C$. Use a scalar analysis for the solution but express the final results in Cartesian vector form.

## SOLUTION

A free-body diagram for the plate is shown in Fig. 6-84b. The bearing at $A$ is a ball bearing; therefore, this reaction is represented by two force components $\mathbf{A}_y$ and $\mathbf{A}_z$. The bearing at $B$ is a thrust bearing; therefore, this reaction is represented by three force components $\mathbf{B}_x$, $\mathbf{B}_y$, and $\mathbf{B}_z$. The force in the cable is represented by tension $\mathbf{T}_C$. The plate is subjected to a general, three-dimensional system of forces; therefore, six equilibrium equations [Eqs. (6-2)] are available for determining the six unknown force components indicated on the free-body diagram. Summing forces in the positive $x$-, $y$-, and $z$-directions and summing moments about the $x$-, $y$-, and $z$-axes in accordance with the right-hand rule yields

$\Sigma F_x = 0$: $B_x = 0$

$$B_x = 0 \qquad \text{(a)}$$

$\Sigma F_y = 0$: $A_y + B_y + 600 = 0$

$$A_y + B_y = -600 \text{ lb} \qquad \text{(b)}$$

$\Sigma F_z = 0$: $A_z + B_z + T_C - 300 - 400 - 500 = 0$

$$A_z + B_z + T_C = 1200 \text{ lb} \qquad \text{(c)}$$

$\Sigma M_x = 0$: $300(3) + 400(3) - 600(2) + 500(1.5) - T_C(3) = 0$

$$T_C = 550 \text{ lb} \qquad \text{(d)}$$

$\Sigma M_y = 0$: $300(3.5) + 400(0.5) + 500(2) - T_C(3.5) - A_z(4) = 0$

$$4A_z + 3.5T_C = 2250 \text{ lb} \cdot \text{ft (e)}$$

$\Sigma M_z = 0$: $A_y(4) + 600(2) = 0$

$$A_y = -300 \text{ lb} \qquad \text{(f)}$$

From Eqs. (b) and (f):

$$A_y + B_y = -300 + B_y = -600 \text{ lb}$$
$$B_y = -300 \text{ lb}$$

From Eqs. (d) and (e):

$$4A_z + 3.5T_C = 4A_z + 3.5(550) = 2250 \text{ lb} \cdot \text{ft}$$
$$A_z = 81.3 \text{ lb}$$

From Eq. (c):

$$A_z + B_z + T_C = 81.3 + B_z + 550 = 1200 \text{ lb}$$
$$B_z = 568.7 \text{ lb}$$

$$\mathbf{A} = -300 \, \mathbf{j} + 81.3 \, \mathbf{k} \text{ lb} \qquad \textbf{Ans.}$$

$$\mathbf{B} = -300 \, \mathbf{j} + 569 \, \mathbf{k} \text{ lb} \qquad \textbf{Ans.}$$

$$\mathbf{T}_C = 550 \, \mathbf{k} \text{ lb} \quad \blacksquare \qquad \textbf{Ans.}$$

## PROBLEMS

### Introductory Problems

**6-102\*** The masses of cartons 1, 2, and 3, which rest on the platform shown in Fig. P6-102, are 300 kg, 100 kg, and 200 kg, respectively. The mass of the platform is 500 kg. Determine the tensions in the three cables $A$, $B$, and $C$ that support the platform.

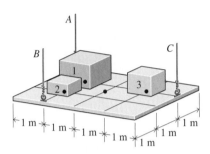

**Figure P6-102**

**6-103\*** Determine the reaction at support $A$ of the pipe system shown in Fig. P6-103 when the force applied to the pipe wrench is 50 lb.

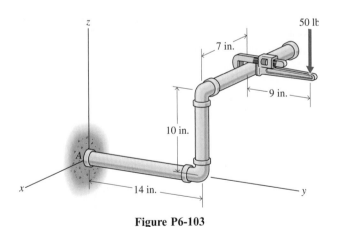

**Figure P6-103**

**6-104** The rectangular plate of uniform thickness shown in Fig. P6-104 has a mass of 500 kg. Determine
(a) The tensions in the three cables supporting the plate.
(b) The deformation of the cable at $A$, which has a diameter of 10-mm, a length of 1.5 m, and a modulus of elasticity of 200 GPa.

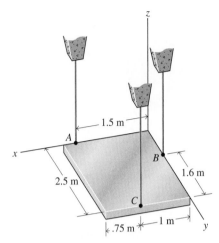

**Figure P6-104**

### Intermediate Problems

**6-105\*** A shaft is loaded through a pulley and a lever (Fig. P6-105) that are fixed to the shaft. Friction between the belt and pulley prevents slipping of the belt. Determine the force $\mathbf{P}$ required for equilibrium and the reactions at supports $A$ and $B$. The support at $A$ is a ball bearing, and the support at $B$ is a thrust bearing. The bearings exert only force reactions on the shaft.

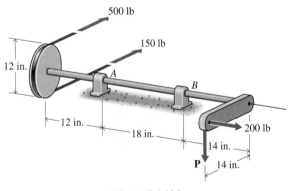

**Figure P6-105**

**6-106\*** The homogeneous door shown in Fig. P6-106 has a mass of 60 kg and is held in the position shown by the rod $AB$. The rod is held in place by smooth horizontal pins at $A$ and $B$. The hinges at $C$ and $D$ are smooth, and the hinge at $C$ can support thrust along its axis. Determine all forces that act on the door.

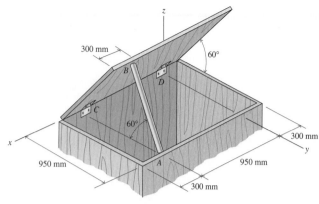

**Figure P6-106**

**6-107** A beam is supported by a ball-and-socket joint and two cables as shown in Fig. P6-107. Determine
(a) The tensions in the two cables.
(b) The reaction at support $A$ (the ball and socket joint).
(c) The axial stress and deformation in each of the cables if they are made of steel ($E = 30{,}000$ ksi) and have $\frac{1}{4}$-in. diameters.

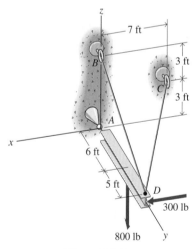

**Figure P6-107**

**Challenging Problems**

**6-108\*** The crankshaft-flywheel arrangement of a one-cylinder engine is shown in Fig. P6-108. A 250 N·m couple $\mathbf{C}$ is delivered to the crankshaft by the flywheel, which is rotating at a constant angular velocity. The support at $A$ is a ball bearing, and the support at $B$ is a

thrust bearing, both of which support only force reactions on the shaft. Determine the magnitudes of the force $\mathbf{P}$ and of the resultant bearing reactions at $A$ and $B$.

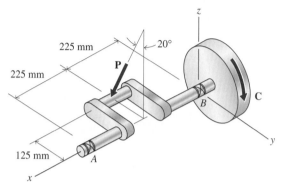

**Figure P6-108**

**6-109** A farmer is using the hand winch shown in Fig. P6-109 to slowly raise a 40-lb bucket of water from a well. In the position shown, force $\mathbf{P}$ is vertical. The bearings at $C$ and $D$ exert only force reactions on the shaft. Bearing $C$ can support thrust loading; bearing $D$ cannot. Determine the magnitude of force $\mathbf{P}$ and the components of the bearing reactions.

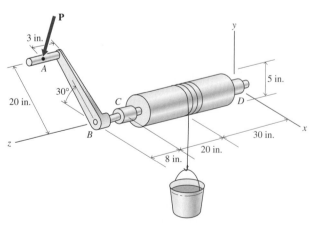

**Figure P6-109**

**6-110** The block $W$ shown in Fig. P6-110 has a mass of 250 kg. Bar $AB$ rests against a smooth vertical wall at end $B$ and is supported at end $A$ with a ball-and-socket joint. The two cables are attached to a point on the bar midway between the ends. Determine

(a) The reactions at supports $A$ and $B$ and the tension in cable $CD$.
(b) The axial stress and deformation in cable $CD$ if it is made of steel ($E$ = 210 GPa) and has an 8-mm diameter.

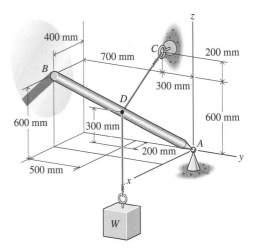

**Figure P6-110**

---

# 6-8 FRICTION

Thus far in this book, two surfaces in contact have been either perfectly smooth or perfectly rough. A perfectly smooth or frictionless surface that exerts only normal forces on bodies is a useful model for a large number of problems. However, frictional forces that act tangent to the surface are present in the contact between all real surfaces. Whether the friction forces are large or small depends on a number of things, including the types of materials in contact.

Friction forces act to oppose the tendency of contacting surfaces to slip relative to one another and can be either good or bad. Without friction it would be impossible to walk, ride a bicycle, drive a car, or pick up objects. In some machine applications, such as brakes and belt drives, a design consideration is to maximize the friction forces. In many other machine applications, however, friction is undesirable. Friction causes energy loss and wears down sliding surfaces in contact. In these cases, a primary design consideration is to minimize the friction forces.

Two main types of friction are encountered in engineering practice—dry friction and fluid friction. As its name suggests, dry friction (or Coulomb friction) describes the tangential component of the contact force that exists when two dry surfaces slide or tend to slide relative to one another. Coulomb friction is the primary concern of this section and will be studied in considerable detail.

Fluid friction describes the tangential component of the contact force that exists between adjacent layers in a fluid that are moving at different velocities relative to each other, as in the thin layer of oil between bearing surfaces. The tangential forces developed between the adjacent fluid layers oppose the relative motion and are dependent primarily on the relative velocity between the two layers. Fluid friction is one of the primary concerns in the study of fluid mechanics and is more properly treated in a course in fluid mechanics.

**Characteristics of Coulomb Friction** To investigate the behavior of frictional forces, consider a simple experiment consisting of a solid block of mass $m$ resting on a rough horizontal surface and acted on by a horizontal

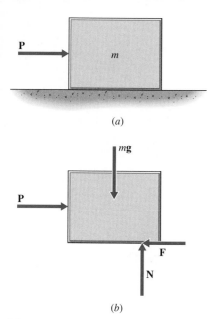

(a)

(b)

**Figure 6-85**

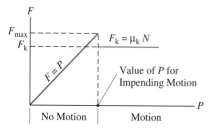

**Figure 6-86**

force **P** (Fig. 6-85). Equilibrium of the block requires a force having both a normal component ($N = mg$) and a horizontal or friction component ($F = P$) acting on the contact surface. When the horizontal force **P** is zero, no horizontal component of force $F$ will be required for equilibrium, and friction will not exert a force on the surface. As the force **P** increases, the friction force $F$ also increases, as shown in the graph of Fig. 6-86. The friction force cannot increase indefinitely, however, and it eventually reaches its maximum value $F_{max}$. The maximum value is also called the *limiting value of static friction*. In other words, $F_{max}$ is the maximum value of the friction force for which static equilibrium exists.

The condition in which the friction force is at its maximum value is called the *condition of impending motion*. That is, if **P** increases beyond the point $P = F_{max}$ then friction can no longer supply the amount of force necessary for equilibrium. Therefore the block will no longer be in equilibrium, but will start moving in the direction of the force **P**. When the block starts moving, the friction force $F$ normally decreases in magnitude by about 20 to 25 percent. From this point on, the block will slide with increasing speed while the friction force (the kinetic friction force, $F_k$) remains approximately constant (see Fig. 6-86).

Repeating the experiment with a second block of mass $m_2 = 2m$ would produce similar results, but the limiting force at which the block starts to move would be observed to be twice as great. Repeating the experiment with two blocks of different sizes but the same mass and material would yield the same limiting force for both blocks. That is, the value of the limiting friction force is proportional to the normal force at the contact surface:

$$F_{max} = \mu_s N \tag{6-12}$$

The constant of proportionality $\mu_s$ is called the *coefficient of static friction*, and it depends on the types of material in contact. However, $\mu_s$ is observed to be relatively independent of both the normal force and the area of contact.

To understand how $\mu_s$ can be independent of the area of contact, one must consider where the friction forces come from. It is generally believed that dry friction results primarily from the roughness between two surfaces and to a lesser extent from attraction between the molecules of the two surfaces.[2] Even two surfaces that are considered to be smooth have small irregularities, as the (idealized) enlargement of the contact surfaces of Fig. 6-87 shows. Therefore, contact between the block and the surface takes place only over a few very small areas of the common surface. The friction force $F$ is then the resultant of the tangential components of the forces acting at each of these tiny contact points, just as

[2]D. Halliday, R. Resnick, and J. Walker, *Fundamentals of Physics*, 4th ed., John Wiley and Sons, Inc., New York, 1993, pp. 133–134.

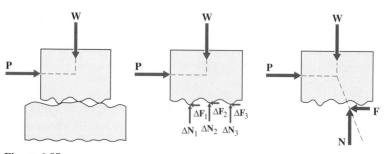

**Figure 6-87**

the normal force $N$ is the resultant of the normal components of the forces acting at each of the contact points. (Normal and tangential here are relative to the overall contact plane and not the individual tiny contact points.) Increasing the number of contact points just means that the normal and frictional components at each point are proportionately smaller, but their sums $F$ and $N$ do not change. Therefore, $\mu_s$ will not change either.

Before going on, it must be noted that the normal force $N$ is the resultant of a distributed force. If the force distribution is uniform, $N$ will act at the center of the surface. In general, however, $N$ will not act at the center of the surface or through the center of the body. Since the actual distribution of forces is generally not known, the location of $N$ must be determined using moment equilibrium.

In many friction problems, it is easily recognized that a body is in no danger of tipping over. Since only force equilibrium is considered, it doesn't matter where the normal force is drawn on the free-body diagram. Even so, the student should not get into the habit of showing the force as always acting through the center of the body.

It must also be noted that friction is a resistive force. That is, friction always acts to oppose motion; it never acts to create motion. Equation (6-12) only tells how much friction is available $F_{\text{avail}} = F_{\max} = \mu_s N$ to prevent motion. No matter how much frictional force is available on a surface, however, the frictional force actually exerted is never greater than that required to satisfy the equations of equilibrium

$$F \le \mu_s N \qquad (6\text{-}13)$$

where the equality holds only at the point of impending motion.

Once the block starts to slip relative to the surface, the friction force will decrease to

$$F = \mu_k N \qquad (6\text{-}14)$$

where $\mu_k$ is called the *coefficient of kinetic friction*. This coefficient is again independent of the normal force and is also independent of the speed of the relative motion—at least for low speeds. Of course the presence of any oil or moisture on the surface can change the problem from one of dry friction, in which the friction force is independent of the speed of the body, to one of fluid friction, in which the friction force is a function of the speed. At higher speeds, the effect of lubrication by an intervening fluid film (such as oil, surface moisture, or even air) can become appreciable.

These results are summarized as follows.

## COULOMB'S LAWS OF DRY FRICTION

The direction of the friction force on a surface is such as to oppose the tendency of one surface to slide relative to the other. It is the relative motion or the impending relative motion of one body relative to another that is important.

The friction force is never greater than just sufficient to prevent motion. For the static equilibrium case in which the two surfaces are stationary with respect to one another, the normal and tangential components of the contact force satisfy

$$F \le \mu_s N$$

where the equality holds for the case of impending motion in which the contacting surfaces are on the verge of sliding relative to each other.

For the case where two contacting surfaces are sliding over each other, the normal and tangential components of the contact force satisfy

$$F = \mu_k N$$

where $\mu_k < \mu_s$.

Of course, Coulomb's laws apply only when $N$ is positive, that is, when the surfaces are being pressed together.

The values of $\mu_s$ and $\mu_k$ must be determined experimentally for each pair of contacting surfaces. Average values of $\mu_s$ for various types of materials are given in Table 6-3. Reported values for $\mu_s$ vary widely, however, depending on the exact nature of the contacting surfaces. Values for $\mu_k$ are generally 20 to 25 percent less than those reported for $\mu_s$. Values for $\mu_k$ are not listed in Table 6-3, since the uncertainty in $\mu_k$ is much larger (by as much as 100 percent in some cases) than the difference between it and $\mu_s$.

Because of the uncertainty in the values of $\mu_s$ and $\mu_k$, Table 6-3 should only be used to get a rough estimate of the magnitude of the friction forces. If more accurate values are needed, experiments should be performed using the actual surfaces being studied.

Since the coefficients of friction are the ratio of two forces, they are dimensionless quantities and can be used with either the SI or U.S. customary system of units.

In many simple friction problems it is convenient to use the resultant of the friction and normal forces rather than their separate components. In the case of the block in Fig. 6-88, this leaves only three forces acting on the block. Moment equilibrium is established simply by making the three forces concurrent, and only force equilibrium need be considered. Since $N$ and $F$ are rectangular components of the resultant **R**, the magnitude and direction of the resultant are given by

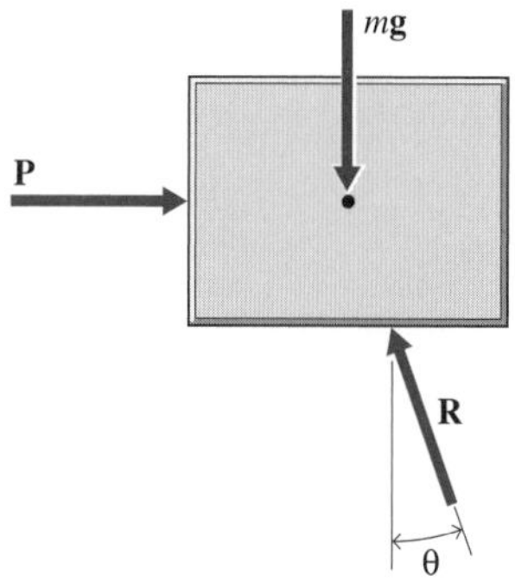

$$R = \sqrt{N^2 + F^2} \qquad \text{and} \qquad \tan \theta = F/N \qquad (6\text{-}15)$$

At the point of impending motion, Eq. (6-15) becomes

$$R = \sqrt{N^2 + F_{max}^2} = \sqrt{N^2 + (\mu_s N)^2} = N\sqrt{1 + \mu_s^2} \qquad (6\text{-}16a)$$

$$\tan \phi_s = \frac{F}{N} = \frac{\mu_s N}{N} = \mu_s \qquad (6\text{-}16b)$$

where $\phi_s$, the angle between the resultant and the normal to the surface, is called the *angle of static friction*. For a given normal force $N$, if the friction force is less than the maximum ($F < \mu_s N$), then the angle of the resultant will be less than the angle of static friction: $\theta < \phi_s$. In no case can the angle of the resultant $\theta$ be greater than $\phi_s$ for a body in equilibrium. A similar relation is obtained in the case of kinetic friction:

$$\tan \phi_k = \mu_k$$

where $\phi_k$ is called the *angle of kinetic friction*.

TABLE 6-3  Coefficients of Friction for Common Surfaces

| Materials | $\mu_s$ |
| --- | --- |
| Metal on metal | 0.5 |
| Wood on metal | 0.5 |
| Wood on wood | 0.4 |
| Leather on wood | 0.4 |
| Rubber on metal | 0.5 |
| Rubber on wood | 0.5 |
| Rubber on pavement | 0.7 |

**Figure 6-88**

When a block sits on an inclined surface and is acted upon only by gravity, the resultant of the normal and friction force must be collinear with the weight, as shown in Fig. 6-89. The angle between the resultant and the normal to the surface can never be greater than the angle of static friction $\phi_s$. Thus, the steepest inclination $\theta$ for which the block will be in equilibrium is equal to the angle of static friction. This angle is called the *angle of repose*.

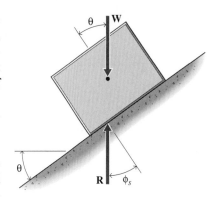

**Figure 6-89**

There are three typical types of friction problems encountered in engineering analysis:

1. **Impending motion is not assumed**. This first case is just the type of equilibrium problem solved in the previous chapters of this book. The required friction force $F_{req}$ ($\neq \mu_s N$) and the normal force $N$ are drawn on the free-body diagram and are determined using force equilibrium. The normal force should not be drawn as acting through the center of the body. Instead, its location is determined using moment equilibrium. The amount of friction required for equilibrium $F_{req}$ and the location of the normal force are then checked against their maximum values. Three possibilities exist.

   a. If the amount of friction required for equilibrium $F_{req}$ is smaller than or equal to the maximum amount of friction available $F_{avail} = F_{max} = \mu_s N$ and the location of the normal force is on the body, then the body is in equilibrium. In this case, the actual friction force supplied by the surface is

   $$F_{actual} = F_{req} < F_{avail} = \mu_s N$$

   That is, the surface supplies just enough friction force (resistance) to keep the body from moving.

   b. If the amount of friction required for equilibrium $F_{req}$ is smaller than or equal to the maximum amount of friction available $F_{avail} = F_{max} = \mu_s N$ but the location of the normal force is not somewhere on the body, then the body is not in equilibrium and will tip over.

   c. If the amount of friction required for equilibrium $F_{req}$ is greater than the maximum amount of friction available $F_{avail} = F_{max} = \mu_s N$, then the body is not in equilibrium and will slide. In this case, the actual friction force supplied by the surface is the kinetic friction force $F_{actual} = \mu_k N$.

2. **Impending slipping is known to occur at all surfaces of contact**. Since impending slipping is known to occur at all surfaces of contact, the magnitude of the friction forces can be shown as $\mu_s N$ on the free-body diagrams. The equations of equilibrium are then written and

   a. If all of the applied forces are given but $\mu_s$ is unknown, the equations of equilibrium can be solved for $N$ and $\mu_s$. This $\mu_s$ is the smallest coefficient of static friction for which the body will be in equilibrium.

   b. If the coefficient of static friction is given but one of the applied forces is unknown, the equations of equilibrium can be solved for $N$ and the unknown applied force.

3. **Impending motion is known to exist but the type of motion or surface of slip is not known**. Since it is not known whether the body tips or slips, the free-body diagrams must be drawn as in Case 1. That is, the friction forces must not be shown as $\mu_s N$ on the free-body diagrams. At this point the three equations of equilibrium will contain more than three unknowns. Assumptions must be made about the type of motion that is about to occur until the number of equations equals the number of unknowns. The

equations are then solved for the remaining unknowns and checked against the assumptions made about slipping or tipping. If $F_{\text{req}}$ comes out greater than $F_{\text{avail}} = \mu_s N$ at some surface or if the location of the normal force is not on the body, then the assumptions must be changed and the problem solved again.

In both of the last two Cases (2b and 3), the friction force is treated as if it is a known force. Care must be taken to be sure that it opposes the tendency of the other forces to cause motion. The result will not include a negative sign to indicate that the friction force is in the wrong direction. If the direction of the friction force is drawn incorrectly, incorrect answers will result. The direction is determined by pretending for a moment that friction does not exist. Then apply the friction force in such a direction as to oppose the motion that would occur in the absence of friction.

## Example Problem 6-25

The pickup truck of Fig. 6-90a is traveling at a constant speed of 50 mi/h and is carrying a 60-lb box in the back. The box projects 1 ft above the cab of the pickup. The wind resistance on the box can be approximated as a uniformly distributed force of 25 lb/ft on the exposed edge of the box. Calculate the minimum coefficient of friction required to keep the box from sliding on the bed of the pickup. Also determine whether or not the box will tip over.

### SOLUTION

The free-body diagram of the box is drawn in Fig. 6-90b. The equilibrium equations for the box

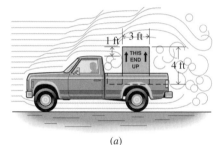

(a)

(b)

**Figure 6-90**

$$+\rightarrow \Sigma F_x = 0: \qquad\qquad\qquad 25(1) - C_f = 0$$

$$+\uparrow \Sigma F_y = 0: \qquad\qquad\qquad C_n - 60 = 0$$

$$+\downdownarrows \Sigma M_A = 0: \qquad 60(1.5) - 25(1)(3.5) - C_n x_C = 0$$

are solved to get

$$C_f = 25.0 \text{ lb} \qquad C_n = 60.0 \text{ lb}$$

$$x_C = 0.042 \text{ ft} = 0.500 \text{ in.}$$

Thus the required coefficient of friction is

$$\mu_s = \frac{C_f}{C_n} = \frac{25.0}{60.0} = 0.417 \qquad\qquad \textbf{Ans.}$$

and since $x_C$ is positive, the box will not tip. **Ans.**

*Note:* The friction and normal force must act on the box; thus, $x_C$ must be a number between 0 and 3 ft. If the solution had given $x_C$ to be negative, then no normal force on the bottom of the box could satisfy moment equilibrium and the box would tip over. ∎

## Example Problem 6-26

The wheels of the refrigerator of Fig. 6-91*a* are stuck and will not turn. The refrigerator weighs 600 N. Assume a coefficient of friction between the wheels and the floor of 0.6 and determine the force necessary to cause the refrigerator to just begin to move (impending motion). Also determine the maximum height *h* at which the force can be applied without causing the refrigerator to tip over.

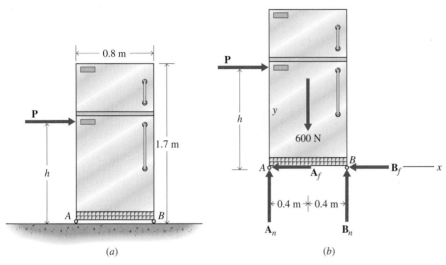

**Figure 6-91**

## SOLUTION

The free-body diagram of the refrigerator is drawn in Fig. 6-91*b*. The equilibrium equations

$$+\rightarrow \Sigma F_x = 0: \qquad\qquad P - A_f - B_f = 0$$
$$+\uparrow \Sigma F_y = 0: \qquad\qquad A_n + B_n - 600 = 0$$
$$+\downdownarrows \Sigma M_B = 0: \qquad 600(0.4) - P(h) - A_n(0.8) = 0$$

give

$$A_n + B_n = 600$$
$$P = A_f + B_f = 0.6(A_n + B_n) = 360 \text{ N} \qquad \textbf{Ans.}$$

and

$$h = \frac{240 - 0.8\, A_n}{360}$$

When $h = 0$, $A_n = B_n = 300$ N, and the wheels share the load of the weight equally. As $h$ increases, $A_n$ gets smaller. However, the force at $A$ cannot be negative, so

$$h < \frac{240}{360} = 0.667 \text{ m} \qquad \textbf{Ans.}$$

Note that at the point of impending tipping, none of the weight is carried by the wheel at $A$; it has all been shifted to the wheel at $B$. ∎

## Example Problem 6-27

A 20-lb homogeneous box has tipped and is resting against a 40-lb homogeneous box (Fig. 6-92a). The coefficient of friction between box $A$ and the floor is 0.7; between box $B$ and the floor, 0.4. Treat the contact surface between the two boxes as smooth and determine whether the boxes are in equilibrium.

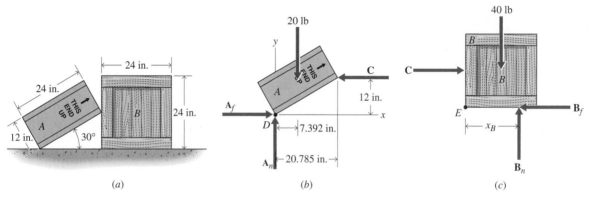

**Figure 6-92**

## SOLUTION

The free-body diagram of box $A$ is drawn in Fig. 6-92b. The equilibrium equations

$$+\rightarrow \Sigma F_x = 0: \qquad A_f - C = 0$$

$$+\uparrow \Sigma F_y = 0: \qquad A_n - 20 = 0$$

$$+\downarrow \Sigma M_D = 0: \qquad C(12) - 20(7.392) = 0$$

are solved to get

$$C = 12.32 \text{ lb} \qquad A_n = 20.00 \text{ lb} \qquad A_f = 12.32 \text{ lb}$$

The friction force available at this surface is

$$F_{\text{max}} = \mu_s A_n = 0.7(20.00) = 14.00 \text{ lb}$$

Since the friction force required (12.32 lb) is less than the friction force available (14.00 lb), box $A$ is in equilibrium.

The free-body diagram of box $B$ is drawn in Fig. 6-92c. The equilibrium equations for this box

$$+\rightarrow \Sigma F_x = 0: \qquad C - B_f = 12.32 - B_f = 0$$

$$+\uparrow \Sigma F_y = 0: \qquad B_n - 40 = 0$$

$$+\downarrow \Sigma M_E = 0: \qquad B_n(x_B) - 12.32(12) - 40(12) = 0$$

give

$$B_f = 12.32 \text{ lb} \qquad B_n = 40.0 \text{ lb} \qquad x_B = 15.696 \text{ in.}$$

The friction force available at this surface is

$$F_{max} = \mu_s B_n = 0.4(40.0) = 16.00 \text{ lb}$$

Again the friction force available (16.00 lb) is greater than the friction force required (12.32 lb) and box $B$ is also in equilibrium.

Thus, both boxes are in equilibrium.    **Ans.**

Note that while the normal force $B_n$ does not act at the center of the crate, it does act on the bottom of the crate since $x_B \leq 24$ in. ■

## Example Problem 6-28

A 500-N weight (Fig. 6-93a) is supported by a lightweight rope wrapped around the inner cylinder of a 1000-N homogeneous drum. The coefficient of friction between the weightless brake arm and the outer cylinder of the drum is 0.40. The force $P$ just prevents motion of the weight. The pins at $A$ and $D$ have diameters of 8 mm. Determine the shearing stress on a cross section of each pin if both pins are in double shear.

### SOLUTION

A free-body diagram of the drum is shown in Fig. 6-93b. The friction force $C_f$ prevents rotation of the drum; therefore, from the equilibrium equation

$$+\downarrow \Sigma M_D = 0: \qquad C_f(200) - 500(100) = 0 \qquad C_f = 250 \text{ N}$$

And since motion is impending,

$$C_f = \mu_s C_n = 0.40 \ C_n = 250 \text{ N} \qquad C_n = 625 \text{ N}$$

From the remaining equilibrium equations:

$$+\rightarrow \Sigma F_x = 0: \qquad D_x - C_f = D_x - 250 = 0 \qquad D_x = 250 \text{ N}$$

$$+\uparrow \Sigma F_y = 0: \qquad D_y - C_n - 500 - 1000 = 0$$

$$D_y - 625 - 500 - 1000 = 0 \qquad D_y = 2125 \text{ N}$$

$$F_D = \sqrt{(D_x)^2 + (D_y)^2} = \sqrt{(250)^2 + (2125)^2} = 2140 \text{ N}$$

The shearing stress on a cross section of the pin at $D$, which is in double shear, is then given by Eq.(4-4) as

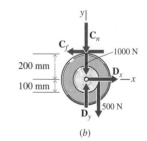

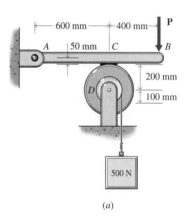

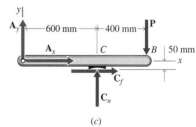

**Figure 6-93**

$$\tau = \frac{F_D}{A_s} = \frac{2140}{2(\pi/4)(0.008)^2} = 21.29(10^6) \text{ N/m}^2 \cong 21.3 \text{ MPa} \qquad \textbf{Ans.}$$

A free-body diagram of the brake arm is shown in Fig. 6-93c. From the equilibrium equation

$$+\downarrow \Sigma M_A = 0: \qquad C_f(50) + C_n(600) - P(1000) = 0$$
$$250(50) + 625(600) - P(1000) = 0 \qquad P = 387.5 \text{ N}$$

From the remaining equilibrium equations

$$+\rightarrow \Sigma F_x = 0: \qquad A_x + C_f = A_x + 250 = 0 \qquad A_x = -250 \text{ N} = 250 \text{ N} \leftarrow$$
$$+\uparrow \Sigma F_y = 0: \qquad A_y + C_n - P = 0$$
$$A_y + 625 - 387.5 = 0 \qquad A_y = -237.5 \text{ N} = 237.5 \text{ N} \downarrow$$
$$F_A = \sqrt{(A_x)^2 + (A_y)^2} = \sqrt{(250)^2 + (237.5)^2} = 344.8 \text{ N}$$

The shearing stress on a cross section of the pin at $A$, which is in double shear, is then given by Eq. (4-4) as

$$\tau = \frac{F_A}{A_s} = \frac{344.8}{2(\pi/4)(0.008)^2} = 3.430(10^6) \text{ N/m}^2 \cong 3.43 \text{ MPa} \quad \blacksquare \qquad \textbf{Ans.}$$

---

# ■ PROBLEMS

### Introductory Problems

**6-111\*** A 20-lb piece of electronic equipment is placed on a wooden skid that weighs 10 lb and rests on a concrete floor (Fig. P6-111). The coefficient of static friction between the skid and the floor is 0.45
(a) Determine the minimum pushing force along the handle necessary to cause the skid to start sliding across the floor (Fig. P6-111a).
(b) Determine the minimum force necessary to start the skid in motion if a pulling force instead of a pushing force is applied to the handle (Fig. P6-111b).

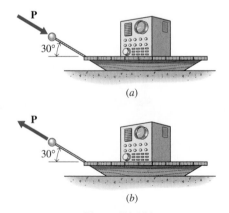

*(a)*

*(b)*

**Figure P6-111**

**6-112\*** A skier is at the point of experiencing impending motion on a ski slope as illustrated in Fig. P6-112. Determine the coefficient of static friction between a ski and the slope if the angle $\theta$ (the angle of repose) is 5°.

**Figure P6-112**

**6-113** The block in Fig. P6-113 weighs 500 lb, and the coefficent of friction between the block and the inclined plane is 0.2. Determine
(a) Whether the block would be in equilibrium for $P = 400$ lb.
(b) The minimum force $P$ to prevent motion.
(c) The maximum force $P$ for which the block is in equilibrium.

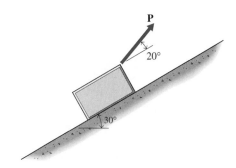

**Figure P6-113**

**6-114** A 75-kg man starts climbing a 5-m-long ladder leaning against a wall (Fig. P6-114). The coefficient of friction is 0.25 at both surfaces. Neglect the weight of the ladder and determine how far up the ladder the man can climb before the ladder starts to slip.

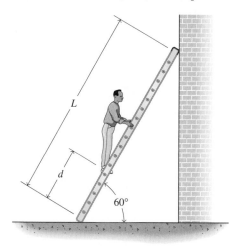

**Figure P6-114**

**6-115\*** A device for lifting rectangular objects such as bricks and concrete blocks is shown in Fig. P6-115. Determine the minimum coefficient of static friction between the contacting surfaces required to make the device work.

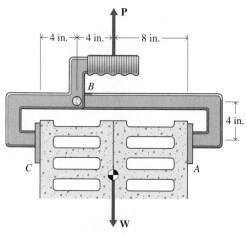

**Figure P6-115**

**6-116\*** A boy is pulling a sled with a box (at constant velocity) up an inclined surface as shown in Fig. P6-116. The mass of the box and sled is 50 kg; the mass of the boy is 40 kg. If the coefficient of kinetic friction between the sled runners and the icy surface is 0.05, determine the minimum coefficient of static friction needed between the boy's shoes and the icy surface.

**Figure P6-116**

**6-117** A 120-lb girl is walking up a 48-lb uniform beam as shown in Fig. P6-117. Determine how far up the beam the girl can walk before the beam starts to slip if
  (a) The coefficient of friction is 0.20 at all surfaces.
  (b) The coefficient of friction at the bottom end of the beam is increased to 0.40 by placing a piece of rubber between the beam and the floor.

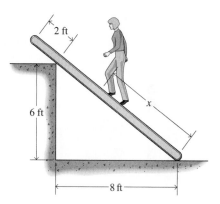

**Figure P6-117**

**Intermediate Problems**

**6-118\*** The broom shown in Fig. P6-118 weighs 8 N and is held up by the two cylinders, which are wedged between the broom handle and the side rails. The coefficient of friction between the broom and cylinders and between the cylinders and side rails is 0.30. The side rails are at an angle of $\theta = 30°$ to the vertical. The weight of the cylinders may be neglected. Determine whether or not this system is in equilibrium. If the system is in equilibrium, determine the force exerted on the broom handle by the rollers.

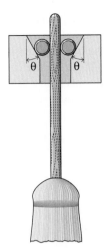

Figure P6-118

**6-119\*** The device shown in Fig. P6-119 is used to raise boxes and crates between floors of a factory. The frame, which slides on the 4-in.-diameter vertical post, weighs 50 lb. If the coefficient of friction between the post and the frame is 0.10, determine the force **P** required to raise a 150-lb box.

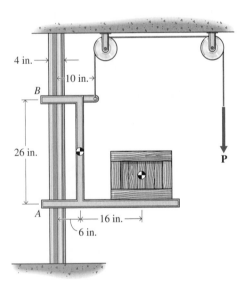

Figure P6-119

**6-120** The automobile shown in Fig. P6-120 has a mass of 1500 kg. The coefficient of friction between the rubber tires and the pavement is 0.70. Determine the maximum incline $\theta$ that the automobile can drive up if the automobile has
(a) A rear-wheel drive.
(b) A front-wheel drive.

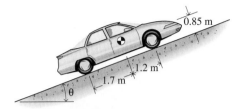

Figure P6-120

**6-121\*** When a drawer is pulled by only one of the handles, it tends to twist and rub as shown (highly exaggerated) in Fig. P6-121, which is a top view of the drawer. The weight of the drawer and its contents is 2 lb and is uniformly distributed. The coefficient of friction between the sides of the drawer and the sides of the dresser is 0.6; between the bottom of the drawer and the side rails the drawer rides on, $\mu_s = 0.1$. Determine the minimum amount of force **P** necessary to pull the drawer out.

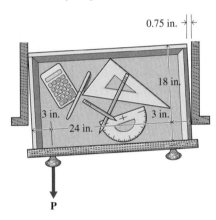

Figure P6-121

**6-122** An ill-fitting window is about 10 mm narrower than its frame (see Fig. P6-122). The window weighs 40 N, and the coefficient of friction between the window and the frame is 0.20. Determine the amount of force **P** that must be applied at the lower corner to keep the window from lowering.

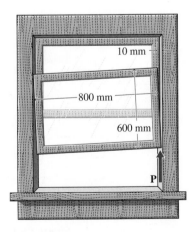

Figure P6-122

**6-123** Determine the minimum coefficient of static friction necessary for the pliers shown in Fig. P6-123 to grip the bolt.

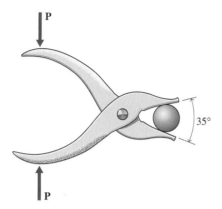

**Figure P6-123**

**6-124\*** A homogeneous book of mass 0.46 kg rests in a bookshelf as shown in Fig. P6-124. The thickness of the book is small compared to the other dimensions shown. The coefficient of friction at all surfaces is 0.4. Determine the minimum angle $\theta$ for which the book is in equilibrium.

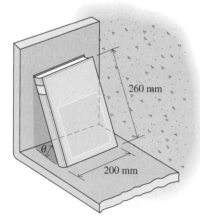

**Figure P6-124**

**Challenging Problems**

**6-125\*** A wedge is used to raise a 350-lb refrigerator onto a platform (Fig. P6-125). The coefficient of friction is 0.2 at all surfaces.
(a) Determine the minimum force **P** needed to insert the wedge.
(b) Determine if the system would still be in equilibrium if **P** = **0**.
(c) If the system is not in equilibrium when **P** = **0**, determine the force necessary to keep the wedge in place, or if the system is in equilibrium when **P** = **0**, determine the force necessary to remove the wedge.

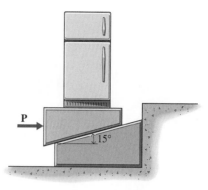

**Figure P6-125**

**6-126\*** A 50-kg uniform plank rests on rough supports at $A$ and $B$ (Fig. P6-126). The coefficient of friction is 0.60 at both surfaces. If a man weighing 800 N pulls on the rope with a force of $P = 400$ N, determine
(a) The minimum and maximum angles $\theta_{min}$ and $\theta_{max}$ for which the system will be in equilibrium.
(b) The minimum coefficient of friction that must exist between the man's shoes and the ground for each of the cases in part $a$.

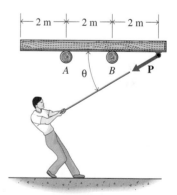

**Figure P6-126**

**6-127** The plunger of a door latch is held in place by a spring as shown in Fig. P6-127. Friction on the sides of the plunger may be ignored. If a force of 2 lb is required to just start closing the door and the coefficient of friction between the plunger and the striker plate is 0.25, determine the force exerted on the plunger by the spring.

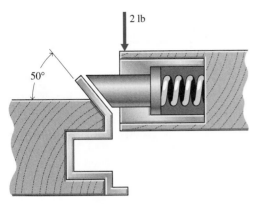

**Figure P6-127**

**6-128** The 25-kg block in Fig. P6-128 is held against the wall by the brake arm. The coefficient of friction between the wall and the block is 0.20; between the block and the brake arm, 0.50. Neglect the weight of the brake arm. Determine

(a) Whether the system would be in equilibrium for $P = 230$ N.

(b) The minimum force $P$ for which the system would be in equilibrium.

(c) The maximum force $P$ for which the system would be in equilibrium.

(d) The shearing stress on a cross section of the pin at $A$ when impending motion of the block is downward if the pin has a 6-mm diameter and is in double shear.

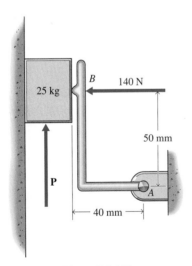

**Figure P6-128**

**6-129** A 250-lb weight is suspended from a lightweight rope wrapped around the inner cylinder of a drum (Fig. P6-129). A brake arm is pressed against the outer cylinder of the drum by a hydraulic cylinder. The coefficient

of friction between the brake arm and the drum is 0.40. Determine

(a) The smallest force in the hydraulic cylinder necessary to prevent motion.

(b) The shearing stress on a cross section of the pin at $A$ when motion is impending if the drum weighs 75 lb and the pin has a $\frac{1}{2}$-in. diameter and is in double shear.

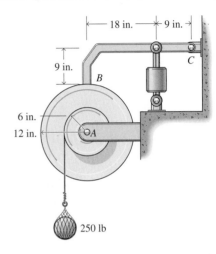

**Figure P6-129**

**6-130\*** The brake shown in Fig. P6-130 is used to control the motion of block $B$. If the mass of block $B$ is 25 kg and the kinetic coefficient of friction between the brake drum and brake pad is 0.30, determine

(a) The force **P** required for a constant-velocity descent.

(b) The shearing stress on cross sections of both pins. Each pin has a diameter of 10 mm and is in double shear.

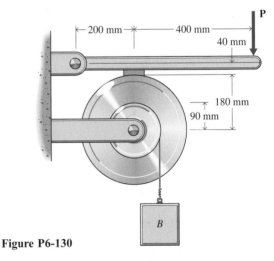

**Figure P6-130**

**6-131** Three identical cylinders are stacked as shown in Fig. P6-131. The cylinders each weigh 22 lb and are 8 in. in diameter. The coefficient of friction is $\mu_s = 0.40$ at all surfaces. Determine the maximum force **P** that the cylinders can support without moving.

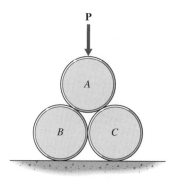

**Figure P6-131**

**Computer Problems**

**6-132** A 10-kg drum rests on a thin, lightweight piece of cardboard as shown in Fig. P6-132. The coefficient of friction is the same at all surfaces.
(a) Plot $P$, the maximum force that may be applied to the cardboard without moving it, as a function of the coefficient of friction $\mu_s$ ($0.05 \leq \mu_s \leq 0.8$).
(b) On the same graph, plot $(A_f)_{actual}$ and $(B_f)_{actual}$, the actual amounts of friction force that act at points $A$ and $B$, and $(A_f)_{avail}$ and $(B_f)_{avail}$, the maximum amounts of friction force available for equilibrium at point $A$ and $B$.
(c) What happens to the system at $\mu_s \cong 0.364$?
(d) What does the solution for $\mu_s > 0.364$ mean?

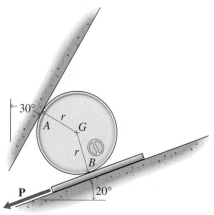

**Figure P6-132**

**6-133** A 150-lb uniformly loaded file cabinet sits on an inclined surface as shown in Fig. P6-133. A horizontal force **P** acts on the cabinet, and the coefficient of friction between the cabinet and the inclined surface is 0.4.
(a) Determine $P_{min}$ and $P_{max}$, the minimum and maximum horizontal forces for which the file cabinet will be in equilibrium.
(b) Calculate and plot $N$ and $F$, the normal and friction forces acting on the bottom of the file cabinet, as functions of $P$ ($P_{min} \leq P \leq P_{max}$).
(c) Calculate and plot $d$, the distance between the line of action of the normal force and the corner $C$, as a function of $P$ ($P_{min} \leq P \leq P_{max}$).

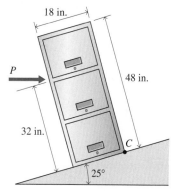

**Figure P6-133**

## 6-9 FLAT BELTS AND V-BELTS

Many types of power machinery rely on belt drives to transfer power from one piece of equipment to another. Without friction, the belts would slip on their pulleys and no power transfer would be possible. Maximum torque is applied to the pulley when the belt is at the point of impending slip.

Although the analysis presented is for flat belts, it also applies to any shape belt as well as circular ropes as long as the only contact between the belt and the

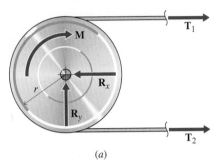

(a)

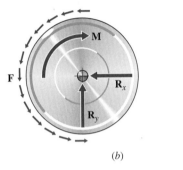

(b)

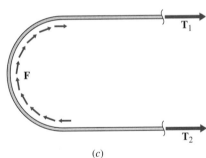

(c)

**Figure 6-94**

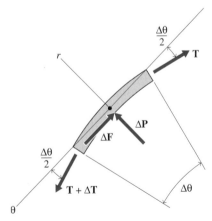

**Figure 6-95**

pulley is on the bottom surface of the belt. This section ends with a brief discussion of V-belts that indicates the kind of modifications required when friction acts on the sides of the belt instead of the bottom.

Figure 6-94a shows a flat belt passing over a circular drum. The tensions in the belt on either side of the drum are $T_1$ and $T_2$, and the bearing reaction is **R**. Friction in the bearing is neglected for this analysis, but a torque **M** is applied to the drum to keep it from rotating. If there is no friction between the belt and the drum $T_1 = T_2$, no torque is required for moment equilibrium, and $\mathbf{M} = \mathbf{0}$. If there is friction between the belt and the drum, however, then the two tensions need not be equal and a torque $M = r(T_2 - T_1)$ is needed to satisfy moment equilibrium. Assuming that $T_2 > T_1$, this means that friction forces must exert a counterclockwise moment on the drum (Fig. 6-94b) and the drum will exert an opposite frictional resistance on the belt (Fig. 6-94c). Because the friction force depends on the normal force and the normal force varies around the drum, care must be taken in adding up the total frictional resistance.

The free-body diagram (Fig. 6-95) of a small segment of the belt includes the friction force $\Delta\mathbf{F}$ and the normal force $\Delta\mathbf{P}$. The tension in the belt increases from **T** on one side of the segment to $\mathbf{T} + \Delta\mathbf{T}$ on the other side. Equilibrium in the radial direction gives

$$+\nwarrow \ \Sigma F_r = 0: \qquad \Delta P - T \sin (\Delta\theta/2) - (T + \Delta T) \sin (\Delta\theta/2) = 0$$

or

$$\Delta P = 2T \sin (\Delta\theta/2) + \Delta T \sin (\Delta\theta/2) \qquad (a)$$

while equilibrium in the circumferential $(\theta-)$ direction gives

$$+\swarrow \ \Sigma F_\theta = 0: \qquad (T + \Delta T) \cos (\Delta\theta/2) - T \cos (\Delta\theta/2) - \Delta F = 0$$

or

$$\Delta T \cos (\Delta\theta/2) = \Delta F \qquad (b)$$

In the limit as $\Delta\theta \rightarrow 0$ the normal force $\Delta P$ on the small segment of the belt must vanish according to Eq. (a). But when the normal force vanishes ($\Delta P \rightarrow 0$), there can be no friction on the belt either ($\Delta F \rightarrow 0$). Therefore, the change in tension across the small segment of the belt must also vanish ($\Delta T \rightarrow 0$) in the limit as $\Delta\theta \rightarrow 0$ according to Eq. (b).

Assuming that slip is impending gives $\Delta F = \mu_s \Delta P$, and Eqs. (a) and (b) can be combined to give

$$\Delta T \cos (\Delta\theta/2) = \mu_s \, 2T \sin (\Delta\theta/2) + \mu_s \, \Delta T \sin (\Delta\theta/2) \qquad (c)$$

which after dividing through by $\Delta\theta$ is

$$\frac{\Delta T}{\Delta\theta} \cos (\Delta\theta/2) = \mu_s T \frac{\sin (\Delta\theta/2)}{(\Delta\theta/2)} + \frac{\mu_s \Delta T \sin (\Delta\theta/2)}{2(\Delta\theta/2)} \qquad (d)$$

Finally, taking the limit as $\Delta\theta \rightarrow 0$ and recalling that $\lim\limits_{\Delta\theta\to0} \dfrac{\Delta T}{\Delta\theta} = \dfrac{dT}{d\theta}$;

$$\lim_{x\to0} \cos x = 1 \qquad \lim_{x\to0} \sin x = x \qquad \lim_{x\to0} \frac{\sin x}{x} = 1$$

gives

$$\frac{dT}{d\theta} = \mu_s T \tag{e}$$

Equation (e) can be rearranged in the form

$$\frac{dT}{T} = \mu_s \, d\theta \tag{6-17}$$

which, since the coefficient of friction is a constant, can be immediately integrated from $\theta_1$, where the tension is $T_1$, to $\theta_2$, where the tension is $T_2$, to get

$$\ln\left(\frac{T_2}{T_1}\right) = \mu_s(\theta_2 - \theta_1) = \mu_s\beta \tag{6-18}$$

or

$$T_2 = T_1 e^{\mu_s\beta} \tag{6-19}$$

where $\beta = \theta_2 - \theta_1$ is the central angle of the drum for which the belt is in contact with the drum. The angle of wrap $\beta$ must be measured in radians and must obviously be positive. Angles greater than $2\pi$ radians are possible and simply mean that the belt is wrapped more than one complete revolution around the drum.

It must be emphasized that Eq. (6-19) assumes impending slip at all points along the belt surface and therefore gives the maximum change in tension that the belt can have. Since the exponential function of a positive value is always greater than 1, Eq. (6-19) gives that $T_2$ (the tension in the belt on the side toward which slip tends to occur) will always be greater than $T_1$ (the tension in the belt on the side away from which slip tends to occur). Of course, if slip is not known to be impending, then Eq. (6-19) does not apply and $T_2$ may be larger or smaller than $T_1$.

V-belts, as shown in Fig. 6-96a, are handled similarly. A view of the belt cross-section (Fig. 6-96b), however, shows that there are now two normal forces and there will also be two frictional forces (acting along the edges of the belt and pointing into the plane of the figure). Equilibrium in the circumferential ($\theta$-) direction now gives

$$\Delta T \cos(\Delta\theta/2) = 2\,\Delta F$$

while equilibrium in the radial direction gives

$$2\,\Delta P \sin(\alpha/2) = 2T\sin(\Delta\theta/2) + \Delta T\sin(\Delta\theta/2)$$

Continuing as above results finally in

$$T_2 = T_1 e^{(\mu_s)_{\text{enh}}\beta} \tag{6-20}$$

in which $(\mu_s)_{\text{enh}} = \left[\dfrac{\mu_s}{\sin(\alpha/2)}\right] > \mu_s$ is an enhanced coefficient of friction. That is, V-belts always give a larger $T_2$ than flat belts for a given coefficient of friction $\mu_s$ and a given angle of wrap $\beta$.

Equations (6-19) and (6-20) can also be used when slipping is actually occurring by replacing the static coefficient of friction $\mu_s$ with the kinetic coefficient of friction $\mu_k$.

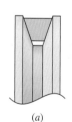

(a)

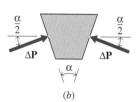

(b)

**Figure 6-96**

**Figure 6-97**

## Example Problem 6-29

A sport utility vehicle is prevented from moving by pulling on a rope that is wrapped $n + \frac{1}{4}$ times around the stump of a tree (Fig. 6-97). The coefficient of friction between the rope and the tree is 0.35 and the force exerted by the vehicle is 750 lb. If it is desired that the force exerted on the rope be no more than 25 lb, determine $n$, the number of times the rope must be wrapped around the tree stump.

### SOLUTION

The angle of wrap of the rope around the stump to hold the vehicle is found from Eq. (6-19). Thus,

$$T_2 = T_1 e^{\mu_s \beta}$$

$$750 = 25 e^{0.35\beta}$$

or

$$\beta = \frac{\ln \dfrac{750}{25}}{0.35} = 9.718 \text{ radians}$$

which is 1.547 times around the tree stump. Any angle less than this will require a resisting force greater than 25 lb, while any angle greater than this will require less force. Thus

$$n = 2 \qquad \text{Ans.}$$

will be sufficient. ∎

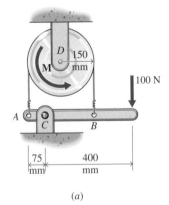

*(a)*

**Figure 6-98**

## Example Problem 6-30

The band brake shown in Fig. 6-98a is used to control the rotation of a drum. The coefficient of friction between the drum and the flat belt is 0.25. The brake arm $AB$ may be considered weightless, but the drum weighs 50 N. The couple **M** applied to the drum causes impending motion of the drum. If the diameters of the pins at $C$ and $D$ are 8 mm, determine the shearing stress on a cross section of each of the pins. Pin $C$ is in single shear and pin $D$ is in double shear.

### SOLUTION

A free-body diagram of the brake arm is shown in Fig. 6-98b. From the equilibrium equation

$$+\circlearrowleft \Sigma M_C = 0: \qquad T_B(225) - T_A(75) - 100(400) = 0$$

The second relationship between $T_A$ and $T_B$ is provided by the belt friction equation [Eq. (6-19)]

$$T_B = T_A e^{\mu_s \beta} = T_A e^{0.25(\pi)} = 2.193 \, T_A$$

Solving yields

$$T_A = 95.60 \text{ N} \qquad T_B = 209.6 \text{ N}$$

From the remaining equilibrium equations

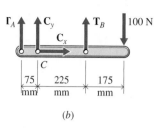

$+\rightarrow \Sigma F_x = 0$:     $C_x = 0$   $C_x = 0$ N

$+\uparrow \Sigma F_y = 0$:     $C_y + T_A + T_B - 100 = 0$

$C_y + 95.60 + 209.6 - 100 = 0$

$C_y = -205.2$ N

(b)

The shearing stress on a cross section of the pin at $C$, which is in single shear, is then given by Eq. (4-4) as

$$\tau = \frac{F_C}{A_S} = \frac{205.2}{(\pi/4)(0.008)^2} = 4.082(10^6) \text{ N/m}^2 \cong 4.08 \text{ MPa} \qquad \textbf{Ans.}$$

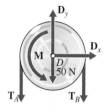

A free-body diagram of the drum is shown in Fig. 6-98c. From the equilibrium equations

(c)

**Figure 6-98**

$+\rightarrow \Sigma F_x = 0$:     $D_x = 0$   $D_x = 0$ N

$+\uparrow \Sigma F_y = 0$:     $D_y - T_A - T_B - 50 = 0$

$D_y - 95.60 - 209.6 - 50 = 0$   $D_y = 355.2$ N

The shearing stress on a cross section of the pin at $D$, which is in double shear, is then given by Eq. (4-4) as

$$\tau = \frac{F_D}{A_s} = \frac{355.2}{2(\pi/4)(0.008)^2} = 3.533(10^6) \text{ N/m}^2 \cong 3.53 \text{ MPa} \qquad \blacksquare \qquad \textbf{Ans.}$$

## PROBLEMS

### Introductory Problems

**6-134\***  A 35-kg child is sitting on a swing suspended by a rope that passes over a tree branch (Fig. P6-134). The coefficient of friction between the rope and the branch (which can be modeled as a flat belt over a drum) is 0.5, and the weight of the rope can be ignored. Determine the minimum force the child must exert on the rope to keep suspended.

**Figure P6-134**

**6-135\*** A rope attached to a 500-lb block passes over a frictionless pulley and is wrapped for one full turn around a fixed post as shown in Fig. P6-135. If the coefficient of friction between the rope and the post is 0.25, determine

(a) The minimum force **P** that must be used to keep the block from falling.

(b) The minimum force **P** that must be used to begin to raise the block.

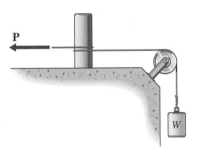

**Figure P6-135**

**6-136** A rope attached to a 220-kg block passes over a fixed drum (Fig. P6-136) If the coefficient of friction between the rope and the drum is 0.30, determine the minimum force **P** that must be used to

(a) Keep the block from falling.

(b) Begin to raise the block.

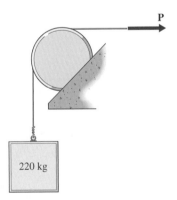

**Figure P6-136**

## Intermediate Problems

**6-137\*** A rotating shaft with a belt-type brake is shown in Fig. P6-137. The static coefficient of friction between the brake drum and the brake belt is 0.20. When a 75-lb force **P** is applied to the brake arm, rotation of the shaft is prevented. Determine the maximum torque that can be resisted by the brake if the shaft is tending to rotate

(a) Counterclockwise.

(b) Clockwise.

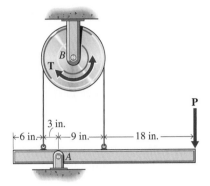

**Figure P6-137**

**6-138\*** The scaffolding of Fig. P6-138 is raised using an electric motor that sits on the scaffolding. Frictionless wheels at the ends of the scaffold restrict horizontal motion. The 250-mm-diameter pulley at the top is jammed and will not rotate. The coefficient of friction between the rope and the pulley is 0.25, and the weight of the motor, scaffold, and supplies is 2500 N. Determine the minimum torque that must be supplied by the motor

(a) To raise the scaffold at a constant rate.

(b) To lower the scaffold at a constant rate.

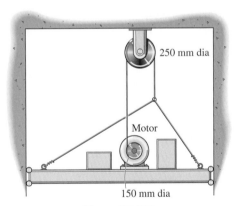

**Figure P6-138**

**6-139** A rope is wrapped one full turn around each of two posts as shown in Fig. P6-139. If the coefficient of friction between the rope and the posts is 0.5, determine

(a) The ratio of $T_A$ to $T_B$.

(b) The ratio of $T_A$ to $T_B$ if only one post is used.

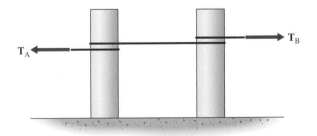

**Figure P6-139**

**Challenging Problems**

**6-140*** A conveyor belt is driven with the 200-mm-diameter multiple-pulley drive shown in Fig. P6-140. Couples $\mathbf{C}_A$ and $\mathbf{C}_B$ are applied to the system at pulleys $A$ and $B$, respectively. The angle of contact between the belt and a pulley is 225° for each 6.5-kg pulley, and the coefficient of friction is 0.30. Determine

(a) The maximum force $T$ that can be developed by the drive and the magnitudes of the input couples $\mathbf{C}_A$ and $\mathbf{C}_B$.

(b) The shearing stress on a cross section of each 25-mm-diameter pin, if each pin is in double shear.

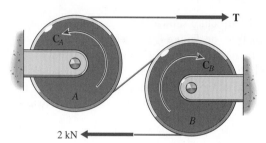

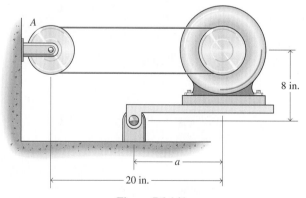

**Figure P6-140**

**6-141*** The electric motor shown in Fig. P6-141 weighs 30 lb and delivers 50 lb · in. of torque to pulley $A$ of a furnace blower by means of a V-belt. The effective diameters of the 36° pulleys are 5 in. The coefficient of friction is 0.30. Determine the minimum distance $a$ to prevent slipping of the belt if the rotation of the motor is clockwise.

**Figure P6-141**

**6-142** Complete the derivation of Eq. (6-20).

**6-143** The band wrench of Fig. P6-143 is used to unscrew an oil filter from a car. (The filter acts as if it were a wheel with a resisting torque of 40 lb · ft.) Neglect the weight of the handle and friction between the end of the metal handle and the metal filter case. Determine

(a) The minimum coefficient of friction between the band and the filter that will prevent slippage.

(b) The shearing stress on a cross section of the pin at $B$ when slipping is impending if the $\frac{1}{4}$-in.-diameter pin is loaded in double shear.

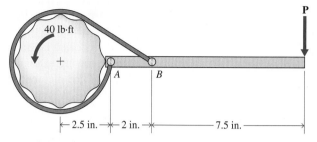

**Figure P6-143**

**6-144*** A uniform belt 2 m long with a mass of 2 kg hangs over a small fixed peg (Fig. P6-144). If the coefficient of friction between the belt and the peg is 0.40, determine the maximum distance $d$ between the two ends for which the belt will not slip off the peg.

**Figure P6-144**

**6-145** The hand brake of Fig. P6-145 is used to control the rotation of a drum. The coefficient of friction between the belt and the drum is $\mu_s = 0.35$, and the weight of the handle may be neglected. If a clockwise torque of 350 lb · ft is applied to the drum, determine the minimum force $\mathbf{P}$ that must be applied to the handle to prevent motion when $a = 4$ in., 8 in., and 12 in.

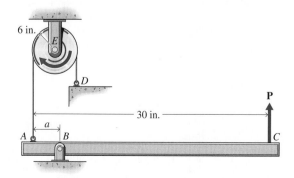

**Figure P6-145**

# 6-10 DESIGN

Design was discussed in Section 4-10, where design was applied to axially loaded members and pins. The methods discussed in section 4-10 can be applied in this chapter to two-force members and pins. Design will be extended to more complicated members and loading situations in later chapters of this book. The following example illustrates the use of design principles to size pins and a two-force member in a pin-connected structure.

## Example Problem 6-31

All members, including pins, of the structure shown in Fig. 6-99a are made of structural steel. All pins are in single shear, failure is by yielding, and the factor of safety is to be 1.5. The yield strength is 36 ksi for normal stress and 18 ksi for shearing stress. Determine the diameter of the rod $CD$ and the diameters of the pins at $A$, $B$, $C$, and $D$ required to support the forces shown. Assume that all other members of the structure are of adequate size to support the applied forces.

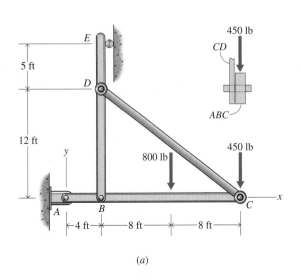

(a)

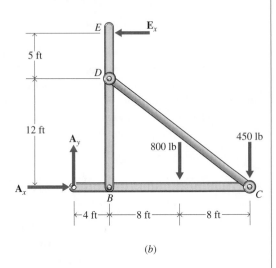

(b)

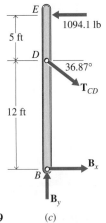

**Figure 6-99**    (c)

## SOLUTION

A free-body diagram of the complete structure is shown in Fig. 6-99b. From the equilibrium equations

$+\circlearrowleft \Sigma M_A = 0$:    $E_x(17) - 800(12) - 450(20) = 0$    $E_x = 1094.1$ lb

$+\rightarrow \Sigma F_x = 0$:    $A_x - E_x = 0$    $A_x = 1094.1$ lb

$+\uparrow \Sigma F_y = 0$:    $A_y - 800 - 450 = 0$    $A_y = 1250.0$ lb

and the magnitude of the force on pin $A$ is

$$F_A = \sqrt{A_x^2 + A_y^2} = \sqrt{(1094.1)^2 + (1250.0)^2} = 1661.2 \text{ lb}$$

A free-body diagram of member *BDE* is shown in Fig. 6-99c. From the equilibrium equations

$+\circlearrowleft \Sigma M_B = 0$:    $1094.1(17) - T_{CD} \cos 36.87°(12) = 0$    $T_{CD} = 1937.5$ lb

$+\rightarrow \Sigma F_x = 0$:    $B_x + T_{CD} \cos 36.87° - 1094.1 = 0$    $B_x = -455.9$ lb

$+\uparrow \Sigma F_y = 0$:    $B_y - T_{CD} \sin 36.87° = 0$    $B_y = 1162.5$ lb

and the magnitude of the force on pin *B* is

$$F_B = \sqrt{B_x^2 + B_y^2} = \sqrt{(-455.9)^2 + (1162.5)^2} = 1248.7 \text{ lb}$$

The force in the rod *CD* and the magnitude of the force on pins *C* and *D* are

$$F_C = F_D = T_{CD} = 1937.5 \text{ lb}$$

The pins and member *CD* may now be sized by applying the failure criterion

$$\text{Strength} \geq (\text{Factor of safety})(\text{Stress})$$

For the pins, strength $= \tau_y$, and for the rod *CD*, strength $= \sigma_y$. Thus, for pin *A*,

$$\tau_y \geq (\text{FS})\left(\frac{F_A}{A}\right) = (\text{FS})\left(\frac{F_A}{\pi d_A^2/4}\right)$$

$$d_A \geq \sqrt{\frac{4(\text{FS})(F_A)}{\pi \tau_y}} = \sqrt{\frac{4(1.5)(1661.2)}{\pi(18,000)}}$$

$$d_A \geq 0.420 \text{ in.} \qquad \qquad \textbf{Ans.}$$

For pin *B*,

$$d_B \geq \sqrt{\frac{4(\text{FS})(F_B)}{\pi \tau_y}} = \sqrt{\frac{4(1.5)(1248.7)}{\pi(18,000)}}$$

$$d_B \geq 0.364 \text{ in.} \qquad \qquad \textbf{Ans.}$$

For pins *C* and *D*,

$$d \geq \sqrt{\frac{4(\text{FS})(T_{CD})}{\pi \tau_y}} = \sqrt{\frac{4(1.5)(1937.5)}{\pi(18,000)}}$$

$$d_C = d_D \geq 0.453 \text{ in.} \qquad \qquad \textbf{Ans.}$$

For rod *CD*,

$$d_{CD} \geq \sqrt{\frac{4(\text{FS})(T_{CD})}{\pi \sigma_y}} = \sqrt{\frac{4(1.5)(1937.5)}{\pi(36,000)}}$$

$$d_{CD} \geq 0.321 \text{ in.} \qquad \qquad \textbf{Ans.}$$

The previous results indicate that $\frac{1}{2}$-in.-diameter pins (a standard size) could be used for all of the connections and either a $\frac{3}{8}$-in.- or a $\frac{1}{2}$-in.-diameter rod could be used for member $CD$.

## PROBLEMS

### Introductory Problems

**6-146\*** The structure shown in Fig. P6-146 consists of a circular tie rod $AB$ and a rigid member $BC$. If the structure is to support a load $P = 40$ kN, determine the required diameters of the pins at $A$, $B$, and $C$, and the required diameter of the tie rod. The tie rod is made of structural steel, and the pins are made of 0.2% C hardened steel. All pins are in double shear. The tie rod is adequately reinforced around the pins so that tensile failure does not occur at the pins. Failure is by yielding, and the factor of safety is 1.3. Take the yield strength in shear to be half the yield strength in tension.

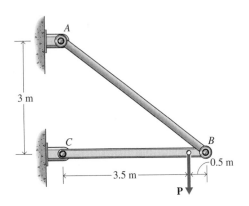

**Figure P6-146**

**6-147\*** A rigid handle is used to twist a 2.5-in.-diameter valve, as shown in Fig. P6-147. A pair of oppositely directed parallel forces $P = 100$ lb is needed to twist the valve. Force is transmitted from the handle to the valve shaft by means of a 1.0-in.-long (into the page) key with a square cross section. The key is to be made of 0.8% C hot-rolled steel. If failure is by fracture, and the factor of safety is to be 2, determine the cross-sectional dimensions of the square key to the nearest $\frac{1}{64}$ in. The yield and ultimate strengths in shear are half the corresponding values in tension.

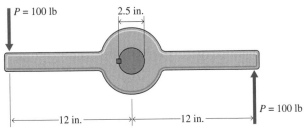

**Figure P6-147**

**6-148** Each member of the truss of Fig. P6-148 has a circular cross section and is made of structural steel. If failure is by yielding and the factor of safety is 2.5, determine the minimum permissible diameter of member $EJ$.

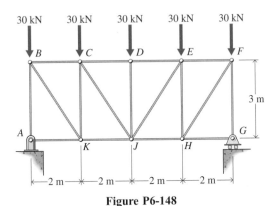

**Figure P6-148**

### Intermediate Problems

**6-149** A pin connected system of levers and bars is used as a toggle for a press to crush cans, as shown in Fig. P6-149. The system is designed to provide a crushing force of 550 lb. The handle may be considered rigid. The pin at $D$ is to be made of 2024-T4 wrought aluminum. If failure is by yielding, and the factor of safety is to be 1.5, determine the required diameter of pin $D$ to the nearest $\frac{1}{32}$ in.

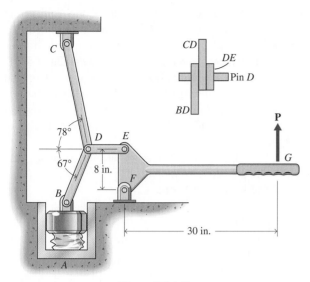

**Figure P6-149**

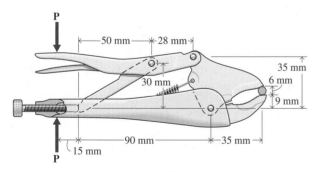

**Figure P6-150**

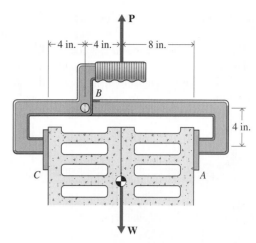

**Figure P6-151**

**6-150\*** A pair of vise grip pliers is shown in Fig. P6-150. All members of the pliers, except for the pins and spring, may be treated as rigid. It is anticipated that the largest applied force on the handles will be $P = 200$ N. If each of the three pins is in double shear and made of 0.4% C hardened steel, determine the required diameters for the pins to the nearest millimeter. In the analysis, neglect the spring force. Failure is by yielding, and the factor of safety is 1.25. The yield strength in shear is half the tensile yield strength.

**6-151** A device for lifting rectangular objects such as bricks and concrete blocks is shown in Fig. P6-151. The coefficients of friction at all vertical contact surfaces are $\mu_s = 0.4$ and $\mu_k = 0.3$. The device is to lift two blocks, each weighing 15 lb. The pin at $B$ is to be made of structural steel with a strength in shear equal to half the strength in tension. For failure by yielding and a factor of safety of 3, determine the minimum permissible diameter of the pin at $B$, which is in double shear.

**Challenging Problems**

**6-152\*** The flat roof of a building is supported by a series of parallel plane trusses spaced 2 m apart (only one such truss is shown in Fig. P6-152). Water of density 1000 kg/m³ may collect on the roof to a depth of 0.2 m. All members of the truss are to be the same size. If the truss members are made of structural steel, failure is by yielding, and the factor of safety is 3.0, determine the smallest-diameter solid circular rod, to the nearest millimeter, that can be used. The members are braced so that buckling does not occur.

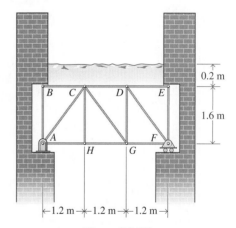

**Figure P6-152**

**6-153** The brake shown in Fig. P6-153 is used to control the motion of block $B$. The weight of the block is 2200 lb, the coefficient of static friction between the brake arm and drum is 0.35, and the coefficient of kinetic friction is 0.30. A force **P** is applied to the rigid brake arm such that the block moves downward at a constant rate. Determine the minimum permissible diameter for the pin at the end of the brake arm (to the nearest $\frac{1}{8}$-in.) if it is in double shear and is made of 0.2% C hardened steel. Failure is by yielding, and the factor of safety is 2.5. The strength in shear is half the strength in tension.

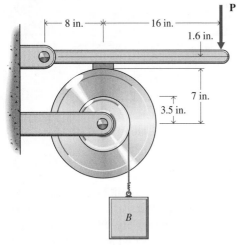

**Figure P6-153**

**6-154** Solid circular bars $AB$ and $BC$ of Fig. P6-154 are pinned at the ends, and the structure is subjected to a load $P = 26$ kN. The angle $\theta$ may vary, but pin $A$ is always directly above pin $C$. Both bars are made of 6061-T6 wrought aluminum alloy. Failure is by yielding, and the factor of safety is 4. In the force analysis assume that the weights of the bars are negligible with respect to the applied loads and that the pins are adequately designed. Determine

(a) The angle $\theta$ for a minimum weight structure.
(b) The diameters of the bars.
(c) The weights of the bars.

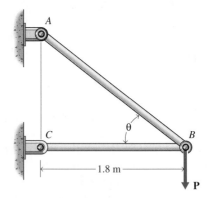

**Figure P6-154**

## 6-11 SUMMARY

Any system of forces acting on a rigid body can be expressed in terms of a resultant force **R** and a resultant couple **C**. Therefore, for a rigid body to be in equilibrium, both the resultant force **R** and the resultant couple **C** must vanish. These two conditions can be expressed by the two vector equations

$$\mathbf{R} = \Sigma F_x \mathbf{i} + \Sigma F_y \mathbf{j} + \Sigma F_z \mathbf{k} = 0$$
$$\mathbf{C} = \Sigma M_x \mathbf{i} + \Sigma M_y \mathbf{j} + \Sigma M_z \mathbf{k} = 0$$

(6-1)

Equations (6-1) can be expressed in scalar form as

$$\Sigma F_x = 0 \qquad \Sigma F_y = 0 \qquad \Sigma F_z = 0$$
$$\Sigma M_x = 0 \qquad \Sigma M_y = 0 \qquad \Sigma M_z = 0$$

(6-2)

Equations (6-2) are the necessary conditions for equilibrium of a rigid body. They are also the sufficient conditions for equilibrium if all of the forces acting on the body can be determined from these equations.

To study the force system acting on a body, it is necessary to identify all forces, both known and unknown, that act on the body. The best way to identify all forces acting on a body is to use the free-body diagram approach. Special care must be exercised when representing the actions of connections and supports on the free-body diagram.

The term *two-dimensional* is used to describe problems in which the forces involved are contained in a plane (say the *xy*-plane) and the axes of all couples are perpendicular to the plane containing the forces. For two-dimensional problems, the equations of equilibrium reduce to

$$\mathbf{R} = \Sigma F_x \mathbf{i} + \Sigma F_y \mathbf{j} = \mathbf{0}$$
$$\mathbf{C} = \Sigma M_A \mathbf{k} = \mathbf{0} \tag{6-3}$$

or in scalar form to

$$\Sigma F_x = 0 \qquad \Sigma F_y = 0 \qquad \Sigma M_A = 0 \tag{6-4}$$

The third equation represents the sum of the moments of all forces about a *z*-axis through any point *A* on or off the body. Equations (6-4) are both the necessary and sufficient conditions for equilibrium of a body subjected to a two-dimensional system of forces. Alternative forms of Eqs. (6-4) are

$$\Sigma F_x = 0 \qquad \Sigma M_A = 0 \qquad \Sigma M_B = 0 \tag{6-5}$$

where points *A* and *B* must have different *x*-coordinates, and

$$\Sigma M_A = 0 \qquad \Sigma M_B = 0 \qquad \Sigma M_C = 0 \tag{6-6}$$

where *A*, *B*, and *C* are any three points not on the same straight line.

Two broad categories of engineering structures were considered in this chapter; namely, frames and trusses. Structures that are not constructed entirely of two-force members are called *frames* or *machines*. While frames and machines may also contain one or more two-force members, they always contain at least one member that is acted upon by forces at more than two points or is acted upon by both forces and moments. The main distinction between frames and machines is that frames are rigid structures while machines are not.

The method of solution for frames and machines consists of taking the structures apart, drawing free-body diagrams of each of the components, and writing the equations of equilibrium for each of the free-body diagrams. Since some of the members of frames and machines are not two-force members, however, the directions of the forces in these members is not known. The analysis of frames and machines will consist of solving the equilibrium equations of a system of rigid bodies.

For machines and nonrigid structures, the structure must be taken apart and analyzed even if the only information desired is the support reactions or the relationship between the external forces (input and output forces) acting on it.

Four main assumptions are made in the analysis of trusses:

1. Truss members are connected only at their ends; no member is continuous through a joint.
2. Members are connected by frictionless pins.
3. The truss structure is loaded only at the joints.
4. The weight of the members may be neglected.

Because of these assumptions, truss members are modeled as two-force members with the forces acting at the ends of the members and directed along the axis of the member.

One method of analysis (the method of joints) for trusses is performed by drawing a free-body diagram for each pin (joint). Application of the vector equilibrium equation, $\Sigma\mathbf{F} = \mathbf{0}$, at each joint yields two algebraic equations that can be solved for two unknowns. The pin forces are solved sequentially, starting from a pin on which only two unknown forces and one or more known forces act. Once these forces are determined, their values can be applied to adjacent joints and treated as known quantities. This process is repeated until all unknown forces have been determined. The method of joints is most often used when the forces in all of the members of a truss are to be determined.

A second method of analysis for trusses is the method of sections. When the method of sections is used, the truss is divided into two parts by passing an imaginary plane or curved section through the members of interest. Free-body diagrams may then be drawn for either or both parts of the truss. Since each part is a rigid body, three independent equations of equilibrium can be written for either part. Therefore, a section that cuts through no more than three members should be used.

It will often happen that a section cutting no more than three members and passing through a given member of interest cannot be found. In such a case it may be necessary to draw a section through a nearby member and solve for the force in that member first. Then the method of joints can be used to find the force in the member of interest. One of the principle advantages of using the method of sections is that the force in a member near the center of a large truss usually can be determined without first obtaining the forces in the rest of the truss.

Tangential forces due to friction are always present at the interface between two contacting bodies, and these forces always act in a direction to oppose the tendency of the contacting surfaces to slip relative to one another. In some types of machine elements, such as bearings and skids, it is desired to minimize friction, while in other types, such as brakes and belt drives, it is desired to maximize friction. Two main types of friction are commonly encountered in engineering practice—dry friction and fluid friction. Dry friction is encountered when the unlubricated surfaces of two solids are in contact and a condition of sliding or tendency to slide exists. Fluid friction develops when adjacent layers in a fluid move at different velocities.

Since friction forces cannot increase without limit, they eventually reach a maximum value $F_{\max}$. The condition when a friction force is at its maximum value is called a *condition of impending motion*. Beyond that point, friction can no longer supply the amount of force required for equilibrium. The value of limiting friction force is proportional to the normal force $N$ at the contact surface. Thus,

$$F_{\max} = \mu_s N \qquad (6\text{-}12)$$

The constant of proportionality $\mu_s$, the coefficient of static friction, depends on the types of material in contact but is independent of both the normal force and the area of contact. The frictional force actually exerted is never greater than that required to satisfy the equation of equilibrium.

$$F \leq \mu_s N \qquad (6\text{-}13)$$

The equality in Eq. (6-13) holds only at the point of impending motion. Thus, friction belongs to a class of forces known as resistances; it only operates to oppose motion, never to produce it. Once a body starts to slip relative to its supporting surface, the friction force will decrease approximately 25 percent to

$$F = \mu_k N \qquad (6\text{-}14)$$

where $\mu_k$ is the coefficient of kinetic friction. This coefficient is again independent of the normal force and is also independent of the speed of the relative motion—at least for low speeds. The presence of any moisture or oil on the surface, however, can change the problem from one of dry friction, in which the friction force is independent of the speed of the body, to one of fluid friction, in which the friction force is directly proportional to the speed.

For some problems, the methods developed in Chapter 4 can be used to determine stress and/or deformation. For example, both normal stress and deformation can be calculated for straight two-force members. For members loaded in a fashion more complex than axially loaded members, the calculation of stress and deformation is more complicated than indicated in Chapter 4. The methods to determine stress and deformation for these members will be developed throughout the remainder of this book. However, shearing stresses in pins and bearing stresses between pins and members may be found using the results of Chapter 4, regardless of the type of loading on the member.

## REVIEW PROBLEMS

**6-155\*** Determine the force **P** required to pull the 250-lb roller over the step shown in Fig. P6-155.

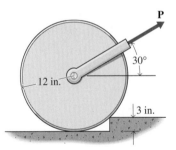

**Figure P6-155**

**6-156\*** A machine is activated by force **T** when a pedal is depressed with a 40-N force, as shown in Fig. P6-156. Determine the magnitude of **T** and the resultant bearing reaction at *B*.

**Figure P6-156**

**6-157\*** The 500-lb block shown in Fig. P6-157 is supported by a ball-and-socket joint at *A*, by a smooth pin at *B*, and by a cable at *C*. Determine the components of the reactions at supports *A* and *B* and the force in the cable at *C*.

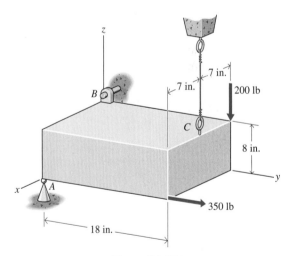

**Figure P6-157**

**6-158** The plate shown in Fig. P6-158 has a mass of 80 kg. The brackets at supports *A* and *B* exert only force reactions on the plate. Each of the brackets can resist a force along the axis of the pins in one direction only. Determine
(a) The reactions at supports *A* and *B* and the tension in the cable.
(b) The change in length of the cable. (Use $E = 200$ GPa and $d = 2.5$ mm.)

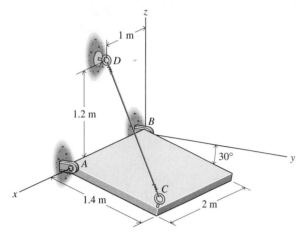

**Figure P6-158**

**6-159** The door to an airplane hangar consists of two uniform sections that are hinged at the middle as shown in Fig. P6-159. The door is raised by means of a cable attached to a bar along the bottom edge of the door. Smooth rollers at the ends of the bar $C$ run in a smooth vertical channel. If the door is 30 ft wide, 15 ft tall, and weighs 1620 lb, determine

(a) The force $P$ when the door opening height $h = 8$ ft.
(b) The resultant hinge forces at $A$ and $B$ when $h = 8$ ft.

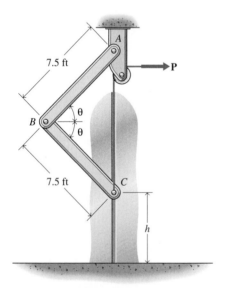

**Figure P6-159**

**6-160\*** Three bars are connected by smooth frictionless pins as shown in Fig. P6-160. Bar $BCDE$ is rigid, bar $AB$ is aluminum ($E = 73$ GPa; $L = 1.5$ m; $d = 30$ mm) and bar $EF$ is steel ($E = 200$ GPa; $L = 1.2$ m; $d = 20$ mm). The 25-mm-diameter pivot pin $D$ is aluminum and is in double shear. When $P = 100$ kN determine

(a) The deformation of bars $AB$ and $EF$.
(b) The shearing stress on a cross section of the pin at $D$.

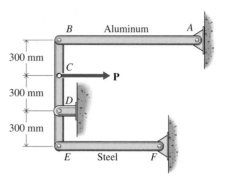

**Figure P6-160**

**6-161** A person is holding a 20-lb object as shown in Fig. P6-161. Determine the force **T** in the biceps muscle and the force **F** of the humerus against the ulna, in terms of the weight **W** of the forearm which acts through $G$. For the position shown, both **T** and **F** act vertically.

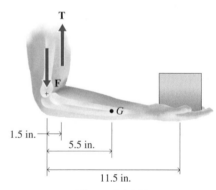

**Figure P6-161**

**6-162** Bodies $A$ and $B$ shown in Fig. P6-162 have masses of 2200 kg and 450 kg, respectively. The coefficient of friction between $B$ and the horizontal surface is 0.2, between $A$ and $B$ it is 0.2, and between $A$ and the fixed surface is 0.5. Determine the force **P** required to cause impending motion of body $B$ to the left.

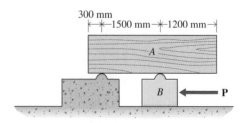

**Figure P6-162**

**6-163*** Three bars are connected by smooth frictionless pins as shown in Fig. P6-163. Bar $BCD$ is rigid, bar $AC$ is aluminum [$E = 12,000$ ksi; $L = 36$ in.; $d = 1$ in.; $\alpha = 12.5(10^{-6})/°F$], and bar $DE$ is steel [$E = 30,000$ ksi; $L = 30$ in.; $d = \frac{1}{2}$ in.; $\alpha = 6.6(10^{-6})/°F$]. The $\frac{5}{8}$-in.-diameter pivot pin $B$ is steel and is in single shear. If the temperature drops by 20°F after the unit is assembled, determine

(a) The normal stresses in bars $AC$ and $DE$.
(b) The change in length of bars $AC$ and $DE$.
(c) The shearing stress on the cross section of the pin at $B$.

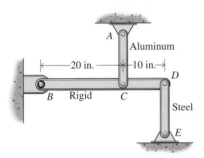

**Figure P6-163**

**6-164*** An adjustable bracket is held in place by a force **P** of magnitude 175 N as shown in Fig. P6-164. Determine the minimum coefficient of friction between the bracket and vertical square member if the bracket is to stay in place.

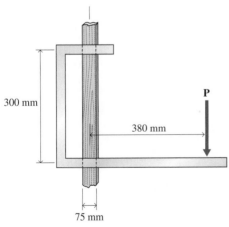

**Figure P6-164**

**6-165** Determine the force in members $CD$, $DE$, and $FG$ of the truss shown in Fig. P6-165.

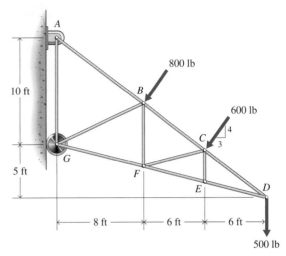

**Figure P6-165**

**6-166** Determine the change in length of members $CF$ and $FG$ of the truss shown in Fig. P6-166. Each member of the truss is made of structural steel, and is circular with a diameter of 25 mm.

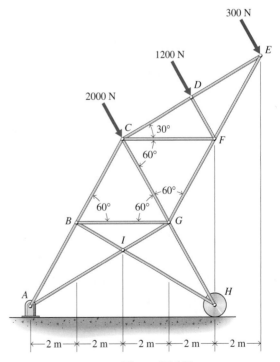

**Figure P6-166**

**6-167** The band brake of Fig. P6-167 is used to control the rotation of a drum. The coefficient of friction between the belt and the drum is 0.25, and the weight of the handle is negligible.

Determine the force **P** that must be applied to the end of the handle to resist a maximum torque of 200 lb · in. if the torque is applied
(a) Clockwise.
(b) Counterclockwise.

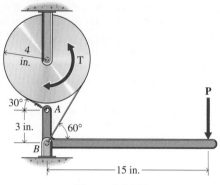

**Figure P6-167**

# TORSIONAL LOADING: SHAFTS

## 7-1 INTRODUCTION

The problem of transmitting a torque (a couple) from one plane to a parallel plane is frequently encountered in the design of machinery. The simplest device for accomplishing this function is a circular shaft such as that connecting an electric motor with a pump, compressor, or other machine. A modified free-body diagram (the weight and bearing reactions are not shown because they contribute no useful information to the torsion problem) of a shaft used to transmit a torque from a driving motor $A$ to a coupling $B$ is shown in Fig. 7-1. The resultant of the electromagnetic forces applied to armature $A$ of the motor is a couple resisted by the resultant of the bolt forces (another couple) acting on the flange coupling $B$. The circular shaft transmits the torque from the armature to the coupling. Typical torsion problems involve determinations of significant stresses in and deformations of shafts.

A segment of the shaft between transverse planes $a–a$ and $b–b$ of Fig. 7-1 will be studied. The complicated stress distributions at the locations of the torque-applying devices are beyond the scope of this elementary treatment of the torsion problem. A free-body diagram of the segment of the shaft between sections $a–a$ and $b–b$ is shown in Fig. 7-2 with the torque applied by the armature indicated on the left end as $T$. The resisting torque $T_r$ at the right end of the segment is the resultant of the differential forces $dF$ acting on the transverse plane $b–b$. The force $dF$ is equal to $\tau_\rho \, dA$, where $\tau_\rho$ is the shearing stress on the transverse plane at a distance $\rho$ from the center of the shaft and $dA$ is a differential area. For circular sections, the shearing stress at a point on any transverse plane is always perpendicular to the radius to the point. If the shaft is in equilibrium, a summation of moments about the axis of the shaft indicates that

$$T = T_r = \int_{\text{area}} \rho \, dF = \int_{\text{area}} \rho \, \tau_\rho \, dA \qquad (7\text{-}1)$$

The law of variation of the shearing stress on the transverse plane ($\tau$ as a function of radial position $\rho$) must be known before the integral of Eq. (7-1) can be evaluated. Thus, the problem of determining the relationship between torque and shearing stress is statically indeterminate. Recalling the procedures developed in Chapter 4, the solution of a statically indeterminate problem requires the use of

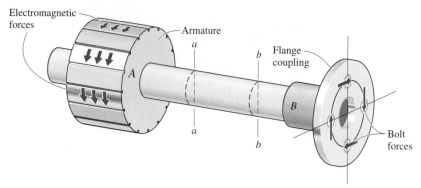

**Figure 7-1**

the equations of equilibrium [Eq. (7-1) for torsion], an analysis of deformation, and the relationship between stress and strain.

In 1784 C. A. Coulomb, a French engineer, developed (experimentally) a relationship between applied torque and angle of twist for circular bars.[1] In a paper published in 1820[1] A. Duleau, another French engineer, derived the same relationship analytically by making the assumption that a plane section before twisting remains plane after twisting and a diameter remains straight. Visual examination of twisted models indicates that these assumptions are correct for circular sections either solid or hollow (provided the hollow section is circular and symmetrical with respect to the axis of the shaft), but incorrect for any other shape. Compare, for example, the distortions of rubber models with circular and rectangular cross sections shown in Fig. 7-3. Figure 7-3*b* shows the circular shaft after loading, and illustrates that plane sections remain plane. For the rectangular shaft, plane sections before loading (Fig. 7-3*c*) become warped after loading (Fig. 7-3*d*).

---

[1]From *History of Strength of Materials*, S. P. Timoshenko, McGraw-Hill, New York, 1953.

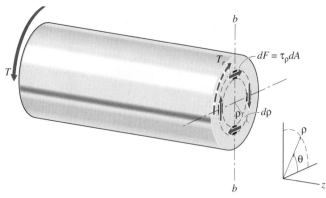

**Figure 7-2**

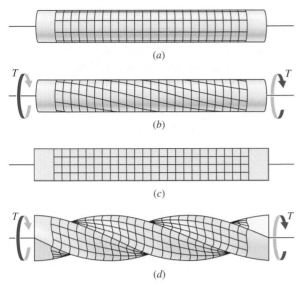

Figure 7-3

---

![segment decoration]

# 7-2 TORSIONAL SHEARING STRAIN

If a plane transverse section before twisting remains plane after twisting and a diameter of the section remains straight, the distortion of the shaft of Fig. 7-2 will be as indicated in Fig. 7-4a, where points $B$ and $D$ on a common radius in a plane move to points $B'$ and $D'$ in the same plane and still on the same radius.

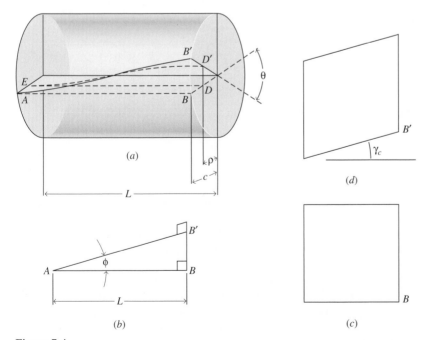

Figure 7-4

The angle $\theta$ is called the *angle of twist*. The surface $ABB'$ of Fig. 7-4a is shown in plan view in Fig. 7-4b, in which a differential element of the material at $B$ (Fig. 7-4c) is distorted due to shearing stress at $B'$ (Fig. 7-4d). Clearly, the angle $\phi$ of Fig. 7-4b is the same as the shearing strain $\gamma_c$ of Fig. 7-4d. Similar figures could be drawn for the surface $EDD'$. It is recommended that the reader review the concept of shearing strain in Section 4-4.

At this point the assumption is made that all longitudinal elements ($AB$, $ED$, etc.) have the same length $L$ (which limits the results to straight shafts of constant diameter). From Fig. 7-4, the shearing strain $\gamma_\rho$ at a distance $\rho$ from the center of the shaft and $\gamma_c$ at the surface of the shaft ($\rho = c$) are related to the angle of twist $\theta$ by

$$\tan \gamma_c = \frac{BB'}{AB} = \frac{c\theta}{L}$$

and

$$\tan \gamma_\rho = \frac{DD'}{ED} = \frac{\rho\theta}{L}$$

or, if the strain is small ($\tan \gamma \cong \sin \gamma \cong \gamma$, $\gamma$ in radians),

$$\gamma_c = \frac{c\theta}{L} \tag{7-2a}$$

and

$$\gamma_\rho = \frac{\rho\theta}{L} \tag{7-2b}$$

Combining Eqs. (7-2a) and (7-2b) gives

$$\theta = \frac{\gamma_c L}{c} = \frac{\gamma_\rho L}{\rho}$$

which indicates that the shearing strain

$$\gamma_\rho = \frac{\gamma_c}{c} \rho \tag{7-3}$$

is zero at the center of the shaft and increases linearly with respect to the distance $\rho$ from the axis of the shaft. Equation (7-3) is the result of the deformation analysis of a circular shaft subjected to torsional loading. This equation can be combined with Eq. (7-1) once the relationship between shearing stress $\tau$ and shearing strain $\gamma$ is known.

Up to this point, no assumptions have been made about the relationship between stress and strain or about the type of material of which the shaft is made. Therefore, Eq. (7-3) is valid for elastic or inelastic action and for homogeneous or heterogeneous materials, provided the strains are not too large ($\tan \gamma \cong \gamma$). Problems in this book will be assumed to satisfy this requirement.

# 7-3 TORSIONAL SHEARING STRESS—THE ELASTIC TORSION FORMULA

If the assumption is now made that Hooke's law applies (the accompanying limitation is that the stresses must be below the proportional limit of the material), the shearing stress $\tau$ is related to the shearing strain $\gamma$ by the expression $\tau = G\gamma$ [Eq. (4-15c)]. Then, multiplying Eq. (7-3) by the shear modulus (modulus of rigidity) $G$ gives

$$\tau_\rho = \frac{\tau_c}{c}\,\rho \qquad (7\text{-}4)$$

When Eq. (7-4) is substituted into Eq. (7-1), the result is

$$T = T_r = \frac{\tau_c}{c}\int \rho^2\,dA = \frac{\tau_\rho}{\rho}\int \rho^2\,dA \qquad (a)$$

The integral in Eq. (a) is called the *polar second moment of area*[2] and is given the symbol $J$. For a solid circular shaft such as is shown in Fig. 7-5a, the polar second moment $J$ is

$$J = \int \rho^2\,dA = \int_0^c \rho^2\,(2\pi\rho\,d\rho) = \frac{\pi c^4}{2} \qquad (7\text{-}5a)$$

For a circular annulus such as shown in Fig. 7-5b, the polar second moment $J$ is

$$J = \int \rho^2\,dA = \int_b^c \rho^2\,(2\pi\rho\,d\rho)$$
$$= \frac{\pi c^4}{2} - \frac{\pi b^4}{2} = \frac{\pi}{2}(r_o^4 - r_i^4) \qquad (7\text{-}5b)$$

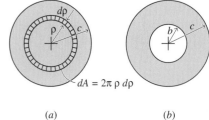

(a)                    (b)

**Figure 7-5**

where $r_o$ and $r_i$ are the outer and inner radii, respectively, of the circular annulus. In terms of the polar second moment $J$, Eq. (a) can be written as

$$T = T_r = \frac{\tau_c J}{c} = \frac{\tau_\rho J}{\rho} \qquad (b)$$

or solving for the unknown shearing stress $\tau$

$$\tau_\rho = \frac{T\rho}{J} \qquad \text{and} \qquad \tau_c = \frac{Tc}{J} \qquad (7\text{-}6)$$

Like the shearing strain $\gamma_\rho$, the shearing stress $\tau_\rho$ is zero at the center of the shaft and increases linearly with respect to the distance $\rho$ from the axis of the shaft. Both the shearing strain $\gamma$ and the shearing stress $\tau$ are maximum when $\rho = c$.

---

[2]Integrals of the type $\int x^2\,dA$ arise often in mechanics and are given the general name "second moments of area." Second moments of area are discussed further in Chapter 8 where they are used to relate stresses to internal forces and moments in beams. Second moments of area are sometimes (improperly) called "moments of inertia," since they are closely related to the moment of inertia integral $\int \rho^2\,dm$, which arises in dynamics.

Equation (7-6) is known as the "elastic torsion formula," in which $\tau_\rho$ is the shearing stress on a transverse plane at a distance $\rho$ from the axis of the shaft and $T$ is the resisting torque (the torque produced on the transverse plane by the shearing stresses). Even though the resisting torque is $T_r$, the equivalent symbol $T$ is used in Eq. (7-6); see Eq. (7-1). Equation (7-6) is valid for both solid and hollow circular shafts. The resisting torque $T_r$ is generally different from the external torques applied to various points along the shaft and must be obtained from a free-body diagram and an equilibrium equation. The procedure for calculating the resisting torque is illustrated in Example Problem 7-1. Note that Eq. (7-6) applies only for linearly elastic action in homogeneous and isotropic materials since Hooke's law $\tau = G\gamma$ was used in its development.

## 7-4 TORSIONAL DISPLACEMENTS

Frequently the amount of twist in a shaft is of paramount importance. Therefore, determination of the angle of twist is a common problem for the machine designer. The fundamental approach to such problems is provided by the following equations

$$\gamma_\rho = \rho \frac{\theta}{L} \qquad \text{or} \qquad \gamma_\rho = \rho \frac{d\theta}{dL} \tag{7-2}$$

$$\tau_\rho = \frac{T\rho}{J} \qquad \text{or} \qquad \tau_c = \frac{Tc}{J} \tag{7-6}$$

$$G = \frac{\tau}{\gamma} \tag{4-15}$$

The second form of Eq. (7-2) is used when the torque or the cross section varies as a function of position along the length of the shaft. Equation (7-2) is valid for both elastic and inelastic action. Equation (7-6) is the elastic torsion formula that provides the shearing stress $\tau_\rho$ on a transverse plane at a distance $\rho$ from the axis of the shaft. Equation (4-15) is Hooke's law for shearing stresses. The last two expressions are limited to stresses below the proportional limit of the material (linearly elastic action). The three equations can be combined to give several different relationships; for example,

$$\theta = \frac{\gamma_\rho L}{\rho} = \frac{\tau_\rho L}{G\rho} \tag{7-7a}$$

or

$$\theta = \frac{TL}{GJ} \tag{7-7b}$$

The angle of twist $\theta$ determined from these expressions is for a length $L$ of shaft, of constant diameter ($J$ = constant), constant material properties ($G$ = constant), and carrying a torque $T$. The resisting torque $T = T_r$ is the torque produced on the transverse plane by the shear stresses and is generally different from the external torques applied to the shaft at various sections by gears, pulleys, or couplings. Ideally, the length of shaft should not include sections too near to (within about one-half shaft diameter of) places where mechanical devices are attached. For practical purposes, however, it is customary to neglect local distortions at all connections and to compute angles as if there were no discontinuities.

If $T$, $G$, or $J$ is not constant along the length of the shaft, Eq. (7-7b) takes the form

$$\theta = \sum_{i=1}^{n} \frac{T_i L_i}{G_i J_i} \qquad (7\text{-}7c)$$

where each term in the summation is for a length $L$ where $T$, $G$, and $J$ are constant. If $T$, $G$, or $J$ is a function of $x$ (the distance along the length of the shaft), the angle of twist is found using

$$\theta = \int_0^L \frac{T \, dx}{GJ} \qquad (7\text{-}7d)$$

## Example Problem 7–1

A steel shaft is used to transmit torque from a motor to operating units in a factory. The torque is input at gear $B$ (see Fig. 7-6a) and is removed at gears $A$, $C$, $D$, and $E$.

(a) Determine the torques transmitted by cross sections (resisting torques) in intervals $AB$, $BC$, $CD$, and $DE$ of the shaft.

(b) Draw a torque diagram for the shaft.

### SOLUTION

(a) The torques transmitted by cross sections, or resisting torques, in intervals $AB$, $BC$, $CD$, and $DE$ of the shaft shown in Fig. 7-6a are obtained by using the four free-body diagrams shown in Figs. 7-6b, c, d, and e. The moment equilibrium equation $\Sigma M = 0$ about the axis of the shaft yields

$+\!\!\downarrow \Sigma M = 0$:    $T_{AB} - 20 = 0$    $T_{AB} = +20.0$ kip · in.    **Ans.**

$+\!\!\downarrow \Sigma M = 0$:    $T_{BC} - 20 + 55 = 0$    $T_{BC} = -35.0$ kip · in.    **Ans.**

$+\!\!\downarrow \Sigma M = 0$:    $T_{CD} - 20 + 55 - 10 = 0$    $T_{CD} = -25.0$ kip · in.    **Ans.**

$+\!\!\downarrow \Sigma M = 0$:    $T_{DE} - 20 + 55 - 10 - 15 = 0$    $T_{DE} = -10.00$ kip · in.    **Ans.**

For all of these calculations, a free-body diagram of the part of the bar to the left of the imaginary cut has been used. A free-body diagram of the part of the bar to the right of the cut would have yielded identical results. In fact, for the determination of $T_{CD}$ and $T_{DE}$, the free-body diagram to the right of the cut would have been more efficient, since fewer torques would have appeared on the diagram.

(b) A torque diagram is a graph in which abscissas represent distances along the shaft and ordinates represent the internal or resisting torques at the corresponding cross sections. Positive torques point outward from the cross section when represented as a vector according to the right-hand rule. A torque

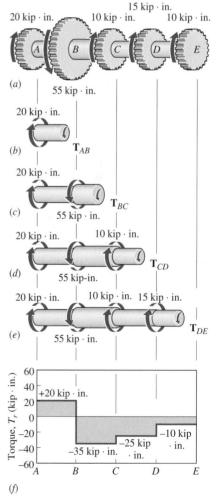

**Figure 7-6**

diagram for the shaft of Fig. 7-6*a*, constructed by using the results from part *a*, is shown in Fig. 7-6*f*. Note in the diagram that the abrupt changes in torque are equal to the applied torques at gears *A*, *B*, *C*, *D*, and *E*. Thus, the torque diagram could have been drawn directly below the sketch of the shaft of Fig. 7-6*a*, without the aid of the free-body diagrams shown in Figs. 7-6*b*, *c*, *d*, and *e*, by using the applied torques at gears *A*, *B*, *C*, *D*, and *E*. However, care must be exercised with the signs used for torque, since Fig. 7-6*f* represents resisting torques. ■

## Example Problem 7-2

A hollow steel shaft with an outside diameter of 400 mm and an inside diameter of 300 mm is subjected to a torque of 300 kN · m as shown in Fig. 7-7. The modulus of rigidity *G* (shear modulus) for the steel is 80 GPa. Determine

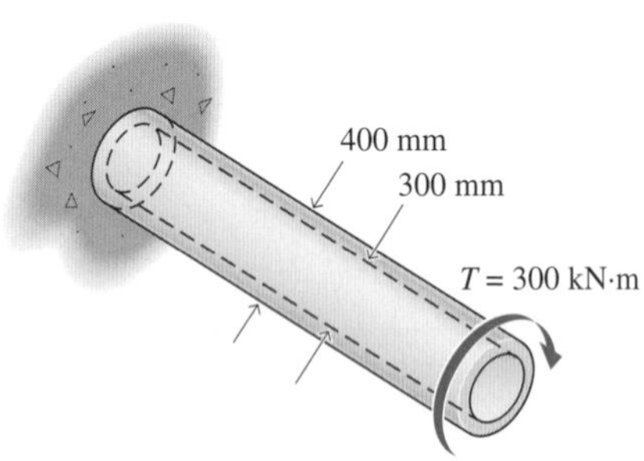

400 mm
300 mm

$T = 300$ kN·m

**Figure 7-7**

(a) The maximum shearing stress in the shaft.

(b) The shearing stress on a transverse cross section at the inside surface of the shaft.

(c) The magnitude of the angle of twist in a 2-m length.

### SOLUTION

Equations for shearing stress and angle of twist in a circular shaft subjected to a torque contain the polar second moment *J* of the cross section, which is given by Eq. (7-5b) as

$$J = \frac{\pi}{2}(r_o^4 - r_i^4) = \frac{\pi}{2}(200^4 - 150^4) = 1718.1(10^6) \text{ mm}^4 = 1718.1(10^{-6}) \text{ m}^4$$

(a) The resisting torque on all cross sections of the shaft is $T_r = T = 300$ kN · m. The maximum shearing stress occurs on a transverse cross section at the outer surface of the shaft and is given by Eq. (7–6) as

$$\tau_{max} = \frac{Tc}{J} = \frac{300(10^3)(200)(10^{-3})}{1718.1(10^{-6})}$$

$$= 34.9(10^6) \text{ N/m}^2 = 34.9 \text{ MPa} \qquad \textbf{Ans.}$$

(b) The shearing stress on a transverse cross section at the inner surface of the shaft is given by Eq. (7-6) as

$$\tau_\rho = \frac{T\rho}{J} = \frac{300(10^3)(150)(10^{-3})}{1718.1(10^{-6})} = 26.2(10^6) \text{ N/m}^2 = 26.2 \text{ MPa} \quad \textbf{Ans.}$$

(c) The angle of twist in a 2-m length is given by Eq. (7–7b) as

$$\theta = \frac{TL}{GJ} = \frac{300(10^3)(2)}{80(10^9)1718.1(10^{-6})} = 0.00437 \text{ rad} \quad \blacksquare \qquad \textbf{Ans.}$$

## Example Problem 7-3

A solid steel shaft 14 ft long has a diameter of 6 in. for 9 ft of its length and a diameter of 4 in. for the remaining 5 ft. The shaft is in equilibrium when subjected to the three torques shown in Fig. 7–8a. The modulus of rigidity (shear modulus) of the steel is 12,000 ksi. Determine

(a) The maximum shearing stress in the shaft.
(b) The rotation of end $B$ of the 6-in. segment with respect to end $A$.
(c) The rotation of end $C$ of the 4-in. segment with respect to end $B$.
(d) The rotation of end $C$ with respect to end $A$.

## SOLUTION

In general, free-body diagrams should be drawn to evaluate the resisting torque in each section of the shaft. Such diagrams are shown in Figs. 7–8b and c. From the free-body diagram of Fig. 7–8b,

$$+ \downarrow \Sigma M_x = 0: \qquad T_6 - 20 + 5 = 0 \qquad T_6 = 15 \text{ kip} \cdot \text{ft} \longrightarrow \downarrow \longrightarrow$$

From the free-body diagram of Fig. 7–8c,

$$+ \downarrow \Sigma M_x = 0: \qquad -T_4 + 5 = 0 \qquad T_4 = 5 \text{ kip} \cdot \text{ft} \longrightarrow \uparrow \longrightarrow$$

A torque diagram, such as the one shown in Fig. 7-8d, provides a pictorial representation of the levels of resisting torque being transmitted by each of the sections and serves as an aid for stress and deformation calculations.

Equations for the shearing stress and angle of twist in a circular shaft subjected to a torque contain the polar second moment $J$ of the cross section, which is given by Eq. (7-5a) as

$$J_4 = \frac{\pi}{2} r^4 = \frac{\pi}{2}(2^4) = 25.13 \text{ in}^4$$

$$J_6 = \frac{\pi}{2} r^4 = \frac{\pi}{2}(3^4) = 127.23 \text{ in}^4$$

(a) The location of the maximum shearing stress is not apparent; hence, the stress must be checked at both sections. The maximum shearing stress on a transverse cross section occurs at the outer surface of the shaft and is given by Eq. (7-6) as

$$\tau_{AB} = \frac{T_{AB} c_{AB}}{J_{AB}} = \frac{15(12)(3)}{127.23} = 4.244 \text{ ksi}$$

$$\tau_{BC} = \frac{T_{BC} c_{BC}}{J_{BC}} = \frac{5(12)(2)}{25.13} = 4.775 \text{ ksi}$$

Therefore,

$$\tau_{max} = \tau_{BC} = 4.775 \text{ ksi} \cong 4.78 \text{ ksi} \qquad \qquad \textbf{Ans.}$$

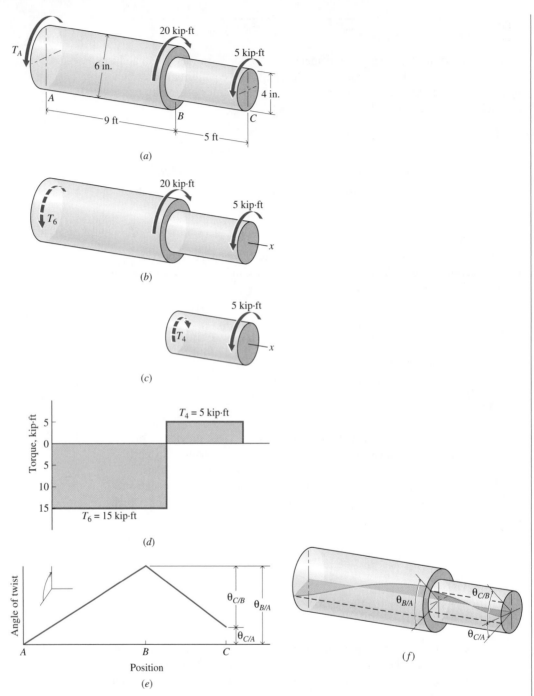

**Figure 7-8**

(b) As the resisting torque of 15 kip · ft is transmitted from section to section in segment *AB* of the shaft, the section at *B* twists relative to the section at *A* by an amount $\theta_{B/A}$ as shown on the angle of twist diagram of Fig. 7-8e. The slope of the angle of twist diagram is constant since the term $T/GJ$ (in Eq. 7-7b) is constant. The rotation of the section at *B* (angle of twist in the 6-in. section) is given by Eq. (7-7b) as

$$\theta_{B/A} = \frac{T_{AB}L_{AB}}{G_{AB}J_{AB}} = \frac{15(12)(9)(12)}{12,000(127.23)} = 0.012733 \text{ rad}$$

$$\cong 0.01273 \text{ rad} - \curvearrowright - \qquad \textbf{Ans.}$$

(c) Similarly for segment $BC$:

$$\theta_{C/B} = \frac{T_{BC}L_{BC}}{G_{BC}J_{BC}} = \frac{5(12)(5)(12)}{12,000(25.13)} = 0.011938 \text{ rad}$$

$$\cong 0.01194 \text{ rad} - \curvearrowleft - \qquad \textbf{Ans.}$$

(d) If there was no resisting torque being transmitted by segment $BC$ it would rotate as a rigid body through angle $\theta_{B/A}$. However, the resisting torque of 5 kip · ft causes the section at $C$ to rotate relative to the section at $B$ by an amount $\theta_{C/B}$ as segment $BC$ deforms, as shown in Fig. 7-8$e$. The resultant of the deformations in the two segments of the shaft is

$$\theta_{C/A} = \theta_{B/A} - \theta_{C/B}$$

$$= 0.012733 - 0.011938 = 0.000795 \text{ rad} - \curvearrowright - \qquad \textbf{Ans.}$$

The distortion for the entire shaft is pictorially shown in Fig. 7-8$f$. ∎

## Example Problem 7–4

A solid steel shaft has a 100-mm diameter for 2 m of its length and a 50-mm diameter for the remaining 1 m of its length, as shown in Fig. 7-9$a$. A 100-mm-long pointer $CD$ is attached to the end of the shaft. The shaft is attached to a rigid support at the left end and is subjected to a 16 kN · m torque at the right end of the 100-mm section and a 4 kN · m torque at the right end of the 50-mm section. The modulus of rigidity $G$ of the steel is 80 GPa. Determine

(a) The maximum shearing stress in the 50-mm section of the shaft.
(b) The maximum shearing stress in the 100-mm section of the shaft.
(c) The rotation of a cross section at $B$ with respect to its no-load position.
(d) The movement of point $D$ with respect to its no-load position.

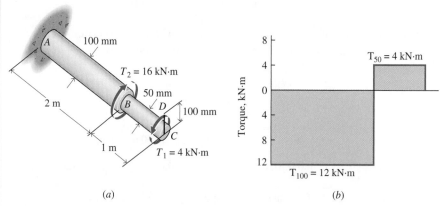

(a)                                                    (b)

Figure 7-9

## SOLUTION

Equations for the shearing stress and angle of twist in a circular shaft subjected to a torque contain the polar second moment $J$ of the cross section, which is given by Eq. (7-5a) as

$$J_{50} = \frac{\pi}{2}r^4 = \frac{\pi}{2}(25^4) = 0.6136(10^6) \text{ mm}^4 = 0.6136(10^{-6}) \text{ m}^4$$

$$J_{100} = \frac{\pi}{2}r^4 = \frac{\pi}{2}(50^4) = 9.817(10^6) \text{ mm}^4 = 9.817(10^{-6}) \text{ m}^4$$

(a) The magnitude of the resisting torque on all cross sections of the shaft in the 50-mm section is $T_r = T_1 = 4$ kN · m (see Fig. 7-9b). The torque diagram of Fig. 7-9b was drawn directly by using the fact that the abrupt changes in the diagram are equal to the applied torques. The maximum shearing stress occurs at the outer surface of the shaft and is given by Eq. (7-6) as

$$\tau_{max} = \frac{T_{BC}c_{BC}}{J_{BC}} = \frac{4(10^3)(25)(10^{-3})}{0.6136(10^{-6})}$$

$$= 163.0(10^6) \text{ N/m}^2 = 163.0 \text{ MPa} \qquad \textbf{Ans.}$$

(b) The magnitude of the resisting torque on all cross sections of the shaft in the 100-mm section is $T_r = T_2 - T_1 = 16 - 4 = 12$ kN · m (see Fig. 7-9b). The maximum shearing stress occurs at the outer surface of the shaft and is given by Eq. (7-6) as

$$\tau_{max} = \frac{T_{AB}c_{AB}}{J_{AB}} = \frac{12(10^3)(50)(10^{-3})}{9.817(10^{-6})}$$

$$= 61.1(10^6) \text{ N/m}^2 = 61.1 \text{ MPa} \qquad \textbf{Ans.}$$

(c) The rotation of the section at $B$ (angle of twist in the 100-mm section) is given by Eq. (7–7b) as

$$\theta_B = \frac{T_{AB}L_{AB}}{G_{AB}J_{AB}} = \frac{12(10^3)(2)}{80(10^9)(9.817)(10^{-6})}$$

$$= 0.03056 \text{ rad} \cong 0.0306 \text{ rad} - \curvearrowright - \qquad \textbf{Ans.}$$

(d) The rotation of the section at $C$ (angle of twist in the 100-mm section minus the angle of twist in the 50-mm section) is given by Eq. (7-7c) as

$$\theta_C = \sum_{i=1}^{n} \frac{T_i L_i}{G_i J_i} = \frac{T_{AB}L_{AB}}{G_{AB}J_{AB}} - \frac{T_{BC}L_{BC}}{G_{BC}J_{BC}}$$

$$= \frac{12(10^3)(2)}{80(10^9)(9.817)(10^{-6})} - \frac{4(10^3)(1)}{80(10^9)(0.6136)(10^{-6})}$$

$$= 0.03056 - 0.08149 = -0.05093 \cong 0.0509 \text{ rad} - \curvearrowleft -$$

The movement of point $D$ with respect to its no-load position is

$$s_D = r\theta_C = 100(0.05093) = 5.09 \text{ mm} - \curvearrowleft - \blacksquare \qquad \textbf{Ans.}$$

## Example Problem 7-5

Two 1.50-in.-diameter steel ($G = 12,000$ ksi) shafts are connected with gears as shown in Fig. 7-10a. End $D$ of shaft $CD$ is fastened to a rigid support that prevents rotation. The diameters of gears $B$ and $C$ are 10 in. and 6 in., respectively. If an input torque of $T_A = 750$ lb · ft is applied at section $A$ of shaft $AB$, determine

(a) The maximum shearing stress on a cross section of shaft $CD$.

(b) The rotation of section $A$ of shaft $AB$ with respect to its no-load position.

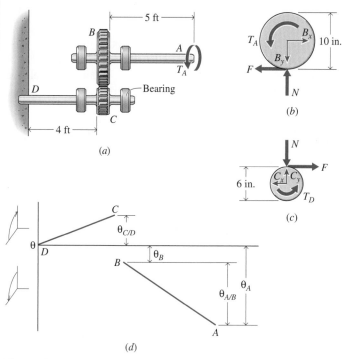

**Figure 7-10**

## SOLUTION

(a) The torque at section $D$ of shaft $CD$ required for equilibrium of the system can be determined from equilibrium considerations for the two shafts. As an aid for these considerations, free-body diagrams for gears $B$ and $C$ are shown in Figs. 7-10b and c, respectively. The input torque $T_A$ in shaft $AB$ is transferred to shaft $CD$ by means of the gear tooth force $F$ shown in the two diagrams. Thus, from a summation of moments about the axis of each of the shafts: For shaft $AB$,

$$T_A - r_B F = 0 \qquad (a)$$

For shaft $CD$,

$$T_D - r_C F = 0 \qquad (b)$$

Since the force $F$ in Eqs. (a) and (b) must be equal,

$$T_D = (r_C/r_B)T_A = (3/5)(750) = 450 \text{ lb} \cdot \text{ft}$$

The magnitude of the resisting torque on all cross sections of the shaft $CD$ is $T_{CD} = 450$ lb · ft. The maximum shearing stress on a transverse cross section of shaft $CD$ occurs at the outside surface of the shaft and is given by Eq. (7-6) as

$$\tau_{max} = \frac{T_{CD}c_{CD}}{J_{CD}} = \frac{450(12)(0.75)}{(\pi/2)(0.75)^4}$$

$$= 8149 \text{ psi} \cong 8150 \text{ psi} \qquad \textbf{Ans.}$$

(b) The quantities required for the determination of the rotation of section $A$ of shaft $AB$ with respect to its no-load position are illustrated on the angle of twist diagram shown in Fig. 7-10d. The rotation of the section at $C$ relative to the section at $D$ in shaft $CD$ is given by Eq. (7-7b) as

$$\theta_{C/D} = \frac{T_{CD}L_{CD}}{G_{CD}J_{CD}} = \frac{450(12)(4)(12)}{12,000,000(\pi/2)(0.75)^4} = 0.04346 \text{ rad} - \curvearrowright -$$

The teeth on gears $B$ and $C$ must move through the same arc length. Therefore,

$$s = r_B\theta_B = r_C\theta_{C/D}$$

from which

$$\theta_B = (r_C/r_B)(\theta_{C/D}) = (3/5)(0.04346) = 0.02608 \text{ rad} - \curvearrowleft -$$

The magnitude of the resisting torque on all cross sections of the shaft $AB$ is $T_{AB} = 750$ lb · ft, and the rotation of the section at $A$ relative to the section at $B$ in shaft $AB$ is given by Eq. (7-7b) as

$$\theta_{A/B} = \frac{T_{AB}L_{AB}}{G_{AB}J_{AB}} = \frac{750(12)(5)(12)}{12,000,000(\pi/2)(0.75)^4} = 0.09054 \text{ rad} - \curvearrowleft -$$

Finally,

$$\theta_A = \theta_B + \theta_{A/B}$$

$$= 0.02608 + 0.09054 = 0.11662 \text{ rad} = 6.68° - \curvearrowleft - \blacksquare \qquad \textbf{Ans}$$

## Example Problem 7-6

The solid circular tapered shaft of Fig. 7-11 is subjected to end torques applied in transverse planes. Determine the magnitude of the angle of twist in terms of $T$, $L$, $G$, and $r$. Assume elastic action and a slight taper.

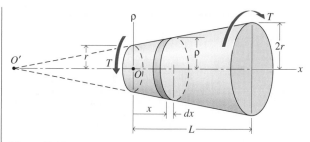

**Figure 7-11**

## SOLUTION

Note that Eq. (7-2) was developed assuming that plane cross sections remain plane and that all longitudinal elements have the same length. Neither of these assumptions is strictly valid for the tapered shaft, but if the taper is slight, the error involved is negligible; therefore, from Eqs. (7-2), (7-6), and (4-15),

$$d\theta = \frac{\gamma}{\rho} dx \qquad \tau = \frac{T\rho}{J} = \frac{T\rho}{\pi\rho^4/2} = \frac{2T}{\pi\rho^3} \qquad \gamma = \frac{\tau}{G}$$

Therefore,

$$d\theta = \frac{\tau}{G\rho} dx = \frac{2T}{G\pi\rho^4} dx$$

The radius $\rho$ can be expressed as a function of $x$; thus,

$$\rho = r + \frac{2r - r}{L} x = \frac{r}{L}(L + x)$$

Substituting this value for $\rho$ into the expression for $d\theta$ gives

$$d\theta = \frac{2TL^4}{G\pi r^4 (L + x)^4} dx$$

Integrating to obtain the angle $\theta$ yields

$$\theta = \frac{2TL^4}{G\pi r^4} \int_0^L \frac{dx}{(L + x)^4} = -\frac{2TL^4}{3G\pi r^4}\left(\frac{1}{8L^3} - \frac{1}{L^3}\right) = \frac{7TL}{12G\pi r^4} \qquad \text{Ans.}$$

Alternatively, the origin of coordinates can be placed at a distance $L$ to the left of $O$ in Fig. 7-11 (point $O'$). The function for $\rho$ then becomes

$$\rho = \frac{r}{L} x$$

and

$$\theta = \frac{2TL^4}{G\pi r^4} \int_L^{2L} \frac{dx}{x^4} = \frac{7TL}{12G\pi r^4} \qquad \blacksquare \qquad \text{Ans.}$$

# Problems

## Introductory Problems

**7-1*** For the steel shaft shown in Fig. P7-1,
(a) Determine the torques transmitted by cross sections in intervals *AB*, *BC*, *CD*, and *DE* of the shaft.
(b) Draw a torque diagram for the shaft.

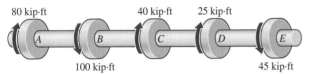

**Figure P7-1**

**7-2*** For the steel shaft shown in Fig. P7-2,
(a) Determine the maximum torque transmitted by any transverse cross section of the shaft.
(b) Draw a torque diagram for the shaft.

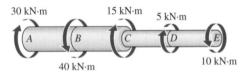

**Figure P7-2**

**7-3** For the steel shaft shown in Fig. P7-3,
(a) Determine the maximum torque transmitted by any transverse cross section of the shaft.
(b) Draw a torque diagram for the shaft.

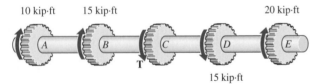

**Figure P7-3**

**7-4** The motor shown in Fig. P7-4 supplies a torque of 500 N · m to the 25-mm-diameter shaft *BCDE*. The torques removed at gears *C*, *D*, and *E* are 100 N · m, 150 N · m, and 250 N · m, respectively.
(a) Determine the torques transmitted by cross sections in intervals *BC*, *CD*, and *DE* of the shaft.
(b) Draw a torque diagram for the shaft.
(c) Determine the maximum shearing stress in the shaft.

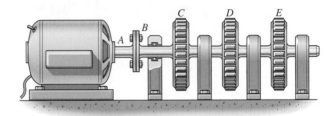

**Figure P7-4**

**7-5*** A solid circular steel shaft 1-in. in diameter is subjected to a torque of 9000 1b · in. The modulus of rigidity (shear modulus) for the steel is 12,000 ksi. Determine
(a) The maximum shearing stress in the shaft.
(b) The magnitude of the angle of twist in a 4-ft length.

**7-6*** A hollow steel shaft has an outside diameter of 150 mm and an inside diameter of 100 mm. The shaft is subjected to a torque of 35 kN · m. The modulus of rigidity (shear modulus) for the steel is 80 GPa. Determine
(a) The shearing stress on a transverse cross section at the outside surface of the shaft.
(b) The shearing stress on a transverse cross section at the inside surface of the shaft.
(c) The magnitude of the angle of twist in a 2.5-m length.
(d) The magnitude of the angle of twist in a 2.5-m long solid shaft that has the same weight as the hollow shaft.

**7-7** A torque of 30,000 lb · in. is supplied to the 3-in.-diameter factory drive shaft of Fig. P7-7 by a belt that drives pulley *A*. A torque of 18,000 lb · in. is taken off by pulley *B* and the remainder by pulley *C*. Shafts *AB* and *BC* are 5 ft and 3.75 ft long, respectively. Both shafts are made of steel (*G* = 12,000 ksi). Determine
(a) The maximum shearing stress in each of the shafts.
(b) The magnitude of the angle of twist of pulley *B* with respect to pulley *A*.
(c) The magnitude of the angle of twist of pulley *C* with respect to pulley *A*.

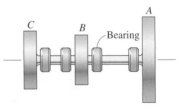

**Figure P7-7**

**7-8** A torque of 10 kN · m is supplied to the steel ($G$ = 80 GPa) factory drive shaft of Fig. P7-8 by a belt that drives pulley $A$. A torque of 6 kN · m is taken off by pulley $B$ and the remainder by pulley $C$. Shafts $AB$ and $AC$ are 2.25 m and 1.60 m long, respectively. If the diameter of shaft $AB$ is 80 mm and the diameter of shaft $AC$ is 65 mm, determine

(a) The maximum shearing stress in each of the shafts.
(b) The angle of twist of pulley $B$ with respect to pulley $A$.
(c) The angle of twist of pulley $C$ with respect to pulley $B$.

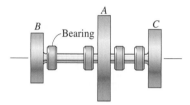

**Figure P7-8**

### Intermediate Problems

**7-9\*** A solid circular aluminum alloy ($G$ = 4000 ksi) shaft with diameters of 2.5 in. and 1.75 in. is subjected to a torque $T$, as shown in Fig. P7-9. The shearing stress is limited to 8000 psi, and the angle of twist in the 7-ft length cannot exceed 0.04 rad. Determine the maximum permissible value of $T$.

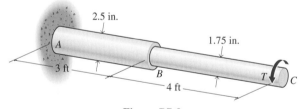

**Figure P7-9**

**7-10\*** The solid circular steel ($G$ = 80 GPa) shaft of Fig. P7-10 has a diameter of 80 mm. If the gears are spaced at 1.25-m intervals, determine

(a) The maximum shearing stress in the shaft.
(b) The rotation of a section at $D$ with respect to a section at $B$.
(c) The rotation of a section at $E$ with respect to a section at $A$.

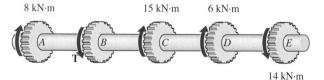

**Figure P7-10**

**7-11** The shaft shown in Fig. P7-11 consists of a brass ($G$ = 5600 ksi) tube $AB$ that is securely connected to a solid stainless steel ($G$ = 12,500 ksi) bar $BC$. Tube $AB$ has an outside diameter of 5 in. and an inside diameter of 2.5 in. Bar $BC$ has a diameter of 3.5 in. Torques $T_1$ and $T_2$ are 100 kip · in. and 40 kip · in., respectively, in the directions shown. Determine

(a) The maximum shearing stress in the shaft.
(b) The rotation of a section at $C$ with respect to its no-load position.

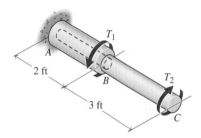

**Figure P7-11**

**7-12** A motor supplies a torque of 5.5 kN · m to the constant-diameter steel ($G$ = 80 GPa) line shaft shown in Fig. P7-12. Three machines are driven by gears $B$, $C$, and $D$ on the shaft, and they require torques of 3 kN · m, 1.5 kN · m, and 1 kN · m, respectively. Determine

(a) The minimum diameter required if the maximum shearing stress in the shaft is limited to 100 MPa.
(b) The rotation of gear $D$ with respect to the coupling at $A$ if the coupling and gears are spaced at 2-m intervals and the shaft diameter is 75 mm.

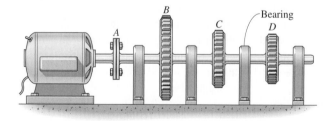

**Figure P7-12**

**7-13\*** The hollow circular steel ($G$ = 12,000 ksi) shaft of Fig. P7-13 is in equilibrium under the torques indicated. Determine

(a) The maximum shearing stress in the shaft.
(b) The rotation of a section at $D$ with respect to a section at $A$.

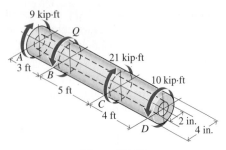

9 kip·ft
*Q*
*A*
3 ft
*B*
21 kip·ft
5 ft
10 kip·ft
*C*
4 ft
*D*
2 in.
4 in.

**Figure P7-13**

**7-14** The solid circular shaft and the hollow tube shown in Fig. P7-14 are both attached to a rigid circular plate at their left ends. A torque $T_A = 2$ kN · m applied to the right end of the shaft is resisted by a torque $T_B$ at the right end of the tube. The shaft is made of steel ($G = 80$ GPa), and the tube is made of an aluminum alloy ($G = 28$ GPa). If the shaft has a diameter of 50 mm and the tube has an outside diameter of 80 mm, determine

(a) The maximum inside diameter that can be used for the tube if the maximum shearing stress in the tube must be limited to 50 MPa.

(b) The maximum inside diameter that can be used for the tube if the rotation of the right end of the shaft with respect to the right end of the tube must be limited to 0.25 rad.

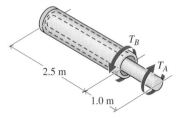

$T_B$
$T_A$
2.5 m
1.0 m

**Figure P7-14**

**7-15** The 4-in-diameter shaft shown in Fig. P7-15 is composed of brass ($G = 5000$ ksi) and steel ($G = 12,000$ ksi) sections rigidly connected. Determine the maximum allowable torque, applied as shown, if the maximum shearing stresses in the brass and the steel are not to exceed 7500 psi and 10,000 psi, respectively, and the distance $AC$, through which the end of the 10-in. pointer $AB$ moves, is not to exceed 0.60 in.

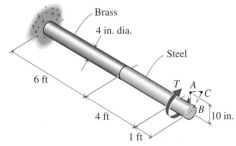

Brass
4 in. dia.
Steel
6 ft
$T$
$A$
$C$
$B$
4 ft
1 ft
10 in.

**Figure P7-15**

**7-16*** A stepped steel ($G = 80$ GPa) shaft has the dimensions and is subjected to the torques shown in Fig. P7-16. Determine

(a) The maximum shearing stress on a section 3 m from the left end of the shaft.

(b) The rotation of the section at the right end of the shaft with respect to its no-load position.

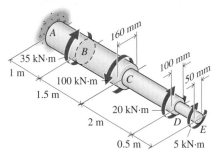

*A*
160 mm
35 kN·m
*B*
1 m
100 kN·m
100 mm
*C*
1.5 m
50 mm
20 kN·m
2 m
*D*
*E*
0.5 m
5 kN·m

**Figure P7-16**

**7-17*** A torque $T$ is applied to the right end of shaft $AB$ of Fig. P7-17. The mean diameter of bevel gear $C$ is twice that of bevel gear $B$. Both shafts are made of steel ($G = 12,000$ ksi). Shaft $AB$ has a diameter of 1.5 in., and shaft $CD$ has a diameter of 2.0 in. If the maximum shearing stress in either shaft must not exceed 15 ksi, determine

(a) The maximum permissible torque $T$.

(b) The rotation of a section at $A$ relative to its no-load position.

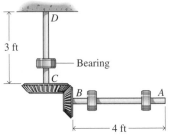

*D*
3 ft
Bearing
*C*
*B*
*A*
4 ft

**Figure P7-17**

### Challenging Problems

**7-18*** Torque is applied to the steel ($G = 80$ GPa) shaft shown in Fig. P7-18 through gear $C$ and is removed through gears $A$ and $B$. If the torque applied to gear $C$ by the motor is 20 kN · m and the torque removed through gear $B$ is 12 kN · m, determine

(a) The minimum permissible diameter for each section of the shaft if the maximum shearing stresses must not exceed 125 MPa.

(b) The minimum permissible uniform diameter for a shaft with $L_1 = 1.5$ m and $L_2 = 1.25$ m if the rotation of gear $A$ relative to gear $C$ must be less than 0.15 rad.

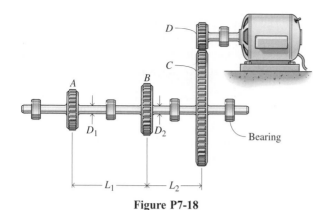

**Figure P7-18**

**7-19\*** A torque of 40 lb · ft is applied through gear *A* to the left end of the gear train shown in Fig. P7-19. The diameters of gears *B* and *C* are 5 in. and 2 in., respectively. If the maximum shearing stresses in the aluminum alloy ($G = 3800$ ksi) shafts *AB* and *CD* are limited to 15 ksi, determine

(a) The minimum permissible diameter for shaft *AB*.
(b) The minimum permissible diameter for shaft *CD*.
(c) The maximum length for shaft *CD* if the rotation of a section at *D* with respect to a section at *C* must not exceed 0.5 rad.

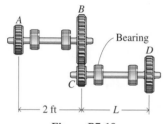

**Figure P7-19**

**7-20** The motor shown in Fig. P7-20 supplies a torque of 45 kN · m to shaft *AB*. Two machines are powered by gears *D* and *E*. The torque delivered by gear *E* to the machine is 8 kN · m. Shafts *AB* and *DCE* are made of steel ($G = 80$ GPa) and have 150-mm and 80-mm diameters, respectively. If the diameters of gears *B* and *C* are 450 mm and 150 mm, respectively, determine

(a) The maximum shearing stress in shaft *AB*.
(b) The maximum shearing stress in shaft *DCE*.
(c) The rotation of gear *E* relative to gear *D*.

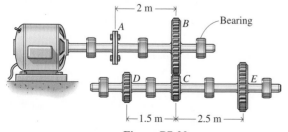

**Figure P7-20**

**7-21** The hollow circular tapered shaft of Fig. P7-21 is subjected to a constant torque *T*. Determine the angle of twist in terms of *T*, *L*, *G*, and *r*.

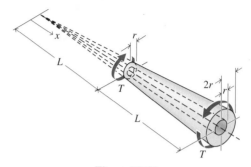

**Figure P7-21**

**7-22\*** The hollow tapered shaft of Fig. P7-22 has a constant wall thickness *t*. Determine the angle of twist for a constant torque *T* in terms of *T*, *L*, *G*, *t*, and *r*. Note that when *t* is small, the approximate expression for the polar second moment of area ($J = r^2 A$, where *A* is the cross-sectional area of the shaft) may be used.

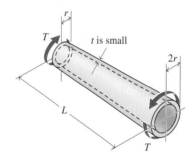

**Figure P7-22**

**7-23\*** The solid cylindrical shaft of Fig. P7-23 is subjected to a uniformly distributed torque of *q* lb · in. per inch of length. Determine, in terms of *q*, *L*, *G*, and *c*, the rotation of the left end caused by the applied torque.

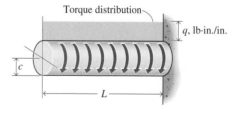

**Figure P7-23**

**7-24** The solid cylindrical shaft of Fig. P7-24 is subjected to a distributed torque that varies linearly from zero at the left end to *q* N · m per meter of length at the right end. Determine, in terms of *q*, *L*, *G*, and *c*, the rotation of the left end caused by the applied torque.

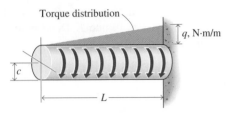

**Figure P7-24**

## Computer Problems

**7-25** A hollow circular steel ($G = 11,000$ ksi) shaft 3 ft long is being designed to transmit a torque $T$ of 3000 lb · ft. The outer radius $r_o$ of the shaft can vary, but the cross-sectional area $A$ of the shaft must remain constant ($A = 3$ in$^2$). Compute and plot
(a) The angle of twist $\theta$ for the 3-ft length as a function of the outer radius $r_o$ (1 in. $\leq r_o \leq 4$ in.).
(b) The maximum shearing stress $\tau_{max}$ in the shaft as a function of the outer radius $r_o$ (1 in. $\leq r_o \leq 4$ in.).

**7-26** A hollow circular brass ($G = 40$ GPa) shaft 2 m long is being designed to transmit a torque $T$ of 7500 N · m. The outer radius $r_o$ of the shaft must be fixed ($r_o = 50$ mm); however, the inner radius $r_i$ of the shaft can vary (0 mm $\leq r_i \leq 45$ mm). Compute and plot

(a) The angle of twist $\theta$ for the 2-m length as a function of the radius ratio $r_i/r_o$ (0 $\leq r_i/r_o \leq 0.9$).
(b) The maximum shearing stress $\tau_{max}$ in the shaft as a function of the radius ratio $r_i/r_o$ (0 $\leq r_i/r_o \leq$ 0.9).

**7-27** The aluminum ($G = 10,600$ ksi) shaft of Fig. P7-27 is 6 ft long, is 4 in. in diameter, and has a 3.5-in.-diameter hole drilled in the end that is attached to the wall. Calculate and plot the angle of twist $\theta_{C/A}$ as a function of the length of the hole $L_{AB}$ (0 ft $\leq L_{AB} \leq 5$ ft) when the 1800 lb · ft torque is applied to the end of the shaft.

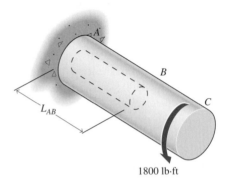

**Figure P7-27**

---

## 7-5 STRESSES ON OBLIQUE PLANES

At this point it is necessary to ascertain whether the transverse plane is a plane of maximum shearing stress and whether there are other significant stresses induced by torsion. For this study, the stresses at point $A$ in the shaft of Fig. 7-12$a$ will be analyzed. Figure 7-12$b$ shows a differential element taken from the shaft at $A$ and the stresses acting on transverse and longitudinal planes. The shearing stress $\tau_{xy}$ can be determined by means of the elastic torsion formula.[3] The equality of shearing stresses on orthogonal planes can be demonstrated by applying the equations of equilibrium to a free-body diagram of the differential element of Fig. 7-12$b$, which has length $dx$, height $dy$, and thickness $dz$. If a shearing force $V_x = \tau_{yx}\, dx\, dz$ is applied to the top surface of the element, the equation $\Sigma F_x = 0$ will require application of an oppositely directed force $V_x$ to the bottom of the element, thus leaving the element subjected to a clockwise couple. This clockwise couple must be balanced by a counterclockwise couple composed of the oppositely directed forces $V_y = \tau_{xy}\, dy\, dz$ applied to the vertical faces of the element. Finally, application of the equation $\Sigma M_z = 0$ yields

$$\tau_{yx}(dx\ dz)\ dy = \tau_{xy}(dy\ dz)\ dx$$

---

[3]The double subscript on the shearing stress is used to designate both the plane on which the stress acts and the direction of the stress. The first subscript indicates the plane (or rather the normal to the plane), and the second subscript indicates the direction of the stress.

from which

$$\tau_{yx} = \tau_{xy} \qquad (7\text{-}8)$$

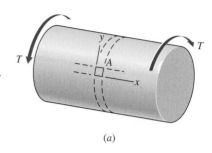

(a)

Therefore, if a shearing stress exists at a point on any plane, a shearing stress of the same magnitude must also exist at this point on an orthogonal plane. This statement is also valid when normal stresses are acting on the planes, since the normal stresses occur in collinear but oppositely directed pairs and thus have zero moment with respect to any axis.

If the equations of equilibrium are applied to the free-body diagram of Fig. 7-12c (which is a wedge-shaped part of the differential element of Fig. 7-12b with $dA$ being the area of the inclined face), the following results are obtained:

$+\nearrow \Sigma F_n = 0$:     $\sigma_n dA - \tau_{xy}(dA \cos \alpha) \sin \alpha - \tau_{xy}(dA \sin \alpha) \cos \alpha = 0$

$+\nwarrow \Sigma F_t = 0$:     $\tau_{nt} dA - \tau_{xy}(dA \cos \alpha) \cos \alpha + \tau_{xy}(dA \sin \alpha) \sin \alpha = 0$

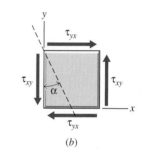

(b)

from which

$$\sigma_n = 2\tau_{xy} \sin \alpha \cos \alpha \qquad (a)$$

$$\tau_{nt} = \tau_{xy}(\cos^2 \alpha - \sin^2 \alpha) \qquad (b)$$

Expressing Eqs. (a) and (b) in terms of the double angle $2\alpha$ yields

$$\sigma_n = \tau_{xy} \sin 2\alpha \qquad (7\text{-}9)$$

$$\tau_{nt} = \tau_{xy} \cos 2\alpha \qquad (7\text{-}10)$$

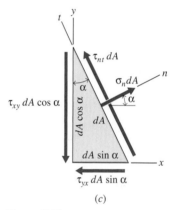

(c)

In the development of Eqs. (7-9) and (7-10) stresses were multiplied by areas to produce forces on the free-body diagram of Fig. 7-12c; forces must always be used when applying the equations of equilibrium. In Eqs. (7-9) and (7-10) $\sigma_n$ is the normal stress on the inclined plane and $\tau_{nt}$ is the shearing stress on the same plane. The shearing stress $\tau_{xy}$ is found using the elastic torsion formula, Eq. (7-6). At a given point of the circular shaft $\tau_{xy}$ is constant, and thus Eqs. (7-9) and (7-10) show that the stresses $\sigma_n$ and $\tau_{nt}$ are functions of the angle of the inclined plane, $\alpha$. The directions of all stresses and the angle $\alpha$ in Eqs. (7-9) and (7-10) and Fig. 7-12c are considered to be positive. The results obtained from Eqs. (7-9) and (7-10) are shown in the graph of Fig. 7-13, from which it is apparent that the maximum shear-

**Figure 7-12**

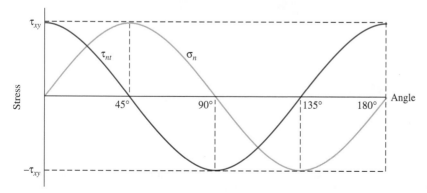

**Figure 7-13**

ing stress occurs on both transverse ($\alpha = 0$) and longitudinal ($\alpha = 90°$) planes. The graph also shows that maximum normal stresses occur on planes oriented at 45° with the axis of the bar and perpendicular to the surface of the bar. On one of these planes ($\alpha = 45°$ on Fig. 7-13), the stress is tension, and on the other ($\alpha = 135°$), the stress is compression. Furthermore, all of these maximum stresses have the same magnitude; hence, the elastic torsion formula gives the magnitude of both the maximum normal stress and the maximum shearing stress at a point in a circular shaft subjected to pure torsion (the only loading is a torque).

Any of the stresses discussed previously may be significant in a given problem. Compare, for example, the failures shown in Fig. 7-14. In Fig. 7-14a, the steel rear axle of a truck split longitudinally. One would also expect this type of failure to occur in a shaft of wood with the grain running longitudinally. In Fig. 7-14b, the compressive stress caused the thin-walled aluminum alloy tube to buckle along one 45° plane, while the tensile stress caused tearing on the other 45° plane. Buckling of thin-walled tubes subjected to torsional loading is a matter of paramount concern to the designer. In Fig. 7-14c, the tensile stresses caused the gray cast iron to fail in tension—typical of any brittle material subjected to torsion. In Fig. 7-14d, the low-carbon steel failed in shear on a plane that is almost transverse—a typical failure for a ductile material. The reason the fracture in Fig. 7-14d did not occur on a transverse plane is that under the large plastic twisting deformation before rupture (note the spiral lines indicating elements originally parallel to the axis of the bar), longitudinal elements were subjected to both torsion and axial tensile loading because the grips of the testing machine would not permit the bar to shorten as the elements were twisted into spirals. This axial tensile stress (not shown in Fig. 7-12) changes the plane of maximum shearing stress from a transverse to an oblique plane (resulting in a warped surface of rupture).[4]

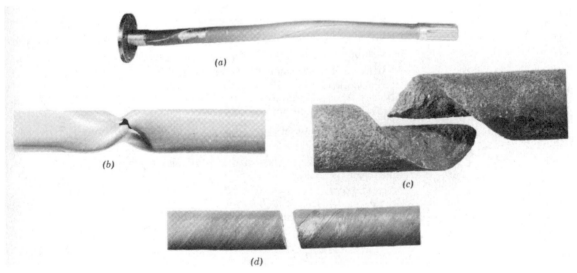

(a)

(b)

(c)

(d)

**Figure 7-14**

[4]The tensile stress is not entirely due to the grips because the plastic deformation of the outer elements of the bar is considerably greater than that of the inner elements. This results in a spiral tensile stress in the outer elements and a similar compressive stress in the inner elements.

## Example Problem 7-7

A cylindrical tube is fabricated by butt-welding a 6-mm-thick steel plate along a spiral seam as shown in Fig. 7-15a. If the maximum compressive stress in the tube must be limited to 80 MPa, determine

(a) The maximum torque $T$ that can be applied to the tube.

(b) The shearing stress parallel to the weld and the normal stress perpendicular to the weld.

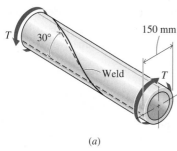

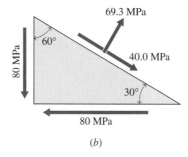

**Figure 7-15**

## SOLUTION

(a) For the cylindrical tube:

$$J = \frac{\pi}{2}(75^4 - 69^4) = 14.096(10^6) \text{ mm}^4 = 14.096(10^{-6}) \text{ m}^4$$

The maximum compressive stress in the tube is given by Eq. (7-9) and Fig. 7-13 as

$$\sigma_n = \tau_{xy} \sin 2\alpha = \frac{Tc}{J} \sin 2(45°)$$

$$80(10^6) = \frac{T_{max}(75)(10^{-3})}{14.096(10^{-6})}$$

Thus,

$$T_{max} = 15.036(10^3) \text{ N} \cdot \text{m} \cong 15.04 \text{ kN} \cdot \text{m} \qquad \textbf{Ans.}$$

(b) The normal stress $\sigma_n$ and the shear stress $\tau_{nt}$ on the weld surface are given by Eqs. (7-9) and (7-10) as

$$\sigma_n = \tau_{xy} \sin 2\alpha = \frac{Tc}{J} \sin 2(\alpha) = \frac{15.036(10^3)(75)(10^{-3})}{14.096(10^{-6})} \sin 2(60°)$$

$$= 69.3 (10^6) \text{ N/m}^2 = 69.3 \text{ MPa T} \qquad \textbf{Ans.}$$

$$\tau_{nt} = \tau_{xy} \cos 2\alpha = \frac{Tc}{J} \cos 2\alpha = \frac{15.036(10^3)(75)(10^{-3})}{14.096(10^{-6})} \cos 2(60°)$$

$$= -40.0(10^6) \text{ N/m}^2 = -40.0 \text{ MPa} \qquad \textbf{Ans.}$$

The minus sign indicates that the direction of $\tau_{nt}$ is opposite to that shown on Fig. 7-12c. The results are shown in Fig. 7-15b. ∎

## Problems

### Introductory Problems

**7-28\***  Determine the maximum torque that can be resisted by a hollow circular shaft having an inside diameter of 50 mm and an outside diameter of 90 mm without exceeding a normal stress of 75 MPa T or a shearing stress of 80 MPa.

**7-29\***  A cylindrical tube is fabricated by butt-welding a 0.075-in. plate with a spiral seam, as shown in Fig. P7-29. A torque $T$ is applied to the tube through a rigid plate. If the outside diameter of the tube is 1.50 in. and $T = 1000$ lb · in., determine
(a) The normal stress perpendicular to the weld and the shear stress parallel to the weld.
(b) The maximum tensile and compressive stresses in the tube.

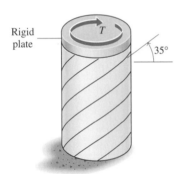

**Figure P7-29**

**7-30**  The hollow circular steel ($G = 80$ GPa) shaft shown in Fig. P7-30 has an outside diameter of 120 mm and an inside diameter of 60 mm. Determine
(a) The maximum compressive stress in the shaft.
(b) The maximum compressive stress in the shaft after the inside diameter is increased to 100 mm.

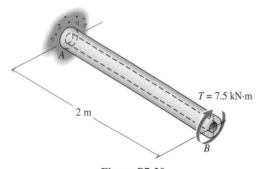

**Figure P7-30**

### Intermediate Problems

**7-31\***  The motor shown in Fig. P7-31 develops a torque $T = 1500$ lb · ft; the output torques from gears $C$ and $D$ are equal. The mean diameters of gears $A$ and $B$ are 12 in. and 4 in., respectively. If the diameter of the motor shaft is 2 in. and the diameter of the power shaft is 1.25 in., determine
(a) The torque in the power shaft between gears $B$ and $C$.
(b) The torque in the power shaft between gears $C$ and $D$.
(c) The maximum tensile and compressive stresses in each shaft.

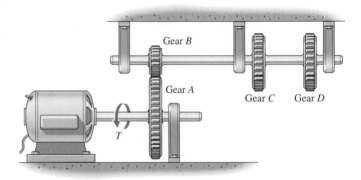

**Figure P7-31**

**7-32**  Five 600-mm-diameter pulleys are keyed to a 40-mm-diameter solid steel ($G = 76$ GPa) shaft, as shown in Fig. P7-32. The pulleys carry belts that are used to drive machinery in a factory. Belt tensions for normal operating conditions are indicated on the figure. Each segment of the shaft is 1.5 m long. Determine
(a) The maximum shearing stress in each segment of the shaft.
(b) The maximum tensile and compressive stresses in the shaft.
(c) The rotation of end $E$ with respect to end $A$.

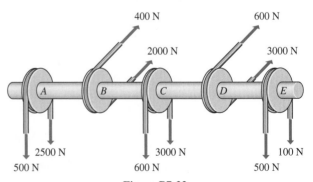

**Figure P7-32**

**Challenging Problems**

**7-33\*** When the two torques are applied to the steel ($G = 12,000$ ksi) shaft of Fig. P7-33, point $A$ moves 0.172 in. in the direction indicated by torque $T_1$. Determine
(a) The torque $T_1$.
(b) The maximum tensile stress in section $BC$ of the shaft.
(c) The maximum compressive stress in section $CD$ of the shaft.

**7-34** A solid circular stepped steel ($G = 80$ GPa) shaft has the dimensions and is subjected to the torques shown in Fig. P7-34. Determine
(a) The maximum tensile stress in section $AB$ of the shaft.
(b) The maximum compressive stress in section $BC$ of the shaft.
(c) The rotation of a section at $C$ with respect to its no-load position.

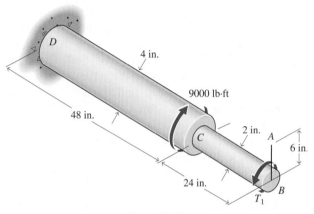

**Figure P7-33**

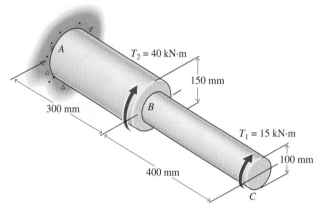

**Figure P7-34**

# 7-6 WORK OF FORCES AND COUPLES

Almost anyone will recognize that one of the most common uses for the circular shaft is the transmission of power; therefore, no discussion of torsion would be complete without including this topic. Power is defined as the time rate of doing work. In everyday life, the word *work* is applied to any form of activity that requires the exertion of muscular or mental effort. In mechanics, however, the term is used in a much more restricted sense. In this section work of a force and work of a couple (such as torque) will be described; the following section will apply the principles to power transmission of a circular shaft.

**Work of a Force** In mechanics, a force does work only when the particle to which the force is applied moves. For example, when a constant force $\mathbf{P}$ is applied to a particle that moves a distance $d$ in a straight line, as shown in Fig. 7-16, the work, $U$, done on the particle by the force $\mathbf{P}$ is defined by the scalar product

$$U = \mathbf{P} \cdot \mathbf{d} = Pd \cos \phi$$
$$= P_x d_x + P_y d_y + P_z d_z \qquad (7\text{-}11)$$

where $\phi$ is the angle between the vectors $\mathbf{P}$ and $\mathbf{d}$. Equation (7-11) is usually interpreted as

> The work done by a force $\mathbf{P}$ is the product of the magnitude of the force ($P$) and the magnitude of the rectangular component of the displacement in the direction of the force ($d \cos \phi$) (see Fig. 7-16).

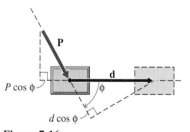

**Figure 7-16**

However, cos $\phi$ can also be associated with the force **P** instead of with the displacement **d**. Then Eq. (7-11) would be interpreted as

> The work done by a force **P** is the product of the magnitude of the displacement ($d$) and the magnitude of the rectangular component of the force in the direction of the displacement ($P \cos \phi$) (see Fig. 7-16).

Since work is defined as the scalar product of two vectors, work is a scalar quantity with only magnitude and algebraic sign. When the sense of the displacement and the sense of the force component in the direction of the displacement are the same ($0 \leq \phi < 90°$), the work done by the force is positive. When the sense of the displacement and the sense of the force component in the direction of the displacement are opposite ($90° < \phi \leq 180°$), the work done by the force is negative. When the direction of the force is perpendicular to the direction of the displacement ($\phi = 90°$), the component of the force in the direction of the displacement is zero and the work done by the force is zero. Of course, the work done by the force is also zero if the displacement is zero, $d = 0$.

Work has the dimensions of force times length. In the SI system of units, this combination of dimensions is called a joule (1 J = 1 N · m).[5] In the U.S. customary system of units, there is no special unit for work. It is expressed simply as lb · ft.

If the force is not constant or if the displacement is not in a straight line, Eq. (7-11) gives the work done by the force only during an infinitesimal part of the displacement, $d\mathbf{r}$ (see Fig. 7-17):

$$dU = \mathbf{P} \cdot d\mathbf{r} = P\, ds \cos \phi = P_t\, ds$$

$$= P_x\, dx + P_y\, dy + P_z\, dz \qquad (7\text{-}12)$$

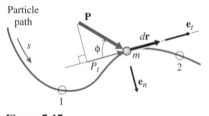

Particle path

**Figure 7-17**

where $d\mathbf{r} = ds\, \mathbf{e}_t = dx\, \mathbf{i} + dy\, \mathbf{j} + dz\, \mathbf{k}$. The total work done by the force as the particle moves from position 1 to position 2 is obtained by integrating Eq. (7-12) along the path of the particle

$$U_{1\to2} = \int_1^2 dU = \int_{s_1}^{s_2} P_t\, ds$$

$$= \int_{x_1}^{x_2} P_x\, dx + \int_{y_1}^{y_2} P_y\, dy + \int_{z_1}^{z_2} P_z\, dz \qquad (7\text{-}13)$$

For the special case in which the force **P** is constant (both in magnitude and direction), the force components can be taken outside the integral signs in Eq. (7-13). Then Eq. (7-13) gives

$$U_{1\to2} = P_x \int_{x_1}^{x_2} dx + P_y \int_{y_1}^{y_2} dy + P_z \int_{z_1}^{z_2} dz$$

$$= P_x\,(x_2 - x_1) + P_y\,(y_2 - y_1) + P_z\,(z_2 - z_1) \qquad (7\text{-}14)$$

---

[5]It may be noted that work and moment of a force have the same dimensions: they are both force times length. However, work and moment are two totally different concepts and the special unit joule should be used only to describe work. The moment of a force must always be expressed as N · m in the SI system of units.

Note that the evaluation of the work done by a constant force depends on the coordinates at the end points of the particle's path but not on the actual path traveled by the particle. For the constant force **P** shown in Fig. 7-18, it doesn't matter if the particle moves along path $a$ from position 1 to position 2, or along path $b$, or along some other path. The work done by the force is always the same. Forces for which the work done is independent of the path are called *conservative forces*.

The weight $W$ of a particle is a particular example of a constant force. When bodies move near the surface of the earth, the force of the earth's gravity is essentially constant ($P_x = 0$, $P_y = 0$, and $P_z = -W$). Therefore, the work done on a particle by its weight is $-W(z_2 - z_1)$. When $z_2 > z_1$, the particle moves upward (opposite the gravitational force), and the work done by gravity is negative. When $z_2 < z_1$, the particle moves downward (in the direction of the gravitational force), and the work done by gravity is positive.

Examples of forces that do work when a body moves from one position to another include the weight of the body, friction between the body and other surfaces, and externally applied loads. Examples of forces that do no work include forces at fixed points ($ds = 0$) and forces acting in a direction perpendicular to the displacement ($\cos \phi = 0$).

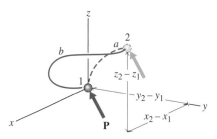

**Figure 7-18**

## Work of a Couple

The work done by a couple is obtained by calculating the work done by each force of the couple separately and adding their works together. For example, consider a couple **C** acting on a rigid body, as shown in Fig. 7-19$a$. During some small time $dt$ the body translates and rotates. If the displacement of point $A$ is $d\mathbf{r} = ds_t\, \mathbf{e}_t$, choose a second point $B$ such that the line $AB$ is perpendicular to $d\mathbf{r}$. Then the motion that takes $A$ to $A'$ will take $B$ to $B'$. This motion may be considered in two parts: first a translation that takes the line $AB$ to $A'\hat{B}$, followed by a rotation $d\theta$ about $A'$ that takes $\hat{B}$ to $B'$ (see Fig. 7-19$b$).

Now represent the couple by a pair of forces of magnitude $P = C/b$ in the direction perpendicular to the line $AB$ (see Fig. 7-19$c$). During the translational part of the motion, one force will do positive work $P\, ds_t$, and the other will do negative work $-P\, ds_t$: therefore, the sum of the work done on the body by the pair of forces during the translational part of the motion is zero. During the rotational part of the motion, $A'$ is a fixed point and the force applied $A'$ does no work. The work done by the force $B$ is $dU = P\, ds_r \cong Pb\, d\theta$, where $d\theta$ is in radians and $C = Pb$ is the magnitude of the moment of the couple. Therefore, when a body is simultaneously translated and rotated, the couple does work only as a result of the rotation.

The total work done by the couple during the differential motion is

$$dU = P\, ds_t - P\, ds_t + Pb\, d\theta = C\, d\theta \tag{7-15}$$

The work is positive if the angular displacement $d\theta$ is in the same direction as the sense of rotation of the couple and negative if the displacement is in the opposite direction. No work is done if the couple is translated or rotated about an axis parallel to the plane of the couple.

The work done on the body by the couple as the body rotates through a finite angle $\Delta\theta = \theta_2 - \theta_1$ is obtained by integrating Eq. (7-15):

$$U_{1 \to 2} = \int_1^2 dU = \int_{\theta_1}^{\theta_2} C\, d\theta \tag{7-16}$$

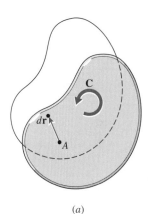

(a)

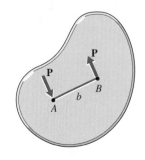

(b)

(c)

**Figure 7-19**

If the couple is constant, then $C$ can be taken outside the integral sign and Eq. (7-16) becomes

$$U_{1 \to 2} = C \int_{\theta_1}^{\theta_2} d\theta = C(\theta_2 - \theta_1) = C \, \Delta\theta \qquad (7\text{-}17)$$

If the body rotates in space, the component of the infinitesimal angular displacement $d\boldsymbol{\theta}$ in the direction of the couple $\mathbf{C}$ is required. For this case, the work done is determined by using the dot product relationship,

$$dU = \mathbf{C} \cdot d\boldsymbol{\theta} = M \, d\theta \cos \phi \qquad (7\text{-}18)$$

where $M$ is the magnitude of the moment of the couple, $d\theta$ is the magnitude of the infinitesimal angular displacement, and $\phi$ is the angle between $\mathbf{C}$ and $d\boldsymbol{\theta}$. For planar motion (in the $xy$-plane), $\mathbf{C} = C \, \mathbf{k}$, $d\boldsymbol{\theta} = d\theta \, \mathbf{k}$, $\mathbf{C} \cdot d\boldsymbol{\theta} = C \, d\theta$, and Eq. (7-18) gives the same result as Eq. (7-17).

Since work is a scalar quantity, the work done on a rigid body by a system of external forces and couples is the algebraic sum of work done by the individual forces and couples.

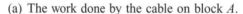

## Example Problem 7-8

A 500-lb block $A$ is held in equilibrium on an inclined surface with a cable and weight system and a force $P$ as shown in Fig. 7-20a. When the force $P$ is removed and the system is disturbed, block $A$ slides slowly down the incline at a constant speed for a distance of 10 ft. The coefficient of friction between block $A$ and the inclined surface is 0.20. Determine

(a) The work done by the cable on block $A$.
(b) The work done by gravity on block $A$.
(c) The work done by the surface of the incline on block $A$.
(d) The work done by all forces on block $A$.
(e) The work done by the cable on block $B$.
(f) The work done by gravity on block $B$.
(g) The work done by all forces on block $B$.

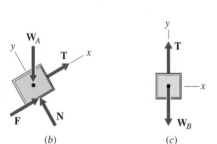

(a)

(b)     (c)

**Figure 7-20**

## SOLUTION

Free-body diagrams for blocks $A$ and $B$ are shown in Figs. 7-20b and c. Four forces act on block $A$: the cable tension $T$, the weight $W_A$, and the normal and frictional forces $N$ and $F$ at the surface of contact between the block and the inclined surface. Two forces act on block $B$: the cable tension $T$ and the weight $W_B$. Since the blocks move at a constant velocity they are in equilibrium and the equilibrium equations applied to block $A$ yield

$$N = W_A \cos 30° = 500 \cos 30° = 433 \text{ lb}$$

$$F = \mu N = 0.20(433) = 86.6 \text{ lb}$$

$$T = W_A \sin 30° - F = 500 \sin 30° - 86.6 = 163.4 \text{ lb}$$

The equilibrium equations applied to block $B$ yield

$$W_B = T = 163.4 \text{ lb}$$

Since all of the forces are constant in both magnitude and direction during the movements of the blocks, the work can be computed by using Eq. (7-14) (in two dimensions so that $z_2 - z_1 = 0$):

$$U_{1\rightarrow2} = P_x(x_2 - x_1) + P_y(y_2 - y_1)$$

Therefore, for block $A$, $x_2 - x_1 = -10$ ft, $y_2 - y_1 = 0$ ft, and

(a) $U_T = 163.4(-10) = -1634$ lb · ft          **Ans.**

(b) $U_W = (-500 \cos 60°)(-10) = 2500$ lb · ft          **Ans.**

Alternatively, the weight force acts vertically downward and block $A$ drops a vertical distance $h = 10 \sin 30°$ so that

$$U_W = W_A h = 500(10 \sin 30°) = 2500 \text{ lb · ft}$$

(c) $U_F = 86.6(-10) = -866$ lb · ft          **Ans.**

   $U_N = 0(-10) + 433(0) = 0$ lb · ft          **Ans.**

(d) $U_{\text{total}} = U_T + U_W + U_F + U_N = -1634 + 2500 - 866 + 0 = 0$          **Ans.**

For block $B$, $x_2 - x_1 = 0$ ft, $y_2 - y_1 = 10$ ft, and

(e) $U_T = 163.4(10) = 1634$ lb · ft          **Ans.**

(f) $U_W = 163.4(-10) = -1634$ lb · ft          **Ans.**

(g) $U_{\text{total}} = U_T + U_W = -1634 + 1634 = 0$ ∎          **Ans.**

---

## Example Problem 7-9

A constant couple

$$\mathbf{C} = 25\,\mathbf{i} + 35\,\mathbf{j} - 50\,\mathbf{k} \text{ N · m}$$

acts on a rigid body. The unit vector associated with the fixed axis of rotation of the body for an infinitesimal angular displacement $d\theta$ is

$$\mathbf{e}_\theta = 0.667\,\mathbf{i} + 0.333\,\mathbf{j} + 0.667\,\mathbf{k}$$

Determine the work done on the body by the couple during an angular displacement of 2.5 radians.

### SOLUTION

The magnitude (moment $M$) of the couple is

$$M = \sqrt{(25)^2 + (35)^2 + (-50)^2} = 65.95 \text{ N · m}$$

The unit vector associated with the couple is

$$\mathbf{e}_C = \frac{+25}{65.95} \mathbf{i} + \frac{+35}{65.95} \mathbf{j} + \frac{-50}{65.95} \mathbf{k} = 0.3791 \mathbf{i} + 0.5307 \mathbf{j} - 0.7582 \mathbf{k}$$

The cosine of the angle between the axis of the couple and the axis of rotation of the body is

$$\cos \alpha = \mathbf{e}_C \cdot \mathbf{e}_\theta$$
$$= (0.3791 \mathbf{i} + 0.5307 \mathbf{j} - 0.7582 \mathbf{k}) \cdot (0.667 \mathbf{i} + 0.333 \mathbf{j} + 0.667 \mathbf{k})$$
$$= -0.07614$$

Therefore,

$$\alpha = 94.37°$$

The work done by the couple during the finite rotation of 2.5 radians can be obtained by using Eq. (7-18). Thus,

$$dU = M \cos \alpha \; d\theta = 65.95(-0.07614) \; d\theta = -5.021 \; d\theta$$

Therefore,

$$U = \int_0^{2.5} dU = \int_0^{2.5} (M \cos \alpha) \; d\theta$$
$$= -5.021 \int_0^{2.5} d\theta = -12.55 \; \text{N} \cdot \text{m} \quad \textbf{Ans.}$$

Alternatively,

$$U = \int_0^{2.5} \mathbf{C} \cdot d\boldsymbol{\theta}$$
$$= \int_0^{2.5} (25 \mathbf{i} + 35 \mathbf{j} - 50 \mathbf{k}) \cdot (0.667 \mathbf{i} + 0.333 \mathbf{j} + 0.667 \mathbf{k}) \; d\theta$$
$$= \int_0^{2.5} -5.02 \; d\theta = -12.55 \; \text{N} \cdot \text{m} \quad \blacksquare \qquad \textbf{Ans.}$$

## Problems

### Introductory Problems

**7-35*** A 175-lb man climbs a flight of stairs 12 ft high. Determine
(a) The work done by the man.
(b) The work done on the man by gravity.

**7-36*** A box with a mass of 600 kg is dragged up an incline 12 m long and 4 m high by using a cable that is parallel to the incline. The force in the cable is 2500 N. Determine the work done on the box
(a) By the cable.
(b) By gravity.

### Intermediate Problems

**7-37*** A 100-lb block is pushed at constant speed for a distance of 20 ft along a level floor by a force **F** that makes

an angle of 35° with the horizontal, as shown in Fig. P7-37. The coefficient of friction between the block and the floor is 0.35. Determine the work done on the block
(a) By the force **F**.
(b) By gravity.
(c) By the floor.

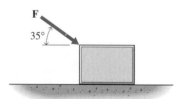

**Figure P7-37**

**7-38** A block with a mass of 100 kg slides down an inclined surface that is 5 m long and makes an angle of 25° with the horizontal. A man pushes horizontally on the block so that it slides down the incline at a constant speed. The coefficient of friction between the block and the inclined surface is 0.15. Determine the work done on the block

(a) By the man.
(b) By gravity.
(c) By the inclined surface.

**Challenging Problems**

**7-39\*** A constant force acting on a particle can be expressed in Cartesian vector form as $\mathbf{F} = (8\ \mathbf{i} - 6\ \mathbf{j} +$ 2 $\mathbf{k}$) lb. Determine the work done by the force on the particle if the displacement of the particle can be expressed in Cartesian vector form as $\mathbf{s} = (5\ \mathbf{i} + 4\ \mathbf{j} + 6\ \mathbf{k})$ ft.

**7-40** A constant couple $\mathbf{C} = (200\ \mathbf{i} + 300\ \mathbf{j} + 350\ \mathbf{k})$ N · m acts on a rigid body. The unit vector associated with the fixed axis of rotation of the body for an infinitesimal angular displacement $d\boldsymbol{\theta}$ is $\mathbf{e}_\theta = 0.250\ \mathbf{i} + 0.350\ \mathbf{j} - 0.903\ \mathbf{k}$. Determine the work done on the body by the couple during an angular displacement of 1.5 radians.

# 7-7 POWER TRANSMISSION BY TORSIONAL SHAFTS

As previously mentioned, power is defined as the time rate of doing work. If a constant torque $T$ acts on a circular shaft (see Fig. 7-21), the work done on the shaft by the torque is given by Eq. (7-17) as $U = T\theta$, where $\theta$ is the angular displacement of the shaft in radians. The time derivative of Eq. (7-17) gives

$$\text{Power} = \frac{dU}{dt} = T\frac{d\theta}{dt} = T\omega \qquad (7\text{-}19)$$

where $dU/dt$ is the power in lb · ft per min (or similar units), $T$ is a constant torque in lb · ft, and $\omega$ is the angular velocity (of the shaft) in radians per minute. All units, of course, may be changed to any other consistent set of units. Since the angular velocity is usually given in revolutions per minute (rpm), the conversion of revolutions to radians will often be necessary. Also, in the U.S. customary system, power is usually given in units of horsepower, and the relation 1 hp = 33,000 lb · ft per min will be found useful. In the SI system, the basic unit of power is the watt (1 watt = 1 N · m/s).

Solution of a power transmission problem is illustrated in the following example problem.

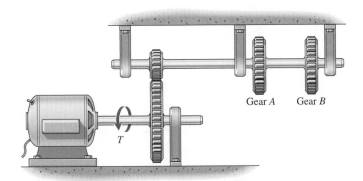

Gear *A*    Gear *B*

*T*

**Figure 7-21**

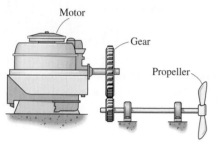

Motor

Gear

Propeller

**Figure 7-22**

## Example Problem 7-10

A diesel engine for a small commercial boat is to operate at 200 rpm and deliver 800 hp through a gearbox with a ratio of 4 to 1 to the propeller shaft as shown in Fig. 7-22. Both the shaft from the engine to the gearbox and the propeller shaft are to be solid and made of heat-treated alloy steel. Determine the minimum permissible diameters for the two shafts if the maximum shearing stress is limited to 20 ksi and the angle of twist in a 10-ft length of the propeller shaft is not to exceed 4°. Neglect power loss in the gearbox and assume (incorrectly because of thrust stresses) that the propeller shaft is subjected to pure torsion.

### SOLUTION

The first step is the determination of the torques to which the shafts are to be subjected. The torques are obtained by using Eq. (7-19):

$$\text{Power} = T\omega = T(2\pi N)$$

$$T_1 = \frac{\text{Power}}{2\pi N} = \frac{33{,}000(\text{hp})}{2\pi N} = \frac{33{,}000(800)}{2\pi(200)} = 21{,}010 \text{ lb} \cdot \text{ft}$$

which is the torque at the crankshaft of the engine. Because the propeller shaft speed is four times that of the crankshaft and power loss in the gearbox is to be neglected, the torque on the propeller shaft is one fourth that on the crankshaft. Thus

$$T_2 = \frac{1}{4} T_1 = \frac{1}{4} (21{,}010) = 5252 \text{ lb} \cdot \text{ft}$$

Note that specifying the power that the shaft must transmit is just another way of specifying the torque that the shaft must withstand. After determining the torque from the given power, the application of the shearing stress equation [(Eq. 7-6)] and the angle of twist equation [(Eq. 7-7)] is the same as in previous examples. Equation (7-6) will be used to determine the shaft sizes needed to satisfy the stress specification. For the main shaft,

$$\frac{J}{c} = \frac{T}{\tau}$$

$$\frac{(\pi/2)c_1^4}{c_1} = \frac{21{,}010(12)}{20(10^3)}$$

which yields

$$c_1^3 = 8.024 \qquad \text{and} \qquad c_1 = 2.002 \text{ in.}$$

Thus, the minimum diameter of the shaft from the engine to the gearbox should be

$$d_m = 2c_1 = 2(2.002) = 4.004 \cong 4.00 \text{ in.} \qquad \textbf{Ans.}$$

The torque on the propeller shaft is one fourth that on the main shaft, and this is the only change in the expression for $c_1^3$; therefore,

$$c_2^3 = 8.024/4 \quad \text{and} \quad c_2 = 1.261 \text{ in.}$$

The propeller shaft size needed to satisfy the distortion specification will be determined from Eq. (7-7b)

$$\frac{\theta}{L} = \frac{T}{GJ}$$

which gives

$$\frac{4(\pi/180)}{10(12)} = \frac{5252(12)}{12(10^6)(\pi c_3^4/2)}$$

$$c_3^4 = 5.747 \quad \text{and} \quad c_3 = 1.548 > 1.261$$

Therefore, the minimum diameter of the propeller shaft must be

$$d_p = 2c_3 = 2(1.548) = 3.096 \cong 3.10 \text{ in.} \quad \blacksquare \qquad \textbf{Ans.}$$

## Problems

### Introductory Problems

**7-41\*** Determine the horsepower that a 6-in.-diameter shaft can transmit at 200 rpm if the maximum shearing stress in the shaft must be limited to 12,000 psi.

**7-42\*** A solid circular steel ($G = 80$ GPa) shaft transmits 200 kW at 180 rpm. Determine the minimum diameter required if the angle of twist in a 2-m length must not exceed 0.025 rad.

**7-43** A steel ($G = 12,000$ ksi) shaft with a 4-in. diameter must not twist more than 0.06 rad in a 20-ft length. Determine the maximum power that the shaft can transmit at 270 rpm.

### Intermediate Problems

**7-44\*** A 3-m-long hollow steel ($G = 80$ GPa) shaft has an outside diameter of 100 mm and an inside diameter of 60 mm. The maximum shearing stress in the shaft is 80 MPa, and the angular velocity is 200 rpm. Determine
(a) The power being transmitted by the shaft.
(b) The magnitude of the angle of twist in the shaft.

**7-45\*** The hydraulic turbines in a water-power plant rotate at 60 rpm and are rated at 20,000 hp with an overload capacity of 25,000 hp. The 30-in.-diameter shaft between the turbine and the generator is made of steel ($G = 12,000$ ksi) and is 20 ft long. Determine
(a) The maximum shearing stress in the shaft at the rated load.
(b) The maximum shearing stress in the shaft at the maximum overload.

**7-46** A solid circular steel ($G = 80$ GPa) shaft 1.5 m long transmits 200 kW at a speed of 400 rpm. If the allowable shearing stress is 70 MPa and the allowable angle of twist is 0.045 rad, determine
(a) The minimum permissible diameter for the shaft.
(b) The speed at which this power can be delivered if the shearing stress is not to exceed 50 MPa in a shaft with a diameter of 75 mm.

### Challenging Problems

**7-47\*** The motor shown in Fig. P7-47 develops 100 hp at a speed of 360 rpm. Gears $A$ and $B$ deliver 40 hp and 60 hp, respectively, to operating units in a factory. If the maximum shearing stress in the shafts must be limited to 12 ksi, determine
(a) The minimum satisfactory diameter for the motor shaft.
(b) The minimum satisfactory diameter for the power shaft.

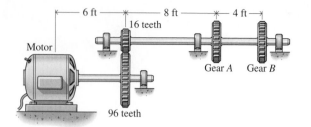

**Figure P7-47**

**7-48\*** A motor delivers 149 kW at 250 rpm to gear $B$ of the factory drive shaft shown in Fig. P7-48. Gears $A$ and $C$ transfer 89 kW and 60 kW, respectively, to operating machinery in the factory. Determine
(a) The maximum shearing stress in shaft $AB$.
(b) The maximum shearing stress in shaft $BC$.
(c) The rotation of gear $C$ with respect to gear $A$ if the shafts are both made of steel ($G = 80$ GPa).

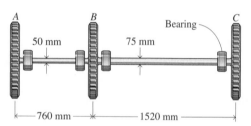

**Figure P7-48**

**7-49** A motor supplies 270 hp at 250 rpm to gear $A$ of the factory drive shaft shown in Fig. P7-49. Gears $B$ and $C$ transfer 170 hp and 100 hp, respectively, to operating machinery in the factory. For an allowable shearing stress of 11 ksi, determine
(a) The minimum permissible diameter $d_1$ for shaft $AB$.
(b) The minimum permissible diameter $d_2$ for shaft $BC$.
(c) The rotation of gear $C$ with respect to gear $A$ if both shafts are made of steel ($G = 11,600$ ksi) and have diameters of 3 in.

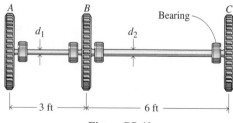

**Figure P7-49**

**7-50** A motor provides 180 kW of power at 400 rpm to the drive shafts shown in Fig. P7-50. The maximum shearing stress in the three solid steel ($G = 80$ GPa) shafts must not exceed 70 MPa. Gears $A$, $B$, and $C$ supply 40 kW, 60 kW, and 80 kW, respectively, to operating units in the plant. Determine
(a) The minimum satisfactory diameter for shaft $D$.
(b) The minimum satisfactory diameter for shaft $E$.
(c) The minimum satisfactory diameter for shaft $F$.

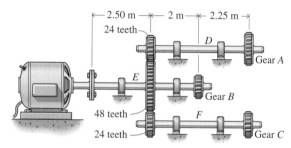

**Figure P7-50**

## Computer Problems

**7-51** A 3-in.-diameter steel drive shaft connects a motor to a machine. The power output of the motor varies with angular speed according to the following data

| $\omega$ (rpm) | $P$ (hp) | $\omega$ (rpm) | $P$ (hp) |
|---|---|---|---|
| 60 | 14.3 | 700 | 103.3 |
| 360 | 71.3 | 940 | 104.7 |
| 450 | 83.5 | 1120 | 90.6 |
| 600 | 96.5 | 1480 | 22.5 |

Calculate and plot the shear stress in the shaft $\tau$ as a function of the angular speed $\omega$.

**7-52** A 75-mm-diameter steel drive shaft connects a motor to a machine. The power output of the motor varies with angular speed according to the following data

| $\omega$ (rpm) | $P$ (kW) | $\omega$ (rpm) | $P$ (kW) |
|---|---|---|---|
| 50 | 15.5 | 750 | 77.2 |
| 100 | 36.0 | 1000 | 78.0 |
| 250 | 61.7 | 1250 | 67.5 |
| 500 | 73.3 | 1500 | 15.0 |

Calculate and plot the shear stress in the shaft $\tau$ as a function of the angular speed $\omega$.

# 7-8 STATICALLY INDETERMINATE MEMBERS

All problems discussed in the preceding sections of this chapter were statically determinate; therefore, only the equations of equilibrium were required to determine the resisting torque at any section. Occasionally, torsionally loaded members are constructed and loaded such that the member is statically indeterminate (the number of independent equilibrium equations is less than the number of unknowns). When this occurs, distortion equations, which involve angles of twist, must be written until the total number of equations agrees with the number of unknowns to be determined. A simplified angle of twist diagram will often be of assistance in obtaining the correct equations. The following examples illustrate the procedures to be followed in solving statically indeterminate torsion problems.

## ▌ Example Problem 7-11

The circular shaft $AC$ of Fig. 7-23a is fixed to rigid walls at $A$ and $C$. The solid section $AB$ is made of annealed bronze ($G_{AB} = 45$ GPa), and the hollow section $BC$ is made of aluminum alloy ($G_{BC} = 28$ GPa). There is no stress in the shaft before the 30 kN · m torque $T$ is applied. Determine the maximum shearing stresses in both the bronze and aluminum portions of the shaft after the torque $T$ is applied.

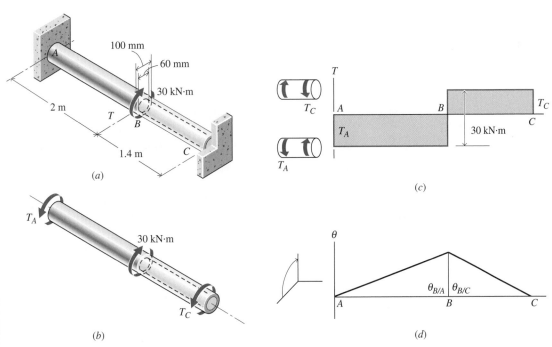

(a)

(b)

(c)

(d)

**Figure 7-23**

### SOLUTION

A free-body diagram of the shaft is shown in Fig. 7-23b. The torques $T_A$ and $T_C$ at the supports are unknown. A summation of moments about the axis of the shaft gives

$$T_A + T_C = 30(10^3) \tag{a}$$

This is the only independent equation of equilibrium relating the two unknown torques $T_A$ and $T_C$; therefore, the problem is statically indeterminate. A second equation can be obtained from the deformation of the shaft, since the left and right portions of the shaft undergo the same angle of twist, as shown in Fig. 7-23d. Thus,

$$\theta_{B/A} = \theta_{B/C}$$

Since Eq. (a) is expressed in terms of $T_A$ and $T_C$, the convenient form of the angle of twist equation for use in this example is Eq. (7-7b). Thus,

$$\frac{T_A L_{AB}}{G_{AB} J_{AB}} = \frac{T_C L_{BC}}{G_{BC} J_{BC}}$$

For the two segments of the shaft the polar second moments of area are

$$J_{AB} = (\pi/2)(50^4) = 9.817(10^6) \text{ mm}^4 = 9.817(10^{-6}) \text{ m}^4$$
$$J_{BC} = (\pi/2)(50^4 - 30^4) = 8.545(10^6) \text{ mm}^4 = 8.545(10^{-6}) \text{ m}^4$$

Therefore,

$$\frac{T_A(2)}{45(10^9)(9.817)(10^{-6})} = \frac{T_C(1.4)}{28(10^9)(8.545)(10^{-6})}$$

from which

$$T_C = 0.7737 \, T_A \tag{b}$$

Solving Eqs. (a) and (b) simultaneously yields

$$T_A = 16,914 \text{ N} \cdot \text{m} = 16.914 \text{ kN} \cdot \text{m}$$
$$T_C = 13,086 \text{ N} \cdot \text{m} = 13.086 \text{ kN} \cdot \text{m}$$

The stresses in the two portions of the shaft can then be obtained by using Eq. (7-6). Thus,

$$\tau_{AB} = \frac{T_A c_{AB}}{J_{AB}} = \frac{16.914(10^3)(50)10^{-3}}{9.817(10^{-6})}$$

$$= 86.1(10^6) \text{ N/m}^2 = 86.1 \text{ MPa} \qquad \textbf{Ans.}$$

$$\tau_{BC} = \frac{T_C c_{BC}}{J_{BC}} = \frac{13.086(10^3)(50)(10^{-3})}{8.545(10^{-6})}$$

$$= 76.6(10^6) \text{ N/m}^2 = 76.6 \text{ MPa} \quad \blacksquare \qquad \textbf{Ans.}$$

## Example Problem 7-12

A hollow circular aluminum alloy ($G_a$ = 4000 ksi) cylinder has a steel ($G_s$ = 11,600 ksi) core as shown in Fig. 7-24a. The steel and aluminum parts are securely connected at the ends. If the shearing stresses in the steel and aluminum must be limited to 14 ksi and 10 ksi, respectively, determine

(a) The maximum torque $T$ that can be applied to the right end of the composite shaft.
(b) The rotation of the right end of the composite shaft when the torque of part (a) is applied.

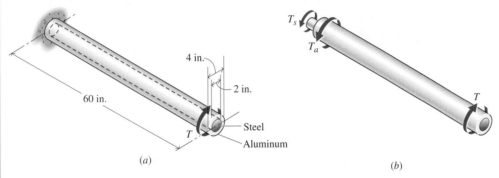

(a)

(b)

**Figure 7-24**

## SOLUTION

A free-body diagram of the shaft is shown in Fig. 7-24b. Since torques and stresses will be related by the elastic torsion formula, which is limited to cross sections of homogeneous material, two unknown torques—the torque in the aluminum $T_a$ and the torque in the steel $T_s$—have been placed on the left end of the shaft. Summing moments with respect to the axis of the shaft yields

$$T_a + T_s = T \tag{a}$$

Since Eq. (a) is the only independent equation of equilibrium, the problem is statically indeterminate. A second equation can be obtained from the deformation of the shaft. The fact that the shaft is nonhomogeneous does not invalidate the assumptions of plane cross sections remaining plane and diameters remaining straight. As a result, strains remain proportional to the distance from the axis of the shaft; however, stresses are not proportional to the radii throughout the entire cross section since $G$ is not single-valued. The steel and aluminum parts of the shaft experience the same angle of twist because of the secure connections at the ends. Thus,

$$\theta_s = \theta_a$$

Since maximum shearing stresses are specified, the convenient form of the angle of twist equation for use in this example is Eq. (7-7a). Thus,

$$\frac{\tau_s L_s}{G_s c_s} = \frac{\tau_a L_a}{G_a c_a}$$

$$\frac{\tau_s(60)}{11.6(10^6)(1)} = \frac{\tau_a(60)}{4.0(10^6)(2)}$$

from which

$$\tau_s = 1.45\tau_a \qquad\qquad\qquad (b)$$

It is obvious from Eq. (b) that the shearing stress in the steel controls; therefore,

$$\tau_s = 14 \text{ ksi} \qquad \text{and} \qquad \tau_a = 14/1.45 = 9.655 \text{ ksi} < 10 \text{ ksi}$$

(a) Once the maximum shearing stresses in the steel and aluminum portions of the shaft are known, Eq. (7-6) can be used to determine the torques transmitted by the two parts of the shaft. Thus,

$$T_s = \frac{\tau_s J_s}{c_s} = \frac{14,000(\pi/2)(1^4)}{1} = 21,990 \text{ lb} \cdot \text{in.}$$

$$T_a = \frac{\tau_a J_a}{c_a} = \frac{9655(\pi/2)(2^4-1^4)}{2} = 113,750 \text{ lb} \cdot \text{in.}$$

From Eq. (a)

$$T = T_a + T_s = 21,990 + 113,750$$

$$= 135,740 \text{ lb} \cdot \text{in.} = 135.7 \text{ kip} \cdot \text{in.} \quad\textbf{Ans.}$$

(b) The rotation of the right end of the shaft with respect to its no-load position can be determined by using either Eq. (7-7a) or Eq. (7-7b). If Eq. (7-7a) is used,

$$\theta = \theta_a = \theta_s = \frac{\tau_s L_s}{G_s c_s} = \frac{14,000(60)}{11.6(10^6)(1)} = 0.0724 \text{ rad} \quad\blacksquare \qquad \textbf{Ans.}$$

---

## ▌ Example Problem 7-13

The torsional assembly shown in Fig. 7-25a consists of a solid bronze ($G_B = 45$ GPa) shaft $CD$ and a hollow aluminum alloy ($G_A = 28$ GPa) shaft $EF$ that has a steel ($G_S = 80$ GPa) core. The ends $C$ and $F$ are fixed to rigid walls, and the steel core of shaft $EF$ is connected to the flange at $E$ so that the aluminum and steel parts act as a unit. The two flanges $D$ and $E$ are bolted together, and the bolt clearance permits flange $D$ to rotate through 0.03 rad before $EF$ carries any of the load. Determine the maximum shearing stress in each of the shaft materials when the torque $T = 54$ kN · m is applied to flange $D$.

### SOLUTION

A free-body diagram for the assembly is shown in Fig. 7-25b. An unknown torque $T_B$ is shown at the left support, and two unknown torques $T_A$ and $T_S$ are

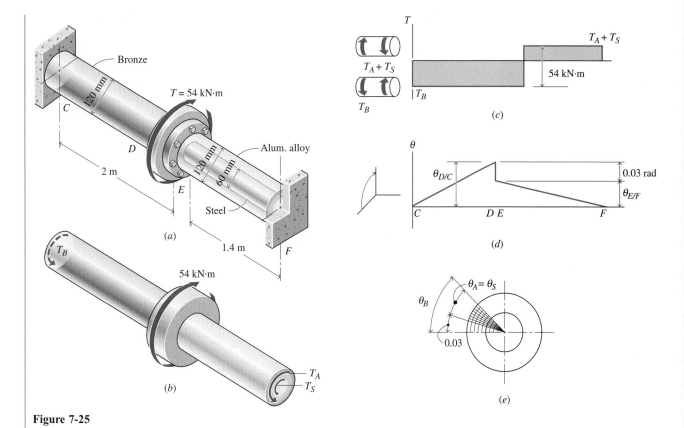

**Figure 7-25**

shown at the right support. Summing moments with respect to the axis of the shaft gives

$$T_B + T_A + T_S = 54(10^3) \qquad (a)$$

Equation (a) is the only independent equilibrium equation that can be written relating the three unknown torques $T_A$, $T_B$, and $T_S$. Since there are three unknown torques and only one equilibrium equation, two distortion equations are needed to solve the problem. Two different types of angle of twist diagrams are shown in Figs. 7-25d and e. Angles of rotation for all cross sections of the shafts are shown in Fig. 7-25d. Relationships between angles $\theta_{D/C}$, $\theta_{E/F}$, and the rotation $\theta = 0.03$ rad permitted by the bolt clearance are required for solution of this example. The same quantities are shown in a polar form of representation in Fig. 7-25e. The angle of twist in the bronze shaft $\theta_B$ is the same as the rotation of coupling $D$ with respect to the support at $C$. The angles of twist in the aluminum and steel shafts $\theta_A$ and $\theta_S$, respectively, are equal and the same as the rotation of coupling $E$ with respect to the support at $F$. Thus, the two distortion equations required for solution of the problem are

$$\theta_B = \theta_A + 0.03 \qquad (b)$$

and

$$\theta_A = \theta_S \qquad (c)$$

Equations (a), (b), and (c) can be written in terms of the same three unknowns (torque, angle, or stress) and solved simultaneously. Since maximum stresses are required, Eqs. (a), (b), and (c) will be written in terms of the maximum stress in each material by using Eqs. (7-6) and (7-7b), in which the polar second moments of area are

$$J_A = (\pi/2)(60^4 - 30^4) = 19.085(10^6) \text{ mm}^4 = 19.085(10^{-6}) \text{ m}^4$$

$$J_B = (\pi/2)(60^4) = 20.36(10^6) \text{ mm}^4 = 20.36(10^{-6}) \text{ m}^4$$

$$J_S = (\pi/2)(30^4) = 1.2723(10^6) \text{ mm}^4 = 1.2723(10^{-6}) \text{ m}^4$$

Thus, from Eq. (a),

$$\frac{\tau_B J_B}{c_B} + \frac{\tau_A J_A}{c_A} + \frac{\tau_S J_S}{c_S} = 54(10^3)$$

$$\frac{\tau_B(20.36)(10^{-6})}{60(10^{-3})} + \frac{\tau_A(19.085)(10^{-6})}{60(10^{-3})} + \frac{\tau_S(1.2723)(10^{-6})}{30(10^{-3})} = 54(10^3)$$

from which

$$339.3\tau_B + 318.1\tau_A + 42.41\tau_S = 54(10^9) \tag{d}$$

Similarly, Eq. (b) may be written

$$\frac{\tau_B L_B}{G_B c_B} = \frac{\tau_A L_A}{G_A c_A} + 0.03$$

$$\frac{\tau_B(2)}{45(10^9)(60)(10^{-3})} = \frac{\tau_A(1.4)}{28(10^9)(60)(10^{-3})} + 0.03$$

from which

$$740.7\tau_B = 833.3\tau_A + 30(10^9) \tag{e}$$

Finally, Eq. (c) can be written

$$\frac{\tau_A L_A}{G_A c_A} = \frac{\tau_S L_S}{G_S c_S}$$

$$\frac{\tau_A(1.4)}{28(10^9)(60)(10^{-3})} = \frac{\tau_S(1.4)}{80(10^9)(30)(10^{-3})}$$

from which

$$\tau_A = 0.700 \, \tau_S \tag{f}$$

Solving Eqs. (d), (e), and (f) simultaneously yields

$$\tau_A = 52.9(10^6) \text{ N/m}^2 = 52.9 \text{ MPa} \qquad \textbf{Ans.}$$

$$\tau_B = 100.0(10^6) \text{ N/m}^2 = 100.0 \text{ MPa} \qquad \textbf{Ans.}$$

$$\tau_S = 75.6(10^6) \text{ N/m}^2 = 75.6 \text{ MPa} \blacksquare \qquad \textbf{Ans.}$$

## Problems

### Introductory Problems

**7-53\*** A hollow circular brass ($G = 5600$ ksi) tube with an outside diameter of 4 in. and an inside diameter of 2 in. is attached at the ends to a solid 2-in.-diameter steel ($G = 12,000$ ksi) core as shown in Fig. P7-53. Determine the maximum shearing stress in the tube when the composite shaft is transmitting a torque of 8 kip · ft.

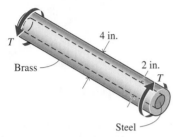

**Figure P7-53**

**7-54\*** A steel ($G = 80$ GPa) tube with an inside diameter of 100 mm and an outside diameter of 125 mm is encased in a Monel ($G = 65$ GPa) tube with an inside diameter of 125 mm and an outside diameter of 175 mm, as shown in Fig. P7-54. The tubes are connected at the ends to form a composite shaft. The shaft is subjected to a torque of 10 kN · m. Determine
(a) The maximum shearing stress in each material.
(b) The angle of twist in a 2-m length.

**Figure P7-54**

**7-55** A 3-in.-diameter cold-rolled steel ($G = 11,600$ ksi) shaft, for which the maximum allowable shearing stress is 15 ksi, exhibited severe corrosion in a certain installation. It is proposed to replace the shaft with one in which an aluminum alloy ($G = 4000$ ksi) tube $\frac{1}{4}$ in. thick is bonded to the outer surface of the cold-rolled steel shaft to produce a composite shaft. If the shearing stress in the aluminum alloy shell is limited to 12 ksi, determine
(a) The maximum torque that the original shaft can transmit.
(b) The maximum torque that the replacement shaft can transmit.

**7-56** The 50-mm-diameter steel ($G = 80$ GPa) shaft shown in Fig. P7-56 is fixed to rigid walls at both ends. When a torque of 4.75 kN · m is applied as shown, determine
(a) The maximum shearing stress in the shaft.
(b) The angle of rotation of the section where the torque is applied with respect to its no load position.

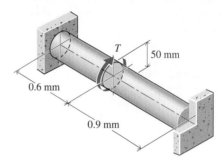

**Figure P7-56**

**7-57\*** Two 3-in.-diameter solid circular steel ($G = 11,600$ ksi) and bronze ($G = 6500$ ksi) shafts are rigidly connected and supported as shown in Fig. P7-57. A torque $T$ is applied at the junction of the two shafts as indicated. The shearing stresses are limited to 18 ksi for the steel and 6 ksi for the bronze. Determine
(a) The maximum torque $T$ that can be applied.
(b) The angle of rotation of the section where the torque is applied with respect to its no-load position.

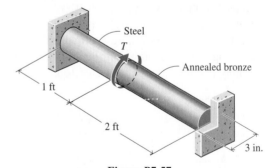

**Figure P7-57**

### Intermediate Problems

**7-58\*** A composite shaft consists of a bronze ($G = 45$ GPa) shell that has an outside diameter of 100 mm bonded to a solid steel ($G = 80$ GPa) core. Determine the diameter of the steel core when the torque resisted by the steel core is equal to the torque resisted by the bronze shell.

**7-59\*** The solid steel ($G = 12,000$ ksi) shaft shown in Fig. P7-59 is fixed to the wall at $C$. The bolt holes in the flange at $A$ have an angular misalignment of 0.0010 rad with respect to the holes in the wall. Determine

(a) The torque, applied at $B$, required to align the bolt holes.

(b) The maximum shearing stress in the shaft after the bolts are inserted and tightened and the torque at $B$ is removed.

(c) The maximum torque that can be applied at section $B$ after the bolts are tightened if the maximum shearing stress in the shaft is not to exceed 10 ksi.

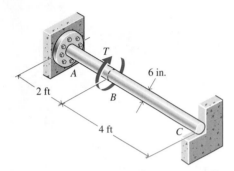

**Figure P7-59**

**7-60** A composite shaft consists of a solid steel ($G = 80$ GPa) core with an outside diameter of 40 mm covered by a brass ($G = 39$ GPa) tube with an inside diameter of 40 mm and a wall thickness of 20 mm, which is in turn covered by an aluminum alloy ($G = 28$ GPa) sleeve with an inside diameter of 80 mm and a wall thickness of 10 mm, as shown in Fig. P7-60. The three materials are bonded so that they act as a unit. Determine

(a) The maximum shearing stress in each material when the assembly is transmitting a torque of 15 kN · m.

(b) The angle of twist in a 3-m length when the assembly is transmitting a torque of 10 kN · m.

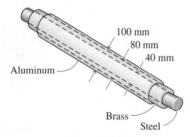

**Figure P7-60**

**7-61** The composite shaft shown in Fig. P7-61 is used as a torsional spring. The solid circular polymer ($G = 150$ ksi) portion of the shaft is encased in and firmly attached to a steel ($G = 12,000$ ksi) sleeve for part of its length. If a torque $T$ of 1.00 kip · in. is being transmitted by the composite shaft, determine

(a) The rotation of a cross section at $C$.

(b) The rotation of a cross section at $C$ if the steel sleeve is assumed to be rigid.

(c) The percent error introduced by assuming the steel sleeve to be rigid.

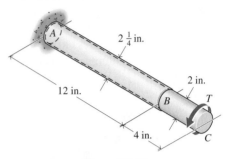

**Figure P7-61**

**7-62\*** A disk and two circular shafts are connected and supported between rigid walls, as shown in Fig. P7-62. Shaft $AB$ is made of brass ($G = 39$ GPa) and has a diameter of 100 mm and a length of 400 mm. Shaft $BC$ is made of Monel ($G = 65$ GPa) and has a diameter of 80 mm and a length of 600 mm. If a torque of 20 kN · m is applied to the disk, determine

(a) The maximum shearing stress in each of the shafts.

(b) The angle of rotation of the disk with respect to its no-load position.

(c) The maximum tensile and compressive stresses in the shaft.

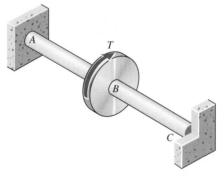

**Figure P7-62**

**Challenging Problems**

**7-63\*** The steel ($G = 12,000$ ksi) shaft shown in Fig. P7-63 will be used to transmit a torque of 1000 lb · in. The hollow portion $AE$ of the shaft is connected to the solid portion $BF$ with two pins, at $C$ and $D$ as shown. If the average shearing stress in the pins must be limited to 25 ksi, determine the minimum satisfactory diameters for each of the pins.

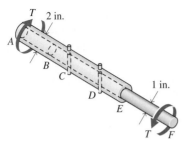

**Figure P7-63**

**7-64\***  A stainless-steel ($G = 86$ GPa) shaft 2.5 m long extends through and is attached to a hollow brass ($G = 39$ GPa) shaft 1.5 m long, as shown in Fig. P7-64. Both shafts are fixed at the wall. When the two torques shown are applied to the shaft, determine

(a) The maximum shearing stress in the steel.
(b) The maximum shearing stress in the brass.
(c) The maximum compressive stress in the brass.
(d) The rotation of the right end of the shaft.

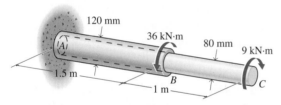

**Figure P7-64**

**7-65**  The shaft shown in Fig. P7-65 consists of a 6-ft hollow steel ($G = 12,000$ ksi) section $AB$ and a 4-ft solid aluminum alloy ($G = 4000$ ksi) section $CD$. The torque $T$ of 40 kip · ft is applied initially only to the steel section $AB$. Section $CD$ is then connected and the torque $T$ is released. When the torque is released, the connection slips 0.010 rad before the aluminum section takes any load. Determine

(a) The maximum shearing stress in the aluminum alloy.
(b) The maximum shearing stress in the steel after the torque $T$ is released.
(c) The final rotation of the collar at $B$ with respect to its no-load position.

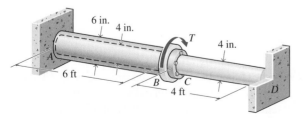

**Figure P7-65**

**7-66**  A torque $T$ of 10 kN · m is applied to the steel ($G = 80$ GPa) shaft shown in Fig. P7-66 without the brass ($G = 40$ GPa) shell. The brass shell is then slipped into place and attached to the steel. After the original torque is released, determine

(a) The maximum shearing stress in the brass shell.
(b) The maximum shearing stress in the steel shaft.
(c) The final rotation of the right end of the steel shaft with respect to the left end.

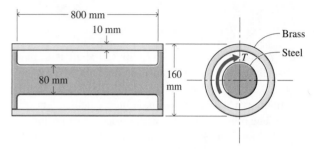

**Figure P7-66**

**7-67**  A hollow steel ($G = 11,600$ ksi) tube with an inside diameter of 2 in., an outside diameter of 2.5 in., and a length of 12 in. is encased with a brass ($G = 5600$ ksi) tube that has an inside diameter of 2.5 in. and an outside diameter of 3.25 in. The brass and steel tubes, while unstressed, are brazed together at one end. A torque is then applied to the other end of the brass tube, which twists one degree with respect to the steel tube. The tubes are then brazed together in that position. Determine the maximum shearing stresses in the steel tube and in the brass tube after the torque is removed from the brass tube.

**Computer Problems**

**7-68**  A composite shaft consists of a 2-m-long solid circular steel ($G_s = 80$ GPa) section securely fastened to a 2-m-long solid circular bronze ($G_b = 40$ GPa) section, as shown in Fig. P7-68. Both ends of the composite shaft are attached to rigid supports. The maximum shearing stress $\tau_{max}$ in the shaft must not exceed 60 MPa, and the rotation $\theta$ of any cross section in the shaft must not exceed 0.04 rad. The ratio of the diameters of the two sections can vary ($\frac{1}{2} \leq d_b/d_s \leq 2$), but the average diameter $(d_b + d_s)/2$ must be 100 mm. Compute and plot

(a) The maximum allowable torque $T$ as a function of the diameter ratio $d_b/d_s$ ($\frac{1}{2} \leq d_b/d_s \leq 2$).
(b) The rotation $\theta$ of a section at $C$ as a function of $d_b/d_s$ ($\frac{1}{2} \leq d_b/d_s \leq 2$).

(c) The maximum shearing stress $\tau_b$ in the bronze shaft as a function of $d_b/d_s$ ($\frac{1}{2} \leq d_b/d_s \leq 2$).

(d) The maximum shearing stress $\tau_s$ in the steel shaft as a function of $d_b/d_s$ ($\frac{1}{2} \leq d_b/d_s \leq 2$).

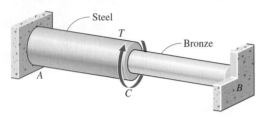

**Figure P7-68**

**7-69**  A hollow steel ($G_s = 12{,}000$ ksi) shaft is stiffened by filling its center with an aluminum alloy ($G_a = 4000$ ksi) shaft as shown in Fig. P7-69. If the steel and aluminum parts rotate as a single unit, calculate and plot

(a) The rotation of end $B$ with respect to its no-load position $\theta_{B/A}$ as a function of the diameter of the aluminum alloy shaft $d_a$ (0 in. $\leq d_a \leq 3.75$ in.).

(b) The shear stress $\tau_a$ in the aluminum alloy shaft at the interface between the two shafts as a function of $d_a$ (0 in. $\leq d_a \leq 3.75$ in.).

(c) The shear stress $\tau_s$ in the steel shaft at the interface between the two shafts as a function of $d_a$ (0 in. $\leq d_a \leq 3.75$ in.).

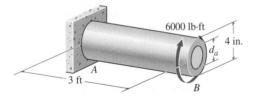

**Figure P7-69**

**7-70**  The solid steel ($G = 80$ GPa) shaft shown in Fig. P7-70 is fixed to the wall at $C$. The flange at $A$ is to be attached to the wall with eight 18-mm-diameter bolts on a 300-mm-diameter circle. However, the bolt holes in the flange have an angular misalignment of 1° with respect to the holes in the wall. If a torque $T$ is applied to the shaft at $B$, calculate and plot

(a) The maximum shearing stress in both sections of the shaft as a function of the torque $T$ ($0 \leq T \leq 60$

kN · m). Assume that the bolts are inserted and tightened as soon as the holes align.

(b) The shearing stress in the bolts as a function of $T$ ($0 \leq T \leq 60$ kN · m).

(c) The rotation $\theta$ of a section at $B$ as a function of $T$ ($0 \leq T \leq 60$ kN · m).

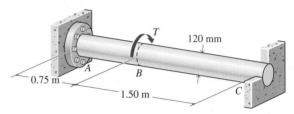

**Figure P7-70**

**7-71**  A 2-in.-diameter solid steel ($G_s = 12{,}000$ ksi, $\tau_{max} = 30$ ksi) shaft and a 3-in.-diameter hollow aluminum alloy ($G_a = 4000$ ksi, $\tau_{max} = 24$ ksi) shaft are fastened together with a $\frac{1}{2}$-in.-diameter brass ($\tau_{max} = 36$ ksi) pin, as shown in Fig. P7-71. Calculate and plot

(a) The maximum shearing stresses, $\tau_a$ in the aluminum shaft and $\tau_s$ in the steel shaft, as a function of the torque $T$ applied to the end of the shaft ($0 \leq T \leq 60$ kip · in.).

(b) The average shearing stresses $\tau_b$ on the cross-sectional area of the brass pin at the interface between the shafts as a function of $T$ ($0 \leq T \leq 60$ kip · in.).

(c) The rotation of end $B$ with respect to its on-load position $\theta_{B/D}$ as a function of $T$ ($0 \leq T \leq 60$ kip · in.).

(d) What is the maximum torque that can be applied to the shaft without exceeding the maximum shearing stresses in either the shaft or the pin?

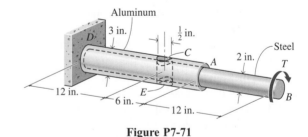

**Figure P7-71**

---

## 7-9 DESIGN

Design, within a limited context, has been discussed previously in Chapters 4 and 6. In those chapters, design was limited to axially loaded members and to pins. This chapter will extend design to solid or hollow circular bars subjected to static torsional loading. Design of these bars will be limited to proportioning a torsionally loaded member to perform a specified function without failure. In this chapter, failure will refer to failure by yielding or failure by fracture.

Furthermore, design will be limited to circular bars made of ductile materials for which the significant failure stress is the shearing stress at yield or fracture. This book will not address the important issue of fatigue loading. That topic is covered in later courses.

For many materials the yield strength in shear is not given in tables such as Tables A-17 and A-18. It will be shown in Chapter 10 that, for ductile materials, the yield strength in shear is either 0.5 or 0.577 times the tensile yield strength, depending upon the criterion selected for failure. For the present, the more conservative value of 0.5 will be used.

## Example Problem 7-14

A solid circular shaft 4 ft long made of 2014-T4 wrought aluminum is subjected to a torsional load of 10,000 lb · in. If failure is by yielding and a factor of safety (FS) of 2 is specified, select a suitable diameter for the shaft, if 2014-T4 wrought aluminum bars are available with diameters in increments of $\frac{1}{8}$ in.

### SOLUTION
The failure criterion is

$$\text{Strength} \geq (\text{Factor of safety})(\text{Stress})$$

For a torsionally loaded circular shaft, stress refers to the shearing stress $\tau = Tc/J$. For a failure mode of yielding, strength is the yield strength in shear $\tau_y$, which for 2014-T4 wrought aluminum is listed in Table A-17 as 24 ksi. Thus,

$$\tau_y \geq \text{FS}\left(\frac{Tc}{J}\right) = \text{FS}\left(\frac{Tr}{\pi r^4/2}\right) = \frac{16(\text{FS})(T)}{\pi d^3}$$

Solving for the diameter,

$$d \geq \left[\frac{16(\text{FS})(T)}{\pi \tau_y}\right]^{1/3} = \left[\frac{16(2)(10,000)}{\pi(24)(10^3)}\right]^{1/3}$$

$$d \geq 1.619 \text{ in.}$$

Since bars are available in $\frac{1}{8}$-in. increments, the smallest permissible bar is

$$d_{\min} = 1\frac{5}{8} \text{ in.} \qquad\qquad \textbf{Ans.}$$

## Example Problem 7-15

A solid circular shaft 2 m long is to transmit 1000 kW at 600 rpm. Failure is by yielding, and the factor of safety is 1.75. If the shaft is made of structural steel, select a suitable diameter for the shaft if bars are available with diameters in increments of 10 mm.

## SOLUTION

Since failure is by yielding, the strength is the yield strength in shear. Table A-17 lists the yield strength in tension as 250 MPa, but does not give a value for the yield strength in shear. According to the discussion in Section 7-9, $\tau_y$ will be taken as $\sigma_y/2$. Thus, $\tau_y = 250/2 = 125$ MPa. Using the results of Example Problem 7-14,

$$d \geq \left[\frac{16(\text{FS})(T)}{\pi\tau_y}\right]^{1/3} \tag{a}$$

The torque may be found using Eq. (7-19)

$$\text{Power} = T\omega = T(2\pi N)$$

or

$$T = \frac{\text{Power}}{2\pi N} \tag{b}$$

Combining Eqs. (a) and (b) gives

$$d \geq \left[\frac{8(\text{FS})(\text{Power})}{\pi^2 N\tau_y}\right]^{1/3}$$

$$d \geq \left[\frac{8(1.75)(1000)(10^3)}{\pi^2(600/60)(125)(10^6)}\right]^{1/3}$$

$$d \geq 0.1043 \text{ m} = 104.3 \text{ mm}$$

Since bars are available in 10-mm increments, the smallest permissible bar is

$$d_{\min} = 110 \text{ mm} \qquad \textbf{Ans.}$$

## Example Problem 7-16

A stepped shaft is subjected to the torques shown in Fig. 7-26. Both segments of the shaft are made of 6061-T6 wrought aluminum. For a factor of safety of 2.0 against failure by yielding, determine the required diameters for the two segments of the shaft.

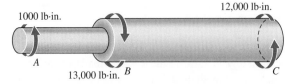

1000 lb·in.         12,000 lb·in.

A

13,000 lb·in.  B      C

**Figure 7-26**

## SOLUTION

For failure by yielding, the significant strength in the shaft is the shearing yield strength. From Table A-17, the shearing yield strength for 6061-T6 wrought aluminum is 26 ksi. Using the results of Example Problem 7-14,

$$d \geq \left[ \frac{16(FS)(T)}{\pi \tau_y} \right]^{1/3}$$

For segment $AB$,

$$d_{AB} \geq \left[ \frac{16(2)(1000)}{\pi(26)(10^3)} \right]^{1/3} = 0.7317 \text{ in.}$$

For segment $BC$,

$$d_{BC} \geq \left[ \frac{16(2)(12,000)}{\pi(26)(10^3)} \right]^{1/3} = 1.675 \text{ in.}$$

The minimum diameters are then

$$d_{AB} = 0.732 \text{ in.} \qquad \textbf{Ans.}$$
$$d_{BC} = 1.675 \text{ in.} \quad \blacksquare \qquad \textbf{Ans.}$$

## Example Problem 7-17

A steel pipe will be used as a shaft to transmit 100 kW at 120 rpm. Failure is by yielding ($\sigma_y = 250$ MPa) and the factor of safety is 1.5. Determine the lightest weight standard steel pipe that can be used for the shaft.

## SOLUTION

The failure criterion

$$\tau_y \geq (FS)\left( \frac{Tc}{J} \right)$$

can be solved for ($J/c$) to yield

$$\frac{J}{c} \geq \frac{FS(T)}{\tau_y} = \frac{FS(Power)}{\tau_y(2\pi N)}$$

Substituting the given numerical values

$$\frac{J}{c} \geq \frac{1.5(100)(10^3)}{125(10^6)(2\pi)(120/60)}$$

or

$$\frac{J}{c} \geq 95.49(10^{-6}) \text{ m}^3 = 95.49(10^3) \text{ mm}^3$$

where

$$\frac{J}{c} = \frac{\pi \left(r_o^4 - r_i^4\right)}{2r_0} \geq 95.49(10^3) \text{ mm}^3 \tag{a}$$

Equation (a) can be satisfied for an infinite number of hollow pipes having different radius ratios. Table A-14 can be used to select a pipe that satisfies the requirement that $J/c \geq 95.49(10^3)$ mm$^3$. However, Table A-14 lists the properties $I$ and $S$, where $I$ is the rectangular second moment of area with respect to a diameter of the pipe, and $S$ is the section modulus, $S = I/c$. Since $J = 2I$ for circular sections (either solid or hollow), the required section modulus for the pipe is

$$S = \frac{I}{c} = \frac{1}{2}\frac{J}{c} \geq \frac{95.49(10^3)}{2} = 47.75(10^3) \text{ mm}^3 \tag{a}$$

The lightest weight standard pipe in Table A-14 with $S \geq 47.75(10^3)$ mm$^3$, is a 102-mm nominal diameter pipe. Any pipe in Table A-14 that has a section modulus greater than $S = 47.75(10^3)$ mm$^3$ would satisfy the stress requirement, but the 102-mm pipe is the lightest, since it has the smallest cross-sectional area. ∎

# Problems

## Introductory Problems

**7-72\*** A motor is to transmit 150 kW to a piece of mechanical equipment. The power is transmitted through a solid structural steel shaft. Failure is by yielding, and the factor of safety is 1.25. The designer has the freedom to operate the motor at 60 rpm or 6000 rpm. For each case, determine the minimum shaft diameter. Shafts are available with diameters in increments of 5 mm. If weight is important, which speed would be used?

**7-73\*** A 3-ft-long steel pipe is subjected to a torque of 1200 lb · ft at each end. The pipe is made of 0.2% C hardened steel, failure is by yielding, and the factor of safety is 1.5. Determine the nominal diameter of the lightest standard-weight steel pipe that can be used for the shaft.

**7-74** A standard-weight structural steel pipe must transmit 150 kW at 60 rpm. The failure mode is yielding and the factor of safety is 1.5.
(a) Select the lightest standard-weight steel pipe that can be used.
(b) If solid structural steel shafts are available with di-

ameters in increments of 10 mm, determine the minimum diameter that can be used.
(c) Compare the weights of the two shafts.

**7-75** A shaft is to transmit 100 hp at 200 rpm. The designer has a variety of solid bars and standard steel pipes to select from. Both the bars and the pipes are made of structural steel, the failure mode is yielding, and the factor of safety is 2.
(a) Select the lightest standard weight steel pipe that can be used.
(b) Select a suitable solid shaft, if they are available with diameters in increments of ⅛ in.
(c) If weight is critical, which shaft should be used?

## Intermediate Problems

**7-76\*** The motor shown in Fig. P7-76 supplies a torque of 1000 N · m to shaft *ABCDE*. The torques removed at *C*, *D*, and *E* are 500 N · m, 300 N · m, and 200 N · m, respectively. The shaft is the same diameter throughout and is made of 0.4% C hot-rolled steel. For a factor of safety of 3 and failure by yielding, select a suitable diameter for the shaft, if shafts are available with diameters in increments of 10 mm.

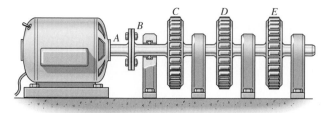

**Figure P7-76**

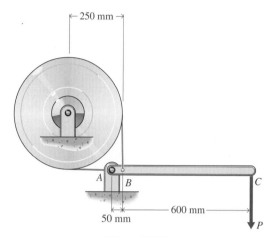

**Figure P7-78**

**7-77** A torque of 30,000 lb · in. is supplied to the factory drive shaft of Fig. P7-77 by a belt that drives pulley *A*. A torque of 10,000 lb · in. is removed by pulley *B* and 20,000 lb · in. by pulley *C*. The shaft is made of structural steel and has a constant diameter over its length. Segment *AB* of the shaft is 3 ft long, and segment *BC* is 4 ft long. Failure is by yielding, and the factor of safety is 2.25. Select a suitable diameter for the shaft, if shafts are available with diameters in increments of $\frac{1}{8}$ in.

**7-79** The motor shown in Fig. P7-79 supplies a torque of 380 lb · ft to shaft *BCD*. The torques removed at gears *C* and *D* are 220 lb · ft and 160 lb · ft, respectively. The shaft *BCD* has a constant diameter and is made of 0.4% C hot-rolled steel, failure is by yielding, and the factor of safety is 2. Determine

(a) The minimum allowable diameter of the shaft, if shafts are available with diameters in increments of $\frac{1}{8}$ in.

(b) The minimum allowable diameter of the bolts used in the coupling, if eight bolts are used, the material is structural steel, the mode of failure is yielding, and the factor of safety is 1.5. The diameter of the bolt circle is $d_1 = 3.5$ in., and the bolts are available with diameters in increments of $\frac{1}{16}$ in.

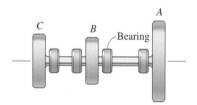

**Figure P7-77**

**Challenging Problems**

**7-78\*** The band brake shown in Fig. P7-78 is part of a hoisting machine. The coefficient of friction between the 500-mm-diameter drum and the flat belt is 0.20. The maximum actuating force *P* that can be applied to the brake arm is 490 N. Rotation of the drum is clockwise. What minimum size shaft should be used to transmit the resisting torque developed by the brake to the machine if the shaft is to be made of 0.4% C hot-rolled steel. The factor of safety is 3 for failure by yielding. Circular steel bars are available with diameters in increments of 5 mm.

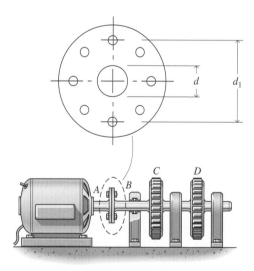

**Figure P7-79**

**7-80**  A shaft used to transmit power is constructed by joining two solid segments of shaft with a collar, as shown in Fig. P7-80. The collar has an inside diameter equal to the diameter of the shaft, and both the collar and the shaft are made of the same material. The collar is securely bonded to the shaft segments. Determine the ratio of the diameters of the collar and shaft such that the splice can transmit the same power as the shaft and at the same maximum shearing stress level. Is the solution dependent on the material selected?

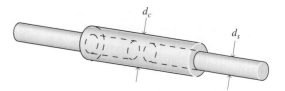

**Figure P7-80**

## 7-10 SUMMARY

The problem of transmitting a torque (a couple) from one plane to a parallel plane is frequently encountered in the design of machinery. The simplest device for accomplishing this function is a circular shaft. If the shaft is in equilibrium, a summation of moments about the axis of the shaft indicates that

$$T = T_r = \int_{\text{area}} \rho \, \tau_\rho \, dA \qquad (7\text{-}1)$$

The law of variation of the shearing stress on the transverse plane ($\tau$ as a function of radial position $\rho$) must be known before the integral of Eq. (7-1) can be evaluated. If the assumption is made that a plane transverse cross section before twisting remains plane after twisting and a diameter of the section remains straight, the distortion of the shaft can be expressed as

$$\gamma_c = \frac{c\theta}{L} \quad \text{and} \quad \gamma_\rho = \frac{\rho\theta}{L} \qquad (7\text{-}2)$$

or

$$\gamma_\rho = \frac{\gamma_c}{c} \rho \qquad (7\text{-}3)$$

The angle $\theta$ is called the angle of twist. Equation (7-3) indicates that the shearing strain is zero at the center of the shaft and increases linearly with respect to the distance $\rho$ from the axis of the shaft. This equation can be combined with Eq. (7-1) once the relationship between shearing stress $\tau$ and shearing strain $\gamma$ is known. Since no assumptions have been made about the relationship between stress and strain or about the type of material of which the shaft is made, Eq. (7-3) is valid for elastic or inelastic action and for homogeneous or heterogeneous materials, provided the strains are not too large (tan $\gamma \cong \gamma$). If the assumption is made that Hooke's law ($\tau = G\gamma$) applies (stresses must be below the proportional limit of the material), Eq. (7-3) can be written

$$\tau_\rho = \frac{\tau_c}{c} \rho \qquad (7\text{-}4)$$

When Eq. (7-4) is substituted into Eq. (7-1), the result is

$$\tau_\rho = \frac{T\rho}{J} \quad \text{and} \quad \tau_c = \frac{Tc}{J} \qquad (7\text{-}6)$$

where $J$ is the polar second moment of the cross-sectional area of the shaft. Equation (7-6) indicates that the shearing stress $\tau_\rho$, like the shearing strain $\gamma_\rho$, is zero at the center of the shaft and increases linearly with respect to the distance $\rho$ from the axis of the shaft. Both the shearing strain $\gamma$ and the shearing stress $\tau$ are maximum when $\rho = c$. Equation (7-6) is known as the elastic torsion formula and is valid for both solid and hollow circular shafts.

Frequently, the amount of twist in a shaft is important. Equations (7-2), (7-6), and Hooke's law ($\tau = G\gamma$) can be combined to give

$$\theta = \frac{\gamma_\rho L}{\rho} = \frac{\tau_\rho L}{G\rho} \quad \text{or} \quad \theta = \frac{TL}{GJ} \qquad \text{(7-7a, b)}$$

The angle of twist determined from these expressions is for a length of shaft of constant diameter ($J = $ constant), constant material properties ($G = $ constant), and carrying a torque $T$. Ideally, the length of shaft should not include sections too near to (within about one-half shaft diameter of) places where mechanical devices (gears, pulleys, or couplings) are attached. For practical purposes, however, it is customary to neglect distortions at connections and to compute angles as if there were no discontinuities.

If $T$, $G$, or $J$ is not constant along the length of the shaft, Eq. (7-7b) takes the form

$$\theta = \sum_{i=1}^{n} \frac{T_i L_i}{G_i J_i} \qquad \text{(7-7c)}$$

where each term in the summation is for a length $L$ where $T$, $G$, and $J$ are constant. If $T$, $G$, or $J$ is a function of $x$ (the distance along the length of the shaft), the angle of twist is found using

$$\theta = \int_0^L \frac{T\,dx}{GJ} \qquad \text{(7-7d)}$$

One of the most common uses for the circular shaft is the transmission of power. Power is defined as the time rate of doing work, and the basic relationship for work done by a constant torque $T$ is $U = T\theta$, where $U$ is work and $\theta$ is the angular displacement of the shaft in radians. The derivative of $U$ with respect to time $t$ gives

$$\frac{dU}{dt} = T\frac{d\theta}{dt} = T\omega \qquad \text{(7-19)}$$

where $dU/dt$ is power, $T$ is a constant torque, and $\omega$ is the angular velocity of the shaft. In the U.S. customary system of units, power is usually given in units of horsepower (1 hp = 33,000 lb · ft per min). In the SI system of units, power is given in watts (N · m/s).

# REVIEW PROBLEMS

**7-81\*** A solid circular steel ($G$ = 12,000 ksi) shaft is loaded and supported as shown in Fig. P7-81. Determine

(a) The maximum shearing stress in the shaft.

(b) The rotation of a section at $B$ with respect to its no-load position.

(c) The rotation of end $C$ with respect to its no-load position.

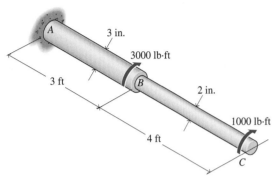

**Figure P7-81**

**7-82\*** A torque of 12.0 kN · m is supplied to the driving gear $B$ of Fig. P7-82 by a motor. Gear $A$ takes off 4.0 kN · m of torque, and the remainder is taken off by gear $C$. Determine the angle of twist of gear $A$ with respect to gear $C$ if both shafts are made of steel ($G$ = 80 GPa) and have diameters of 75 mm.

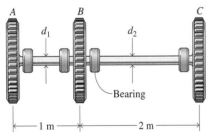

**Figure P7-82**

**7-83** An aluminum alloy ($G$ = 3800 ksi) tube will be used to transmit a torque in a control mechanism. The tube has an outside diameter of 1.25 in. and a wall thickness of 0.065 in. Because of the tendency of thin sections to buckle, the maximum compressive stress in the tube must be limited to 8000 psi. Determine

(a) The maximum torque that can be applied.

(b) The angle of twist in a 3-ft length when a torque of 1000 lb · in. is applied.

**7-84** A 2-m-long hollow brass ($G$ = 39 GPa) shaft has an outside diameter of 50 mm. The maximum shearing stress in the shaft must be limited to 50 MPa, and the angle of twist in the 2-m length must be limited to 2°. If the shaft will rotate at 1500

rpm and transmit 50 kW, determine the maximum inside diameter that can be used for the shaft.

**7-85\*** The inner surface of the aluminum alloy ($G$ = 4000 ksi) sleeve $A$ and the outer surface of the steel ($G$ = 12,000 ksi) shaft $B$ of Fig. P7-85 are smooth. Both the sleeve and the shaft are rigidly fixed to the wall at $D$. The 0.500-in.-diameter pin $C$ fills a hole drilled completely through a diameter of the sleeve and shaft. If the average shearing stress on the cross-sectional area of the pin at the interface between the shaft and the sleeve must not exceed 5000 psi, determine

(a) The maximum torque $T$ that can be applied to the right end of the steel shaft $B$.

(b) The maximum shearing stress in the aluminum alloy sleeve $A$ when the torque of part $a$ is applied.

(c) The rotation of the right end of the shaft when the maximum torque $T$ is applied.

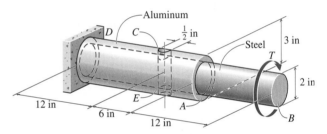

**Figure P7-85**

**7-86\*** The 100-mm-diameter segment $ABC$ of the shaft shown in Fig. P7-86 is initially not connected to the 60-mm-diameter segment $CD$. Torque $T_B$ = 15 kN · m is applied at section $B$, and a secure connection between the two segments is then made at $C$, after which the torque $T_B$ is removed. Determine the resulting maximum shearing stress in segment $CD$ after torque $T_B$ is removed. The moduli of rigidity are 40 GPa for $ABC$ and 80 GPa for $CD$.

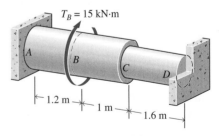

**Figure P7-86**

**7-87** The motor shown in Fig. P7-87 delivers 200 hp at 300 rpm to a piece of equipment at $B$. The shaft is made from 0.4% C hot-rolled steel, which is available with diameters in increments of $1/8$ in. Determine the shaft diameter required if failure is by yielding and the factor of safety is 1.5.

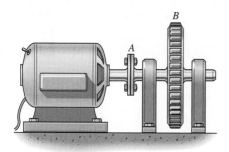

**Figure P7-87**

**7-88** The 160-mm-diameter steel ($G = 80$ GPa) shaft shown in Fig. P7-88 has a 100-mm-diameter bronze ($G = 40$ GPa) core inserted in 3 m of the right end and securely bonded to the steel. Determine

(a) The maximum shearing stress in each of the materials.

(b) The rotation of the free end of the shaft.

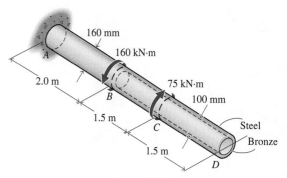

**Figure P7-88**

# FLEXURAL LOADING: STRESSES IN BEAMS

## 8-1 INTRODUCTION

A member subjected to loads applied transverse to the long dimension of the member and which cause the member to bend is known as a *beam*. The beam, or flexural member, is frequently encountered in structures and machines, and its elementary stress analysis constitutes one of the more interesting facets of mechanics of materials. Figure 8-1 is a photograph of an I-beam, *AB*, simply supported in a testing machine and loaded at the one-third points. Figure 8-2 depicts the shape (exaggerated) of the beam when loaded.

Before proceeding with a discussion of stress analysis for flexural members, it may be well to classify some of the various types of beams and loadings encountered in practice. Beams are frequently classified on the basis of their supports or reactions. A beam supported by a pin, roller, or smooth surface at the ends and having one span is called a *simple beam* (Fig. 8-3a). A simple support (a pin or roller) will develop a reaction normal to the beam but will not produce a couple. If either or both ends of the beam project beyond the supports, it is called a *simple beam with overhang* (Fig. 8-3b). A beam with more than two simple supports is a *continuous beam* (Fig. 8-3c). A *cantilever beam* is one in which

**Figure 8-1**

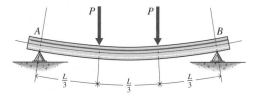

**Figure 8-2**

one end is built into a wall or other support so that the built-in end cannot move transversely or rotate (Fig. 8-3*d*). The built-in end is said to be *fixed* if no rotation occurs and *restrained* if a limited amount of rotation occurs. The supports shown in Figs. 8-3*d*, *e*, and *f* represent fixed ends unless otherwise stated. The beams in Figs. 8-3*d*, *e*, and *f* are, in order, a cantilever beam, a beam fixed (or restrained) at the left end and simply supported near the other end (which has an overhang), and a beam fixed (or restrained) at both ends.

    Cantilever beams and simple beams have only two reactions (two forces or one force and a couple), and these reactions can be obtained from a free-body diagram of the beam by applying the equations of equilibrium. Such beams are said to be statically determinate because the reactions can be obtained from the equations of equilibrium. Beams with more than two reaction components are called statically indeterminate because there are not enough equations of equilibrium to determine the reactions.

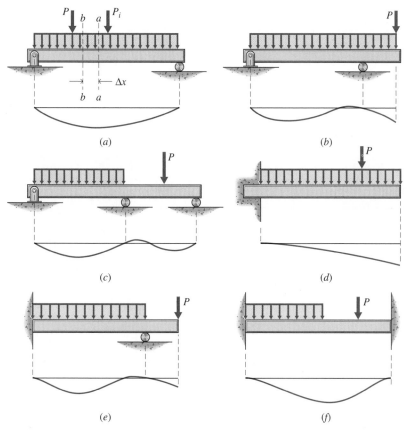

**Figure 8-3**

All of the beams shown in Fig. 8-3 are subjected to both uniformly distributed loads and concentrated loads, and, although shown as horizontal, they may have any orientation. Distributed loads will be shown on the side of the beam on which they are acting; that is, if drawn on the bottom of the beam, the load is pushing upward and if drawn on the right side of a vertical beam, the load is pushing to the left. Deflection curves (greatly exaggerated) are shown beneath the beams to assist in visualizing the shapes of the loaded beams.

A free-body diagram of the portion of the beam of Fig. 8-3$a$ between the left end and plane $a$–$a$ is shown in Fig. 8-4$a$. A study of this diagram shows that a transverse force $V_r$ and a couple $M_r$ at the cut section and a force $R$ (a reaction) at the left support are needed to maintain equilibrium. The force $V_r$ is the resultant force due to the shearing stresses acting on the cut section (on plane $a$–$a$) and is called the *resisting shear*. The couple $M_r$ is the resultant moment due to the normal stresses acting on the cut section (on plane $a$–$a$) and is called the *resisting moment*. The magnitudes and senses of $V_r$ and $M_r$ are obtained from the equations of equilibrium $\Sigma F_y = 0$ and $\Sigma M_O = 0$ where $O$ is any axis perpendicular to the $xy$-plane. The reaction $R$ must be evaluated from a free body of the entire beam.

The normal and shearing stresses $\sigma$ and $\tau$ on plane $a$–$a$ are related to the resisting moment $M_r$ and the shear $V_r$ by the equations

$$V_r = -\int_{\text{Area}} \tau \, dA \tag{8-1a}$$

$$M_r = -\int_{\text{Area}} y \, \sigma \, dA \tag{8-1b}$$

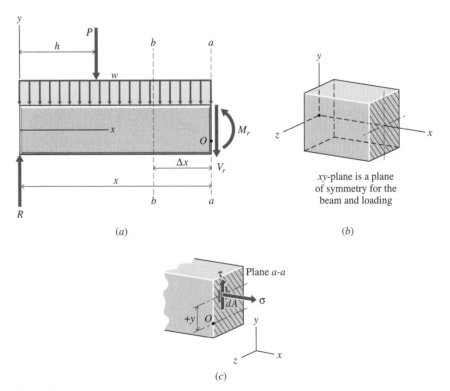

(a)

(b)

xy-plane is a plane
of symmetry for the
beam and loading

(c)

**Figure 8-4**

The resisting shear and moment ($V_r$ and $M_r$), as shown on Fig. 8-4a, will be defined as positive quantities later in Section 8-6. The normal and shear stresses ($\sigma$ and $\tau$), as shown on Fig. 8-4c, are defined as positive stresses. The minus signs in Eqs. (8-1a) and (8-1b) are required to bring these two definitions into agreement. It is obvious from Eqs. (8-1) that the laws of variation of the normal and shearing stresses must be known before the integrals can be evaluated. Thus, the problem is statically indeterminate. For the present, the shearing stresses will be ignored while the normal stresses are studied.

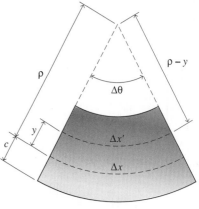

## 8-2 FLEXURAL STRAINS

**Figure 8-5**

A segment of the beam of Fig. 8-4a, between planes a–a and b–b, is shown in Fig. 8-5 with the distortion greatly exaggerated. When Fig. 8-5 was drawn, the assumption was made that a plane section before bending remains a plane after bending. For this to be strictly true, it is necessary that the beam be bent only with couples (no shear on transverse planes). Also, the beam must be so proportioned that it will not buckle and the loads applied so that no twisting occurs (this last limitation will be satisfied if the loads are applied in a plane of symmetry—a sufficient though not a necessary condition). When a beam is bent only with couples, the deformed shape of all longitudinal elements (also referred to as "fibers") is an arc of a circle.

Precise experimental measurements indicate that at some distance c above the bottom of the beam, longitudinal elements undergo no change in length. The curved surface formed by these elements (at radius $\rho$ in Fig. 8-5) is referred to as the *neutral surface of the beam*, and the intersection of this surface with any cross section is called the *neutral axis of the section*. All elements (fibers) on one side of the neutral surface are compressed, and those on the opposite side are elongated. As shown in Fig. 8-5, the fibers above the neutral surface of the beam of Fig. 8-4a (on the same side as the center of curvature) are compressed and the fibers below the neutral surface (on the side opposite the center of curvature) are elongated.

Finally, the assumption is made that all longitudinal elements have the same initial length. This assumption imposes the restriction that the beam be initially straight and of constant cross section; however, in practice, considerable deviation from these last restrictions is often tolerated.

The longitudinal strain $\epsilon_x$ experienced by a longitudinal element that is located a distance y from the neutral surface of the beam is determined by using the definition of normal strain as expressed by Eq. (4-11). Thus,

$$\epsilon_x = \frac{\delta}{L} = \frac{L_f - L_i}{L_i}$$

where $L_f$ is the final length of the fiber after the beam is loaded and $L_i$ is the initial length of the fiber before the beam is loaded. From the geometry of the beam segment shown in Fig. 8-5,

$$\epsilon_x = \frac{\Delta x' - \Delta x}{\Delta x} = \frac{(\rho - y)(\Delta\theta) - \rho(\Delta\theta)}{\rho(\Delta\theta)} = -\frac{1}{\rho}y \qquad (8\text{-}2)$$

Equation (8-2) indicates that the strain developed in a fiber is directly proportional to the distance y of the fiber from the neutral surface of the beam. This

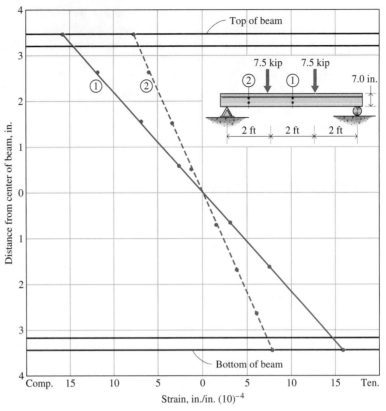

**Figure 8-6**

variation can be demonstrated experimentally by means of strain gages attached to a beam, as shown in Fig. 8-1. The strains, as measured by gages on two different sections, are plotted against the vertical position of the gages on the beam in Fig. 8-6. Curve 1 represents strains on a section at the center of the beam where pure bending occurs (no transverse shear), and curve 2 shows strains at a section near one end of the beam where both flexural (normal) stresses and transverse shearing stresses exist. These curves are both straight lines within the limits of the accuracy of the measuring equipment.[1] Note that Eq. (8-2) is valid for elastic or inelastic action so long as the beam does not twist or buckle and the transverse shearing stresses are small. Problems in this book will be assumed to satisfy these restrictions.

## 8-3 FLEXURAL STRESSES

With the acceptance of the premise that the longitudinal strain $\epsilon_x$ is proportional to the distance of the fiber from the neutral surface of the beam, the law of variation of the normal stress $\sigma = \sigma_x$ on the transverse plane can be determined by using a tensile–compressive stress–strain diagram for the material used in fabri-

---

[1]A more exact analysis using principles developed in the theory of elasticity indicates that curve 2 should be curved slightly. *Note*: Other experiments indicate that a plane section of an initially curved beam will also remain plane after bending and that the deformations will still be proportional to the distance of the fiber from the neutral surface. The strain, however, will not be proportional to this distance, since each deformation must be divided by a different original length.

cating the beam. For most real materials, the tension and compression stress–strain diagrams are identical in the linearly elastic range. Although the diagrams may differ somewhat in the inelastic range, the differences can be neglected for most real problems. For beam problems in this book, the compressive stress–strain diagram will be assumed to be identical to the tensile diagram.

For the special case of linearly elastic action, the relationship between stress $\sigma_x$ and strain $\epsilon_x$ is given by Hooke's law, Eq. (4-15a) (since the state of stress is uniaxial, $\sigma_y = \sigma_z = 0$) as

$$\sigma_x = E\epsilon_x \tag{a}$$

Substituting Eq. (8-2) into Eq. (a) yields

$$\sigma_x = E\epsilon_x = -\frac{E}{\rho}y \tag{8-3}$$

Equation (8-3) shows that the normal stress $\sigma_x$ on the transverse cross section of the beam varies linearly with distance $y$ from the neutral surface. Also, since plane cross sections remain plane, the normal stress $\sigma_x$ is uniformly distributed in the $z$-direction (see Fig. 8-4c).

With the law of variation of flexural stress known, Fig. 8-4 can now be redrawn as shown in Fig. 8-7. The forces $F_C$ and $F_T$ are the resultants of the compressive and tensile flexural stresses, respectively. Since the sum of the forces in the $x$-direction must be zero, $F_C$ is equal to $F_T$; hence, they form a couple of magnitude $M_r$.

The resisting moment $M_r$ developed by the normal stresses in a typical beam with loading in a plane of symmetry but of arbitrary cross section, such as the one shown in Fig. 8-8, is given by Eq. (8-1) as

$$M_r = -\int_A y\,dF = -\int_A y\sigma_x\,dA \tag{b}$$

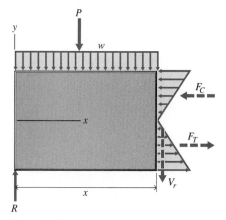

**Figure 8-7**

The minus sign in Eq. (b) arises because the resisting moment is assumed to be counterclockwise (Fig. 8-4a) and a tension stress at location $y$ (positive $y$ is above the neutral surface) will create a clockwise moment. Since $y$ is measured from the neutral surface, it is first necessary to locate this surface by means of the equilibrium equation $\Sigma F_x = 0$, which gives

$$\int_A dF = \int_A \sigma_x\,dA = 0 \tag{c}$$

Substituting Eq. (8-3) into Eq. (c) yields

$$\int_A \sigma_x\,dA = \int_A -\frac{E}{\rho}y\,dA$$

$$= -\frac{E}{\rho}\int_A y\,dA = -\frac{E}{\rho}y_C A = 0 \tag{8-4}$$

where $y_C$ is the distance from the neutral axis (NA) to the centroidal axis ($c$–$c$) of the cross section (Fig. 8-8) that is perpendicular to the plane of bending. Since neither $(E/\rho)$ nor $A$ are zero, $y_C$ must equal zero. Thus, for flexural loading and linearly elastic action, the neutral axis passes through the centroid of the cross section.

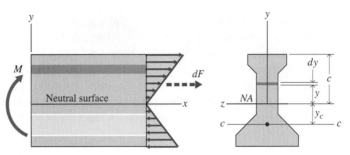

**Figure 8-8**

Since the normal stress $\sigma_x$ varies linearly with distance $y$ from the neutral surface, the maximum normal stress $\sigma_{max}$ on the cross section can be written as

$$\sigma_{max} = -\frac{E}{\rho} c \qquad (8\text{-}5)$$

where $c$ is the distance to the surface of the beam (top or bottom) farthest from the neutral surface. If the quantity $(E/\rho)$ is eliminated from Eqs. (8-3) and (8-5), a useful relationship between the maximum stress $\sigma_{max}$ on a transverse cross section and the stress $\sigma_x$ at an arbitrary distance $y$ from the neutral surface is obtained. Thus,

$$\sigma_x = \frac{y}{c} \sigma_{max} = \frac{y}{c} \sigma_c \qquad (8\text{-}6)$$

in which $\sigma_{max} = \sigma_c$ (equals $\sigma_x$ evaluated at $y = c$).

Substitution of Eq. (8-6) into Eq. (8-1b) yields

$$M_r = -\int_A y\, \sigma_x\, dA = -\frac{\sigma_c}{c} \int_A y^2\, dA \qquad (8\text{-}7)$$

The integral $\int y^2 dA$ is called the *second moment of area*.[2] The second moments of area of several common shapes are given in Table 8-1. Second moments of more complex areas can usually be derived (without integration) from combinations of these simple shapes as shown in the next section. After a discussion of second moments of area, attention will be focused on the application of Eq. (8-7) to calculate the normal stress acting on a transverse section of a beam.

---

[2]In Chapter 5, the centroid for an area was located by evaluating an integral of the form $\int_A x\, dA$, which is called the *first moment of the area with respect to the y-axis*. By analogy, the integral $\int_A x^2\, dA$ is called the *second moment of the area with respect to the y-axis*. In the analysis of the angular motion of rigid bodies, an integral of the form $\int_m r^2\, dm$ occurs in which $dm$ represents an element of mass and $r$ represents the distance from the element to some axis. Euler gave the name "moment of inertia" to these integrals. Because of the similarity between the two types of integrals, both have become widely known as moments of inertia. In this book, integrals involving areas will be referred to as second moments of area and less frequently as area moments of inertia. Integrals involving masses will be referred to as mass moments of inertia or simply moments of inertia.

# 8-4 SECOND MOMENTS OF AREAS

The second moment of an area with respect to an axis in the plane of the area will be denoted by the symbol $I$. The particular axis about which the second moment is taken will be denoted by subscripts. Thus, the second moments of the area $A$ shown in Fig. 8-9 with respect to $x$- and $y$-axes in the plane of the area are

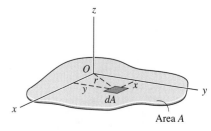

**Figure 8-9**

$$I_x = \int_A y^2 \, dA \quad \text{and} \quad I_y = \int_A x^2 \, dA \qquad (8\text{-}8)$$

where $y$ is the distance of the area $dA$ from the $x$-axis and $x$ is the distance of the area $dA$ from the $y$-axis. The quantities $I_x$ and $I_y$ are sometimes referred to as rectangular second moments[3] of the area $A$.

The second moment of an area can be visualized as the sum of a number of terms each consisting of an area multiplied by a distance squared. Thus, the dimensions of a second moment are a length raised to the fourth power ($L^4$). Common units are $\text{mm}^4$ and $\text{in}^4$. Also, the sign of each term summed to obtain the second moment is positive, since either a positive or negative distance squared is positive. Therefore, the second moment of an area is always positive.

**Radius of Gyration** Since the second moment of an area has the dimensions of length to the fourth power, it can be expressed as the area $A$ multiplied by a length $k$ squared. That is,

$$I_x = \int_A y^2 \, dA = A k_x^2 \quad \text{or} \quad k_x = \sqrt{\frac{I_x}{A}} \qquad (8\text{-}9a)$$

$$I_y = \int_A x^2 \, dA = A k_y^2 \quad \text{or} \quad k_y = \sqrt{\frac{I_y}{A}} \qquad (8\text{-}9b)$$

The distance $k$ is called the *radius of gyration*. The subscript denotes the axis about which the second moment of area is taken. The radius of gyration does not identify a physical point on the area $A$. Instead, it can be visualized as the distance from the axis to a point where a concentrated area of the same size could be placed and would have the same second moment of area with respect to the given axis. Radius of gyration will be discussed in more detail in Chapter 11.

## Parallel-Axis Theorem for Second Moments of Area

When the second moment of an area has been determined with respect to a given axis, the second moment with respect to a parallel axis can be obtained by means of the parallel-axis theorem (also known as the transfer formula). If one of the

---

[3]The second moment of an area with respect to an axis perpendicular to the plane of the area is denoted by the symbol $J$. For example, the second moment of the area $A$ shown in Fig. 8-9 with respect to a $z$-axis that is perpendicular to the plane of the area at the origin $O$ of the $xy$-coordinate system is

$$J_z = \int_A r^2 \, dA = \int_A (x^2 + y^2) \, dA = \int_A x^2 \, dA + \int_A y^2 \, dA = I_y + I_x$$

The quantity $J_z$ is known as the polar second moment of the area $A$ and was used in Chapter 7 in the calculation of the stress in circular shafts transmitting torques.

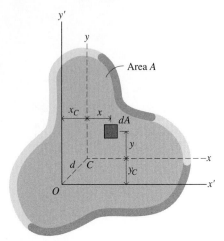

**Figure 8-10**

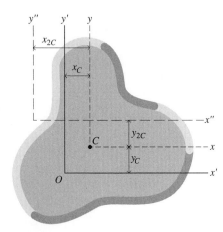

**Figure 8-11**

axes (say the $x$-axis) passes through the centroid $C$ of the area as shown in Fig. 8-10, the second moment of the area about a parallel $x'$-axis is

$$I_{x'} = \int_A (y + y_C)^2 \, dA$$

$$= \int_A y^2 \, dA + 2 \, y_C \int_A y \, dA + y_C^2 \int_A dA \qquad \text{(a)}$$

where $y_C$ has been taken outside the integral signs because it is the same for every element of area $dA$. The integral in the first term is the second moment of the area with respect to the (centroidal) $x$-axis $\int_A y^2 \, dA = I_{xC}$, the integral in the last term is the total area $\int_A dA = A$, and the integral $\int_A y \, dA$ is the first moment of the area with respect to the $x$-axis. Since the $x$-axis passes through the centroid $C$ of the area, the first moment is zero (the middle term vanishes) and Eq. (a) becomes

$$I_{x'} = I_{xC} + y_C^2 A \qquad \text{(8-10)}$$

The parallel-axis theorem, Eq. (8-10), states that the second moment of an area with respect to any axis in the plane of the area is equal to the second moment of the area with respect to a parallel axis through the centroid of the area added to the product of the area and the square of the perpendicular distance between the two axes. The theorem also indicates that the second moment of an area with respect to an axis through the centroid of the area is less than that for any parallel axis, since

$$I_{xC} = I_{x'} - y_C^2 A$$

and $I_{xC}$, $y_C^2 A$, and $I_{x'}$ are all positive. As a point of caution, note that the parallel-axis theorem is valid only for transfers to or from a centroidal axis. That is, if $x''$ is a second axis parallel to the $x'$-axis and $y_{2C}$ is the distance between $x''$ and the centroidal axis (Fig. 8-11), then

$$I_{x''} = I_{xC} + y_{2C}^2 A = \left(I_{x'} - y_C^2 A\right) + y_{2C}^2 A$$

$$= I_{x'} + \left(y_{2C}^2 - y_C^2\right) A \neq I_{x'} + (y_{2C} - y_C)^2 A$$

**Second Moments of Composite Areas** Frequently in engineering practice, an irregular area $A$ will be encountered that can be broken up into a series of simple areas $A_1, A_2, A_3, \ldots, A_n$ for which the second-moment integrals have been evaluated and tabulated. The second moment of the irregular area (the composite area) with respect to any axis is equal to the sum of the second moments of the separate parts of the area with respect to the specified axis

$$I_x = \int_A y^2 \, dA$$

$$= \int_{A_1} y^2 \, dA + \int_{A_2} y^2 \, dA + \int_{A_3} y^2 \, dA + \cdots + \int_{A_n} y^2 \, dA$$

$$= I_{x1} + I_{x2} + I_{x3} + \cdots + I_{xn}$$

Table 8-1 contains a listing of the values of the integrals for frequently encountered shapes such as rectangles, triangles, circles, and semicircles. Tables listing

TABLE 8-1 Second Moments of Plane Areas

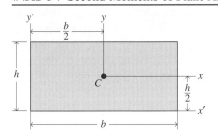

$$I_x = \frac{bh^3}{12}$$

$$I_{x'} = \frac{bh^3}{3}$$

$$A = bh$$

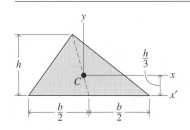

$$I_x = \frac{bh^3}{36}$$

$$I_{x'} = \frac{bh^3}{12}$$

$$A = \frac{1}{2}bh$$

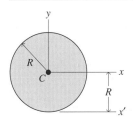

$$I_x = \frac{\pi R^4}{4}$$

$$I_{x'} = \frac{5\pi R^4}{4}$$

$$A = \pi R^2$$

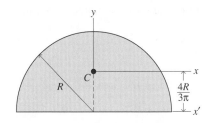

$$I_x = \frac{\pi R^4}{8} - \frac{8R^4}{9\pi}$$

$$I_{x'} = \frac{\pi R^4}{8}$$

$$I_y = \frac{\pi R^4}{8}$$

$$A = \frac{1}{2}\pi R^2$$

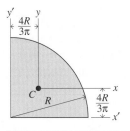

$$I_x = \frac{\pi R^4}{16} - \frac{4R^4}{9\pi}$$

$$I_{x'} = \frac{\pi R^4}{16}$$

$$A = \frac{1}{4}\pi R^2$$

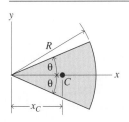

$$I_x = \frac{R^4}{4}\left(\theta - \frac{1}{2}\sin 2\theta\right)$$

$$x_C = \frac{2}{3}\frac{R\sin\theta}{\theta}$$

$$I_y = \frac{R^4}{4}\left(\theta + \frac{1}{2}\sin 2\theta\right)$$

$$A = \theta R^2$$

**Figure 8-12**

second moments of area and other properties for common structural shapes can be found in engineering handbooks. Properties of selected structural shapes are listed in Tables A1 through A16 of Appendix A for use in solving problems.

When an area such as a hole is removed from a larger area, its second moment must be subtracted from the second moment of the larger area to obtain the resulting second moment. For example, for the case of a square plate with a circular hole (Fig. 8-12)

$$I_{\blacksquare} = I_{\square} + I_{\boxdot}$$

Rearranging gives

$$I_{\square} = I_{\blacksquare} - I_{\boxdot}$$

Of course, all second moments must be evaluated with respect to the same axis before they are added or subtracted. If necessary, the parallel-axis theorem can be used to relate the second moments tabulated in Table 8-1 to the axes required in particular problems.

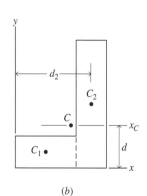

**Figure 8-13**

## Example Problem 8-1

A beam is constructed by gluing a 2- × 6-in. wooden plank 10 ft long to a second 2- × 6-in. wooden plank, also 10 ft long, forming the cross section shown in Fig. 8-13a. Determine the second moment of the cross-sectional area with respect to

(a) The $x$-axis.

(b) The $y$-axis.

(c) The $x_C$-axis, which passes through the centroid of the area and is parallel to the $x$-axis.

### SOLUTION

(a)   As shown in Fig. 8-13b, the cross-sectional area can be divided into two simple rectangles. Since both rectangles have an edge along the $x$-axis, their second moments of area are just $bh^3/3$ (see Table 8-1) where $b$ is the base and $h$ the height of the rectangle. Therefore, the second moment of area of the entire area with respect to the $x$-axis is

$$I_x = I_{x1} + I_{x2} = \frac{1}{3}(4)(2)^3 + \frac{1}{3}(2)(8)^3 = 352 \text{ in}^4 \qquad \textbf{Ans.}$$

(b)   Using the same division of areas as in part (a), the first area has an edge along the $y$-axis and its second moment of area is given by $bh^3/3$. However, the parallel-axis theorem is needed for the second rectangle, since neither the centroid of the rectangle nor either edge is along the $y$-axis. Therefore, the second moment of area of the entire area with respect to the $y$-axis is

$$I_y = I_{y1} + I_{y2} = I_{y1} + \left[ I_{yC2} + d_2^2 A_2 \right]$$

$$= \frac{1}{3}(2)(4)^3 + \left[\frac{1}{12}(8)(2)^3 + (5)^2(16)\right]$$

$$= 448 \text{ in}^4 \qquad \textbf{Ans.}$$

where $I_{yC2} = \frac{1}{12}bh^3$ from Table 8-1

(c) The centroid will be located using the principle of moments as applied to areas and the same division of areas as above

$$d(8 + 16) = 1(8) + 4(16) \qquad d = 3 \text{ in.}$$

Then, using the parallel axis theorem for both rectangles, the second moment of area of the entire area with respect to the $x_C$-axis is

$$I_{xC} = I_{xC1} + I_{xC2}$$

$$= \left[\frac{1}{12}(4)(2)^3 + (3-1)^2(8)\right]$$

$$+ \left[\frac{1}{12}(2)(8)^3 + (4-3)^2(16)\right] = 136.0 \text{ in}^4 \qquad \textbf{Ans.}$$

Note that, since the $x_C$-axis passes through the centroid of the entire area, the second moments of area $I_x$ and $I_{xC}$ are related by the parallel axis theorem

$$I_x = I_{xC} + Ad^2$$

$$= 136.0 + 24(3)^2 = 352 \text{ in}^4$$

as in part (a). ∎

## Example Problem 8-2

Determine the second moment of the shaded area shown in Fig. 8-14a with respect to

(a) The $x$-axis.

(b) The $y$-axis.

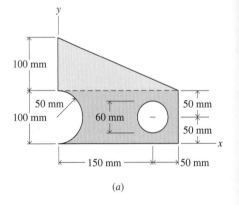

(a)

### SOLUTION

As shown in Fig. 8-14b, the shaded area can be divided into a 100- × 200-mm rectangle (1) with a 60-mm-diameter circle (2) and a 100-mm-diameter half circle (3) removed and a 100- × 200-mm triangle (4). The second moments for these four areas, with respect to the $x$- and $y$-axes, can be obtained by using information from Table 8-1, as follows.

(a) For the rectangle (shape 1),

$$I_{x1} = \frac{bh^3}{3} = \frac{200(100^3)}{3} = 66.667(10^6) \text{ mm}^4$$

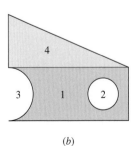

(b)

**Figure 8-14**

For the circle (shape 2),

$$I_{x2} = I_{xC} + y_C^2 A = \frac{\pi R^4}{4} + y_C^2(\pi R^2)$$

$$= \frac{\pi(30^4)}{4} + (50^2)(\pi)(30^2) = 7.705(10^6) \text{ mm}^4$$

For the half circle (shape 3),

$$I_{x3} = I_{xC} + y_C^2 A = \frac{\pi R^4}{8} + y_C^2 \left[\frac{\pi R^2}{2}\right]$$

$$= \frac{\pi(50^4)}{8} + (50)^2 \left[\frac{\pi(50)^2}{2}\right] = 12.272(10^6) \text{ mm}^4$$

For the triangle (shape 4),

$$I_{x4} = I_{xC} + y_C^2 A = \frac{bh^3}{36} + y_C^2 \left[\frac{bh}{2}\right]$$

$$= \frac{200(100^3)}{36} + \left[100 + \frac{100}{3}\right]^2 \left[\frac{200(100)}{2}\right] = 183.333(10^6) \text{ mm}^4$$

For the composite area,

$$I_x = I_{x1} - I_{x2} - I_{x3} + I_{x4}$$

$$= 66.667(10^6) - 7.705(10^6) - 12.272(10^6) + 183.333(10^6)$$

$$= 230.023(10^6) \cong 230(10^6) \text{ mm}^4$$ **Ans.**

(b) For the rectangle (shape 1),

$$I_{y1} = \frac{b^3 h}{3} = \frac{200^3(100)}{3} = 266.667(10^6) \text{ mm}^4$$

For the circle (shape 2),

$$I_{y2} = I_{yC} + x_C^2 A = \frac{\pi R^4}{4} + x_C^2(\pi R^2)$$

$$= \frac{\pi(30^4)}{4} + (150^2)(\pi)(30^2) = 64.253(10^6) \text{ mm}^4$$

For the half circle (shape 3),

$$I_{y3} = \frac{\pi R^4}{8} = \frac{\pi(50^4)}{8} = 2.454(10^6) \text{ mm}^4$$

For the triangle (shape 4),

$$I_{y4} = \frac{bh^3}{12} = \frac{100(200^3)}{12} = 66.667(10^6) \text{ mm}^4$$

For the composite area,

$$I_y = I_{y1} - I_{y2} - I_{y3} + I_{y4}$$
$$= 266.667(10^6) - 64.253(10^6) - 2.454(10^6) + 66.667(10^6)$$
$$= 266.627(10^6) \cong 267(10^6) \text{ mm}^4 \quad \blacksquare \qquad \textbf{Ans.}$$

## Example Problem 8-3

A column with the cross section shown in Fig. 8-15 is constructed from a W24 × 84 wide-flange section and a C12 × 30 channel. Determine the second moments and radii of gyration of the cross-sectional area with respect to horizontal and vertical axes through the centroid of the cross section.

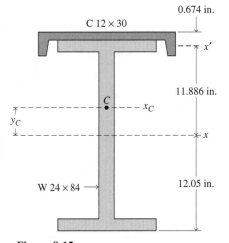

**Figure 8-15**

### SOLUTION

Properties and dimensions for the structural shapes can be obtained from Tables A-1 and A-5 of Appendix A. In Fig. 8-15, the x-axis passes through the centroid of the wide-flange section and the $x'$-axis passes through the centroid of the channel. The centroidal $x_C$-axis for the composite section can be located by using the principle of moments as applied to areas.

The total area $A_T$ for the composite section is

$$A_T = A_{WF} + A_{CH} = 24.7 + 8.82 = 33.52 \text{ in}^2$$

The moment of the composite area about the x-axis is

$$M_x = A_{WF}(y_C)_{WF} + A_{CH}(y_C)_{CH} = 24.7(0) + 8.82(11.886) = 104.835 \text{ in}^3$$

The distance $y_C$ from the x-axis to the centroid of the composite section is

$$y_C = \frac{M_x}{A_T} = \frac{104.835}{33.52} = 3.128 \text{ in.}$$

The second moment $I_{xCWF}$ for the wide-flange section about the centroidal $x_C$-axis of the composite section is

$$I_{xCWF} = I_{xWF} + (y_C)^2_{WF}A_{WF} = 2370 + (3.128)^2(24.7) = 2611.7 \text{ in}^4$$

The second moment $I_{xCCH}$ for the channel about the centroidal $x_C$-axis of the composite section is

$$I_{xCCH} = I_{x'CH} + [(y_C)_{CH} - (y_C)]^2 A_{CH}$$
$$= 5.14 + (11.886 - 3.128)^2(8.82) = 681.7 \text{ in}^4$$

For the composite area:

$$I_{xC} = I_{xCWF} + I_{xCCH}$$
$$= 2611.7 + 681.7 = 3293.4 \cong 3290 \text{ in}^4 \qquad \textbf{Ans.}$$

The $y$-axis passes through the centroid of each of the areas; therefore, the second moment $I_{yC}$ for the composite section is

$$I_{yC} = I_{yCWF} + I_{yCCH}$$
$$= 94.4 + 162 = 256.4 \cong 256 \text{ in}^4 \qquad \textbf{Ans.}$$

The radius of gyration about the $x_C$-axis for the composite section is

$$k_{xC} = \left[\frac{I_{xC}}{A_T}\right]^{1/2} = \left[\frac{3293.4}{33.52}\right]^{1/2} = 9.912 \cong 9.91 \text{ in.} \qquad \textbf{Ans.}$$

The radius of gyration about the $y_C$-axis for the composite section is

$$k_{yC} = \left[\frac{I_{yC}}{A_T}\right]^{1/2} = \left[\frac{256.4}{33.52}\right]^{1/2} = 2.766 \cong 2.77 \text{ in.} \ \blacksquare \qquad \textbf{Ans.}$$

# PROBLEMS

## Introductory Problems

**8-1\*** Determine the second moment of the shaded area shown in Fig. P8-1 with respect to
(a) The $x$-axis.    (b) The $y$-axis.

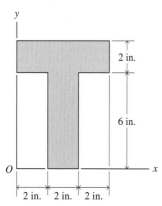

**Figure P8-1**

**8-2\*** Determine the second moments of the shaded area shown in Fig. P8-2 with respect to $x$- (horizontal) and $y$- (vertical) axes through the centroid of the area.

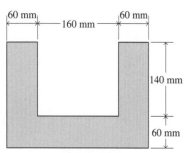

**Figure P8-2**

**8-3** Determine the second moments of the shaded area shown in Fig. P8-3 with respect to $x$- (horizontal) and $y$- (vertical) axes through the centroid of the area.

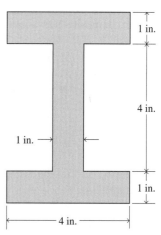

**Figure P8-3**

## Intermediate Problems

**8-4\*** Four C305 × 45 channels are welded together to form the cross section shown in Fig. P8-4. Determine the second moments of the area with respect to $x$- (horizontal) and $y$- (vertical) axes through the centroid of the area.

**Figure P8-4**

**8-5\*** Two 10- × 1-in. steel plates are welded to the flanges of an S18 × 70 I-beam as shown in Fig. P8-5. Determine the second moments of the area with respect to $x$- (horizontal) and $y$- (vertical) axes through the centroid of the area.

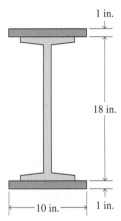

**Figure P8-5**

**8-6** Determine the second moment of the shaded area shown in Fig. P8-6 with respect to
(a) The $x$-axis.    (b) The $y$-axis.

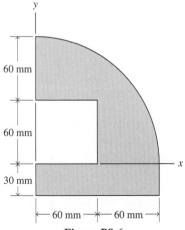

**Figure P8-6**

**Challenging Problems**

**8-7\*** Determine the second moments of the shaded area shown in Fig. P8-7 with respect to
(a) The $x$- and $y$-axes shown on the figure.
(b) The $x$- and $y$-axes through the centroid of the area.

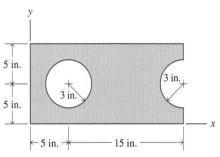

**Figure P8-7**

**8-8** Determine the second moments of the shaded area shown in Fig. P8-8 with respect to
(a) The $x$- and $y$-axes shown on the figure.
(b) The $x$- and $y$-axes through the centroid of the area.

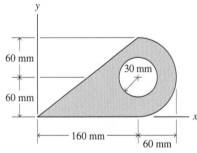

**Figure P8-8**

**8-9** Determine the second moments of the shaded area shown in Fig. P8-9 with respect to the $x$- and $y$-axes.

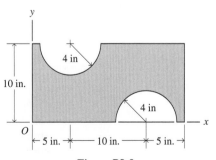

**Figure P8-9**

## 8-5 THE ELASTIC FLEXURE FORMULA

When the integral $\int_A y^2 \, dA$ in Eq. (8-7) is replaced by the symbol $I$, the elastic flexure formula is obtained by combining Eqs. (8-6) and (8-7) as

$$\sigma_x = -\frac{M_r y}{I} = -\frac{My}{I} \tag{8-11}$$

where $\sigma_x$ is the normal stress on a transverse plane, at a distance $y$ from the neutral surface, $M_r = M$ is the resisting moment of the section, and $I$ is the second moment of area of the transverse section with respect to the neutral axis. Recall that the neutral axis passes through the centroid of the area.

At any section of the beam, the flexural stress, or normal stress, will be maximum (have the greatest magnitude) at the surface farthest from the neutral axis ($y = c$), and Eq. (8-11) becomes

$$\sigma_{\max} = \frac{M_r c}{I} = \frac{M_r}{S} = \frac{M}{S} \tag{8-12}$$

where $S = I/c$ is called the *section modulus of the beam*. Although the section modulus can be readily calculated for a given section, values are often included in tables to simplify calculations. Observe that for a given area, $S$ becomes larger as the shape is altered to concentrate more of the area as far as possible from the neutral axis. Commercial rolled shapes such as I- and WF-beams and the various built-up sections are intended to optimize the area–section modulus relation.

Thus far, the discussion of flexural behavior has been limited to structural members with symmetric cross sections that are loaded in a plane of symmetry. Many other shapes are subjected to flexural loadings, and methods are needed to determine stress distributions in these nonsymmetric shapes. The flexure formula [Eq. (8-11)] provides a means for relating the resisting moment $M_r$ at a section of a beam to the normal stress at a point on the transverse cross section. Further insight into the applicability of the flexure formula to nonsymmetric sections can be gained by considering the requirements for equilibrium when the applied moment $M$ does not have a component about the $y$-axis of the cross section. Thus, from Eq. (8-6) and the equilibrium equation $\Sigma M_y = 0$,

$$\int_A z \, \sigma_x \, dA = \int_A z \, \frac{\sigma_c}{c} y \, dA = \frac{\sigma_c}{c} \int_A zy \, dA = \frac{\sigma_c}{c} I_{yz} = 0 \tag{a}$$

The quantity $I_{yz}$ is commonly known as the *mixed second moment* of the cross-sectional area with respect to the centroidal $y$- and $z$-axes. Obviously, Eq. (a) can be satisfied only if $I_{yz} = 0$. For symmetric cross sections, $I_{yz} = 0$ when the $y$- and $z$-axes coincide with the axes of symmetry. For nonsymmetric cross sections, $I_{yz} = 0$ when the $y$- and $z$-axes are centroidal principal axes for the cross section. Thus, the flexure formula is valid for any cross section, provided $y$ is measured along a principal direction, $I$ is a principal second moment of area, and $M$ is a moment about a principal axis. Problems of this type will not be discussed in this book.

# Example Problem 8-4

A timber beam consists of four 2- × 8-in. planks fastened together to form a box section 8 in. wide × 12 in. deep, as shown in Fig. 8-16a. If the resisting moment at the section is $M_r = -200(10^3)$ lb · in., determine

(a) The flexural stress at point $A$ of the cross section.
(b) The flexural stress at point $B$ of the cross section.
(c) The flexural stress at point $C$ of the cross section.
(d) The flexural stress at point $D$ of the cross section.

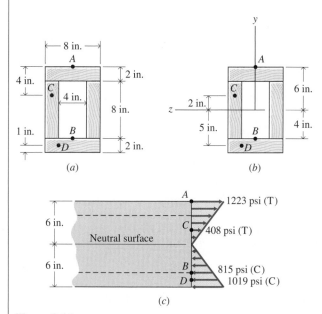

Figure 8-16

## SOLUTION

The resisting moment $M_r$ is assumed to act in the vertical plane of symmetry of the cross section. The neutral axis passes through the centroid of the cross section and is perpendicular to the plane of the resisting moment. As a result of symmetry, the centroid is at the geometric center of the cross section. The second moment of area for the cross section with respect to the neutral axis is found by subtracting the second moment of area for the hollow part of the section (4 × 8 in.) from the second moment of area for the solid part of the section (8 × 12 in.). The equation listed in Table 8-1 for the second moment of area of a rectangular section is $I = bh^3/12$. Thus, for the cross section shown in Fig. 8-16a

$$I = \frac{b_s h_s^3}{12} - \frac{b_h h_h^3}{12} = \frac{8(12)^3}{12} - \frac{4(8)^3}{12} = 981.3 \text{ in}^4$$

The distance $y$, positive upward, from the neutral surface is shown in Fig. 8-16b. Thus,

(a) The flexural (normal) stress at $A$ is

$$\sigma_A = -\frac{M_r y_A}{I} = -\frac{(-200)(10^3)(+6)}{981.3}$$

$$= +1222.9 \text{ lb/in}^2 \cong 1223 \text{ psi T} \qquad \textbf{Ans.}$$

Since the resisting moment $M_r$ is negative, the flexural stress $\sigma$ has the same sign as $y$. When $y$ is positive (the upper portion of the beam), the stress is positive (tension). When $y$ is negative (the lower portion of the beam), the stress is negative (compression).

(b) The flexural (normal) stress at $B$ is

$$\sigma_B = -\frac{M_r y_B}{I} = -\frac{(-200)(10^3)(-4)}{981.3}$$

$$= -815 \text{ lb/in}^2 = 815 \text{ psi C} \qquad \textbf{Ans.}$$

(c) The flexural (normal) stress at $C$ is

$$\sigma_C = -\frac{M_r y_C}{I} = -\frac{(-200)(10^3)(+2)}{981.3}$$

$$= +408 \text{ lb/in}^2 = 408 \text{ psi T} \qquad \textbf{Ans.}$$

(d) The flexural (normal) stress at $D$ is

$$\sigma_D = -\frac{M_r y_D}{I} = -\frac{(-200)(10^3)(-5)}{981.3}$$

$$= -1019 \text{ lb/in}^2 = 1019 \text{ psi C} \qquad \textbf{Ans.}$$

The flexural stress varies linearly with the $y$-coordinate and is constant with respect to the $z$-coordinate (Fig. 8-16b). The variation of flexural stress over the depth of the beam is shown in Fig. 8-16c.

Alternatively, from Eq. (8-11), note on a given cross section that

$$\sigma_x = -\frac{M_r y}{I} \qquad \text{or} \qquad \frac{\sigma_x}{y} = -\frac{M_r}{I} = \text{constant}$$

Therefore, if the stress is known at one point on the cross section it can be determined at any other point without knowing either the resisting moment $M_r$ or the second moment of area $I$ for the cross section. Thus,

$$\frac{\sigma_A}{y_A} = \frac{\sigma_B}{y_B} = \frac{\sigma_C}{y_C} = \frac{\sigma_D}{y_D}$$

(b) $\qquad \sigma_B = \frac{y_B}{y_A}\sigma_A = \frac{-4}{+6}(+1222.9) = -815 = 815 \text{ psi C} \qquad \textbf{Ans.}$

(c) $\qquad \sigma_C = \frac{y_C}{y_A}\sigma_A = \frac{+2}{+6}(+1222.9) = +408 = 408 \text{ psi T} \qquad \textbf{Ans.}$

(d) $\qquad \sigma_D = \frac{y_D}{y_A}\sigma_A = \frac{-5}{+6}(+1222.9) = -1019 = 1019 \text{ psi C} \qquad \blacksquare \qquad \textbf{Ans.}$

## Example Problem 8-5

The maximum flexural stress at a certain section in a beam with a rectangular cross section 100 mm wide × 200 mm deep (Fig. 8-17) is 15 MPa. Determine

(a) The resisting moment $M_r$ developed at the section.
(b) The percentage decrease in $M_r$ if the dotted central portion of the cross section shown in Fig. 8-17 is removed.

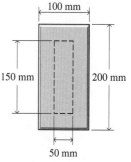

**Figure 8-17**

## SOLUTION

The normal stress $\sigma_x$ at a distance $y$ from the neutral surface on a transverse cross section of a beam is given by Eq. (8-11) as

$$\sigma_x = -\frac{M_r y}{I} \qquad \text{or} \qquad M_r = -\frac{\sigma_x I}{y}$$

(a) As a result of symmetry, the centroid is the geometric center of the cross section. The neutral axis is the horizontal axis passing through the centroid. For the original cross section,

$$I = I_{NA} = \frac{100(200)^3}{12} = 66.67(10^6) \text{ mm}^4 = 66.67(10^{-6}) \text{ m}^4$$

$$|M_r| = \frac{\sigma_x I}{c} = \frac{15(10^6)(66.67)(10^{-6})}{100(10^{-3})}$$

$$= 10.00(10^3) \text{ N} \cdot \text{m} = 10.00 \text{ kN} \cdot \text{m} \qquad \textbf{Ans.}$$

(b) For the modified cross section,

$$I = I_{NA} = \frac{100(200)^3}{12} - \frac{50(150)^3}{12} = 52.60(10^6) \text{ mm}^4 = 52.60(10^{-6}) \text{ m}^4$$

$$|M_r| = \frac{\sigma_x I}{c} = \frac{15(10^6)(52.60)(10^{-6})}{100(10^{-3})}$$

$$= 7.89(10^3) \text{ N} \cdot \text{m} = 7.89 \text{ kN} \cdot \text{m}$$

The percent decrease is

$$D = \frac{10.00 - 7.89}{10.00}(100) = 21.1\% \qquad \textbf{Ans.}$$

Note that removing 38% of the area of the beam (near the neutral axis) resulted in only a 21% reduction in $M_r$. If the same amount of area had been removed from the top and bottom of the beam, the reduction in $M_r$ would have been 61%! ∎

## Example Problem 8-6

A beam has the cross section shown in Fig. 8-18a. On a section where the resisting moment is $M_r = -75$ kN $\cdot$ m, determine

(a) The maximum tensile flexural stress.

(b) The maximum compressive flexural stress.

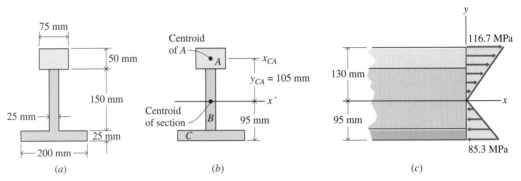

Figure 8-18

### SOLUTION

The neutral axis is horizontal and passes through the centroid of the cross section. The centroid is located using the principle of moments as applied to areas. The total area $A$ of the cross section is

$$A = 200(25) + 150(25) + 50(75) = 12{,}500 \text{ mm}^2$$

The moment of the area $M_A$ about the bottom edge of the cross section is

$$M_A = 12.5[200(25)] + 100[150(25)] + 200[50(75)] = 1{,}187{,}500 \text{ mm}^3$$

The distance $d$ from the bottom edge of the cross section to the centroid is

$$d = \frac{M_A}{A} = \frac{1{,}187{,}500}{12{,}500} = 95 \text{ mm}$$

Since the centroid of the cross section is not located at the centroid of any of the individual parts (three rectangles), the second moment of the cross section with respect to the horizontal centroidal axis (neutral axis) is found using the parallel-axis theorem. For part $A$ shown in Fig. 8-18b,

$$I_{x'A} = I_{xCA} + A_A y_{CA}^2 = \frac{75(50)^3}{12} + 75(50)(105)^2 = 42.13(10^6) \text{ mm}^4$$

Similarly, for parts $B$ and $C$ of Fig. 8-18b,

$$I_{x'B} = I_{xCB} + A_B y_{CB}^2 = \frac{25(150)^3}{12} + 25(150)(5)^2 = 7.13(10^6) \text{ mm}^4$$

$$I_{x'C} = I_{xCC} + A_C y_{CC}^2 = \frac{200(25)^3}{12} + 200(25)(-82.5)^2 = 34.29(10^6) \text{ mm}^4$$

Adding the second moments of area for the three parts gives the second moment of area for the cross section as

$$I_{x'} = I_{x'A} + I_{x'B} + I_{x'C} = 42.13(10^6) + 7.13(10^6) + 34.29(10^6)$$

$$= 83.55(10^6) \text{ mm}^4 = 83.55(10^{-6}) \text{ m}^4$$

(a) Since the resisting moment $M_r$ is negative, the top portion of the beam is in tension ($\sigma$ positive) and the bottom portion of the beam is in compression ($\sigma$ negative). The flexural stress varies linearly with distance from the neutral axis. Therefore, the maximum tensile flexural stress occurs at the top surface of the beam

$$\sigma_{\max T} = -\frac{M_r y_t}{I} = -\frac{(-75)(10^3)(130)(10^{-3})}{83.55(10^{-6})}$$

$$= +116.7(10^6) \text{ N/m}^2 = 116.7 \text{ MPa T} \qquad \textbf{Ans.}$$

(b) The maximum compressive flexural stress occurs at the bottom surface of the beam

$$\sigma_{\max C} = -\frac{M_r y_b}{I} = -\frac{(-75)(10^3)(-95)(10^{-3})}{83.55(10^{-6})}$$

$$= -85.3(10^6) \text{ N/m}^2 = 85.3 \text{ MPa C} \qquad \textbf{Ans.}$$

The linear variation of flexural stress over the depth of the beam is shown in Fig. 8-18c. ∎

## PROBLEMS

### Introductory Problems

**8-10\*** The maximum flexural stress on a transverse plane in a beam with a rectangular cross section 150 mm wide × 300 mm deep is 15 MPa. Determine the resisting moment $M_r$ transmitted by the plane.

**8-11\*** The maximum flexural stress on a transverse plane in a beam with a rectangular cross section 4 in. wide × 8 in. deep is 1500 psi. Determine the resisting moment $M_r$ transmitted by the plane.

**8-12\*** Determine the maximum flexural stress required to produce a resisting moment $M_r$ of −15 kN · m if the beam has the cross section shown in Fig. P8-12.

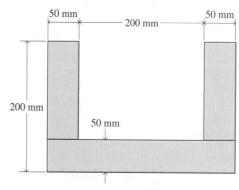

**Figure P8-12**

**8-13** Determine the maximum flexural stress required to produce a resisting moment $M_r$ of $+5000$ lb · ft if the beam has the cross section shown in Fig. P8-13.

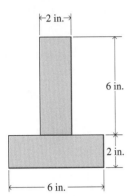

**Figure P8-13**

**8-14** A timber beam consists of three 50- × 200-mm planks fastened together to form an I-beam 200 mm wide × 300 mm deep, as shown in Fig. P8-14. If the flexural stress at point $A$ of the cross section is 7.5 MPa C, determine the flexural stress
(a) At point $B$ of the cross section.
(b) At point $C$ of the cross section.
(c) At point $D$ of the cross section.

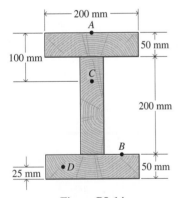

**Figure P8-14**

**8-15\*** A timber beam is made of three 2- × 6-in. planks fastened together to form an I-beam 6 in. wide × 10 in. deep. If the maximum flexural stress must not exceed 1200 psi, determine the maximum resisting moment $M_r$ that the beam can support.

**8-16** The beam of Fig. P8-16 is made of material that has a yield strength of 250 MPa. Determine the maximum resisting moment that the beam can support if yielding must be avoided.

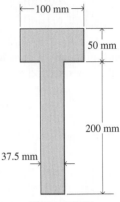

**Figure P8-16**

**8-17** The maximum flexural stress on a transverse cross section of a beam with a 2- × 4-in. rectangular cross section must not exceed 10 ksi. Determine the maximum resisting moment $M_r$ that the beam can support if the neutral axis is
(a) Parallel to the 4-in. side.
(b) Parallel to the 2-in. side.

**Intermediate Problems**

**8-18\*** A hardened steel ($E = 210$ GPa) bar with a 50-mm square cross section is subjected to a flexural form of loading that produces a flexural strain of $+1200$ $\mu$m/m at a point on the top surface of the beam. Determine
(a) The maximum flexural stress at the point.
(b) The resisting moment $M_r$ developed in the beam on a transverse cross section through the point.

**8-19\*** A steel ($E = 29,000$ ksi) bar with a rectangular cross section is bent over a rigid mandrel ($R = 10$ in.) as shown in Fig. P8-19. If the maximum flexural stress in the bar is not to exceed the yield strength ($\sigma_y = 36$ ksi) of the steel, determine the maximum allowable thickness $h$ for the bar.

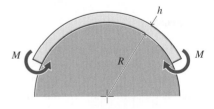

**Figure P8-19**

**8-20** An aluminum alloy ($E = 73$ GPa) bar with a rectangular cross section is bent over a rigid mandrel as shown in Fig. P8-19. The thickness $h$ of the bar is 25 mm. If the maximum flexural stress in the bar must be limited to 200 MPa, determine the minimum allowable radius $R$ for the mandrel.

**8-21** The load carrying capacity of an S24 × 80 American Standard beam (see Appendix A for dimensions) is to be increased by fastening two 8- × ¾-in. plates to the flanges of the beam, as shown in Fig. P8-21. The maximum flexural stress in both the original and modified beams must be limited to 15 ksi. Determine
(a) The maximum resisting moment that the original beam can support.
(b) The maximum resisting moment that the modified beam can support.

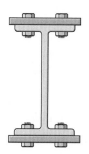

**Figure P8-21**

**8-22\*** Two L102 × 102 × 12.7-mm structural steel angles (see Appendix A for dimensions) are attached back to back to form a T-section, as shown in Fig. P8-22. Determine the maximum resisting moment $M_r$ that can be supported by the beam if the maximum flexural stress must be limited to 125 MPa.

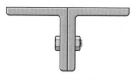

**Figure P8-22**

**8-23\*** An I-beam is fabricated by welding two 16- × 2 in. flange plates to a 24- × 1-in. web plate. The beam is loaded in the plane of symmetry parallel to the web. On a section where the resisting moment $M_r = 1400$ kip · ft, determine the maximum flexural stress.

**8-24\*** Determine the maximum tensile and compressive flexural stresses on a section where the resisting moment $M_r = -4$ kN · m if the beam has the cross section shown in Fig. P8-24. The beam is loaded in the plane of symmetry parallel to the web.

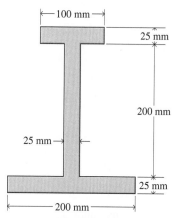

**Figure P8-24**

**8-25** Determine the maximum tensile and compressive flexural stresses on a section where the resisting moment $M_r = +30,000$ lb · in. if the beam has the cross section shown in Fig. P8-25. The beam is loaded in the vertical plane of symmetry.

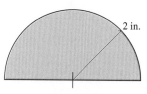

**Figure P8-25**

**8-26** Determine the maximum resisting moment $M_r$ that can be supported by a beam having the cross section shown in Fig. P8-26 if the maximum flexural stress must be limited to 120 MPa.

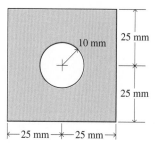

**Figure P8-26**

**Challenging Problems**

**8-27\*** A beam has the cross section shown in Fig. P8-27. On a section where the resisting moment is −30 kip · ft, determine
(a) The maximum tensile flexural stress.
(b) The maximum compressive flexural stress.

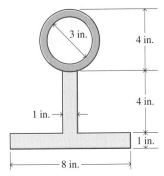

**Figure P8-27**

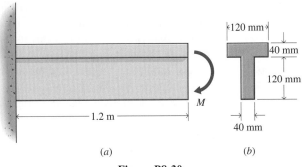

(a)　　　　　　　　(b)

**Figure P8-30**

**8-28\*** Determine the maximum flexural stress on a section where the resisting moment $M_r = +100$ kN · m if the beam has the cross section shown in Fig. P8-28. The beam is loaded in the vertical plane of symmetry.

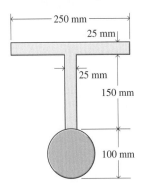

**Figure P8-28**

**8-29** The cantilever beam shown in Fig. P8-29 is subjected to a moment $M = 20,000$ lb · in. at its free end. The beam has a 2- × 2-in. cross section. Determine the maximum tensile flexural stress in the beam.

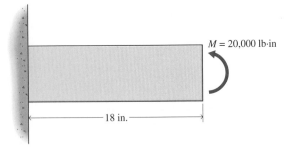

**Figure P8-29**

**8-30** The cantilever beam shown in Fig. P8-30a is subjected to a moment $M$ at its free end. The cross section of the beam is shown in Fig. P8-30b. If the allowable stresses are 110 MPa T and 170 MPa C, determine the maximum moment that can be applied to the beam.

**8-31\*** A steel pipe with an outside diameter of 4 in. and an inside diameter of 3 in. is simply supported at its ends and carries two concentrated loads, as shown in Fig. P8-31. On a section 5 ft from the right support, determine the flexural stress
(a) At point $A$ (top, outside) on the cross section.
(b) At point $B$ (bottom, inside) on the cross section.

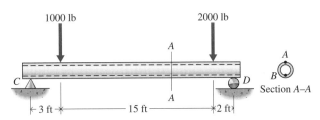

**Figure P8-31**

**8-32** A beam has the cross section shown in Fig. P8-32 and is loaded in the vertical plane of symmetry. On a section where the resisting moment $M_r = +50$ kN · m, determine
(a) The maximum tensile flexural stress.
(b) The maximum compressive flexural stress.

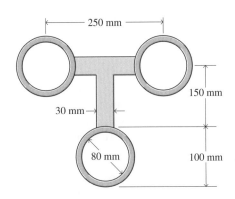

**Figure P8-32**

## Computer Problems

**8-33** A beam with a hollow circular cross section (see Fig. P8-33) is being designed to support a maximum moment of 100 kip · in. If the maximum flexural stress in the beam must not exceed 40 ksi and the maximum diameter of the beam must not exceed 5 in., prepare a design curve that shows the acceptable values of the inside diameter $d_i$ as a function of the outside diameter $d_o$ (2.5 in. $\leq d_o \leq$ 5 in.).

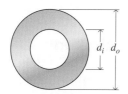

**Figure P8-33**

**8-34** A beam with a solid rectangular cross section (see Fig. P8-34) is being designed to support a maximum moment of 4 kN · m. If the maximum flexural stress in the beam must not exceed 50 MPa and the width of the beam must not exceed 75 mm, prepare a design curve that shows the acceptable values of depth $h$ as a function of the width $b$ (0 mm $\leq b \leq$ 75 mm).

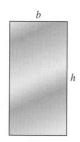

**Figure P8-34**

**8-35** The load carrying capacity of an S24 × 80 American standard beam (see Appendix A for dimensions) is to be increased by fastening plates to the flanges of the beam, as shown in Fig. P8-35. The maximum flexural stress in both the original beam and the modified beams must be limited to 20 ksi. If the width of the plates is the same as the width of the flanges, compute and plot the percent increase in the maximum resisting moment that the modified beam can support compared to the original beam as a function of the thickness of the plates $t$ (0 $\leq t \leq 2t_{\text{flange}}$).

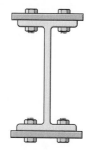

**Figure P8-35**

**8-36** A beam with a solid rectangular cross section (150 × 260 mm) is being redesigned as an I-beam by removing material from the sides of the beam (Fig. P8-36). If the maximum flexural stress in the beam must not exceed 150 MPa and the thickness of the web must not be less than 10 mm, compute and plot

(a) The percent decrease in weight (cross-sectional area) of the modified beam compared to the original beam as a function of the thickness of the web $w$ (10 mm $\leq w \leq$ 150 mm).

(b) The percent decrease in maximum resisting moment that the modified beam can support compared to the original beam as a function of the thickness of the web $w$ (10 mm $\leq w \leq$ 150 mm).

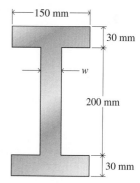

**Figure P8-36**

# 8-6 SHEAR FORCES AND BENDING MOMENTS IN BEAMS

The method for determining flexural stresses outlined in Section 8-5 is adequate if one wishes to determine the flexural stresses on any specified transverse cross section of the beam. However, if the maximum flexural stress is required in a beam subjected to a loading that produces a resisting moment that varies with position along the beam, it is desirable to have a method for determining the maximum resisting moment. Similarly, the maximum transverse shearing stress will occur at a section where the resisting shear ($V_r$ of Fig. 8-4$a$) is maximum, and a method for determining such sections is likewise desirable.

In the following section equations for the resisting shear $V_r$ and the resisting moment $M_r$ will be obtained using equilibrium equations. In Section 8-7, relationships will be developed between loads applied to the beam and the resisting shear $V_r$ at a section and between the resisting shear $V_r$ at a section and the resisting moment $M_r$ at that section.

## Shear Force and Bending Moment: An Equilibrium Approach

When the equilibrium equation $\Sigma F_y = 0$ is applied to the free-body diagram of Fig. 8-4$a$, the result can be written as

$$R - wx - P = V_r \qquad \text{or} \qquad V = V_r$$

where $V = R - wx - P$ is the resultant of the external transverse forces acting on the part of the beam to either side of a transverse section and is called the transverse shear or just the shear at the section.

As seen from the definitions of $V$ and $V_r$, the shear $V$ is equal in magnitude and opposite in sense to the resisting shear $V_r$. Since these shear forces are always equal in magnitude, they are frequently treated as though they were identical. For simplicity the symbol $V$ will be used henceforth to represent both the transverse shear and the resisting shear. The sign convention to be selected will have sufficient generality to apply to both quantities.

The resultant of the flexural stresses on any transverse section has been shown to be a couple (if only transverse loads are considered) and has been designated as $M_r$. When the equilibrium equation $\Sigma M_O = 0$ (where $O$ is any axis parallel to the neutral axis of the section) is applied to the free-body diagram of Fig. 8-4$a$, the result can be written as

$$Rx - \frac{wx^2}{2} - P(x - h) = M_r \qquad \text{or} \qquad M = M_r$$

where $M = Rx - \dfrac{wx^2}{2} - P(x - h)$ is the algebraic sum of the moments of the external forces acting on the part of the beam to either side of the transverse section, with respect to an axis in the section and is called the bending moment or just the moment at the section. The axis $O$ is usually taken to be the neutral axis of the cross section.

As seen from the definitions of $M$ and $M_r$, the bending moment $M$ is equal in magnitude and opposite in sense to the resisting moment $M_r$. Since these moments are always equal in magnitude, they are frequently treated as though they were identical. For simplicity the symbol $M$ will be used henceforth to represent both the bending moment and the resisting moment. The sign convention for moments will have sufficient generality to apply to both quantities.

The bending moment $M$ and the transverse shear $V$ are not normally shown on the free-body diagram. It is customary procedure to show each external force individually, as previously indicated in Fig. 8-4. The variation of $V$ and $M$ (actually $V_r$ and $M_r$) along the beam can be shown conveniently by means of equations or by means of shear and bending moment diagrams (graphs of $V$ and $M$ as functions of $x$).

A sign convention is necessary for the correct interpretation of results obtained from equations or diagrams for shear and moment. The following convention will give consistent results regardless of whether one proceeds from left to right or from right to left along the beam. By definition, the shear at a section is positive when the portion of the beam to the left of the section (for a horizontal beam) tends to move upward with respect to the portion to the right of the section, as shown in Fig. 8-19a. Then for a positive shear force, the transverse shear pushes up on the portion of the beam to the left of the section while the resisting shear pushes downward on the section, as shown in Fig. 8-19b. The converse applies when the portion of the beam is taken to the right of the section. Also, by definition, the bending moment in a horizontal beam is positive at sections for which the top of the beam is in compression and the bottom is in tension, as shown in Fig. 8-19c. Then for a positive bending moment, the bending moment acting on the portion of the beam to the left of the section must act clockwise while the resisting moment on the section must act counterclockwise, as shown in Fig. 8-19d. Observe that the signs of the terms in the preceding equations for $V$ and $M$ agree with these definitions (sign conventions).

Since $M$ and $V$ vary with $x$ (Fig. 8-4a), they are functions of $x$, and equations for $M$ and $V$ can be obtained from free-body diagrams of portions of the beam. The procedure can be summarized as follows:

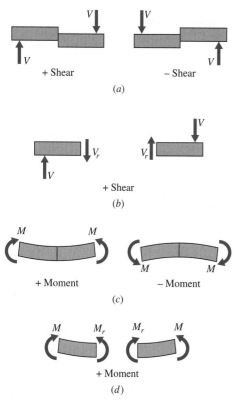

**Figure 8-19**

- The beam is sectioned at an arbitrary location $x$.
- A free-body diagram is drawn for the portion of the beam to the left (or for the portion to the right) of the transverse cross section. The shear force and bending moment must be drawn according to the sign convention established earlier.
- Equilibrium equations $\Sigma F_y = 0$ and $\Sigma M_{\text{cut}} = 0$ are written from the free-body diagrams and solved to get the shear force and bending moment at the location $x$. The equations obtained for $V(x)$ and $M(x)$ will be valid for a range of $x$ for which the nature of the loading does not change. That is, if the beam is sectioned in a region in which there are no concentrated or distributed loads, the equations will be valid for the entire region $x_L < x < x_R$ where $x_L$ is the location of the nearest load to the left of the section and $x_R$ is the location of the nearest load to the right of the section. No matter where in the region $x_L < x < x_R$ that the beam is sectioned, the free-body diagram will look exactly the same and the equations will be exactly the same.
- The process must be repeated for each different segment of the beam.

Example Problem 8-7 will illustrate the procedure to calculate the shear force and the bending moment for any section of a beam.

## Example Problem 8-7

A beam is loaded and supported as shown in Fig. 8-20$a$. Write equations for the shear force $V$ and the bending moment $M$ for any section of the beam

(a) In the interval $AB$.

(b) In the interval $BC$.

(c) In the interval $CD$.

### SOLUTION

A free-body diagram, or load diagram, for the beam is shown in Fig. 8-20$b$. The reactions $R_1$ and $R_2$ shown on the load diagram are determined from the equations of equilibrium.

$$+\curvearrowright \Sigma M_D = 0: \qquad R_1(8) - 400(6)(3) - 2000(2) = 0$$

$$+\curvearrowleft \Sigma M_A = 0: \qquad R_2(8) - 400(6)(5) - 2000(6) = 0$$

Solving yields

$$R_1 = 1400 \text{ lb} \qquad \text{and} \qquad R_2 = 3000 \text{ lb}$$

(a) A free-body diagram of a portion of the beam from the left end to any section between $A$ and $B$ is shown in Fig. 8-20$c$. The bending moment $M$ (which is the same as the resisting moment $M_r$) and the transverse shear $V$ (which is the same as the resisting shear $V_r$) are shown as positive quantities. As discussed in the preceding section, the subscript $r$ has been dropped from

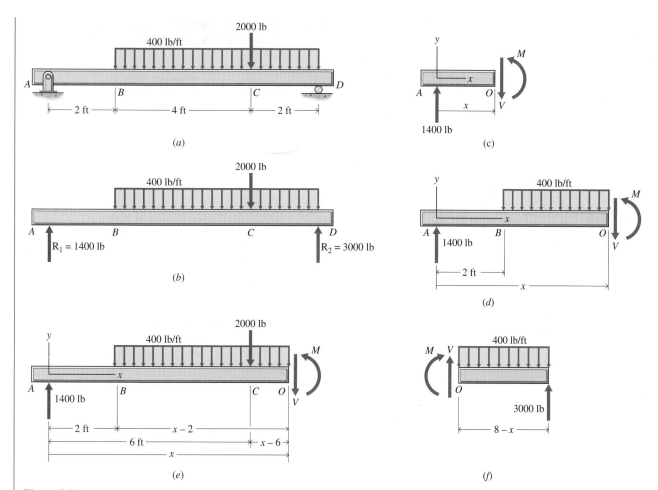

**Figure 8-20**

the resisting shear and the resisting moment. From the definition of $V$, or from the equilibrium equation $\Sigma F_y = 0$,

$$+\uparrow \Sigma F_y = 0: \qquad 1400 - V = 0$$

Solving yields

$$V = 1400 \text{ lb} \qquad (0 < x < 2 \text{ ft}) \qquad \textbf{Ans.}$$

From the definition of $M$, or from the equilibrium equation $\Sigma M_O = 0$,

$$+\downarrow \Sigma M_O = 0: \qquad -1400(x) + M = 0$$

Solving yields

$$M = 1400x \text{ lb} \cdot \text{ft} \qquad (0 < x < 2 \text{ ft}) \qquad \textbf{Ans.}$$

Thus, in the interval $AB$ $(0 < x < 2 \text{ ft})$, the shear force $V$ is constant and the bending moment $M$ varies linearly with $x$.

(b) A free-body diagram of a portion of the beam from the left end to any section between $B$ and $C$ is shown in Fig. 8-20d. The equations of equilibrium for this portion of the beam are

$$+\uparrow \; \Sigma F_y = 0: \qquad 1400 - 400(x - 2) - V = 0$$

$$+\downarrow \Sigma M_O = 0: \qquad -1400(x) + 400(x - 2)\left(\frac{x - 2}{2}\right) + M = 0$$

Solving yields

$$V = -400x + 2200 \text{ lb} \qquad (2 \text{ ft} < x < 6 \text{ ft}) \qquad \textbf{Ans.}$$

$$M = -200x^2 + 2200x - 800 \text{ lb} \cdot \text{ft} \qquad (2 \text{ ft} < x < 6 \text{ ft}) \qquad \textbf{Ans.}$$

In the interval $BC$ (2 ft $< x <$ 6 ft), the shear force $V$ varies linearly with $x$ and the bending moment $M$ varies with the square of $x$.

(c) A free-body diagram of a portion of the beam from the left end to any section between $C$ and $D$ is shown in Fig. 8-20e. The equations of equilibrium for this portion of the beam are

$$+\uparrow \; \Sigma F_y = 0: \qquad 1400 - 400\,(x - 2) - 2000 - V = 0$$

$$+\downarrow \Sigma M_O = 0: \qquad -1400(x) + 400(x - 2)\left(\frac{x - 2}{2}\right)$$
$$+ 2000(x - 6) + M = 0$$

Solving yields

$$V = -400x + 200 \text{ lb} \qquad (6 \text{ ft} < x < 8 \text{ ft}) \qquad \textbf{Ans.}$$

$$M = -200x^2 + 200x + 11,200 \text{ lb} \cdot \text{ft} \qquad (6 \text{ ft} < x < 8 \text{ ft}) \qquad \textbf{Ans.}$$

In the interval $CD$ (6 ft $< x <$ 8 ft), the shear force $V$ varies linearly with $x$ and the bending moment $M$ varies with the square of $x$.

An alternate free-body diagram for interval $CD$ is shown in Fig. 8-20f. The equations of equilibrium for this portion of the beam are

$$+\uparrow \; \Sigma F_y = 0: \qquad V - 400(8 - x) + 3000 = 0$$

$$+\downarrow \Sigma M_O = 0: \qquad -M - 400(8 - x)\left(\frac{8 - x}{2}\right) + 3000(8 - x) = 0$$

Solving yields

$$V = -400x + 200 \text{ lb} \qquad (6 \text{ ft} < x < 8 \text{ ft}) \qquad \textbf{Ans.}$$

$$M = -200x^2 + 200x + 11,200 \text{ lb} \cdot \text{ft} \qquad (6 \text{ ft} < x < 8 \text{ ft}) \qquad \textbf{Ans.}$$

which are identical to the results obtained using the free-body diagram shown in Fig. 8-20e. This is to be expected, since the values of $V$ and $M$ in Figs. 8-20e and 8-20f are equal in magnitude and opposite in direction, according to Newton's third law. ∎

## Example Problem 8-8

An S152 × 19 steel beam is loaded and supported as shown in Fig. 8-21a. On a section 3 m to the right of $A$, determine

(a) The flexural stress at a point 25 mm below the top of the beam.
(b) The maximum flexural stress on the section.

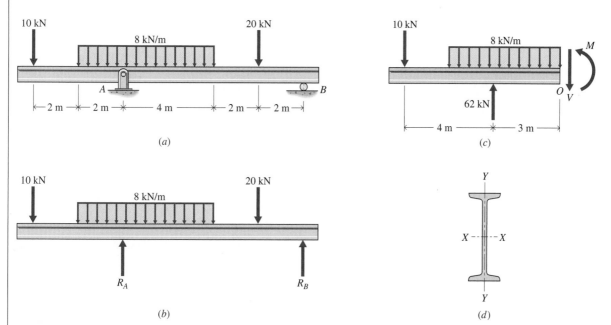

Figure 8-21

## SOLUTION

The reaction at $A$ is found using the free-body diagram shown in Fig. 8-21b and the equation of equilibrium $\Sigma M_B = 0$. Thus,

$$+\!\!\!\downarrow \Sigma M_B = 0: \qquad 20(2) + 8(6)(7) + 10(12) - R_A(8) = 0$$

from which

$$R_A = 62 \text{ kN}$$

The flexural stresses are required at a specific section of the beam; therefore, bending moment equations for the complete beam are not needed. On a section 3 m to the right of $A$, the bending moment can be found by using the free-body diagram shown in Fig. 8-21c and the equation of equilibrium $\Sigma M_O = 0$. Thus,

$$+\!\!\!\downarrow \Sigma M_O = 0: \qquad M + 8(5)(2.5) - 62(3) + 10(7) = 0$$

from which

$$M = +16 \text{ kN} \cdot \text{m}$$

The cross section for the beam is shown in Fig. 8-21d. The X–X-axis is the neutral axis. For an S152 × 19 section (see Appendix A).

$$I = 9.20(10^6) \text{ mm}^4 = 9.20(10^{-6}) \text{ m}^4$$

$$S = 121(10^3) \text{ mm}^3 = 121(10^{-6}) \text{ m}^3$$

$$d = 152.4 \text{ mm} = 0.1524 \text{ m}$$

(a) At a point 25 mm below the top of the beam,

$$y = \frac{d}{2} - 25 = \frac{152.4}{2} - 25 = 51.2 \text{ mm} = 0.0512 \text{ m}$$

$$\sigma = -\frac{My}{I} = -\frac{16(10^3)(0.0512)}{9.20(10^{-6})}$$

$$= -89.04(10^6) \text{ N/m}^2 \cong 89.0 \text{ MPa C} \qquad \textbf{Ans.}$$

(b) The maximum flexural stress is

$$\sigma_{max} = \frac{M}{S} = \frac{16(10^3)}{121(10^{-6})} = 132.2(10^6) \text{ N/m}^2$$

$$= 132.2 \text{ MPa T (on the bottom)} \qquad \textbf{Ans.}$$

$$= 132.2 \text{ MPa C (on the top)} \blacksquare \qquad \textbf{Ans.}$$

# PROBLEMS

**Introductory Problems**

**8-37\*** For the cantilever beam shown in Fig. P8-37, write equations for the shear force $V$ and the bending moment $M$ for any section of the beam in the interval $0 < x < 4$ ft. Use the coordinate system shown.

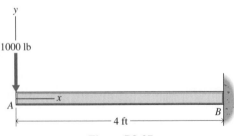

**Figure P8-37**

**8-38\*** For the cantilever beam shown in Fig. P8-38, write equations for the shear force $V$ and the bending moment $M$ for any section of the beam in the interval $0 < x < 2$ m. Use the coordinate system shown.

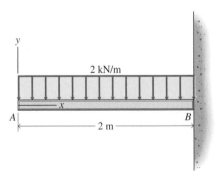

**Figure P8-38**

**8-39** A beam is loaded and supported as shown in Fig. P8-39. Using the coordinate axes shown, write equations for the shear force $V$ and the bending moment $M$ for any section of the beam in the interval 4 ft $< x < 8$ ft.

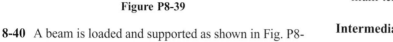

**Figure P8-39**

**8-40**  A beam is loaded and supported as shown in Fig. P8-40. Using the coordinate axes shown, write equations for the shear force $V$ and the bending moment $M$ for any section of the beam in the interval $0 < x < 4$ m.

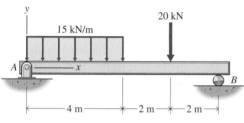

**Figure P8-40**

**8-41**  A beam is loaded and supported as shown in Fig. P8-41. Using the coordinate axes shown, write equations for the shear force $V$ and the bending moment $M$ for any section of the beam in the interval $0 < x < 10$ ft.

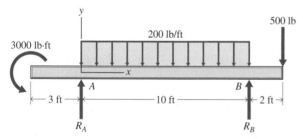

**Figure P8-41**

**8-42\***  A beam is loaded and supported as shown in Fig. P8-42. Using the coordinate axes shown, write equations for the shear force $V$ and the bending moment $M$ for any section of the beam in the interval $0 < x < L$.

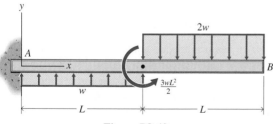

**Figure P8-42**

**8-43\***  The beam shown in Fig. P8-39 has a solid rectangular cross section that is 3 in. wide and 8 in. deep. Determine the maximum tensile flexural stress on a section at $x = 10$ ft.

**8-44**  The beam shown in Fig. P8-40 is an S254 × 52 steel section. On a section at $x = 5$ m, determine the maximum tensile and compressive flexural stresses.

### Intermediate Problems

**8-45\***  A beam is loaded and supported as shown in Fig. P8-45. Using the coordinate axes shown, write equations for the shear force $V$ and the bending moment $M$ for any section of the beam.
(a) In the interval $-3$ ft $< x < 0$.
(b) In the interval $0 < x < 2$ ft.
(c) In the interval $2$ ft $< x < 8$ ft.
(d) In the interval $8$ ft $< x < 10$ ft.

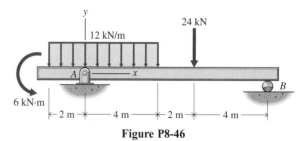

**Figure P8-45**

**8-46\***  A beam is loaded and supported as shown in Fig. P8-46. Using the coordinate axes shown, write equations for the shear force $V$ and the bending moment $M$ for any section of the beam
(a) In the interval $-2$ m $< x < 0$.
(b) In the interval $0 < x < 4$ m.
(c) In the interval $4$ m $< x < 6$ m.
(d) In the interval $6$ m $< x < 10$ m.

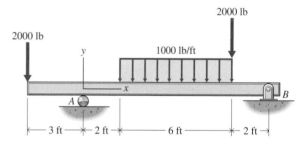

**Figure P8-46**

**8-47**  A beam is loaded and supported as shown in Fig. P8-47. Using the coordinate axes shown,
(a) Write equations for the shear force $V$ and the bending moment $M$ for any section of the beam.

(b) Determine the magnitudes and locations of the maximum shear force and the maximum bending moment in the beam.

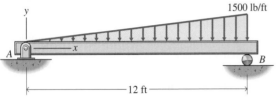

**Figure P8-47**

**8-48** A beam is loaded and supported as shown in Fig. P8-48. Using the coordinate axes shown,
(a) Write equations for the shear force $V$ and the bending moment $M$ for any section of the beam.
(b) Determine the magnitudes and locations of the maximum shear force and the maximum bending moment in the beam.

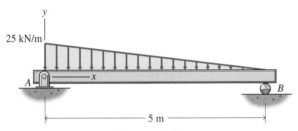

**Figure P8-48**

**8-49** An S8 × 23 steel beam (see Appendix A) is loaded and supported as shown in Fig. P8-49.
(a) Using the coordinate axes shown, write equations for the shear force $V$ and the bending moment $M$ for any section of the beam in the interval 6 ft < $x$ < 8 ft.
(b) Determine the flexural stress at a point 1 in. below the top of the beam on a section at $x$ = 3 ft.
(c) Determine the maximum flexural stress on a section at $x$ = 3 ft.

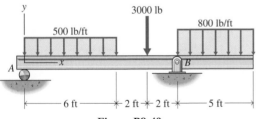

**Figure P8-49**

**8-50\*** An S178 × 30 steel beam (see Appendix A) is loaded and supported as shown in Fig. P8-50.
(a) Using the coordinate axes shown, write equations for the shear force $V$ and the bending moment $M$ for

any section of the beam in the interval 0.5 m < $x$ < 2.5 m.
(b) Determine the flexural stress at a point 15 mm above the bottom of the beam on a section at $x$ = 1.5 m.
(c) Determine the maximum flexural stress on a section at $x$ = 1.5 m.

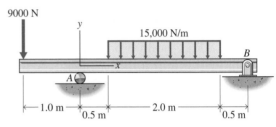

**Figure P8-50**

**8-51\*** A beam is loaded and supported as shown in Fig. P8-51. On a section 6 ft to the right of support $A$, determine the maximum tensile and compressive flexural stresses if the load $P$ = 3000 lb.

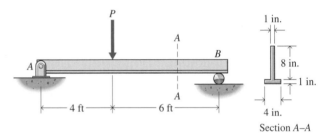

**Figure P8-51**

**8-52** A beam is loaded and supported as shown in Fig. P8-52. On a section 3 m to the right of support $A$, determine the maximum tensile and compressive flexural stresses if $w$ = 3.5 kN/m.

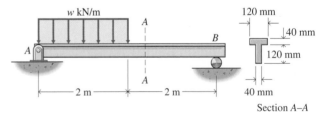

**Figure P8-52**

**Challenging Problems**

**8-53\*** A beam is loaded and supported as shown in Fig. P8-53. Using the coordinate axes shown,
(a) Write equations for the shear force $V$ and the bending moment $M$ for any section of the beam.
(b) Determine the magnitudes and locations of the maximum shear force and the maximum bending moment in the beam.

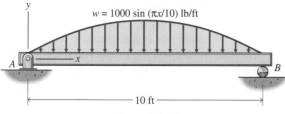

**Figure P8-53**

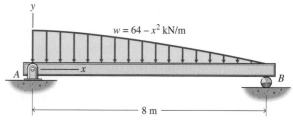

**Figure P8-56**

**8-54\*** A beam is loaded and supported as shown in Fig. P8-54. Using the coordinate axes shown,
(a) Write equations for the shear force $V$ and the bending moment $M$ for any section of the beam.
(b) Determine the magnitudes and locations of the maximum shear force and the maximum bending moment in the beam.

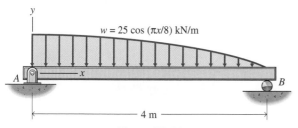

**Figure P8-54**

**8-55** A beam is loaded and supported as shown in Fig. P8-55. Using the coordinate axes shown,
(a) Write equations for the shear force $V$ and bending moment $M$ for any section of the beam.
(b) Determine the magnitudes and locations of the maximum shear force and the maximum bending moment in the beam.

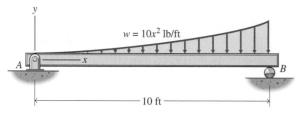

**Figure P8-55**

**8-56** A beam is loaded and supported as shown in Fig. P8-56. Using the coordinate axes shown,
(a) Write equations for the shear force $V$ and the bending moment $M$ for any section of the beam.
(b) Determine the magnitudes and locations of the maximum shear force and the maximum bending moment in the beam.

**8-57\*** Two C10 × 15.3 steel channels (see Appendix A) are placed back to back to form a 10-in.-deep beam, as shown in Fig. P8-57. The beam is 10 ft long and is simply supported at its ends. The beam carries a uniformly distributed load of 1000 lb/ft over its entire length and a concentrated load $P$ at the center of the span, as shown in Fig. P8-57. If the maximum permissible flexural stress at the center of the span is 16,000 psi, determine the maximum permissible value of the concentrated load $P$.

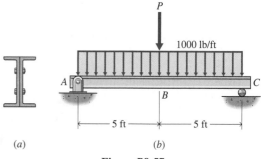

*(a)* *(b)*

**Figure P8-57**

**8-58** Two 50- × 200-mm structural timbers are used to fabricate a beam with an inverted T cross section, as shown in Fig. P8-58. The beam is simply supported at the ends and is 4 m long. If the maximum flexural stress must be limited to 30 MPa, determine
(a) The maximum moment that can be resisted by the beam.
(b) The largest concentrated load that can be supported at the center of the span.
(c) The largest uniformly distributed load (over the entire span) that can be supported by the beam.

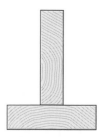

**Figure P8-58**

**Computer Problems**

**8-59**  The supports for the beam shown in Fig. P8-59 are symmetrically located. If the distance $d$ from the supports to the ends of the beam is adjustable, compute and plot

(a) The bending moment $M(x)$ in the beam for $d = 2$ ft, 3 ft, and 4 ft as a function of $x$ (0 ft $\leq x \leq$ 15 ft).

(b) The maximum bending moments $(M_{max})_{AB}$ in segment $AB$ and $(M_{max})_{BC}$ in segment $BC$ as functions of $d$ (0 ft $\leq d \leq$ 6 ft).

(c) What value of $d$ gives the smallest $M_{max}$ in the beam?

**8-60**  An overhead crane consists of a carriage that moves along a beam as shown in Fig. P8-60. If the carriage moves slowly along the beam, compute and plot

(a) The bending moment $M(x)$ in the beam for $b = 1.2$ m, 2 m, and 2.8 m as a function of $x$ (0 m $\leq x \leq$ 5 m).

(b) The bending moments $M_B$ under the left wheel and $M_C$ under the right wheel as functions of $b$ (0.3 m $\leq b \leq$ 4.7 m).

(c) What is the largest bending moment in the beam? When and where does it occur?

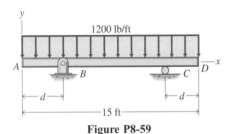

**Figure P8-59**

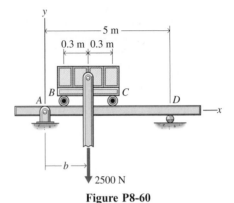

**Figure P8-60**

# 8-7 LOAD, SHEAR FORCE, AND BENDING MOMENT RELATIONSHIPS

The equilibrium approach is a fairly simple and straightforward method of obtaining equations for the shear force and bending moment in a beam. However, if the loading on the beam is complex, the equilibrium approach can require several sections and several free-body diagrams. An alternative approach is to derive mathematical relationships between the loads acting on the beam and the shear forces in the beam, and relationships between the shear forces and bending moments in the beam.

Consider the beam loaded and supported as shown in Fig. 8-22a. At some location $x$, the beam is acted on by a distributed load $w$, a concentrated load $P$,

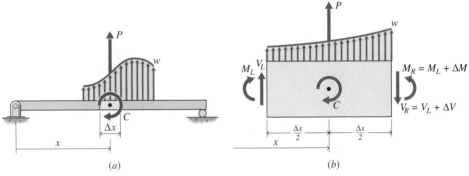

**Figure 8-22**

and a concentrated couple $C$. A free-body diagram of a segment of the beam centered at the location $x$ is shown in Fig. 8-22$b$. The upward direction is considered positive for the applied loads $w$ and $P$; the shear forces and the bending moments are shown in the positive directions of the sign convention established earlier; and $\Delta x$, $\Delta V$, and $\Delta M$ may be large or small. The element must be in equilibrium, and force equilibrium gives

$$+ \uparrow \Sigma F_y = 0: \qquad V_L + w_{\text{avg}}\Delta x + P - (V_L + \Delta V) = 0$$

from which

$$\Delta V = P + w_{\text{avg}}\Delta x \qquad\qquad (8\text{-}13a)$$

Four important relationships are obtained from Eq. (8-13a).

1. If the concentrated force $P$ and the distributed force $w$ are both zero in some region of the beam ($\Delta x$ large or small), then

$$\Delta V = 0 \qquad \text{or} \qquad V_L = V_R \qquad\qquad (8\text{-}13b)$$

That is, in any segment of a beam where there are no loads, the shear force is constant.

2. If the concentrated load $P$ is not zero, then in the limit as $\Delta x \to 0$,

$$\Delta V = P \qquad \text{or} \qquad V_R = V_L + P \qquad\qquad (8\text{-}13c)$$

That is, across any concentrated load $P$, the shear force graph (shear force versus $x$) jumps by the amount of the concentrated load. Furthermore, moving from left to right along the beam, the shear force graph jumps in the direction of the concentrated load.

3. If the concentrated load $P$ is zero, then in the limit as $\Delta x \to 0$,

$$\Delta V = w_{\text{avg}}\Delta x \to 0$$

and the shear force is a continuous function at $x$. Dividing through by $\Delta x$ then gives

$$\lim_{\Delta x \to 0} \frac{\Delta V}{\Delta x} = \frac{dV}{dx} = w \qquad\qquad (8\text{-}13d)$$

That is, the slope of the shear force graph at any location (section) $x$ in the beam is equal to the intensity of loading at that section of the beam. Moving from left to right along the beam, if the distributed force is upward, then the slope of the shear force graph ($dV/dx = w$) is positive and the shear force graph is increasing (moving upward). If the distributed force is zero, then the slope of the shear force graph ($dV/dx = 0$) is zero and the shear force is constant.

4. In any region of the beam in which Eq. (8-13d) is valid (any region in which there are no concentrated loads), the equation can be integrated between definite limits to obtain

$$V_2 - V_1 = \int_{V_1}^{V_2} dV = \int_{x_1}^{x_2} w \, dx \qquad\qquad (8\text{-}13e)$$

That is, for any segment of the beam acted on by a distributed load $w$ and no concentrated load ($P = 0$), the change in shear force between sections at $x_1$ and $x_2$ is equal to the area under the load diagram between the two sections.

Similarly, applying moment equilibrium to the free-body diagram of Fig. 8-22$b$ gives

$$+ \circlearrowleft \ \Sigma M_{center} = 0: \qquad (M_L + \Delta M) - M_L - C - V_L \frac{\Delta x}{2}$$

$$-(V_L + \Delta V) \frac{\Delta x}{2} + a(w_{avg} \Delta x) = 0$$

or

$$\Delta M = C + V_L \, \Delta x + \Delta V \frac{\Delta x}{2} - a \, (w_{avg} \, \Delta x) \qquad (8\text{-}14a)$$

in which $\dfrac{-\Delta x}{2} < a < \dfrac{\Delta x}{2}$ and in the limit as $\Delta x \to 0$, $a \to 0$, and $w_{avg} \to w$. Three more important relationships are obtained from Eq. (8-14a).

5. If the concentrated couple $C$ is not zero, then in the limit as $\Delta x \to 0$,

$$\Delta M = C \qquad \text{or} \qquad M_R = M_L + C \qquad (8\text{-}14b)$$

That is, across any concentrated couple $C$, the bending moment graph (bending moment versus $x$) jumps by the amount of the concentrated couple. Furthermore, moving from left to right along the beam, the bending moment graph jumps upward for a clockwise concentrated couple and jumps downward for a counterclockwise concentrated couple.

6. If the concentrated couple $C$ and the concentrated load $P$ are both zero,[4] then in the limit as $\Delta x \to 0$,

$$\Delta M = V_L \Delta x + \Delta V \frac{\Delta x}{2} - a(w_{avg} \Delta x) \to 0$$

and the bending moment is a continuous function of $x$. Dividing through by $\Delta x$ then gives

$$\lim_{\Delta x \to 0} \frac{\Delta M}{\Delta x} = \frac{dM}{dx} = V \qquad (8\text{-}14c)$$

That is, the slope of the bending moment graph at any location (section) $x$ in the beam is equal to the value of the shear force at that section of the beam.

---

[4]If the concentrated force $P$ is not zero, the bending moment will still be continuous at $x$ but it will not be continuously differentiable at $x$. For a point slightly to the left of $P$, $dM/dx = V_L$ while for a point slightly to the right of $P$, $dM/dx = V_R$.

Moving from left to right along the beam, if the shear force is positive, then $dM/dx = V$ is positive and the bending moment graph is increasing.

7. In any region of the beam in which Eq. (8-14c) is valid (any region in which there are no concentrated loads or couples), the equation can be integrated between definite limits to obtain

$$M_2 - M_1 = \int_{M_1}^{M_2} dM = \int_{x_1}^{x_2} V \, dx \qquad (8\text{-}14d)$$

That is, for any segment of the beam in which the shear force is continuous ($C = P = 0$), the change in bending moment between sections at $x_1$ and $x_2$ is equal to the area under the shear force graph between the two sections.

Note that Eqs. (8-13) and (8-14) were derived with the $x$-axis positive to the right, the applied loads positive upward, and the shear force and bending moment with signs as indicated in Fig. 8-19. If one or more of these assumptions are changed, the algebraic signs in Eqs. (8-13) and (8-14) may need to be altered.

Equations (8-13) and (8-14) can be used to draw the shear and bending moment graphs and to compute values of shear force and bending moment at various sections along a beam.

## Shear Force and Bending Moment Diagrams
Shear force and bending moment diagrams provide a convenient method for obtaining maximum values of shear force and bending moment. A shear force diagram is a graph in which abscissas represent distances along the beam and ordinates represent the transverse shear force at the corresponding sections. A bending moment diagram is a graph in which abscissas represent distances along the beam and ordinates represent the bending moment at the corresponding sections.

Shear force and bending moment diagrams can be drawn by calculating values of shear force and bending moment at various sections along the beam and plotting enough points to obtain a smooth curve. Such a procedure is rather time-consuming; therefore, other more rapid methods will be developed using the load, shear force, and bending moment relationships developed in this section.

A convenient arrangement for constructing shear force and bending moment diagrams is to draw a free-body diagram of the entire beam and construct shear force and bending moment diagrams directly below. Two methods of procedure are used.

The first method consists of writing algebraic equations for the shear force $V$ and the bending moment $M$, and constructing curves from the equations. This method has the disadvantage that unless the load is uniformly distributed or varies according to a known equation along the entire beam, no single elementary expression can be written for the shear force $V$ or the bending moment $M$ that applies to the entire length of the beam. Instead, it is necessary to divide the beam into intervals bounded by the abrupt changes in the loading. An origin should be selected, positive directions should be shown for the coordinate axes, and the limits of the abscissa (usually $x$) should be indicated for each interval.

Complete shear force and bending moment diagrams should indicate values of shear force and bending moment at each section where the load changes abruptly and at sections where they are maximum or minimum (negative maximum values). Sections where the shear force and bending moment are zero should also be located.

The second method consists of drawing the shear force diagram from the load diagram and the bending moment diagram from the shear force diagram by using Eqs. (8-13) and (8-14). This latter method, though it may not produce a precise curve, is less time-consuming than the first and does provide the information usually required.

When all loads and reactions are known, the shear force and bending moment at the ends of the beam can be determined by inspection. Both shear force and bending moment are zero at the free end of a beam unless a force or a couple is applied there, in which case, the shear force is the same as the force and the bending moment the same as the couple. At a simply supported or pinned end, the shear force must equal the end reaction and the bending moment must be zero. At a built-in or fixed end, the reactions are the shear force and the bending moment values.

Once a starting point for the shear force diagram is established, the diagram can be sketched by using the definition of shear force and the fact that the slope of the shear force diagram can be obtained from the load diagram. When positive directions are chosen as upward and to the right, a positive distributed load (acting upward) produces a positive slope on the shear force diagram. Similarly, a negative load (acting downward) produces a negative slope on the shear force diagram. A concentrated force produces an abrupt change in shear force. The change in shear force between any two sections is given by the area under the load diagram between the same two sections. The change in shear force at a concentrated force is equal to the concentrated force.

A bending moment diagram is drawn from the shear force diagram in the same manner. The slope at any point on the bending moment diagram is given by the shear force at the corresponding point on the shear force diagram, a positive shear force produces a positive slope, and a negative shear force produces a negative slope, when upward and to the right are positive. The change in the bending moment between any two sections is given by the area under the shear force diagram between the two sections. A concentrated couple applied to a beam at any section will cause the bending moment at the section to change abruptly by an amount equal to the moment of the couple.

The choice of which method to use depends on the type of information needed. If only the maximum values of shear force or bending moment are needed, then the second method usually gives these values more easily than the first. If equations of the bending moment are needed (they will be needed in Chapter 9 for finding the deflected shape of the beam), then the equilibrium approach must be used.

Example Problems 8-9 and 8-10 illustrate the two methods for drawing shear force and bending moment diagrams.

## ■ Example Problem 8-9

A beam is loaded and supported as shown in Fig. 8-23a.

(a) Write equations for the shear force and the bending moment for any section of the beam in the interval *AB*.

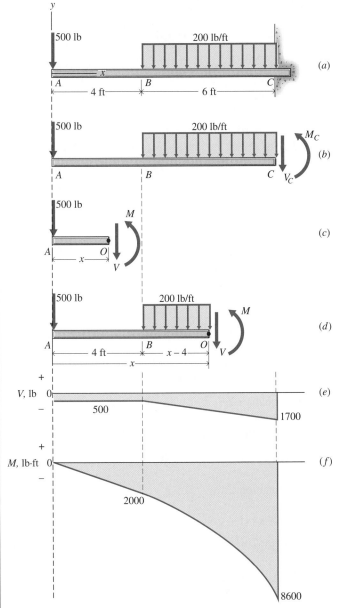

**Figure 8-23**

(b) Write equations for the shear force and the bending moment for any section of the beam in the interval $BC$.

(c) Draw complete shear force and bending moment diagrams for the beam.

## SOLUTION

A free-body diagram for the beam is shown in Fig. 8-23$b$. It is not necessary to compute the reactions on a cantilever beam in order to write shear force and bending moment equations or to draw shear force and bending moment diagrams; however, the reactions provide a convenient check. Thus,

$$+\uparrow \ \Sigma F_y = 0: \qquad\qquad -500 - 200(6) - V_C = 0$$

$$+\downarrow \ \Sigma M_C = 0: \qquad\qquad 500(10) + 200(6)(3) + M_C = 0$$

from which

$$V_C = -1700 \ \text{lb} \qquad \text{and} \qquad M_C = -8600 \ \text{lb} \cdot \text{ft}$$

(a) A free-body diagram of a portion of the beam from the left end to any section between $A$ and $B$ is shown in Fig. 8-23c. The resisting shear force $V$ and the resisting moment $M$ are shown as positive values.

From the equilibrium equation $\Sigma F_y = 0$,

$$+\uparrow \ \Sigma F_y = 0: \qquad\qquad -500 - V = 0$$

$$V = -500 \ \text{lb} \qquad 0 < x < 4 \ \text{ft} \qquad\qquad \textbf{Ans.}$$

From the equilibrium equation $\Sigma M_O = 0$,

$$+\downarrow \ \Sigma M_O = 0: \qquad\qquad 500(x) + M = 0$$

$$M = -500x \ \text{lb} \cdot \text{ft} \qquad 0 < x < 4 \ \text{ft} \qquad\qquad \textbf{Ans.}$$

(b) A free-body diagram of a portion of the beam from the left end to any section between $B$ and $C$ is shown in Fig. 8-23d.

From the equilibrium equation $\Sigma F_y = 0$,

$$+\uparrow \ \Sigma F_y = 0: \qquad\qquad -500 - 200(x - 4) - V = 0$$

$$V = 300 - 200x \ \text{lb} \qquad 4 < x < 10 \ \text{ft} \qquad\qquad \textbf{Ans.}$$

From the equilibrium equation $\Sigma M_O = 0$,

$$+\downarrow \ \Sigma M_O = 0: \qquad\qquad 500(x) + 200(x - 4)(x - 4)/2 + M = 0$$

$$M = -100x^2 + 300x - 1600 \ \text{lb} \cdot \text{ft} \qquad 4 < x < 10 \ \text{ft} \qquad\qquad \textbf{Ans.}$$

(c) The equations for $V$ can be plotted in the appropriate intervals to give the shear force diagram of Fig. 8-23e. Likewise, the equations for $M$ can be plotted to give the bending moment diagram of Fig. 8-23f.

The shear force diagram can also be drawn using the load diagram of Fig. 8-23b and the relationships (1) through (4) developed earlier in this section. There is no need to write shear force and bending moment equations using this approach.

Using the results developed in (2), the shear force diagram jumps at the concentrated force at $A$; the magnitude of the jump is the magnitude of the concentrated force, 500 lb, and the jump is in the direction of the concentrated force, that is, downward.

Between $A$ and $B$ there is no distributed load, and the relationships in (1) and (3) indicate that the slope of the shear force diagram is zero, that is, $dV/dx = w = 0$. Therefore, the diagram between $A$ and $B$ is horizontal at a value of $-500$ lb.

The distributed load between $B$ and $C$ is constant with a value of $-200$ lb/ft. The results developed in (3) show that the slope of the shear force diagram between $B$ and $C$ has a negative slope of $-200$ lb/ft and is constant. Thus the diagram is a straight line. The value of the shear force at $C$ is found using the results of (4). The change in shear force between $B$ and $C$ is equal to the area under the load diagram between $B$ and $C$. That is, $\Delta V = (6)(-200)$, and since the shear force at $B$ is $-500$ lb, the shear force at $C$ is

$$V_C = V_B + \Delta V = -500 + (6)(-200) = -1700 \text{ lb}$$

This checks with the value of the shear force at $C$ from the load diagram of Fig. 8-23$b$. The complete shear force diagram is shown in Fig. 8-23$e$.

The bending moment diagram can be drawn using the shear force diagram and the results developed in (5) through (7). At $A$ the bending moment is zero. Between $A$ and $B$, the results developed in (6) show that the slope of the bending moment diagram is constant, that is, $dM/dx = V = -500$ lb · ft/ft. Therefore, the bending moment diagram between $A$ and $B$ is a straight line. The value of the bending moment at $B$ is found using the results of (7). The change in bending moment between $A$ and $B$ is equal to the area under the shear force diagram between $A$ and $B$. That is, $\Delta M = (4)(-500)$ lb · ft, and since the bending moment at $A$ is zero,

$$M_B = M_A + \Delta M = 0 + (4)(-500) = -2000 \text{ lb} \cdot \text{ft}$$

The results of (6) indicate that between $B$ and $C$ the slope of the bending moment diagram is $dM/dx = V$. In this region $V$ is negative and the magnitude of $V$ gets larger and larger. Therefore, the slope of the bending moment diagram is negative and is increasing in magnitude as $V$ gets larger. The shape of the bending moment diagram in this region is shown in Fig. 8-23$f$.

The value of the bending moment at $C$ is found using the relationship developed in (7). The change in bending moment between $B$ and $C$ is equal to the area under the shear force diagram between $B$ and $C$. That is, $\Delta M = (-500 - 1700)(6)/2 = -6600$ lb · ft, and since the bending moment at $B$ is $-2000$ lb · ft,

$$M_C = M_B + \Delta M = -2000 - 6600 = -8600 \text{ lb} \cdot \text{ft}$$

This checks with the value of the concentrated moment at $C$ in Fig. 8-23$b$ and the results developed in (5). The complete bending moment diagram is shown in Fig. 8-23$f$. ∎

## Example Problem 8-10

A beam is loaded and supported as shown in Fig. 8-24$a$.

(a) Write equations for the shear force and the bending moment for any section of the beam in the interval $CD$.

(b) Draw complete shear force and bending moment diagrams for the beam.

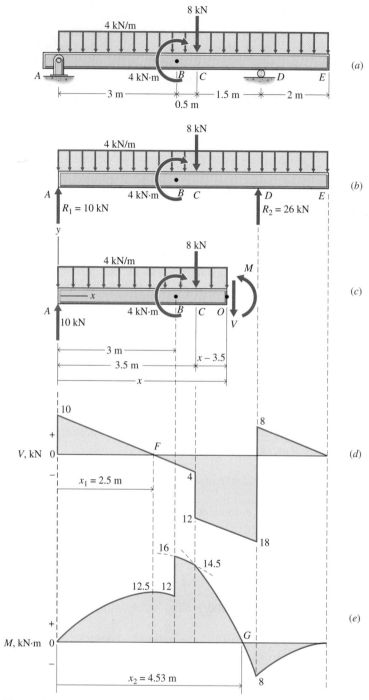

**Figure 8-24**

## SOLUTION

The reactions are determined by using the free-body diagram of the beam shown in Fig. 8-24b. From the equilibrium equations,

$$+\uparrow \Sigma M_D = 0: \qquad R_1(5) + 4 - 4(7)(1.5) - 8(1.5) = 0$$

$$+\downarrow \Sigma M_A = 0: \qquad R_2(5) - 4 - 4(7)(3.5) - 8(3.5) = 0$$

from which

$$R_1 = 10 \text{ kN} \quad \text{and} \quad R_2 = 26 \text{ kN}$$

(a) A free-body diagram of a portion of the beam from the left end to any section between $C$ and $D$ is shown in Fig. 8-24c.

From the equilibrium equation $\Sigma F_y = 0$,

$$+\uparrow \Sigma F_y = 0: \quad 10 - 4(x) - 8 - V = 0$$
$$V = -4x + 2 \text{ kN} \quad 3.5 < x < 5 \text{ m} \qquad \textbf{Ans.}$$

From the equilibrium equation $\Sigma M_O = 0$,

$$+\circlearrowleft \Sigma M_O = 0: \quad -10(x) + 4(x)(x/2) - 4 + 8(x - 3.5) + M = 0$$
$$M = -2x^2 + 2x + 32 \text{ kN} \cdot \text{m} \quad 3.5 < x < 5 \text{ m} \qquad \textbf{Ans.}$$

(b) The equations for $V$ and $M$ in the other intervals can be written in the same manner, and the shear force and the bending moment diagrams (Figs. 8-24d and e) can be obtained by plotting these equations. In this example, the shear force diagram will be drawn directly from the load diagram (Fig. 8-24b). The shear force just to the right of $A$ is 10 kN. From $A$ to $C$ the shear force decreases at a constant rate of 4 kN/m. Thus, the shear force just to the left of $C$ is

$$V_C = V_A + \Delta V = 10 - 4(3.5) = -4 \text{ kN}$$

The concentrated downward load at $C$ causes the shear force to change suddenly from $-4$ kN just to the left of $C$ to $-12$ kN just to the right of $C$. From $C$ to $D$ the shear force continues to decrease at a constant rate of 4 kN/m. Thus, the shear force just to the left of $D$ is

$$V_D = V_C + \Delta V = -12 - 4(1.5) = -18 \text{ kN}$$

The reaction at $D$ causes the shear force to change suddenly from $-18$ kN just to the left of $D$ to $+8$ kN just to the right of $D$. From $D$ to $E$ the shear force continues to decrease at a constant rate of 4 kN/m. Thus, the shear force at $E$ is

$$V_E = V_D + \Delta V = +8 - 4(2) = 0$$

Since the distributed load is uniform over the entire beam, the slope of the shear force diagram is constant. Points of zero shear, such as point $F$ in Fig. 8-24d, are located from the geometry of the shear force diagram. For example, the slope of the shear force diagram is 4 kN/m. Therefore,

$$x_1 = 10/4 = 2.5 \text{ m}$$

The moment is zero at $A$, and the slope of the bending moment diagram (equal to the shear force) is 10 kN $\cdot$ m/m. From $A$ to $C$ the shear force and, hence, the slope of the bending moment diagram decreases uniformly to zero at $F$ and to $-4$ at $C$. The abrupt change of shear force at $C$ indicates a sudden change of slope of the bending moment diagram; thus, the two parts of the bending moment diagram at $C$ are not tangent. The slope of the diagram changes from $-12$ at $C$ to $-18$ just to the left of $D$. From $D$ to $E$ the slope changes from $+8$ to zero. The change of moment from $A$ to $F$ is equal to the area under the shear force diagram from $A$ to $F$; therefore,

$$M_F = M_A + \Delta M = 0 + 10(2.5)/2 = 12.5 \text{ kN} \cdot \text{m}$$

Similarly,

$$V_B = V_A + \Delta V = 10 - 4(3) = -2 \text{ kN}$$

$$M_B = M_F + \Delta M = 12.5 - 2(0.5)/2 = 12.0 \text{ kN} \cdot \text{m}$$

The 4 kN $\cdot$ m concentrated couple is applied at $B$. Since the bending moment just to the left of $B$ is positive and the couple contributes an additional positive moment to sections to the right of $B$, the bending moment changes abruptly from $+12$ kN $\cdot$ m to $+16$ kN $\cdot$ m. In a similar manner bending moments at $C$, $D$, and $E$ are determined from the shear force diagram areas as

$$M_C = M_B + \Delta M = 16.0 - (4 + 2)(0.5)/2 = 14.5 \text{ kN} \cdot \text{m}$$

$$M_D = M_C + \Delta M = 14.5 - (12 + 18)(1.5)/2 = -8.0 \text{ kN} \cdot \text{m}$$

$$M_E = M_D + \Delta M = -8.0 + 8(2)/2 = 0$$

There is no moment applied at end $E$ of the beam, and if the bending moment at $E$ is not zero, it indicates that an error has occurred. The point $G$, where the bending moment is zero, can be determined by setting the expression for the bending moment from part $a$ equal to zero and solving for $x$. The result is

$$x_2 = 0.5 \pm \sqrt{16.25} = 4.531 \text{ m} \cong 4.53 \text{ m} \qquad \textbf{Ans.}$$

Note in this example that maximum and minimum bending moments may occur at sections where the shear force diagram passes through zero. In general, the shear force diagram may pass through zero at a number of points along the beam, and each such crossing indicates a point of possible maximum bending moment (in engineering, the bending moment with the largest absolute value is the maximum bending moment). It should be emphasized that the shear force diagram does not indicate the presence of abrupt discontinuities in the bending moment curve; hence, the maximum bending moment may occur where a couple is applied to the beam, rather than where the shear force diagram passes through zero. All possibilities should be examined to determine the maximum bending moment.

Since the flexural stress is zero at sections where the bending moment is zero, if a beam must be spliced, the splice should be located at or near such a section. ∎

## PROBLEMS

*Note*: In problems involving rolled shapes (see Appendix A), the beam is so oriented that the maximum section modulus applies.

### Introductory Problems

**8-61\*** Draw complete shear force and bending moment diagrams for the beam shown in Fig. P8-61.

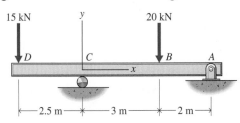

**Figure P8-61**

**8-62\*** Draw complete shear force and bending moment diagrams for the beam shown in Fig. P8-62.

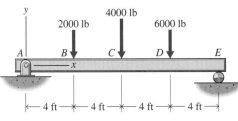

**Figure P8-62**

**8-63** Draw complete shear force and bending moment diagrams for the beam shown in Fig. P8-63.

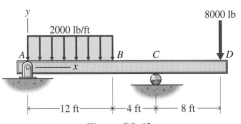

**Figure P8-63**

**8-64** Draw complete shear force and bending moment diagrams for the beam shown in Fig. P8-64.

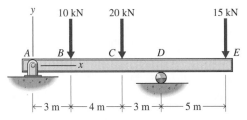

**Figure P8-64**

**8-65** Draw complete shear force and bending moment diagrams for the beam shown in Fig. P8-65.

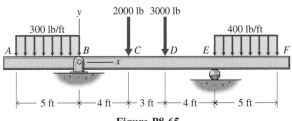

**Figure P8-65**

**8-66\*** Draw complete shear force and bending moment diagrams for the beam shown in Fig. P8-66.

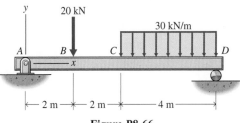

**Figure P8-66**

**8-67\*** Draw complete shear force and bending moment diagrams for the beam shown in Fig. P8-67.

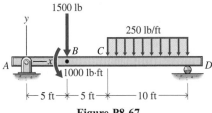

**Figure P8-67**

**8-68** Draw complete shear force and bending moment diagrams for the beam shown in Fig. P8-68.

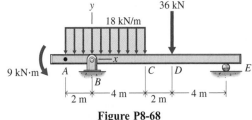

**Figure P8-68**

### Intermediate Problems

**8-69\*** The beam shown in Fig. P8-69*a* has the cross section shown in Fig. P8-69*b*. Determine the maximum tensile and compressive flexural stresses in the beam.

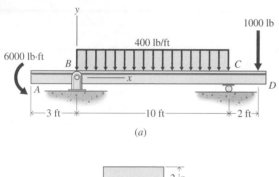

(a)

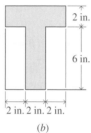

(b)

**Figure P8-69**

**8-70\*** A W102 × 19 wide-flange beam is loaded and supported as shown in Fig. P8-70. Determine the maximum tensile and compressive flexural stresses in the beam.

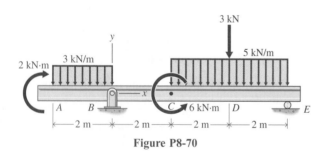

**Figure P8-70**

**8-71** A beam is loaded and supported as shown in Fig. P8-71a. Two 10- × 1-in. steel plates are welded to the flanges of an S18 × 70 American standard I-beam to form the cross section shown in Fig. P8-71b. Determine the maximum tensile and compressive flexural stresses in the beam.

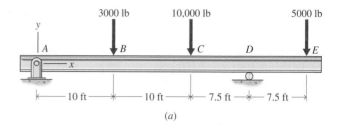

(a)

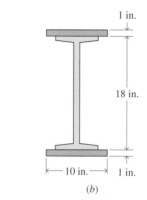

(b)

**Figure P8-71**

**8-72** A beam is loaded and supported as shown in Fig. P8-72a. Two 250- × 25-mm steel plates and two C254 × 45 channels are welded together to form the cross section shown in Fig. P8-72b. Determine the maximum tensile and compressive flexural stresses in the beam.

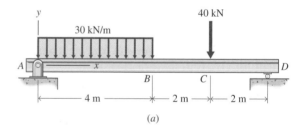

(a)

(b)

**Figure P8-72**

**8-73\*** A WT8 × 25 structural steel T-section is loaded and supported as shown in Fig. P8-73. Determine the maximum tensile and compressive flexural stresses in the beam.

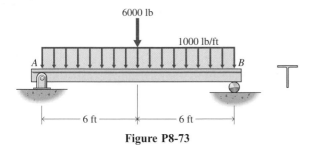

**Figure P8-73**

**8-74**  A 203- × 203-mm nominal size structural timber (see Appendix A) is supported by two brick columns, as shown in Fig. P8-74. Assume that the brick columns transmit only vertical forces to the timber beam. The beam supports the roof of a building through three timber columns. Columns *A* and *C* each transmit forces of 8 kN to the beam; column *B* transmits a force of 10 kN.

(a) Draw complete shear force and bending moment diagrams for the beam.

(b) Determine the maximum tensile and compressive flexural stresses in the beam.

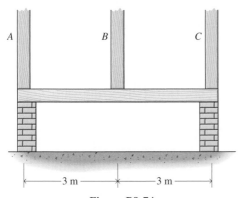

**Figure P8-74**

## Challenging Problems

**8-75\***  Draw complete shear force and bending moment diagrams for the beam shown in Fig. P8-75.

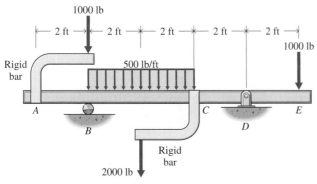

**Figure P8-75**

**8-76\***  An S457 × 81 American standard steel beam (see Appendix A) is loaded and supported as shown in Fig. P8-76. The two segments of the beam are connected with a smooth pin at *D*.

(a) Draw complete shear force and bending moment diagrams for the beam.

(b) Determine the maximum tensile and compressive flexural stresses in the beam.

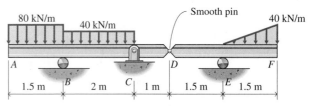

**Figure P8-76**

**8-77**  An S15 × 50 American standard steel beam (see Appendix A) is loaded and supported as shown in Fig. P8-77. The total length of the beam is 15 ft. If the flexural stress is limited to 15 ksi, determine the maximum permissible value for the distributed load *w*.

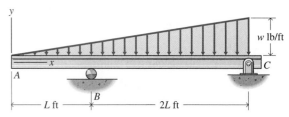

**Figure P8-77**

**8-78**  Draw complete shear force and bending moment diagrams for segments *AB* and *CD* of the structure shown in Fig. P8-78.

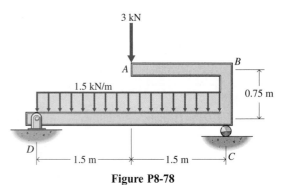

**Figure P8-78**

**8-79\***  Three members are connected with smooth pins to form the frame shown in Fig. P8-79. The weights of the members are negligible. Member *CD* has a 3- × 3-in. cross section, and member *BE* has a 1- × 1-in. cross section. The pin at $A$ has a $1/2$-in. diameter and is in double shear.

(a) Draw complete shear force and bending moment diagrams for member $CD$.

(b) Determine the maximum tensile and compressive flexural stresses in the member $CD$.

(c) Determine the normal stress in member $BE$.

(d) Determine the shearing stress in the pin at $A$.

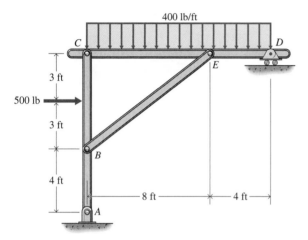

Figure P8-79

**8-80*** A body with a mass of 1500 kg is supported by a roller on an I-beam, as shown in Fig. P8-80. The roller moves slowly along the beam, thereby causing the shear force $V$ and the bending moment $M$ to be functions of the position $b$.

(a) Draw complete shear force and bending moment diagrams when the roller is at position $b$.

(b) Determine the position of the roller when the bending moment is maximum.

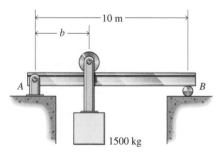

Figure P8-80

**8-81** Member $AB$ supports a 55-lb sign, as shown in Fig. P8-81. Determine

(a) The maximum tensile flexural stress in the $1/2$-in. nominal diameter standard steel pipe $AB$ (see Appendix A).

(b) The normal stress in the $3/16$-in.-diameter wire $BC$.

(c) The shearing stress in the $1/4$-in.-diameter pin at $A$, which is in double shear.

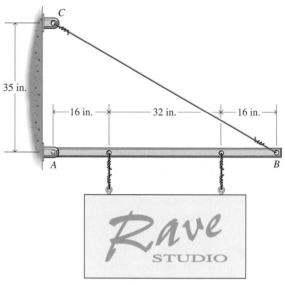

Figure P8-81

**8-82** Two beams $AD$ and $EH$ are spliced, as shown in Fig. P8-82. Draw complete shear force and bending moment diagrams for the beams $AD$, $CF$, and $EH$.

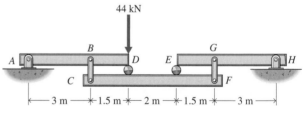

Figure P8-82

**8-83** A tractor is moving slowly over a bridge, as shown in Fig. P8-83. The forces exerted on one beam of the bridge by the tractor are 4050 lb by the rear wheels and 1010 lb by the front wheels. Determine the position $b$ of the tractor for which the bending moment in the beam is maximum.

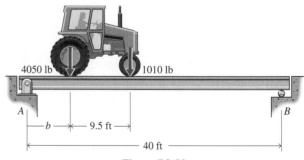

Figure P8-83

# 8-8 SHEARING STRESSES IN BEAMS

The discussion of shearing stresses in beams was delayed while flexural stresses were studied in Section 8-3. This procedure seems to be in keeping with the historical record on the study of beam stresses. From the time of Coulomb's paper, which contained the correct theory of the distribution of flexural stresses, approximately seventy years elapsed before the Russian engineer D. J. Jourawski (1821–1891), while designing timber railroad bridges in 1844–1850, developed the elementary shear stress theory used today. In 1856, Saint-Venant developed a rigorous solution for shearing stresses in beams; however, the elementary solution of Jourawski is the one in general use today by engineers and architects because it yields adequate results and is much easier to apply. The method requires use of the elastic flexure formula in its development; therefore, the formula developed is limited to elastic action. The shearing stress evaluation discussed in this section is used

1. For timber beams, because of their longitudinal plane of low shear resistance.
2. In design codes for shear stresses in the webs of I-beams.

If one constructs a beam by stacking flat slabs one on top of another without fastening them together, and then loads this beam in a direction normal to the surface of the slabs, the resulting deformation will appear somewhat like that in Fig. 8-25a. This same type of deformation can be observed by taking a pack of cards and bending them, and noting the relative motion

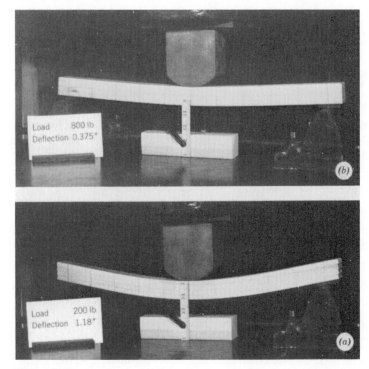

**Figure 8-25**

of the ends of the cards with respect to each other. The fact that a solid beam does not exhibit this relative movement of longitudinal elements (see Fig. 8-25b, in which the beam is identical to that of Fig. 8-25a except that the layers are glued together) indicates the presence of shearing stresses on longitudinal planes. Evaluation of these shearing stresses will be determined by means of equilibrium and the free-body diagrams of a short portion of a beam with a rectangular cross section.

A segment of a beam of length $\Delta x$ is shown in Fig. 8-26a. Acting on transverse sections of the beam are the forces due to normal stresses (the distribution of normal stresses is shown in Fig. 8-26a) and the resultant shear forces due to shearing stresses ($V_L$ and $V_R$ in Fig. 8-26a). The normal stresses vary linearly with distance $y$ from the neutral surface and are constant in the $z$-direction. Thus, Fig. 8-26a is redrawn as shown in Fig. 8-26b. A horizontal cut of length $\Delta x$ is made at a distance $y_1$ above (or below) the neutral surface, and a segment of the beam between $y_1$ and $c$ (distance from the neutral surface to the top, or bottom, of the beam) is shown in Fig. 8-26c. A shear force $V_H$ acts over an area of length $\Delta x$ and thickness $t$. Forces $F_1$ and $F_2$ are due to normal stresses, and are given by $F = \int dF = \int \sigma \, dA = \int \sigma (t \, dy)$ where $dA = t \, dy$ as shown in Fig. 8-26d. The normal (flexural) stress $\sigma$ at a distance $y$ from the neutral surface is given by the expression $\sigma = -My/I$. Therefore, the resultant normal force $F_1$ on the left end of the segment from $y_1$ to the top of the beam is[5]

$$F_1 = -\frac{M}{I} \int_A y \, dA = -\frac{M}{I} \int_{y_1}^{c} y \, (t \, dy)$$

Similarly, the resultant force $F_2$ on the right side of the element is

$$F_2 = -\frac{(M + \Delta M)}{I} \int_{y_1}^{c} y \, (t \, dy)$$

These forces are shown on the free-body diagram of Fig. 8-26c. Also shown on the free-body diagram of Fig. 8-26c are the resultants of the vertical shear stresses on the left $V_L'$ and right $V_R'$ sides of the element and the resultant of the horizontal shear stress $V_H$ on the bottom of the element. A summation of forces in the horizontal direction yields

$$V_H = F_2 - F_1 = -\frac{\Delta M}{I} \int_{y_1}^{c} y \, (t \, dy)$$

The average shearing stress $\tau_{\text{avg}}$ is the horizontal shear force $V_H$ divided by the horizontal shear area $A_s = t \, \Delta x$ between sections $A$ and $B$. Thus,

$$\tau_{\text{avg}} = \frac{V_H}{A_s} = -\frac{\Delta M}{It \, \Delta x} \int_{y_1}^{c} y \, (t \, dy)$$

---

[5]If $M$ is positive, then the normal stress above the neutral axis (where $y$ is positive) will be negative (compression) and the force $F_1$ will be a compressive force. If $M$ is negative, then the normal stress above the neutral axis will be positive (tension) and the force $F_1$ will be a tensile force as drawn.

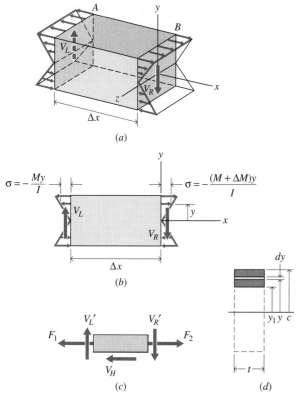

**Figure 8-26**

In the limit as $\Delta x \to 0$

$$\tau = \lim_{\Delta x \to 0} \frac{\Delta M}{\Delta x} \left(-\frac{1}{It}\right) \int_{y_1}^{c} ty \, dy = \frac{dM}{dx} \left(-\frac{1}{It}\right) \int_{y_1}^{c} ty \, dy \qquad (a)$$

The shear force $V$ at the beam section where the stress is to be evaluated is given by Eq. (8-14c) as $V = dM/dx$. The integral of Eq. (a) is the first moment of the portion of the cross-sectional area between the transverse line where the stress is to be evaluated ($y = y_1$) and the extreme fiber of the beam ($y = c$). This integral is designated $Q$, and when values of $V$ and $Q$ are substituted into Eq. (a), the formula for the horizontal (or longitudinal) shearing stress becomes

$$\tau_H = -\frac{VQ}{It} \qquad (b)$$

The minus sign in Eq. (b) is needed to satisfy Eq. (8-1a) and is consistent with the sign convention for shearing stresses (Fig. 8-4c). At each point in the beam, the horizontal (longitudinal) and vertical (transverse) shearing stresses have the same magnitude ($\tau_{xy} = \tau_{yx}$); hence, Eq. (b) also gives the vertical shearing stress at a point in a beam (averaged across the width).[6] For the balance of this chap-

---

[6]If the shear force $V$ is positive (downward on section $B$), then the horizontal shear stress will be negative (to the right on the bottom of the element) and the vertical shear stress will also be negative (downward on section $B$—in the same direction as the shear force).

ter, magnitudes of $V$ and $Q$ will be used to determine the magnitude of the shearing stress $\tau$ and Eq. (b) will be written as

$$\tau = \frac{VQ}{It} \tag{8-15}$$

The sense of the stress $\tau$ will be determined from the sense of the shear force $V$ on transverse planes and from $\tau_{xy} = \tau_{yx}$ on longitudinal planes.

Because the flexure formula was used in the derivation of Eq. (8-15), it is subject to the same assumptions and limitations as the flexure formula. Although the stress given by Eq. (8-15) is associated with a particular point in a beam, it is averaged across the thickness $t$ and hence is accurate only if $t$ is not too great.

The variation of shearing stress on a transverse cross section of a beam will be demonstrated by using the rectangular cross section shown in Fig. 8-27$a$. The transverse shearing stress at any point of the section at a distance $y_1$ from the neutral axis is from Fig. 8-27$b$ and Eq. (8-15),

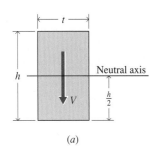

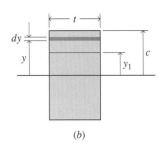

$(a)$

$(b)$

$$\tau = \frac{VQ}{It} = \frac{V}{It}\int_A y\, dA = \frac{V}{It}\int_{y_1}^{c} y\, t\, dy$$

$$= \frac{V}{I}\int_{y_1}^{h/2} y\, dy = \frac{V}{2I}\left[\left(\frac{h}{2}\right)^2 - y_1^2\right] \tag{c}$$

Equation (c) indicates that the transverse shearing stress on a rectangular cross section has a parabolic distribution, as shown in Fig. 8-27$c$. The shearing stress acts in the direction of the shear force $V$ that produces the stress. The maximum shearing stress occurs when $y_1 = 0$ (at the neutral axis) and has a magnitude

$$\tau_{\max} = \frac{Vh^2}{8I} = \frac{Vh^2}{8(th^3/12)} = \frac{3}{2}\frac{V}{th} = \frac{3}{2}\frac{V}{A} \tag{8-16}$$

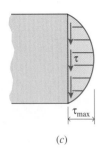

$(c)$

**Figure 8-27**

where $A$ is the area of the transverse cross section.

The maximum shearing stress given by Eq. (8-16) exists on both the transverse plane and the longitudinal plane along the neutral surface. This equation is useful in the design of timber beams with rectangular cross sections because timber has a low shearing strength parallel to the grain.

Equation (8-16) is valid for a rectangular beam and should not be used for other sections. For a rectangular section the maximum shearing stress is 1.5 times the average shearing stress ($\tau_{\text{avg}} = V/A$). For a rectangular section having a depth twice the width, the maximum stress as computed by Saint-Venant's more rigorous method is about 3% greater than that given by Eq. (8-16). If the beam is square, the error is about 12%. If the width is four times the depth, the error is almost 100%, from which one must conclude that, if Eq. (8-15) were applied to a point in the flange of an I-beam or T-section, the result would be worthless. Furthermore, if Eq. (8-15) is applied to sections where the sides of the beam are not parallel, such as a triangular section, the average transverse shearing stress is subject to additional error because the variation of transverse shearing stress is greater when the sides are not parallel.

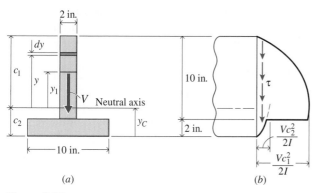

**Figure 8-28**

A second illustration of the variation of shearing stress on a transverse section of a beam will be demonstrated by using the inverted T-shaped beam shown in Fig. 8-28a. For this section,

$$y_C = \frac{2(10)(7) + 10(2)(1)}{2(10) + 10(2)} = 4 \text{ in.}$$

$$c_1 = 8 \text{ in.} \quad \text{and} \quad c_2 = 4 \text{ in.}$$

Note that in Eq. (8-15), $V$ and $I$ are constant for any section, and only $Q$ and $t$ vary for different points in the section. The transverse shearing stress at any point in the stem of the section a distance $y_1$ from the neutral axis is from Fig. 8-28a and Eq. (8-15),

$$\tau = \frac{VQ}{It} = \frac{V}{It} \int_{y_1}^{c_1} yt \, dy$$

$$= \frac{V}{2I}(c_1^2 - y_1^2) = \frac{V}{2I}(8^2 - y_1^2) \quad (-2 < y_1 < 8)$$

An expression for the average shearing stress in the flange can be written in a similar manner and is

$$\tau = \frac{V}{2I}(c_2^2 - y_1^2) = \frac{V}{2I}(4^2 - y_1^2) \quad (-4 < y_1 < -2)$$

These are parabolic equations for the theoretical stress distribution, and the results are shown in Fig. 8-28b. The diagram has a discontinuity at the junction of the flange and stem because the thickness of the section changes abruptly. The distribution in the flange is fictitious because the stress at the top of the flange must be zero (a free surface). From Fig. 8-28b and Eq. (8-15), one may conclude that, in general, the maximum[7] longitudinal and transverse shearing stress occurs

---

[7]In this book the term *maximum*, as applied to a longitudinal and transverse shearing stress, will mean the average stress across the thickness $t$ at a point where such average has the maximum value.

at the neutral surface at a section where the transverse shear $V$ is maximum. There may be exceptions such as a beam with a cross section in the form of a Greek cross with $Q/t$ at the neutral surface less than the value some distance from the neutral surface.

Another example of importance is the determination of the shearing stress in an I-beam. Consider the W203 × 22 section [$I = 20.0(10^6)$ mm$^4$] shown in Fig. 8-29, and let the shear force $V$ on the section be 37.5 kN. Equation (8-15) may be used to calculate the shearing stress at various distances $y_1$ from the neutral axis of the beam. For example, at the neutral axis ($y_1 = 0$)

$$Q_{NA} = 102(8)(99) + 95(6.2)(47.5) = 108.76(10^3) \text{ mm}^3 = 108.76(10^{-6}) \text{ m}^3$$

$$\tau_{NA} = \frac{VQ_{NA}}{It_w} = \frac{37.5(10^3)(108.76)(10^{-6})}{20.0(10^{-6})(6.2)(10^{-3})} = 32.89(10^6) \text{ N/m}^2 \cong 32.9 \text{ MPa}$$

Similarly, at the junction between the web and the flange with $t = 6.2$ mm (in the web),

$$Q_J = 102(8)(99) = 80.78(10^3) \text{ mm}^3 = 80.78(10^{-6}) \text{ m}^3$$

$$\tau_J = \frac{VQ_J}{It_w} = \frac{37.5(10^3)(80.78)(10^{-6})}{20.0(10^{-6})(6.2)(10^{-3})} = 24.43(10^6) \text{ N/m}^2 \cong 24.4 \text{ MPa}$$

The shearing stress distribution for the complete cross section of the beam is shown in Fig. 8-29$b$. Equation (8-15) gives $\tau = 1.485$ MPa at the junction of the web and flange with $t = 102$ mm (in the flange). However, this result is incorrect because the bottom surface of the top flange (or the top surface of the bottom flange) is a free surface and thus $\tau = 0$. A similar result was obtained for the inverted T-section, shown in Fig. 8-28. More advanced methods of the theory of elasticity must be used to derive a correct solution.

The variation of shearing stress over the depth of the web is small, and the shearing stresses in the flanges are small compared to those in the web. As a result, the majority of the shear force $V$ is carried by the web. In the design of

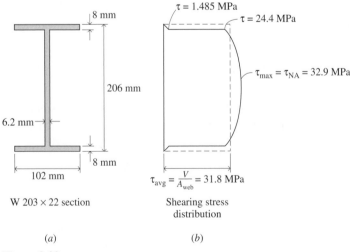

W 203 × 22 section

(a)

$\tau = 1.485$ MPa

$\tau = 24.4$ MPa

$\tau_{max} = \tau_{NA} = 32.9$ MPa

$\tau_{avg} = \dfrac{V}{A_{web}} = 31.8$ MPa

Shearing stress distribution

(b)

**Figure 8-29**

I-beams, the maximum shearing stress calculated by using Eq. (8-15) is approximated by dividing the shear force $V$ by the area of the web, that is,

$$\tau_{\text{avg}} = \frac{V}{A_{\text{web}}} = \frac{V}{t_w(d - 2t_f)} \qquad (8\text{-}17)$$

where $t_w$ is the thickness of the web, $d$ is the depth of the beam, and $t_f$ is the thickness of the flange. For the example being considered,

$$\tau_{\text{avg}} = \frac{V}{A_{\text{web}}} = \frac{37.5(10^3)}{6.2(10^{-3})[206-2(8)](10^{-3})}$$

$$= 31.8(10^6)\text{N/m}^2 = 31.8 \text{ MPa}$$

The maximum and average shearing stresses differ by approximately 3%. For commercial I-beams, the maximum difference is approximately 10%. Equation (8-17) should only be used to calculate shearing stresses in I-beams; it should not be used for T-, rectangular, or other sections. Equation (8-17) is specified in design codes for I-beams.

As a final example, consider a beam with a circular cross section. This type of beam is important in the transmission of power; for example, a shaft between a motor and a piece of equipment. Bending loads are induced in shafts by forces at gears, bearings, and pulleys.

Consider a beam with a solid circular cross section subjected to a shear force $V$, as shown in Fig. 8-30a. According to Eq. (8-15), a shear force $V$ causes a shearing stress $\tau_{xy}$ in the direction of $V$, as shown in Fig. 8-30b. This shearing stress at point $A$ can be resolved into normal ($n$) and tangential ($t$) components, as shown in Fig. 8-30c. However, Eq. (7-8) requires that $\tau_{xn} = \tau_{nx}$ (where $\tau_{nx}$ is the shearing stress in the $x$-direction on a plane with outward normal in the $n$-direction). The outside surface of the shaft (beam) is a free surface; therefore, $\tau_{nx} = \tau_{xn} = 0$, which indicates that any shearing stress at point $A$ must be tangent to the surface of the shaft (beam) and not in the direction of the shear force $V$ as required by Eq. (8-15). At the neutral axis, the shearing stress $\tau_{xt}$ is in the direction of shear force $V$ and Eq. (8-15) gives

$$\tau_{\text{max}} = \tau_{NA} = \frac{VQ_{NA}}{It_{NA}} = \frac{V(\pi r^2/2)(4r/3\pi)}{(\pi r^4/4)(2r)} = \frac{4V}{3\pi r^2} = \frac{4}{3}\frac{V}{A} \qquad (8\text{-}18)$$

where $A$ is the cross-sectional area of the beam. The maximum shearing stress given by Eq. (8-18) differs from the maximum shearing stress given by the mathematical theory of elasticity[8] by approximately 5%.

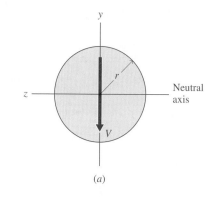

(a)

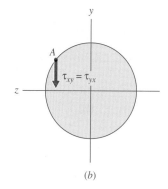

(b)

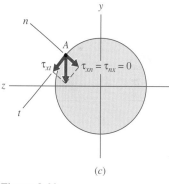

(c)

**Figure 8-30**

---

## Example Problem 8-11

A W254 × 33 wide-flange beam is loaded and supported as shown in Fig. 8-31a. At section $A$–$A$ of the beam, determine the flexural and shearing stresses

(a) At point $a$ on the top surface of the flange.

---

[8]*The Mathematical Theory of Elasticity*, 4th ed., A. E. H. Love, Dover Publications, New York, 1994.

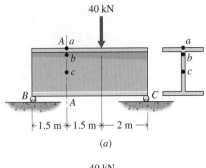

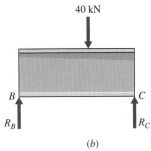

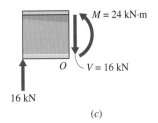

**Figure 8-31**

(b) At point $b$ in the web at the junction of the web and the flange.

(c) At point $c$ on the neutral axis.

## SOLUTION

A free-body diagram of the beam is shown in Fig. 8-31$b$. The reactions at supports $B$ and $C$ are determined by using the equilibrium equation $\Sigma M = 0$. Thus,

$$+\circlearrowleft \Sigma M_C = 0: \qquad 40(2) - R_B(5) = 0$$

$$+\circlearrowleft \Sigma M_B = 0: \qquad R_C(5) - 40(3) = 0$$

from which

$$R_B = 16 \text{ kN} \qquad \text{and} \qquad R_C = 24 \text{ kN}$$

A free-body diagram of the part of the beam to the left of section $A$–$A$ is shown in Fig. 8-31$c$. The shear force $V$ and bending moment $M$ transmitted by section $A$–$A$ are determined by using the equilibrium equations $\Sigma F_y = 0$ and $\Sigma M_O = 0$. Thus,

$$+\uparrow \Sigma F_y = 0: \qquad 16 - V = 0$$

$$+\circlearrowleft \Sigma M_O = 0: \qquad M - 16(1.5) = 0$$

from which

$$V = 16 \text{ kN} \qquad \text{and} \qquad M = 24 \text{ kN} \cdot \text{m}$$

(a) For a W254 $\times$ 33 section:

$$I = 49.1(10^6) \text{ mm}^4, \ d = 258 \text{ mm}, \ w_f = 146 \text{ mm}, \ t_f = 9.1 \text{ mm}, \ t_w = 6.1 \text{ mm}$$

The flexural stress at point $a$ is given by Eq. (8-11) as

$$\sigma_a = -\frac{My}{I} = -\frac{24(10^3)(129)(10^{-3})}{49.1(10^{-6})}$$

$$= -63.05(10^6)\text{N/m}^2 \cong 63.1 \text{ MPa C} \qquad \text{Ans.}$$

The shearing stress at point $a$ is given by Eq. (8-15). At point $a$, $Q = 0$, and thus

$$\tau = 0 \qquad \text{Ans.}$$

(b) At point $b$, $y = 129 - 9.1 = 119.9$ mm

$$\sigma_b = -\frac{My}{I} = -\frac{24(10^3)(119.9)(10^{-3})}{49.1(10^{-6})}$$

$$= -58.61(10^6)\text{N/m}^2 \cong 58.6 \text{ MPa C} \qquad \text{Ans.}$$

At point $b$, the first moment $Q_b$ of the area above the point is

$$Q_b = y_c A = 124.45(146)(9.1) = 165.34(10^3) \text{ mm}^3 = 165.34(10^{-6}) \text{ m}^3$$

$$\tau_b = \frac{VQ_b}{I_{NA}t_w} = \frac{16(10^3)(165.34)(10^{-6})}{49.1(10^{-6})(6.1)(10^{-3})}$$

$$= 8.832(10^6) \text{ N/m}^2 \cong 8.83 \text{ MPa} \qquad \qquad \textbf{Ans.}$$

(c) Since point $c$ is on the neutral axis, the flexural stress is zero. The shearing stress is maximum, and an approximate value is given by Eq. (8-17) as

$$\tau_{max} = \frac{V}{A_{web}} = \frac{16(10^3)}{6.1(10^{-3})(239.8)(10^{-3})}$$

$$= 10.938(10^6) \text{ N/m}^2 \cong 10.94 \text{ MPa} \qquad \qquad \textbf{Ans.}$$

The shearing stress at point $c$ can also be obtained using Eq. (8-15). At point $c$, the first moment $Q_c$ of the area above the point is

$$Q_c = y_C A = 124.45(146)(9.1) + 59.95(119.9)(6.1)$$

$$= 209.2(10^3) \text{ mm}^3 = 209.2(10^{-6}) \text{ m}^3$$

$$\tau_c = \frac{VQ_c}{I_{NA}t_w} = \frac{16(10^3)(209.2)(10^{-6})}{49.1(10^{-6})(6.1)(10^{-3})} = 11.176(10^6) \text{ N/m}^2 \cong 11.18 \text{ MPa}$$

The percent difference between the two values of shearing stress is

$$D = \frac{11.176 - 10.938}{11.176}(100) = 2.13\% \quad \blacksquare$$

## Example Problem 8-12

A beam is simply supported and carries a concentrated load of 1800 lb at the center of a 15-ft span, as shown in Fig. 8-32a. If the beam has the T cross section shown in Fig. 8-32b, determine

(a) The average shearing stress on a horizontal plane 4 in. above the bottom of the beam and 6 ft from the left support.
(b) The maximum transverse shearing stress in the beam.
(c) The average shearing stress in the joint between the flange and the stem at a section 6 ft from the left support.
(d) The force transmitted from the flange to the stem by the glue in a 12-in. length of the joint centered 6 ft from the left support.
(e) The maximum tensile flexural stress in the beam.

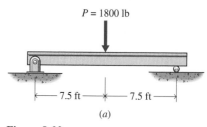

$P = 1800$ lb

— 7.5 ft — ✳ — 7.5 ft —

(a)

**Figure 8-32**

## SOLUTION

The second moment of the cross-sectional area about the neutral axis is

$$I_{NA} = \frac{1}{12}(2)(10^3) + 2(10)(3)^2 + \frac{1}{12}(10)(2)^3 + 10(2)(3)^2 = 533.3 \text{ in}^4$$

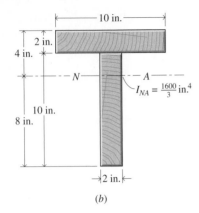

(b)

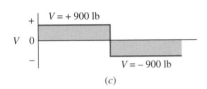

(c)

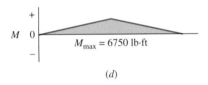

(d)

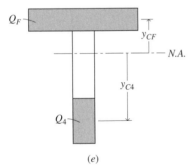

(e)

**Figure 8-32**

(a) The shear force $V$ on a cross section 6 ft from the left support is $+900$ lb, as shown on Fig. 8-32c. The first moment $Q_4$ for the bottom 4 in. of the stem (see Fig. 8-32e) is

$$Q_4 = y_{C4}A_4 = 6(2)(4) = 48 \text{ in}^3$$

The average shearing stress on a horizontal plane 4 in. above the bottom of the beam is then given by Eq. (8-15) as

$$\tau_4 = \frac{VQ_4}{I_{NA}t_4} = \frac{900(48)}{533.3(2)} = 40.50 \text{ psi} \cong 40.5 \text{ psi} \qquad \textbf{Ans.}$$

(b) The maximum transverse shearing stress in the beam will occur at the neutral axis on the cross section supporting the largest shear force $V$. The first moment $Q_{NA}$ for the part of the stem below the neutral axis is

$$Q_{NA} = 4(2)(8) = 64 \text{ in}^3$$

In the computation of $Q$, it is immaterial whether one takes the area above or below a transverse line. For example, $Q_{NA}$ for the area above the neutral axis is $Q_{\text{flange}} + Q_{\text{stem}} = 3(10)(2) + 1(2)(2) = 64 \text{ in}^3$, which is the same as that for the part of the stem below the neutral axis. Since the shear force $V$ equals 900 lb on all cross sections of the beam,

$$\tau_{\max} = \frac{VQ_{NA}}{I_{NA}t_S} = \frac{900(64)}{533.3(2)} = 54.00 \text{ psi} \cong 54.0 \text{ psi} \qquad \textbf{Ans.}$$

(c) The first moment $Q_F$ for the flange of the beam (see Fig. 8-32e) is

$$Q_F = y_{CF}A_F = 3(10)(2) = 60 \text{ in}^3$$

The average shearing stress on the horizontal plane at the joint between the flange and the stem is then given by Eq. (8-15) as

$$\tau_J = \frac{VQ_F}{I_{NA}t_S} = \frac{900(60)}{533.3(2)} = 50.63 \text{ psi} \cong 50.6 \text{ psi} \qquad \textbf{Ans.}$$

(d) The force transmitted from the flange to the stem by the glue is

$$V_g = \tau_J A_J = 50.63(12)(2) = 1215.1 \text{ lb} \cong 1215 \text{ lb} \qquad \textbf{Ans.}$$

(e) The maximum tensile flexural stress in the beam will occur in a fiber at the bottom of the beam because the resisting moment at all cross sections of the beam is positive. The largest moment occurs at midspan, as shown in Fig. 8-32d. The flexural stress is given by Eq. (8-11) as

$$\sigma_{\max} = \frac{M_{\max}c}{I} = \frac{6750(12)(8)}{533.3} = 1215.1 \text{ psi} \cong 1215 \text{ psi T} \quad \blacksquare \qquad \textbf{Ans.}$$

## PROBLEMS

### Introductory Problems

**8-84\*** A timber beam is loaded and supported as shown in Fig. P8-84. At section $A–A$ of the beam, determine the shearing stresses on horizontal planes that pass through points $a$, $b$, and $c$ of the cross section.

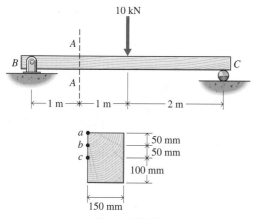

**Figure P8-84**

**8-85\*** The transverse shear $V$ at a certain section of a timber beam is 7500 lb. If the beam has the cross section shown in Fig. P8-85, determine
(a) The horizontal shearing stress in the glued joint 2 in. below the top of the beam.
(b) The transverse shearing stress at a point 3 in. below the top of the beam.
(c) The magnitude and location of the maximum transverse shearing stress on the cross section.

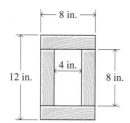

**Figure P8-85**

**8-86** A W254 × 89 structural steel wide-flange beam (see Appendix A) is loaded and supported as shown in Fig. P8-86. Determine the maximum transverse shearing stress at section $A–A$ of the beam
(a) Using Eq. (8-15).
(b) Using Eq. (8-17).

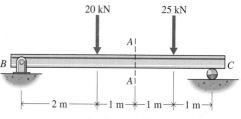

**Figure P8-86**

**8-87** A WT7 × 34 structural steel T-section (see Appendix A) is loaded and supported as shown in Fig. P8-87. Determine the maximum transverse shearing stress at section $A–A$ of the beam.

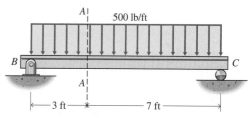

**Figure P8-87**

**8-88\*** A timber beam 4 m long is simply supported at its ends and carries a uniformly distributed load $w$ of 8 kN/m over its entire length. If the beam has the cross section shown in Fig. P8-88, determine
(a) The maximum horizontal shearing stress in the glued joints between the web and flanges of the beam.
(b) The maximum horizontal shearing stress in the beam.

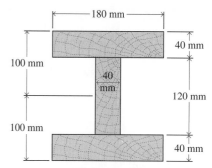

**Figure P8-88**

**8-89** A W10 × 30 structural steel wide-flange beam (see Appendix A) is loaded and supported as shown in Fig. P8-89. At section $A–A$ of the beam, determine
(a) The maximum transverse shearing stress due to the 5000-lb load.

(b) The maximum transverse shearing stress due to the 5000-lb load plus the weight of the beam.

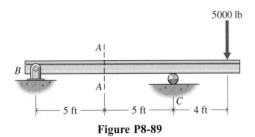

**Figure P8-89**

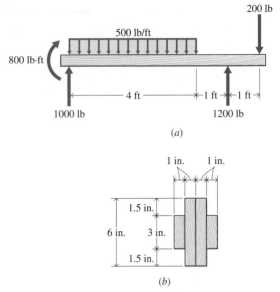

(a)

(b)

**Figure P8-91**

## Intermediate Problems

**8-90\*** A W203 × 60 structural steel wide-flange section (see Appendix A) is used for the cantilever beam shown in Fig. P8-90. Determine the maximum flexural and transverse shearing stresses in the beam and state where they occur.

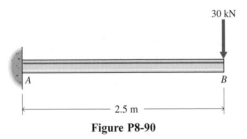

**Figure P8-90**

**8-91\*** The beam shown in Fig. P8-91a is composed of two 1- × 6-in. and two 1- × 3-in. hard maple boards that are glued together as shown in Fig. P8-91b. Determine the magnitude and location of
(a) The maximum tensile flexural stress in the beam.
(b) The maximum horizontal shearing stress in the beam.

**8-92** The beam shown in Fig. P8-92a is composed of three pieces of timber that are glued together as shown in Fig. P8-92b. Determine
(a) The maximum horizontal shearing stress in the glued joints.
(b) The maximum horizontal shearing stress in the wood.
(c) The maximum tensile and compressive flexural stresses in the beam.

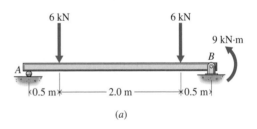

(a)

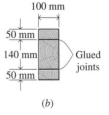

(b)

**Figure P8-92**

**8-93** The lintel beam *AB* shown in Fig. P8-93*a* has a rectangular cross section as shown in Fig. P8-93*b* and is used to support a brick wall over a door opening. The brick wall is assumed to produce a triangular load distribution. The total load carried by the beam is 500 lb. If the beam is simply supported at the ends, determine the maximum flexural and transverse shearing stresses in the beam and state where they occur.

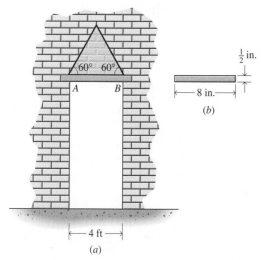

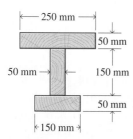

**Figure P8-93**

**8-94** A timber beam is simply supported and carries a uniformly distributed load of 4 kN/m over the full length of the beam. If the beam has the cross section shown in Fig. P8-94 and a span of 6 m, determine

(a) The horizontal shearing stress in the glued joint 50 mm below the top of the beam and 1 m from the left support.

(b) The horizontal shearing stress in the glued joint 50 mm above the bottom of the beam and $\frac{1}{2}$ m from the left support.

(c) The maximum horizontal shearing stress in the beam.

(d) The maximum tensile flexural stress in the beam.

**Figure P8-94**

**8-95\*** The timber beam shown in Fig. P8-95*a* is fabricated by gluing two 1- × 5-in. and two 1- × 4-in. boards together as shown in Fig. P8-95*b*. Determine

(a) The maximum horizontal shearing stress in the glued joints.

(b) The maximum horizontal shearing stress in the wood.

(c) The maximum tensile and compressive flexural stresses in the beam.

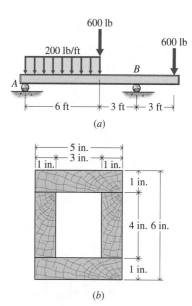

**Figure P8-95**

## Challenging Problems

**8-96\*** A cantilever beam is used to support a concentrated load of 20 kN at the end of the beam. The beam is fabricated by bolting two C457 × 86 steel channels (see Appendix A) back to back to form the H-section shown in Fig. P8-96. If the pairs of bolts are spaced at 300-mm intervals along the beam, determine

(a) The shear force carried by each of the bolts.

(b) The bolt diameter required if the shear and bearing stresses for the bolts must be limited to 60 MPa and 125 MPa, respectively.

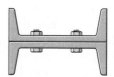

**Figure P8-96**

**8-97\*** A timber beam is fabricated from one 2- × 8-in. and two 2- × 6-in. pieces of lumber to form the cross section shown in Fig. P8-97. The flanges of the beam are fastened to the web with nails that can safely transmit a shear force of 100 lb. If the beam is simply supported and carries a 1000-lb load at the center of a 12-ft span, determine
(a) The shear force transferred by the nails from the flange to the web in a 12-in. length of the beam.
(b) The spacing required for the nails.

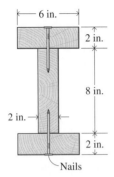

**Figure P8-97**

**8-98** A box beam will be fabricated by bolting two 15- × 260-mm steel plates to two C305 × 45 steel channels (see Appendix A), as shown in Fig. P8-98. The beam will be simply supported at the ends and will carry a concentrated load of 125 kN at the center of a 5-m span. Determine the bolt spacing required if the bolts have a diameter of 20 mm and an allowable shearing stress of 150 MPa.

**Figure P8-98**

**8-99** A W21 × 101 structural steel wide-flange section (see Appendix A) is simply supported at its ends and carries a concentrated load at the center of a 20-ft span. The concentrated load must be increased to 125 kip, which requires that the beam be strengthened. It has been decided that two ³/₄- × 16-in. steel plates will be bolted to the flanges, as shown in Fig. P8-99. Determine the bolt spacing required if the bolts have a diameter of ³/₄ in. and an allowable shearing stress of 17.5 ksi.

**Figure P8-99**

**8-100** A W356 × 122 structural steel wide-flange section (see Appendix A) has a C381 × 74 channel bolted to the top flange, as shown in Fig. P8-100. The beam is simply supported at its ends and carries a concentrated load of 96 kN at the center of an 8-m span. If the pairs of bolts are spaced at 500-mm intervals along the beam, determine
(a) The shear force carried by each of the bolts.
(b) The bolt diameter required if the shear and bearing stresses for the bolts must be limited to 60 MPa and 125 MPa, respectively.

**Figure P8-100**

### Computer Problems

**8-101** A timber beam is simply supported and carries a uniformly distributed load $w$ of 360 lb/ft over its entire 18-ft span (Fig. P8-101a). If the beam has the cross section shown in Fig. P8-101b, compute and plot the vertical shearing stress $\tau$ as a function of distance $y$ from the neutral axis for a cross section 2 ft from the left end of the beam.

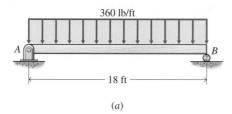

(a)

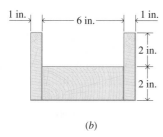

(b)

**Figure P8-101**

**8-102** The transverse shear force $V$ at a certain section of a timber beam is 18 kN. If the beam has the cross section shown in Fig. P8-102, compute and plot the vertical shearing stress $\tau$ as a function of distance $y$ ($-150$ mm $< y <$ 150 mm) from the neutral axis.

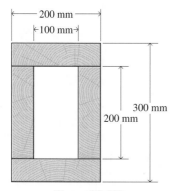

**Figure P8-102**

**8-103** The transverse shear force $V$ at a certain section of a timber beam is 2500 lb. If the beam has the cross section shown in Fig. P8-103, compute and plot the vertical shearing stress $\tau$ as a function of distance $y$ ($-3$ in. $< y <$ 3 in.) from the neutral axis.

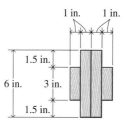

**Figure P8-103**

**8-104** The beam shown in Fig. P8-104a is fabricated by gluing two pieces of timber together to form the cross section shown in Fig. P8-104b. Compute and plot the vertical shearing stress $\tau$ as a function of distance $y$ from the neutral axis for a cross section 0.5 m from the left end of the beam.

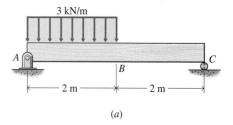

(a)

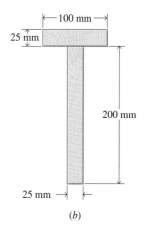

(b)

**Figure P8-104**

# 8-9 DESIGN

Again in this chapter, design will be limited to proportioning a member (in this case, a beam) to perform a specified function without exceeding specified levels of stress. Failure by excessive deformation (excessive elastic deflection) will be discussed in Chapter 9. Failure by yielding or failure by fracture, which results from excessive normal (flexural) stresses or shearing stresses, must be considered when designing beams. The elastic flexure formula and the shearing stress formula developed in Sections 8-5 and 8-8 and used to calculate flexural stresses and transverse (or longitudinal) shearing stresses in beams are

$$\sigma_{max} = \frac{Mc}{I} = \frac{M}{S} \quad \text{and} \quad \tau = \frac{VQ}{It}$$

To determine maximum normal and shearing stresses in beams, sections must be located where $M$ and $V$ are maximum (critical sections). A shear force diagram is used to locate the critical section of the beam where $V$ is maximum. The critical section for flexure, the section where $M$ is maximum, is found from a bending moment diagram. In general, the absolute maximum value of $M$ is used for design purposes. However, care must be exercised for beams with cross sections that are nonsymmetrical with respect to the neutral axis (such as T-beams) and beams made of materials with different properties in tension and compression. In these cases, both the largest positive and the largest negative values of $M$ must be considered.

Beam design consists of finding a cross-sectional shape so that flexural and shearing stresses do not exceed permissible values, called allowable values. For a safe design

$$\text{Strength} \geq (\text{Factor of safety})(\text{Stress}) \tag{a}$$

where strength is a material property and stress is computed using either Eq. (8-12) or (8-15). Equation (a) may be written

$$\frac{\text{Strength}}{\text{Factor of safety}} \geq \text{Stress} \tag{b}$$

where (Strength/Factor of safety) is the allowable stress. If the symbol $\sigma_{all}$ is used for allowable stress, Eq. (b) may be written for flexure as

$$\sigma_{all} \geq \frac{Mc}{I} = \frac{M}{S} \tag{8-19}$$

and

$$\tau_{all} \geq \frac{VQ}{It} \tag{8-20}$$

for shear.

Experience indicates that beam design is usually governed by flexural stresses. Thus, a beam is usually designed for flexure using Eq. (8-19), and then checked for shearing stress using Eq. (8-20). If the shearing stress is less than the allowable shearing stress, this procedure is adequate. If the allowable shearing stress has been exceeded, the beam is redesigned and the process is repeated. However, the shearing stress (longitudinal) may be the controlling factor for beams made of timber. The examples that follow illustrate the procedures for designing beams. Design problems involving combined loading will be presented in Chapter 10.

Other design factors, such as local web yielding, web crippling, and side-sway web buckling for steel beams may be found in *Steel Structures Design and Behavior*.[9] For wood beams, additional design factors such as load duration, moisture content, temperature, and beam stability are discussed in *Design of Wood Structures*.[10]

## Example Problem 8-13

An air-dried Douglas fir beam of rectangular cross section is to support the load shown in Fig. 8-33a. If the allowable flexural stresses in tension and compression are 1200 psi and the allowable shearing stress is 100 psi, determine the lightest weight standard structural timber that can be used.

### SOLUTION

First, load (free-body), shear force, and bending moment diagrams are drawn, as shown in Figs. 8-33b, c, and d, respectively. From these diagrams it is determined that the maximum shear force is 900 lb and the maximum bending moment is 3375 lb · ft. Using the maximum bending moment in Eq. (8-19) yields the required minimum section modulus for the beam as

$$S \geq \frac{M_{max}}{\sigma_{all}} = \frac{3375(12)}{1200} = 33.75 \text{ in}^3$$

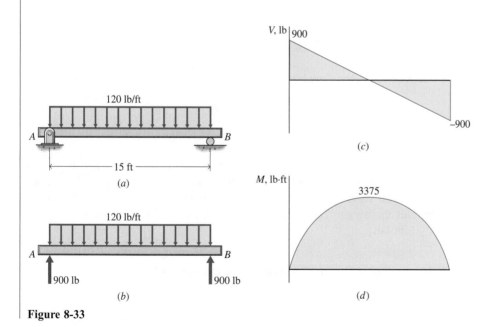

**Figure 8-33**

[9]*Steel Structures Design and Behavior*, 4th ed., C. G. Salmon and J. F. Johnson, Harper Collins, New York, 1996.

[10]*Design of Wood Structures*, 3rd ed., D. E. Breyer, McGraw-Hill, New York, 1993.

Table A-15 in Appendix A contains a listing of standard structural timbers together with values for their section modulus $S$. Note that the properties listed are for dressed (actual size) timbers. Since the lightest-weight beam is wanted, a beam with $S \geq 33.75$ in$^3$ and the smallest weight per unit length is a 2- × 12-in. nominal size timber. For this beam, the actual values of $S$ and $I$ are 35.8 in$^3$ and 206 in$^4$, respectively. Equation (8-20) can now be used to see whether or not the allowable shearing stress requirement is met. The maximum shearing stress for a beam with a rectangular cross section occurs at the neutral axis and is given by

$$\tau_{max} = \frac{VQ}{It} = 1.5\frac{V}{A}$$

where $V$ is the absolute value of the maximum shear force and $A$ is the cross-sectional area of the beam. Thus,

$$\tau_{max} = 1.5\frac{V}{A} = 1.5\frac{900}{(1.625)(11.5)} = 72.24 \text{ psi} < 100 \text{ psi}$$

Equation (8-20) is satisfied, since $\tau_{all} \geq \tau_{max}$ or 100 psi $>$ 72.24 psi. The maximum shearing stress is within the allowable limit; therefore, the beam selected is satisfactory. However, the analysis assumed that the beam was weightless, whereas the beam selected weighs 5.19 lb/ft. The bending moment diagram for the uniformly distributed weight of the beam is similar to Fig. 8-33$d$, and $M_{max}$ for the weight is $M_{max} = 146$ lb · ft. Adding the maximum bending moments for the applied loading and the beam weight gives

$$M_{max} = 3375 + 146 = 3521 \text{ lb} \cdot \text{ft}$$

The required section modulus then becomes

$$S \geq \frac{M_{max}}{\sigma_{all}} = \frac{3521(12)}{1200} = 35.21 \text{ in}^3$$

If the value of $S$ (35.21 in$^3$ for this example) had been greater than $S$ for the original beam selected (35.8 in$^3$ for this example), the entire procedure would be repeated based on the section modulus for the beam with the applied loading plus the weight. Thus, the procedure is a trial-and-error process.

Similarly,

$$V_{max} = 900 + \frac{1}{2}(5.19)(15) = 939 \text{ lb}$$

Thus,

$$\tau_{max} = 1.5\frac{V}{A} = 1.5\frac{939}{(1.625)(11.5)} = 75.37 \text{ psi} < 100 \text{ psi}$$

The original beam selected had a section modulus of $S = 35.8$ in$^3$. Thus, the 2- × 12-in. nominal beam is satisfactory. ∎

# Example Problem 8-14

Select the lightest wide-flange beam that can be used to support the load shown in Fig. 8-34a. The allowable flexural stresses in tension and compression are 160 MPa and the allowable shearing stress is 82 MPa.

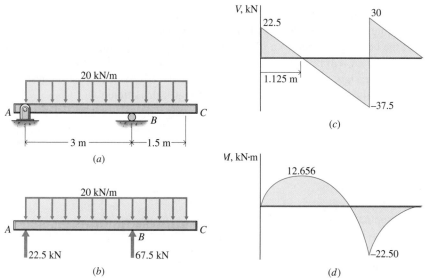

**Figure 8-34**

## SOLUTION

Load (free-body), shear force, and bending moment diagrams for the beam are shown in Figs. 8-34b, c, and d, respectively. From these diagrams it is determined that the maximum shear force is 37.5 kN and the maximum bending moment is 22.50 kN · m. Using the maximum bending moment in Eq. (8-19) yields the required section modulus for the beam as

$$S \geq \frac{M_{max}}{\sigma_{all}} = \frac{22.50(10^3)}{160(10^6)} = 140.63(10^{-6}) \text{ m}^3 = 140.63(10^3) \text{ mm}^3$$

Table A-2 in Appendix A contains a listing of the properties of wide-flange sections and from this listing it is determined that the lightest section with $S \geq 140.63(10^3)$ mm$^3$ (with respect to the $x$–$x$-axis), is a W203 × 22 section.

When designing wide-flange beams or American Standard Beams, it is often assumed that the entire shear load is carried by the web of the beam and that it is uniformly distributed (see Section 8-8). Thus,

$$\tau_{avg} = \frac{V}{A_{web}} = \frac{37.5(10^3)}{6.2(10^{-3})[206-2(8)](10^{-3})}$$

$$= 31.8(10^6) \text{ N/m}^2 = 31.8 \text{ MPa}$$

Since $\tau_{avg}$ is less than $\tau_{all} = 82$ MPa, the W203 × 22 section is satisfactory.

As in the previous example, the weight of the beam should be considered. The mass per unit length for this beam is 22 kg/m, and thus the weight per unit length is $W = mg = 22(9.81) = 215.8$ N/m, or 0.2158 kN/m. The weight per unit length of the beam is about 1.1% of the applied force per unit length. The actual section modulus $S = 193(10^3)$ mm$^3$ is about 27% higher than the required minimum value of $S_{min} = 140.63 \,(10^3)$ mm$^3$. Thus, the maximum flexural stress is less than the allowable value even when the weight of the beam is considered and the W203 × 22 wide-flange section is acceptable.

Since the cross section is symmetric, the maximum tensile stress and the maximum compressive stress both occur on the section where the magnitude of the bending moment is a maximum. For a negative bending moment, the top of the beam will be in tension and the bottom of the beam will be in compression. The sign of the shear force is completely arbitrary, and the largest (absolute value) shear force is used in the design calculations. ■

# PROBLEMS

## Introductory Problems

**8-105\*** A 20-ft-long simply supported timber beam is loaded with 1800-lb concentrated loads applied 6 ft from each support. The allowable flexural stress is 1900 psi, and the allowable shearing stress is 90 psi. Select the lightest standard structural timber that can be used to support the loading.

**8-106\*** A simply supported timber beam 5 m long is loaded with 6700-N concentrated loads applied 2 m from each support. If the allowable flexural stress is 9 MPa and the allowable shearing stress is 0.6 MPa, select the lightest standard structural timber that can be used to support the loading.

**8-107** The lever shown in Fig. P8-107 is used to lift a 600-lb rock. Select a standard steel pipe to perform the task. The allowable flexural stress is 20 ksi. Neglect the effects of shear.

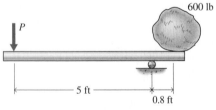

**Figure P8-107**

## Intermediate Problems

**8-108\*** A structural steel beam is subjected to the loading shown in Fig. P8-108. The allowable flexural stress is 152 MPa, and the allowable shearing stress is 100 MPa. Select the lightest American Standard beam that can be used to support the loading.

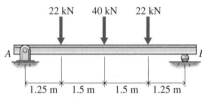

**Figure P8-108**

**8-109** A 16-ft-long simply supported beam is loaded with a uniform load of 4000 lb/ft over its entire length. If the allowable flexural stress is 22 ksi and the allowable shearing stress is 14.5 ksi, select the lightest structural steel wide-flange beam that can be used to support the loading.

**8-110** Select the lightest wide-flange beam that can be used to support the loading shown in Fig. P8-110. The allowable flexural stress is 152 MPa, and the allowable shearing stress is 100 MPa.

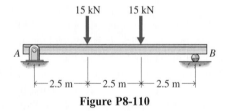

**Figure P8-110**

## Challenging Problems

**8-111\*** The floor framing plan for a residential dwelling is shown in Fig. P8-111. The floor decking is to be supported by 2-in. nominal width joists spaced 16 in. apart. Each joist is to span 12 ft and is simply supported at the ends. The floor decking is subjected to a uniform loading of 60 lb/ft$^2$, which includes the live load plus an allowance for the dead load of the flooring system. The joists are made of construction grade Douglas fir with an allowable flexural stress of 1200 psi and an allowable shearing stress of 120 psi. Determine the required nominal depth of the joists.

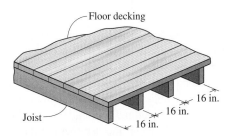

Figure P8-111

**8-112**  A 15-kN load is supported by a roller on an I-beam, as shown in Fig. P8-112. The roller moves slowly along the beam, thereby causing the shear force and bending moment to be functions of $b$. Select the lightest permissible American Standard beam to support the loading. The allowable flexural stress is 152 MPa, and the allowable shearing stress is 100 MPa. Neglect the weight of the beam.

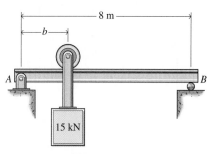

Figure P8-112

**8-113**  A carriage moves slowly along a simply supported I-beam, as shown in Fig. P8-113. Select the lightest permissible wide-flange beam to support the loading, if the allowable flexural stress is 24 ksi and the allowable shearing stress is 14.5 ksi. Neglect the weight of the beam. Note that the shear force and bending moment are functions of $b$, the position of the left-hand wheel.

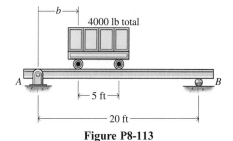

Figure P8-113

## 8-10 SUMMARY

A member subjected to loads applied transverse to the long dimension of the member and that cause the member to bend is known as a *beam*. A beam supported by pins, rollers, or smooth surfaces at the ends is called a *simple beam*. A simple support will develop a reaction normal to the beam but will not produce a couple. A cantilever beam has one end built into a wall or other support. The built-in end is said to be fixed if no rotation occurs and restrained if a limited amount of rotation occurs.

Cantilever beams and simple beams have only two reactions (two forces or one force and a couple), and these reactions can be obtained from a free-body diagram of the beam by applying the equations of equilibrium. Such beams are said to be statically determinate. Beams with more than two reaction components are called statically indeterminate since there are not enough equations of equilibrium to determine the reactions.

A free-body diagram of a portion of a beam with a cross section exposed by an imaginary cut shows that a transverse force $V_r$ and a couple $M_r$ at the cut section are needed to maintain equilibrium. The force $V_r$ is the resultant force due to the shearing stresses. The couple $M_r$ is the resultant couple due to the normal stresses. The magnitudes and senses of $V_r$ and $M_r$ are obtained from the equations of equilibrium $\Sigma F_y = 0$ and $\Sigma M_O = 0$, where $O$ is any axis perpendicular to the $xy$-plane.

The normal and shearing stresses $\sigma$ and $\tau$ on a transverse plane of a beam are related to the resisting moment $M_r$ and the shear force $V_r$ by the equations

$$V_r = -\int_{\text{Area}} \tau \, dA \qquad (8\text{-}1a)$$

$$M_r = -\int_{\text{Area}} y \, \sigma \, dA \qquad (8\text{-}1b)$$

It is obvious from Eqs. (8-1) that the laws of variation of the normal and shearing stresses must be known before the integrals can be evaluated.

The variation of normal stress on a plane is obtained by assuming that a plane section before bending remains a plane after bending. For this to be strictly true, it is necessary that the beam be bent only with couples. When a beam is bent with couples, the deformed shape of all longitudinal elements (also referred to as fibers) is an arc of a circle. Precise experimental measurements indicate that at some distance $c$ above the bottom of the beam, longitudinal elements undergo no change in length. The curved surface formed by these elements is referred to as the *neutral surface of the beam*, and the intersection of this surface with any cross section is called the *neutral axis of the section*. All elements (fibers) on one side of the neutral surface are compressed, while those on the opposite side are elongated. As a result, the normal strain at any point on the plane can be expressed as

$$\epsilon_x = -\frac{1}{\rho} y \qquad (8\text{-}2)$$

Equation (8-2) indicates that the strain in a fiber is proportional to the distance of the fiber from the neutral surface of the beam. Equation (8-2) is valid for elastic or inelastic action so long as the transverse shearing stresses are small.

Since the longitudinal strain $\epsilon_x$ is proportional to the distance of the fiber from the neutral surface of the beam, the normal stress $\sigma_x$ on the plane (for linearly elastic action) is given by Hooke's law as

$$\sigma_x = E\epsilon_x = -\frac{E}{\rho} y \qquad (8\text{-}3)$$

Substituting Eq. (8-3) into Eq. (8-1) yields

$$M_r = -\int_A y \, \sigma_x \, dA = -\frac{E}{\rho} \int_A y^2 \, dA$$

The integral $\int y^2 \, dA$ is called the second moment of area. When the integral $\int_A y^2 dA$ is replaced by the symbol $I$, the elastic flexure formula is obtained as

$$\sigma_x = -\frac{M_r y}{I} \qquad (8\text{-}11)$$

where $\sigma_x$ is the flexural stress at a distance $y$ from the neutral surface and on a transverse plane, $M_r$ is the resisting moment of the section, and $I$ is the second moment of area of the transverse section with respect to the neutral axis.

At any section of the beam, the flexural stress will be maximum (have the greatest magnitude) at the surface farthest from the neutral axis ($y = c$), and Eq. (8-11) becomes

$$\sigma_{max} = \frac{M_r c}{I} = \frac{M_r}{S} \qquad (8\text{-}12)$$

where $S = I/c$ is called the *section modulus of the beam.*

If the maximum flexural stress is required in a beam subjected to a loading that produces a bending moment that varies with position along the beam, it is desirable to have a method for determining the maximum moment. Similarly, the maximum transverse shearing stress will occur at a section where the resisting shear is maximum. Shear force and bending moment diagrams provide a method for obtaining maximum values of shear and moment. A shear force diagram is a graph in which abscissas represent distances along the beam and ordinates represent the transverse shear at the corresponding sections. A bending moment diagram is a graph in which abscissas represent distances along the beam and ordinates represent the bending moment at the corresponding sections. Four simple relationships developed by using equilibrium considerations that are used to construct shear force and bending moment diagrams are

**3.**
$$\frac{dV}{dx} = w \qquad (8\text{-}13d)$$

That is, the slope of the shear force diagram at any location $x$ in the beam is equal to the intensity of loading at the section of the beam.

**4.**
$$V_2 - V_1 = \int_{x_1}^{x_2} w \, dx \qquad (8\text{-}13e)$$

That is, for any segment of the beam acted on by a distributed load $w$ and no concentrated load ($P = 0$), the change in shear force between sections at $x_1$ and $x_2$ is equal to the area under the load diagram between the two sections.

**6.**
$$\frac{dM}{dx} = V \qquad (8\text{-}14c)$$

That is, the slope of the bending moment diagram at any location $x$ in the beam is equal to the value of the shear force at that section of the beam.

**7.**
$$M_2 - M_1 = \int_{x_1}^{x_2} V \, dx \qquad (8\text{-}14d)$$

That is, for any segment of the beam in which the shear force is continuous ($C = P = 0$), the change in bending moment between sections at $x_1$ and $x_2$ is equal to the area under the shear diagram between the two sections.

At each point in a beam, the horizontal (longitudinal) and vertical (transverse) shearing stresses have the same magnitude and are given by the expression

$$\tau = \frac{VQ}{It} \tag{8-15}$$

where $Q$ is the first moment of the portion of the area of the cross section between the transverse line where the stress is to be evaluated and the extreme fiber of the beam. The sense of stress $\tau$ is the same as the sense of the shear force $V$ on the transverse plane. Because the flexure formula was used in the derivation of Eq. (8-15), it is subject to the same limitations as the flexure formula.

# REVIEW PROBLEMS

**8-114\*** A beam has the cross section shown in Fig. P8-114. If the flexural stress at point $A$ is 14 MPa T, determine
(a) The maximum flexural stress on the section.
(b) The resisting moment $M_r$ at the section.

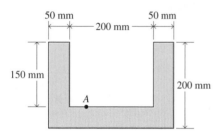

**Figure P8-114**

**8-115\*** A T-beam has the cross section shown in Fig. P8-115. Determine the maximum tensile and compressive flexural stresses on a cross section of the beam where the resisting moment being transmitted is 74 kip · ft.

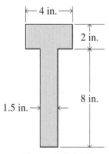

**Figure P8-115**

**8-116** Determine the percentage of the resisting moment $M_r$ carried by the flanges of a W838 × 226 wide-flange beam (see Appendix A for dimensions).

**8-117** The maximum flexural stress on the cross section of the beam shown in Fig. P8-117 is 8000 psi C. Determine
(a) The resisting moment being transmitted by the section.
(b) The magnitude of the flexural force carried by the flange.

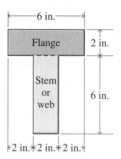

**Figure P8-117**

**8-118** A beam is loaded and supported as shown in Fig. P8-118.
(a) Draw complete shear force and bending moment diagrams for the beam.
(b) Using the coordinate axes shown, write equations for the shear force and bending moment for any section of the beam in the interval $0 < x < 4$ m.

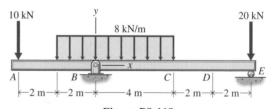

**Figure P8-118**

**8-119\*** A beam is loaded and supported as shown in Fig. P8-119.
(a) Draw complete shear force and bending moment diagrams for the beam.
(b) Using the coordinate axes shown, write equations for the shear force and bending moment for any section of the beam in the interval $0 < x < 10$ ft.

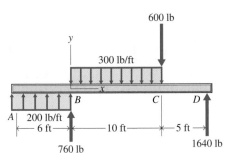

**Figure P8-119**

**8-120\*** Select the lightest pair of structural-steel angles (see Appendix A) that may be used for the beam of Fig. P8-120 if the maximum flexural stress must be limited to 60 MPa. The angles will be fastened back to back to form a T-section.

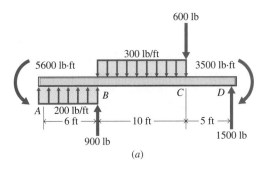

(a)

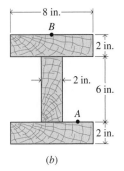

(b)

**Figure P8-121**

**8-122** Select the lightest steel wide-flange or American standard beam (see Appendix A) that may be used for the beam of Fig. P8-122 if the maximum flexural stress must be limited to 75 MPa.

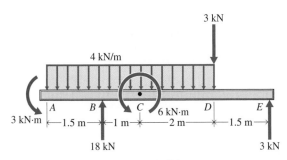

**Figure P8-120**

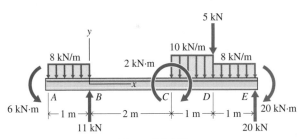

**Figure P8-122**

**8-123\*** The beam shown in Fig. P8-123a has the cross section shown in Fig. P8-123b. Determine the maximum tensile and compressive flexural stresses in the beam.

**8-121** A timber beam is loaded and supported as shown in Fig. P8-121a. The beam has the cross section shown in Fig. P8-121b. Determine

(a) The flexural stresses at points $A$ and $B$ on a transverse cross section 1 ft from the left end of the beam.

(b) The maximum tensile and compressive flexural stresses in the beam.

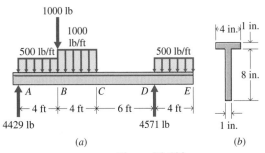

(a)

(b)

**Figure P8-123**

**8-124\***  A WT305 $\times$ 70 structural tee (see Appendix A) is loaded and supported as a beam (with the flange on top) as shown in Fig. P8-124. Determine
(a) The maximum tensile flexural stress in the beam.
(b) The maximum compressive flexural stress in the beam.
(c) The maximum vertical shearing stress in the beam.
(d) The vertical shearing stress at a point in the stem just below the flange on the cross section where the maximum vertical shearing stress occurs.

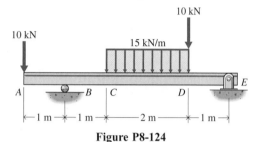

**Figure P8-124**

**8-125**  A WT12 $\times$ 52 structural tee (see Appendix A) is loaded and supported as a beam (with the flange on the bottom), as shown in Fig. P8-125. Determine
(a) The maximum tensile flexural stress in the beam.
(b) The maximum compressive flexural stress in the beam.
(c) The maximum vertical shearing stress in the beam.
(d) The vertical shearing stress at a point in the stem just above the flange on the cross section where the maximum vertical shearing stress occurs.

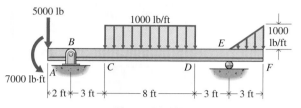

**Figure P8-125**

**8-126**  An extruded aluminum alloy beam has the cross section shown in Fig. P8-126. All parts of the section are 5 mm thick. When a constant shear force $V = 4450$ N is being supported by the beam, determine
(a) The shearing stresses at points $A$ and $B$ of the cross section.
(b) The maximum horizontal shearing stress in the beam.

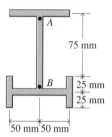

**Figure P8-126**

# FLEXURAL LOADING: BEAM DEFLECTIONS

## 9-1 INTRODUCTION

Important relations between applied load and stress (flexural and shear) in a beam were presented in Chapter 8. A beam design, however, is frequently not complete until the amount of deflection has been determined for the specified load. Failure to control beam deflections within proper limits in building construction is frequently reflected by the development of cracks in plastered walls and ceilings. Beams in many machines must deflect just the right amount for gears or other parts to make proper contact. In innumerable instances the requirements for a beam involve a given load-carrying capacity with a specified maximum deflection.

The deflection of a beam depends on the stiffness of the material and the dimensions of the beam as well as on the applied loads and supports. Three common methods for calculating beam deflections owing to flexural stresses are presented here: (1) the integration method, (2) the singularity function method, and (3) the superposition method.

## 9-2 THE DIFFERENTIAL EQUATION OF THE ELASTIC CURVE

When a straight beam is loaded and the action is elastic, the centroidal axis of the beam is a curve defined as the elastic curve. In regions of constant bending moment, the elastic curve is an arc of a circle of radius $\rho$ as indicated in Fig. 9-1. Since the portion $AB$ of the beam is bent only with couples, plane sections $A$ and $B$ remain plane and the deformation of the fibers (elongation and contraction) is proportional to the distance from the neutral surface, which is unchanged in length. From Fig. 9-1,

$$\theta = \frac{L}{\rho} = \frac{L + \delta}{\rho + c}$$

from which

$$\frac{c}{\rho} = \frac{\delta}{L} = \epsilon = \frac{\sigma}{E} = \frac{Mc}{EI} \tag{9-1}$$

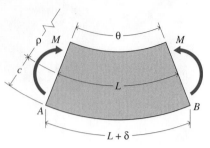

**Figure 9-1**

Therefore,

$$\frac{1}{\rho} = \frac{M}{EI} \tag{9-2}$$

which relates the radius of curvature $\rho$ of the neutral surface of the beam to the bending moment $M$, the modulus of elasticity $E$ of the material, and the second moment $I$ of the cross-sectional area.

Equation (9-2) for the curvature, $1/\rho$, of the elastic curve is useful only when the bending moment is constant for the interval of the beam involved. For most beams, the bending moment is a function of position along the beam and a more general expression is required.

The curvature from the calculus (see any standard calculus textbook) is

$$\frac{1}{\rho} = \frac{d^2y/dx^2}{[1 + (dy/dx)^2]^{3/2}}$$

For most beams the slope $dy/dx$ is very small, and its square can be neglected in comparison to unity. With this approximation,

$$\frac{1}{\rho} = \frac{d^2y}{dx^2}$$

and Eq. (9-2) becomes

$$EI\frac{d^2y}{dx^2} = M(x) \tag{9-3}$$

which is the differential equation for the elastic curve of a beam where the moment $M$ is a function of $x$, $M(x)$.

The differential equation of the elastic curve [Eq. (9-3)] can also be obtained from the geometry of the bent beam as shown in Fig. 9-2, where it is evident that $dy/dx = \tan \theta \cong \theta$ for small angles and that $d^2y/dx^2 = d\theta/dx$. Again from Fig. 9-2,

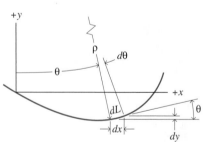

**Figure 9-2**

$$d\theta = \frac{dL}{\rho} \cong \frac{dx}{\rho}$$

for small angles, Therefore,

$$\frac{d^2y}{dx^2} = \frac{d\theta}{dx} = \frac{1}{\rho} = \frac{M}{EI}$$

or

$$EI\frac{d^2y}{dx^2} = M \tag{9-3}$$

The sign convention for bending moments established in Section 8-6 will be used for Eq. (9-3). Both $E$ and $I$ are always positive; therefore, the signs of the bending moment and the second derivative must be consistent. With the coordinate axes shown in Fig. 9-3, the slope changes from positive to negative in the interval from $A$ to $B$; therefore, the second derivative is negative, which agrees with the sign convention for the moment established in Section 8-6. For the interval $BC$, both $d^2y/dx^2$ and $M$ are positive.

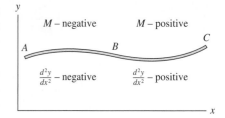

**Figure 9-3**

Figure 9-3 also reveals that the signs of the bending moment and the second derivative are also consistent when the origin of the coordinate system is selected at the right end of the beam with $x$ positive to the left and $y$ positive upward.

Equations (8-13d), (8-14c), and (9-3) provide a means for correlating the successive derivatives of the deflection $y$ of the elastic curve with the physical quantities that they represent in beam action. They are

$$\text{deflection} = y$$

$$\text{slope} = \frac{dy}{dx}$$

$$\text{moment} = EI\frac{d^2y}{dx^2} \text{ [from Eq. (9-3)]}$$

$$\text{shear} = \frac{dM}{dx} \text{ [from Eq. (8-14c)]} = EI\frac{d^3y}{dx^3} \text{ (for } EI \text{ constant)}$$

$$\text{load} = \frac{dV}{dx} \text{ [from Eq. (8-13d)]} = EI\frac{d^4y}{dx^4} \text{ (for } EI \text{ constant)}$$

where the signs are as given in Section 8-5.

Before developing specific methods for calculating beam deflections, it is advisable to consider the assumptions used in the development of the basic relation, Eq. (9-3). All of the limitations that apply to the flexure formula apply to the calculation of deflections because the flexure formula was used in the derivation of Eq. (9-3). It is further assumed that

1. The square of the slope of the beam is negligible compared to unity.
2. The beam deflection due to shearing stresses is negligible (a plane section is assumed to remain plane).
3. The values of $E$ and $I$ remain constant for any interval along the beam. In case either of them varies and can be expressed as a function of the distance $x$ along the beam, a solution of Eq. (9-3) that takes this variation into account may be possible.

## 9-3 DEFLECTIONS BY INTEGRATION

Whenever the assumptions of the previous section are essentially correct and the bending moment can be readily expressed as an integrable function of $x$, Eq. (9-3) can be solved for the deflection $y$ of the elastic curve of a beam at any point $x$ along the beam. The constants of integration can be evaluated from the applicable boundary or matching conditions.

A *boundary condition* is defined as a known set of values for $x$ and $y$, or $x$ and $dy/dx$, at a specific location along the beam. One boundary condition can be used to determine one and only one constant of integration. For example, a

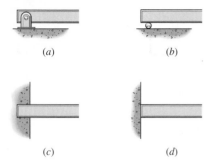

(a)          (b)

(c)          (d)

**Figure 9-4**

pin or roller at any point in a beam (Figs. 9-4a and b) represents a simple support at which the beam cannot deflect (unless otherwise stated in the problem) but can rotate. At a fixed end, as represented by Figs. 9-4c and d, the beam can neither deflect nor rotate unless otherwise stated. Thus, a boundary condition is $y = 0$ at the pin support of Fig. 9-4a and $y = 0$ at the roller support of Fig. 9-4b. At the fixed support shown in Fig. 9-4c (or d), the boundary conditions are $y = 0$ and $dy/dx = 0$.

Many beams are subjected to abrupt changes in loading along the beam, such as concentrated loads, reactions, or even distinct changes in the amount of uniformly distributed load. Because the expressions for the bending moment on the left and right of any abrupt change in load are different functions of $x$, it is impossible to write a single equation for the bending moment in terms of ordinary algebraic functions that is valid for the entire length of the beam. This can be resolved by writing separate bending moment equations for each interval of the beam. Although the intervals are bounded by abrupt changes in load, the beam is continuous at such locations; therefore, the slope and the deflection at the junction of adjacent intervals must match. A *matching condition* is defined as the equality of slope or deflection, as determined at the junction of two intervals from the elastic curve equations for both intervals. One matching condition (for example, at $x$ equals $L/3$, $y$ from the left equation equals $y$ from the right equation) can be used to determine one and only one constant of integration.

The procedure for obtaining beam deflections when matching conditions are required is lengthy and tedious. A method is presented in Section 9-5 in which singularity functions are used to write a single equation for the bending moment that is valid for the entire length of the beam; this eliminates the need for matching conditions and, accordingly, reduces the labor involved.

Calculating the deflection of a beam by the integration method, that is, integrating Eq. (9-3) twice, involves four definite steps, and the following sequence for these steps is strongly recommended.

1. Select the interval or intervals of the beam to be used; next, place a set of coordinate axes on the beam with the origin at one end of an interval and then indicate the range of values of $x$ in each interval. For example, two adjacent intervals might be

$$0 \leq x \leq L/3 \quad \text{and} \quad L/3 \leq x \leq L$$

2. List the available boundary and matching conditions (where two or more adjacent intervals are used) for each interval selected. Remember that two conditions are required to evaluate the two constants of integration for each interval used.

3. Express the bending moment as a function of $x$ for each interval selected and equate it to $EI(d^2y/dx^2)$.

4. Solve the differential equation or equations from step 3 and evaluate all constants of integration. Check the resulting equations for dimensional homogeneity. Calculate the deflection at specific points when required.

The following examples illustrate the use of the integration method for calculating beam deflections.

## Example Problem 9-1

A beam is loaded and supported as shown in Fig. 9-5a. Determine

(a) The equation of the elastic curve.
(b) The deflection at the left end of the beam.
(c) The slope at the left end of the beam.

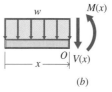

(a)

### SOLUTION

(a) The beam is sectioned at position $x$, and a free-body diagram is drawn for the segment of the beam to the left of the section, as shown in Fig. 9-5b. The notations $V(x)$ and $M(x)$ indicate that the shear force and bending moment are functions of $x$. Both $V(x)$ and $M(x)$ are taken as positive quantities. Summing moments about a horizontal axis in the plane of the section eliminates the shear force and yields

(b)

**Figure 9-5**

$$+ \circlearrowleft \Sigma M_O = 0: \qquad M(x) + \frac{wx^2}{2} = 0$$

from which

$$M(x) = -\frac{wx^2}{2} \qquad 0 \le x \le L$$

Equation (9-3) then gives

$$EI \frac{d^2y}{dx^2} = -\frac{wx^2}{2} \qquad 0 \le x \le L$$

Successive integration gives

$$EI \frac{dy}{dx} = -\frac{wx^3}{6} + C_1 \qquad\qquad \text{(a)}$$

and

$$EIy = -\frac{wx^4}{24} + C_1 x + C_2 \qquad\qquad \text{(b)}$$

where $C_1$ and $C_2$ are constants of integration to be determined using the boundary conditions. Boundary conditions are simply conditions that must be satisfied by a function [in this case $y(x)$] at the boundaries or ends of the interval. The cantilevered connection at the wall provides whatever force is necessary to prevent vertical motion ($y = 0$ at $x = L$) and whatever moment is necessary to prevent rotation ($dy/dx = \theta = 0$ at $x = L$). Therefore, the available boundary conditions are

$$\frac{dy}{dx} = 0 \qquad \text{when} \qquad x = L \qquad\qquad \text{(c)}$$

$$y = 0 \qquad \text{when} \qquad x = L \qquad\qquad \text{(d)}$$

Substituting Eq. (c) into Eq. (a) and Eq. (d) into Eq. (b) gives

$$0 = -\frac{wL^3}{6} + C_1 \qquad C_1 = \frac{wL^3}{6}$$

and

$$0 = -\frac{wL^4}{24} + \frac{wL^3}{6}(L) + C_2 \qquad C_2 = -\frac{wL^4}{8}$$

Substituting values of $C_1$ and $C_2$ into Eq. (b) gives the equation of the elastic curve for the beam. Thus,

$$y = -\frac{w}{24EI}(x^4 - 4L^3x + 3L^4) \qquad 0 \le x \le L \qquad \textbf{Ans.}$$

Note that the elastic curve equation is dimensionally homogeneous.

(b) At the left end of the beam, $x = 0$; therefore,

$$y = -\frac{w}{24EI}(0 - 0 + 3L^4) = -\frac{wL^4}{8EI} = \frac{wL^4}{8EI} \downarrow \qquad \textbf{Ans.}$$

The negative sign indicates that the deflection is downward.

(c) The slope of the elastic curve is given by Eq. (a) as

$$\frac{dy}{dx} = -\frac{w}{6EI}(x^3 - L^3) \qquad 0 \le x \le L$$

from which the slope at the left end of the beam, where $x = 0$, is

$$\frac{dy}{dx} = -\frac{w}{6EI}(0 - L^3) = +\frac{wL^3}{6EI} = \frac{wL^3}{6EI} \; \measuredangle \qquad \textbf{Ans.}$$

The plus sign indicates that the slope of the beam at the left end is positive (upward and to the right). ∎

## Example Problem 9-2

For the beam loaded and supported as shown in Fig. 9-6a, determine

(a) The equation of the elastic curve for the interval between the supports.
(b) The deflection midway between the supports.
(c) The point of maximum deflection between the supports.
(d) The maximum deflection in the interval between the supports.

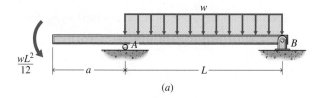

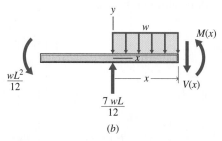

**Figure 9-6**

## SOLUTION

(a) From a free-body diagram of the beam and the equation $\Sigma M_B = 0$,

$$+\circlearrowleft \Sigma M_B = 0: \qquad R_A(L) - \frac{wL^2}{12} - wL\frac{L}{2} = 0$$

$$R_A = +\frac{7wL}{12} = \frac{7wL}{12} \uparrow$$

As indicated in Fig. 9-6b, the origin of coordinates is selected at the left support, and the interval to be used is $0 \leq x \leq L$. The two required boundary conditions are $y = 0$ when $x = 0$ and $y = 0$ when $x = L$. (The origin of coordinates is often placed at the left support to simplify the determination of the constants of integration. Since the deflection is zero at a support, this choice makes $y = 0$ at $x = 0$, which often makes $C_2 = 0$.)

From the free-body diagram of the portion of the beam shown in Fig. 9-6b, Eq. (9-3) yields

$$EI\frac{d^2y}{dx^2} = M(x) = \frac{7wL}{12}x - \frac{wL^2}{12} - wx\left(\frac{x}{2}\right) \qquad 0 \leq x \leq L$$

Successive integration gives

$$EI\frac{dy}{dx} = \frac{7wL}{24}x^2 - \frac{wL^2}{12}x - \frac{w}{6}x^3 + C_1$$

and

$$EIy = \frac{7wL}{72}x^3 - \frac{wL^2}{24}x^2 - \frac{w}{24}x^4 + C_1x + C_2$$

where $C_1$ and $C_2$ are constants of integration to be determined using the boundary conditions. Substitution of the boundary condition $y = 0$ when $x = 0$ gives

$$C_2 = 0$$

Substitution of the remaining boundary condition $y = 0$ when $x = L$ gives

$$C_1 = -\frac{wL^3}{72}$$

Therefore, the elastic curve equation is

$$y = -\frac{w}{72EI}(3x^4 - 7Lx^3 + 3L^2x^2 + L^3x) \qquad 0 \le x \le L \qquad \textbf{Ans.}$$

(b) The deflection midway between the supports is obtained by substituting $x = L/2$ into the elastic curve equation. Thus,

$$y = -\frac{w}{72EI}\left[3\left(\frac{L}{2}\right)^4 - 7L\left(\frac{L}{2}\right)^3 + 3L^2\left(\frac{L}{2}\right)^2 + L^3\left(\frac{L}{2}\right)\right]$$

from which

$$y = -\frac{wL^4}{128EI} = \frac{wL^4}{128EI}\downarrow \qquad \textbf{Ans.}$$

(c) The maximum deflection occurs where the slope $dy/dx$ is zero

$$\frac{dy}{dx} = -\frac{w}{72EI}(12x^3 - 21Lx^2 + 6L^2x + L^3) = 0 \qquad 0 \le x \le L$$

from which

$$12x^3 - 21Lx^2 + 6L^2x + L^3 = 0$$

This cubic equation has three roots: $x = -0.1162L$, $x = +0.541L$, and $x = +1.325L$. Only the middle root has physical significance in this problem, since $0 \le x \le L$. Therefore, the point of maximum deflection is at

$$x = 0.541L \text{ to the right of the left support} \qquad \textbf{Ans.}$$

(d) The maximum deflection can readily be obtained by substituting $0.541L$ for $x$ in the elastic curve equation. The result is

$$y = -\frac{7.88(10^{-3})wL^4}{EI} = \frac{7.88(10^{-3})wL^4}{EI}\downarrow \quad \blacksquare \qquad \textbf{Ans.}$$

## Example Problem 9-3

For the beam loaded and supported as shown in Fig. 9-7a, determine the deflection of the right end.

### SOLUTION

From the free-body diagram of Fig. 9-7b, the equations of equilibrium give the support reactions

$+\uparrow \Sigma F_y = 0$:     $V_A - w(L/3) = 0$     $V_A = wL/3 \uparrow$

$+\circlearrowleft \Sigma M_A = 0$:     $M_A - (5L/6)[w(L/3)] = 0$     $M_A = 5wL^2/18 \circlearrowleft$

Therefore, with the sign conventions established in Section 8-6, the shear force and bending moment at $x = 0$ are

$$V(0) = +\frac{wL}{3} \quad \text{and} \quad M(0) = -\frac{5wL^2}{18}$$

In Fig. 9-7b there are two intervals to be considered, namely, the loaded portion of the beam and the unloaded portion of the beam. The loaded portion of the beam must be used because it contains the point where the deflection is required. However, a quick check reveals the absence of boundary conditions in this interval. It therefore becomes necessary to use both intervals as well as matching and boundary conditions; hence, the origin of coordinates is selected at the left end of the beam, as shown. The intervals are

$$0 \le x \le 2L/3 \quad \text{and} \quad 2L/3 \le x \le L$$

The available boundary conditions are

$$\frac{dy}{dx} = 0 \quad \text{when} \quad x = 0$$

$$y = 0 \quad \text{when} \quad x = 0$$

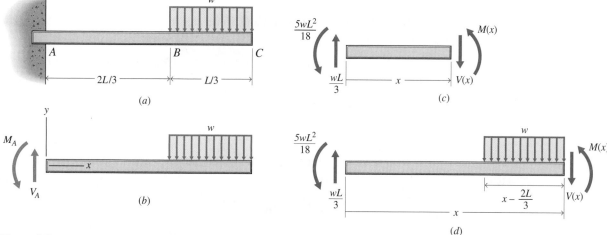

(a)    (b)    (c)    (d)

**Figure 9-7**

The available matching conditions are when $x = 2L/3$

$$\frac{dy}{dx} \text{ from the left equation} = \frac{dy}{dx} \text{ from the right equation}$$

$$y \text{ from the left equation} = y \text{ from the right equation}$$

where the left equation is for the interval $0 \le x \le 2L/3$ and the right equation is for the interval $2L/3 \le x \le L$. Four conditions (two boundary and two matching) are sufficient for the evaluation of the four constants of integration (two in each of the two elastic curve differential equations); therefore, the problem can be solved in the following manner.

From the free-body diagram of Fig. 9-7c where the beam is sectioned in the unloaded interval, Eq. (9-3) yields for $0 \le x \le 2L/3$

$$EI \frac{d^2y}{dx^2} = M(x) = \frac{wL}{3} x - \frac{5wL^2}{18} \tag{a}$$

From the free-body diagram of Fig. 9-7d where the beam is sectioned in the loaded interval, Eq. (9-3) yields for $2L/3 \le x \le L$

$$EI \frac{d^2y}{dx^2} = M(x) = \frac{wL}{3} x - \frac{5wL^2}{18} - w \left( x - \frac{2L}{3} \right) \left( \frac{x - 2L/3}{2} \right) \tag{b}$$

Integration of Equations (a) and (b) gives

$$EI \frac{dy}{dx} = \frac{wL}{6} x^2 - \frac{5wL^2}{18} x + C_1 \qquad\qquad 0 \le x \le 2L/3 \quad \text{(c)}$$

$$EI \frac{dy}{dx} = \frac{wL}{6} x^2 - \frac{5wL^2}{18} x - \frac{w}{6} \left( x - \frac{2L}{3} \right)^3 + C_3 \qquad 2L/3 \le x \le L \quad \text{(d)}$$

Substitution of the boundary condition $\frac{dy}{dx} = 0$ when $x = 0$ into Eq. (c) gives

$$C_1 = 0$$

Since the beam has a continuous slope at $x = 2L/3$

$$\frac{dy}{dx} [\text{from Eq. (c) at } x = 2L/3] = \frac{dy}{dx} [\text{from Eq. (d) at } x = 2L/3]$$

which gives

$$C_3 = 0$$

Integration of the resulting differential equations gives

$$EIy = \frac{wL}{18} x^3 - \frac{5wL^2}{36} x^2 + C_2 \qquad\qquad 0 \le x \le 2L/3$$

$$EIy = \frac{wL}{18} x^3 - \frac{5wL^2}{36} x^2 - \frac{w}{24} \left( x - \frac{2L}{3} \right)^4 + C_4 \qquad 2L/3 \le x \le L$$

Use of the remaining boundary condition yields $C_2 = 0$. Use of the final matching condition yields $C_4 = 0$.

The deflection of the right end of the beam can now be obtained from the elastic curve equation for the right interval by replacing $x$ by its value $L$. The result is

$$EIy = \frac{wL}{18}(L^3) - \frac{5wL^2}{36}(L^2) - \frac{w}{24}\left(L - \frac{2L}{3}\right)^4$$

from which

$$y = -\frac{163}{1944}\frac{wL^4}{EI} = \frac{163}{1944}\frac{wL^4}{EI}\downarrow\ \blacksquare \qquad\qquad \textbf{Ans.}$$

## Example Problem 9-4

Determine the deflection at the left end of the cantilever beam with variable width shown in Fig. 9-8.

### SOLUTION

Since the width $w$ of the beam varies linearly with respect to position $x$ along the length of the beam, the second moment of area $I$ of the cross section also varies linearly with respect to position $x$. Thus, at position $x$,

$$I = \frac{wh^3}{12} = \frac{(bx/L)h^3}{12} = \left(\frac{bh^3}{12}\right)\left(\frac{x}{L}\right) = I_L\left(\frac{x}{L}\right)$$

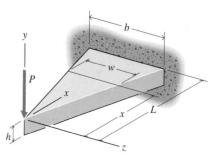

**Figure 9-8**

where $I_L = bh^3/12$ is the second moment of area of the cross section of the beam at the support. This variation in second moment of area $I$ must be included in the integration process used to determine the equation of the elastic curve for the beam. The equation of the elastic curve for the beam is obtained by successive integration from Eq. (9-3). Thus,

$$E\frac{d^2y}{dx^2} = \frac{M(x)}{I} = -\frac{Px}{I} = -\frac{Px}{I_L(x/L)} = -\frac{PL}{I_L}$$

Successive integration gives

$$EI_L\frac{dy}{dx} = -PLx + C_1$$

$$EI_Ly = -\frac{PLx^2}{2} + C_1x + C_2$$

From the boundary conditions,

$$\frac{dy}{dx} = 0 \qquad \text{when} \qquad x = L, \qquad C_1 = PL^2$$

$$y = 0 \qquad \text{when} \qquad x = L, \qquad C_2 = -\frac{PL^3}{2}$$

Thus, the equation of the elastic curve is

$$y = -\frac{PL}{EI_L}\left(\frac{x^2}{2} - Lx + \frac{L^2}{2}\right) \qquad 0 \le x \le L$$

At the left end of the beam where $x = 0$, the deflection is

$$y = -\frac{PL^3}{2EI_L} = \frac{6PL^3}{Ebh^3} \downarrow \qquad\qquad \textbf{Ans.}$$

It is also interesting to note how the maximum flexural stress varies along the length of the beam. From Eq. (8-12),

$$\sigma_{max} = \frac{Mc}{I} = \frac{Px\left(\frac{h}{2}\right)}{\left(\frac{bh^3}{12}\right)\left(\frac{x}{L}\right)} = \frac{6PL}{bh^2} \qquad\qquad (a)$$

Equation (a) indicates that the maximum flexural stress does not depend on position $x$ but is constant along the entire length of the beam. This type of beam is frequently referred to as a *constant stress beam.* ■

## PROBLEMS

**Introductory Problems**

**9-1*** A beam is loaded and supported as shown in Fig. P9-1. Determine
(a) The equation of the elastic curve. Use the designated axes.
(b) The deflection at the left end of the beam.
(c) The slope at the left end of the beam.

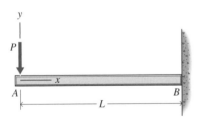

**Figure P9-1**

**9-2*** A beam is loaded and supported as shown in Fig. P9-2. Determine
(a) The equation of the elastic curve. Use the designated axes.
(b) The deflection at the right end of the beam.
(c) The slope at the right end of the beam.

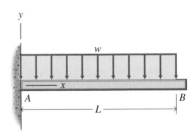

**Figure P9-2**

**9-3** A beam is loaded and supported as shown in Fig. P9-3. Determine
(a) The equation of the elastic curve. Use the designated axes.
(b) The deflection midway between the supports.
(c) The slope at the left end of the beam.

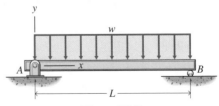

**Figure P9-3**

**9-4** A beam is loaded and supported as shown in Fig. P9-4. Determine
(a) The equation of the elastic curve. Use the designated axes.
(b) The deflection at the left end of the beam.
(c) The slope at the left end of the beam.

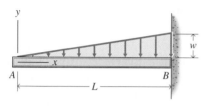

**Figure P9-4**

**9-5** A beam is loaded and supported as shown in Fig. P9-5. Determine
(a) The equation of the elastic curve. Use the designated axes.
(b) The deflection midway between the supports.
(c) The slope at the right end of the beam.

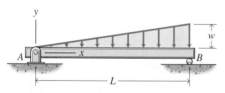

**Figure P9-5**

**9-6\*** For the steel beam [$E = 200$ GPa and $I = 32.0(10^6)$ mm$^4$] shown in Fig. P9-6, determine the deflection at a section midway between the supports.

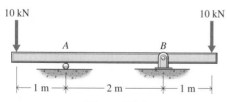

**Figure P9-6**

**9-7\*** For the steel beam ($E = 30,000$ ksi and $I = 32.1$ in$^4$) shown in Fig. P9-7, determine the deflection at a section midway between the supports.

**Figure P9-7**

**9-8** A cantilever beam is fixed at the left end and carries a uniformly distributed load $w$ over the full length of the beam. In addition, the right end is subjected to a moment of $+3wL^2/8$, as shown in Fig. P9-8. Determine the maximum deflection in the beam if $I = 2.5(10^6)$ mm$^4$, $E = 210$ GPa, $L = 3$ m, and $w = 1500$ N/m.

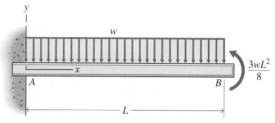

**Figure P9-8**

### Intermediate Problems

**9-9\*** A beam is loaded and supported as shown in Fig. P9-9. Determine
(a) The equation of the elastic curve. Use the designated axes.
(b) The slope at the left end of the beam.
(c) The deflection midway between the supports.

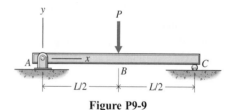

**Figure P9-9**

**9-10\*** A 100- × 300-mm timber having a modulus of elasticity of 8 GPa is loaded and supported as shown in Fig. P9-10. Determine
(a) The deflection at the 7-kN load.
(b) The deflection at the free end of the beam.

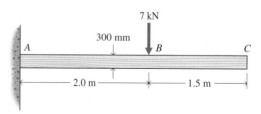

**Figure P9-10**

**9-11** A 130-lb boy is standing on a 1.6- × 12-in. wood ($E = 1400$ ksi) diving board, as shown in Fig. P9-11. If length $AB$ is 2 ft and length $BC$ is 5 ft, determine the maximum deflection in the diving board.

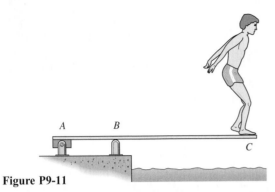

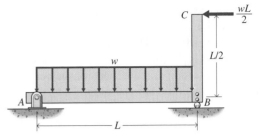

**Figure P9-11**

**Figure P9-15**

**9-12\*** A timber beam 150 mm wide × 300 mm deep is loaded and supported as shown in Fig. P9-12. The modulus of elasticity of the timber is 10 GPa. A pointer is attached to the right end of the beam. Determine
(a) The deflection of the right end of the pointer.
(b) The maximum deflection of the beam.

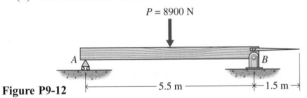

**Figure P9-12**

**9-13** A beam is loaded and supported as shown in Fig. P9-13. Determine
(a) The equation of the elastic curve. Use the designated axes.
(b) The slope at the right end of the beam.

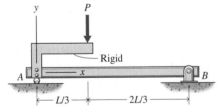

**Figure P9-13**

**9-14** A beam is loaded and supported as shown in Fig. P9-14. Determine
(a) The equation of the elastic curve. Use the designated axes.
(b) The deflection at the left end of the beam.

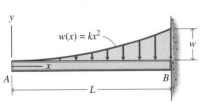

**Figure P9-14**

**9-15\*** Determine the deflection midway between the supports for beam $AB$ of Fig. P9-15. Segment $BC$ of the beam is rigid.

**9-16** The beam shown in Fig. P9-16 is a W203 × 60 structural steel ($E = 200$ GPa) wide-flange section (see Appendix A). Determine
(a) The equation of the elastic curve for the region of the beam between the supports. Use the designated axes.
(b) The deflection midway between the supports if $w = 3.5$ kN/m and $L = 5$ m.

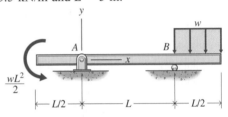

**Figure P9-16**

**Challenging Problems**

**9-17\*** A beam $AB$ is loaded and supported as shown in Fig. P9-17. The load $P$ is applied through a collar, which can be positioned on the load bar $DE$ at any location in the interval $L/4 < a < 3L/4$. Determine
(a) The equation of the elastic curve for beam $AB$.
(b) The location of load $P$ for maximum deflection at end $B$.
(c) The location of load $P$ for zero deflection at end $B$.

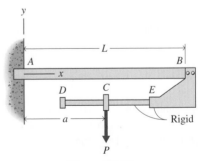

**Figure P9-17**

**9-18\*** The cantilever beam $ABC$ shown in Fig. P9-18 has a second moment of area $I_{AB} = 2I$ in the interval $AB$ and a second moment of area $I_{BC} = I$ in the interval $BC$. Determine
(a) The deflection at section $B$.
(b) The deflection at section $C$.

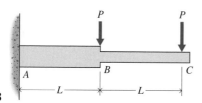

**Figure P9-18**

**9-19** The simply supported beam $ABCD$ shown in Fig. P9-19 has a second moment of area $I_{BC} = 2I$ in the center section $BC$ and a second moment of area $I_{AB} = I_{CD} = I$ in the other two sections near the supports. Determine
(a) The deflection at section $B$.
(b) The maximum deflection in the beam.

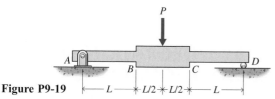

**Figure P9-19**

**9-20** The cantilever beam $ABC$ shown in Fig. P9-20 has a second moment of area $I_{AB} = 4I$ in the interval $AB$ and a second moment of area $I_{BC} = I$ in the interval $BC$. Determine
(a) The deflection at section $B$.
(b) The deflection at section $C$.

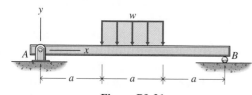

**Figure P9-20**

**9-21\*** A beam is loaded and supported as shown in Fig. P9-21. Determine the deflection midway between the supports.

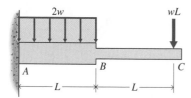

**Figure P9-21**

**9-22** A beam is loaded and supported as shown in Fig. P9-22. Determine the deflection at the left end of the distributed load.

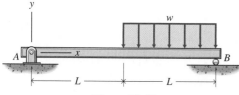

**Figure P9-22**

**9-23** A beam is loaded and supported as shown in Fig. P9-23. Determine
(a) The slope at the left end of the beam.
(b) The maximum deflection between the supports.

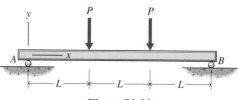

**Figure P9-23**

**9-24\*** The cantilever beam $ABC$ shown in Fig. P9-24 has a second moment of area $I_{AB} = 4I$ in the interval $AB$ and a second moment of area $I_{BC} = I$ in the interval $BC$. Determine
(a) The deflection at section $B$.
(b) The deflection at section $C$.

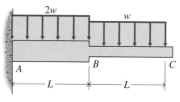

**Figure P9-24**

**Computer Problems**

**9-25** A 160-lb diver walks slowly onto a diving board. The diving board is a wood ($E = 1800$ ksi) plank 10 ft long, 18 in. wide, and 2 in. thick, and is modeled as the cantilever beam shown in Fig. P9-25. For the diver at positions $a = nL/5$ ($n = 1, 2, \ldots, 5$), compute and plot the deflection curve for the diving board (plot $y$ as a function of $x$ for $0 \le x \le 10$ ft).

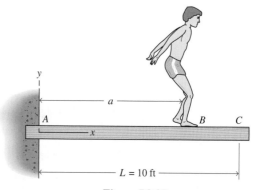

**Figure P9-25**

**9-26** A diving board consists of a wood ($E = 12$ GPa) plank that is pinned at the left end and rests on a movable support as shown in Fig. P9-26. The board is 3 m long, 500 mm wide, and 80 mm thick. If a 70-kg diver stands at the end of the board,

(a) Compute and plot the deflection curve for the beam (plot $y$ as a function of $x$ for $0 \le x \le 3$ m) for the right support at positions $b = 0.5$ m, 1.0 m, and 1.5 m.

(b) If the stiffness of the board is defined as the ratio of the diver's weight to the deflection at the end of the board $k = (W/y)$, compute and plot the stiffness of the board as a function of $b$ for $0 \le b \le 1.5$ m. Does the stiffness depend on the weight of the diver?

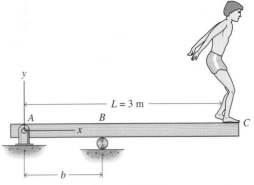

**Figure P9-26**

---

# 9-4 DEFLECTIONS BY INTEGRATION OF SHEAR-FORCE OR LOAD EQUATIONS

In Section 9-3, the equation of the elastic curve was obtained by integrating Eq. (9-3) and applying the appropriate boundary conditions to evaluate the two constants of integration. In a similar manner, the equation of the elastic curve can be obtained from load and shear-force equations. The differential equations that relate deflection $y$ to load $w(x)$ or deflection $y$ to shear-force $V(x)$ are obtained by substituting Eq. (8-14c) or (8-13d), respectively, into Eq. (9-3). Thus,

$$EI \, \frac{d^2y}{dx^2} = M(x) \tag{9-3}$$

$$EI \, \frac{d^3y}{dx^3} = V(x) \tag{9-4}$$

$$EI \, \frac{d^4y}{dx^4} = w(x) \tag{9-5}$$

When Eq. (9-4) or (9-5) is used to obtain the equation of the elastic curve, either three or four integrations will be required instead of the two integrations required with Eq. (9-3). These additional integrations will introduce additional constants of integration. The boundary conditions, however, now include conditions on the shear forces and bending moments, in addition to the conditions on slopes and deflections. Use of a particular differential equation is usually made on the basis of mathematical convenience or personal preference. In those instances when the expression for the load is easier to write than the expression for the moment, Eq. (9-5) would be preferred over Eq. (9-3). The following examples illustrate the use of Eq. (9-5) for calculating beam deflections.

## ■ Example Problem 9-5

A beam is loaded and supported as shown in Fig. 9-9. Determine

(a) The equation of the elastic curve.

(b) The maximum deflection of the beam.

## SOLUTION

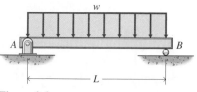

**Figure 9-9**

(a) Since the equation for the load distribution [$w(x) = w = $ constant] is given, Eq. (9-5) will be used to determine the equation of the elastic curve. In Section 8-7 (see Fig. 8-22), the upward direction was considered positive for a distributed load $w$; therefore, Eq. (9-5) is written as (where the origin of the coordinate axes is at $A$)

$$EI \frac{d^4 y}{dx^4} = w(x) = -w \qquad 0 \le x \le L$$

Successive integration gives

$$EI \frac{d^3 y}{dx^3} = V(x) = -wx + C_1$$

$$EI \frac{d^2 y}{dx^2} = M(x) = -\frac{wx^2}{2} + C_1 x + C_2$$

$$EI \frac{dy}{dx} = -\frac{wx^3}{6} + C_1 \frac{x^2}{2} + C_2 x + C_3$$

$$EIy = -\frac{wx^4}{24} + C_1 \frac{x^3}{6} + C_2 \frac{x^2}{2} + C_3 x + C_4$$

The four constants of integration are determined by applying the boundary conditions. Thus,

At $x = 0$, $y = 0$;    therefore, $C_4 = 0$

At $x = 0$, $M = 0$;    therefore, $C_2 = 0$

At $x = L$, $M = 0$;    therefore, $C_1 = \dfrac{wL}{2}$

At $x = L$, $y = 0$;    therefore, $C_3 = -\dfrac{wL^3}{24}$

Thus,

$$y = \frac{w}{24EI}[-x^4 + 2Lx^3 - L^3 x] \qquad 0 \le x \le L \qquad \textbf{Ans.}$$

The constant $C_1$ could also have been determined from a boundary condition involving the shear force $V$. For example, the shear force jumps upward by $wL/2$ across the left support. Therefore, at $x = 0$ the shear force equation gives $V = -w(0) + C_1 = wL/2$ and the first constant of integration is $C_1 = wL/2$.

(b) The maximum deflection occurs at $x = L/2$ (because of symmetry), which when substituted into the equation of the elastic curve gives

$$y_{max} = -\frac{5wL^4}{384EI} = \frac{5wL^4}{384EI} \downarrow \blacksquare \qquad \textbf{Ans.}$$

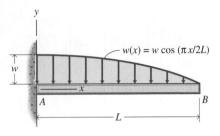

**Figure 9-10**

# Example Problem 9-6

A beam is loaded and supported as shown in Fig. 9-10. Determine

(a) The equation of the elastic curve.

(b) The deflection at the right end of the beam.

(c) The support reactions $V_A$ and $M_A$ at the left end of the beam.

## SOLUTION

(a) Since the equation for the load distribution is given and the moment equation is not easy to write, Eq. (9-5) will be used to determine the deflections. In Section 8-7 (see Fig. 8-22), the upward direction was considered positive for a distributed load $w$; therefore, Eq. (9-5) is written as

$$EI \frac{d^4y}{dx^4} = w(x) = -w \cos \frac{\pi x}{2L} \qquad 0 \le x \le L$$

Successive integration gives

$$EI \frac{d^3y}{dx^3} = V(x) = -\frac{2wL}{\pi} \sin \frac{\pi x}{2L} + C_1$$

$$EI \frac{d^2y}{dx^2} = M(x) = \frac{4wL^2}{\pi^2} \cos \frac{\pi x}{2L} + C_1 x + C_2$$

$$EI \frac{dy}{dx} = \frac{8wL^3}{\pi^3} \sin \frac{\pi x}{2L} + C_1 \frac{x^2}{2} + C_2 x + C_3$$

$$EIy = -\frac{16wL^4}{\pi^4} \cos \frac{\pi x}{2L} + C_1 \frac{x^3}{6} + C_2 \frac{x^2}{2} + C_3 x + C_4$$

The four constants of integration are determined by applying the boundary conditions. Thus,

At $x = 0$, $y = 0$;     therefore, $C_4 = \dfrac{16wL^4}{\pi^4}$

At $x = 0$, $\dfrac{dy}{dx} = 0$;     therefore, $C_3 = 0$

At $x = L$, $V = 0$;     therefore, $C_1 = \dfrac{2wL}{\pi}$

At $x = L$, $M = 0$;     therefore, $C_2 = -\dfrac{2wL^2}{\pi}$

Thus,

$$y = -\frac{w}{3\pi^4 EI} \left[ 48L^4 \cos \frac{\pi x}{2L} - \pi^3 Lx^3 + 3\pi^3 L^2 x^2 - 48L^4 \right] \qquad \textbf{Ans.}$$

where $0 \le x \le L$.

(b) The deflection at the right end of the beam is

$$y_B = y_{x=L} = -\frac{w}{3\pi^4 EI}(-\pi^3 L^4 + 3\pi^3 L^4 - 48L^4)$$

$$= -\frac{(2\pi^3 - 48)wL^4}{(3\pi^4 EI)} = -0.04795wL^4/EI \qquad \textbf{Ans.}$$

(c) The shear force $V(x)$ and bending moment $M(x)$ at any distance $x$ from the support are

$$V(x) = \frac{2wL}{\pi}\left[1 - \sin\frac{\pi x}{2L}\right]$$

$$M(x) = \frac{2wL}{\pi^2}\left[2L \cos\frac{\pi x}{2L} + \pi x - \pi L\right]$$

Thus, the support reactions at the left end of the beam are

$$V_A = V_{x=0} = \frac{2wL}{\pi} \qquad \textbf{Ans.}$$

$$M_A = M_{x=0} = -\frac{2(\pi - 2)wL^2}{\pi^2} \quad \blacksquare \qquad \textbf{Ans.}$$

# PROBLEMS

## Introductory Problems

**9-27\***  For the beam and loading shown in Fig. P9-27, determine
(a) The equation of the elastic curve.
(b) The maximum deflection for the beam.

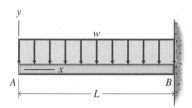

**Figure P9-27**

**9-28**  For the beam and loading shown in Fig. P9-28, determine
(a) The equation of the elastic curve.
(b) The deflection midway between the supports.

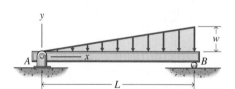

**Figure P9-28**

## Intermediate Problems

**9-29\***  A beam is loaded and supported as shown in Fig. P9-29. Determine
(a) The equation of the elastic curve.
(b) The deflection at the left end of the beam.
(c) The support reactions $V_B$ and $M_B$.

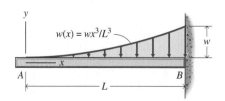

**Figure P9-29**

**9-30** A beam is loaded and supported as shown in Fig. P9-30. Determine

(a) The equation of the elastic curve.

(b) The deflection midway between the supports.

(c) The maximum deflection of the beam.

(d) The support reactions $R_A$ and $R_B$.

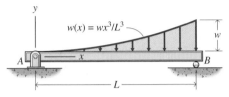

**Figure P9-30**

**Challenging Problems**

**9-31\*** A beam is loaded and supported as shown in Fig. P9-31. Determine

(a) The equation of the elastic curve.

(b) The deflection at the left end of the beam.

(c) The support reactions $V_B$ and $M_B$.

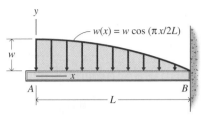

**Figure P9-31**

**9-32** A beam is loaded and supported as shown in Fig. P9-32. Determine

(a) The equation of the elastic curve.

(b) The deflection midway between the supports.

(c) The maximum deflection of the beam.

(d) The slope at the left end of the beam.

(e) The support reactions $R_A$ and $R_B$.

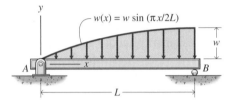

**Figure P9-32**

# 9-5 SINGULARITY FUNCTIONS

The double integration method of Section 9-3 becomes tedious and time-consuming when several intervals and several sets of matching conditions are required. The labor involved in solving problems of this type, however, can be diminished by making use of singularity functions following the method developed in 1862 by the German mathematician A. Clebsch (1833–1872).[1]

Singularity functions are closely related to the unit step function used by the British physicist O. Heaviside (1850–1925) to analyze the transient response of electrical circuits. Singularity functions will be used here for writing one bending moment equation that applies in all intervals along a beam, thus eliminating the need for matching conditions.

To illustrate the use of singularity functions, consider the beam loaded as shown in Fig. 9-11. The terms $R_L$ and $R_R$ represent support reactions at the left and right supports, respectively. The moment equations at the four designated sections are

$$M_1 = R_L x \qquad\qquad\qquad 0 < x < x_1$$

$$M_2 = R_L x - P(x - x_1) \qquad\qquad x_1 < x < x_2$$

$$M_3 = R_L x - P(x - x_1) + M_A \qquad x_2 < x < x_3$$

$$M_4 = R_L x - P(x - x_1) + M_A - \frac{w}{2}(x - x_3)^2 \qquad x_3 < x < L$$

---

[1]For a rather complete history of the Clebsch method and the numerous extensions thereof, see "Clebsch's Method for Beam Deflections," Walter D. Pilkey, *Journal of Engineering Education*, January 1964, p. 170.

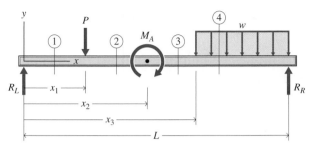

**Figure 9-11**

Note that the origin of the coordinate system is at the left end of the beam, with positive $x$ to the right. Each time a section "jumps over" a discontinuity in load (at $x_1$, $x_2$, and $x_3$), an additional term appears in the moment equation. For example, in section 2 the moment equation involves all loads to the left of the section; the same is true for section 3 and section 4.

First, consider the moment equation

$$M_2 = R_L x - P(x - x_1) \qquad x_1 < x < x_2$$

The moment equation for both $M_1$ and $M_2$ can be represented by a single moment equation

$$M = R_L x - P\langle x - x_1 \rangle^1 \qquad 0 < x < x_2$$

where $\langle x - x_1 \rangle^1$ is ignored when $x < x_1$ and replaced by parentheses $(x - x_1)^1$ when $x \geq x_1$. That is, the bracket $\langle x - x_1 \rangle^1 = 0$ when $x < x_1$ and $\langle x - x_1 \rangle^1 = (x - x_1)^1$ when $x \geq x_1$. Therefore

$$M = M_1 = R_L x \qquad 0 < x < x_1$$
$$M = M_2 = R_L x - P(x - x_1) \qquad x_1 < x < x_2$$

The exponent one in $\langle x - x_1 \rangle^1$ has the same meaning as the exponent one (which is not usually written) in $(x - x_1)$.

For a section at 3, the moment equation can be written

$$M = M_3 = R_L x - P\langle x - x_1 \rangle^1 + M_A \langle x - x_2 \rangle^0 \qquad 0 < x < x_3$$

For $x < x_2$, the term $\langle x - x_2 \rangle^0$ is zero, and for $x \geq x_2$ the term $\langle x - x_2 \rangle^0 = (x - x_2)^0 = 1$, that is, the zero power of the $\langle x - x_2 \rangle^0$ term is unity for $x \geq x_2$.

Proceeding in a similar fashion, the moment equation for the entire beam can be written using a single expression

$$M = M_4 = R_L x - P\langle x - x_1 \rangle^1 + M_A \langle x - x_2 \rangle^0 - \frac{w}{2}\langle x - x_3 \rangle^2 \qquad 0 < x < L$$

The terms $\langle x - x_1 \rangle^1$, $\langle x - x_2 \rangle^0$, and $\langle x - x_3 \rangle^2$ are called *singularity functions.*

A singularity function of $x$ is written as $\langle x - x_0 \rangle^n$, where $n$ is any integer (positive or negative) including zero, and $x_0$ is a constant equal to the value of $x$ at the initial boundary of a specific interval along a beam. The brackets $\langle \, \rangle$ are

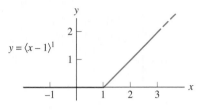

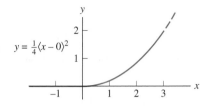

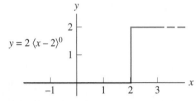

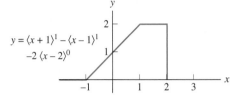

**Figure 9-12**

replaced by parentheses ( ) when $x \geq x_0$ and by zero when $x < x_0$. Selected properties of singularity functions required for beam-deflection problems are listed here for emphasis and ready reference.

$$\langle x - x_0 \rangle^n = \begin{cases} (x - x_0)^n & \text{when } n > 0 \text{ and } x \geq x_0 \\ 0 & \text{when } n > 0 \text{ and } x < x_0 \end{cases}$$

$$\langle x - x_0 \rangle^0 = \begin{cases} 1 & \text{when } x \geq x_0 \\ 0 & \text{when } x < x_0 \end{cases}$$

$$\int \langle x - x_0 \rangle^n \, dx = \frac{1}{n+1} \langle x - x_0 \rangle^{n+1} + C \qquad \text{when } n \geq 0$$

$$\frac{d}{dx} \langle x - x_0 \rangle^n = n \langle x - x_0 \rangle^{n-1} \qquad \text{when } n \geq 1$$

Several examples of singularity functions are shown in Fig. 9-12.

Distributed loadings that are sectionally continuous (the distributed load cannot be represented by a single function of $x$ for all values of $x$) are readily obtained by superposition, as illustrated in the following examples.

A special word of caution is warranted for distributed loadings. For the beam shown in Fig. 9-11 the distributed load is extended to the right end of the beam. If the distributed load does not extend to the right end of the beam (as in Fig. 9-13), the distributed load should be replaced with an equivalent loading in which each load extends to the right end of the beam (as in Fig. 9-14). Example Problems 9-8 and 9-9 illustrate how to handle such distributed loadings.

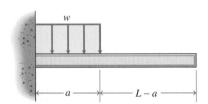

**Figure 9-13**

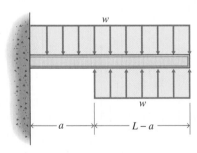

**Figure 9-14**

## Example Problem 9-7

For the beam loaded and supported as shown in Fig. 9-15a, determine the deflection of the right end of the beam.

### SOLUTION

A free-body diagram of the entire beam is shown in Fig. 9-15b. The equations of equilibrium written for this free-body diagram yield the reactions at the fixed support

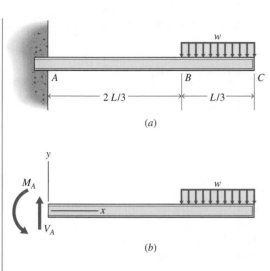

(a)

(b)

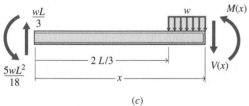

(c)

**Figure 9-15**

$+\uparrow \Sigma F_y = 0:$ $\qquad$ $V_A - w\left(\dfrac{L}{3}\right) = 0$ $\qquad$ $V_A = \dfrac{wL}{3} \uparrow$

$+\circlearrowleft \Sigma M_A = 0:$ $\qquad$ $M_A - w\left(\dfrac{L}{3}\right)\left(\dfrac{5L}{6}\right) = 0$ $\qquad$ $M_A = \dfrac{5wL^2}{18} \circlearrowright$

The free-body diagram of Fig. 9-15c is constructed by sectioning the beam in the rightmost interval and keeping the portion of the beam to the left of the section. This ensures that the resulting moment equation will be valid for the entire beam. If the cut had been made between $x = 0$ and $x = 2L/3$, then the distributed load between $x = 2L/3$ and $x = L$ would not appear on the free-body diagram or in the moment equation and the moment equation could not possibly be valid for points of the beam under the distributed load. Using moment equilibrium and summing moments about a point on the cut section

$+\circlearrowleft \Sigma M_{\text{cut}} = 0:$ $\quad$ $M(x) + \dfrac{5wL^2}{18} - \dfrac{wL}{3}x + \left[ w\left\langle x - \dfrac{2L}{3}\right\rangle^1 \right] \dfrac{\left\langle x - \dfrac{2L}{3}\right\rangle^1}{2} = 0$

yields the moment equation, for $0 \le x \le L$

$$M(x) = -\dfrac{5wL^2}{18} + \dfrac{wLx}{3} - \dfrac{w}{2}\left\langle x - \dfrac{2L}{3}\right\rangle^2 \qquad \text{(a)}$$

Substituting Eq. (a) into Eq. (9-3) yields the differential equation of the elastic curve

$$EI\frac{d^2y}{dx^2} = -\frac{5wL^2}{18} + \frac{wLx}{3} - \frac{w}{2}\left\langle x - \frac{2L}{3}\right\rangle^2 \qquad \text{(b)}$$

The boundary conditions, in the interval $0 \le x \le L$, are $dy/dx = 0$ when $x = 0$ and $y = 0$ when $x = 0$ using the axes shown in Fig. 9-15b. The first integration of Eq. (b) gives

$$EI\frac{dy}{dx} = -\frac{5wL^2x}{18} + \frac{wLx^2}{6} - \frac{w}{6}\left\langle x - \frac{2L}{3}\right\rangle^3 + C_1 \qquad \text{(c)}$$

Using the boundary condition $dy/dx = 0$ when $x = 0$ gives $C_1 = 0$ because the term in the brackets is zero when $x < 2L/3$. Integrating again gives

$$EIy = -\frac{5wL^2x^2}{36} + \frac{wLx^3}{18} - \frac{w}{24}\left\langle x - \frac{2L}{3}\right\rangle^4 + C_2 \qquad \text{(d)}$$

Using the boundary condition $y = 0$ when $x = 0$ gives $C_2 = 0$ because the term in the brackets is zero when $x < 2L/3$.

The deflection at the right end of the beam is obtained by substituting $x = L$ in the elastic curve equation [Eq. (d)],

$$EIy = -\frac{5wL^2(L)^2}{36} + \frac{wL(L)^3}{18} - \frac{w}{24}\left(L - \frac{2L}{3}\right)^4$$

or

$$y = -\frac{163}{1944}\frac{wL^4}{EI} = \frac{163}{1944}\frac{wL^4}{EI}\downarrow \qquad \textbf{Ans.}$$

This result agrees with that of Example Problem 9-3. ∎

## Example Problem 9-8

Determine the deflection at the left end of the beam of Fig. 9-16a.

### SOLUTION

A free-body diagram of the entire beam is shown in Fig. 9-16b. Moment equilibrium about support $B$ yields the support reaction at $A$

$$+\,\circlearrowleft\;\Sigma M_B = 0: \qquad R_L L - w\left(\frac{L}{2}\right)\left(\frac{7L}{4}\right) + \frac{wL^2}{2} = 0 \qquad R_L = \frac{3wL}{8}\uparrow$$

Before the moment can be written in terms of singularity functions that are valid for the full length of the beam, the distributed load must be replaced with an equivalent loading as shown in the free-body diagram of Fig. 9-16c. Between $x = -L/2$ and $x = +L$ the downward distributed load on the top of the beam

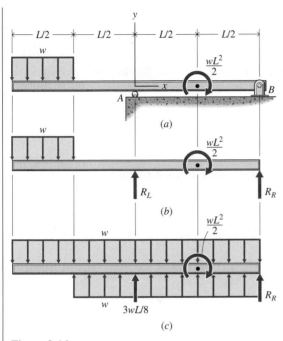

**Figure 9-16**

and the upward distributed load on the bottom of the beam cancel, leaving just the original downward distributed load over the first quarter of the beam. When the expression for the bending moment obtained from the free-body diagram of Fig. 9-16c is substituted in Eq. (9-3), the result is

$$EI\frac{d^2y}{dx^2} = -\frac{w}{2}(x + L)^2 + \frac{w}{2}\left\langle x + \frac{L}{2}\right\rangle^2 + \frac{3wL}{8}\langle x - 0\rangle^1 + \frac{wL^2}{2}\left\langle x - \frac{L}{2}\right\rangle^0$$

where the first term after the equals sign represents the distributed load on the top of the beam and the second term represents the distributed load on the bottom of the beam. The effect of the two terms is to terminate the distributed load at $x = -L/2$. The boundary conditions are $y = 0$ when $x = 0$, and $y = 0$ when $x = L$. Two integrations of the moment equation give

$$EI\frac{dy}{dx} = -\frac{w}{6}(x + L)^3 + \frac{w}{6}\left\langle x + \frac{L}{2}\right\rangle^3 + \frac{3wL}{16}\langle x - 0\rangle^2 + \frac{wL^2}{2}\left\langle x - \frac{L}{2}\right\rangle^1 + C_1$$

and

$$EIy = -\frac{w}{24}(x + L)^4 + \frac{w}{24}\left\langle x + \frac{L}{2}\right\rangle^4 + \frac{wL}{16}\langle x - 0\rangle^3$$

$$+ \frac{wL^2}{4}\left\langle x - \frac{L}{2}\right\rangle^2 + C_1x + C_2$$

The first boundary condition, $y = 0$ when $x = 0$, gives

$$0 = -\frac{wL^4}{24} + \frac{wL^4}{384} + 0 + 0 + 0 + C_2 \qquad C_2 = +\frac{5}{128}wL^4$$

Note that even though the origin of coordinates is at the left support (so $y = 0$ at $x = 0$), the constant of integration $C_2$ is not zero. At $x = 0$, $(x + L)^4 = L^4$; $\langle x + \frac{L}{2} \rangle^4 = \frac{L^4}{16}$; $\langle x - 0 \rangle^3 = 0^3 = 0$; and $\langle x - \frac{L}{2} \rangle^2 = \langle -\frac{L}{2} \rangle^2 = 0$ (because a singularity function is zero whenever its argument is negative). The second boundary condition, $y = 0$ when $x = L$, gives

$$0 = -\frac{16wL^4}{24} + \frac{81wL^4}{384} + \frac{wL^4}{16} + \frac{wL^4}{16} + C_1 L + \frac{5wL^4}{128} \qquad C_1 = +\frac{7}{24} wL^3$$

The deflection at the left end is obtained by substituting $x = -L$ in the elastic curve equation. The result is

$$EIy = 0 + 0 + 0 + 0 + \frac{7}{24} wL^3 (-L) + \frac{5}{128} wL^4 = -\frac{97}{384} wL^4$$

Thus,

$$y = -\frac{97}{384} \frac{wL^4}{EI} = \frac{97}{384} \frac{wL^4}{EI} \downarrow \quad \blacksquare \qquad \text{Ans.}$$

---

## ▌ Example Problem 9-9

Use singularity functions to write a single equation for the bending moment at any section of the beam shown in Fig. 9-17a.

### SOLUTION

The loading on the beam of Fig. 9-17a can be considered a combination of the loadings shown in Figs. 9-17b, c, d, and e, where downward-acting loads are shown on the top of the beam and upward-acting loads on the bottom of the beam. The initial distributed load is a linear function of $x$, $w = ax + b$. The constants $a$ and $b$ must be chosen so that $w = 0$ when $x = x_1$ ($0 = ax_1 + b$), and $w = w_1$ when $x = x_2$, ($w_1 = ax_2 + b$). Together, these two conditions give

$$a = \frac{w_1}{x_2 - x_1} \qquad \text{and} \qquad b = -\frac{w_1 x_1}{x_2 - x_1}$$

so that

$$w = \frac{w_1 \langle x - x_1 \rangle}{x_2 - x_1}$$

The slope of the second triangular load has to be the same as the first—it just starts later:

$$w = \frac{w_1 \langle x - x_2 \rangle}{x_2 - x_1}$$

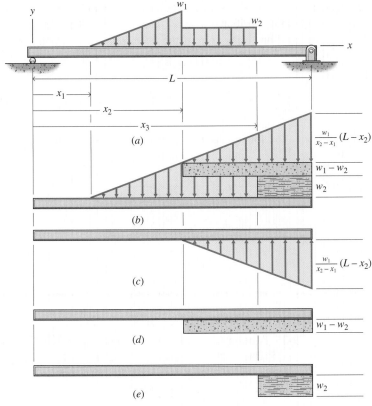

**Figure 9-17**

The moment of the linearly varying load on the top of the beam (Fig. 9-17$b$) relative to a point on a section at $x$ is

$$\left[\frac{1}{2}\frac{w_1\langle x-x_1\rangle}{x_2-x_1}\langle x-x_1\rangle\right]\frac{\langle x-x_1\rangle}{3}=\frac{w_1}{6(x_2-x_1)}\langle x-x_1\rangle^3 \qquad (a)$$

This term will affect the moment equation for all values of $x \geq x_1$. The moment of the linearly varying load on the bottom of the beam (Fig. 9-17$c$) relative to a point on a section at $x$ is

$$-\left[\frac{1}{2}\frac{w_1\langle x-x_2\rangle}{x_2-x_1}\langle x-x_2\rangle\right]\frac{\langle x-x_2\rangle}{3}=-\frac{w_1}{6(x_2-x_1)}\langle x-x_2\rangle^3 \qquad (b)$$

This term will affect the moment equation for all values of $x \geq x_2$. Together, the two linearly varying loads act to create a constant load of magnitude $w_1$ for values of $x \geq x_2$. The magnitude of this constant distributed load is reduced from $w_1$ to $w_2$ by the uniform distributed load of magnitude $w_1 - w_2$ at $x = x_2$ shown in Fig. 9-17$d$. The moment of this uniform load on the bottom of the beam relative to a point on a section at $x$ is

$$-\left[(w_1-w_2)\langle x-x_2\rangle\right]\frac{\langle x-x_2\rangle}{2}=-\frac{w_1-w_2}{2}\langle x-x_2\rangle^2 \qquad (c)$$

Finally, the constant distributed load of magnitude $w_2$ is terminated at $x = x_3$ by introducing the constant distributed load $w_2$ shown in Fig. 9-17e. The moment of this uniform load on the bottom of the beam relative to a point on a section at $x$ is

$$-[w_2 \langle x - x_3 \rangle] \frac{\langle x - x_3 \rangle}{2} = -\frac{w_2}{2} \langle x - x_3 \rangle^2 \qquad \text{(d)}$$

The moment equation for the beam is then written in terms of singularity functions as

$$M(x) = R_L x - \frac{w_1}{6(x_2 - x_1)} \langle x - x_1 \rangle^3 + \frac{w_1}{6(x_2 - x_1)} \langle x - x_2 \rangle^3$$

$$+ \frac{w_1 - w_2}{2} \langle x - x_2 \rangle^2 + \frac{w_2}{2} \langle x - x_3 \rangle^2 \ \blacksquare \qquad \text{Ans.}$$

---

## PROBLEMS

### Introductory Problems

**9-33\*** A cantilever beam is loaded and supported as shown in Fig. P9-33. Use singularity functions to determine the deflection
(a) At a distance $x = L$ from the support.
(b) At the right end of the beam.

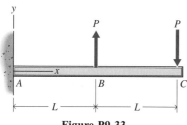

**Figure P9-33**

**9-34\*** A cantilever beam is loaded and supported as shown in Fig. P9-34. Use singularity functions to determine the deflection
(a) At a distance $x = L$ from the support.
(b) At the right end of the beam.

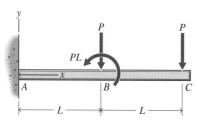

**Figure P9-34**

**9-35** A beam is loaded and supported as shown in Fig. P9-35. Use singularity functions to determine the deflection
(a) At a distance $x = L$ from the left support.
(b) At the middle of the span.

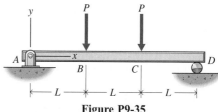

**Figure P9-35**

**9-36** A beam is loaded and supported as shown in Fig. P9-36. Use singularity functions to determine the deflection
(a) At the right end of the beam.
(b) At a section midway between the supports.

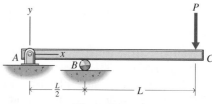

**Figure P9-36**

**9-37** A cantilever beam is loaded and supported as shown in Fig. P9-37. Use singularity functions to determine the deflection
(a) At the point of application of the concentrated load $3P$.
(b) At the right end of the beam.

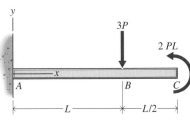

**Figure P9-37**

**9-38\*** A beam is loaded and supported as shown in Fig. P9-38. Use singularity functions to determine the deflection
(a) At the left end of the beam.
(b) At a point midway between the supports.

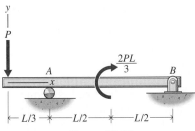

**Figure P9-38**

### Intermediate Problems

**9-39\*** Use singularity functions to determine the deflection at the left end of the cantilever beam shown in Fig. P9-39.

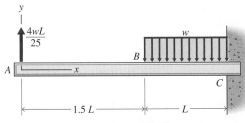

**Figure P9-39**

**9-40\*** A beam is loaded and supported as shown in Fig. P9-40. Use singularity functions to determine the deflection
(a) At the left end of the distributed load.
(b) At a section midway between the supports.

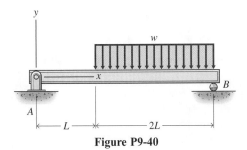

**Figure P9-40**

**9-41** A beam is loaded and supported as shown in Fig. P9-41. Use singularity functions to determine
(a) The deflection midway between the supports.
(b) The maximum deflection of the beam.

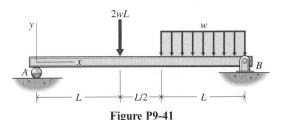

**Figure P9-41**

**9-42** A beam is loaded and supported as shown in Fig. P9-42. Use singularity functions to determine
(a) The deflection at the right end of the beam.
(b) The deflection midway between the supports.

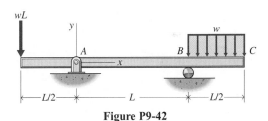

**Figure P9-42**

### Challenging Problems

**9-43\*** Use singularity functions to determine the deflection at the right end of the cantilever beam shown in Fig. P9-43.

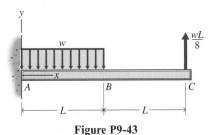

**Figure P9-43**

**9-44\*** A beam is loaded and supported as shown in Fig. P9-44. Use singularity functions to determine the deflection
(a) At the left end of the distributed load.
(b) At a section midway between the supports.

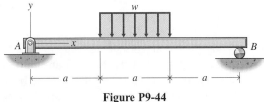

**Figure P9-44**

**9-45** A beam is loaded and supported as shown in Fig. P9-45. Use singularity functions to determine
(a) The deflection midway between the supports.
(b) The maximum deflection of the beam.

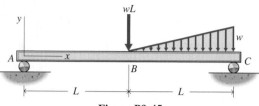

**Figure P9-45**

**9-46** A beam is loaded and supported as shown in Fig. P9-46. Use singularity functions to determine
(a) The deflection at the middle of the span.
(b) The maximum deflection in the beam.

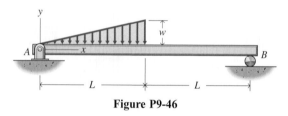

**Figure P9-46**

## Computer Problems

**9-47** A 160-lb diver walks slowly onto a diving board. The diving board is a wood ($E = 1800$ ksi) plank 10 ft long, 18 in. wide, and 2 in. thick, and is modeled as the cantilever beam shown in Fig. P9-47. For the diver at positions $a = nL/5$ ($n = 1, 2, \ldots, 5$), compute and plot the deflection curve for the diving board (plot $y$ as a function of $x$ for $0 \le x \le 10$ ft).

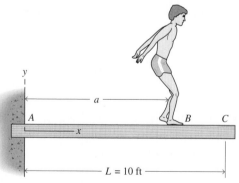

**Figure P9-47**

**9-48** A diving board consists of a wood ($E = 12$ GPa) plank, which is pinned at the left end and rests on a movable support as shown in Fig. P9-48. The board is 3 m long, 500 mm wide, and 80 mm thick. If a 70-kg diver stands at the end of the board,

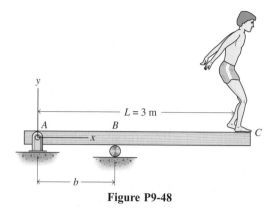

**Figure P9-48**

(a) Compute and plot the deflection curve for the beam (plot $y$ as a function of $x$ for $0 \le x \le 3$ m) for the right support at positions $b = 0.5$ m, 1.0 m, and 1.5 m.
(b) If the stiffness of the board is defined as the ratio of the diver's weight to the deflection at the end of the board $k = (W/y)$, compute and plot the stiffness of the board as a function of $b$ for $0 \le b \le 1.5$ m. Does the stiffness depend on the weight of the diver?

**9-49** A bridge over a small stream on a golf course consists of a wood deck (weighing 40 lb/ft) laid over two wood ($E = 1800$ ksi) beams. Each of the beams is 15 ft long, 4 in. wide, and $h$ in. high; each beam carries half of the loading on the bridge. It is desired that the bridge support the weight of a loaded golf cart (total weight of 840 lb) with a maximum deflection of no more than 2 in. Model the beams and the loading as shown in Fig. P9-49 and
(a) Determine the minimum depth $h_{min}$ of the beams that will support the given weight.
(b) If $h = 5$ in., compute and plot the deflection curve for the beam (plot $y$ as a function of $x$ for $0 \le x \le 15$ ft) for the cart at positions $a = 3$ ft, 6 ft, and 9 ft.

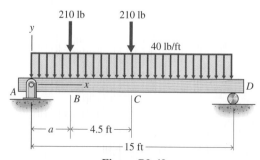

**Figure P9-49**

**9-50**  The gangplank between a fishing boat and a dock consists of a wood ($E = 12$ GPa) plank 3 m long, 300 mm wide, and 50 mm thick. If the plank is modeled as the simply supported beam shown in Fig. P9-50, compute and plot the deflection curve for the gangplank (plot $y$ as a function of $x$ for $0 \le x \le 3$ m) as a 75-kg man walks across the plank. Plot curves for the man at positions $a = nL/7$ ($n = 1, 2, \dots, 5$).

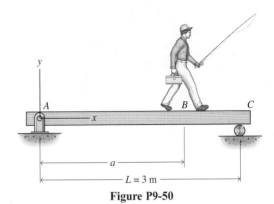

Figure P9-50

# 9-6 DEFLECTIONS BY SUPERPOSITION

The method of superposition is based on the fact that the resultant effect of several loads acting on a member simultaneously is the sum of the contributions from each of the loads applied individually. The results for the separate loads are frequently available from previous work or easily determined by previous methods. In such instances, the superposition method becomes a powerful concept or tool for finding stresses, deflections, and the like. The method is applicable in all cases in which a linear relation exists between the stresses or deflections and the applied loads.

To show that beam deflections can be accurately determined by the method of superposition, consider the cantilever beam of Fig. 9-18 with loads $Q$, $w$, and $P$. To determine the deflection at any point of this beam by the double integration method, it is necessary to express the bending moment in terms of the applied loads. For each interval along the beam, the value of $M$ is the algebraic sum of the moments due to the separate loads. After two successive integrations, the solution for the deflection at any point will still be the algebraic sum of the contributions for each applied load. Furthermore, for any given value of $x$, the relation between applied load and resulting deflection will be linear. It is evident, therefore, that the deflection of a beam is the sum of the deflections produced by the individual loads. Once the deflections produced by a few typical individual loads have been determined by one of the methods already presented, the superposition method provides a means of rapidly solving a wide variety of more complicated problems by various combinations of known results. As more data becomes available, a wider range of problems can be solved by superposition.

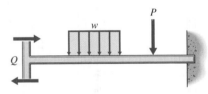

Figure 9-18

The data in Appendix A (Table A-19) are provided for use in mastering the superposition method. No attempt is made to give a large number of results because such data are readily available in various handbooks. The data given and the illustrative examples are for the purpose of making the concept and methods clear.

## Example Problem 9-10

A 16-ft long, simply supported beam carries a uniformly distributed load of 500 lb/ft and a concentrated load of 1000 lb, as shown in Fig. 9-19$a$. The beam is rough sawn (4 in. wide $\times$ 8 in. tall, $I = 170.67$ in$^4$) out of air-dried Douglas fir ($E = 1900$ ksi). Determine the deflection $y_c$ at the center of the beam.

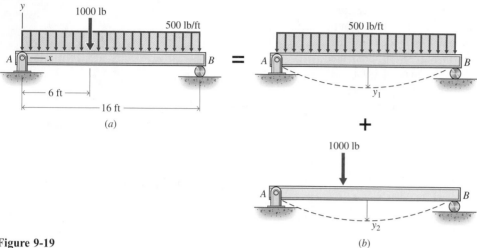

**Figure 9-19**

## SOLUTION

The deflection at the center of the beam consists of two parts, $y_1$ due to the distributed load and $y_2$ due to the concentrated load. As shown in Fig. 9-19b, the original beam with two loads can be replaced by two beams, each carrying one of the two loads.

From case 7 of Table A-19 (Appendix A), the deflection at the center of the beam due to the distributed load, with $L = 16$ ft $= 192$ in., is given by

$$y_1 = -\frac{5wL^4}{384EI} = -\frac{5(500/12)(192)^4}{384(1.9)(10^6)(170.67)} = -2.2736 \text{ in.}$$

Similarly, from case 5 of Table A-19 (in which $b = 6$ ft $= 72$ in. is the shorter distance between the concentrated load and the end of the beam), the deflection at the center of the beam due to the concentrated load is given by

$$y_2 = -\frac{Pb(3L^2 - 4b^2)}{48EI} = -\frac{1000(72)[3(192)^2 - 4(72)^2]}{48(1.9)(10^6)(170.67)} = -0.4156 \text{ in.}$$

Consequently, the total deflection of the center of the beam is

$$y_c = y_1 + y_2 = (-2.2736) + (-0.4156)$$
$$= -2.6892 \text{ in.} \cong 2.69 \text{ in.} \downarrow \blacksquare \qquad \textbf{Ans.}$$

## Example Problem 9-11

A 5-m-long cantilever beam carries a uniformly distributed load of 7.5 kN/m and a concentrated load of 25 kN, as shown in Fig. 9-20a. The steel ($E = 28$ GPa) beam is a wide-flange section [$I = 500(10^{-6})$ m⁴]. Determine the deflection at the right end of the beam.

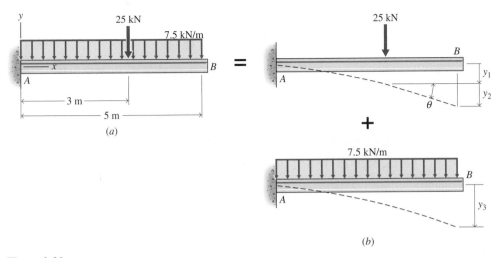

**Figure 9-20**

## SOLUTION

As shown in Fig. 9-20b, the cantilever beam with two loads can be replaced by two beams, each carrying one of the two loads. The elastic curve for the concentrated load is shown greatly exaggerated in Fig. 9-20b. The deflection at the right end is given as $y_1 + y_2$, where $y_1$ is the deflection at the location of the concentrated load and $y_2$ is the additional deflection of the unloaded 2 m. From case 1 of Table A-19 (Appendix A),

$$y_1 = -\frac{PL^3}{3EI} = -\frac{25(10^3)(3)^3}{3(28)(10^9)(500)(10^{-6})} = -0.016071 \text{ m}$$

From the concentrated load to the end of the beam, the slope is constant. Again from case 1 of Table A-19, the constant slope of the beam is

$$\theta = -\frac{PL^2}{2EI} = -\frac{25(10^3)(3)^2}{2(28)(10^9)(500)(10^{-6})} = -0.008036 \text{ rad}$$

and the deflection $y_2$ is given by

$$y_2 = \theta L = -0.008036(2) = -0.016071 \text{ m}$$

The correct expression for $y_2$ should be $\tan \theta = y_2/L$. However, since the deflection (and angle of deflection) is small, $\tan \theta \cong \theta$ and $y_2 \cong \theta L$. Consequently, the total deflection of the right end of the beam due to the concentrated load is

$$y_1 + y_2 = (-0.016071) + (-0.016071) = -0.03214 \text{ m} = 32.14 \text{ mm} \downarrow$$

The elastic curve for the distributed load is also shown (greatly exaggerated) in Fig. 9-20b. From case 2 of Table A-19 (Appendix A), the deflection of the right end of the beam due to the distributed load is

$$y_3 = -\frac{wL^4}{8EI} = -\frac{7.5(10^3)(5)^4}{8(28)(10^9)(500)(10^{-6})} = -0.04185 \text{ m} = 41.85 \text{ mm} \downarrow$$

Finally, the total deflection of the right end of the beam due to both loads is

$$y = (y_1 + y_2) + y_3 = (-32.14) + (-41.85)$$

$$= -73.99 \text{ mm} \cong 74.0 \text{ mm} \downarrow \quad \blacksquare \qquad \textbf{Ans.}$$

## Example Problem 9-12

For the beam in Fig.9-21a, determine the maximum deflection when $E$ is $12(10^6)$ psi and $I$ is 81 in⁴.

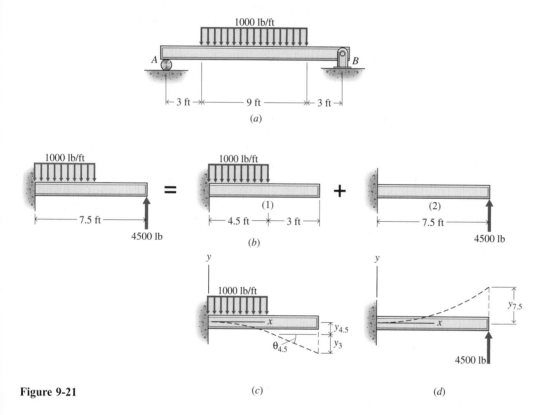

**Figure 9-21**

## SOLUTION

From symmetry and the equilibrium equation $\Sigma F_y = 0$, the two support reactions are equal and are each equal to 4500 lb upward. Because of the symmetrical loading, the slope of the beam is zero at the center of the span, and the left half of the beam could be replaced with a cantilevered support without changing any of the stresses, forces, or moments in the right half of the beam. Therefore, the right half of the beam can be considered a cantilever beam with two loads—a downward distributed load over the first 4.5 ft and an upward concentrated load of 4500 lb at the end. As shown in Fig. 9-21b, for the right half, the cantilever with two loads can be replaced by two beams (designated 1 and 2), each carrying one of the two loads.

The elastic curve (exaggerated) for part 1, as shown in Fig. 9-21c, gives the deflection at the right end as $y_{4.5} + y_3$, where $y_{4.5}$ is the deflection at the end of the uniformly distributed load and $y_3$ is the additional deflection of the unloaded 3 ft. From case 2 of Table A-19 of Appendix A,

$$y_{4.5} = -\frac{wL^4}{8EI} = -\frac{(1000/12)[4.5(12)]^4}{8(12)(10^6)(81)} = -0.09113 \text{ in.}$$

Similarly,

$$\theta_{4.5} = -\frac{wL^3}{6EI} = -\frac{(1000/12)[4.5(12)]^3}{6(12)(10^6)(81)} = -0.002250 \text{ rad}$$

from which

$$y_3 = \theta L = (-0.002250)(3)(12) = -0.0810 \text{ in.}$$

Consequently, the total deflection of the right end for part 1 is

$$y_1 = y_{7.5} = y_{4.5} + y_3 = (-0.09113) + (-0.0810)$$
$$= -0.17213 \text{ in.} = 0.17213 \text{ in.} \downarrow$$

The elastic curve (exaggerated) for part 2 is shown in Fig. 9-21d. From case 1 of Table A-19 of Appendix A,

$$y_2 = y_{7.5} = +\frac{PL^3}{3EI} = +\frac{4500[7.5(12)]^3}{3(12)(10^6)(81)}$$
$$= +1.1250 \text{ in.} = 1.1250 \text{ in.} \uparrow$$

The algebraic sum of the deflections for parts 1 and 2 is

$$y_R = y_1 + y_2 = (-0.17213) + (1.1250)$$
$$= +0.9529 \text{ in.} \cong 0.953 \text{ in.} \uparrow$$

which means that the right end of the beam is 0.953 in. above the center. Obviously, the right end does not move, and the maximum deflection is at the center and is

$$y_{max} = 0.953 \text{ in.} \downarrow \blacksquare \qquad \text{**Ans.**}$$

## Example Problem 9-13

A beam is loaded and supported as shown in Fig. 9-22a. Use the method of superposition to determine the deflection

(a) At a point midway between the supports.

(b) At the right end of the beam.

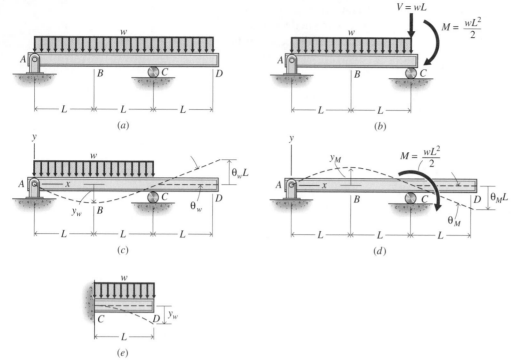

**Figure 9-22**

## SOLUTION

(a) The deflection at a point midway between the supports is determined by using the beam shown in Fig. 9-22*b*. The support reactions at *A* and *C* are unchanged if the distributed load on the overhang *CD* is replaced with an equivalent force-couple at *C*. Also, the shear force and bending moment at points between the supports are unchanged if the distributed load on the overhang *CD* is replaced with an equivalent force-couple at *C*. Therefore, the deflection and slope at points between the supports will be unchanged if the distributed load on the overhang *CD* is replaced with an equivalent force-couple at *C*.

The effects of the overhang *CD* on span *AC* of the beam can be represented by a shear force $V = wL$ and a moment $M = wL^2/2$. Since the shear force $V$ does not contribute to the deflection at any point in span *AC* of the beam, the deflection at the middle of the span, as shown in Figs. 9-22*c* and *d*, can be expressed as

$$y_B = y_w + y_M$$

The deflections $y_w$ and $y_M$ are listed in cases 7 and 8, of Table A-19 Appendix A, respectively. Thus,

$$y_B = -\frac{5w(2L)^4}{384EI} + \frac{(wL^2/2)(2L)^2}{16EI} = -\frac{wL^4}{12EI} = \frac{wL^4}{12EI} \downarrow \qquad \textbf{Ans.}$$

(b) The deflection at the right end of the beam is produced by the combined effects of the distributed load on the overhang and the rotation of the cross section of the beam at support $C$, as shown in Figs. 9-22c, d, and e. Thus,

$$y_D = \theta_w L + \theta_M L + y_w$$

The angles $\theta_w$ and $\theta_M$ and the deflection $y_w$ are listed in cases 7, 8, and 2 of Table A-19 of Appendix A, respectively. Thus,

$$y_D = \frac{w(2L)^3(L)}{24EI} - \frac{(wL^2/2)(2L)(L)}{3EI} - \frac{wL^4}{8EI} = -\frac{wL^4}{8EI} = \frac{wL^4}{8EI} \downarrow \quad \blacksquare \quad \textbf{Ans.}$$

# PROBLEMS

**Introductory Problems**

**9-51*** Determine the deflection at the right end of the cantilever beam shown in Fig. P9-51.

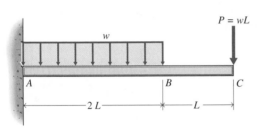

**Figure P9-51**

**9-52*** Determine the deflection at the right end of the cantilever beam shown in Fig. P9-52.

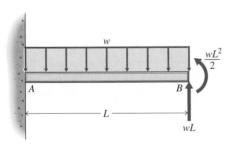

**Figure P9-52**

**9-53** For the cantilever beam shown in Fig. P9-53, determine
(a) The slope and deflection at section $B$.
(b) The slope and deflection at section $C$.

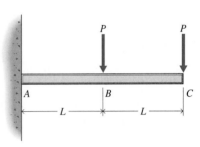

**Figure P9-53**

**9-54** Determine the deflection at a point midway between the supports of the beam shown in Fig. P9-54 when $P = 13.5$ kN, $L = 3$ m, $I = 80(10^6)$ mm$^4$, and $E = 200$ GPa.

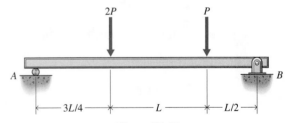

**Figure P9-54**

**9-55** Determine the deflection at the right end of the cantilever beam shown in Fig. P9-55.

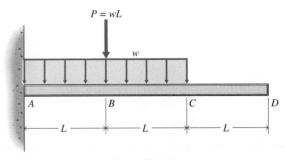

**Figure P9-55**

**9-56*** For the beam shown in Fig. P9-56, determine
(a) The deflection at a point midway between the supports.
(b) The deflection at the right end of the beam.

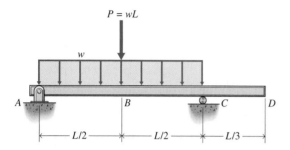

$P = wL$

**Figure P9-56**

### Intermediate Problems

**9-57*** Member $AB$ of Fig. P9-57 is the flexural member of a scale that is used to weigh food in a microwave oven. Determine the deflection of point $C$ when $W = 5$ lb, $L = 2$ in., and $EI = 100$ lb · in².

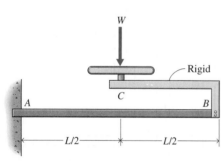

**Figure P9-57**

**9-58*** Determine the deflection at a point midway between the supports of the beam shown in Fig. P9-58.

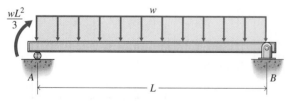

**Figure P9-58**

**9-59** Determine the deflection at a point midway between the supports of the beam shown in Fig. P9-59.

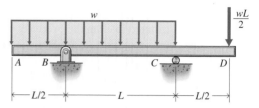

**Figure P9-59**

**9-60*** Determine the deflection at the free end of the cantilever beam shown in Fig. P9-60 when $w = 7$ kN/m, $L = 1.8$ m, $I = 130(10^{-6})$ m⁴, and $E = 200$ GPa.

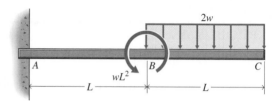

**Figure P9-60**

**9-61** Determine the deflection at the right end of the cantilever beam shown in Fig. P9-61.

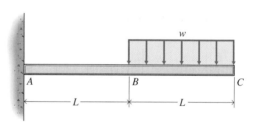

**Figure P9-61**

**9-62** Determine the deflection at the right end of the cantilever beam shown in Fig. P9-62.

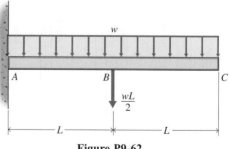

**Figure P9-62**

### Challenging Problems

**9-63*** Determine the midspan deflection of beam $AC$ of Fig. P9-63 if both beams have the same flexural rigidity.

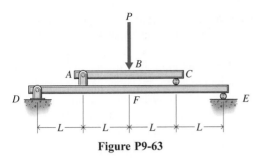

**Figure P9-63**

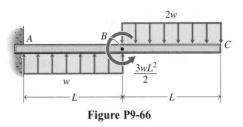

**Figure P9-66**

**9-64*** Determine the deflection at the right end of the beam shown in Fig. P9-64.

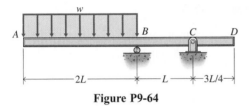

**Figure P9-64**

**9-65** Determine the deflection at the free end of the cantilever beam shown in Fig. P9-65.

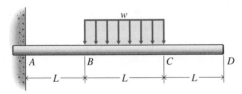

**Figure P9-65**

**9-66** Determine the deflection at the free end of the cantilever beam shown in Fig. P9-66 when $w = 7.5$ kN/m, $L = 3$ m, $I = 180(10^6)$ mm$^4$, and $E = 200$ GPa.

**9-67*** Determine the deflection at the free end of the cantilever beam shown in Fig. P9-67.

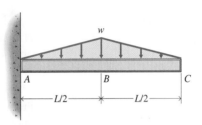

**Figure P9-67**

**9-68** Determine the deflection at the right end of the cantilever beam shown in Fig. P9-68.

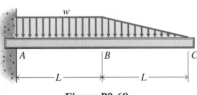

**Figure P9-68**

# 9-7 STATICALLY INDETERMINATE BEAMS: THE INTEGRATION METHOD

A beam, subjected only to transverse loads, with more than two reaction components is statically indeterminate because the equations of equilibrium are not sufficient to determine all the reactions. In such cases the geometry of the deformation of the loaded beam is used to obtain the additional relations needed for an evaluation of the reactions (or other unknown forces). For problems involving elastic action, each additional constraint on a beam provides additional information concerning slopes or deflections. Such information, when used with the appropriate slope or deflection equations, yields expressions that supplement the independent equations of equilibrium.

For statically determinate beams, known slopes and deflections were used to obtain boundary and matching conditions, from which the constants of integration in the elastic curve equation could be evaluated. For statically indeterminate beams, the procedures are identical. However, the moment equations will contain reactions or loads that cannot be evaluated from the available equations of equilibrium, and one additional boundary condition is needed for the evaluation of each such unknown. For example, if a beam is subjected to a force system for which there are two independent equilibrium equations and if there are four unknown reactions or loads on the beam, two boundary or matching conditions are needed in addition to those necessary for the determination of the constants of integration. These extra boundary conditions, when substituted in the appropriate elastic curve equations (slope or deflection), will yield the necessary additional equations. The following examples illustrate the method.

## Example Problem 9-14

A beam is loaded and supported as shown in Fig. 9-23a. Determine the reactions at $A$ and $B$.

### SOLUTION

From the free-body diagram of Fig. 9-23b it is seen that there are three unknown reaction components ($M_A$, $V_A$, and $R_B$) and that only two independent equations of equilibrium are available. The additional unknown requires the use of the elastic curve equation, for which one extra boundary condition is required in addition to the two required for the constants of integration. Because three boundary conditions are available in the interval between the supports, only one elastic curve equation needs to be written. The origin of coordinates is arbitrarily placed at the wall and for the interval $0 \le x \le L$, the boundary condi-

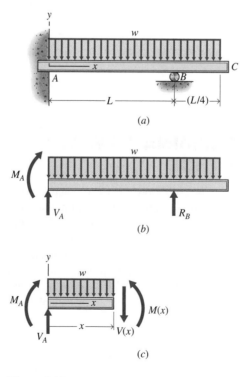

(a)

(b)

(c)

**Figure 9-23**

tions are as follows: when $x = 0$, $dy/dx = 0$; when $x = 0$, $y = 0$; and when $x = L$, $y = 0$. From Fig. 9-23c and Eq. (9-3),

$$EI \frac{d^2y}{dx^2} = M(x) = V_A x + M_A - \frac{wx^2}{2} \qquad 0 \leq x \leq L$$

Integration gives

$$EI \frac{dy}{dx} = \frac{V_A x^2}{2} + M_A x - \frac{wx^3}{6} + C_1$$

The first boundary condition $dy/dx = 0$ when $x = 0$ gives $C_1 = 0$. A second integration yields

$$EIy = \frac{V_A x^3}{6} + \frac{M_A x^2}{2} - \frac{wx^4}{24} + C_2$$

The second boundary condition $y = 0$ when $x = 0$ gives $C_2 = 0$, and the last boundary condition $y = 0$ when $x = L$ gives

$$0 = \frac{V_A L^3}{6} + \frac{M_A L^2}{2} - \frac{wL^4}{24}$$

which reduces to

$$4V_A L + 12M_A = wL^2 \qquad\qquad \text{(a)}$$

The equation of equilibrium $\Sigma M_B = 0$ for the free-body diagram of Fig. 9-23b yields

$$+\curvearrowright \Sigma M_B = 0: \qquad V_A L + M_A - w\left(\frac{5L}{4}\right)\left(\frac{3L}{8}\right) = 0$$

which reduces to

$$32V_A L + 32M_A = 15wL^2 \qquad\qquad \text{(b)}$$

Simultaneous solution of Eqs. (a) and (b) gives

$$M_A = -\frac{7wL^2}{64} = \frac{7wL^2}{64} \curvearrowleft \qquad\qquad \textbf{Ans.}$$

and

$$V_A = +\frac{37wL}{64} = \frac{37wL}{64} \uparrow \qquad\qquad \textbf{Ans.}$$

Finally, the equation $\Sigma F_y = 0$ for Fig. 9-23b gives

$$+\uparrow \Sigma F_y = 0: \qquad R_B + V_A - w(5L/4) = 0$$

from which

$$R_B = +\frac{43wL}{64} = \frac{43wL}{64} \uparrow \qquad \textbf{Ans.}$$

An alternate solution would be to place the origin of coordinates at the right support and write the moment equation for the interval $0 \le x \le L$. This equation would involve only one unknown, the reaction $R_B$. Upon integration and evaluation of the constants, the third boundary condition would directly yield the value of $R_B$. The two independent equilibrium equations could then be used to evaluate $M_A$ and $V_A$. ∎

## Example Problem 9-15

A beam is loaded and supported as shown in Fig. 9-24a. Determine

(a) The reactions at supports $A$ and $B$.
(b) The deflection at the middle of the span.

### SOLUTION

(a) There are three unknown reaction components ($M_B$, $V_B$, and $R_A$) on the free-body diagram of Fig. 9-24b. Because there are three unknown reaction components and only two independent equations of equilibrium, the problem is statically indeterminate and a deformation equation will be needed in order to solve for the three unknowns. Additional complications arise because two moment equations are needed, one in the interval $0 \le x \le L$ and another in the interval $L \le x \le 2L$. Integrating each moment equation twice results in four constants of integration, along with the unknown reaction $R_A$. To solve the resulting equations requires three boundary conditions (one at $A$ and two at $B$) and two matching conditions at the point where the moment is applied. The lengthy algebraic computations can be simplified if one uses singularity functions. The moment equation is

$$EI\frac{d^2y}{dx^2} = M(x) = R_A x - M\langle x - L\rangle^0 \qquad 0 \le x \le 2L$$

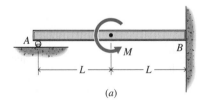

(a)

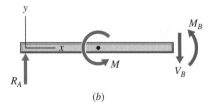

(b)

**Figure 9-24**

which upon integration yields

$$EI\frac{dy}{dx} = \frac{R_A x^2}{2} - M\langle x - L\rangle^1 + C_1$$

$$EIy = \frac{R_A x^3}{6} - \frac{M}{2}\langle x - L\rangle^2 + C_1 x + C_2$$

Using the boundary condition $y = 0$ when $x = 0$ gives $C_2 = 0$. The boundary condition $dy/dx = 0$ when $x = 2L$ gives

$$2R_A L^2 - ML + C_1 = 0 \qquad\qquad\text{(a)}$$

and the boundary condition $y = 0$ when $x = 2L$ gives

$$\frac{4R_A L^3}{3} - \frac{ML^2}{2} + 2C_1 L = 0 \qquad\qquad\text{(b)}$$

Solving Eqs. (a) and (b) simultaneously yields

$$C_1 = -\frac{ML}{8} \quad\text{and}\quad R_A = +\frac{9M}{16L} = \frac{9M}{16L}\uparrow \qquad\qquad\textbf{Ans.}$$

The remaining two unknowns are found using the equations of equilibrium and the free-body diagram shown in Fig. 9-24b.

$$+\uparrow \Sigma F_y = 0: \qquad R_A - V_B = \frac{9M}{16L} - V_B = 0$$

$$+\curvearrowleft \Sigma M_B = 0: \qquad -R_A(2L) + M + M_B = -\frac{9M}{16L}(2L) + M + M_B = 0$$

The reactions at $B$ obtained from these two equations are

$$V_B = +\frac{9M}{16L} = \frac{9M}{16L}\downarrow \quad\text{and}\quad M_B = +\frac{M}{8} = \frac{M}{8}\curvearrowleft \qquad\qquad\textbf{Ans.}$$

(b) Setting $x = L$ in the elastic curve equation

$$EIy = \frac{R_A x^3}{6} - \frac{M}{2}\langle x - L\rangle^2 + C_1 x + C_2$$

yields

$$EIy = \frac{1}{6}\left(\frac{9M}{16L}\right)(L)^3 - \frac{1}{2}(M)(0)^2 + \left(-\frac{ML}{8}\right)(L) + 0$$

from which the deflection at the middle of the span is

$$y = -\frac{ML^2}{32EI} = \frac{ML^2}{32EI}\downarrow \quad\blacksquare \qquad\qquad\textbf{Ans.}$$

# PROBLEMS

*Note:* Use the integration method (with singularity functions if needed) to solve the following problems.

## Introductory Problems

**9-69*** A beam is loaded and supported as shown in Fig. P9-69. Determine
(a) The reactions at supports $A$ and $B$.
(b) The deflection at the middle of the span.

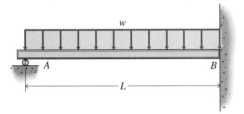

**Figure P9-69**

**9-70*** When a moment $M$ is applied to the left end of the beam shown in Fig. P9-70, the slope at the left end of the beam is zero. Determine the magnitude of the moment $M$.

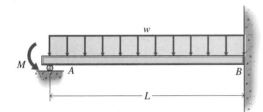

**Figure P9-70**

**9-71** A beam is loaded and supported as shown in Fig. P9-71. Determine the magnitude of the moment $M$ required to make the slope at the left end of the beam zero.

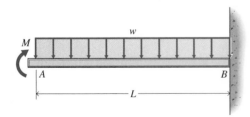

**Figure P9-71**

**9-72** When a moment $M$ is applied to the left end of the cantilever beam shown in Fig. P9-72, the slope at the left end of the beam is zero. Determine the magnitude of the moment $M$.

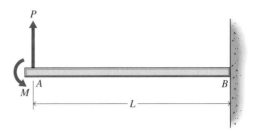

**Figure P9-72**

**9-73*** When a load $P$ is applied to the right end of the cantilever beam shown in Fig. P9-73, the deflection at the right end of the beam is zero. Determine the magnitude of the load $P$.

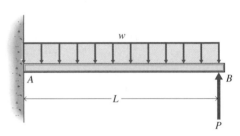

**Figure P9-73**

## Intermediate Problems

**9-74*** A beam is loaded and supported as shown in Fig. P9-74. Determine
(a) The reactions at supports $A$ and $B$.
(b) The maximum deflection in the beam.

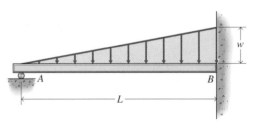

**Figure P9-74**

**9-75*** A beam is loaded and supported as shown in Fig. P9-75. Determine
(a) The reactions at supports $A$ and $B$.
(b) The maximum deflection in the beam.

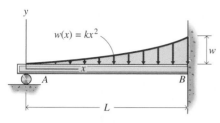

**Figure P9-75**

**9-76** A beam is loaded and supported as shown in Fig. P9-76. Determine the reactions at supports $A$ and $B$.

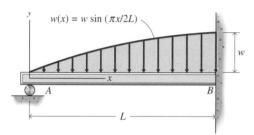

**Figure P9-76**

**9-77** A beam is loaded and supported as shown in Fig. P9-77. Determine
(a) The reactions at supports $A$ and $B$.
(b) The deflection at the middle of the span.

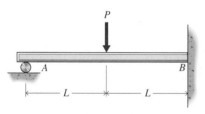

**Figure P9-77**

**9-78\*** A beam is loaded and supported as shown in Fig. P9-78. Determine
(a) The reactions at supports $A$ and $B$.
(b) The deflection at the middle of the span.

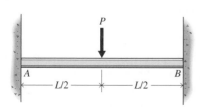

**Figure P9-78**

**9-79** A beam is loaded and supported as shown in Fig. P9-79. Determine the reactions at supports $B$, $C$, and $D$.

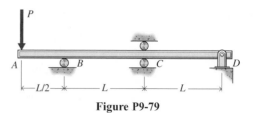

**Figure P9-79**

## Challenging Problems

**9-80\*** A beam is loaded and supported as shown in Fig. P9-80. Determine
(a) The reactions at supports $A$, $B$, and $C$.
(b) The bending moment over the middle support.
(c) The deflection at the middle of span $BC$.

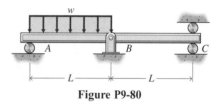

**Figure P9-80**

**9-81\*** A beam is loaded and supported as shown in Fig. P9-81. Determine
(a) The reactions at supports $A$, $C$ and $D$.
(b) The deflection at the middle of span $AC$.

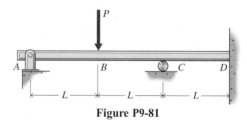

**Figure P9-81**

**9-82** A beam is loaded and supported as shown in Fig. P9-82. Determine
(a) The reactions at supports $A$, $B$, and $C$.
(b) The bending moment over the center support.
(c) The maximum bending moment in the beam.

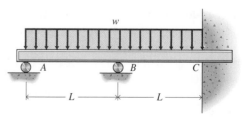

**Figure P9-82**

**9-83** A beam is loaded and supported as shown in Fig. P9-83. Determine

(a) The reactions at supports $A$, $B$, and $C$.

(b) The bending moment over the center support.

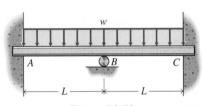

**Figure P9-83**

**9-84*** A beam is loaded and supported as shown in Fig. P9-84. Determine

(a) The reactions at supports $A$ and $D$.

(b) The deflection at $B$ if $E = 200$ GPa and $I = 350(10^6)$ mm$^4$.

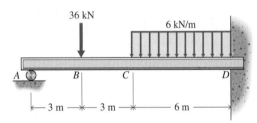

**Figure P9-84**

**9-85** A beam is loaded and supported as shown in Fig. P9-85. Determine

(a) The reactions at supports $A$, $B$, and $D$.

(b) The deflection at $C$ if $E = 30,000$ ksi and $I = 26$ in$^4$.

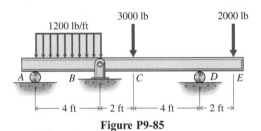

**Figure P9-85**

## Computer Problems

**9-86** A 10-m-long beam carries a uniformly distributed load and is supported as shown in Fig. P9-86. Horizontal reactions at the supports may be neglected. If the beam is constructed of wood ($EI = 1500$ kN·m$^2$), compute and plot

(a) The deflection curves for the beam (plot $y$ as a function of $x$ for $0 \leq x \leq 10$ m) for center support locations of $b = 2$ m, 4 m, and 7 m.

(b) The bending moment distribution along the beam (plot $M$ as a function of $x$ for $0 \leq x \leq 10$ m) for center support locations of $b = 2$ m, 4 m, and 7 m.

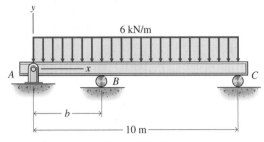

**Figure P9-86**

**9-87** A concentrated load $P = 3900$ lb moves slowly across the beam shown in Fig. P9-87. Horizontal reactions at the supports may be neglected. The 16-ft-long beam is a W4 × 13 structural steel section. Compute and plot

(a) The deflection curves for the beam (plot $y$ as a function of $x$ for $0 \leq x \leq 16$ ft) for load locations of $b = 3$ ft, 7 ft, and 11 ft.

(b) The bending moment distribution along the beam (plot $M$ as a function of $x$ for $0 \leq x \leq 16$ ft) for load locations of $b = 3$ ft, 7 ft, and 11 ft.

(c) The maximum flexural stresses in the beam as a function of $b$ for 1 ft $\leq b \leq 8$ ft. What range of $b$ would be acceptable for this beam?

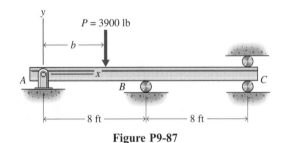

**Figure P9-87**

**9-88** A 6-m-long WT178 × 51 structural steel section is used for the cantilever beam shown in Fig. P9-88. The flange is at the top of the beam. If a roller support is added to the beam at $B$, compute and plot

(a) The deflection curves for the beam (plot $y$ as a function of $x$ for $0 \leq x \leq 6$ m) for roller support locations of $b = 3$ m, 4 m, and 5 m.

(b) The bending moment distribution along the beam (plot $M$ as a function of $x$ for $0 \leq x \leq 6$ m) for roller support locations of $b = 3$ m, 4 m, and 5m.

(c) The maximum tensile and compressive flexural stresses in the beam as a function of $b$ for 1 m $\leq b \leq 6$ m. What range of $b$ would be acceptable for this beam?

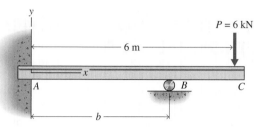

**Figure P9-88**

**9-89** A 20-ft-long W8 × 24 structural steel section is used for the two-span beam shown in Fig. P9-89. The beam supports a uniformly distributed load $w$ of 1800 lb/ft. If the support at $B$ settles with time until it provides no resistance to deflection of the beam,

(a) Determine $b_{max}$, the maximum amount that support $B$ will settle.

(b) Compute and plot the deflection curves for the beam (plot $y$ as a function of $x$ for $0 \le x \le 20$ ft) for support $B$ initially ($b = 0$) and settled ($b = b_{max}/3$, $b = 2b_{max}/3$, and $b = b_{max}$).

(c) Compute and plot the bending moment distribution along the beam (plot $M$ as a function of $x$ for $0 \le x \le 20$ ft) for support $B$ initially ($b = 0$) and settled ($b = b_{max}/3$, $b = 2b_{max}/3$, and $b = b_{max}$).

(d) Compute and plot the maximum flexural stresses in the beam as a function of $b$ for $0 \le b \le b_{max}$. What range of $b$ would be acceptable for this beam?

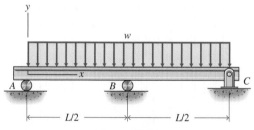

**Figure P9-89**

**9-90** A 6-m-long S457 × 81 structural steel section is used for the two-span beam shown in Fig. P9-90. The beam supports a uniformly distributed load $w$ of 100 kN/m. If the support at $A$ settles with time until it provides no resistance to deflection of the beam.

(a) Determine $a_{max}$, the maximum amount that support $A$ will settle.

(b) Compute and plot the deflection curves for the beam (plot $y$ as a function of $x$ for $0 \le x \le 6$ m) for support $A$ initially ($a = 0$) and settled ($a = a_{max}/3$, $a = 2a_{max}/3$, and $a = a_{max}$).

(c) Compute and plot the bending moment distribution along the beam (plot $M$ as a function of $x$ for $0 \le x \le 6$ m) for support $A$ initially ($a = 0$) and settled ($a = a_{max}/3$, $a = 2a_{max}/3$, and $a = a_{max}$).

(d) Compute and plot the maximum flexural stresses in the beam as a function of $a$ for $0 \le a \le a_{max}$. What range of $a$ would be acceptable for this beam?

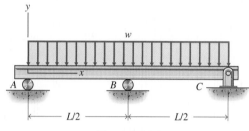

**Figure P9-90**

**9-91** A 20-ft-long S7 × 20 structural steel section is used for the cantilever beam shown in Fig. P9-91. If the support at $B$ settles with time until it provides no resistance to deflection of the beam,

(a) Determine $b_{max}$, the maximum amount that support $B$ will settle.

(b) Compute and plot the deflection curves for the beam (plot $y$ as a function of $x$ for $0 \le x \le 20$ ft) for support $B$ initially ($b = 0$) and settled ($b = b_{max}/3$, $b = 2b_{max}/3$, and $b = b_{max}$).

(c) Compute and plot the bending moment distribution along the beam (plot $M$ as a function of $x$ for $0 \le x \le 20$ ft) for support $B$ initially ($b = 0$) and settled ($b = b_{max}/3$, $b = 2b_{max}/3$, and $b = b_{max}$).

(d) Compute and plot the maximum flexural stresses in the beam as a function of $b$ for $0 \le b \le b_{max}$. What range of $b$ would be acceptable for this beam?

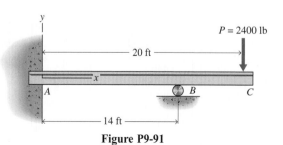

**Figure P9-91**

# 9-8 STATICALLY INDETERMINATE BEAMS: THE SUPERPOSITION METHOD

The concept (discussed in Section 9-6) that a slope or deflection due to several loads in the algebraic sum of the slopes or deflections due to each of the loads acting individually is frequently used to provide the deformation equations needed to supplement the equilibrium equations in the solution of statically indeterminate beam problems. To provide the necessary deformation equations, selected restraints are removed and replaced by unknown loads (forces and couples); the deformation diagrams corresponding to individual loads (both known and unknown) are sketched; and the component deflections or slopes are summed to produce the known configuration. The following examples illustrate the use of superposition for this purpose.

## Example Problem 9-16

A steel ($E = 30,000$ ksi) beam 20 ft long is simply supported at the ends and at the midpoint, as shown in Fig. 9-25a. Determine the reactions at supports $A$, $B$, and $C$. The second moment of area of the cross section with respect to the neutral axis is 100 in$^4$.

### SOLUTION

With three unknown support reactions and only two equations of equilibrium available, the beam is statically indeterminate. The center support exerts whatever force is necessary to prevent the center of the beam from settling. Therefore, the problem can be reformulated. "If the center support $C$ were removed, what upward force at $C$ would be needed to make the deflection at $C$ equal to zero?" The resulting simply supported beam is equivalent to two beams with individual loads, as shown in Fig. 9-25b. The resulting deflection at the midpoint of the beam is the deflection $y_R$ due to the reaction force $R_C$, plus the deflection $y_w$ due to the uniform distributed load $w$, that is,

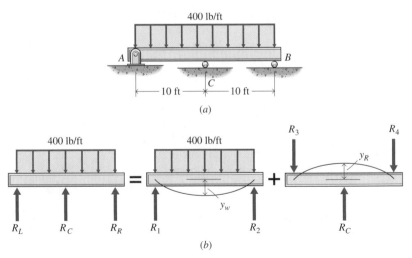

(a)

(b)

**Figure 9-25**

$$y = y_R + y_w = 0 \qquad\qquad\text{(a)}$$

The deflection $y_w$ can be obtained from case 7 of Table A-19 in Appendix A, and is

$$y_w = -\frac{5wL^4}{384EI} = -\frac{5(400/12)[20(12)]^4}{384(30)(10^6)(100)} = -0.4800 \text{ in.}$$

The deflection $y_R$ can be obtained in terms of $R_C$ from case 6 of Table A-19 in Appendix A, and is

$$y_R = \frac{R_C L^3}{48EI} = \frac{R_C[20(12)]^3}{48(30)(10^6)(100)} = 96(10^{-6})R_C \text{ in.}$$

When these values are substituted in Eq. (a) the result is

$$y = 96(10^{-6})R_C - 0.4800 = 0$$

from which

$$R_C = +5000 \text{ lb} = 5000 \text{ lb} \uparrow \qquad\qquad \textbf{Ans.}$$

The equilibrium equation $\Sigma F_y = 0$ and symmetry give

$$R_L = R_R = \frac{1}{2}[400(20) - 5000] = +1500 \text{ lb} = 1500 \text{ lb} \uparrow \qquad \textbf{Ans.}$$

The arithmetic will frequently be simplified in beam deflection problems if expressions for the deflections are substituted in the deflection equation in symbol form. In this example, Eq. (a) becomes

$$\frac{R_C L^3}{48EI} - \frac{5wL^4}{384EI} = 0$$

which reduces to

$$R_C = \frac{5wL}{8}$$

or

$$R_C = \frac{5(400)(20)}{8} = 5000 \text{ lb} \uparrow \quad \blacksquare$$

## ▌ Example Problem 9-17

A beam is loaded and supported as shown in Fig. 9-26a. Determine the reactions at supports $A$ and $B$.

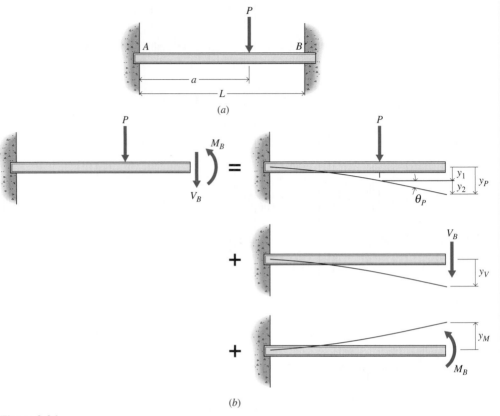

**Figure 9-26**

## SOLUTION

There are four unknown support reactions (a shear and moment at each end), and only two equations of equilibrium are available. Therefore, the beam is statically indeterminate and two deformation equations are needed to complete the solution for the support reactions. The built-in support at the right end exerts whatever force and moment are necessary to prevent the right end of the beam from settling or rotating. Therefore, the problem can be reformulated: "If the right support $B$ were removed, what force and moment at $B$ would be needed to make the deflection and slope of the beam at $B$ equal to zero?" The resulting cantilever beam is equivalent to three beams with individual loads, as shown in Fig. 9-26$b$. Note that the unknown shear force and bending moment at the right end are both shown according to the sign convention established earlier so that the algebraic sign of the result will be correct.

From the geometry of the constrained beam, the resultant slope and the resultant deflection at the right end are both zero. The slope and deflection at the end of each of the three replacement beams can be obtained from the expressions in Table A-19 of Appendix A. Thus, the first beam with load $P$ (see case 1 of Table A-19) has a constant slope from $P$ to the end of the beam, which is

$$\theta_P = -\frac{Pa^2}{2EI}$$

The deflection $y_P$ at the end is made up of two parts; $y_1$ for a beam of length

$a$, and $y_2$ the added deflection of the tangent segment (straight line) from $P$ to the end of the beam. This deflection is

$$y_P = y_1 + y_2 = -\frac{Pa^3}{3EI} + (L - a)\theta_P$$

$$= -\frac{Pa^3}{3EI} + (L - a)\left(-\frac{Pa^2}{2EI}\right) = \frac{Pa^3}{6EI} - \frac{Pa^2L}{2EI}$$

The slope and deflection at the end of the beam due to the shear $V_B$ (also from case 1 of Table A-19) are

$$\theta_V = -\frac{V_B L^2}{2EI} \quad \text{and} \quad y_V = -\frac{V_B L^3}{3EI}$$

Finally, the slope and deflection at the right end of the beam due to $M_B$ (see case 4 of Table A-19) are

$$\theta_M = \frac{M_B L}{EI} \quad \text{and} \quad y_M = \frac{M_B L^2}{2EI}$$

Since the resultant slope is zero,

$$\theta_P + \theta_V + \theta_M = -\frac{Pa^2}{2EI} - \frac{V_B L^2}{2EI} + \frac{M_B L}{EI} = 0$$

Similarly,

$$y_P + y_V + y_M = \frac{Pa^3}{6EI} - \frac{Pa^2L}{2EI} - \frac{V_B L^3}{3EI} + \frac{M_B L^2}{2EI} = 0$$

The simultaneous solution of these two equations yields

$$M_B = -\frac{Pa^2(L - a)}{L^2} \quad \text{and} \quad V_B = -\frac{Pa^2(3L - 2a)}{L^3} \quad \textbf{Ans.}$$

When the equations of equilibrium are applied to a free-body diagram of the entire beam, the shear and moment at the left end are found to be

$$M_A = -\frac{Pa(L - a)^2}{L^2} \quad \text{and} \quad V_A = \frac{P(L^3 - 3a^2L + 2a^3)}{L^3} \quad ■ \quad \textbf{Ans.}$$

# PROBLEMS

## Introductory Problems

**9-92\*** A beam is loaded and supported as shown in Fig. P9-92. Determine the magnitude of the load $P$ required to make the slope at the left end of the beam zero.

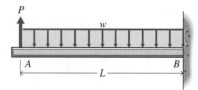

**Figure P9-92**

**9-93\*** A beam is loaded and supported as shown in Fig. P9-93. Determine the reactions at supports $A$ and $B$.

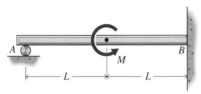

**Figure P9-93**

**9-94** A beam is loaded and supported as shown in Fig. P9-94. Determine the reactions at supports $A$ and $B$.

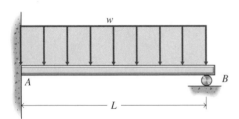

**Figure P9-94**

**9-95** A beam is loaded and supported as shown in Fig. P9-95. Determine the reactions at supports $A$ and $B$.

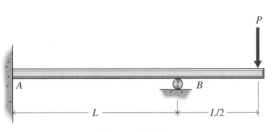

**Figure P9-95**

**9-96\*** A beam is loaded and supported as shown in Fig. P9-96. Determine the reactions at supports $A$, $B$, and $C$.

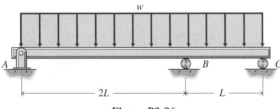

**Figure P9-96**

**Intermediate Problems**

**9-97\*** A beam is loaded and supported as shown in Fig. P9-97. When the load $P$ is applied, the slope at the right end of the beam is zero. Determine
(a) The magnitude of the load $P$.
(b) The reactions at supports $A$ and $B$.

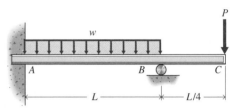

**Figure P9-97**

**9-98\*** A beam is loaded and supported as shown in Fig. P9-98. Determine
(a) The reactions at supports $A$ and $C$.
(b) The deflection at the right end of the distributed load.

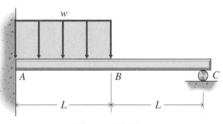

**Figure P9-98**

**9-99** A beam is loaded and supported as shown in Fig. P9-99. Determine the magnitude of the moment $M$ required to make
(a) The slope at the right end of the beam zero.
(b) The deflection at the right end of the beam zero.

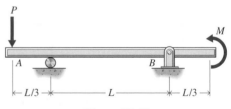

**Figure P9-99**

**9-100** Draw complete shear force and bending moment diagrams for the beam shown in Fig. P9-100.

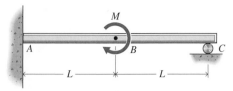

**Figure P9-100**

**9-101** Draw complete shear force and bending moment diagrams for the beam shown in Fig. P9-101.

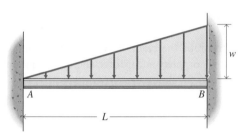

**Figure P9-101**

**Challenging Problems**

**9-102\*** Two beams are loaded and supported as shown in Fig. P9-102. Determine the reactions at supports $A$, $B$, and $D$. Both beams have the same flexural rigidity.

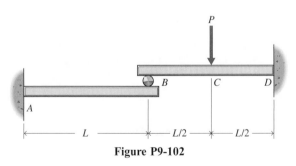

**Figure P9-102**

**9-103\*** A steel ($E = 29,000$ ksi and $I = 120$ in$^4$) beam is loaded and supported as shown in Fig. P9-103. The post $BD$ is a 6- $\times$ 6-in. timber ($E = 1500$ ksi) that is braced to prevent buckling. Determine the load carried by the post if it is unstressed before the 530 lb/ft distributed load is applied.

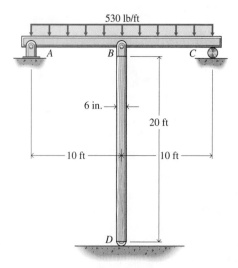

**Figure P9-103**

**9-104** A uniformly distributed load of 7 kN/m is supported by two 100- $\times$ 100-mm timber ($E = 8.5$ GPa) beams arranged as shown in Fig. P9-104. Beam $AB$ is fixed at the wall and beam $CD$ is simply supported. Before the load is applied, the beams are in contact at $B$, but the reaction at $B$ is zero. After the 7 kN/m distributed load is applied, determine
(a) The maximum flexural stress in each beam.
(b) The maximum longitudinal shearing stress in each beam.

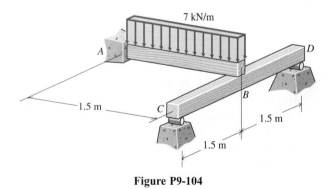

**Figure P9-104**

**9-105** The 4-in.- wide $\times$ 6-in.- deep timber ($E = 1200$ ksi) beam shown in Fig. P9-105 is fixed at the left end and supported at the right end with a tie rod that has a cross-sectional area of 0.125 in$^2$. Determine the tension in the tie rod if it is unstressed before the load is applied to the timber beam and
(a) The tie rod is made of steel ($E = 30,000$ ksi).
(b) The tie rod is made of aluminum alloy ($E = 10,000$ ksi).

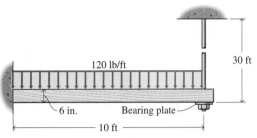

**Figure P9-105**

**9-106*** In Fig. P9-106, the aluminum alloy tie rod passes through a hole in the aluminum alloy cantilever beam and through the coil spring positioned on the end of the tie rod. Before loading, there is a clearance of 2.5 mm between the bottom of the beam and the top of the spring. The cross-sectional area of the tie rod is 100 mm², the second moment of area of the cross section of the beam with respect to the neutral axis is $40(10^6)$ mm⁴, the modulus of elasticity of the aluminum alloy is 70 GPa, and the spring modulus is 1000 kN/m. Determine the axial stress in the tie rod when $M = 9$ kN · m and $w = 90$ kN/m.

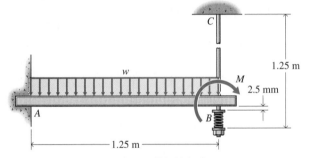

**Figure P9-106**

**9-107*** Two steel ($E = 29,000$ ksi) beams support a 1600-lb concentrated load, as shown in Fig. P9-107. In the unloaded condition, beam $AB$ touches but exerts no force on beam $CD$. Beam $AB$ is an S4 × 9.5 American standard section, and beam $CD$ is an S5 × 14.75 American standard section (see Appendix A). Determine
(a) The maximum flexural stress in each beam.
(b) The maximum transverse shearing stress in each beam.

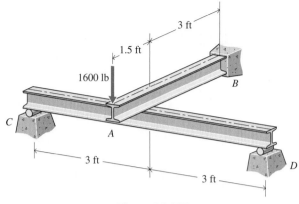

**Figure P9-107**

**9-108** A beam is loaded and supported as shown in Fig. P9-108.
(a) Determine the reactions at supports $A$ and $C$.
(b) Draw complete shear force and bending moment diagrams for the beam.

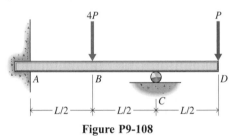

**Figure P9-108**

**9-109** A beam is loaded and supported as shown in Fig. P9-109.
(a) Determine the reactions at supports $A$, $B$, and $C$.
(b) Draw complete shear force and bending moment diagrams for the beam.

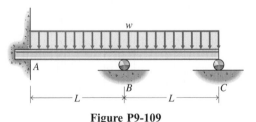

**Figure P9-109**

## 9-9 DESIGN

In Chapter 8 when the design of beams was discussed, either flexural strength or shear strength was the controlling parameter. In this chapter, an additional parameter, deflection, will be introduced. Thus, the design of a beam may be based on flexural stress, shearing stress, or deflection. The design procedure is similar to that presented in Chapter 8. The beam is first designed based on flexural stress and then checked for shearing stress and deflection. If the shearing stress and deflection are within allowable limits, the design is satisfactory. If either the shearing stress or the deflection is greater than the allowable value, the beam must be redesigned until all of the allowable limits are satisfied. Clearly, this is a trial-and-error process. The following examples illustrate the procedures for designing beams where allowable limits are given for flexural stress, shearing stress, and deflection.

## Example Problem 9-18

An air-dried Douglas fir timber ($E = 13$ GPa) beam is loaded as shown in Fig. 9-27a. If the allowable flexural stress is 8 MPa, the allowable shearing stress is 0.7 MPa, and the allowable deflection is 14 mm, determine the lightest-weight standard structural timber that can be used for the beam.

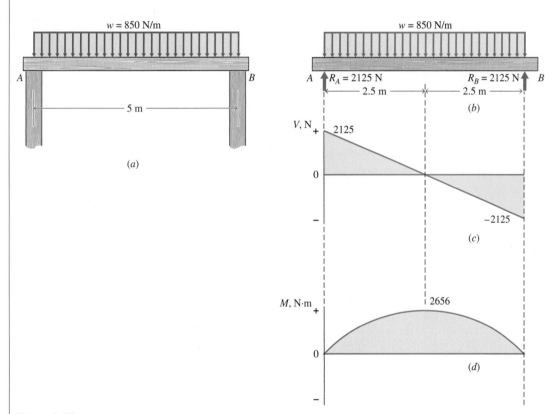

**Figure 9-27**

### SOLUTION

Load (free-body), shear force, and bending moment diagrams for the beam of Fig. 9-27a are shown in Figs. 9-27b, c, and d, respectively. Since the load is uniformly distributed over the full length of the beam,

$$R_A = R_B = \frac{1}{2}(850)(5) = 2125 \text{ N}$$

$$V_{\text{max}} = 2125 \text{ N} \qquad M_{\text{max}} = \frac{1}{2}(2125)(2.5) = 2656 \text{ N} \cdot \text{m}$$

The minimum section modulus needed to satisfy the allowable value of the flexural stress is given by Eq. (8-12) as

$$S \geq \frac{M_{max}}{\sigma_{all}} = \frac{2656}{8(10^6)} = 332.0(10^{-6}) \text{ m}^3 = 332.0(10^3) \text{ mm}^3$$

The lightest-weight standard structural timber listed in Table A-16 with $S \geq 332.0(10^3)$ mm$^3$ is a timber with nominal dimensions of 51 × 254 mm. Some properties of this timber that will be needed later are

$$\text{Mass/unit length} = 6.38 \text{ kg/m}$$
$$\text{Area} = 9.88(10^3) \text{ mm}^2 = 9.88(10^{-3}) \text{ m}^2$$
$$I = 48.3(10^6) \text{ mm}^4 = 48.3(10^{-6}) \text{ m}^4$$
$$S = 400(10^3) \text{ mm}^3 = 400(10^{-6}) \text{ m}^3$$

For a rectangular section, the area is $A = bh$, the second moment of area is $I = bh^3/12$ and the maximum shearing stress occurs at the neutral axis. Then,

$$Q = \frac{h}{4}\left(b\frac{h}{2}\right) = \frac{bh^2}{8}$$

and the maximum shearing stress is

$$\tau = \frac{VQ}{It} = \frac{V(bh^2/8)}{(bh^3/12)b} = 1.5\frac{V}{bh} = 1.5\frac{V}{A} = 1.5\frac{2125}{9.88(10^{-3})}$$
$$= 0.3226(10^6) \text{ N/m}^2 = 0.3226 \text{ MPa} < 0.7 \text{ MPa}$$

Thus, the shearing stress requirement is satisfied.

For a simply supported beam with a uniformly distributed load, case 7 of Table A-19 gives the maximum deflection as

$$|y_{max}| = \frac{5wL^4}{384EI} = \frac{5(850)(5)^4}{384(13)(10^9)(48.3)(10^{-6})}$$
$$= 11.017(10^{-3}) \text{ m} = 11.017 \text{ mm} < 14 \text{ mm}$$

Therefore, the deflection requirement is satisfied. The 51- × 254-mm standard structural timber satisfies the requirements for flexural stress, shearing stress, and deflection; however, the analysis thus far neglected the weight of the beam. The timber beam weighs $(6.38)(9.81) = 62.59$ N/m. Adding this uniformly distributed load to the applied load gives a uniformly distributed load $w = 850 + 62.6 = 912.6$ N/m. For this loading, the maximum shear force is 2282 N and the maximum bending moment is 2852 N · m. The section modulus needed to satisfy the flexural stress requirement is

$$S \geq \frac{M_{max}}{\sigma_{all}} = \frac{2852}{8(10^6)} = 356.5(10^{-6}) \text{ m}^3 = 356.5(10^3) \text{ mm}^3 < 400(10^3) \text{ mm}^3$$

Thus, the 51- × 254-mm timber satisfies the flexural stress requirement. The maximum shearing stress and the maximum deflection with the beam weight included are

$$\tau_{max} = 1.5 \frac{V_{max}}{A} = 1.5 \frac{2282}{9.88(10^{-3})}$$

$$= 0.3465(10^6) \text{ N/m}^2 = 0.3465 \text{ MPa} < 0.7 \text{ MPa}$$

$$|y_{max}| = \frac{5wL^4}{384EI} = \frac{5(912.6)(5)^4}{384(13)(10^9)(48.3)(10^{-6})}$$

$$= 11.828(10^{-3}) \text{ m} = 11.828 \text{ mm} < 14 \text{ mm}$$

Thus, the 51- × 254-mm standard structural timber satisfies all requirements with the weight of the beam included. ∎

## Example Problem 9-19

A structural steel ($E = 29,000$ ksi) beam is loaded as shown in Fig. 9-28a. If the allowable flexural stress is 24,000 psi, the allowable shearing stress is 14,000 psi, and the allowable deflection midway between supports $A$ and $B$ is 0.5 in., determine the lightest American Standard section (S-shape) that can be used for the beam.

### SOLUTION

Load (free-body), shear force, and bending moment diagrams for the beam of Fig. 9-28a are shown in Figs. 9-28b, c, and d, respectively. The equilibrium equations yield

$+\circlearrowleft \Sigma M_B = 0$:     $R_A(16) - 500(16)(8) - 1000(6) = 0$     $R_A = 4375$ lb

$+\uparrow \Sigma F_y = 0$:     $4375 - 500(16) - 1000 + R_B = 0$     $R_B = 4625$ lb

From the shear force and bending moment diagrams,

$$V_{max} = 4625 \text{ lb}$$

$$M_{max} = \frac{1}{2}(4375)(8.75) = 19,141 \text{ lb} \cdot \text{ft}$$

The minimum section modulus needed to satisfy the allowable value of the flexural stress is given by Eq. (8-12) as

$$S \geq \frac{M_{max}}{\sigma_{all}} = \frac{19,141(12)}{24,000} = 9.571 \text{ in}^3$$

The lightest weight American Standard beam (S-shape) listed in Table A-3 with $S \geq 9.571 \text{ in}^3$ is an S7 × 15.3 section. For this section,

$$S = 10.5 \text{ in}^3 \qquad t_{web} = 0.252 \text{ in.}$$

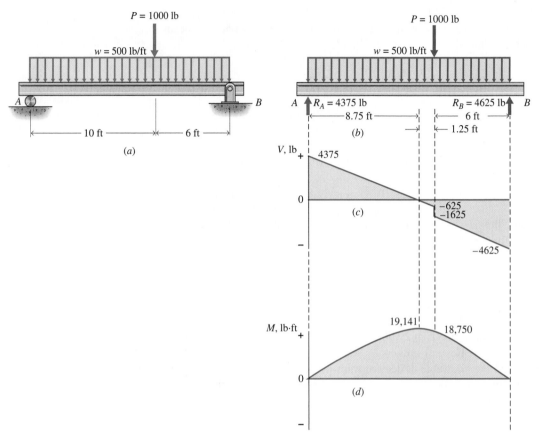

**Figure 9-28**

$$d = 7.00 \text{ in.} \qquad I = 36.7 \text{ in}^4$$

$$t_f = 0.392 \text{ in.}$$

The average value of the shearing stress in the web is

$$\tau_{avg} = \frac{V_{max}}{A_{web}} = \frac{4625}{0.252[7.00 - 2(0.392)]} = 2953 \text{ psi} << 14,000 \text{ psi}$$

Thus, the shearing stress requirement is satisfied, because the maximum shearing stress in the web of an American Standard beam is only slightly larger than the average shearing stress.

The deflection at midspan is found by using the method of superposition and Table A-19. The given loading (Fig. 9-29a) is equivalent to the two loads, parts (1) and (2), as shown in Fig. 9-29b. For part (1), the midspan deflection is given as case 7 of Table A-19. It is

$$y_1 = -\frac{5wL^4}{384EI} = -\frac{5(500/12)[16(12)]^4}{384(29)(10^6)(36.7)} = -0.6927 \text{ in.}$$

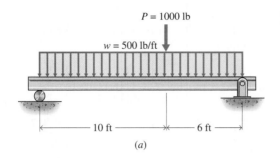

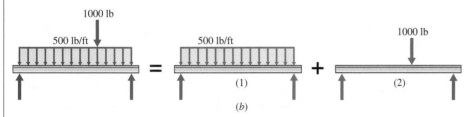

**Figure 9-29**

For part (2), the midspan deflection is given as case 5 of Table A-19. It is

$$y_2 = -\frac{Pb(3L^2 - 4b^2)}{48EI} = -\frac{1000(72)[3(192)^2 - 4(72)^2]}{48(29)(10^6)(36.7)} = -0.12664 \text{ in.}$$

The midspan deflection is the algebraic sum of $y_1$ and $y_2$ and is

$$|y_{\text{mid}}| = 0.6927 + 0.12664 = 0.8193 \text{ in.} > 0.5 \text{ in.}$$

Since $|y_{\text{mid}}|$ is greater than $y_{\text{all}}$, a new section must be selected with sufficient $I$ to satisfy the deflection requirement, Thus,

$$|y_{\text{mid}}| \geq \frac{5wL^4}{384EI} + \frac{Pb(3L^2 - 4b^2)}{48EI}$$

Solving for $I$ yields

$$I \geq \frac{5wL^4}{384E|y_{\text{mid}}|} + \frac{Pb(3L^2 - 4b^2)}{48E|y_{\text{mid}}|}$$

$$= \frac{5(500/12)(192)^4}{384(29)(10^6)(0.5)} + \frac{1000(72)[3(192)^2 - 4(72)^2]}{48(29)(10^6)(0.5)} = 60.14 \text{ in}^4$$

The lightest S-shape in Table A-3 with $I \geq 60.14$ in$^4$ is an S8 × 23 beam with $I = 64.9$ in$^4$ and $S = 16.2$ in$^3$. The S8 × 23 beam satisfies the requirements for flexural stress, shearing stress, and deflection.

Consider now the effect of the beam's weight on the deflection. The S8 × 23 beam weighs 23 lb/ft. Adding the weight of the beam to the 500 lb/ft applied distributed load gives a uniformly distributed load $w = 500 + 23 =$

523 lb/ft, resulting in maximum values of shear force and bending moment of 4809 lb and 19,870 lb · ft, respectively. The value of $I$ required as a result of this increase in load is

$$I \geq \frac{5wL^4}{384E|y_{mid}|} + \frac{Pb(3L^2 - 4b^2)}{48E|y_{mid}|}$$

$$= \frac{5(523/12)(192)^4}{384(29)(10^6)(0.5)} + \frac{1000(72)[3(192)^2 - 4(72)^2]}{48(29)(10^6)(0.5)} = 62.49 \text{ in}^4$$

For the S8 × 23 beam, $I = 64.9 \text{ in}^4 > 62.49 \text{ in}^4$; therefore, the addition of the weight of the beam does not change the selection of the beam. An S8 × 23 beam satisfies all requirements of flexural stress, shearing stress, and deflection. ■

# PROBLEMS

### Introductory Problems

**9-110\*** A 3-m simply supported beam is loaded with a uniformly distributed load of 2.6 kN/m over its entire length. The beam is made of air-dried Douglas fir ($E = 13$ GPa) with an allowable flexural stress of 8 MPa and an allowable shearing stress of 0.7 MPa. The maximum deflection at the center of the span must not exceed 10 mm. Select the lightest standard structural timber that can be used for the beam.

**9-111** A portion of a pedestrian walkway along the side of a bridge is shown in Fig. P9-111. Cantilever beams support the loading, one of which is shown. The beams are select structural eastern hemlock ($E = 1200$ ksi) with allowable flexural and shearing stresses of 1300 psi and 80 psi, respectively. The allowable deflection is 0.2 in. Select the lightest standard structural timber that can be used for the beams.

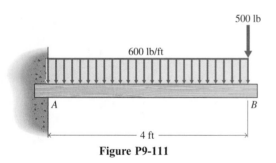

**Figure P9-111**

### Intermediate Problems

**9-112\*** A solid circular shaft made of ASTM A36 steel ($E = 200$ GPa) is supported by bearings spaced 1.5 m apart. The shaft is to support a 4-kN load perpendicular

to the shaft; the load may be placed at any point between the bearings. The allowable flexural stress is 152 MPa, the allowable shearing stress is 100 MPa, and the allowable deflection is 5 mm. If shafts are available with diameters in increments of 5 mm, determine the smallest-diameter shaft that can be used to support the load. Neglect the weight of the shaft.

**9-113** A simply supported structural steel ($E = 29,000$ ksi) beam has a span of 24 ft and carries a uniformly distributed load of 1200 lb/ft. The beam has an allowable flexural stress of 24 ksi, an allowable shearing stress of 14 ksi, and an allowable deflection of $1/360$ of the span. Select the lightest American standard (S-shape) beam that can be used to support the loading.

### Challenging Problems

**9-114\*** The simply supported structural steel ($E = 200$ GPa) beam shown in Fig. P9-114 has an allowable flexural stress of 165 MPa, an allowable shearing stress of 100 MPa, and an allowable deflection of $1/360$ of the span. Select the lightest wide-flange beam that can be used to support the loading shown in the figure.

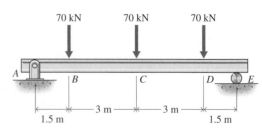

**Figure P9-114**

**9-115** The simply supported beam shown in Fig. P9-115 is made of air-dried Douglas fir ($E = 1900$ ksi) with an allowable flexural stress of 1900 psi, an allowable shearing stress of 85 psi, and an allowable deflection of $\frac{1}{360}$ of the span. Select the lightest standard structural timber that can be used to support the loads shown in the figure.

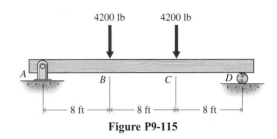

**Figure P9-115**

# 9-10 SUMMARY

A beam design is frequently not complete until the amount of deflection has been determined for the specified load. The deflection of a beam depends on the stiffness of the material and the dimensions of the beam as well as on the applied loads and type of supports.

When a straight beam is loaded and the action is elastic, the centroidal axis of the beam is a curve defined as the elastic curve. In regions of constant bending moment, the elastic curve is an arc of a circle of radius $\rho$. Since the beam is bent only with couples, plane sections remain plane and the deformation of the fibers (elongation and contraction) is proportional to the distance from the neutral surface, which is unchanged in length. Thus

$$\frac{c}{\rho} = \frac{\delta}{L} = \epsilon = \frac{\sigma}{E} = \frac{Mc}{EI} \tag{9-1}$$

Therefore,

$$\frac{1}{\rho} = \frac{M}{EI} \tag{9-2}$$

which relates the radius of curvature of the neutral surface of the beam to the bending moment $M$, the stiffness of the material $E$, and the second moment $I$ of the cross-sectional area of the beam with respect to the neutral axis.

Equation (9-2) for the curvature of the elastic curve is useful only when the bending moment is constant for the interval of the beam involved. For most beams, the bending moment is a function of position along the beam, and a more general expression is required. The curvature from the calculus (see any standard calculus textbook) is

$$\frac{1}{\rho} = \frac{d^2y/dx^2}{[1 + (dy/dx)^2]^{3/2}}$$

For most beams the slope $dy/dx$ is very small, and its square can be neglected in comparison to unity. With this approximation,

$$\frac{1}{\rho} = \frac{d^2y}{dx^2}$$

and Eq. (9-2) becomes

$$EI \frac{d^2y}{dx^2} = M(x) \tag{9-3}$$

which is the differential equation for the elastic curve of a beam where the moment $M$ is a function of $x$.

Whenever the bending moment can be readily expressed as an integrable function of $x$, Eq. (9-3) can be solved for the deflection $y$ of the elastic curve of a beam at any point $x$ along the beam. The constants of integration are evaluated from the applicable boundary conditions.

The integration method becomes tedious and time-consuming when several intervals and several sets of matching conditions are required. The labor involved in solving problems of this type, however, can be diminished by making use of singularity functions. Singularity functions are used to write one bending moment equation that applies in all intervals along a beam, thus eliminating the need for matching conditions.

The method of superposition for determining beam deflections is based on the fact that the resultant effect of several loads acting simultaneously on a member is the sum of the contributions from each of the loads applied individually. The results for the separate loads are listed in Table A-19 of Appendix A.

A beam, subjected only to transverse loads, with more than two reaction components is statically indeterminate because the equations of equilibrium are not sufficient to determine all the reactions. The additional relations needed for an evaluation of reactions (or other unknown forces) are obtained from deformation (slope or deflection) equations.

# REVIEW PROBLEMS

**9-116\*** The boards for a concrete form are to be bent to a circular curve of radius 5 m. What maximum thickness can be used if the stress is not to exceed 15 MPa? The modulus of elasticity for the wood is 10 GPa.

**9-117\*** During fabrication of a laminated timber arch, one of the 10-in.-wide $\times$ 1-in.-thick Douglas fir ($E$ = 1900 ksi) planks is bent to a radius of 12 ft. Determine the maximum flexural stress developed in the plank.

**9-118** A beam is loaded and supported as shown in Fig. P9-118. Determine

(a) The equation of the elastic curve, using the designated axes.

(b) The slope at the right end of the beam.

(c) The deflection at the right end of the beam.

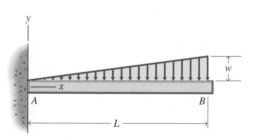

**Figure P9-118**

**9-119** Select the lightest structural steel ($E$ = 29,000 ksi) wide-flange or American standard beam (Appendix A) that can be used for the beam shown in Fig. P9-119 if the maximum flexural stress must not exceed 10 ksi and the maximum deflection must not exceed 0.200 in. when $L$ = 8 ft and $w$ = 2000 lb/ft.

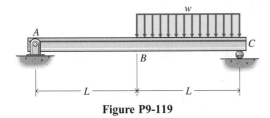

**Figure P9-119**

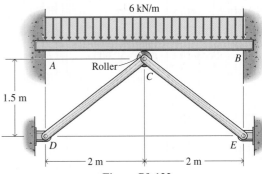

**Figure P9-122**

**9-120\*** A cantilever beam is fabricated by bolting two structural steel ($E = 200$ GPa) C127 × 10 channel sections together, as shown in Fig. P9-120. Determine
(a) The deflection at the right end of the beam.
(b) The maximum tensile flexural stress in the beam.

**9-123\*** In Fig. P9-123, beam $AC$ is made of brass and beam $BC$ is made of steel. The second moment of the cross-sectional area of beam $BC$ with respect to its neutral axis is twice that of beam $AC$, and the modulus of elasticity of the steel is twice that of the brass. The reaction at $C$ is zero before the load $w$ is applied. Determine
(a) The reaction at $C$ on beam $BC$.
(b) The reactions at supports $A$ and $B$.

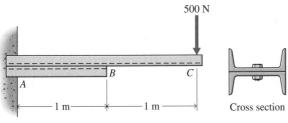

**Figure P9-120**

**9-121** An aluminum beam ($E = 10,000$ ksi and $I = 48$ in⁴) is loaded and supported as shown in Fig. P9-121. Determine
(a) The deflection at the left end of the beam.
(b) The deflection at the middle of span $BD$.

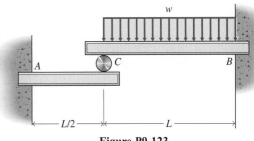

**Figure P9-123**

**9-124** A beam is loaded and supported as shown in Fig. P9-124. Determine
(a) The reactions at supports $A$ and $D$.
(b) The deflection at $B$ if $E = 200$ GPa and $I = 350(10^6)$ mm⁴.

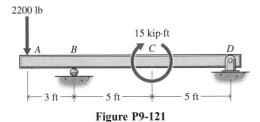

**Figure P9-121**

**9-122** The steel ($E = 200$ GPa) beam $AB$ of Fig. P9-122 is fixed at ends $A$ and $B$ and supported at the center by the pin-connected timber ($E = 10$ GPa) struts $CD$ and $CE$. The cross-sectional area of each strut is 6400 mm², and the second moment of the cross-sectional area of the beam with respect to its neutral axis is $25(10^6)$ mm⁴. Determine the force in each strut after the 6 kN/m distributed load is applied to the beam.

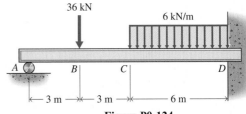

**Figure P9-124**

**9-125** A beam is loaded and supported as shown in Fig. P9-125. Determine

(a) The reactions at supports $A$, $B$, and $D$.

(b) The deflection at $C$ if $E = 30,000$ ksi and $I = 26$ in$^4$.

**9-126\*** A beam is loaded and supported as shown in Fig. P9-126. Determine

(a) The reactions at supports $A$, $B$, and $C$.

(b) The deflection midway between supports $B$ and $C$.

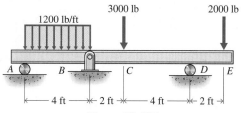

**Figure P9-125**

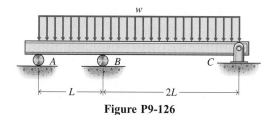

**Figure P9-126**

# COMBINED STATIC LOADING

## 10-1 INTRODUCTION

In previous chapters, formulas were developed for determining normal and shearing stresses on specific planes in axially loaded bars, circular shafts, and beams. For example, the normal stress at a point on a transverse cross section of a beam can be determined by using the flexure formula [Eq. (8-11)]. The shearing stress at the same point on the cross section can be determined by using the shearing stress formula [Eq. (8-15)]. In this chapter, methods will be developed for determining the normal and shearing stresses on other planes through the point of interest.

For the case of an axially loaded bar, expressions were developed in Section 4-3 for determining the normal stress [Eq. (4-7)] and shearing stress [Eq. (4-8)] on inclined planes through the bar. This analysis indicated that maximum normal stresses occur on transverse planes and that maximum shearing stresses occur on planes inclined at 45° to the axis of the bar. Similarly, for the case of pure torsion in a circular shaft, maximum shearing stresses occur on transverse planes and maximum tensile and compressive stresses occur on planes inclined at 45° to the axis of the shaft. For the case of a beam subjected to arbitrary transverse loading, both normal and shearing stresses develop on a transverse cross section. The stresses on a specified inclined plane can be determined by using the free-body diagram method of Section 7-5; however, this method is not suitable for the determination of maximum normal and maximum shearing stresses, which are often required. A more general approach for determining stresses on inclined planes will be developed to handle this type of analysis.

## 10-2 STRESS AT A GENERAL POINT IN AN ARBITRARILY LOADED MEMBER

The nature of the internal force distribution at an arbitrary interior point $O$ of a body of arbitrary shape that is in equilibrium under the action of a system of applied forces and support reactions $F_1, F_2, F_3, \ldots, F_i$ can be studied by exposing an interior plane through $O$, as shown in Fig. 10-1a. The force distribution required on such an interior plane to maintain equilibrium of the isolated part of the body will not be uniform, in general; however, any distributed force acting on a small area $\Delta A$ surrounding a point of interest $O$ can be replaced by a stati-

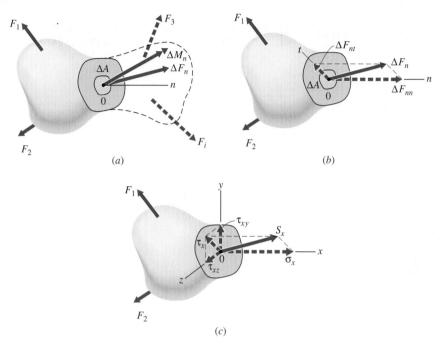

**Figure 10-1**

cally equivalent resultant force $\Delta F_n$ through $O$ and a couple $\Delta M_n$. The subscript $n$ indicates that the resultant force and couple are associated with a particular plane through $O$—namely, the one having an outward normal in the $n$-direction at $O$. For any other plane through $O$ the values of $\Delta F$ and $\Delta M$ could be different. Note that the line of action of $\Delta F_n$ or $\Delta M_n$ may not coincide with the direction of $n$. If the resultant force $\Delta F_n$ is divided by the area $\Delta A$, an average force per unit area (average resultant stress) is obtained. As the area $\Delta A$ is made smaller and smaller, the force distribution becomes more and more uniform and the couple $\Delta M_n$ vanishes. In the limit $\Delta A \to 0$, a quantity known as the stress vector[1] or resultant stress, $S_n$, is obtained. Thus,

$$S_n = \lim_{\Delta A \to 0} \frac{\Delta F_n}{\Delta A}$$

In Section 4-3 it was pointed out that materials respond differently to components of the stress vector. In particular, components normal ($n$) and tangent ($t$) to the internal plane are important. As shown in Fig. 10-1$b$, the resultant force $\Delta F_n$ can be resolved into components $\Delta F_{nn}$ normal to the plane and $\Delta F_{nt}$ tangent to the plane. A normal stress $\sigma_n$ and a shearing stress $\tau_n$ are then defined as

$$\sigma_n = \lim_{\Delta A \to 0} \frac{\Delta F_{nn}}{\Delta A}$$

and

$$\tau_n = \lim_{\Delta A \to 0} \frac{\Delta F_{nt}}{\Delta A}$$

---

[1]The component of a tensor on a plane is a vector; therefore, on a particular plane, the stresses can be treated as vectors.

For purposes of analysis it is convenient to reference stresses to some coordinate system. In a Cartesian coordinate system, the stresses on planes having outward normals in the $x$-, $y$-, and $z$-directions are usually chosen. Consider the plane having an outward normal in the $x$-direction (see Fig. 10-1$c$). In this case the normal and shear stresses on the plane will be $\sigma_x$ and $\tau_x$, respectively. Since $\tau_x$, in general, will not coincide with the $y$- or $z$-axes, it must be resolved into the components $\tau_{xy}$ and $\tau_{xz}$, as shown in Fig. 10-1$c$.

Unfortunately, the state of stress at a point in a material is not completely defined by these three components of the stress vector, since the stress vector itself depends on the orientation of the plane with which it is associated. An infinite number of planes can be passed through the point, resulting in an infinite number of stress vectors being associated with the point. Fortunately it can be shown[2] that the specification of stresses on three mutually perpendicular planes is sufficient to completely describe the state of stress at the point. The rectangular components of stress vectors on planes having outward normals in the coordinate directions are shown in Fig. 10-2. The six faces of the small element are denoted by the directions of their outward normals so that the positive $x$-face is the one whose outward normal is in the direction of the positive $x$-axis. The coordinate axes $x$, $y$, and $z$ are arranged as a right-hand system.

The sign convention for stresses is as follows:

Normal stresses are indicated by the symbol $\sigma$ and a single subscript to indicate the plane (actually the outward normal to the plane) on which the stress acts. Normal stresses are positive if they point in the direction of the outward normal. Thus, normal stresses are positive if tensile and negative if compressive.

Shearing stresses are denoted by the symbol $\tau$ followed by two subscripts; the first subscript designates the normal to the plane on which the stress acts and the second designates the coordinate axis to which the stress is parallel. Thus $\tau_{xz}$ is the shearing stress on a plane with outward normal in the $x$-direction. The stress acts parallel to the $z$-axis. A positive shearing stress points in the positive direction of the coordinate axis of the second

---

[2]See Section 2-11 of *Mechanics of Materials*, 5th ed., W. F. Riley, L. D. Sturges, and D. H. Morris, John Wiley and Sons, New York, 1999.

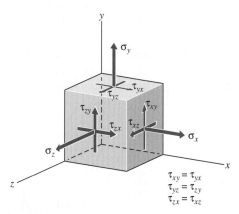

**Figure 10-2**

subscript if it acts on a surface with an outward normal in the positive direction. Conversely, if the outward normal of the surface is in the negative direction, then the positive shearing stress points in the negative direction of the coordinate axis of the second subscript. The stresses shown on the element in Fig. 10-2 are all positive.

Summarizing, the resultant stress vector $S_n$ is the result of the action of one part of the body on another part of the body, and it is a measure of the force intensity over a section of the body. The components of the stress vector, $\sigma$ and $\tau$, depend on the location of the point $O$ as well as the orientation of the plane through the point. Of the nine components of stress, $\sigma_x$, $\sigma_y$, $\sigma_z$, $\tau_{xy}$, $\tau_{yx}$, $\tau_{yz}$, $\tau_{zy}$, $\tau_{zx}$, and $\tau_{xz}$, only six are independent because moment equilibrium requires that $\tau_{xy} = \tau_{yx}$, $\tau_{yz} = \tau_{zy}$, and $\tau_{zx} = \tau_{xz}$ (see Section 7-5).

# 10-3 TWO-DIMENSIONAL OR PLANE STRESS

Considerable insight into the nature of stress distributions can be gained by considering a state of stress known as two-dimensional or plane stress. For this case, two parallel faces of the small element shown in Fig. 10-2 are assumed to be free of stress. For purposes of analysis, let these faces be perpendicular to the $z$-axis. Thus,

$$\sigma_z = \tau_{zx} = \tau_{zy} = 0$$

From Eq. (7-8), however, this also implies that

$$\tau_{xz} = \tau_{yz} = 0$$

A state of plane stress occurs naturally at points on the outside surface of a body where the $z$-components of force are zero. A state of plane stress also occurs at points within thin plates where the $z$-dimension of the body is small and the $z$-components of force are zero.

For plane stress analysis, then, the only components of stress present are $\sigma_x$, $\sigma_y$, and $\tau_{xy} = \tau_{yx}$. For convenience, this state of stress is represented by the two-dimensional sketch shown in Fig.10-3. However, the three-dimensional element of which the two-dimensional sketch is a plane projection should be kept in mind at all times.

Normal and shearing stresses on an arbitrary plane, such as plane $A$–$A$ in Fig. 10-3, can be obtained by using a free-body diagram and the equations of equilibrium. The solution to the following example problem illustrates this method of approach for plane stress problems.

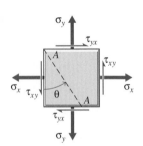

**Figure 10-3**

# Example Problem 10-1

At a point on the outside surface of a thin-walled pressure vessel (see Fig. 10-4a), stresses are 8000 psi T and zero shear on a horizontal plane and 4000 psi C and zero shear on a vertical plane, as shown in Figs. 10-4b and c. Determine

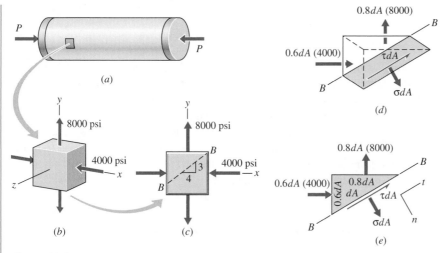

**Figure 10-4**

the stresses at this point on plane $B-B$ having a slope of 3 vertical to 4 horizontal.

## SOLUTION

The differential blocks of Figs. 10-4$b$ and $c$ are drawn as an aid to visualization of the data. A free-body diagram of a wedge-shaped element defined by the three given planes and subjected to ordinary force vectors is shown in Fig. 10-4$d$. Since stresses are not vectors, stress components cannot be used in the equations of equilibrium. The stresses must first be multiplied by the areas on which they act to get force vectors. It is these force vectors that are shown on the free-body diagram and that must be used in the equations of equilibrium.

The shaded area of Fig. 10-4$d$ indicates the plane on which the stresses to be evaluated are acting. This area is arbitrarily assigned the magnitude $dA$; therefore, the corresponding areas of the horizontal and vertical faces of the element are 0.8 $dA$ and 0.6 $dA$, respectively. For practical purposes, the two-dimensional free-body diagram of Fig. 10-4$e$ is usually adequate.

Summing forces in the $n$-direction yields

$$+\searrow \Sigma F_n = 0: \qquad \sigma \, dA + [(0.6 \, dA)(4000)](0.6)$$
$$- [(0.8 \, dA)(8000)](0.8) = 0 \qquad (a)$$

from which

$$\sigma = +3680 \text{ psi} = 3680 \text{ psi T} \qquad \textbf{Ans.}$$

In Eq. (a), the stress $\sigma_x = 4000$ psi is multiplied by the area $dA_x = 0.6 \, dA$ to get the force $dF_x = \sigma_x \, dA_x = (0.6 \, dA)(4000)$. The component of this force in the $n$-direction is $(dF_x)_n = (\sigma_x \, dA_x)(0.6) = [(4000)(0.6 \, dA)](0.6)$.

Summing forces in the $t$-direction yields

$$+\nearrow \Sigma F_t = 0: \qquad \tau \, dA + [(0.6 \, dA)(4000)](0.8)$$
$$+ [(0.8 \, dA)(8000)](0.6) = 0 \qquad (b)$$

from which

$$\tau = -5760 \text{ psi} = 5760 \text{ psi } \swarrow$$   **Ans.**

Normal stresses should always be designated as tension or compression. The presence of shearing stresses on the horizontal and vertical planes, had there been any, would merely have required two more forces on the free-body diagram: one parallel to the vertical face and one parallel to the horizontal face. The magnitudes of the shearing stresses (not the shearing forces) must be the same on any two orthogonal planes [Eq. (7-8)], and the vectors used to represent the shearing forces must be directed as in Fig. 10-3 or both vectors reversed, depending on the given data. ■

# PROBLEMS

### Introductory Problems

**10-1\*** At a point in a thin plate, there are normal stresses of 20 ksi T on a vertical plane and 10 ksi C on a horizontal plane, as shown in Fig. P10-1. Determine the normal and shear stresses at this point on the inclined plane $A-B$ shown in the figure.

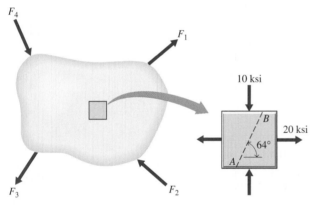

**Figure P10-1**

**10-2\*** At a point in a stressed body, there are normal stresses of 95 MPa T on a vertical plane and 125 MPa T on a horizontal plane, as shown in Fig. P10-2. Determine the normal and shear stresses at this point on the inclined plane $A-B$ shown in the figure.

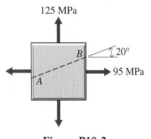

**Figure P10-2**

**10-3** The stresses shown in Fig. P10-3 act at a point on the surface of a circular shaft that is subjected to a twisting moment $M$ as shown. Determine the normal and shear stresses at this point on the inclined plane $A-B$ shown in the figure.

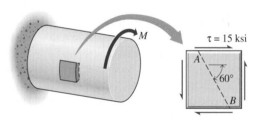

**Figure P10-3**

### Intermediate Problems

**10-4\*** The stresses shown in Fig. P10-4 act at a point on the surface of a cantilever beam. Determine the normal and shear stresses at this point on the inclined plane $A-B$ shown in the figure.

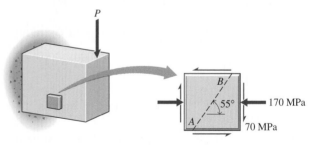

**Figure P10-4**

**10-5** The stresses shown in Fig. P10-5 act at a point on the surface of a thin-walled pressure vessel that is subjected to an internal pressure, an axial load $P$, and a torque $T$. Determine the normal and shear stresses at this point on the inclined plane $A-B$ shown in the figure.

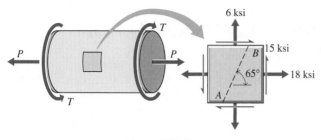

**Figure P10-5**

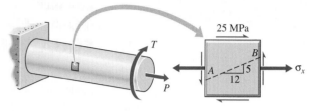

**Figure P10-8**

**10-6** The stresses shown in Fig. P10-6 act at a point in a stressed body. Determine the normal and shear stresses at this point on the inclined plane $A-B$ shown in the figure.

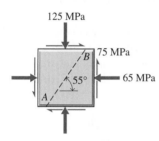

**Figure P10-6**

**10-9** The thin-walled cylindrical pressure vessel shown in Fig. P10-9 was constructed by wrapping a thin steel plate into a helix that forms an angle $\theta = 35°$ with respect to a transverse plane through the cylinder, and butt-welding the resulting seam. In a thin-walled cylindrical pressure vessel, the normal stress $\sigma_y$ on a horizontal plane through a point on the surface of the vessel is twice as large as the normal stress $\sigma_x$ on a vertical plane through the point and the shear stresses on both the horizontal and vertical planes are zero. If the stresses in the weld material on the plane of the weld must be limited to 10 ksi in tension and 7 ksi in shear, determine the maximum normal stress $\sigma_x$ permitted in the vessel.

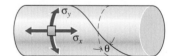

**Figure P10-9**

**Computer Problems**

**Challenging Problems**

**10-7\*** At a point in a machine component, the normal and shear stresses on an inclined plane are 4800 psi T and 1500 psi, respectively, as shown in Fig. P10-7. The normal stress on a vertical plane through the point is zero. Determine
(a) The shear stresses on horizontal and vertical planes.
(b) The normal stress on a horizontal plane.

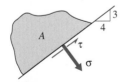

**Figure P10-7**

**10-8** The stresses on horizontal and vertical planes at a point on the outside surface of a solid circular bar subjected to an axial load $P$ and a torsional load $T$ are shown in Fig. P10-8. The normal stress on the inclined plane $A-B$ is 15 MPa T. Determine
(a) The normal stress $\sigma_x$ on the vertical plane.
(b) The magnitude and direction of the shear stress on the inclined plane $A-B$.

**10-10** The thin-walled cylindrical pressure vessel shown in Fig. P10-9 was constructed by wrapping a thin steel plate into a helix that forms an angle $\theta$ with respect to a transverse plane through the cylinder and butt-welding the resulting seam. The normal stresses $\sigma_x$ and $\sigma_y$ on vertical and horizontal planes through a point on the surface of the vessel are 50 MPa and 100 MPa, respectively, and the shear stresses are zero. Prepare a plot showing the variation of normal and shear stresses on the plane of the weld as the angle $\theta$ is varied from $0°$ to $180°$.

**10-11** In a solid circular shaft subjected to a torsional form of loading, the normal stresses $\sigma_x$ and $\sigma_y$ on vertical and horizontal planes through a point on the surface of the vessel are zero and the shear stresses are 15 ksi, as shown in Fig. P10-11. Prepare a plot showing the variation of normal and shear stresses on plane $A-A$ through the point as the angle $\theta$ is varied from $0°$ to $180°$.

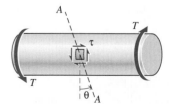

**Figure P10-11**

## 10-4 THE STRESS TRANSFORMATION EQUATIONS FOR PLANE STRESS

Equations relating the normal and shearing stresses $\sigma_n$ and $\tau_{nt}$ on an arbitrary plane (whose normal is oriented at an angle $\theta$ with respect to a reference $x$-axis) through a point and the known stresses $\sigma_x$, $\sigma_y$, and $\tau_{xy} = \tau_{yx}$ on the reference planes can be developed using the free-body diagram method. Consider the plane stress situation indicated in Fig. 10-5$a$ (all stresses are positive), where the dotted line $A$–$A$ represents any plane through the point (all planes are perpendicular to the plane of zero stress—the plane of the paper). In the following derivation, a counterclockwise angle $\theta$ is positive, where $\theta$ is measured from the positive $x$-axis to the positive $n$-axis.

Figure 10-5$b$ is a free-body diagram of a wedge-shaped element in which the areas of the faces are $dA$ for the inclined face (plane $A$–$A$), $dA \cos \theta$ for the vertical face, and $dA \sin \theta$ for the horizontal face. The $n$-axis is perpendicular to the inclined face; the $t$-axis is parallel to the inclined face. The positive direction for the $n$-axis is outward from the surface (it is the outward normal), and the positive direction for the $t$-axis is 90° counterclockwise from the $n$-axis. The $xy$- and $nt$-axes shown in Fig. 10-5 are positive. The stresses shown in Fig. 10-5$a$ are multiplied by the areas over which they act, resulting in the free-body diagram shown in Fig. 10-5$b$. Summing forces in the $n$-direction gives

$$+\nearrow \Sigma F_n = 0: \qquad \sigma_n \, dA - \sigma_x(dA \cos \theta) \cos \theta - \sigma_y(dA \sin \theta) \sin \theta$$
$$- \tau_{yx}(dA \sin \theta) \cos \theta - \tau_{xy}(dA \cos \theta) \sin \theta = 0$$

from which, since $\tau_{yx} = \tau_{xy}$,

$$\sigma_n = \sigma_x \cos^2 \theta + \sigma_y \sin^2 \theta + 2\tau_{xy} \sin \theta \cos \theta \qquad (10\text{-}1a)$$

or, in terms of the double angle,

$$\sigma_n = \frac{\sigma_x(1 + \cos 2\theta)}{2} + \frac{\sigma_y(1 - \cos 2\theta)}{2} + \frac{2\tau_{xy}(\sin 2\theta)}{2}$$

$$= \frac{\sigma_x + \sigma_y}{2} + \frac{\sigma_x - \sigma_y}{2} \cos 2\theta + \tau_{xy} \sin 2\theta \qquad (10\text{-}1b)$$

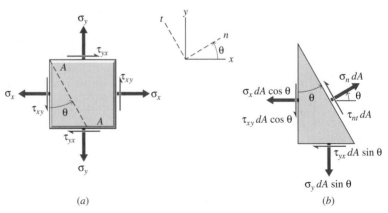

(a)

(b)

**Figure 10-5**

Summing forces in the $t$-direction gives

$$+\nwarrow \Sigma F_t = 0: \qquad \tau_{nt}\,dA + \sigma_x(dA\cos\theta)\sin\theta - \sigma_y(dA\sin\theta)\cos\theta$$

$$-\tau_{xy}(dA\cos\theta)\cos\theta + \tau_{yx}(dA\sin\theta)\sin\theta = 0$$

from which

$$\tau_{nt} = -(\sigma_x - \sigma_y)\sin\theta\cos\theta + \tau_{xy}(\cos^2\theta - \sin^2\theta) \qquad (10\text{-}2a)$$

or, in terms of the double angle,

$$\tau_{nt} = -\frac{\sigma_x - \sigma_y}{2}\sin 2\theta + \tau_{xy}\cos 2\theta \qquad (10\text{-}2b)$$

Equations (10-1) and (10-2) provide a means for determining normal and shear stresses on any plane whose outward normal is perpendicular to the $z$-axis and is oriented at an angle $\theta$ with respect to the reference $x$-axis. When these equations are used, the sign conventions used in their development must be rigorously followed; otherwise, erroneous results will be obtained.

These sign conventions can be summarized as follows:

1. Tensile normal stresses are positive; compressive normal stresses are negative. All of the normal stresses shown on Fig. 10-5 are positive. The sign of a normal stress is independent of the coordinate system being used.

2. A shearing stress is positive if it points in the positive direction of the coordinate axis of the second subscript when it is acting on a surface whose outward normal is in a positive direction of the coordinate axis of the first subscript. Similarly, if the outward normal of the surface is in a negative direction, then a positive shearing stress points in the negative direction of the coordinate axis of the second subscript. All of the shearing stresses shown on Fig. 10-5 are positive. Shearing stresses pointing in the opposite directions would be negative. The sign of a shearing stress depends on the coordinate system being used.

3. An angle measured counterclockwise from the reference positive $x$-axis is positive. Conversely, angles measured clockwise from the reference $x$-axis are negative.

4. The $(n, t, z)$ axes have the same order as the $(x, y, z)$ axes. Both sets of axes form a right-hand coordinate system.

## ▌ Example Problem 10-2

At a point in a structural member subjected to plane stress there are normal and shearing stresses on horizontal and vertical planes through the point, as shown in Fig. 10-6a. Use the stress transformation equations to determine:

(a) The normal and shearing stresses on plane $AB$.

(b) The normal and shearing stresses on plane $CD$, which is perpendicular to plane $AB$.

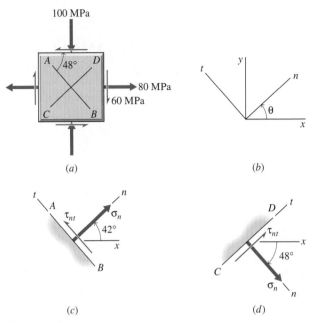

**Figure 10-6**

## SOLUTION

First define the $x$–$y$ and $n$–$t$ (if needed) directions if they have not been specified. On the basis of the axes shown in Fig. 10-6$b$ and the established sign conventions, $\sigma_x$ is positive, whereas $\sigma_y$ and $\tau_{xy}$ are negative. Thus, the given values for use in Eqs. (10-1) and (10-2) are

$$\sigma_x = +80 \text{ MPa}, \qquad \sigma_y = -100 \text{ MPa}, \qquad \text{and} \qquad \tau_{xy} = -60 \text{ MPa}$$

(a) The angle $\theta$ for plane $AB$ is $+42°$, as shown in Fig. 10-6$c$. Thus, from Eqs. (10-1) and (10-2),

$$\sigma_n = \sigma_x \cos^2 \theta + \sigma_y \sin^2 \theta + 2\tau_{xy} \sin \theta \cos \theta$$
$$= 80 \cos^2(+42°) + (-100) \sin^2(+42°)$$
$$\quad + 2(-60) \sin(+42°) \cos(+42°)$$
$$= -60.26 \text{ MPa} \cong 60.3 \text{ MPa C} \qquad \textbf{Ans.}$$

$$\tau_{nt} = -(\sigma_x - \sigma_y) \sin \theta \cos \theta + \tau_{xy}(\cos^2 \theta - \sin^2 \theta)$$
$$= -[80 - (-100)] \sin(+42°) \cos(+42°)$$
$$\quad +(-60)[\cos^2(+42°) - \sin^2(+42°)]$$
$$= -95.78 \text{ MPa} \cong -95.8 \text{ MPa} \qquad \textbf{Ans.}$$

(b) The angle $\theta$ for plane $CD$ is $-48°$ (or $+312°$), as shown in Fig. 10-6$d$. Thus, from Eqs. (10-1) and (10-2),

$$\sigma_n = \sigma_x \cos^2 \theta + \sigma_y \sin^2 \theta + 2\tau_{xy} \sin \theta \cos \theta$$
$$= 80 \cos^2(-48°) + (-100) \sin^2(-48°)$$

$$+ 2(-60) \sin(-48°) \cos(-48°)$$
$$= +40.26 \text{ MPa} \cong 40.3 \text{ MPa (T)} \qquad \textbf{Ans.}$$

$$\tau_{nt} = -(\sigma_x - \sigma_y) \sin\theta \cos\theta + \tau_{xy}(\cos^2\theta - \sin^2\theta)$$
$$\tau_{nt} = -[80 - (-100)] \sin(-48°) \cos(-48°)$$
$$+(-60)[\cos^2(-48°) - \sin^2(-48°)]$$
$$= +95.78 \text{ MPa} \cong +95.8 \text{ MPa} \qquad \textbf{Ans.}$$

Since planes $AB$ and $CD$ are orthogonal, the shearing stresses on the two planes must be equal in magnitude to satisfy Eq. (7-8). Also, one of the stresses must tend to produce a clockwise rotation of the element while the other tends to produce a counterclockwise rotation. With the coordinate systems shown in Figs. 10-6c and 10-6d, this means that the shearing stresses calculated using Eq. (10-2) will have opposite signs on the two planes. ∎

## PROBLEMS

*Note*: Normal and shear stresses on horizontal and vertical planes through a point in a structural member subjected to plane stress are shown in the following figures. Use the stress transformation equations for plane stress [Eqs. (10-1) and (10-2)] to solve for the normal and shear stresses on the inclined plane $A$–$B$.

### Introductory Problems
**10-12\***  At the point shown in Fig. P10-12.

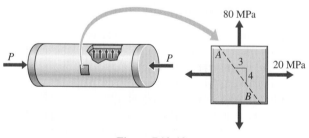

**Figure P10-12**

**10-13\***  At the point shown in Fig. P10-13.

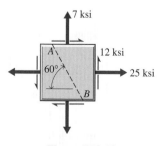

**Figure P10-13**

**10-14**  At the point shown in Fig. P10-14.

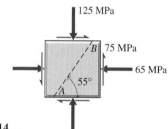

**Figure P10-14**

**10-15**  At the point shown in Fig. P10-15.

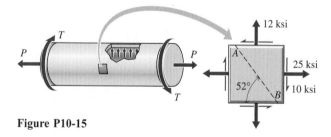

**Figure P10-15**

**10-16\***  At the point shown in Fig. P10-16.

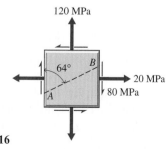

**Figure P10-16**

## Intermediate Problems

**10-17\*** The stresses shown in Fig. P10-17*a* act at a point on the free surface of a stressed body. Determine
(a) The normal and shear stresses at this point on the inclined plane *A–B* shown in the figure.
(b) The normal stresses $\sigma_n$ and $\sigma_t$ and the shear stress $\tau_{nt}$ at this point if they act on the rotated stress element shown in Fig. P10-17*b*.

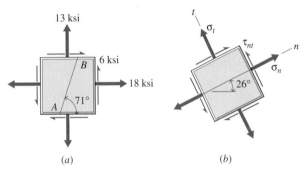

(a)                                    (b)

**Figure P10-17**

**10-18\*** The stresses shown in Fig. P10-18*a* act at a point on the free surface of a stressed body. Determine
(a) The normal and shear stresses at this point on the inclined plane *AB* shown in the figure.
(b) The normal stresses $\sigma_n$ and $\sigma_t$ and the shear stress $\tau_{nt}$ at this point if they act on the rotated stress element shown in Fig. P10-18*b*.

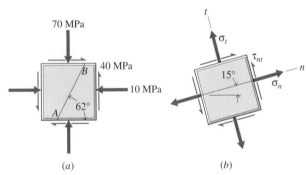

(a)                                    (b)

**Figure P10-18**

**10-19** The stresses shown in Fig. P10-19 act at a point in a structural member. Determine the normal stresses $\sigma_n$ and $\sigma_t$ and the shear stress $\tau_{nt}$ at this point.

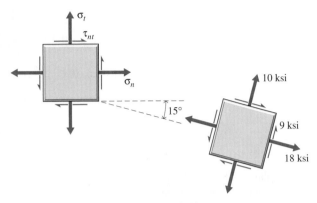

**Figure P10-19**

**10-20** The stresses shown in Fig. P10-20 act at a point in a structural member. Determine the normal stresses $\sigma_n$ and $\sigma_t$ and the shear stress $\tau_{nt}$ at this point.

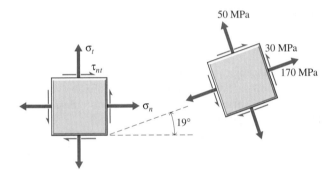

**Figure P10-20**

## Challenging Problems

**10-21\*** The stresses on horizontal and vertical planes at a point on the outside surface of a solid circular shaft subjected to an axial load *P* and a torque *T* are shown in Fig. P10-21. The normal stress on plane *A–A* at this point is 8000 psi T. Determine
(a) The magnitude of the shearing stresses $\tau_h$ and $\tau_v$.
(b) The magnitude and direction of the shearing stress on the inclined plane *A–A*.

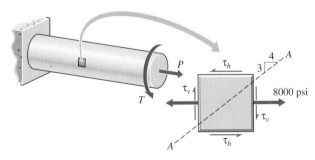

**Figure P10-21**

**10-22\*** The stresses on horizontal and vertical planes at a point are shown in Fig. P10-22. The normal stress on plane *AB* is 15 MPa T. Determine

(a) The normal stress $\sigma_x$ on the vertical plane.

(b) The magnitude and direction of the shearing stress on the inclined plane *A–B*.

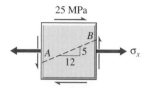

25 MPa

**Figure P10-22**

**10-23** At a point in a structural member, there are stresses on horizontal and vertical planes, as shown in Fig. P10-23. The magnitude of the compressive stress $\sigma_c$ is three times the magnitude of the tensile stress $\sigma_t$. Specifications require that the shearing stress on plane *AB* not exceed 4000 psi and the normal stress on plane *AB* not exceed 7800 psi. Determine the maximum value of stress $\sigma_c$ that will satisfy the specifications.

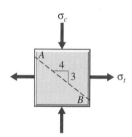

**Figure P10-23**

**10-24** At a point in a stressed body, the stresses on two perpendicular planes are as shown in Fig. P10-24. Determine

(a) The stresses on plane *a–a*.

(b) The stresses on horizontal and vertical planes at the point.

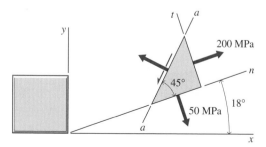

**Figure P10-24**

**10-25** At a point in a stressed body, the stresses on two perpendicular planes are as shown in Fig. P10-25. Determine

(a) The stresses on plane *a–a*.

(b) The stresses on horizontal and vertical planes at the point.

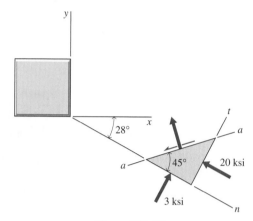

**Figure P10-25**

## Computer Problems

**10-26** The stresses shown in Fig. P10-26 act at a point on the free surface of a stressed body.

(a) Calculate and graph the normal stress $\sigma_n$ and the shearing stress $\tau_{nt}$ on the inclined plane *AB* as a function of the angle $\theta$ ($0° \le \theta \le 180°$).

(b) For what angle $\theta$ is the normal stress a maximum? A minimum? What is the value of the shear stress on these planes?

(c) For what angle $\theta$ is the shear stress a maximum? A minimum? What is the value of the normal stress on these planes?

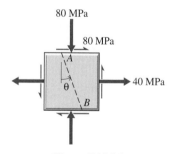

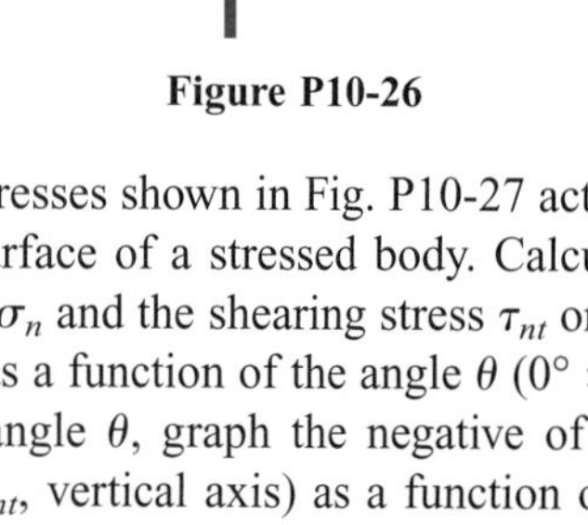

**Figure P10-26**

**10-27** The stresses shown in Fig. P10-27 act at a point on the free surface of a stressed body. Calculate the normal stress $\sigma_n$ and the shearing stress $\tau_{nt}$ on the inclined plane *AB* as a function of the angle $\theta$ ($0° \le \theta \le 180°$). For each angle $\theta$, graph the negative of the shearing stress ($-\tau_{nt}$, vertical axis) as a function of the normal stress ($\sigma_n$, horizontal axis). On your graph, clearly identify the points associated with the angles $\theta = 0°$, 30°, 45°, 60°, 90°, 120°, 135°, 150°, and 180°.

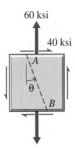

**Figure P10-27**

# 10-5 PRINCIPAL STRESSES AND MAXIMUM SHEARING STRESS—PLANE STRESS

The transformation equations for plane stress [Eqs. (10-1) and (10-2)] provide a means for determining the normal stress $\sigma_n$ and the shearing stress $\tau_{nt}$ on different planes through a point in a stressed body. As an example, consider the state of stress shown in Fig. 10-7a, which acts at a point on the free surface of a machine component or structural member. As the element is rotated through an angle $\theta$ about an axis perpendicular to the free surface, the normal stress $\sigma_n$ and the shearing stress $\tau_{nt}$ on the different planes vary continuously as shown in Fig. 10-7b. For design purposes, critical stresses at the point are usually the maximum tensile stress and the maximum shearing stress.

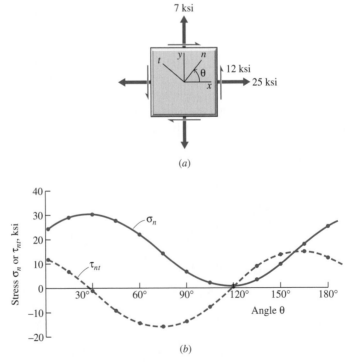

**Figure 10-7**

For a bar under axial load, the planes on which maximum normal stresses and maximum shearing stresses act are known from the results of Section 4-3. For more complicated forms of loading, these stresses can be determined by plotting curves similar to those shown in Fig. 10-7b for each different state of stress encountered, but this process is time-consuming and inefficient. Therefore, more general methods for finding the critical stresses have been developed.

The transformation equations for plane stress developed previously are as follows:

For normal stress $\sigma_n$,

$$\sigma_n = \frac{\sigma_x + \sigma_y}{2} + \frac{\sigma_x - \sigma_y}{2}\cos 2\theta + \tau_{xy}\sin 2\theta \tag{10-1b}$$

For shear stress $\tau_{nt}$,

$$\tau_{nt} = -\frac{\sigma_x - \sigma_y}{2}\sin 2\theta + \tau_{xy}\cos 2\theta \tag{10-2b}$$

Maximum and minimum values of $\sigma_n$ occur at values of $\theta$ for which $d\sigma_n/d\theta$ is equal to zero. Differentiation of $\sigma_n$ with respect to $\theta$ yields

$$\frac{d\sigma_n}{d\theta} = -(\sigma_x - \sigma_y)\sin 2\theta + 2\tau_{xy}\cos 2\theta \tag{a}$$

Setting Eq. (a) equal to zero and solving gives

$$\tan 2\theta_p = \frac{2\tau_{xy}}{\sigma_x - \sigma_y} \tag{10-3}$$

Note that the expression for $d\sigma_n/d\theta$ from Eq. (a) is numerically twice the value of the expression for $\tau_{nt}$ from Eq. (10-2b). Consequently, the shearing stress is zero on planes experiencing maximum and minimum values of normal stress. Planes free of shear stress are known as *principal planes*. Normal stresses occurring on principal planes are known as *principal stresses*. The values of $\theta_p$ from Eq. (10-3) give the orientations of two principal planes. A third principal plane for the plane stress state has an outward normal in the $z$-direction. For a given set of values of $\sigma_x$, $\sigma_y$, and $\tau_{xy}$, there are two values of $2\theta_p$ differing by 180° and, consequently, two values of $\theta_p$ that are 90° apart. This proves that the principal planes are normal to each other.

When $\tau_{xy}$ and $(\sigma_x - \sigma_y)$ have the same sign, $\tan 2\theta_p$ is positive and one value of $\theta_p$ is between 0° and 90°, with the other value 180° greater, as shown in Fig. 10-8. Consequently, one value of $\theta_p$ is between 0° and 45°, and the other one is 90° greater. In the first case, both $\sin 2\theta_p$ and $\cos 2\theta_p$ are positive, and in the second case both are negative. When these functions of $2\theta_p$ are substituted into Eq. (10-1b), two in-plane principal stresses $\sigma_{p1}$ and $\sigma_{p2}$ are found to be

$$\sigma_{p1,\,p2} = \frac{\sigma_x + \sigma_y}{2} \pm \sqrt{\left(\frac{\sigma_x - \sigma_y}{2}\right)^2 + \tau_{xy}^2} \tag{10-4}$$

Equation (10-4) gives the two principal stresses in the $xy$-plane, and the third one is $\sigma_{p3} = \sigma_z = 0$. Equation (10-3) gives the angles $\theta_p$ and $\theta_p + 90°$ between the $x$- (or $y$-) plane and the mutually perpendicular planes on which the principal

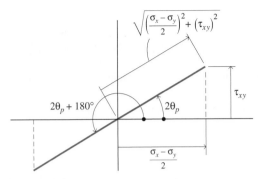

**Figure 10-8**

stresses act. When $\tan 2\theta_p$ is positive, $\theta_p$ is positive and the rotation is counterclockwise from the $x$- and $y$-planes to the planes on which the two principal stresses act. When $\tan 2\theta_p$ is negative, the rotation is clockwise. Note that one value of $\theta_p$ will always be between positive and negative $45°$ (inclusive) with the other value $90°$ greater. The numerically greater principal stress will act on the plane that makes an angle of $45°$ or less with the plane of the numerically larger of the two given normal stresses. This statement can be confirmed by substituting the value of $\theta_p$ from Eq. (10-3) into Eq. (10-1b) to obtain the corresponding principal stress.

Note that if one or both of the principal stresses from Eq. (10-4) is negative, the algebraic maximum stress can have a smaller absolute value than the "minimum" stress.

The maximum in-plane shearing stress $\tau_p$ occurs on planes located by values of $\theta$ where $d\tau_{nt}/d\theta$ is equal to zero. Differentiation of Eq. (10-2b) yields

$$\frac{d\tau_{nt}}{d\theta} = -(\sigma_x - \sigma_y)\cos 2\theta - 2\tau_{xy}\sin 2\theta$$

When $d\tau_{nt}/d\theta$ is equated to zero, the value of $\theta_\tau$ is given by the expression

$$\tan 2\theta_\tau = -\frac{(\sigma_x - \sigma_y)}{2\tau_{xy}} \tag{b}$$

where $\theta_\tau$ locates the planes of maximum in-plane shearing stress. Comparison of Eq. (b) and Eq. (10-3) reveals that the two tangents are negative reciprocals. Therefore, the two angles $2\theta_p$ and $2\theta_\tau$ differ by $90°$, and $\theta_p$ and $\theta_\tau$ are $45°$ apart. This means that the planes on which the maximum in-plane shearing stresses occur are $45°$ from the principal planes. The maximum in-plane shearing stresses are found by substituting values of angle functions obtained from Eq. (b) in Eq. (10-2b). The results are

$$\tau_p = \pm\sqrt{\left(\frac{\sigma_x - \sigma_y}{2}\right)^2 + \tau_{xy}^2} \tag{10-5}$$

Equation (10-5) has the same magnitude as the second term of Eq. (10-4).

A useful relation between the principal stresses and the maximum in-plane shearing stress is obtained from Eqs. (10-4) and (10-5) by subtracting the values for the two in-plane principal stresses and substituting the value of the radical from Eq. (10-5). The result is

$$\tau_p = (\sigma_{p1} - \sigma_{p2})/2 \qquad (10\text{-}6)$$

or, in words, the maximum value of $\tau_{nt}$ ($\tau_p$) is equal in magnitude to half the difference between the two in-plane principal stresses.

In general, when stresses act in three directions it can be shown[3] that there are three orthogonal planes on which the shearing stress is zero. These planes are known as the principal planes, and the stresses acting on them (the principal stresses) will have three values: one maximum, one minimum, and a third stress between the other two. The maximum shearing stress $\tau_{max}$ on any plane that could be passed through the point is half the difference between the maximum and minimum principal stresses and acts on planes that bisect the angles between the planes of the maximum and minimum normal stresses.

$$\tau_{max} = \frac{\sigma_{max} - \sigma_{min}}{2} \qquad (10\text{-}7)$$

When a state of plane stress exists, one of the principal stresses is zero. If the values of $\sigma_{p1}$ and $\sigma_{p2}$ from Eq. (10-4) have the same sign, then the third principal stress, $\sigma_{p3}$ equals zero, will be either the maximum or the minimum normal stress. Thus, the maximum shearing stress may be

$$(\sigma_{p1} - \sigma_{p2})/2 \qquad (\sigma_{p1} - 0)/2 \qquad \text{or} \qquad (0 - \sigma_{p2})/2$$

depending on the relative magnitudes and signs of the principal stresses. These three possibilities are illustrated in Fig. 10-9, in which one of the two orthogonal planes on which the maximum shearing stress acts is shaded for each example.

The direction of the maximum shearing stress can be determined by drawing a wedge-shaped block with two sides parallel to the planes having the maximum and minimum principal stresses, and with the third side at an angle $45°$

---

[3]See Section 2-11 of *Mechanics of Materials*, 5th ed., W. F. Riley, L. D. Sturges, and D. H. Morris, John Wiley and Sons, New York, 1999.

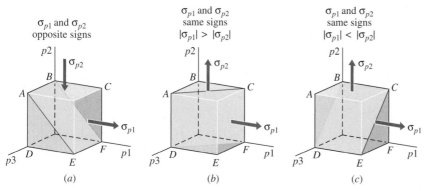

**Figure 10-9**

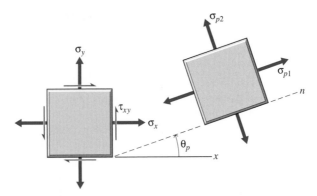

**Figure 10-10**

with the other two sides. The direction of the maximum shearing stress must oppose the larger of the two principal stresses.

Another useful relation between the principal stresses and the normal stresses on the orthogonal planes shown in Fig. 10-10 is obtained by adding the values for the two principal stresses, as given by Eq. (10-4). Thus,

$$\sigma_{p1} + \sigma_{p2} = \sigma_x + \sigma_y \qquad (10\text{-}8)$$

or, in words, for plane stress, the sum of the normal stresses on any two orthogonal planes through a point in a body is a constant or invariant.

In the preceding discussion, "maximum" and "minimum" stresses were considered algebraic quantities, and it has already been pointed out that the minimum algebraic stress may have a larger magnitude than the maximum stress. However, in the application to engineering problems (which includes the problems in this book), the term *maximum* will always refer to the largest absolute value (largest magnitude).

Application of the formulas and procedures developed in this section are illustrated by the following examples.

## Example Problem 10-3

At a point in a structural member subjected to plane stress there are normal and shearing stresses on horizontal and vertical planes through the point, as shown in Fig. 10-11a.

(a) Determine the principal stresses and the maximum shearing stress at the point.

(b) Locate the planes on which these stresses act and show the stresses on a complete sketch.

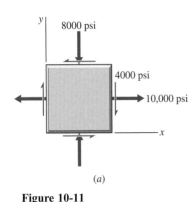

(a)

**Figure 10-11**

### SOLUTION

(a) On the basis of the axes shown in Fig. 10-11a and the established sign conventions, $\sigma_x$ is positive, whereas $\sigma_y$ and $\tau_{xy}$ are negative. For use in Eqs. (10-3) and (10-4), the given values are

$$\sigma_x = +10{,}000 \text{ psi} \qquad \sigma_y = -8000 \text{ psi} \qquad \tau_{xy} = -4000 \text{ psi}$$

When these values are substituted in Eq. (10-4), the principal stresses are found to be

$$\sigma_{p1, p2} = \frac{\sigma_x + \sigma_y}{2} \pm \sqrt{\left(\frac{\sigma_x - \sigma_y}{2}\right)^2 + \tau_{xy}^2}$$

$$= \frac{10,000 + (-8000)}{2} \pm \sqrt{\left(\frac{10,000 - (-8000)}{2}\right)^2 + (-4000)^2}$$

$$= 1000 \pm 9849$$

$$\sigma_{p1} = 1000 + 9849 = 10,849 \text{ psi} \cong 10,850 \text{ psi T} \qquad \textbf{Ans.}$$

$$\sigma_{p2} = 1000 - 9849 = -8849 \text{ psi} \cong 8850 \text{ psi C} \qquad \textbf{Ans.}$$

$$\sigma_{p3} = \sigma_z = 0 \qquad \textbf{Ans.}$$

Since $\sigma_{p1}$ and $\sigma_{p2}$ are of opposite sign, the maximum shearing stress is given by Eq. (10-7) as

$$\tau_{max} = \frac{\sigma_{max} - \sigma_{min}}{2}$$

$$= \frac{10,849 - (-8849)}{2} = 9849 \text{ psi} \cong 9850 \text{ psi} \qquad \textbf{Ans.}$$

The sum of the normal stresses on any two orthogonal planes is a constant for plane stress:

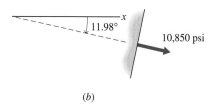

$$\sigma_x + \sigma_y = 10,000 + (-8000) = 2000 \text{ psi}$$

$$= \sigma_{p1} + \sigma_{p2} = 10,850 + (-8850)$$

(b)

Also, the normal stress on the planes of maximum shear stress is equal to the average of the normal stresses

$$\sigma_n = \frac{\sigma_x + \sigma_y}{2} = \frac{\sigma_{p1} + \sigma_{p2}}{2} = 1000 \text{ psi}$$

(b) When the given data are substituted in Eq. (10-3), the results are

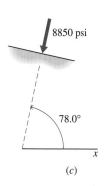

$$\tan 2\theta_p = \frac{2\tau_{xy}}{\sigma_x - \sigma_y} = \frac{2(-4000)}{10,000 - (-8000)} = -0.4444$$

from which

$$2\theta_p = -23.96°, + 156.04°, \ldots$$

and

$$\theta_p = -11.98°, +78.02°, \ldots$$

$$\cong 11.98° \downarrow \text{ and } 78.0° \uparrow \qquad \textbf{Ans.}$$

(c)

When $\theta = -11.98°$, Eq. (10-1a) gives

$$\sigma_n = \sigma_x \cos^2 \theta + \sigma_y \sin^2 \theta + 2\tau_{xy} \sin \theta \cos \theta$$

$$= 10,000 \cos^2(-11.98°) + (-8000) \sin^2(-11.98°)$$
$$+ 2(-4000) \sin(-11.98°) \cos(-11.98°)$$
$$= \sigma_{p1} = 10,849 \text{ psi} \cong 10,850 \text{ psi T}$$

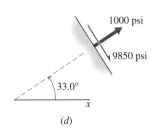

(d)

The result is shown in Fig. 10-11b.
When $\theta = +78.02°$, Eq. (10-1a) gives

$$\sigma_n = \sigma_x \cos^2 \theta + \sigma_y \sin^2 \theta + 2\tau_{xy} \sin \theta \cos \theta$$
$$= 10,000 \cos^2(+78.02°) + (-8000) \sin^2(+78.02°)$$
$$+ 2(-4000) \sin(+78.02°) \cos(+78.02°)$$
$$= \sigma_{p2} = -8849 \text{ psi} \cong 8850 \text{ psi C}$$

The result is shown in Fig. 10-11c.

The maximum in-plane shear stress occurs on a surface oriented at 45° to the principal directions. When $\theta = -11.98° + 45° = +33.02°$, Eqs. (10-1a) and (10-2a) give

$$\sigma_n = \sigma_x \cos^2 \theta + \sigma_y \sin^2 \theta + 2\tau_{xy} \sin \theta \cos \theta$$
$$= 10,000 \cos^2(+33.02°) + (-8000) \sin^2(+33.02°)$$
$$+ 2(-4000) \sin(+33.02°) \cos(+33.02°)$$
$$= +999.6 \text{ psi} \cong 1000 \text{ psi T}$$

and

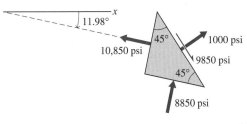

(e)

$$\tau_{nt} = -(\sigma_x - \sigma_y) \sin \theta \cos \theta + \tau_{xy}[\cos^2 \theta - \sin^2 \theta]$$
$$= -[10,000 - (-8000)] \sin(+33.02°) \cos(+33.02°)$$
$$+ (-4000) [\cos^2(+33.02°) - \sin^2(+33.02°)]$$
$$= -9849 \text{ psi} \cong -9850 \text{ psi}$$

The result is shown in Fig. 10-11d.
All three of these results can be combined in the triangular stress block shown in Fig. 10-11e. ∎

---

## Example Problem 10-4

At a point on the outside surface of a thin-walled pressure vessel there are normal and shearing stresses on horizontal and vertical planes through the point, as shown in Figs. 10-12a and b.

(a) Determine the principal stresses and the maximum shearing stress at the point.
(b) Locate the planes on which these stresses act and show the stresses on a complete sketch.

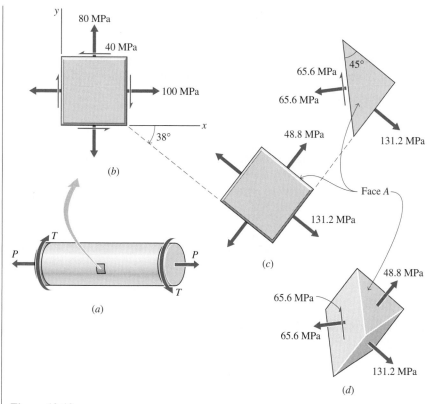

**Figure 10-12**

## SOLUTION

(a) On the basis of the axes shown in Fig. 10-12b and the established sign conventions, $\sigma_x$ is positive, $\sigma_y$ is positive, and $\tau_{xy}$ is negative. For use in Eqs. (10-3) and (10-4), the given values are

$$\sigma_x = +100 \text{ MPa} \qquad \sigma_y = +80 \text{ MPa} \qquad \tau_{xy} = -40 \text{ MPa}$$

When these values are substituted in Eq. (10-4), the principal stresses are found to be

$$\sigma_{p1,\,p2} = \frac{\sigma_x + \sigma_y}{2} \pm \sqrt{\left(\frac{\sigma_x - \sigma_y}{2}\right)^2 + \tau_{xy}^2}$$

$$= \frac{100 + 80}{2} \pm \sqrt{\left(\frac{100 - 80}{2}\right)^2 + (-40)^2}$$

$$= 90 \pm 41.23$$

$$\sigma_{p1} = 90 + 41.23 = +131.23 \text{ MPa} \cong 131.2 \text{ MPa T} \qquad \textbf{Ans.}$$

$$\sigma_{p2} = 90 - 41.23 = +48.77 \text{ MPa} \cong 48.8 \text{ MPa T} \qquad \textbf{Ans.}$$

$$\sigma_{p3} = \sigma_z = 0 \qquad \textbf{Ans.}$$

The maximum shear stress in the $xy$-plane (the 12-plane) is

$$\tau_p = \frac{\sigma_{p1} - \sigma_{p2}}{2} = 41.2 \text{ MPa}$$

and occurs on the plane midway between the principal planes 1 and 2. The normal stress on this surface is the average of the normal stresses

$$\sigma_n = \frac{\sigma_{p1} + \sigma_{p2}}{2} = 90 \text{ MPa}$$

However, the difference between the principal stresses $\sigma_{p1}$ and $\sigma_{p3}$ is bigger than the difference between the principal stresses $\sigma_{p1}$ and $\sigma_{p2}$. Therefore, the maximum shear stress occurs on the surface midway between the principal planes 1 and 3 and has a magnitude of

$$\tau_{max} = \frac{\sigma_{max} - \sigma_{min}}{2} = \frac{\sigma_{p1} - \sigma_{p3}}{2}$$

$$= \frac{131.23 - 0}{2} = 65.61 \text{ MPa} \cong 65.6 \text{ MPa} \qquad \textbf{Ans.}$$

The normal stress on this surface is

$$\sigma_n = \frac{\sigma_{p1} + \sigma_{p3}}{2} = 65.6 \text{ MPa}$$

(b) When the given data are substituted in Eq. (10-3), the results are

$$\tan 2\theta_p = \frac{2\tau_{xy}}{\sigma_x - \sigma_y} = \frac{2(-40)}{100 - 80} = -4.000$$

from which

$$2\theta_p = -75.96°, + 104.04°, \ldots$$

and

$$\theta_p = -37.98°, + 52.02°, \ldots$$
$$\cong 38.0° \downarrow \text{ and } 52.0° \uparrow \qquad \textbf{Ans.}$$

When $\theta = -37.98°$, Eq. (10-1a) gives

$$\sigma_n = \sigma_x \cos^2 \theta + \sigma_y \sin^2 \theta + 2\tau_{xy} \sin \theta \cos \theta$$
$$= 100 \cos^2(-37.98°) + 80 \sin^2(-37.98°)$$
$$+ 2(-40) \sin(-37.98°) \cos(-37.98°)$$
$$= \sigma_{p1} = +131.23 \text{ MPa} \cong 131.2 \text{ MPa T}$$

When $\theta = +52.02°$, Eq. (10-1a) gives

$$\sigma_n = \sigma_x \cos^2 \theta + \sigma_y \sin^2 \theta + 2\tau_{xy} \sin \theta \cos \theta$$

$$= 100 \cos^2(+52.02°) + 80 \sin^2(+52.02°)$$
$$+ 2(-40) \sin(+52.02°) \cos(+52.02°)$$
$$= \sigma_{p2} = +48.77 \text{ MPa} \cong 48.8 \text{ MPa T}$$

The maximum shearing stress occurs on the plane making an angle of 45° with the planes of maximum and minimum normal stress—in this case, 131.2 MPa and zero. The complete sketch is given in Fig. 10-12c, where the upper (wedge-shaped) block is the orthographic projection of the lower block. The three-dimensional wedge (Fig. 10-12d) is presented as an aid to the visualization of Fig. 10-12c. ■

## PROBLEMS

*Note*: Normal and shear stresses on horizontal and vertical planes through a point in a structural member subjected to plane stress are shown in the following figures. Determine and show on a sketch the principal stresses and the maximum shear stress at the point.

### Introductory Problems

**10-28\*** At the point shown in Fig. P10-28.

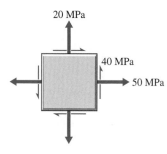

**Figure P10-28**

**10-29\*** At the point shown in Fig. P10-29.

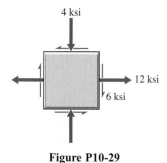

**Figure P10-29**

**10-30** At the point shown in Fig. P10-30.

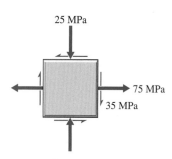

**Figure P10-30**

**10-31\*** At the point shown in Fig. P10-31.

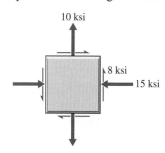

**Figure P10-31**

**10-32** At the point shown in Fig. P10-32.

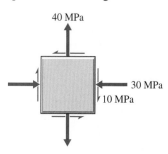

**Figure P10-32**

**Intermediate Problems**

**10-33*** At the point shown in Fig. P10-33.

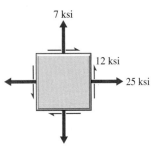

**Figure P10-33**

**10-34*** At the point shown in Fig. P10-34.

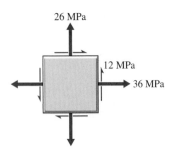

**Figure P10-34**

**10-35** At the point shown in Fig. P10-35.

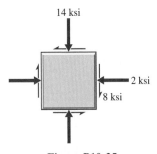

**Figure P10-35**

**10-36*** At the point shown in Fig. P10-36.

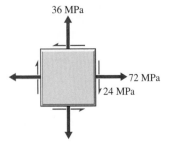

**Figure P10-36**

**10-37** At the point shown in Fig. P10-37.

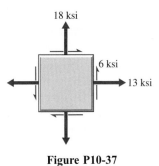

**Figure P10-37**

**Challenging Problems**

**10-38*** The stresses shown in Fig. P10-38 act at a point on the free surface of a stressed body. The magnitude of the maximum shear stress at the point is 125 MPa. Determine the principal stresses, the magnitude of the unknown shear stress on the vertical plane, and the angle $\theta_p$ between the x-axis and the maximum tensile stress at the point.

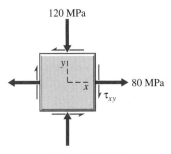

**Figure P10-38**

**10-39*** The principal compressive stress on a vertical plane through a point in a wooden block is equal to four times the principal compressive stress on a horizontal plane, as shown in Fig. P10-39. The plane of the grain is 30° clockwise from the vertical plane. If the normal and shear stresses on the plane of the grain must not exceed 300 psi C and 125 psi shear, determine the maximum allowable compressive stress on the horizontal plane.

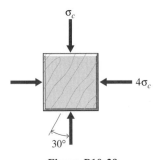

**Figure P10-39**

**10-40** At a point on the free surface of a stressed body, a normal stress of 64 MPa C and an unknown positive shear stress exist on a horizontal plane. One principal stress at the point is 8 MPa C. The maximum shear stress at the point has a magnitude of 95 MPa. Determine the unknown stresses on the horizontal and vertical planes shown in Fig. P10-40 and the unknown principal stress at the point.

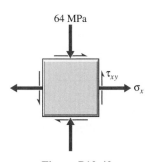

**Figure P10-40**

**10-41** At a point on the free surface of a stressed body, the normal stresses are 20 ksi T on a vertical plane and 30 ksi C on a horizontal plane, as shown in Fig. P10-41. An unknown negative shear stress acts on the horizontal and vertical planes. The maximum shear stress at the point has a magnitude of 32 ksi. Determine the principal stresses and the shear stress on the vertical plane at the point.

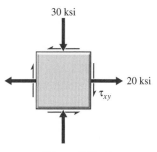

**Figure P10-41**

**10-42** At a point on the free surface of a stressed body, a normal stress of 75 MPa T and an unknown negative shear stress exist on a horizontal plane, as shown in Fig. P10-42. One principal stress at the point is 200 MPa T. The maximum in-plane shear stress at the point has a magnitude of 85 MPa. Determine the unknown stresses on the vertical plane, the unknown principal stress, and the maximum shear stress at the point.

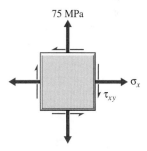

**Figure P10-42**

# 10-6 MOHR'S CIRCLE FOR PLANE STRESS

The German engineer Otto Mohr (1835–1918) developed a useful pictorial or graphic interpretation of the transformation equations for plane stress. This representation, commonly called Mohr's circle, involves the construction of a circle in such a manner that the coordinates of each point on the circle represent the normal and shearing stresses on one plane through the stressed point, and the angular position of the radius to the point gives the orientation of the plane. The proof that normal and shearing components of stress on an arbitrary plane through a point can be represented as points on a circle follows from Eqs. (10-1) and (10-2). Recall Eqs. (10-1b) and (10-2b),

$$\sigma_n - \frac{\sigma_x + \sigma_y}{2} = \frac{\sigma_x - \sigma_y}{2}\cos 2\theta + \tau_{xy}\sin 2\theta$$

$$\tau_{nt} = -\frac{\sigma_x - \sigma_y}{2}\sin 2\theta + \tau_{xy}\cos 2\theta$$

Squaring both equations, adding, and simplifying yields

$$\left(\sigma_n - \frac{\sigma_x + \sigma_y}{2}\right)^2 + \tau_{nt}^2 = \left(\frac{\sigma_x - \sigma_y}{2}\right)^2 + \tau_{xy}^2$$

This is the equation of a circle in terms of the variables $\sigma_n$ and $\tau_{nt}$. The circle is centered on the $\sigma$-axis at a distance $(\sigma_x + \sigma_y)/2$ from the $\tau$-axis, and the radius of the circle is given by

$$R = \sqrt{\left(\frac{\sigma_x - \sigma_y}{2}\right)^2 + \tau_{xy}^2}$$

Normal stresses are plotted as horizontal coordinates, with tensile stresses (positive) plotted to the right of the origin and compressive stresses (negative) plotted to the left. Shearing stresses are plotted as vertical coordinates, with those tending to produce a clockwise rotation of the stress element plotted above the $\sigma$-axis, and those tending to produce a counterclockwise rotation of the stress element plotted below the $\sigma$-axis. The method for interpreting the sign to be associated with a particular shear stress value obtained from a Mohr's circle analysis will be illustrated in the discussion and example problems that follow.

Mohr's circle for any point subjected to plane stress can be drawn when stresses on two mutually perpendicular planes through the point are known. Consider, for example, the stress situation of Fig. 10-13a with $\sigma_x$ greater than $\sigma_y$ and plot on Fig. 10-13b the points representing the given stresses. The coordinates of point $V$ of Fig. 10-13b are the stresses on the vertical plane through the stressed point of Fig. 10-13a, and point $H$ is determined by the stresses on the horizontal

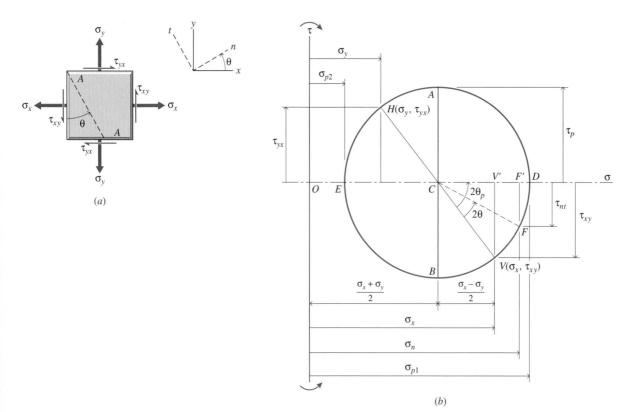

(a)

(b)

**Figure 10-13**

plane through the point. Because $\tau_{yx} = \tau_{xy}$, point $C$, the center of the circle, is on the $\sigma$-axis. Line $CV$ on Mohr's circle represents the plane (the vertical plane of Fig. 10-13a) through the stressed point from which the angle $\theta$ is measured. The coordinates of each point on the circle represent $\sigma_n$ and $\tau_{nt}$ for one particular plane through the stressed point, the abscissa representing $\sigma_n$ and the ordinate representing $\tau_{nt}$. To demonstrate this statement, draw any radius $CF$ in Fig. 10-13b at an angle $2\theta$ counterclockwise from radius $CV$. From the figure, it is apparent that

$$OF' = OC + CF \cos(2\theta_p - 2\theta)$$

and since $CF$ equals $CV$, this equation reduces to

$$OF' = OC + CV \cos 2\theta_p \cos 2\theta + CV \sin 2\theta_p \sin 2\theta$$

Referring to Fig. 10-13b, note that

$$CV \cos 2\theta_p = CV' = (\sigma_x - \sigma_y)/2$$
$$CV \sin 2\theta_p = VV' = \tau_{xy}$$
$$OC = (\sigma_x + \sigma_y)/2 = \sigma_{\text{avg}}$$

Therefore,

$$OF' = OC + CV' \cos 2\theta + VV' \sin 2\theta$$
$$= \frac{\sigma_x + \sigma_y}{2} + \frac{\sigma_x - \sigma_y}{2} \cos 2\theta + \tau_{xy} \sin 2\theta$$

This expression is identical to Eq. (10-1b). Therefore, $OF'$ is equal to $\sigma_n$. In a similar manner,

$$F'F = CF \sin(2\theta_p - 2\theta)$$
$$= CV \sin 2\theta_p \cos 2\theta - CV \cos 2\theta_p \sin 2\theta$$
$$= V'V \cos 2\theta - CV' \sin 2\theta$$
$$= \tau_{xy} \cos 2\theta - \frac{\sigma_x - \sigma_y}{2} \sin 2\theta$$

This expression is identical to Eq. (10-2b). Therefore, $F'F$ is equal to $\tau_{nt}$.

Since the horizontal coordinate of each point on the circle represents the normal stress $\sigma_n$ on some plane through the point, the maximum normal stress at the point is represented by $OD$, and its value is

$$\sigma_{p1} = OD = OC + CD = OC + CV$$
$$= \frac{\sigma_x + \sigma_y}{2} + \sqrt{\left(\frac{\sigma_x - \sigma_y}{2}\right)^2 + \tau_{xy}^2}$$

which agrees with Eq. (10-4).

Likewise, the vertical coordinate of each point on the circle represents the shearing stress $\tau_{nt}$ (called the *in-plane* shearing stress) on some plane through the point, which means that the maximum in-plane shearing stresses at the point are represented by $CA$ and $CB$, and their value is

$$\tau_p = CA = CB = \sqrt{\left(\frac{\sigma_x - \sigma_y}{2}\right)^2 + \tau_{xy}^2}$$

which agrees with Eq. (10-5). If the two nonzero principal stresses have the same sign, the maximum shearing stress at the point will not be in the plane of the applied stresses.

The angle $2\theta_p$ from $CV$ to $CD$ is counterclockwise or positive, and its tangent is

$$\tan 2\theta_p = \frac{\tau_{xy}}{(\sigma_x - \sigma_y)/2}$$

which is Eq. (10-3). From the derivation of Eq. (10-4), the angle between the vertical plane and one of the principal planes was $\theta_p$. In obtaining the same equation from Mohr's circle, the angle between the radii representing these same two planes is $2\theta_p$. In other words, all angles on Mohr's circle are twice the corresponding angles for the actual stressed body. The angle from the vertical plane to the horizontal plane in Fig. 10-13a is 90°, but in Fig. 10-13b, the angle between line $CV$ (which represents the vertical plane) and line $CH$ (which represents the horizontal plane) on Mohr's circle is 180°.

The results obtained from Mohr's circle have been shown to be identical with the equations derived from the free-body diagram of Fig. 10-5. Thus, Mohr's circle provides an extremely useful aid for the visualization and solution of stresses on various planes through a point in a stressed body in terms of the stresses on two mutually perpendicular planes through the point. Although Mohr's circle can be drawn to scale and used to obtain values of stresses and angles by direct measurements on the figure, it is probably more useful as a pictorial aid to the analyst who is performing analytical determinations of stresses and their directions at the point.

When the state of stress at a point is specified by means of a sketch of a small element, the procedure for drawing and using Mohr's circle to obtain specific stress information can be briefly summarized as follows:

1. Choose a set of $x$–$y$ reference axes.
2. Identify the stresses $\sigma_x$, $\sigma_y$ and $\tau_{xy} = \tau_{yx}$, and list them with the proper sign.
3. Draw a set of $\sigma$–$\tau$ coordinate axes with $\sigma$ and $\tau$ positive to the right and upward, respectively.
4. Plot the point $(\sigma_x, -\tau_{xy})$ and label it point $V$ (vertical plane).
5. Plot the point $(\sigma_y, \tau_{yx})$ and label it point $H$ (horizontal plane).
6. Draw a line between $V$ and $H$. This establishes the center $C$ and the radius $R$ of Mohr's circle.
7. Draw the circle.
8. An extension of the radius between $C$ and $V$ can be identified as the $x$-axis or the reference line for angle measurements (i.e., $\theta = 0°$).

By plotting points $V$ and $H$ as $(\sigma_x, -\tau_{xy})$ and $(\sigma_y, \tau_{yx})$ respectively, shear stresses that tend to rotate the stress element clockwise will plot above the $\sigma$-axis, while those tending to rotate the element counterclockwise will plot below the $\sigma$-axis. The use of a negative sign with one of the shearing stresses ($\tau_{xy}$ or $\tau_{yx}$) is required for plotting purposes, since for a given state of stress, the shearing stresses ($\tau_{xy} = \tau_{yx}$) have only one sign (both are positive or both are negative). The use of the negative sign at point $V$ on Mohr's circle brings the direction of angular measurements $2\theta$ on Mohr's circle into agreement with the direction of angular measurements $\theta$ on the stress element.

Once the circle has been drawn, the normal and shearing stresses on an arbitrary inclined plane $A-A$ having an outward normal $n$ that is oriented at an angle $\theta$ with respect to the reference $x$-axis (see Fig. 10-13a) can be obtained from the coordinates of point $F$ (see Fig. 10-13b) on the circle that is located at angular position $2\theta$ from the reference axis through point $V$. The coordinates of point $F$ must be interpreted as stresses $\sigma_n$ and $-\tau_{nt}$. Other points on Mohr's circle that provide stresses of interest are

1. Point $D$, which provides the principal stress $\sigma_{p1}$.
2. Point $E$, which provides the principal stress $\sigma_{p2}$.
3. Point $A$, which provides the maximum in-plane shearing stress $-\tau_p$ and the accompanying normal stress $\sigma_{avg}$ that acts on the plane.

A negative sign must be used when interpreting shearing stresses $\tau_{nt}$ and $\tau_p$ obtained from the circle since a shearing stress tending to produce a clockwise rotation of the stress element is a negative shearing stress when a right-hand $n-t$ coordinate system is used.

Problems of the type presented in Sections 10-4 and 10-5 can readily be solved by this semigraphic method, as illustrated in the following examples.

## Example Problem 10-5

At a point in a structural member subjected to plane stress there are normal and shearing stresses on horizontal and vertical planes through the point, as shown in Fig. 10-14a. Determine and show on a sketch:

(a) The principal and maximum shearing stresses at the point.
(b) The normal and shearing stresses on plane $A-A$ through the point.

### SOLUTION

Mohr's circle (see Fig. 10-14b) is constructed from the given data by plotting point $V$ (representing the stresses on the vertical plane) at $(8, -4)$ because the stresses on the vertical plane are 8 ksi T and 4 ksi (counterclockwise) shear. Likewise, point $H$ (representing the stresses on the horizontal plane) has the coordinates $(-6, 4)$. Draw line $HV$, which is a diameter of Mohr's circle, and note that the center of the circle is at $(1, 0)$. The radius of the circle is

$$CV = \sqrt{7^2 + 4^2} = 8.062 \text{ ksi}$$

(a) The principal stresses and the maximum shearing stress at the point are

$$\sigma_{p1} = OD = 1 + 8.062 = +9.062 \text{ ksi} \cong 9.06 \text{ ksi T} \qquad \textbf{Ans.}$$

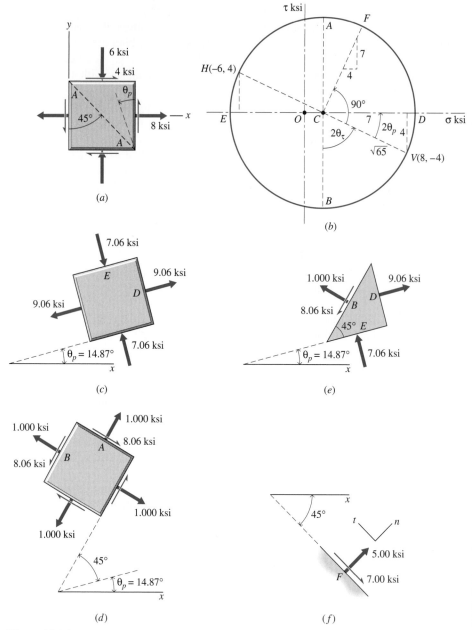

**Figure 10-14**

$$\sigma_{p2} = OE = 1 - 8.062 = -7.062 \text{ ksi} \cong 7.06 \text{ ksi C} \qquad \textbf{Ans.}$$

$$\sigma_{p3} = \sigma_z = 0 \qquad \textbf{Ans.}$$

Since $\sigma_{p1}$ and $\sigma_{p2}$ have opposite signs, the maximum shearing stress is

$$\tau_p = \tau_{max} = CA = CB = 8.062 \text{ ksi} \cong 8.06 \text{ ksi} \qquad \textbf{Ans.}$$

The principal planes are represented by lines $CD$ and $CE$, where

$$\tan 2\theta_p = 4/7 = 0.5714$$

which gives

$$2\theta_p = +29.74° \quad \text{or} \quad \theta_p = 14.87° \nwarrow$$

Since the angle $2\theta_p$ is counterclockwise, the principal planes are counterclockwise from the vertical and horizontal planes of the stress block, as shown in Fig. 10-14c. To determine which principal stress acts on which plane, note that as the radius of the circle rotates counterclockwise, the end of the radius $CV$ moves from $V$ to $D$, indicating that as the initially vertical plane rotates through 14.87°, the stresses change to 9.06 ksi T (normal) and 0 ksi (shear). Note also that the end $H$ or radius $CH$ moves to $E$, indicating that as the initially horizontal plane rotates through 14.87°, the stresses change to 7.06 ksi C (normal) and 0 ksi (shear). These stresses are shown on the sketch of Fig. 10-14c.

Point $A$ represents the state of stress on a surface rotated 45° counterclockwise from the surface represented by point $D$. On this surface, the shear stress is the maximum it can be (in the $xy$-plane) $\tau_{nt} = \tau_p = \tau_{max} = 8.06$ ksi, and the normal stress is the average value $\sigma_n = (\sigma_x + \sigma_y)/2 = \sigma_C = 1.000$ ksi (the normal stress at the center of the circle). Likewise, point $B$ represents the state of stress on a surface rotated 45° counterclockwise from the surface represented by point $E$. The stresses on this surface are also $\sigma_n = \sigma_C = 1.000$ ksi and $\tau_{nt} = \tau_p = 8.06$ ksi. These stresses are shown on the sketch of Fig. 10-14d.

Note that any two orthogonal surfaces of Fig. 10-14c are sufficient to completely specify the principal stresses. Also, only one of the surfaces of Fig. 10-14d is required to completely specify the maximum shear stress. Therefore, these two separate sketches can be combined as shown in Fig. 10-14e.

(b) Plane $A-A$ is 45° counterclockwise from the vertical plane; therefore, the corresponding radius of Mohr's circle is $2\theta = 90°$ counterclockwise from the line $CV$, and is shown as $CF$ on Fig. 10-14b. The coordinates of point $F$ are seen to be (5, 7), which means that the stresses on plane $A-A$ are 5.00 ksi T and 7.00 ksi shear in a direction to produce a clockwise couple on a stress element having plane $A-A$ as one of its faces. This shear stress would be defined as a negative shear stress because to produce a clockwise couple, it must be directed in the negative $t$-direction associated with plane $A-A$ (see Fig. 10-14f). ■

## Example Problem 10-6

At a point in a structural member subjected to plane stress there are normal and shearing stresses on horizontal and vertical planes through the point, as shown in Fig. 10-15a. Determine, and show on a sketch, the principal stresses and the maximum shearing stress at the point.

### SOLUTION

Mohr's circle (see Fig. 10-15b) is constructed from the given data by plotting point $V$ (representing the stresses on the vertical plane) at (72, 24) and point $H$ (representing the stresses on the horizontal plane) at (36, −24). Line $HV$ be-

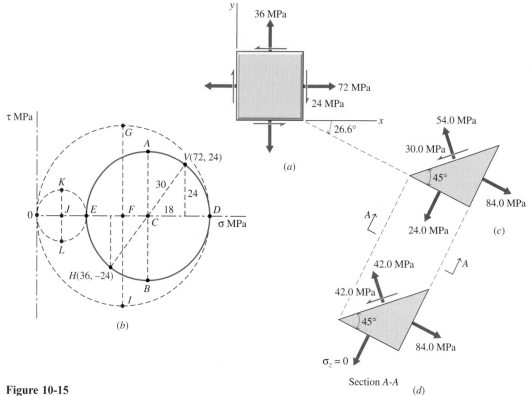

**Figure 10-15**

tween the two points is a diameter of Mohr's circle. The circle is centered at point $C$ (54, 0) and has a radius $CV$ equal to 30 MPa. The principal stresses at the point are

$$\sigma_{p1} = OD = 54 + 30 = 84.0 \text{ MPa T} \qquad \textbf{Ans.}$$

$$\sigma_{p2} = OE = 54 - 30 = 24.0 \text{ MPa T} \qquad \textbf{Ans.}$$

$$\sigma_{p3} = \sigma_z = 0 \qquad \textbf{Ans.}$$

The principal planes are represented by lines $CD$ and $CE$, where

$$\tan 2\theta_p = -24/18 = -1.3333$$

which gives

$$2\theta_p = -53.13° = 53.13°\downarrow$$
$$\theta_p = -26.57° = 26.57°\downarrow$$

Since $\sigma_{p1}$ and $\sigma_{p2}$ have the same sign, the maximum in-plane shearing stress $\tau_p$ is not the maximum shearing stress $\tau_{\text{max}}$ at the point. The maximum shearing stress at the point is represented on Mohr's circle by drawing an additional circle (shown dashed in Fig. 10-15b), which has line $OD$ as a diameter. This circle is centered at point $F$ (42, 0) and has a radius $FG$ equal to 42 MPa. This circle represents the combinations of normal and shearing stresses existing on planes obtained by rotating the element about the principal axis associated with the principal stress $\sigma_{p2}$. A third Mohr's circle (shown dotted in Fig. 10-15b) has line OE as a diameter. This circle is centered at point J (12, 0) and has a radius JK equal to 12 MPa. This circle represents the combinations of normal and shearing stresses existing on planes obtained by rotating the element about the principal axis associated with the principal stress $\sigma_{p1}$. Thus,

$$\tau_p = CA = 30.0 \text{ MPa}$$

$$\tau_{\max} = FG = 42.0 \text{ MPa} \qquad\qquad \textbf{Ans.}$$

The principal stresses $\sigma_{p1}$, $\sigma_{p2}$, and $\sigma_z = \sigma_{p3} = 0$, the maximum in-plane shearing stress $\tau_p$, and the maximum shearing stress $\tau_{\max}$ at the point are all shown on Figs. 10-15c and 10-15d. ∎

---

## PROBLEMS

### Introductory Problems

**10-43\*** The stresses shown in Fig. P10-43 act at a point on the free surface of a stressed body. Use Mohr's circle to determine the normal and shear stresses at this point on the inclined plane A−B shown in the figure.

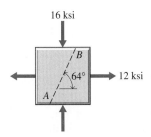

**Figure P10-43**

**10-44** The stresses shown in Fig. P10-44 act at a point on the free surface of a stressed body. Use Mohr's circle to determine the normal and shear stresses at this point on the inclined plane A−B shown in the figure.

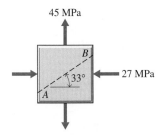

**Figure P10-44**

### Intermediate Problems

**10-45\*** At a point in a structural member subjected to plane stress there are normal and shear stresses on horizontal and vertical planes through the point, as shown in Fig. P10-45. Use Mohr's circle to determine the principal stresses and the maximum shear stress at the point.

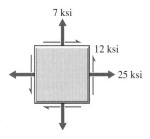

**Figure P10-45**

**10-46\*** At a point in a structural member subjected to plane stress there are normal and shear stresses on horizontal and vertical planes through the point, as shown in Fig. P10-46. Use Mohr's circle to determine the principal stresses and the maximum shear stress at the point.

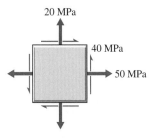

**Figure P10-46**

**10-47** At a point in a structural member subjected to plane stress there are normal and shear stresses on horizontal and vertical planes through the point, as shown in Fig. P10-47. Use Mohr's circle to determine the principal stresses and the maximum shear stress at the point.

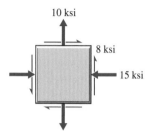

**Figure P10-47**

**10-48** At a point in a structural member subjected to plane stress there are normal and shear stresses on horizontal and vertical planes through the point, as shown in Fig. P10-48. Use Mohr's circle to determine the principal stresses and the maximum shear stress at the point.

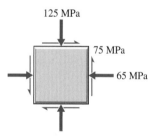

**Figure P10-48**

**Challenging Problems**

**10-49** At a point in a structural member subjected to plane stress there are normal and shear stresses on horizontal and vertical planes through the point, as shown in Fig. P10-49. Use Mohr's circle to determine
(a) The principal stresses and the maximum shear stress at the point.
(b) The normal and shear stresses on the inclined plane A–B shown in the figure.

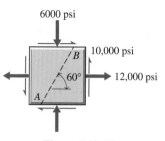

**Figure P10-49**

**10-50\*** At a point in a structural member subjected to plane stress there are normal and shear stresses on horizontal and vertical planes through the point, as shown in Fig. P10-50. Use Mohr's circle to determine
(a) The principal stresses and the maximum shear stress at the point.
(b) The normal and shear stresses on the inclined plane A–B shown in the figure.

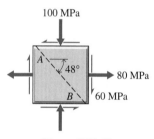

**Figure P10-50**

**10-51** At a point in a structural member subjected to plane stress there are normal and shear stresses on horizontal and vertical planes through the point, as shown in Fig. P10-51. Use Mohr's circle to determine
(a) The principal stresses and the maximum shear stress at the point.
(b) The normal and shear stresses on the inclined plane A–B shown in the figure.

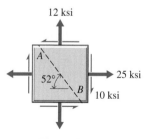

**Figure P10-51**

# 10-7 TWO-DIMENSIONAL OR PLANE STRAIN

The material of Section 4-4 serves to convey the concept of strain as a unit deformation, but it is inadequate for other than uniaxial loading. The extension of the concept to biaxial loading is essential because of the role played by strain in experimental methods of stress evaluation. In many practical problems involving the design of machine elements, the shape and loading are too complicated to permit stress determination solely by mathematical analysis; hence, this technique is supplemented by laboratory measurements.

Strains can be measured by several methods but, except for the simplest cases, stresses cannot be obtained directly. Consequently, the usual procedure in experimental stress analysis is to measure strains and calculate the state of stress by using stress–strain equations as illustrated later in Section 10-11.

The complete state of strain at an arbitrary point $P$ in a body under load can be determined by considering the deformation associated with a small volume of material surrounding the point. For convenience the volume is normally assumed to have the shape of a rectangular parallelepiped with its faces oriented perpendicular to the reference $x$-, $y$-, and $z$-axes in the undeformed state, as shown in Fig. 10-16a. Since the element of volume is very small, deformations are assumed to be uniform; therefore, parallel planes remain plane and parallel and straight lines remain straight in the deformed element, as shown in Fig. 10-16b. The final size of the deformed element is determined by the lengths of the three edges $dx'$, $dy'$, and $dz'$. The distorted shape of the element is determined by the angles $\theta'_{xy}$, $\theta'_{yz}$, and $\theta'_{zx}$ between faces.

The Cartesian components of strain at the point can be expressed in terms of the deformations by using the definitions of normal and shearing strain presented in Section 4-4. These are the strain components associated with the Cartesian components of stress discussed in Section 10-3 and shown in Fig. 10-2. Thus,

$$\epsilon_x = \frac{dx' - dx}{dx} = \frac{d\delta_x}{dx} \qquad \gamma_{xy} = \frac{\pi}{2} - \theta'_{xy} \qquad (10\text{-}9a)$$

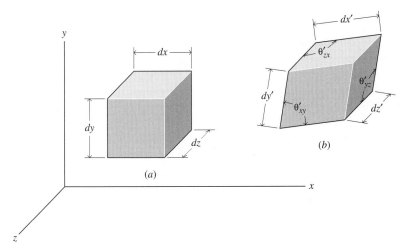

**Figure 10-16**

$$\epsilon_y = \frac{dy' - dy}{dy} = \frac{d\delta_y}{dy} \qquad \gamma_{yz} = \frac{\pi}{2} - \theta'_{yz}$$

(10-9a) cont'd

$$\epsilon_z = \frac{dz' - dz}{dz} = \frac{d\delta_z}{dz} \qquad \gamma_{zx} = \frac{\pi}{2} - \theta'_{zx}$$

In a similar manner, the normal strain component associated with a line oriented in an arbitrary $n$-direction and the shearing strain component associated with two arbitrary orthogonal lines oriented in the $n$- and $t$-directions in the undeformed element are given by

$$\epsilon_n = \frac{dn' - dn}{dn} = \frac{d\delta_n}{dn} \qquad \gamma_{nt} = \frac{\pi}{2} - \theta'_{nt} \qquad (10\text{-}9b)$$

Alternative forms of Eq. (10-9), which will be useful in later developments, are

$$dx' = (1 + \epsilon_x)dx \qquad \theta'_{xy} = \frac{\pi}{2} - \gamma_{xy}$$

$$dy' = (1 + \epsilon_y)dy \qquad \theta'_{yz} = \frac{\pi}{2} - \gamma_{yz}$$

(10-10)

$$dz' = (1 + \epsilon_z)dz \qquad \theta'_{zx} = \frac{\pi}{2} - \gamma_{zx}$$

$$dn' = (1 + \epsilon_n)dn \qquad \theta'_{nt} = \frac{\pi}{2} - \gamma_{nt}$$

# 10-8 THE STRAIN TRANSFORMATION EQUATIONS FOR PLANE STRAIN

The method of relating the components of strain associated with a Cartesian co-ordinate system to the normal and shearing strains associated with other orthogonal directions will be illustrated by considering the two-dimensional or plane strain case. If the $xy$-plane is taken as the reference plane, then for conditions of plane strain $\epsilon_z = \gamma_{zx} = \gamma_{zy} = 0$.[4]

Consider Fig. 10-17$a$, in which the shaded rectangle represents a small un-strained element of material having the configuration of a rectangular paral-lelepiped. The sides of the rectangle are along directions for which the strains $\epsilon_x$, $\epsilon_y$, and $\gamma_{xy}$ are known. The proportion of the rectangle is chosen so that the di-agonal $OB$ points in the direction $n$ for which the strain $\epsilon_n$ is to be determined. When the body is subjected to a system of loads, the element assumes the shape indicated in Fig. 10-17$b$. The dimensions of the deformed element are given in terms of strains obtained from Eq. (10-10).

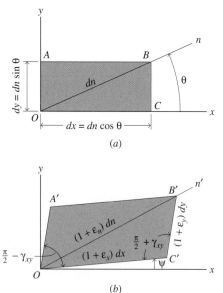

(a)

(b)

**Figure 10-17**

---

[4]Note that although there is no strain in the $z$-direction, $\epsilon_z = 0$, there must be a stress in the $z$-direc-tion, $\sigma_z \neq 0$. That is, a stress is required in the $z$-direction to prevent deformation in the $z$-direction.

**Normal Strain $\epsilon_n$.** An expression for the normal strain $\epsilon_n$ in the $n$-direction can be obtained by applying the law of cosines to the triangle $OC'B'$ shown in Fig. 10-17b. Thus,

$$(OB')^2 = (OC')^2 + (C'B')^2 - 2(OC')(C'B') \cos\left(\frac{\pi}{2} + \gamma_{xy}\right)$$

or in terms of strains

$$[(1 + \epsilon_n)dn]^2 = [(1 + \epsilon_x)dx]^2 + [(1 + \epsilon_y)dy]^2$$
$$- 2[(1 + \epsilon_x)dx][(1 + \epsilon_y)dy][-\sin\gamma_{xy}] \qquad \text{(a)}$$

Substituting $dx = dn\cos\theta$ and $dy = dn\sin\theta$ (see Fig. 10-17a) into Eq. (a) yields

$$(1 + \epsilon_n)^2(dn)^2 = (1 + \epsilon_x)^2(dn)^2(\cos^2\theta) + (1 + \epsilon_y)^2(dn)^2(\sin^2\theta)$$
$$+ 2(dn)^2(\sin\theta)(\cos\theta)(1 + \epsilon_x)(1 + \epsilon_y)(\sin\gamma_{xy}) \qquad \text{(b)}$$

Since the strains are small, it follows that $\epsilon^2 << \epsilon$, $\sin\gamma \approx \gamma$, and so forth; hence, all second-degree terms such as $\epsilon^2$, $\gamma\epsilon$, and the like, can be neglected as Eq. (b) is expanded to become

$$1 + 2\epsilon_n = (1 + 2\epsilon_x)\cos^2\theta + (1 + 2\epsilon_y)\sin^2\theta + 2\gamma_{xy}\sin\theta\cos\theta$$

which reduces to

$$\epsilon_n = \epsilon_x \cos^2\theta + \epsilon_y \sin^2\theta + \gamma_{xy}\sin\theta\cos\theta \qquad \text{(10-11a)}$$

or in terms of the double angle

$$\epsilon_n = \frac{\epsilon_x + \epsilon_y}{2} + \frac{\epsilon_x - \epsilon_y}{2}\cos 2\theta + \frac{\gamma_{xy}}{2}\sin 2\theta \qquad \text{(10-11b)}$$

**Shearing Strain $\gamma_{nt}$.** The shearing strain $\gamma_{nt}$ measures the amount by which the right angle between the $n$- and $t$-directions decreases as the material deforms. As the material deforms, the $n$-direction rotates counterclockwise through an angle $\phi_n$ as shown in Fig. 10-18. Applying the law of sines to triangle $OC'B'$

$$\frac{OB'}{\sin\angle OC'B'} = \frac{B'C'}{\sin\angle B'OC'}$$

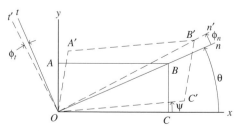

**Figure 10-18**

which gives

$$B'C' \sin \angle OC'B' = OB' \sin \angle B'OC'$$

or in terms of the strains,

$$(1 + \epsilon_y) \, dy \sin\left(\frac{\pi}{2} + \gamma_{xy}\right) = (1 + \epsilon_n) \, dn \sin[\theta + (\phi_n - \psi)] \qquad \text{(c)}$$

Since the strains ($\epsilon_x$, $\epsilon_y$, $\gamma_{xy}$, and $\epsilon_n$) and the angles $\phi_n$ and $\psi$ are all small,

$$\sin\left(\frac{\pi}{2} + \gamma_{xy}\right) = \cos \gamma_{xy} \cong 1$$

$$\sin[\theta + (\phi_n - \psi)] = \sin\theta \cos(\phi_n - \psi) + \cos\theta \sin(\phi_n - \psi)$$

$$\cong \sin\theta + (\phi_n - \psi) \cos\theta$$

and Eq. (c) can be written

$$(1 + \epsilon_y) \, dy \cong (1 + \epsilon_n) \, dn \, [\sin\theta + (\phi_n - \psi) \cos\theta] \qquad \text{(d)}$$

where $dy = dn \sin\theta$ (see Fig. 10-17a). Therefore, Eq. (d) can be reduced to

$$(\epsilon_y - \epsilon_n) \sin\theta \cong (\phi_n - \psi) \cos\theta + \epsilon_n(\phi_n - \psi) \cos\theta$$

$$\cong (\phi_n - \psi) \cos\theta \qquad \text{(e)}$$

since $\epsilon_n$, $\phi_n$, and $\psi$ are all small. Substituting Eq. (10-11a) into Eq. (e) and solving for $\phi_n$ yields

$$\phi_n = -(\epsilon_x - \epsilon_y) \sin\theta \cos\theta - \gamma_{xy} \sin^2\theta + \psi \qquad \text{(f)}$$

Equation (f) gives the counterclockwise rotation of a line that makes an angle of $\theta$ with the $x$-axis initially. Thinking of $\phi_n$ as $\phi(\theta)$, the counterclockwise rotation of the $t$-axis can be written

$$\phi_t = \phi\left(\theta + \frac{\pi}{2}\right)$$

$$= -(\epsilon_x - \epsilon_y) \sin\left(\theta + \frac{\pi}{2}\right) \cos\left(\theta + \frac{\pi}{2}\right) - \gamma_{xy} \sin^2\left(\theta + \frac{\pi}{2}\right) + \psi$$

$$= (\epsilon_x - \epsilon_y) \cos\theta \sin\theta - \gamma_{xy} \cos^2\theta + \psi \qquad \text{(g)}$$

Finally, the shearing strain $\gamma_{nt}$ is the decrease in the right angle between the $n$- and $t$-directions or the difference between the rotations $\phi_n$ and $\phi_t$. Therefore,

$$\gamma_{nt} = \phi_n - \phi_t$$

$$= -2(\epsilon_x - \epsilon_y) \sin\theta \cos\theta + \gamma_{xy}(\cos^2\theta - \sin^2\theta) \qquad \text{(10-12a)}$$

or in terms of the double angle

$$\gamma_{nt} = -(\epsilon_x - \epsilon_y) \sin 2\theta + \gamma_{xy} \cos 2\theta \qquad \text{(10-12b)}$$

Equations (10-11) and (10-12) provide a means for determining the normal strain $\epsilon_n$ associated with a line oriented in an arbitrary $n$-direction in the $xy$-plane and the shearing strain $\gamma_{nt}$ associated with any two orthogonal lines oriented in the $n$- and $t$-directions in the $xy$-plane when the strains $\epsilon_x$, $\epsilon_y$, and $\gamma_{xy}$ associated with the coordinate directions are known. When these equations are used, the sign conventions used in their development must be rigorously followed. The sign conventions used are as follows:

**1.** Tensile strains are positive; compressive strains are negative.
**2.** Shearing strains that decrease the angle between the two lines at the origin of coordinates are positive.
**3.** Angles measured counterclockwise from the reference $x$-axis are positive.
**4.** The $(n, t, z)$ axes have the same order as the $(x, y, z)$ axes. Both sets of axes form a right-hand coordinate system.

## ▌ **Example Problem 10-7**

The strain components at a point are $\epsilon_x = +800 \ \mu\text{m/m}$, $\epsilon_y = -1000 \ \mu\text{m/m}$, and $\gamma_{xy} = -600 \ \mu\text{rad}$. Determine the strain components $\epsilon_n$, $\epsilon_t$, and $\gamma_{nt}$ if the $xy$- and $nt$-axes are oriented as shown in Fig. 10-19.

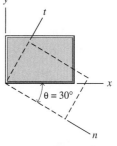

### SOLUTION

The $n$-axis is located at an angle $\theta_n = -30°$ with respect to the $x$-axis; therefore, the strain $\epsilon_n$ is given by Eq. (10-11a) as

$$\epsilon_n = \epsilon_x \cos^2 \theta_n + \epsilon_y \sin^2 \theta_n + \gamma_{xy} \sin \theta_n \cos \theta_n$$

$$= 800 \cos^2(-30°) + (-1000) \sin^2(-30°)$$

$$\qquad +(-600) \sin(-30°) \cos(-30°)$$

$$= +610 \ \mu\text{m/m} \qquad\qquad\qquad\qquad \textbf{Ans.}$$

**Figure 10-19**

The $t$-axis is located at an angle $\theta_t = +60°$ with respect to the $x$-axis; therefore, the strain $\epsilon_t$ is given by Eq. (10-11a) as

$$\epsilon_t = \epsilon_x \cos^2 \theta_t + \epsilon_y \sin^2 \theta_t + \gamma_{xy} \sin \theta_t \cos \theta_t$$

$$= 800 \cos^2(+60°) + (-1000) \sin^2(+60°)$$

$$\qquad +(-600) \sin(+60°) \cos(+60°)$$

$$= -810 \ \mu\text{m/m} \qquad\qquad\qquad\qquad \textbf{Ans.}$$

In a similar manner, the shearing strain $\gamma_{nt}$ is given by Eq. (10-12a) as

$$\gamma_{nt} = -2(\epsilon_x - \epsilon_y) \sin \theta_n \cos \theta_n + \gamma_{xy} (\cos^2 \theta_n - \sin^2 \theta_n)$$

$$= -2[800 - (-1000)] \sin(-30°) \cos(-30°)$$

$$\qquad +(-600) [\cos^2(-30°) - \sin^2(-30°)]$$

$$= 1258.8 \ \mu\text{rad} \cong 1259 \ \mu\text{rad} \qquad\qquad \textbf{Ans.}$$

As a check of the small angle assumptions, note that when $\gamma_{nt} = 0.0012588$ rad, $\sin \gamma_{nt} = 0.00125900 \cong \gamma_{nt}$, and $\cos \gamma_{nt} = 0.9999992 \cong 1$. ∎

## PROBLEMS

### Introductory Problems

**10-52\*** The thin rectangular plate shown in Fig. P10-52 is uniformly deformed such that $\epsilon_x = -2000 \ \mu m/m$, $\epsilon_y = -1500 \ \mu m/m$, and $\gamma_{xy} = +1250 \ \mu rad$. Determine the normal strain $\epsilon_n$ in the plate.

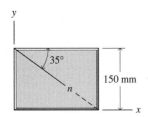

**Figure P10-52**

**10-53\*** The thin rectangular plate shown in Fig. P10-53 is uniformly deformed such that $\epsilon_x = +880 \ \mu in./in.$, $\epsilon_y = +960 \ \mu in./in.$, and $\gamma_{xy} = -750 \ \mu rad$. Determine the normal strain $\epsilon_{AC}$ along diagonal $AC$ of the plate.

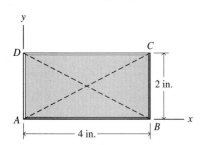

**Figure P10-53**

**10-54** The thin rectangular plate shown in Fig. P10-54 is uniformly deformed such that $\epsilon_x = +1500 \ \mu m/m$, $\epsilon_y = -1250 \ \mu m/m$, and $\gamma_{xy} = +1000 \ \mu rad$. Determine the normal strain $\epsilon_{BD}$ along diagonal $BD$ of the plate.

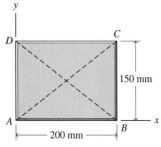

**Figure P10-54**

### Intermediate Problems

In Problems 10-55 through 10-62 the strain components $\epsilon_x$, $\epsilon_y$, and $\gamma_{xy}$ are given for a point in a body subjected to plane strain. Determine the strain components $\epsilon_n$, $\epsilon_t$, and $\gamma_{nt}$ at the point if the $nt$-axes are rotated with respect to the $xy$-axes by the amount and in the direction indicated by the angle $\theta$ shown in Fig. P10-55 or P10-56.

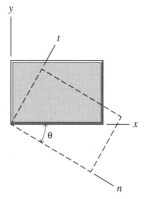

**Figure P10-55**

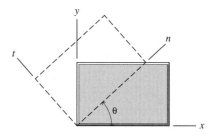

**Figure P10-56**

| Problem | Figure | $\epsilon_x$ | $\epsilon_y$ | $\gamma_{xy}$ | $\theta$ |
|---|---|---|---|---|---|
| 10-55\* | P10-55 | $-800\mu$ | $+640\mu$ | $-960\mu$ | 42° |
| 10-56\* | P10-56 | $+720\mu$ | $-480\mu$ | $+360\mu$ | 30° |
| 10-57 | P10-55 | $+900\mu$ | $+650\mu$ | $-300\mu$ | 32° |
| 10-58 | P10-56 | $+800\mu$ | $-950\mu$ | $-800\mu$ | 42° |
| 10-59\* | P10-55 | $+750\mu$ | $+360\mu$ | $-300\mu$ | 52° |
| 10-60\* | P10-56 | $-100\mu$ | $-700\mu$ | $+400\mu$ | 28° |
| 10-61 | P10-55 | $+850\mu$ | $+450\mu$ | $+600\mu$ | 20° |
| 10-62 | P10-56 | $+1200\mu$ | $+200\mu$ | $-1600\mu$ | 64° |

## Challenging Problems

**10-63\*** The thin rectangular plate shown in Fig. P10-63 is uniformly deformed such that $\epsilon_n = +1575$ $\mu$in./in., $\epsilon_t = +1350$ $\mu$in./in., and $\epsilon_x = +1250$ $\mu$in./in. Determine

(a) The shearing strain $\gamma_{nt}$ in the plate.
(b) The normal strain $\epsilon_y$ in the plate.

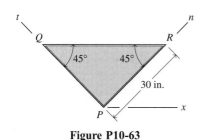

**Figure P10-63**

**10-64** The thin rectangular plate shown in Fig. P10-64 is uniformly deformed such that $\epsilon_{AB} = -1200$ $\mu$m/m, $\epsilon_{BD} = +750$ $\mu$m/m, and $\epsilon_{AD} = -600$ $\mu$m/m. Determine the normal strain $\epsilon_y$ and the shearing strain $\gamma_{xy}$ in the plate.

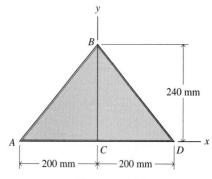

**Figure P10-64**

# 10-9 PRINCIPAL STRAINS AND MAXIMUM SHEARING STRAIN

The similarity between Eqs. (10-11) and (10-12) for plane strain and Eqs. (10-1) and (10-2) for plane stress indicates that all the equations developed for plane stress can be applied to plane strain by substituting $\epsilon_x$ for $\sigma_x$, $\epsilon_y$ for $\sigma_y$, and $\gamma_{xy}/2$ for $\tau_{xy}$. Thus, from Eqs. (10-3), (10-4), and (10-5), expressions are obtained for determining in-plane principal directions, in-plane principal strains, and the maximum in-plane shearing strain. Thus,

$$\tan 2\theta_p = \frac{\gamma_{xy}}{\epsilon_x - \epsilon_y} \tag{10-13}$$

$$\epsilon_{p1}, \epsilon_{p2} = \frac{\epsilon_x + \epsilon_y}{2} \pm \sqrt{\left(\frac{\epsilon_x - \epsilon_y}{2}\right)^2 + \left(\frac{\gamma_{xy}}{2}\right)^2} \tag{10-14}$$

$$\gamma_p = 2 \sqrt{\left(\frac{\epsilon_x - \epsilon_y}{2}\right)^2 + \left(\frac{\gamma_{xy}}{2}\right)^2} \tag{10-15}$$

In the previous equations, normal strains that are tensile and shearing strains that decrease the angle between the faces of the element at the origin of coordinates (see Fig. 10-17) are positive.

When a state of plane strain exists, Eq. (10-14) gives the two in-plane principal strains while the third principal strain is $\epsilon_{p3} = \epsilon_z = 0$.[5] An examination of Eqs. (10-14) and (10-15) indicates that the maximum in-plane shearing strain is the difference between the in-plane principal strains, but this may not be the

---

[5]Note that although there is no strain in the z-direction, $\epsilon_{p3} = \epsilon_z = 0$, there must be a stress in the z-direction, $\sigma_{p3} = \sigma_z \neq 0$. That is, a stress is required in the z-direction to prevent deformation in the z-direction.

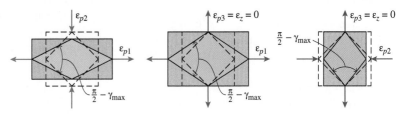

**Figure 10-20**

maximum shearing strain at the point. The maximum shearing strain at the point may be $(\epsilon_{p1} - \epsilon_{p2})$, $(\epsilon_{p1} - 0)$, or $(0 - \epsilon_{p2})$, depending on the relative magnitudes and signs of the principal strains. The lines associated with the maximum shearing strain bisect the angles between lines experiencing maximum and minimum normal strains. The three possibilities are illustrated in Fig. 10-20.

## Example Problem 10-8

The strain components at a point in a body under a state of plane strain are $\epsilon_x = +1200\ \mu$, $\epsilon_y = -600\ \mu$, and $\gamma_{xy} = +900\ \mu$. Determine the principal strains and the maximum shearing strain at the point. Show the principal strain deformations and the maximum shearing strain distortion on a sketch.

### SOLUTION
When the given data are substituted in Eqs. (10-13), (10-14), and (10-15), they yield the in-plane principal strains at the point, their orientations, and the maximum in-plane shearing strain. Thus,

$$\tan 2\theta_p = \frac{\gamma_{xy}}{\epsilon_x - \epsilon_y} = \frac{900}{1200 - (-600)}$$

$$(\epsilon_{p1}, \epsilon_{p2}) = \frac{\epsilon_x + \epsilon_y}{2} \pm \sqrt{\left(\frac{\epsilon_x - \epsilon_y}{2}\right)^2 + \left(\frac{\gamma_{xy}}{2}\right)^2}$$

$$(\epsilon_{p1}, \epsilon_{p2})\,(10^6) = \frac{1200 + (-600)}{2} \pm \sqrt{\left(\frac{1200 - (-600)}{2}\right)^2 + \left(\frac{900}{2}\right)^2}$$

$$\frac{\gamma_p}{2} = \sqrt{\left(\frac{\epsilon_x - \epsilon_y}{2}\right)^2 + \left(\frac{\gamma_{xy}}{2}\right)^2}$$

$$\gamma_p\,(10^6) = 2\sqrt{\left(\frac{(1200 - (-600)}{2}\right)^2 + \left(\frac{900}{2}\right)^2}$$

from which

$$\epsilon_{p1} = +1306\mu \qquad \qquad \textbf{Ans.}$$

$$\epsilon_{p2} = -706\mu \qquad \qquad \textbf{Ans.}$$

$$\epsilon_{p3} = 0$$

$$\theta_p = 13.28°$$

$$\gamma_p = +2010\mu = \gamma_{max}$$

The required sketch is given in Fig. 10-21. ∎

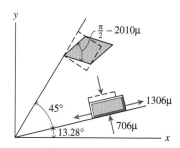

**Figure 10-21**

## PROBLEMS

In Problems 10-65 through 10-76 the strain components $\epsilon_x$, $\epsilon_y$, and $\gamma_{xy}$ are given for a point in a body subjected to plane strain. Determine the principal strains and the maximum shearing strain at the point. Show the principal strain deformations and the maximum shearing strain distortion on a sketch.

| Problem | $\epsilon_x$ | $\epsilon_y$ | $\gamma_{xy}$ |
|---|---|---|---|
| 10-65* | $+600\mu$ | $-200\mu$ | $+500\mu$ |
| 10-66* | $+960\mu$ | $-320\mu$ | $-480\mu$ |
| 10-67 | $+900\mu$ | $-300\mu$ | $-420\mu$ |
| 10-68 | $-900\mu$ | $+600\mu$ | $+480\mu$ |
| 10-69* | $-750\mu$ | $+1000\mu$ | $-250\mu$ |
| 10-70* | $+750\mu$ | $-410\mu$ | $+360\mu$ |
| 10-71 | $+720\mu$ | $+520\mu$ | $+480\mu$ |
| 10-72 | $-540\mu$ | $-980\mu$ | $-560\mu$ |
| 10-73* | $+864\mu$ | $+432\mu$ | $-288\mu$ |
| 10-74* | $+650\mu$ | $+900\mu$ | $+300\mu$ |
| 10-75 | $-325\mu$ | $-625\mu$ | $+380\mu$ |
| 10-76 | $+900\mu$ | $+650\mu$ | $+600\mu$ |

# 10-10 MOHR'S CIRCLE FOR PLANE STRAIN

The pictorial or graphic representation of Eqs. (10-1) and (10-2), known as Mohr's circle for stress, can be used with Eqs. (10-11) and (10-12) to yield a Mohr's circle for strain. The equation for the strain circle obtained from the equation for the stress circle by using a change in variables is

$$\left(\epsilon_n - \frac{\epsilon_x + \epsilon_y}{2}\right)^2 + \left(\frac{\gamma_{nt}}{2}\right)^2 = \left(\frac{\epsilon_x - \epsilon_y}{2}\right)^2 + \left(\frac{\gamma_{xy}}{2}\right)^2$$

The variables in this equation are $\epsilon_n$ and $\gamma_{nt}/2$. The circle is centered on the $\epsilon$ axis at a distance $(\epsilon_x + \epsilon_y)/2$ from the origin and has a radius

$$R = \sqrt{\left(\frac{\epsilon_x - \epsilon_y}{2}\right)^2 + \left(\frac{\gamma_{xy}}{2}\right)^2}$$

Mohr's circle for the strains of Fig. 10-17 (with $\epsilon_x > \epsilon_y$) is given in Fig. 10-22. It is apparent that the sign convention for shearing strain needs to be extended to cover the construction of Mohr's circle. Observe that for positive shearing strain (indicated in Fig. 10-17), the edge of the element parallel to the x-axis tends to

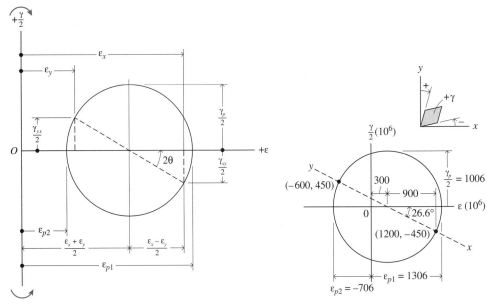

**Figure 10-22**                               **Figure 10-23**

rotate counterclockwise while the edge parallel to the $y$-axis tends to rotate clockwise. For Mohr's circle construction, the clockwise rotation will be designated positive and the counterclockwise designated negative. This is consistent with the sign convention for shearing stresses given in Section 10-4. The Mohr's circle solution for Example Problem 10-8 is shown in Fig. 10-23.

## PROBLEMS

In Problems 10-77 through 10-88 certain strains and angles are given for a point in a body subjected to plane strain. Use Mohr's circle to determine the unknown quantities for each problem and prepare a sketch showing the angle $\theta_p$, the principal strain deformations, and the maximum shearing strain distortions. In some problems there may be more than one possible value of $\theta_p$ depending on the sign of $\gamma_{xy}$.

### Introductory Problems

| Problem | $\epsilon_x$ | $\epsilon_y$ | $\gamma_{xy}$ | $\epsilon_{p1}$ | $\epsilon_{p2}$ | $\gamma_p$ | $\gamma_{max}$ | $\theta_p$ |
|---------|------|------|------|--------|--------|------|-------|--------|
| 10-77* | — | — | — | $+600\mu$ | $-400\mu$ | — | — | $+18.43°$ |
| 10-78* | — | — | — | $+785\mu$ | $-945\mu$ | — | — | $+16.85°$ |
| 10-79 | — | — | — | $+708\mu$ | $-104\mu$ | — | — | $-34.10°$ |
| 10-80 | — | — | — | $-114\mu$ | $-903\mu$ | — | — | $+19.26°$ |

### Intermediate Problems

| Problem | $\epsilon_x$ | $\epsilon_y$ | $\gamma_{xy}$ | $\epsilon_{p1}$ | $\epsilon_{p2}$ | $\gamma_p$ | $\gamma_{max}$ | $\theta_p$ |
|---------|------|------|------|------|------|------|------|------|
| 10-81* | $+950\mu$ | $-225\mu$ | $+275\mu$ | — | — | — | — | — |
| 10-82* | $+900\mu$ | $-333\mu$ | $+982\mu$ | — | — | — | — | — |
| 10-83 | $-750\mu$ | $-390\mu$ | $+900\mu$ | — | — | — | — | — |
| 10-84 | $+600\mu$ | $+480\mu$ | $-480\mu$ | — | — | — | — | — |

**Challenging Problems**

| Problem | $\epsilon_x$ | $\epsilon_y$ | $\gamma_{xy}$ | $\epsilon_{p1}$ | $\epsilon_{p2}$ | $\gamma_p$ | $\gamma_{max}$ | $\theta_p$ |
|---------|------|------|------|------|------|------|------|------|
| 10-85* | $-680\mu$ | $+320\mu$ | — | $+414\mu$ | — | — | — | — |
| 10-86* | $+450\mu$ | $+150\mu$ | — | $+780\mu$ | — | — | — | — |
| 10-87 | $+360\mu$ | $+750\mu$ | — | — | $+197\mu$ | — | — | — |
| 10-88 | $-300\mu$ | $+600\mu$ | — | — | $-522\mu$ | — | — | — |

# 10-11 GENERALIZED HOOKE'S LAW

Hooke's law [see Eq. (4-15)] can be extended to include the biaxial (see Fig. 10-3) and triaxial (see Fig. 10-2) states of stress often encountered in engineering practice. Consider Fig. 10-24, which shows a differential element of material subjected to a biaxial state of normal stress. Shearing stresses have not been shown on the faces of the element, since they produce distortion of the element (angle changes) but do not produce changes in the lengths of the sides of the element, which would contribute to the normal strains. The deformations of the element in the directions of the normal stresses, for a combined loading, can be determined by computing the deformations resulting from the individual stresses separately and adding the values obtained algebraically. This procedure is based on the principle of superposition, which states that the effects of separate loadings can be added algebraically if two conditions are satisfied:

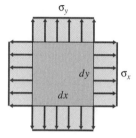

**Figure 10-24**

1. Each effect is linearly related to the load that produced it.
2. The effect of the first load does not significantly change the effect of the second load.

Condition 1 is satisfied if the stresses do not exceed the proportional limit for the material. Condition 2 is satisfied if the deformations are small so that the small changes in the areas of the faces of the element do not produce significant changes in the stresses.

The deformations of the element of Fig. 10-24, associated with the stresses $\sigma_x$ and $\sigma_y$, are shown in Fig. 10-25. The shaded square in Fig. 10-25a indicates the original or unstrained configuration of the element. Under the action of the tensile stress $\sigma_x$, the element experiences a tensile strain of $\sigma_x/E$ in the x-direction and a compressive strain of $v\sigma_x/E$ in the y-direction. These strains cause the element to stretch an amount $(\sigma_x/E)\,dx$ in the x-direction and to contract an amount $(v\sigma_x/E)\,dy$ in the y-direction to the configuration indicated in Fig. 10-25b (the deformations are greatly exaggerated). Then, under the action of the tensile stress $\sigma_y$ superimposed on the stress $\sigma_x$, the element experiences a tensile strain of $\sigma_y/E$ in the y-direction and a compressive strain of $v\sigma_y/E$ in the x-direction. These strains cause the element to stretch an amount $(\sigma_y/E)\,dy$ in the y-direction and to contract an amount $(v\sigma_y/E)\,dx$ in the x-direction, as shown in Fig. 10-25c. If the material is isotropic, Young's modulus has the same value for all directions and the final deformation in the x-direction is (Fig. 10-25d)

$$d\delta_x = \epsilon_x\,dx = \frac{\sigma_x}{E}dx - v\frac{\sigma_y}{E}dx$$

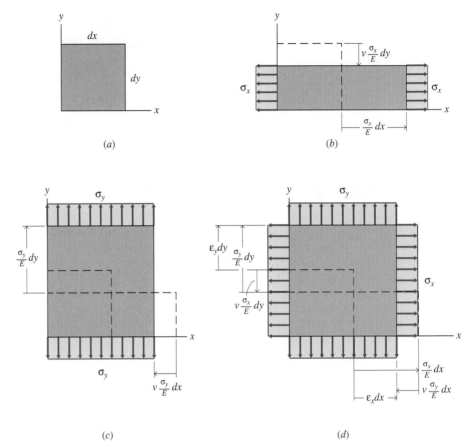

**Figure 10-25**

and the three normal strains are

$$\epsilon_x = \frac{1}{E}(\sigma_x - v\sigma_y)$$

$$\epsilon_y = \frac{1}{E}(\sigma_y - v\sigma_x) \tag{10-16}$$

$$\epsilon_z = -\frac{v}{E}(\sigma_x + \sigma_y)$$

The analysis above is readily extended to triaxial principal stresses, and the expressions for strain become

$$\epsilon_x = \frac{1}{E}[\sigma_x - v(\sigma_y + \sigma_z)]$$

$$\epsilon_y = \frac{1}{E}[\sigma_y - v(\sigma_x + \sigma_z)] \tag{10-17}$$

$$\epsilon_z = \frac{1}{E}[\sigma_z - v(\sigma_x + \sigma_y)]$$

In these expressions, tensile stresses and strains are considered positive, and compressive stresses and strains are considered negative.

When Eqs. (10-16) are solved for the stresses in terms of the strains, they give

$$\sigma_x = \frac{E}{1 - v^2}(\epsilon_x + v\epsilon_y)$$

$$\sigma_y = \frac{E}{1 - v^2}(\epsilon_y + v\epsilon_x) \tag{10-18}$$

Equations (10-18) can be used to calculate normal stresses from measured or computed normal strains. When Eqs. (10-17) are solved for stresses in terms of strains, they give

$$\sigma_x = \frac{E}{(1 + v)(1 - 2v)}[(1 - v)\epsilon_x + v(\epsilon_y + \epsilon_z)]$$

$$\sigma_y = \frac{E}{(1 + v)(1 - 2v)}[(1 - v)\epsilon_y + v(\epsilon_z + \epsilon_x)] \tag{10-19}$$

$$\sigma_z = \frac{E}{(1 + v)(1 - 2v)}[(1 - v)\epsilon_z + v(\epsilon_x + \epsilon_y)]$$

Torsion test specimens are used to study material behavior under pure shear, and it is observed that a shearing stress produces only a single corresponding shearing strain. Thus Hooke's law extended to shearing stresses is simply

$$\tau = G\gamma \tag{10-20}$$

Equations (10-19) and (10-20) seem to indicate that three elastic constants—$E$, $v$, and $G$—are required to determine the deformations (and strains) in a material resulting from an arbitrary state of stress. In fact, only two of these constants need be determined experimentally for a given material.

The relationship between the elastic constants $E$, $v$, and $G$ can be determined by considering the stresses and strains produced by an axial tensile load in a bar of the material. Equations (4-7) and (4-8) indicate that both normal and shearing stresses are produced on different inclined planes through the bar by the axial tensile load. For the discussion that follows, consider the stresses that develop on the faces of the two small square elements shown in Fig. 10-26. From Eqs. (4-7) and (4-8),

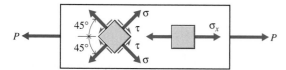

**Figure 10-26**

$$\sigma_x = \frac{P}{2A}(1 + \cos 0°) = \frac{P}{A}$$

$$\sigma = \frac{P}{2A}(1 + \cos 90°) = \frac{P}{2A} = \frac{1}{2}\sigma_x \tag{a}$$

$$\tau = \frac{P}{2A}\sin 90° = \frac{P}{2A} = \frac{1}{2}\sigma_x$$

The strains produced by these stresses are given by Eqs. (10-16) and (10-20) as

$$\epsilon_x = \frac{\sigma_x}{E} \qquad \epsilon_y = -\frac{\nu\sigma_x}{E} \qquad \gamma = \frac{\tau}{G} = \frac{\sigma_x}{2G} \tag{b}$$

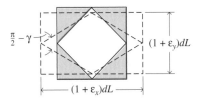

**Figure 10-27**

Under the action of the axial tensile load, the two square elements shown in Fig. 10-26 deform into the shapes shown in Fig. 10-27. In this figure, the deformations of the two elements have been superimposed to show their relationships. The relationship between the shearing strain $\gamma$ and the two normal strains $\epsilon_x$ and $\epsilon_y$ is determined from the geometry of Fig. 10-27. Thus,

$$\tan\left(\frac{\pi}{4} - \frac{\gamma}{2}\right) = \frac{(1 + \epsilon_y)(dL/2)}{(1 + \epsilon_x)(dL/2)} \tag{c}$$

The tangent of the difference between two angles can be expressed as

$$\frac{\tan\dfrac{\pi}{4} - \tan\dfrac{\gamma}{2}}{1 + \tan\dfrac{\pi}{4}\tan\dfrac{\gamma}{2}} = \frac{1 + \epsilon_y}{1 + \epsilon_x} = \frac{1 - \nu\epsilon_x}{1 + \epsilon_x} \tag{d}$$

Since all of the strains are small, Eq. (d) reduces to

$$\frac{1 - \dfrac{\gamma}{2}}{1 + \dfrac{\gamma}{2}} = \frac{1 - \nu\epsilon_x}{1 + \epsilon_x} \tag{e}$$

Solving for the shearing strain $\gamma$ yields

$$\gamma = \frac{(1 + \nu)\epsilon_x}{1 + (1/2)(1 - \nu)(\epsilon_x)} \approx (1 + \nu)\epsilon_x \tag{f}$$

Finally, if Eqs. (b) are substituted into Eq. (f),

$$\frac{\tau}{G} = \frac{\sigma_x}{2G} = (1 + \nu)\frac{\sigma_x}{E} \tag{g}$$

Solving Eq. (g) for $G$ yields the desired relationship between $G$, $E$, and $\nu$, Thus,

$$G = \frac{E}{2(1 + \nu)} \tag{10-21}$$

If Eq. (10-21) is substituted into Eq. (10-20), an alternate form of generalized Hooke's law for shearing stress and strain in isotropic materials is obtained. Thus

$$\tau_{xy} = G\gamma_{xy} = \frac{E}{2(1 + v)} \gamma_{xy}$$

$$\tau_{yz} = G\gamma_{yz} = \frac{E}{2(1 + v)} \gamma_{yz} \qquad\qquad (10\text{-}22)$$

$$\tau_{zx} = G\gamma_{zx} = \frac{E}{2(1 + v)} \gamma_{zx}$$

Equations (10-16) thru (10-22) are widely used for experimental stress determinations. The following example illustrates the method of application.

## Example Problem 10-9

At a point on the surface of an alloy steel ($E = 210$ GPa and $v = 0.30$) machine part subjected to a biaxial state of stress, the measured strains were $\epsilon_x = +1394$ $\mu$m/m, $\epsilon_y = -660$ $\mu$m/m, and $\gamma_{xy} = 2054$ $\mu$rad. Determine

(a) The stress components $\sigma_x$, $\sigma_y$, and $\tau_{xy}$ at the point.
(b) The principal stresses and the maximum shear stress at the point. Locate the planes on which these stresses act and show the stresses on a complete sketch.

### SOLUTION

(a) The normal stresses $\sigma_x$ and $\sigma_y$ are obtained by using Eqs. (10-18). Thus,

$$\sigma_x = \frac{E}{1 - v^2} (\epsilon_x + v\epsilon_y)$$

$$= \frac{210(10^9)}{1 - (0.30)^2} [1394 + 0.30(-660)](10^{-6})$$

$$= +276.0(10^6) \text{ N/m}^2 = 276 \text{ MPa T} \qquad\qquad \textbf{Ans.}$$

$$\sigma_y = \frac{E}{1 - v^2}(\epsilon_y + v\epsilon_x)$$

$$= \frac{210(10^9)}{1 - (0.30)^2} [-660 + 0.30(1394)](10^{-6})$$

$$= -55.80(10^6) \text{ N/m}^2 = 55.8 \text{ MPa C} \qquad\qquad \textbf{Ans.}$$

The shear stress $\tau_{xy}$ is obtained by using Eqs. (10-22). Thus,

$$\tau_{xy} = \frac{E}{2(1 + v)} \gamma_{xy}$$

$$= \frac{210(10^9)}{2(1 + 0.30)} (2054)(10^{-6})$$

$$= 165.90(10^6) \text{ N/m}^2 = 165.9 \text{ MPa} \qquad\qquad \textbf{Ans.}$$

(b) When the values for $\sigma_x$, $\sigma_y$, and $\tau_{xy}$ are substituted in Eq. (10-4), the principal stresses are found to be

$$\sigma_{p1,\,p2} = \frac{\sigma_x + \sigma_y}{2} \pm \sqrt{\left(\frac{\sigma_x - \sigma_y}{2}\right)^2 + \tau_{xy}^2}$$

$$= \frac{276.0 - 55.80}{2} \pm \sqrt{\left(\frac{276.0 + 55.80}{2}\right)^2 + (165.90)^2}$$

$$= 110.10 \pm 234.62$$

$$\sigma_{p1} = 110.10 + 234.62 = 344.72 \text{ MPa} \cong 345 \text{ MPa T} \qquad \textbf{Ans.}$$

$$\sigma_{p2} = 110.10 - 234.62 = -124.52 \text{ MPa} \cong 124.5 \text{ MPa C} \qquad \textbf{Ans.}$$

A state of plane stress exists on the surface of the machine part; therefore,

$$\sigma_{p3} = \sigma_z = 0 \qquad \textbf{Ans.}$$

Since $\sigma_{p1}$ and $\sigma_{p2}$ have opposite signs, the maximum shearing stress is given by Eq. (10-6) as

$$\tau_{\max} = \tau_p = \frac{1}{2}(\sigma_{p1} - \sigma_{p2})$$

$$= \frac{1}{2}(344.72 + 124.52) = 234.62 \text{ MPa} \cong 235 \text{ MPa} \qquad \textbf{Ans.}$$

The in-plane principal stress directions are given by Eq. (10-3) as

$$\tan 2\theta_p = \frac{2\tau_{xy}}{\sigma_x - \sigma_y} = \frac{2(165.90)}{276.0 + 55.80} = 1.0000$$

$$2\theta_p = 45.00° \qquad \theta_p = 22.5° \qquad \textbf{Ans.}$$

The required sketch is shown in Fig. 10-28. The direction of $\tau_{\max} = 235$ MPa on Fig. 10-28 is to oppose the larger of the principal stresses $\sigma_{p1} = 345$ MPa and $\sigma_{p2} = 124.5$ MPa. As observed in Example Problems 10-3 and 10-4, the normal stress on the surface of maximum shear stress is the average of the principal stresses

$$\sigma_n = \frac{344.72 + (-124.52)}{2} = 110.1 \text{ MPa.} \quad \blacksquare$$

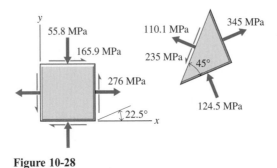

**Figure 10-28**

# PROBLEMS

## Introductory Problems

**10-89\*** At a point on the surface of an aluminum alloy ($E = 10,000$ ksi and $G = 3800$ ksi) machine part subjected to a biaxial state of stress, the measured strains were $\epsilon_x = +900\ \mu\text{in./in.}$, $\epsilon_y = -300\ \mu\text{in./in.}$, and $\gamma_{xy} = -400\ \mu\text{rad}$. Determine the stresses $\sigma_x$, $\sigma_y$, and $\tau_{xy}$ at the point.

**10-90\*** At a point on the surface of a structural steel ($E = 200$ GPa and $G = 76$ GPa) machine part subjected to a biaxial state of stress, the measured strains were $\epsilon_x = +750\ \mu\text{m/m}$, $\epsilon_y = +350\ \mu\text{m/m}$, and $\gamma_{xy} = -560\ \mu\text{rad}$. Determine the stresses $\sigma_x$, $\sigma_y$, and $\tau_{xy}$ at the point.

**10-91** At a point on the surface of a titanium alloy ($E = 14,000$ ksi and $G = 5300$ ksi) machine part subjected to a biaxial state of stress, the measured strains were $\epsilon_x = +1250\ \mu\text{in./in.}$, $\epsilon_y = +600\ \mu\text{in./in.}$, and $\gamma_{xy} = +650\ \mu\text{rad}$. Determine the stresses $\sigma_x$, $\sigma_y$, and $\tau_{xy}$ at the point.

## Intermediate Problems

**10-92\*** At a point on the surface of a stainless steel ($E = 190$ GPa and $G = 76$ GPa) machine part subjected to a biaxial state of stress, the measured strains were $\epsilon_x = +1175\ \mu\text{m/m}$, $\epsilon_y = -1250\ \mu\text{m/m}$, and $\gamma_{xy} = +850\ \mu\text{rad}$. Determine the stresses $\sigma_x$, $\sigma_y$, and $\tau_{xy}$ at the point.

**10-93** Determine the state of strain that corresponds to the following state of stress at a point in a steel ($E = 30,000$ ksi and $v = 0.30$) machine part: $\sigma_x = 15,000$ psi, $\sigma_y = 5000$ psi, $\sigma_z = 7500$ psi, $\tau_{xy} = 5500$ psi, $\tau_{yz} = 4750$ psi, and $\tau_{zx} = 3200$ psi.

**10-94** Determine the state of strain that corresponds to the following state of stress at a point in an aluminum alloy ($E = 73$ GPa and $v = 0.33$) machine part: $\sigma_x = 120$ MPa, $\sigma_y = -85$ MPa, $\sigma_z = 45$ MPa, $\tau_{xy} = 35$ MPa, $\tau_{yz} = 48$ MPa, and $\tau_{zx} = 76$ MPa.

## Challenging Problems

In Problems 10-95 through 10-97 the strain components $\epsilon_x$, $\epsilon_y$, and $\gamma_{xy}$ are given for a point on the free surface of a machine component. Determine the principal stresses and the maximum shear stress at the point. Locate the planes on which these stresses act and show the stresses on a complete sketch.

| Problem | $\epsilon_x$ | $\epsilon_y$ | $\gamma_{xy}$ | $E$ | $v$ |
|---|---|---|---|---|---|
| 10-95\* | $+900\mu$ | $-300\mu$ | $-400\mu$ | 10,000 ksi | 0.30 |
| 10-96 | $+750\mu$ | $+350\mu$ | $-560\mu$ | 200 GPa | 0.30 |
| 10-97 | $+1250\mu$ | $+600\mu$ | $+650\mu$ | 14,000 ksi | 0.32 |

# 10-12 STRAIN MEASUREMENT AND ROSETTE ANALYSIS

Electrical resistance strain gages provide accurate measurements of normal strain. The gage may consist of a length of 0.001-in.-diameter wire arranged as shown in Fig. 10-29a and cemented between two pieces of paper, or it may be an etched foil conductor mounted on epoxy or polyimide backing (see Fig. 10-29b). The wire or foil gage is cemented to the material for which the strain is to be determined. As the material is strained, the wires are lengthened or shortened; this changes the electrical resistance of the gage. The change in resistance can be measured and calibrated to provide a normal strain. Shearing strains are more difficult to measure directly than normal strains and are often obtained by measuring normal strains in two or three different directions. The shearing strain $\gamma_{xy}$ can be computed from normal strain data by using Eq. (10-11a). For example, consider the most general case of three arbitrary normal strain measurements as shown in Fig. 10-30. From Eq. (10-11a)

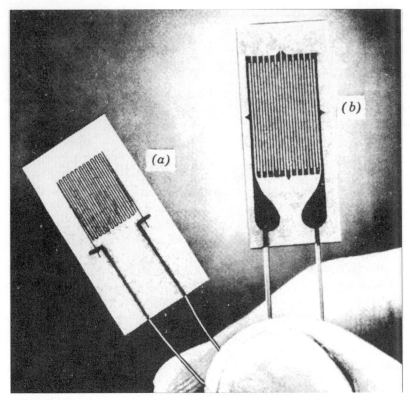

**Figure 10-29**

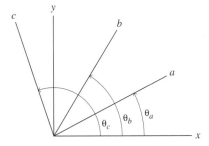

**Figure 10-30**

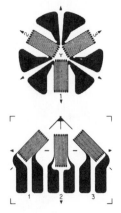

**Figure 10-31**

$$\epsilon_a = \epsilon_x \cos^2 \theta_a + \epsilon_y \sin^2\theta_a + \gamma_{xy} \sin \theta_a \cos \theta_a$$

$$\epsilon_b = \epsilon_x \cos^2 \theta_b + \epsilon_y \sin^2\theta_b + \gamma_{xy} \sin \theta_b \cos \theta_b$$

$$\epsilon_c = \epsilon_x \cos^2 \theta_c + \epsilon_y \sin^2\theta_c + \gamma_{xy} \sin \theta_c \cos \theta_c$$

From the measured value of $\epsilon_a$, $\epsilon_b$, and $\epsilon_c$ and a knowledge of the gage orientations $\theta_a$, $\theta_b$, and $\theta_c$ with respect to the reference $x$-axis, the values of $\epsilon_x$, $\epsilon_y$, and $\gamma_{xy}$ can be determined by simultaneous solution of the three equations. In practice the angles $\theta_a$, $\theta_b$, and $\theta_c$ are selected to simplify the calculations. Multiple-element strain gages used for this type of measurement are known as *strain rosettes*. Two rosette configurations marketed commercially are shown in Fig. 10-31. Once $\epsilon_x$, $\epsilon_y$, and $\gamma_{xy}$ have been determined, Eqs. (10-13), (10-14), and (10-15) or the corresponding Mohr's circle can be used to determine the in-plane principal strains, their orientations, and the maximum in-plane shearing strain at the point.

The principal strain $\epsilon_z = \epsilon_{p3}$ can be determined from the measured data. From Eqs. (10-16) and (10-18), which represent Hooke's law for the case of plane stress,

$$\epsilon_z = -\frac{v}{E}(\sigma_x + \sigma_y) = -\frac{v}{E}\left[\frac{E}{1 - v^2}\right][(\epsilon_x + v\epsilon_y) + (\epsilon_y + v\epsilon_x)]$$

$$= -\frac{v}{1 - v}(\epsilon_x + \epsilon_y) \tag{10-23}$$

This out-of-plane principal strain is important since the maximum shearing strain at the point may be $(\epsilon_{p1} - \epsilon_{p2})$, $(\epsilon_{p1} - \epsilon_{p3})$, or $(\epsilon_{p3} - \epsilon_{p2})$, depending on the relative magnitudes and signs of the principal strains at the point.

The following example illustrates the application of Eqs. (10-11) through (10-23) and Mohr's circle to principal stress and strain and maximum shearing stress and strain determinations under conditions of plane stress.

## Example Problem 10-10

A strain rosette, composed of three electrical resistance strain gages making angles of $0°$, $60°$, and $120°$ with the x-axis (see Fig. 10-32a), was mounted on the free surface of a steel ($E = 30,000$ ksi and $v = 0.30$) machine component. Under load, the following strains were measured:

$$\epsilon_a = +1000 \ \mu\text{in./in.} \qquad \epsilon_b = +750 \ \mu\text{in./in.} \qquad \epsilon_c = -650 \ \mu\text{in./in.}$$

(a) Determine the principal strains and the maximum shearing strain at the point. Show the directions of the in-plane principal strains on a sketch.

(b) Determine the principal stresses and the maximum shearing stress at the point. Show the principal stresses and the maximum shearing stress on a sketch.

## SOLUTION

(a) When the given data are substituted into Eq. (10-11a), the following equations are obtained:

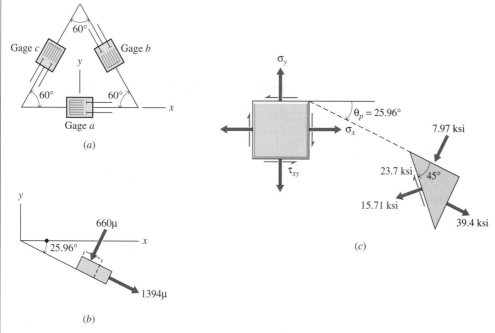

(a)

(b)

(c)

**Figure 10-32**

$$\epsilon_a = \epsilon_0 = \epsilon_x = +1000 \ \mu\text{in./in. (measured)}$$

$$\epsilon_c = \epsilon_{60} = -650 \ \mu = 1000 \ \mu \cos^2 60° + \epsilon_y \sin^2 60° \\ + \gamma_{xy} \sin 60° \cos 60°$$

$$\epsilon_b = \epsilon_{120} = +750 \ \mu = 1000 \ \mu \cos^2 120° + \epsilon_y \sin^2 120° \\ + \gamma_{xy} \sin 120° \cos 120°$$

from which

$$\epsilon_y = -266.7 \ \mu\text{in./in.} \qquad \gamma_{xy} = -1616.6 \ \mu\text{rad}$$

The in-plane principal strains $\epsilon_{p1}$ and $\epsilon_{p2}$ are given by Eq. (10-14) as

$$(\epsilon_{p1}, \epsilon_{p2}) = \frac{\epsilon_x + \epsilon_y}{2} \pm \sqrt{\left(\frac{\epsilon_x - \epsilon_y}{2}\right)^2 + \left(\frac{\gamma_{xy}}{2}\right)^2}$$

$$(\epsilon_{p1}, \epsilon_{p2})(10^6) = \frac{1000 + (-266.7)}{2} \\ \pm \sqrt{\left(\frac{1000-(-266.7)}{2}\right)^2 + \left(\frac{-1616.6}{2}\right)^2} \\ = 366.7 \pm 1026.9$$

$$\epsilon_{p1} = (366.7 + 1026.9)(10^{-6}) = 1393.6(10^{-6}) \cong 1394 \ \mu\text{in./in.} \qquad \textbf{Ans.}$$

$$\epsilon_{p2} = (366.7 - 1026.9)(10^{-6}) = -660.2(10^{-6}) \cong -660 \ \mu\text{in./in.} \qquad \textbf{Ans.}$$

The principal strain perpendicular to the surface, $\epsilon_{p3} = \epsilon_z$ is given by Eq. (10-23) as

$$\epsilon_{p3} = \epsilon_z = -\frac{\nu}{1-\nu}(\epsilon_x + \epsilon_y)$$

$$= -\frac{0.3}{1-0.3}(1000 - 266.7)(10^{-6})$$

$$= -314.3(10^{-6}) \cong -314 \ \mu\text{in./in.} \qquad \textbf{Ans.}$$

Since $\epsilon_{p3}$ is less compressive than the in-plane principal strain $\epsilon_{p2}$, the maximum shearing strain at the point is the maximum in-plane shearing strain $\gamma_p$. Thus,

$$\gamma_{max} = \gamma_p = \pm(1394 \ \mu + 660 \ \mu) = \pm 2054 \ \mu \cong \pm 2050 \ \mu\text{rad} \qquad \textbf{Ans.}$$

The in-plane principal strain directions are obtained using Eq. (10-13)

$$\tan 2\theta_p = \frac{\gamma_{xy}}{\epsilon_x - \epsilon_y} = \frac{(-1616.6)}{1000 - (-266.7)} = -1.2763$$

$$\theta_p = -25.96° = 25.96° \ \circlearrowright \qquad \textbf{Ans.}$$

The required sketch is given in Fig. 10-32b.

(b) The principal stresses and the maximum shearing stress at the point can be obtained from the strain results of part a by using Eqs. (10-18) and (10-22). Thus,

$$\sigma_{max} = \frac{30(10^6)}{1 - (0.30)^2}[1394 + 0.30(-660)](10^{-6})$$

$$= +39,429 \text{ psi} \cong 39.4 \text{ ksi T} \qquad \textbf{Ans.}$$

$$\sigma_{min} = \frac{30(10^6)}{1 - (0.30)^2}[-660 + 0.30(1394)](10^{-6})$$

$$= -7971 \text{ psi} \cong 7.97 \text{ ksi C} \qquad \textbf{Ans.}$$

$$\tau_{max} = \frac{30(10^6)}{2(1 + 0.30)}(2054)(10^{-6}) = 23.7(10^3) \text{ psi} = 23.7 \text{ ksi} \qquad \textbf{Ans.}$$

Note that the value for $\tau_{max}$ checks with that given by Eq. (10-7); that is,

$$\tau_{max} = \frac{\sigma_{p1} - \sigma_{p2}}{2} = \frac{39,429 - (-7971)}{2} = 23.7(10^3) \text{ psi} = 23.7 \text{ ksi} \quad \textbf{Ans.}$$

The directions of the principal stresses will be the same as those for the principal strains, $\theta_p = -25.96°$. The required sketch is shown in Fig. 10-32c.

## ALTERNATE SOLUTION

The Cartesian components of strain obtained from the original rosette data were

$$\epsilon_x = +1000 \ \mu\text{in./in.} \qquad \epsilon_y = -266.7 \ \mu\text{in./in.} \qquad \gamma_{xy} = -1616.6 \ \mu\text{rad}$$

The Cartesian components of stress at the point can be obtained from these strains by using Eqs. (10-18) and (10-22). Thus,

$$\sigma_x = \frac{30(10^6)}{1 - (0.30)^2}[1000 + 0.30(-266.7)](10^{-6})$$

$$= 30,330 \text{ psi}$$

$$\sigma_y = \frac{30(10^6)}{1 - (0.30)^2}[(-266.7) + 0.30(1000)](10^{-6})$$

$$= 1097.8 \text{ psi}$$

$$\tau_{xy} = \frac{30(10^6)}{2(1 + 0.30)}(-1616.6)(10^{-6}) = -18,653 \text{ psi}$$

Then the principal stresses and the maximum in-plane shear stress are obtained using Eqs. (10-4) and (10-5).

$$(\sigma_{p1}, \sigma_{p2}) = \frac{\sigma_x + \sigma_y}{2} \pm \sqrt{\left(\frac{\sigma_x - \sigma_y}{2}\right)^2 + (\tau_{xy})^2}$$

$$= \frac{30,330 + 1097.8}{2} \pm \sqrt{\left(\frac{30,330 - 1097.8}{2}\right)^2 + (-18,653)^2}$$

$$= 15,714 \pm 23,697 \text{ psi}$$

$$\sigma_{p1} = 15,714 + 23,697 = 39,411 \text{ psi} \cong 39.4 \text{ ksi T} \qquad \textbf{Ans.}$$

$$\sigma_{p2} = 15,714 - 23,697 = -7983 \text{ psi} \cong 7.98 \text{ ksi C} \qquad \textbf{Ans.}$$

$$\tau_p = \sqrt{\left(\frac{\sigma_x - \sigma_y}{2}\right)^2 + (\tau_{xy})^2} = \sqrt{\left(\frac{30,330 - 1097.8}{2}\right)^2 + (-18,653)^2}$$

$$\tau_p = \tau_{max} = 23,697 \text{ psi} = 23.7 \text{ ksi} \qquad \textbf{Ans.}$$

The principal directions are obtained using Eq. (10-3)

$$\tan 2\theta_p = \frac{2(-18,653)}{30,330 - 1097.8} = -1.2762$$

$$\theta_p = -25.96° = 25.96° \circlearrowright \qquad \textbf{Ans.}$$

Any differences in numerical values found in part $b$ of the solution and in the alternate solution is due to rounding off. ∎

## PROBLEMS

In the following problems, when the material is named but the properties are not given, refer to Tables A-17 and A-18 in Appendix A for the properties. Use Eq. (10-21) to determine Poisson's ratio when data from Tables A-17 and A-18 are used.

### Introductory Problems

**10-98*** At a point on the free surface of an aluminum alloy ($E = 73$ GPa and $v = 0.33$) machine part, the strain rosette shown in Fig. P10-98 was used to obtain the following normal strain data: $\epsilon_a = +780$ $\mu$m/m, $\epsilon_b = +345$ $\mu$m/m, and $\epsilon_c = -332$ $\mu$m/m. Determine
(a) The strain components $\epsilon_x$, $\epsilon_y$, and $\gamma_{xy}$ at the point.
(b) The principal strains and the maximum shearing strain at the point.

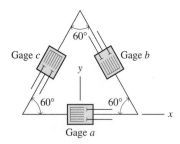

**Figure P10-98**

**10-99*** At a point on the free surface of a steel ($E = 30,000$ ksi and $v = 0.30$) machine part, the strain rosette shown in Fig. P10-99 was used to obtain the following normal strain data: $\epsilon_a = +750$ $\mu$in./in., $\epsilon_b = -125$ $\mu$in./in., and $\epsilon_c = -250$ $\mu$in./in. Determine
(a) The strain components $\epsilon_x$, $\epsilon_y$, and $\gamma_{xy}$ at the point.
(b) The principal strains and the maximum shearing strain at the point.

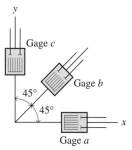

**Figure P10-99**

**10-100** At a point on the free surface of a steel ($E = 200$ GPa and $v = 0.30$) machine part, the strain rosette shown in Fig. P10-100 was used to obtain the following normal strain data: $\epsilon_a = -555$ $\mu$m/m, $\epsilon_b = +925$ $\mu$m/m, and $\epsilon_c$ $+740$ $\mu$m/m. Determine
(a) The strain components $\epsilon_x$, $\epsilon_y$, and $\gamma_{xy}$ at the point.
(b) The principal strains and the maximum shearing strain at the point.

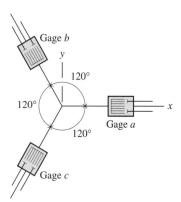

**Figure P10-100**

**Intermediate Problems**

**10-101\*** The strain rosette shown in Fig. P10-101 was used to obtain normal strain data at a point on the free surface of a structural steel machine part. The gage readings were $\epsilon_a = -350$ $\mu$in./in., $\epsilon_b = +1000$ $\mu$in./in., and $\epsilon_c = +550$ $\mu$in./in. Determine
(a) The strain components $\epsilon_x$, $\epsilon_y$, and $\gamma_{xy}$ at the point.
(b) The stress components $\sigma_x$, $\sigma_y$, and $\tau_{xy}$ at the point.
(c) The principal stresses and the maximum shearing stress at the point. Express the stresses in U.S. customary units.

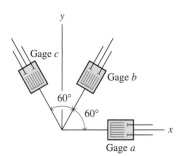

**Figure P10-101**

**10-102\*** The strain rosette shown in Fig. P10-102 was used to obtain normal strain data at a point on the free surface of a 2024-T4 aluminum alloy machine part. The gage readings were $\epsilon_a = +525$ $\mu$m/m, $\epsilon_b = +450$ $\mu$m/m, and $\epsilon_c = +1425$ $\mu$m/m. Determine
(a) The strain components $\epsilon_x$, $\epsilon_y$, and $\gamma_{xy}$ at the point.
(b) The stress components $\sigma_x$, $\sigma_y$, and $\tau_{xy}$ at the point.
(c) The principal stresses and the maximum shearing stress at the point. Express the stresses in SI units.

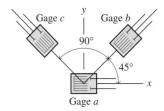

**Figure P10-102**

**10-103** The strain rosette shown in Fig. P10-103 was used to obtain normal strain data at a point on the free surface of an 18-8 cold-rolled stainless steel machine part. The gage readings were $\epsilon_a = +665$ $\mu$in./in., $\epsilon_b = +390$ $\mu$in./in., and $\epsilon_c = +870$ $\mu$in./in. Determine
(a) The strain components $\epsilon_x$, $\epsilon_y$, and $\gamma_{xy}$ at the point.
(b) The stress components $\sigma_x$, $\sigma_y$, and $\tau_{xy}$ at the point.

(c) The principal stresses and the maximum shearing stress at the point. Express the stresses in U.S. customary units.

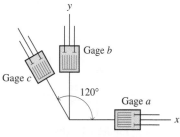

**Figure P10-103**

**Challenging Problems**

**10-104\*** At a point on the free surface of an aluminum alloy ($E = 73$ GPa and $v = 0.33$) machine part, the strain rosette shown in Fig. P10-104 was used to obtain the following normal strain data: $\epsilon_a = +875$ $\mu$m/m, $\epsilon_b = +700$ $\mu$m/m, and $\epsilon_c = -650$ $\mu$m/m. Determine
(a) The stress components $\sigma_x$, $\sigma_y$, and $\tau_{xy}$ at the point.
(b) The principal strains and the maximum shearing strain at the point. Prepare a sketch showing all of these strains.
(c) The principal stresses and the maximum shearing stress at the point. Prepare a sketch showing all of these stresses.

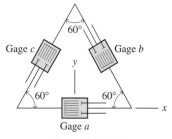

**Figure P10-104**

**10-105** At a point on the free surface of an aluminum alloy ($E = 10,600$ ksi and $v = 0.33$) machine part, the strain rosette shown in Fig. P10-105 was used to obtain the following normal strain data: $\epsilon_a = +800$ $\mu$in./in., $\epsilon_b = +950$ $\mu$in./in., and $\epsilon_c = +600$ $\mu$in./in. Determine
(a) The stress components $\sigma_x$, $\sigma_y$, and $\tau_{xy}$ at the point.
(b) The principal strains and the maximum shearing strain at the point. Prepare a sketch showing all of these strains.

(c) The principal stresses and the maximum shearing stress at the point. Prepare a sketch showing all of these stresses.

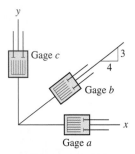

**Figure P10-105**

**10-106** At a point on the free surface of a steel ($E = 200$ GPa and $v = 0.30$) machine part, the strain rosette shown in Fig. P10-106 was used to obtain the following normal strain data: $\epsilon_a = +875$ $\mu$m/m, $\epsilon_b = +700$ $\mu$m/m, and $\epsilon_c = +350$ $\mu$m/m. Determine

(a) The stress components $\sigma_x$, $\sigma_y$, and $\tau_{xy}$ at the point.
(b) The principal strains and the maximum shearing strain at the point. Prepare a sketch showing all of these strains.
(c) The principal stresses and the maximum shearing stress at the point. Prepare a sketch showing all of these stresses.

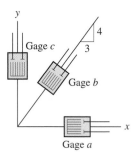

**Figure P10-106**

# 10-13 THIN-WALLED PRESSURE VESSELS

A pressure vessel is described as thin-walled when the ratio of the wall thickness to the radius of the vessel is so small that the distribution of normal stress on a plane perpendicular to the surface of the vessel is essentially uniform throughout the thickness of the vessel. Actually, this stress varies from a maximum value at the inside surface to a minimum value at the outside surface of the vessel, but it can be shown that if the ratio of the wall thickness to the inner radius of the vessel is less than 0.1, the maximum normal stress is less than 5% greater than the average stress. Boilers, gas storage tanks, pipelines, metal tires, and hoops are normally analyzed as thin-walled elements. Gun barrels, certain high-pressure vessels in the chemical processing industry, and cylinders and piping for heavy hydraulic presses need to be treated as thick-walled vessels.

Problems involving thin-walled vessels subjected to liquid (or gas) pressure $p$ are readily solved with the aid of free-body diagrams of sections of the vessels together with the fluid contained therein. In the following subsections, spherical and cylindrical pressure vessels are considered.

**Spherical Pressure Vessels.** A typical thin-walled spherical pressure vessel used for gas storage is shown in Fig. 10-33. If the weights of the gas and vessel are negligible (a common situation), symmetry of loading and geometry requires that stresses on sections that pass through the center of the sphere be equal. Thus, on the small element shown in Fig. 10-34$a$,

$$\sigma_x = \sigma_y = \sigma_n$$

Furthermore, there are no shearing stresses on any of these planes since there are no loads to induce them. The normal stress component in a sphere is known as a *meridional* or *axial stress* and is commonly denoted as $\sigma_m$ or $\sigma_a$.

**Figure 10-33** Hortonsphere for gas storage in Superior, Wisconsin. (Courtesy of Chicago Bridge and Iron Co., Chicago, Ill.)

The free-body diagram shown in Fig. 10-34$b$ can be used to evaluate the stress $\sigma_x = \sigma_y = \sigma_n = \sigma_a$ in terms of the pressure $p$, and the inside radius $r$ and thickness $t$ of the spherical vessel. The force $R$ is the resultant of the internal forces that act on the cross-sectional area of the sphere that is exposed by passing a plane through the center of the sphere. The force $P$ is the resultant of the fluid forces acting on

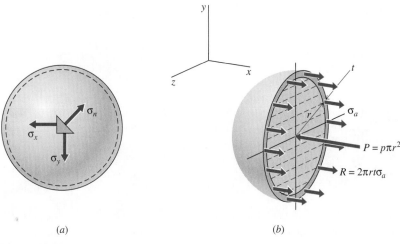

$(a)$ $\qquad\qquad\qquad$ $(b)$

**Figure 10-34**

the fluid remaining within the hemisphere. From a summation of forces in the $x$-direction,

$$R - P = 0$$

or

$$2\pi r t \sigma_a = p\pi r^2$$

from which

$$\sigma_a = \frac{pr}{2t} \tag{10-24}$$

**Cylindrical Pressure Vessels.** A typical thin-walled cylindrical pressure vessel used for gas storage is shown in Fig. 10-35. Normal stresses, such as those shown on the small element of Fig. 10-36$a$, are easy to evaluate by using appropriate free-body diagrams. The normal stress component on a transverse plane is known as an *axial* or *meridional stress* and is commonly denoted as $\sigma_a$ or $\sigma_m$. The normal stress component on a longitudinal plane is known as a *hoop, tangential*, or *circumferential stress* and is denoted as $\sigma_h$, $\sigma_t$, or $\sigma_c$. There are no shearing stresses on transverse or longitudinal planes.

The free-body diagram used for axial stress determinations is similar to Fig. 10-34$b$, which was used for the sphere, and the results are the same. The free-body diagram used for the hoop stress determination is shown in Fig. 10-36$b$. The force $P_x$ is the resultant of the fluid forces acting on the fluid remaining within the portion of the cylinder isolated by the longitudinal plane and two transverse planes. The forces $Q$ are the resultant of the internal forces on the cross-sectional area exposed by the longitudinal plane containing the axis of the cylinder. From a summation of forces in the $x$-direction,

$$2Q - P_x = 0$$

**Figure 10-35** Cylindrical tanks for gas storage in Northlake, Illinois. (Courtesy of Chicago Bridge and Iron Co., Chicago, Ill.)

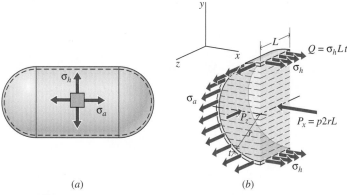

(a)        (b)

**Figure 10-36**

or

$$2\sigma_h Lt = p2rL$$

from which

$$\sigma_h = \frac{pr}{t} \qquad (10\text{-}25a)$$

Also [see Eq. (10-24)],

$$\sigma_a = \frac{pr}{2t} \qquad (10\text{-}25b)$$

The previous analysis of the stresses in a cylindrical vessel subjected to uniform internal pressure indicates that the stress on a longitudinal plane is twice the stress on a transverse plane. Consequently, a longitudinal joint needs to be twice as strong as a transverse (or girth) joint.

## Example Problem 10-11

A cylindrical pressure vessel with an inside diameter of 1.50 m is constructed by wrapping a 15-mm-thick steel plate into a spiral and butt-welding the mating edges of the plate, as shown in Fig. 10-37a. The butt-welded seams form an angle of 30° with a transverse plane through the cylinder. Determine the normal stress $\sigma_n$ perpendicular to the weld and the shearing stress $\tau_{nt}$ parallel to the weld when the internal pressure in the vessel is 1500 kPa.

### SOLUTION

The longitudinal (axial) and hoop directions are principal directions, and surfaces perpendicular to these directions are free of shear stress. However, the weld seam (and any other surface not aligned with the axial or hoop directions) will be subjected to both a normal stress and a shear stress.

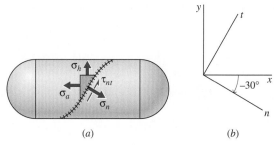

(a)                                (b)

**Figure 10-37**

The hoop stress $\sigma_h$ and the axial stress $\sigma_a$ in the cylinder can be determined by using Eqs. (10-25a) and (10-25b). Thus,

$$\sigma_h = \frac{pr}{t} = \frac{1500(10^3)(0.75)}{0.015} = 75.0(10^6) \text{ N/m}^2 = 75.0 \text{ MPa}$$

$$\sigma_a = \frac{pr}{2t} = \frac{1500(10^3)(0.75)}{2(0.015)} = 37.5(10^6) \text{ N/m}^2 = 37.5 \text{ MPa}$$

The normal stress $\sigma_n$ perpendicular to the weld and the shearing stress $\tau_{nt}$ parallel to the weld can be determined by using the stress transformation equations, Eqs. (10-1a) and (10-2a). The stresses for use in these equations are:

$$\sigma_x = \sigma_a = +37.5 \text{ MPa}$$

$$\sigma_y = \sigma_h = +75.0 \text{ MPa}$$

$$\tau_{xy} = 0 \text{ MPa}$$

The angle $\theta$ for the plane parallel to the weld is $-30°$, as shown in Fig. 10-37$b$. Thus, from Eqs. (10-1a) and (10-2a),

$$\sigma_n = \sigma_x \cos^2\theta + \sigma_y \sin^2\theta + 2\tau_{xy}\sin\theta\cos\theta$$
$$= 37.5 \cos^2(-30°) + 75.0 \sin^2(-30°) + 0$$
$$= 46.9 \text{ MPa T} \qquad\qquad\qquad\qquad \textbf{Ans.}$$

$$\tau_{nt} = -(\sigma_x - \sigma_y)\sin\theta\cos\theta + \tau_{xy}(\cos^2\theta - \sin^2\theta)$$
$$= -(37.5 - 75.0)\sin(-30°)\cos(-30°) + 0$$
$$= -16.24 \text{ MPa} \qquad\qquad\qquad\qquad \textbf{Ans.}$$

The minus sign indicates that the direction of the shearing stress $\tau_{nt}$ is opposite to that shown in Fig. 10-37$a$.

# PROBLEMS

### Introductory Problems

**10-107\*** A spherical gas storage tank, similar to the one shown in Fig. 10-33, has a diameter of 35 ft and a wall thickness of $7/8$ in. Determine the maximum normal stress in the tank if the gas pressure is 100 psi.

**10-108\*** Determine the maximum normal stress in a 300-mm-diameter basketball that has a 2-mm wall thickness after it has been inflated to a pressure of 100 kPa.

**10-109** A steel pipe with an inside diameter of 10 in. will be used to transmit steam under a pressure of 800 psi. If the hoop stress in the pipe must be limited to 10 ksi because of a longitudinal weld in the pipe, determine the minimum satisfactory thickness for the pipe.

**10-110** A cylindrical propane tank, similar to the ones shown in Fig. 10-35, has an outside diameter of 3.25 m and a wall thickness of 22 mm. If the allowable hoop stress is 100 MPa and the allowable axial stress is 45 MPa, determine the maximum internal pressure that can be applied to the tank.

**10-111\*** A cylindrical boiler with hemispherical ends has an outside diameter of 6 ft. If the hoop and axial stresses in the boiler must be limited to 15 ksi and 8 ksi, respectively, determine the thickness of steel plate required if the internal pressure in the boiler will be 200 psi.

### Intermediate Problems

**10-112\*** The cylindrical tank shown in Fig. P10-112 is 20 m in diameter, is made of structural steel, and will be used to store stove oil ($\rho = 850$ kg/m$^3$). Determine the minimum wall thickness required if the allowable stress is 80 MPa.

**Figure P10-112**

**10-113** A standpipe 12 ft in diameter and 50 ft tall is being constructed for use as a storage tank for water ($\gamma = 62.4$ lb/ft$^3$). Determine the minimum thickness of steel plate that can be used if the hoop stress in the standpipe must be limited to 5000 psi.

**10-114** A cylindrical pressure vessel is fabricated by butt-welding 20-mm plate with a spiral seam, as shown in Fig. P10-114. The pressure in the tank is 2800 kPa. Determine
(a) The normal stress perpendicular to the weld.
(b) The shearing stress parallel to the weld.
(c) The maximum shearing stress at a point on the outside surface of the vessel.
(d) The maximum shearing stress at a point on the inside surface of the vessel.

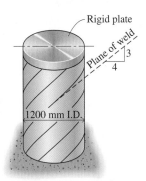

**Figure P10-114**

**10-115** A steel boiler 3 ft in diameter is welded using a spiral seam that makes an angle of 30° with respect to a transverse plane of the boiler, as shown in Fig. P10-115. For an internal pressure of 140 psi and a wall thickness of 2 in., determine
(a) The normal stress perpendicular to the weld.
(b) The shearing stress parallel to the weld.
(c) The maximum shearing stress in the boiler.

**Figure P10-115**

**Challenging Problems**

**10-116\*** A spherical pressure vessel 3 m in diameter is being designed to withstand a maximum internal pressure of 500 kPa. The material being used in its construction has an allowable stress of 105 MPa, a modulus of elasticity of 210 GPa, and a Poisson's ratio of 0.30. Determine

(a) The minimum satisfactory wall thickness.
(b) The circumferential normal strain at maximum pressure when the wall thickness of part *a* is used.
(c) The change in diameter of the pressure vessel at maximum pressure when the wall thickness of part *a* is used.

**10-117\*** A cylindrical tank, similar to the one shown in Fig. 10-35, has an outside diameter of 8 ft and a wall thickness of ¾ in. The tank is subjected to an internal pressure of 150 psi.

(a) Determine the axial and hoop stresses in the tank.
(b) Prepare a plot, similar to Fig. 4-20, showing the variation of normal and shearing stresses on planes through a point in the wall of the tank as the angle $\theta$, measured counterclockwise from a transverse plane, varies from 0° to 90°.

**10-118** The strains measured on the outside surface of the cylindrical pressure vessel shown in Fig. P10-118 are $\epsilon_1 = +619\ \mu m/m$ and $\epsilon_2 = +330\ \mu m/m$. The angle $\theta = 30°$, the outside diameter of the vessel is 510 mm, and the wall thickness is 3 mm. The vessel is made of 0.4% carbon hot-rolled steel (see Appendix A for properties). Determine

(a) The stresses $\sigma_1$ and $\sigma_2$ in the vessel.
(b) The internal pressure applied to the vessel.

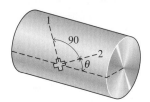

**Figure P10-118**

# 10-14 COMBINED AXIAL, TORSIONAL, AND FLEXURAL LOADS

In numerous industrial situations, machine members are subjected to combinations of the three general types of loads studied so far. When a flexural load is combined with torsional and axial loads, it is frequently difficult to locate the point or points where the most severe stresses occur. The following review of summary statements may be helpful in this regard.

The longitudinal and transverse shearing stresses in a beam are maximum where $Q$ is maximum—usually at the centroidal axis of the section where $V$ is maximum [see Eq. (8-15)]. The flexural stress is maximum at the greatest distance from the centroidal axis on the section where $M$ is maximum [see Eq. (8-12)]. The torsional shearing stress is maximum at the surface of a shaft at the section where $T$ is maximum [see Eq. (7-6)]. With these facts in mind, it is normally possible to locate one or more possible points of high stress. Even so, it may be necessary to calculate the various stresses at more than one point of a member before locating the most severely stressed point. For stresses below the proportional limit of the material, the superposition method can be used to combine stresses on any given plane at any specific point of a loaded member.

After the stresses on a pair of mutually perpendicular planes at a specific point are determined, the methods of Section 10-5 and 10-6 can be used to find the principal stresses and the maximum shearing stress at the point.

The following Example Problems illustrate the procedure for the solution of elastic combined stress problems.

## Example Problem 10-12

The solid 100-mm diameter shaft shown in Fig. 10-38a is subjected to an axial compressive force $P = 200$ kN and a torque $T = 30$ kN · m. For point A on the outside surface of the shaft, determine

(a)  The x- and y-components of stress.
(b)  The principal stresses and the maximum shearing stress at the point.

## SOLUTION

(a)  Since point A is on the outside surface (a free surface) of the shaft, a state of plane stress exists at the point. The coordinate system is selected as shown in Fig. 10-38b. This is the same coordinate system for which the equations in this chapter were developed. Passing a transverse plane through point A and isolating the segment of the shaft to the right of point A results in the free-body diagram shown in Fig. 10-38c. The internal forces on the transverse cross section are an axial compressive force $P_A$ of 200 kN and a torque $T_A$ of 30 kN · m. The directions of the stresses on the left face of the element are in accordance with the directions of the forces that produce the stresses, that is, in the directions of the internal forces. These stresses are shown in Fig. 10-38d. The magnitudes of the stresses are determined as follows.

$$\sigma_x = \frac{P_A}{A} = \frac{200(10^3)}{(\pi/4)(0.100)^2} = 25.46(10^6)\text{N/m}^2 = 25.46 \text{ MPa}$$

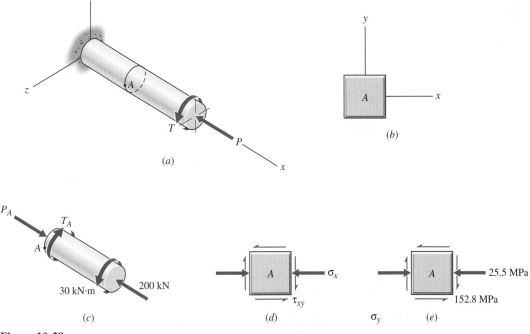

*(a)*

*(b)*

*(c)*      *(d)*      *(e)*

**Figure 10-38**

$$\tau_{xy} = \frac{T_A c}{J} = \frac{30(10^3)(0.050)}{(\pi/2)(0.050)^4} = 152.79(10^6)\text{N/m}^2 = 152.79 \text{ MPa}$$

The $x$- and $y$-components of stress on the element at $A$ are

$$\sigma_x = 25.46 \text{ MPa} \cong 25.5 \text{ MPa C} \qquad \textbf{Ans.}$$

$$\tau_{xy} = 152.79 \text{ MPa} \cong 152.8 \text{ MPa} \qquad \textbf{Ans.}$$

There is no normal stress in the $y$-direction (the circumferential direction), since there is no force to cause such a stress. The shear stresses on the four sides of the element have the same magnitude. The shear stresses that meet at a corner either both point toward the corner or both point away from the corner. These stresses are shown on the element of Fig. 10-38e. Note that point $A$ could be anywhere on the outside surface along the length of the shaft that is not in the vicinity of the fixed or loaded ends.

(b) Equation (10-4) is used to calculate the principal stresses. The stresses for use in this equation are

$$\sigma_x = -25.46 \text{ MPa} \qquad \sigma_y = 0 \qquad \tau_{xy} = -152.79 \text{ MPa}$$

Substituting these stress components into Eq. (10-4) yields

$$\sigma_{p1,\,p2} = \frac{\sigma_x + \sigma_y}{2} \pm \sqrt{\left(\frac{\sigma_x - \sigma_y}{2}\right)^2 + \tau_{xy}^2}$$

$$= \frac{-25.46 + 0}{2} \pm \sqrt{\left(\frac{-25.46 - 0}{2}\right)^2 + (-152.79)^2}$$

$$= -12.73 \pm 153.32$$

Thus, the principal stresses are

$$\sigma_{p1} = -12.73 + 153.32 = 140.59 \text{ MPa} \cong 140.6 \text{ MPa T} \qquad \textbf{Ans.}$$

$$\sigma_{p2} = -12.73 - 153.32 = -166.05 \text{ MPa} \cong 166.1 \text{ MPa C} \qquad \textbf{Ans.}$$

$$\sigma_{p3} = \sigma_z = 0 \qquad \textbf{Ans.}$$

Since the two in-plane principal stresses are of opposite sign and $\sigma_{p3} = \sigma_z = 0$,

$$\tau_{max} = \frac{\sigma_{max} - \sigma_{min}}{2}$$

$$= \frac{140.59 - (-166.05)}{2}$$

$$= 153.32 \text{ MPa} \cong 153.3 \text{ MPa} \quad \blacksquare \qquad \textbf{Ans.}$$

## Example Problem 10-13

The cast-iron frame of a small press is shaped as shown in Fig. 10-39a. The cross section $a$–$a$ of the frame is shown in Fig. 10-39b. For a load $Q$ of 16 kip, and assuming linearly elastic action, determine

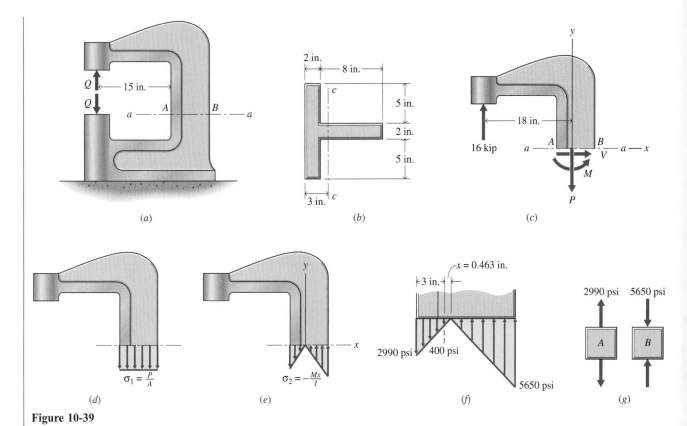

**Figure 10-39**

(a) The normal stress distribution on section $a–a$.

(b) The location of the neutral axis (line of zero stress).

(c) The principal stresses for the critical points on section $a–a$.

## SOLUTION

(a) The frame is sectioned at $a–a$, and a free-body diagram for the part of the frame above the section is shown in Fig. 10-39c. Since the load acts in a plane of symmetry, there are three independent equations of equilibrium. The internal forces at the section are found using the equations of equilibrium as follows:

$$+\rightarrow \Sigma F_x = 0: \qquad\qquad V = 0 \qquad V = 0$$

$$+\uparrow \Sigma F_y = 0: \qquad\qquad 16 - P = 0 \qquad P = 16 \text{ kip}$$

$$+\downarrow \Sigma M_{c-c} = 0: \qquad M - 16(18) = 0 \qquad M = 288 \text{ kip} \cdot \text{in.}$$

Thus, there are two internal forces on the section, an axial force $P$, which produces a constant normal stress $\sigma_1 = P/A$ over the section, and a bending moment $M$, which produces a linear variation of normal stress $\sigma_2 = -Mx/I$ over the section. The cross-sectional area $A$ and the second moment $I$ of the cross-sectional area with respect to the centroidal axis $c–c$ are

$$A = 12(2) + 8(2) = 40 \text{ in}^2$$

$$I = \frac{1}{12}(12)(2)^3 + 12(2)(2)^2 + \frac{1}{12}(2)(8)^3 + 2(8)(3)^2 = 333.3 \text{ in}^4$$

The stresses $\sigma_1$ and $\sigma_2$ due to the internal forces are

$$\sigma_1 = \frac{P}{A} = \frac{16{,}000}{40} = +400 \text{ psi} = 400 \text{ psi T}$$

The maximum tensile flexural stress occurs at the left edge of section $a$–$a$ (point $A$) and is

$$\sigma_{2A} = -\frac{Mx_A}{I} = -\frac{288(10^3)(-3)}{333.3} = +2592 \text{ psi} = 2592 \text{ psi T}$$

The maximum compressive flexural stress occurs at the right edge of section $a$–$a$ (point $B$) and is

$$\sigma_{2B} = -\frac{Mx_B}{I} = -\frac{288(10^3)(+7)}{333.3} = -6049 \text{ psi} = 6049 \text{ psi C}$$

The distributions of stresses $\sigma_1$ and $\sigma_2$ are shown in Figs. 10-39$d$ and $e$, respectively. Superimposing the normal stresses at points $A$ and $B$ gives

$$\sigma_A = \sigma_{1A} + \sigma_{2A} = 400 + 2592 = +2992 \text{ psi} \cong 2990 \text{ psi T}$$

$$\sigma_B = \sigma_{1B} + \sigma_{2B} = 400 - 6049 = -5649 \text{ psi} \cong 5650 \text{ psi C}$$

The distribution of normal stress on section $a$–$a$ is shown in Fig. 10-39$f$.

(b) The normal stress due to the axial load is constant across the section, and the normal stress due to bending varies from zero at the centroidal axis ($c$–$c$ on Fig. 10-38$b$) to a maximum at the edges of the "beam." The location of the neutral axis (the line of zero stress) is determined from

$$\sigma_y = \sigma_1 + \sigma_2 = \frac{P}{A} - \frac{Mx}{I} = \frac{16{,}000}{40} - \frac{288(10^3)x}{333.3} = 0$$

from which

$$x = \frac{16{,}000(333.3)}{40(288)(10^3)} = 0.463 \text{ in.} \qquad\qquad \textbf{Ans.}$$

That is, the neutral axis is 0.463 in. to the right of the centroidal axis of the cross section, at which point the compressive flexural stress just balances the axial tensile stress of 400 psi.

(c) With no shearing stresses on section $a$–$a$, the normal stresses are principal stresses, and the critical points, as observed from the stress distribution of Fig. 10-39$f$, are the left and right edges of the section.
The principal stresses for the right edge are

$$\sigma_p = 5650 \text{ psi C} \qquad \text{and} \qquad 0 \qquad\qquad \textbf{Ans.}$$

The principal stresses for the left edge are

$$\sigma_p = 2990 \text{ psi T} \quad \text{and} \quad 0 \qquad \textbf{Ans.}$$

The principal stresses are shown on elements $A$ and $B$ of Fig. 10-39g. ∎

## Example Problem 10-14

A 100-mm-diameter shaft is loaded and supported, as shown in Fig. 10-40a. Determine

(a) The normal and shearing stresses at points $A$, $B$, $C$, and $D$ on a section at the wall. Neglect stress concentrations.

(b) The principal stresses and maximum shearing stresses at points $A$, $B$, $C$, and $D$ of section $ABCD$.

## SOLUTION

A free-body diagram of the part of the shaft to the left of section $ABCD$ is shown in Fig. 10-40b. The internal forces acting on section $ABCD$ are shear forces $V_y$ and $V_z$, an axial force $P$, a torque $T = M_x$, and bending moments $M_y$ and $M_z$. The six equations of equilibrium used to determine these internal forces are

$$\Sigma F_x = 0: \qquad P - 150 = 0 \qquad P = 150 \text{ kN}$$

$$\Sigma F_y = 0: \qquad V_y - 5 = 0 \qquad V_y = 5 \text{ kN}$$

$$\Sigma F_z = 0: \qquad V_z = 0 \qquad V_z = 0$$

$$\Sigma M_x = 0: \qquad T + 5(0.600) = 0 \qquad T = -3 \text{ kN} \cdot \text{m}$$

$$\Sigma M_y = 0: \qquad M_y = 0 \qquad M_y = 0$$

$$\Sigma M_z = 0: \qquad M_z + 5(0.750) = 0 \qquad M_z = -3.75 \text{ kN} \cdot \text{m}$$

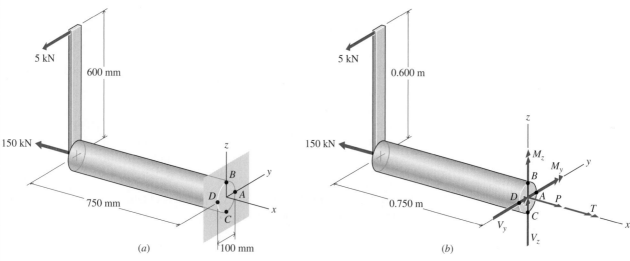

(a)    (b)

**Figure 10-40**

Thus, there are four internal forces on the section: (1) an axial force $P$ that produces a constant tensile stress over the section, (2) a shear force $V_y$ that produces shearing stresses at points $B$ and $C$ of the section but zero shearing stresses at points $A$ and $D$, (3) a torque $T$ that produces the same shearing stress at all surface points of the shaft, and (4) a bending moment $M_z$ that produces a tensile stress at point $A$ and an equal compressive stress at point $D$. The cross-sectional area $A$, the first moment $Q$, the second moment $I$, and the polar second moment $J$ are

$$A = \frac{\pi}{4}(d)^2 = \frac{\pi}{4}(100)^2 = 7854 \text{ mm}^2 = 7854(10^{-6}) \text{ m}^2$$

$$Q = \frac{\pi}{2}(r)^2\left(\frac{4r}{3\pi}\right) = \frac{2}{3}(r)^3 = \frac{2}{3}(50)^3 = 83.33(10^3) \text{ mm}^3 = 83.33(10^{-6}) \text{ m}^3$$

$$I = \frac{\pi}{4}(r)^4 = \frac{\pi}{4}(50)^4 = 4.909(10^6) \text{ mm}^4 = 4.909(10^{-6}) \text{ m}^4$$

$$J = \frac{\pi}{2}(r)^4 = \frac{\pi}{2}(50)^4 = 9.817(10^6) \text{ mm}^4 = 9.817(10^{-6}) \text{ m}^4$$

(a) The magnitude of the normal stress produced by axial force $P$ is

$$\sigma_1 = \frac{P}{A} = \frac{150(10^3)}{7854(10^{-6})} = 19.099(10^6)\text{N/m}^2 = 19.099 \text{ MPa}$$

The magnitude of the normal stress produced by moment $M_z$ is

$$\sigma_2 = \frac{M_z c}{I} = \frac{3.75(10^3)(50)(10^{-3})}{4.909(10^{-6})} = 38.195(10^6) \text{ N/m}^2 = 38.195 \text{ MPa}$$

The magnitude of the shearing stress produced by torque $T$ is

$$\tau_1 = \frac{Tc}{J} = \frac{3(10^3)(50)(10^{-3})}{9.817(10^{-6})} = 15.280(10^6) \text{ N/m}^2 = 15.280 \text{ MPa}$$

The magnitude of the shearing stress produced by shear force $V_y$ is

$$\tau_2 = \frac{V_y Q}{It} = \frac{5(10^3)(83.33)(10^{-6})}{4.909(10^{-6})(100)(10^{-3})} = 0.8487(10^6) \text{ N/m}^2 = 0.8487 \text{ MPa}$$

These component stresses are shown in Fig. 10-41 for points $A$, $B$, $C$, and $D$. The state of stress at each point is plane stress. Therefore, the superimposed states of stress for points $A$, $B$, $C$, and $D$ can be represented on two-dimensional stress elements, as shown in Figs. 10-42$a$, $b$, $c$, and $d$, respectively.                                                                          **Ans.**

(b) The principal stresses at each of the points is obtained by using Eq. (10-4). Thus, for point $A$ shown in Fig. 10-42$a$,

$$\sigma_{p1, p2} = \frac{\sigma_x + \sigma_z}{2} \pm \sqrt{\left(\frac{\sigma_x - \sigma_z}{2}\right)^2 + \tau_{xz}^2}$$

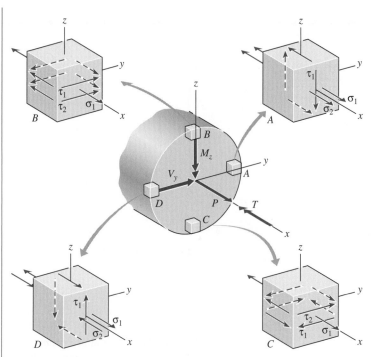

**Figure 10-41**

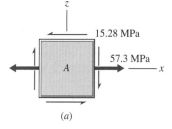

(a)

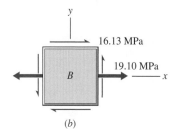

(b)

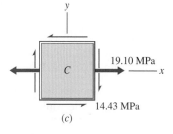

(c)

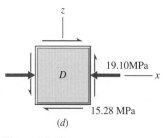

(d)

**Figure 10-42**

$$= \frac{57.3 + 0}{2} \pm \sqrt{\left(\frac{57.3 - 0}{2}\right)^2 + (-15.28)^2}$$

$$= 28.65 \pm 32.47$$

$$\sigma_{p1} = 28.65 + 32.47 = +61.12 \text{ MPa} \cong 61.1 \text{ MPa T}$$

$$\sigma_{p2} = 28.65 - 32.47 = -3.820 \text{ MPa} \cong 3.82 \text{ MPa C}$$

$$\sigma_{p3} = \sigma_y = 0$$

Since $\sigma_{p1}$ and $\sigma_{p2}$ have opposite signs, the maximum shearing stress is given by Eq. (10-7) as

$$\tau_{\max} = \frac{\sigma_{\max} - \sigma_{\min}}{2} = \frac{61.12 - (-3.820)}{2} = 32.47 \text{ MPa} \cong 32.5 \text{ MPa}$$

Proceeding in a similar fashion for the remaining points yields the principal stresses and the maximum shearing stress as

| Point | $\sigma_{p1}$ | $\sigma_{p2}$ | $\tau_{\max}$ | |
|-------|---------------|---------------|---------------|------|
| A | 61.6 MPa T | 3.82 MPa C | 32.5 MPa | **Ans.** |
| B | 28.3 MPa T | 9.20 MPa C | 18.75 MPa | **Ans.** |
| C | 26.9 MPa T | 7.75 MPa C | 17.30 MPa | **Ans.** |
| D | 8.47 MPa T | 27.6 MPa C | 18.02 MPa | **Ans.** |

This example problem illustrates that it may be necessary to determine stresses at a number of points in order to locate the most severely stressed point. ∎

# PROBLEMS

## Introductory Problems

**10-119\*** A 4-in-diameter shaft is subjected to both a torque of 30 kip · in. and an axial tensile load of 50 kip, as shown in Fig. P10-119. Determine the principal stresses and the maximum shearing stress at point $A$ on the surface of the shaft.

**Figure P10-119**

**10-120\*** A hollow shaft with an outside diameter of 400 mm and an inside diameter of 300 mm is subjected to both a torque of 350 kN · m and an axial tensile load of 1500 kN, as shown in Fig. P10-120. Determine the principal stresses and the maximum shearing stress at a point on the outside surface of the shaft.

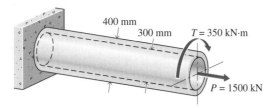

**Figure P10-120**

**10-121** A 2-in-diameter shaft is used in an aircraft engine to transmit 360 hp at 1500 rpm to a propeller that develops a thrust of 2800 lb. Determine the principal stresses and the maximum shearing stress produced at any point on the outside surface of the shaft.

**10-122** A 60-mm-diameter shaft must transmit a torque of unknown magnitude while it is supporting an axial tensile load of 150 kN. Determine the maximum allowable value for the torque if the tensile principal stress at a point on the outside surface of the shaft must not exceed 125 MPa.

**10-123\*** The T-section shown in Fig. P10-123 is used as a short post to support a compressive load $P = 150$ kip. The load is applied on the centerline of the stem at a distance $e = 2$ in. from the centroid of the cross section. Determine the normal stresses at points $C$ and $D$ on section $AB$.

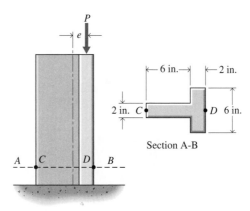

**Figure P10-123**

**10-124\*** The cross section of the straight vertical portion of the coil-loading hook shown in Fig. P10-124a is shown in Fig. P10-124b. The horizontal distance from the line of action of the applied load to the inside face $CD$ of the cross section is 600 mm. Determine the maximum tensile and compressive stresses on section $CDEF$ for a 40-kN load.

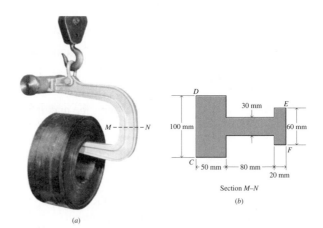

**Figure P10-124**

**10-125** The estimated maximum total force $P$ to be exerted on the C-clamp shown in Fig. P10-125 is 450 lb. If the normal stress on section $A$–$A$ is not to exceed 16,000 psi, determine the minimum allowable value for the dimension $h$ of the cross section.

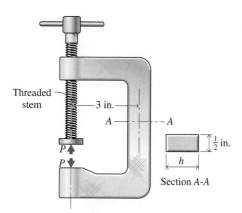

Figure P10-125

## Intermediate Problems

**10-126\*** A 150-mm-diameter shaft will be used to support the axial load and torques shown in Fig. P10-126. Determine the principal stresses and the maximum shearing stress at point *A* on the surface of the shaft.

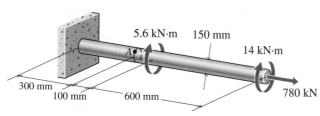

Figure P10-126

**10-127\*** The thin-walled cylindrical pressure vessel shown in Fig. P10-127 has an inside diameter of 24 in. and a wall thickness of $^{1}/_{2}$ in. The vessel is subjected to an internal pressure of 250 psi. In addition, a torque of 150 kip · ft is applied to the vessel through rigid plates on the ends of the vessel. Determine the maximum normal and shearing stresses at a point on the outside surface of the vessel.

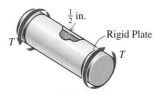

Figure P10-127

**10-128** A steel shaft is loaded and supported as shown in Fig. P10-128. If the maximum shearing stress in the shaft must not exceed 55 MPa and the maximum ten-

sile stress in the shaft must not exceed 83 MPa, determine the maximum torque *T* that can be applied to the shaft.

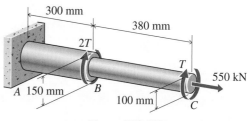

Figure P10-128

**10-129** A 30-lb force *P* is applied to the brake pedal of an automobile as shown in Fig. P10-129. Force *Q* is applied to the brake cylinder. Determine the maximum tensile and compressive normal stresses on section *a–a*, which is midway between points *A* and *B*. Section *a–a* may be modeled as a $^{3}/_{16} \times$ 1-in. rectangle.

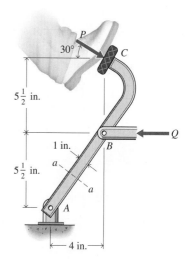

Figure P10-129

**10-130\*** An automobile engine with a mass of 360 kg is supported by an engine hoist, as shown in Fig. P10-130. Determine the maximum tensile and compressive normal stresses on section *a–a* if member *ABC* is a hollow square 100- $\times$ 100-mm section with a wall thickness of 20 mm.

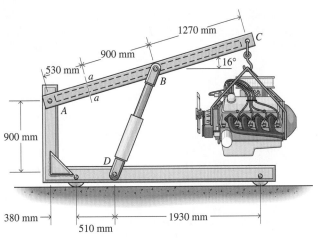

Figure P10-130

**10-131\*** A short post supports a vertical force $P = 9600$ lb and a horizontal force $H = 800$ lb, as shown in Fig. P10-131. Determine the vertical normal stresses at corners $A$, $B$, $C$, and $D$ of the post. Neglect stress concentrations.

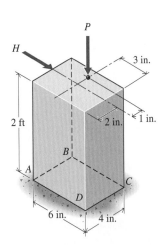

Figure P10-131

**10-132** Determine the vertical normal stresses at points $A$, $B$, $C$, and $D$ of the rectangular post shown in Fig. P10-132. Neglect stress concentrations.

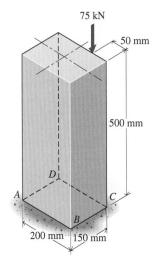

Figure P10-132

**10-133** Determine the principal stresses and the maximum shearing stress at points on the top and bottom of section $a$–$a$ of the pipe system shown in Fig. P10-133. The pipe has an outside diameter of 1 in. and a wall thickness of $\frac{1}{8}$ in.

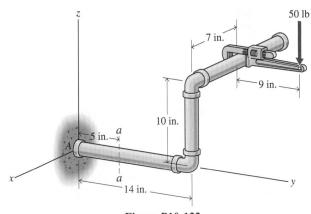

Figure P10-133

**10-134** The output from a strain gage located on the bottom surface of the hat section shown in Fig. P10-134 will be used to indicate the magnitude of the load $P$ applied to the section. The hat section is made of aluminum alloy ($E = 73$ GPa and $v = \frac{1}{3}$ and is 24 mm wide. When the maximum load $P = 500$ N is applied to the section, the strain gage should read $\epsilon = +1000$ $\mu$m/m. Plot a curve showing the combinations of thickness $t$ and height $h$ that will satisfy the specification. Limit the range of $h$ from 0 to 50 mm.

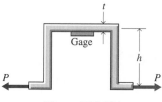

**Figure P10-134**

**10-135*** A solid shaft 4 in. in diameter is acted on by forces $P$ and $Q$, as shown in Fig. P10-135. Determine the principal stresses and the maximum shearing stress at point $A$ on the surface of the shaft.

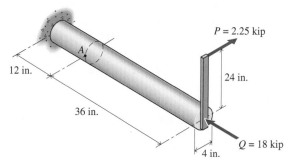

**Figure P10-135**

**Challenging Problems**

**10-136*** A thin-walled cylindrical pressure vessel with an inside diameter of 1.2 m is fabricated by butt-welding 15-mm plate with a spiral seam as shown in Fig. P10-136. The pressure in the tank is 2500 kPa. Additional loads are applied to the cylinder through a rigid end plate as shown in Fig. P10-136. Determine

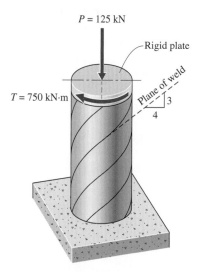

**Figure P10-136**

(a) The normal and shearing stresses on the plane of the weld at a point on the outside surface of the tank.
(b) The principal stresses and the maximum shearing stress at a point on the inside surface of the tank.

**10-137*** A 1-in-diameter steel ($E = 30,000$ ksi and $v = 0.30$) bar is subjected to a tensile load $P$ and a torque $T$, as shown in Fig. P10-137. Determine the axial load $P$ and the torque $T$ if the strains indicated by gages $a$ and $b$ on the bar are $\epsilon_a = +1084$ $\mu$in./in. and $\epsilon_b = -754$ $\mu$in./in.

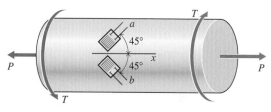

**Figure P10-137**

**10-138** A bag of potatoes is resting on a chair, as shown in Fig. P10-138. The force exerted by the potatoes on the frame at one side of the chair is equivalent to horizontal and vertical forces of 24 N and 84 N, respectively, at $E$ and a force of 28 N perpendicular to member $BH$ at $G$. Determine the maximum tensile and compressive normal stresses on a section midway between pins $C$ and $F$. Member $BH$ has a 10- $\times$ 30-mm cross section; the 30-mm dimension lies in the plane of the page.

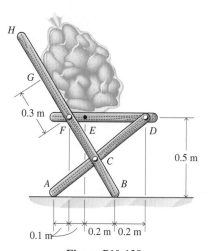

**Figure P10-138**

**10-139** Three strain gages are mounted on a 1-in.-diameter aluminum alloy rod ($E = 10,000$ ksi and $v = \frac{1}{3}$), as shown in Fig. P10-139. When loads $P$ and $Q$ are applied to the rod, they produce longitudinal strains $\epsilon_A = +550$ $\mu$in./in., $\epsilon_B = +400$ $\mu$in./in., and $\epsilon_C = -300$ $\mu$in./in. Determine the magnitudes of loads $P$ and $Q$ and the location $x$ of load $P$.

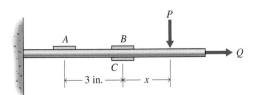

**Figure P10-139**

**10-140*** A steel shaft 120 mm in diameter is supported in flexible bearings at its ends. Two pulleys each 500 mm in diameter are keyed to the shaft. The pulleys carry belts that produce the forces shown in Fig. P10-140. Determine the principal stresses and the maximum shearing stress at point $A$ on the top surface of the shaft.

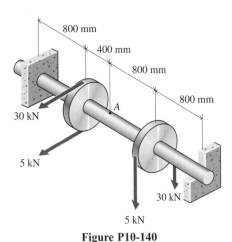

**Figure P10-140**

**10-141*** A 120-lb girl is walking up a uniform 2- $\times$ 12-in. beam of negligible weight, as shown in Fig. P10-141. The coefficient of friction is 0.20 at all surfaces. When the beam begins to slip, determine the maximum tensile and compressive normal stresses on a section at a distance $x/2$ from the right end of the beam.

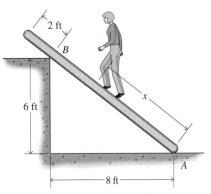

**Figure P10-141**

**10-142** Four strain gages are mounted at 90° intervals around the circumference of a 100-mm-diameter steel ($E = 200$ GPa and $v = 0.30$) shaft, as shown in Fig. P10-142. All of the gages are oriented at an angle of 45° with respect to a line on the surface of the shaft that is parallel to the axis of the shaft. Determine the torque $T$, shear force $V$, and bending moment $M$ at the cross section where the gages are located if $\epsilon_A = +450$ $\mu$m/m, $\epsilon_B = +325$ $\mu$m/m, $\epsilon_C = +550$ $\mu$m/m, and $\epsilon_D = +675$ $\mu$m/m.

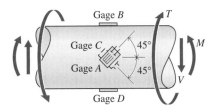

**Figure P10-142**

**10-143** A 4-in.-diameter solid circular steel shaft is loaded and supported as shown in Fig. P10-143. Because of discontinuities at points $A$ and $B$, the normal and shearing stresses at these points must be limited to 13 ksi T and 8.5 ksi, respectively. Determine the maximum permissible value for the load $P$.

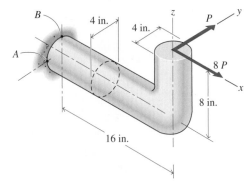

**Figure P10-143**

**10-144** The frame shown in Fig. P10-144 is constructed of 100- × 100-mm timbers. Determine and show on sketches the principal and maximum shearing stresses at points $G$ and $H$.

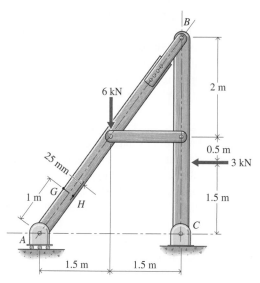

**Figure P10-144**

**Computer Problems**

**10-145** An aircraft engine is designed to transmit 360 hp at 1500 rpm to a propeller that develops a thrust of 2800 lb. Calculate and plot the axial normal stress $\sigma_x$, the torsional shear stress $\tau_{xy}$, the principal normal stress $\sigma_{p1}$, and the maximum shearing stress $\tau_p$ at a point on the outside surface of the propeller shaft as a function of the shaft diameter $d$ (0.75 in. $\leq d \leq$ 2.50 in.).

**10-146** A 120-mm-diameter shaft is used to transmit an axial load of 250 kN and a torque of 20 kN · m. It is proposed to replace the solid shaft with a hollow shaft having the same weight (the same cross-sectional area). Calculate and plot the axial normal stress $\sigma_x$, the torsional shear stress $\tau_{xy}$, the principal normal stress $\sigma_{p1}$, and the maximum shearing stress $\tau_p$ at a point on the outside surface of the propeller shaft as a function of the outside diameter $d_o$ (120 mm $\leq d_o \leq$ 300 mm) of the shaft.

**10-147** As the C-clamp shown in Fig. P10-147 is tightened, the web of the C-section is subjected to flexural stresses as well as axial tensile stresses. If the web of the section has a width of 1 in. and a thickness of ¼ in., and the clamp has a capacity of $d = 3$ in.,

(a) Compute and plot the maximum tensile and compressive stresses on section $a - a$ as a function of the clamp force $P$ (0 lb $\leq P \leq$ 500 lb).

(b) Determine $a$, the distance from the inside edge of the web to the point where the normal stress is zero. Is $a$ a function of $P$?

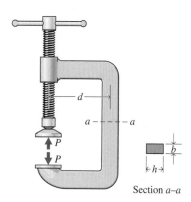

Section $a$–$a$

**Figure P10-147**

**10-148** A 1200-mm-diameter cylindrical pressure tank is fabricated by butt-welding 20-mm plate with a spiral seam as shown in Fig. P10-148. An axial load of 130 kN is applied to the end of the tank through a rigid bearing plate. If the seam angle is $\theta = 37°$, compute and plot the normal stress perpendicular to the weld and the shearing stress parallel to the weld as functions of the internal pressure $p$ (10 kPa $\leq p \leq$ 2800 kPa).

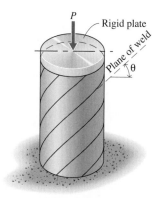

**Figure P10-148**

# 10-15 DESIGN: DUCTILE MATERIALS

A tension test of an axially loaded member is easy to conduct, and the results, for many types of materials, are well known. When such a member fails, the failure occurs at a specific principal (axial) stress, a definite axial strain, a maximum shearing stress of half the axial stress, and a specific amount of strain energy per unit volume of stressed material. Since all of these limits are reached simultaneously for an axial load, it makes no difference which criterion (stress, strain, or energy) is used for predicting failure in another axially loaded member of the same material. Thus, there is only one value of stress and only one value of strength.

For an element subjected to biaxial or triaxial loading, however, the situation is more complicated because the limits of normal stress, normal strain, shearing stress, and strain energy existing at failure for an axial load are not all reached simultaneously. In other words, the cause of failure, in general, is unknown. In such cases, it becomes important to determine the best criterion for predicting failure, because test results are difficult to obtain and the combinations of loads are endless. Three theories have been proposed for predicting failure of various types of material subjected to many combinations of loads. Unfortunately, none of the theories agree with test data for all types of materials and combinations of loads. Three of the more common theories of failure for ductile materials are presented and briefly explained in the following sections.

**Maximum-Normal-Stress Theory**[6] The maximum-normal-stress theory predicts failure of a specimen subjected to any combination of loads when the maximum normal stress at any point in the specimen reaches the axial failure stress as determined by an axial tensile or compressive test of the same material.

The maximum-normal-stress theory is presented graphically in Fig. 10-43b for an element subjected to biaxial principal stresses in the $p1$ and $p2$ directions, as shown in Fig. 10-43a. The limiting stress $\sigma_f$ is the failure stress for this material when loaded axially and is assumed equal in tension and compression. Any combination of biaxial principal stresses $\sigma_{p1}$ and $\sigma_{p2}$ represented by a point inside the square of Fig. 10-43b is safe according to this theory, whereas any combination of stresses represented by a point outside of the square will cause failure of the element on the basis of this theory.

Once the principal stresses have been found, they may be ordered $\sigma_{p1} > \sigma_{p2} > \sigma_{p3}$. For a state of biaxial stress (plane stress), one of the principal stresses is zero (assumed to be $\sigma_{p3}$ in Fig. 10-43a). If the principal stress with the largest magnitude is tension, the maximum-normal-stress theory predicts failure when

$$\sigma_{p1} = \sigma_f \tag{10-26a}$$

where $\sigma_f$ is the failure stress for uniaxial tensile loading. When the principal stress with the largest magnitude is compression, the maximum-normal-stress theory predicts failure when

$$\sigma_{p3} = -\sigma_f \tag{10-26b}$$

---

[6]Often called Rankine's theory after W. J. M. Rankine (1820–1872), an eminent engineering educator in England.

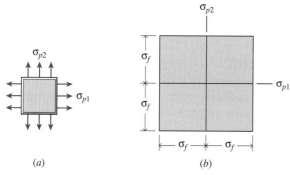

(a)

(b)

**Figure 10-43**

where $\sigma_f$ is the failure stress for uniaxial compressive loading. This failure theory will be compared to experimental data for several materials later in this section.

## Maximum-Shear-Stress Theory[7]

The maximum-shear-stress theory predicts failure of a specimen subjected to any combination of loads when the maximum shear stress at any point reaches the failure stress $\tau_f$ equal to $\sigma_f/2$, as determined by an axial tension or axial compression test of the same material. For ductile materials the shearing elastic limit, as determined from a torsion test (pure shear), is greater than half the tensile elastic limit (with an average value of $\tau_f$ about 0.57 $\sigma_f$). This means that the maximum-shear-stress theory errs on the conservative side by being based on the limit obtained from an axial test.

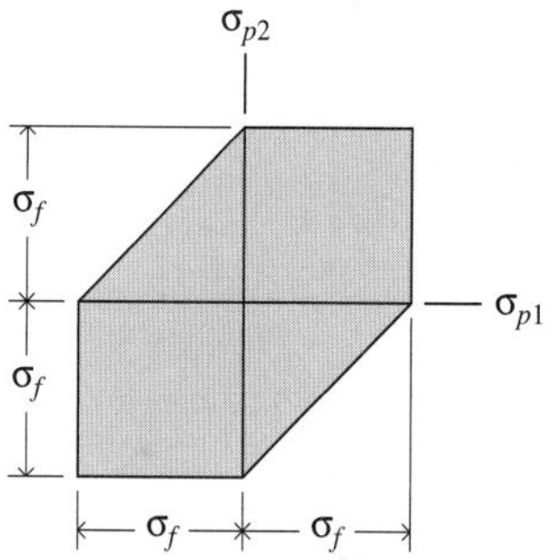

The maximum-shear-stress theory is presented graphically in Fig. 10-44 for an element subjected to biaxial ($\sigma_{p3}$ is equal to zero) principal stresses, as shown in Fig. 10-43a. In the first and third quadrants, $\sigma_{p1}$ and $\sigma_{p2}$ have the same sign, and the maximum shearing stress is half the numerically larger principal stress $\sigma_{p1}$ or $\sigma_{p3}$, as explained in Section 10-5. According to the maximum-shear-stress theory, failure occurs when

$$\tau_{\max} = \tau_f$$

where $\tau_f$ is the failure stress determined from a uniaxial tension or compression test, $\tau_f = \sigma_f/2$. In quadrants one or three

$$\tau_{\max} = \frac{\sigma_{\max} - \sigma_{\min}}{2}$$

$$= \frac{\sigma_{p1}}{2} \quad \text{(First quadrant)} \qquad (10\text{-}27a)$$

$$= \frac{\sigma_{p3}}{2} \quad \text{(Third quadrant)}$$

**Figure 10-44**

---

[7]Sometimes called Coulomb's theory because it was originally stated by him in 1773. More frequently called Guests's theory or law because of the work of J. J. Guest in England in 1900.

where $\sigma_{p1} > \sigma_{p2} > \sigma_{p3}$. The maximum-shear-stress theory predicts failure in the first and third quadrants when

$$\sigma_{\max} = \sigma_f \tag{10-27b}$$

and is the same as the maximum-normal-stress theory of failure.

In the second and fourth quadrants, where $\sigma_{p1}$ and $\sigma_{p2}$ are of an opposite sign, the maximum shearing stress is half the arithmetical sum of the two principal stresses. In the fourth quadrant, the equation of the boundary, or limit stress, line is

$$\sigma_{p1} - \sigma_{p2} = \sigma_f \tag{10-27c}$$

and in the second quadrant the relation is

$$\sigma_{p1} - \sigma_{p2} = -\sigma_f \tag{10-27d}$$

A comparison of this theory with experimental data will be presented later in this section.

### Maximum-Distortion-Energy Theory[8]  The maximum-distortion-energy theory predicts failure of a specimen subjected to any combination of loads when the distortion component of the strain energy intensity of any portion of the stressed member reaches the failure value of the distortion component of the strain energy intensity as determined from an axial tension or compression test of the same material. This theory assumes that the portion of the strain energy producing volume change is ineffective in causing failure by yielding. Supporting evidence comes from experiments showing that homogeneous materials can withstand very high hydrostatic stresses without yielding. The portion of the strain energy producing the element's change of shape is assumed to be completely responsible for the failure of the material by inelastic action.

The strain energy of distortion is most readily computed by determining the total strain energy of the stressed material and subtracting the strain energy corresponding to the volume change. Strain energy per unit volume (strain energy intensity) is the area under the stress-strain curve, $u = \sigma\epsilon/2$ for the linear region of the $\sigma$–$\epsilon$ curve. This expression can also be written

$$u = \frac{\sigma\epsilon}{2} = \frac{\sigma^2}{2E}$$

where $u$ is the strain energy per unit volume (strain energy intensity) and $\sigma$ and $\epsilon$ are the slowly applied axial stress and strain. This equation assumes that the stress does not exceed the proportional limit.

When an elastic element is subjected to triaxial loading, the stresses can be resolved into three principal stresses such as $\sigma_{p1}$, $\sigma_{p2}$, and $\sigma_{p3}$, where $p1$, $p2$, and $p3$ are the principal axes. These stresses will be accompanied by three principal strains related to the stresses by Eq. (10-17) of Section 10-11. If it is assumed that

[8]Frequently called the Huber–Hencky–von Mises theory, because it was proposed by M. T. Huber of Poland in 1904 and independently by R. von Mises of Germany in 1913. The theory was further developed by H. Hencky and von Mises in Germany and the United States.

the loads are applied simultaneously and gradually, the stresses and strains will increase in the same manner. The total strain energy per unit volume is the sum of the energies produced by each of the stresses (energy is a scalar quantity and can be added algebraically regardless of the directions of the individual stresses); thus,

$$u = \frac{1}{2}(\sigma_{p1}\epsilon_{p1} + \sigma_{p2}\epsilon_{p2} + \sigma_{p3}\epsilon_{p3})$$

When the strains are expressed in terms of the stresses, this equation becomes

$$u = \frac{1}{2E}[\sigma_{p1}^2 + \sigma_{p2}^2 + \sigma_{p3}^2 - 2v(\sigma_{p1}\sigma_{p2} + \sigma_{p2}\sigma_{p3} + \sigma_{p3}\sigma_{p1})]$$

The strain energy can be resolved into two components $u_v$ and $u_d$, resulting from a volume change and a distortion, respectively, by considering the principal stresses to be made up of two sets of stresses, as indicated in Figs. 10-45a, b, and c. The state of stress in Fig.10-45c will result in distortion only (no volume change) if the sum of the three normal strains is zero. That is,

$$E[\epsilon_{p1} + \epsilon_{p2} + \epsilon_{p3}]_d = [(\sigma_{p1} - p) - v(\sigma_{p2} + \sigma_{p3} - 2p)]$$
$$+ [(\sigma_{p2} - p) - v(\sigma_{p3} + \sigma_{p1} - 2p)]$$
$$+ [(\sigma_{p3} - p) - v(\sigma_{p1} + \sigma_{p2} - 2p)] = 0$$

which reduces to

$$(1 - 2v)(\sigma_{p1} + \sigma_{p2} + \sigma_{p3} - 3p) = 0$$

Therefore,

$$p = \frac{1}{3}(\sigma_{p1} + \sigma_{p2} + \sigma_{p3})$$

The three normal strains due to $p$ are, from Eq. (10-17),

$$\epsilon_v = (1 - 2v)\frac{p}{E}$$

and the energy resulting from the hydrostatic stress (the volume change) is

$$u_v = 3\left(\frac{p\epsilon_v}{2}\right) = \frac{3}{2}\frac{1 - 2v}{E}p^2$$

$$= \frac{1 - 2v}{6E}[\sigma_{p1} + \sigma_{p2} + \sigma_{p3}]^2$$

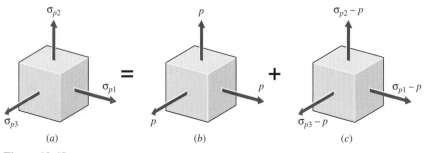

(a)        (b)        (c)

**Figure 10-45**

The energy resulting from the distortion (change of shape) is

$$u_d = u - u_v$$

$$= \frac{1}{6E}[3(\sigma_{p1}^2 + \sigma_{p2}^2 + \sigma_{p3}^2) - 6v(\sigma_{p1}\sigma_{p2} + \sigma_{p2}\sigma_{p3} + \sigma_{p3}\sigma_{p1})$$

$$-(1 - 2v)(\sigma_{p1} + \sigma_{p2} + \sigma_{p3})^2]$$

When the third term in the brackets is expanded, the expression can be rearranged to give

$$u_d = \frac{1 + v}{6E}[(\sigma_{p1}^2 - 2\sigma_{p1}\sigma_{p2} + \sigma_{p2}^2) + (\sigma_{p2}^2 - 2\sigma_{p2}\sigma_{p3} + \sigma_{p3}^2)$$

$$+(\sigma_{p3}^2 - 2\sigma_{p3}\sigma_{p1} + \sigma_{p1}^2)]$$

$$= \frac{1 + v}{6E}[(\sigma_{p1} - \sigma_{p2})^2 + (\sigma_{p2} - \sigma_{p3})^2 + (\sigma_{p3} - \sigma_{p1})^2] \qquad (a)$$

The maximum-distortion-energy theory of failure assumes that inelastic action will occur whenever the energy given by Eq. (a) exceeds the limiting value obtained from a tensile test. For this test, only one of the principal stresses will be nonzero. If this stress is called $\sigma_f$, the value of $u_d$ becomes

$$(u_d)_f = \frac{1 + v}{3E}\sigma_f^2$$

and when this value is substituted in Eq. (a), it becomes

$$2\sigma_f^2 = (\sigma_{p1} - \sigma_{p2})^2 + (\sigma_{p2} - \sigma_{p3})^2 + (\sigma_{p3} - \sigma_{p1})^2 \qquad (b)$$

for failure by yielding.

When a state of plane stress exists, assuming $\sigma_{p3}$ equals zero, Eq. (b) becomes

$$\sigma_{p1}^2 - \sigma_{p1}\sigma_{p2} + \sigma_{p2}^2 = \sigma_f^2 \qquad (10\text{-}28)$$

This last expression is the equation of an ellipse with its major axis along the line $\sigma_{p1}$ equals $\sigma_{p2}$, as shown in Fig. 10-46. A comparison of this theory with experimental data will be presented later in this section.

The graphic representation of Figs. 10-43, 10-44, and Fig. 10-46 are superimposed in Fig. 10-47 for convenient comparison of the different theories. The

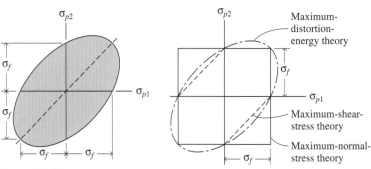

**Figure 10-46**          **Figure 10-47**

failure theories are compared to experimental data in Fig. 10-48.[9] In the first quadrant, failure is predicted most closely by the maximum-distortion-energy theory. The maximum-normal-stress and maximum-shear-stress theories are identical in the first quadrant; both theories are conservative because the experimental data lie outside the prediction envelope.

In the fourth quadrant, the principal stresses have opposite signs. The best correlation between theory and experimental data is for the maximum-distortion-energy theory. In this quadrant, the maximum-shear-stress theory is conservative because the experimental data lie outside the prediction envelope. The prediction envelope for the maximum-normal-stress theory lies outside the experimental data in the fourth quadrant; therefore, the theory overpredicts failure stresses and is unsafe.

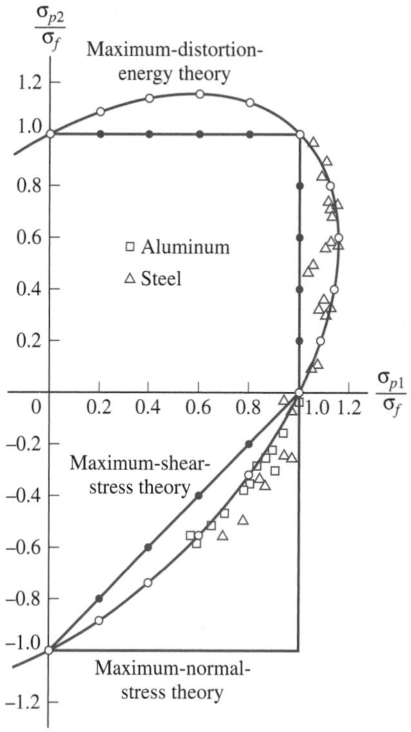

**Figure 10-48**

## Example Problem 10-15

At a point in a structural member subjected to plane stress, the state of stress is as shown in Fig. 10-49a. Determine which, if any, of the theories of failure will predict failure by yielding for this state of stress. The yield strength of the material in tension and compression is 36 ksi.

### SOLUTION

In order to apply the theories of failure, the principal stresses must first be determined. The given stresses for use in Eq. (10-4) are

$$\sigma_x = +10{,}000 \text{ psi} \qquad \sigma_y = -8000 \text{ psi} \qquad \tau_{xy} = -4000 \text{ psi}$$

When these values are substituted in Eq. (10-4), the principal stresses are found to be

$$\sigma_{p1,\,p3} = \frac{\sigma_x + \sigma_y}{2} \pm \sqrt{\left(\frac{\sigma_x - \sigma_y}{2}\right)^2 + \tau_{xy}^2}$$

$$= \frac{10{,}000 + (-8000)}{2} \pm \sqrt{\left(\frac{10{,}000 - (-8000)}{2}\right)^2 + (-4000)^2}$$

$$= 1000 \pm 9849$$

$$\sigma_{p1} = 1000 + 9849 = 10{,}849 \text{ psi} \cong 10{,}850 \text{ psi T}$$

$$\sigma_{p2} = \sigma_z = 0$$

$$\sigma_{p3} = 1000 - 9849 = -8849 \text{ psi} \cong 8850 \text{ psi C}$$

where the principal stresses have been ordered such that $\sigma_{p1} > \sigma_{p2} > \sigma_{p3}$. Thus, the largest principal stress is 10,850 psi tension and the smallest principal stress is 8850 psi compression.

---

[9]*Handbook of Mechanics, Materials, and Structures*, Alexander Blake, Editor, Wiley-Interscience, New York, 1985.

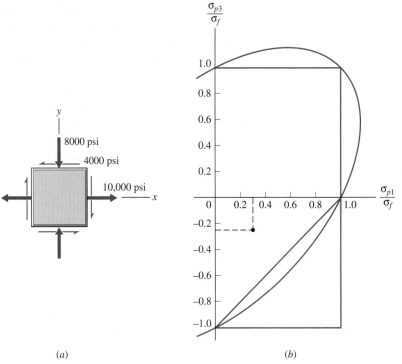

(a)                                   (b)

**Figure 10-49**

The maximum-normal-stress theory predicts that failure will *not* occur if

$$\sigma_{p1} < \sigma_f$$

Since

$$\sigma_{p1} = 10{,}850 \text{ psi} < \sigma_f = 36{,}000 \text{ psi}$$

failure will not occur according to the maximum-normal-stress theory.

The maximum-shear-stress theory of failure predicts that the state of stress shown in Fig. 10-49*a* is safe if

$$\tau_{\max} < \tau_f$$

Since the principal stresses have opposite signs,

$$\tau_{\max} = \frac{\sigma_{\max} - \sigma_{\min}}{2} = \frac{10{,}850 - (-8850)}{2} = 9850 \text{ psi}$$

The failure stress (yield strength) is 36,000 psi; therefore,

$$\tau_f = \frac{\sigma_f}{2} = \frac{36{,}000}{2} = 18{,}000 \text{ psi}$$

Since

$$\tau_{\max} = 9850 \text{ psi} < \tau_f = 18{,}000 \text{ psi}$$

failure will not occur according to the maximum-shear-stress theory.

The maximum-distortion-energy theory of failure states that the state of stress shown in Fig. 10-49a is safe if

$$\sigma_{p1}^2 - \sigma_{p1}\sigma_{p3} + \sigma_{p3}^2 < \sigma_f^2$$
$$(10,850)^2 - 10,850(-8850) + (-8850)^2 = (17,090)^2 < (36,000)^2$$

Thus, failure will not occur according to the maximum-distortion-energy theory of failure.

Alternatively, the ratios $\sigma_{p1}/\sigma_f = 10,850/36,000 = 0.301$ and $\sigma_{p3}/\sigma_f = -8850/36,000 = -0.246$ can be plotted on Fig. 10-48, and compared to the failure envelopes. The result is shown in Fig. 10-49b. Clearly, the state of stress shown in Fig. 10-49a is within all of the failure envelopes; therefore, yielding of the structural member will not occur. ■

## Example Problem 10-16

The solid circular shaft of Fig. 10-50a has a proportional limit of 64 ksi. Determine the value of the load $R$ for failure by yielding as predicted by each of the theories of failure. Assume that point $A$ is the most severely stressed point.

### SOLUTION

Passing a section through point $A$, drawing a free-body diagram of the portion of the body to the right of the section (see Fig. 10-50b), and applying the equations of equilibrium yields the following internal reactions if $R$ is expressed in kip

$$P = 12R \text{ kip} \qquad T = 24R \text{ kip} \cdot \text{in}$$
$$V = R \text{ kip} \qquad M = 36R \text{ kip} \cdot \text{in}$$

The shearing stress at $A$ due to the shear force $V$ is zero because $Q = 0$. The flexural stress at $A$ is

$$\sigma_1 = \frac{Mc}{I} = \frac{36R(2)}{\pi(4)^4/64} = \frac{18R}{\pi} \text{ ksi T}$$

The axial stress at $A$ is

$$\sigma_2 = \frac{P}{A} = \frac{12R}{\pi(4)^2/4} = \frac{3R}{\pi} \text{ ksi T}$$

The sum of the two tensile normal stresses at $A$ becomes

$$\sigma = \sigma_1 + \sigma_2 = \frac{21R}{\pi} \text{ ksi T}$$

The torsional shearing stress at $A$ is

$$\tau = \frac{Tc}{J} = \frac{24R(2)}{\pi(4)^4/32} = \frac{6R}{\pi} \text{ ksi}$$

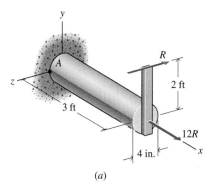

(a)

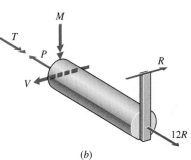

(b)

**Figure 10-50**

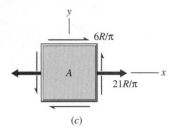

(c)

The stresses on an element at $A$ are shown in Fig. 10-50c. Substituting the stresses $\sigma_x = 21R/\pi$ ksi, $\sigma_y = 0$, and $\tau_{xy} = 6R/\pi$ ksi into the principal stress equation [Eq. (10-4)] gives

$$\sigma_{p1,\,p3} = \frac{\sigma_x + \sigma_y}{2} \pm \sqrt{\left(\frac{\sigma_x - \sigma_y}{2}\right)^2 + \tau_{xy}^2}$$

$$= \frac{21R/\pi + 0}{2} \pm \sqrt{\left(\frac{21R/\pi - 0}{2}\right)^2 + (6R/\pi)^2}$$

$$= \frac{10.5R}{\pi} \pm \frac{12.09R}{\pi}$$

Ordering the principal stresses such that $\sigma_{p1} > \sigma_{p2} > \sigma_{p3}$ gives

$$\sigma_{p1} = \frac{22.59R}{\pi} \text{ ksi} \qquad \sigma_{p2} = 0 \qquad \sigma_{p3} = -\frac{1.59R}{\pi} \text{ ksi}$$

Since the principal stresses have opposite signs,

$$\tau_{max} = \frac{\sigma_{max} - \sigma_{min}}{2} = \frac{22.59R/\pi - (-1.59R/\pi)}{2} = \frac{12.09R}{\pi} \text{ ksi}$$

According to the maximum-normal-stress theory

$$\sigma_{p1} = \frac{22.59R}{\pi} = \sigma_f = 64 \text{ ksi}$$

from which

$$R = \frac{64\pi}{22.59} = 8.90 \text{ kip} \qquad \textbf{Ans.}$$

According to the maximum-shear-stress theory

$$\tau_{max} = \frac{12.09R}{\pi} = \tau_f = \frac{\sigma_f}{2} = \frac{64}{2} \text{ ksi}$$

from which

$$R = \frac{32\pi}{12.09} = 8.315 \text{ kip} \cong 8.32 \text{ kip} \qquad \textbf{Ans.}$$

According to the maximum-distortion-energy theory

$$\sigma_{p1}^2 - \sigma_{p1}\sigma_{p3} + \sigma_{p3}^2 = \sigma_f^2$$

$$\left(\frac{22.59R}{\pi}\right)^2 - \frac{22.59R}{\pi}\left(-\frac{1.59R}{\pi}\right) + \left(-\frac{1.59R}{\pi}\right)^2 = (64)^2$$

which gives

$$R = \frac{64\pi}{23.43} = 8.581 \text{ kip} \cong 8.58 \text{ kip} \qquad \textbf{Ans.}$$

In this example, the maximum-shear-stress theory gives the most conservative result. This is in agreement with the results shown in Fig. 10-48. It would not be wise to use the result from the maximum-normal-stress theory, since it is not safe when the principal stresses have opposite signs (see Fig. 10-48). ∎

## Example Problem 10-17

A thin-walled cylindrical pressure vessel is constructed of a ductile material with a yield strength of 36 ksi. The internal diameter of the vessel is 4 ft, and the internal pressure in the vessel is 100 psi. Determine the minimum required wall thickness based on the maximum-shear-stress theory of failure, if the factor of safety is 2.

### SOLUTION

A point on the outside of the vessel is subjected to a biaxial state of stress, whereas a point on the inside is subjected to a triaxial state of stress. In either case, the hoop stress is given by Eq. (10-25a) as

$$\sigma_h = \frac{pr}{t} = \frac{100(2)(12)}{t} = \frac{2400}{t}$$

and the axial stress is

$$\sigma_a = \frac{\sigma_h}{2} = \frac{1200}{t}$$

Neither the inside point nor the outside point are subjected to shearing stresses, thus $\sigma_h$ and $\sigma_a$ are principal stresses. For a point on the outside surface of the vessel

$$\sigma_{p1} = \sigma_h \qquad \sigma_{p2} = \sigma_a \qquad \sigma_{p3} = 0$$

where the principal stresses are ordered $\sigma_{p1} > \sigma_{p2} > \sigma_{p3}$. A point on the inside surface of the vessel has principal stresses

$$\sigma_{p1} = \sigma_h \qquad \sigma_{p2} = \sigma_a \qquad \sigma_{p3} = -p = -100 \text{ psi}$$

The maximum shearing stress is

$$\tau_{max} = \frac{\sigma_{max} - \sigma_{min}}{2}$$

Therefore, for the point on the outside surface

$$\tau_{max} = \frac{\sigma_h - 0}{2} = \frac{1200}{t} \text{ psi} \qquad (a)$$

while for the point on the inside surface

$$\tau_{max} = \frac{\sigma_h - (-100)}{2} = \frac{1200}{t} + 50 \text{ psi} \qquad (b)$$

The maximum-shear-stress theory of failure is

$$\tau_{max} \leq \tau_f = \frac{\sigma_f}{2}$$

Since a factor of safety is involved

$$\tau_{max} \leq \frac{\tau_f}{FS} = \frac{\sigma_f}{2(FS)} = \frac{36,000}{2(2)} = 9000 \text{ psi} \qquad (c)$$

For the point on the outside surface of the vessel, Eqs. (a) and (c) give

$$\frac{1200}{t} \leq 9000$$

$$t \geq 0.1333 \text{ in.}$$

For the point on the inside surface of the vessel, Eqs. (b) and (c) give

$$\frac{1200}{t} + 50 \leq 9000$$

$$t \geq 0.1341 \text{ in.}$$

Therefore, the minimum required wall thickness is

$$t = 0.1341 \text{ in.} \qquad \blacksquare \qquad \textbf{Ans.}$$

# PROBLEMS

### Introductory Problems

**10-149\*** A machine component fabricated from a material with a proportional limit in tension and compression of 60 ksi is subjected to a biaxial state of stress. The principal stresses are 30 ksi T and 50 ksi C. Determine which, if any, of the theories will predict failure by yielding for this state of stress.

**10-150** The state of stress at a point on the surface of a machine component is shown in Fig. P10-150. If the yield strength of the material is 250 MPa, determine which, if any, of the theories will predict failure by yielding for this state of stress.

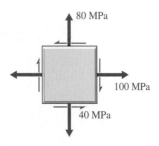

**Figure P10-150**

**10-151** At a point on the free surface of an alloy steel machine component the principal stresses are 45 ksi T and 25 ksi C. What minimum proportional limit is required according to each of the theories if failure by yielding is to be avoided?

**10-152\*** At a point on the free surface of an aluminum alloy machine component the principal stresses are 125 MPa T and 190 MPa C. What minimum proportional limit is required according to each of the theories if failure by yielding is to be avoided?

**10-153\*** A material with a proportional limit of 36 ksi in tension and compression is subjected to a biaxial state of stress. The principal stresses are 18 ksi T, 16 ksi T, and $\sigma_z = 0$. Determine the factor of safety with respect to failure by yielding according to each of the theories of failure.

**10-154** A material with a proportional limit of 180 MPa in tension and compression is subjected to a biaxial state of stress. The principal stresses are 90 MPa T, 80 MPa C, and $\sigma_z = 0$. Determine the factor of safety with respect to failure by yielding according to each of the theories of failure.

**10-155** A material with a yield strength in tension and compression of 60 ksi is subjected to the biaxial state of stress shown in Fig. P10-155. Determine the factor of safety with respect to failure by yielding according to each of the theories of failure.

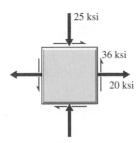

**Figure P10-155**

**10-156*** A point on the free surface of a machine component is subjected to the state of stress shown in Fig. P10-156. If the yield strength of the material is 250 MPa, determine the factor of safety with respect to failure by yielding according to each of the theories of failure.

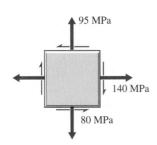

**Figure P10-156**

**Intermediate Problems**

**10-157*** A thin-walled cylindrical pressure vessel is capped at the ends and is subjected to an internal pressure. The inside diameter of the vessel is 5 ft, and the wall thickness is 1.5 in. The vessel is made of steel with tensile and compressive yield strengths of 36 ksi. Determine the internal pressure required to initiate yielding according to
(a) The maximum-shear-stress theory of failure.
(b) The maximum-distortion-energy theory of failure.

**10-158*** The solid circular shaft shown in Fig. P10-158 is subjected to a torque $T$. The yield strength of the material in tension and compression is 400 MPa. Determine the largest permissible value of the torque $T$ according to

(a) The maximum-shear-stress theory of failure.
(b) The maximum-distortion-energy theory of failure.

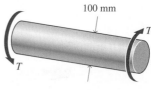

**Figure P10-158**

**10-159** The shaft shown in Fig. P10-159 is made of steel having a proportional limit of 60 ksi in tension or compression. If a factor of safety of 2.5 with respect to failure by yielding is specified, determine the maximum permissible value for the axial load $P$ according to the maximum-shear-stress theory of failure.

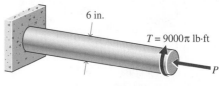

**Figure P10-159**

**10-160** A shaft, similar to the one shown in Fig. P10-159, has a 150-mm diameter and is made of steel having a yield strength in tension and compression of 360 MPa. The applied loads are $P = 2200$ kN and $T = 38$ kN · m. If failure is by yielding, determine the factor of safety according to
(a) The maximum-shear-stress theory of failure.
(b) The maximum-distortion-energy theory of failure.

**10-161** The yield strength $\sigma_y$ of a ductile material may be determined from a tension test (Fig. P10-161a). Using the three theories of failure discussed in Section 10-15, show that the yield strength $\tau_y$ of a member determined from a torsion test (Fig. P10-161b) is predicted to be
(a) $\tau_y = \sigma_y$ using the maximum-normal-stress theory of failure.
(b) $\tau_y = 0.5\sigma_y$ using the maximum-shear-stress theory of failure.
(c) $\tau_y = 0.577\sigma_y$ using the maximum-distortion-energy theory of failure.

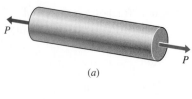

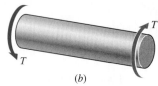

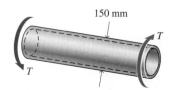

**Figure P10-161**

**10-162*** The hollow steel ($\sigma_y$ = 250 MPa) shaft shown in Fig. P10-162 is subjected to a torque $T$ = 40 kN · m. The factor of safety with respect to failure by yielding is 1.5. Determine the maximum permissible inside diameter for the shaft according to the maximum-shear-stress theory of failure.

150 mm

**Figure P10-162**

**10-163*** A solid circular steel ($\sigma_y$ = 36 ksi) shaft is subjected to an axial tensile load $P$ = 10 kip and a bending moment $M$ = 5 kip · in. Determine the minimum permissible diameter for the shaft according to the maximum-shear-stress theory, if the factor of safety for failure by yielding is 2.0.

**10-164** A solid circular shaft has a diameter of 100 mm and is made of a material with a yield strength of 360 MPa in tension and compression. The shaft is subjected to a bending moment $M$ = 12 kN · m and a torque $T$ = 20 kN · m. Determine the factor of safety with respect to failure by yielding according to the maximum-shear-stress theory of failure.

**Challenging Problems**

**10-165** The solid circular shaft shown in Fig. P10-165 is subjected to an axial tensile force $P$, a bending moment $M$, and a torque $T$. The shaft is made from a ductile material with a yield strength $\sigma_y$. Show that the minimum diameter $d$ of the shaft may be found according to the maximum-shear-stress theory of failure from the equation

$$\frac{\sigma_y}{2\,\text{FS}} = \frac{2}{\pi d^3}\sqrt{(Pd + 8M)^2 + (8T)^2}$$

where FS is the factor of safety.

**Figure P10-165**

**10-166** A solid circular shaft has a diameter $d$ and is subjected to a bending moment $M$ and a torque $T$. The shaft is made from a ductile material with a yield strength $\sigma_y$. Show that the minimum diameter $d$ of the shaft may be found according to the maximum-shear-stress theory of failure from the equation

$$d = \left[\frac{32\,\text{FS}}{\pi\sigma_y}\sqrt{(M)^2 + (T)^2}\right]^{1/3}$$

where FS is the factor of safety.

**10-167*** The shaft shown in Fig. P10-167 is made of an aluminum alloy that has a proportional limit of 48 ksi in tension or compression. If a factor of safety of 2.0 with respect to failure by yielding is specified, determine the maximum permissible value for the load $R$ according to
(a) The maximum-shear-stress theory of failure.
(b) The maximum-distortion-energy theory of failure.

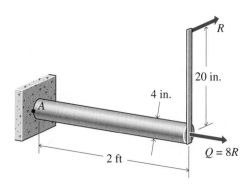

**Figure P10-167**

**10-168*** The shaft shown in Fig. P10-168 is made of steel having a proportional limit of 360 MPa in tension or compression. If a factor of safety of 2.0 with respect to failure by yielding is specified, determine the maximum permissible diameter $D$ according to
(a) The maximum-shear-stress theory of failure.
(b) The maximum-distortion-energy theory of failure.

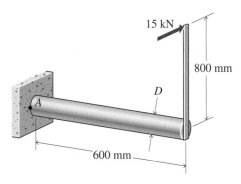

**Figure P10-168**

**10-169**  A steel shaft 4-in. in diameter is supported in flexible bearings at its ends. Two pulleys, each 24 in. in diameter, are keyed to the shaft. The pulleys carry belts that are loaded, as shown in Fig. P10-169. The steel has a proportional limit of 40 ksi in tension and compression and 23 ksi in shear. If a factor of safety of 2.5 with

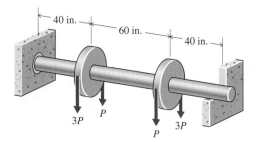

**Figure P10-169**

respect to failure by yielding is specified, determine the maximum permissible belt tension $P$ according to
(a) The maximum-shear-stress theory of failure.
(b) The maximum-distortion-energy theory of failure.

**10-170\***  A structural steel pipe used to transmit steam has an inside diameter of 300 mm. If the steam pressure is 5.5 MPa and a factor of safety of 4 with respect to failure by yielding is specified, determine the minimum permissible wall thickness of the pipe according to the maximum-shear-stress theory of failure. The wall of the pipe does not carry any axial load.

**10-171**  A solid circular shaft acting as a cantilever beam is subjected to the loading shown in Fig. P10-171. The material is 2024-T4 wrought aluminum with a yield strength in tension and compression of 48 ksi. If the factor of safety for failure by yielding is 2.5, determine the minimum permissible diameter of the shaft according to the maximum-shear-stress theory of failure. Neglect the effects of transverse shear.

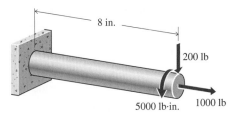

**Figure P10-171**

# 10-16 DESIGN: BRITTLE MATERIALS

Brittle materials, unlike ductile materials, do not yield; therefore, failure is by fracture and the critical stress is the fracture stress (or the ultimate strength). The ultimate strength in compression is greater than the ultimate strength in tension. The ultimate shear strength of a brittle material is approximately equal to the ultimate tensile strength; this is not the case for ductile materials. Of the theories proposed to predict fracture of brittle materials, only two will be presented in this text: the Coulomb–Mohr theory and the maximum-normal-stress theory.

## Coulomb–Mohr and Maximum-Normal-Stress Theories
The tension test and the compression test are the basis for the Coulomb–Mohr theory. A Mohr's circle for each of these tests is shown in Fig. 10-51a, where $\sigma_{ut}$ is the ultimate tensile strength and $\sigma_{uc}$ is the ultimate compressive strength. According to the Coulomb–Mohr theory, failure (fracture) occurs for any state of stress whose Mohr's circle is tangent to the envelope of the two circles shown in Fig. 10-51a (the envelope is the line that is tangent to the two circles, as shown in Fig. 10-51b). Mohr's circle for an arbitrary state of stress is shown dashed in Fig. 10-51b, where $\sigma_{p1}$ and $\sigma_{p3}$ are the principal stresses (the

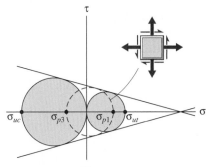

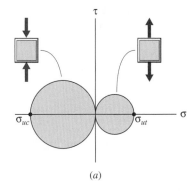

**Figure 10-51**    (b)

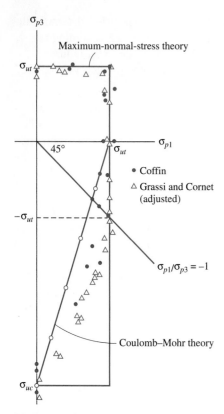

**Figure 10-52**

principal stresses have been ordered $\sigma_{p1} > \sigma_{p2} > \sigma_{p3}$). When Mohr's circle (dashed) for any state of stress is tangent to the envelope of Fig. 10-51$b$, failure occurs. The principal stresses $\sigma_{p1}$ and $\sigma_{p3}$, the ultimate tensile strength $\sigma_{ut}$, and the ultimate compressive strength $\sigma_{uc}$ are related by the equation

$$\frac{\sigma_{p1}}{\sigma_{ut}} - \frac{\sigma_{p3}}{\sigma_{uc}} = 1 \qquad (10\text{-}29)$$

where $\sigma_{uc}$ is the magnitude of the compressive strength and $\sigma_{p1}$ and $\sigma_{p3}$ carry algebraic signs. A state of stress is safe when

$$\frac{\sigma_{p1}}{\sigma_{ut}} - \frac{\sigma_{p3}}{\sigma_{uc}} \leq 1 \qquad (10\text{-}30)$$

Comparison between theory and experimental data (for gray cast iron) is shown in Fig. 10-52, which is adapted from Blake.[10] In the first quadrant the principal stresses have the same sign, and the maximum-normal-stress theory and the Coulomb–Mohr theory are identical. Note that the failure envelope for the maximum-normal-stress theory is not square as was the case for ductile materials where the tensile and compressive yield strengths were equal (see Fig. 10-43$b$). The two theories give different predictions of failure in the fourth quadrant, where the principal stresses have opposite signs. In the fourth quadrant, the maximum-normal-stress theory is unsafe because the data lie inside the failure envelope. The Coulomb–Mohr theory is conservative because the data lie outside the failure envelope.

The line for pure torsion, $\sigma_{p1} = -\sigma_{p3}$, is also shown in Fig. 10-52. This line gives two predictions for the ultimate shear strength, that is, where the line intersects the envelopes for the two failure theories. As previously mentioned, one of the characteristics of a brittle material is that the ultimate shear strength is approximately equal to the ultimate tensile strength. This is predicted by the maximum-normal-stress theory, whereas the Coulomb–Mohr theory predicts a value somewhat less.

## Example Problem 10-18

At the critical point in a machine component, the principal stresses (plane stress) are as shown in Fig. 10-53. The failure strengths for the material are $\sigma_{ut} = 2.5$ ksi and $\sigma_{uc} = 16$ ksi. Use the Coulomb–Mohr theory to determine if this state of stress is safe.

### SOLUTION

The state of stress is safe if the stresses satisfy Eq. (10-30).

$$\frac{\sigma_{p1}}{\sigma_{ut}} - \frac{\sigma_{p3}}{\sigma_{uc}} \leq 1$$

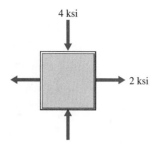

**Figure 10-53**

---

[10]*Handbook of Mechanics, Materials, and Structures*, Alexander Blake, Editor, Wiley-Interscience, New York, 1985.

Substituting the principal stresses from Fig. 10-53 (after noting that $\sigma_{p3}$ is compressive and that $\sigma_{uc}$ is the magnitude of the ultimate compressive strength) into Eq. (10-30) yields

$$\frac{2}{2.5} - \frac{(-4)}{16} = 1.05 > 1$$

Thus, the state of stress shown in Fig. 10-53 is unsafe. ■ **Ans.**

## Example Problem 10-19

A 200-mm-diameter solid circular shaft is subjected to a torque $T$, as shown in Fig. 10-54a. The shaft is made of a material with an ultimate tensile strength of 620 MPa and an ultimate compressive strength of 820 MPa. Determine the maximum permissible value for the torque $T$ according to the Coulomb–Mohr theory of failure.

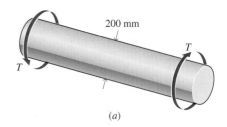

(a)

### SOLUTION

The state of stress for a shaft subjected to a torsional load $T$ is shown in Fig. 10-54b. The shearing stress $\tau$ is given by Eq. (7-6) as

$$\tau = \frac{Tc}{J} = \frac{T(0.100)}{\pi(0.200)^4/32} = 636.6T$$

The principal stresses are

$$\sigma_{p1} = -\sigma_{p3} = \tau = 636.6T$$

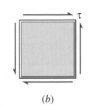

(b)

**Figure 10-54**

The maximum permissible value of $T$ according to the Coulomb–Mohr theory is given by Eq. (10-29) as

$$\frac{\sigma_{p1}}{\sigma_{ut}} - \frac{\sigma_{p3}}{\sigma_{uc}} = 1$$

or

$$\frac{636.6T}{620(10^6)} - \frac{(-636.6T)}{820(10^6)} = 1$$

Solving for $T$ yields

$$T = 0.5546(10^6) \text{ N} \cdot \text{m} \cong 555 \text{ kN} \cdot \text{m} \quad ■$$

**Ans.**

# PROBLEMS

## Introductory Problems

**10-172\*** Two states of stress are shown in Fig. P10-172. The failure strengths for the material are 152 MPa in tension and 572 MPa in compression. Use the Coulomb–Mohr theory to determine if these states of stress are safe.

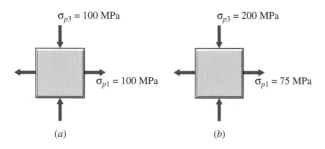

*(a)*            *(b)*

**Figure P10-172**

**10-173\*** The state of stress at the critical point in a machine component is shown in Fig. P10-173. The failure strengths for the material are 26 ksi in tension and 97 ksi in compression. Use the Coulomb–Mohr theory to determine if this state of stress is safe.

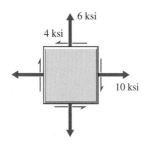

**Figure P10-173**

**10-174** The state of stress at the critical point in a machine component made of cast iron is shown in Fig. P10-174. The failure strengths for this material are 214 MPa in tension and 750 MPa in compression. Use the Coulomb–Mohr theory to determine if this state of stress is safe.

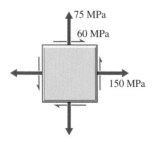

**Figure P10-174**

## Intermediate Problems

**10-175\*** The solid circular cast iron shaft shown in Fig. P10-175 is subjected to a torque $T = 240$ kip · in. and a bending moment $M = 110$ kip · in. The failure strengths for this material are 43 ksi in tension and 140 ksi in compression. Use the Coulomb–Mohr theory to determine the minimum permissible diameter for this shaft.

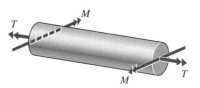

**Figure P10-175**

**10-176\*** A solid circular cast iron shaft is subjected to the loads shown in Fig. P10-176. The failure strengths for this material are 214 MPa in tension and 750 MPa in compression. Use the Coulomb–Mohr theory to determine the minimum permissible diameter for the shaft. Neglect the effects of transverse shear.

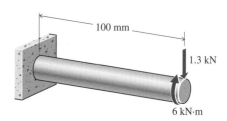

**Figure P10-176**

**10-177** A thin-walled cylindrical pressure vessel has an inside diameter of 12 in. and a wall thickness of 0.20 in. The vessel is made of a material with $\sigma_{ut} = 40$ ksi and $\sigma_{uc} = 50$ ksi. Use the Coulomb–Mohr theory to determine the maximum internal pressure that the vessel can safely support.

## Challenging Problems

**10-178\*** The C-clamp shown in Fig. P10-178 is made of a gray cast iron that has an ultimate tensile strength of 180 MPa and an ultimate compressive strength of 670 MPa. Use the Coulomb–Mohr theory to determine the maximum load $P$ that can be applied safely to the clamp.

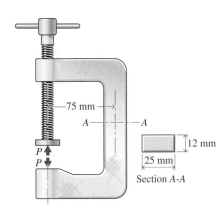

**Figure P10-178**

**10-179** A solid circular gray cast iron shaft is loaded as shown in Fig. P10-179. The failure strengths for this type of cast iron are 36.5 ksi in tension and 124 ksi in compression. Use the Coulomb–Mohr theory to determine the minimum permissible diameter for the shaft. Neglect the effects of transverse shear.

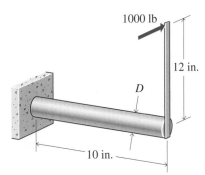

**Figure P10-179**

**10-180** A 100-mm-diameter solid circular shaft is loaded as shown in Fig. P10-180. The shaft is made of a gray cast iron that has an ultimate tensile strength of 150 MPa and an ultimate compressive strength of 570 MPa. Use the Coulomb–Mohr theory to determine if the shaft can safely support the loading shown.

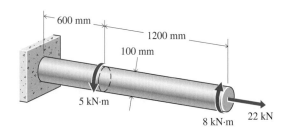

**Figure P10-180**

## 10-17 SUMMARY

In previous chapters, formulas were developed for determining normal and shearing stresses on specific planes in axially loaded bars, circular shafts, and beams. In this chapter, methods were developed for determining normal and shearing stresses on other planes through a point of interest.

Equations relating the normal and shearing stresses $\sigma_n$ and $\tau_{nt}$ on an arbitrary plane (oriented at an angle $\theta$ with respect to a reference $x$-axis) through a point and the known stresses $\sigma_x$, $\sigma_y$, and $\tau_{xy} = \tau_{yx}$ on the reference planes were developed using the free-body diagram method introduced earlier for the axially loaded bars, circular shafts, and beams. The results are

$$\sigma_n = \sigma_x \cos^2 \theta + \sigma_y \sin^2 \theta + 2\tau_{xy} \sin \theta \cos \theta \qquad (10\text{-}1a)$$

$$\tau_{nt} = -(\sigma_x - \sigma_y) \sin \theta \cos \theta + \tau_{xy}(\cos^2 \theta - \sin^2 \theta) \qquad (10\text{-}2a)$$

Equations (10-1) and (10-2) provide a means for determining the normal stress $\sigma_n$ and the shearing stress $\tau_{nt}$ on different planes through a point in a stressed body. For design purposes, critical stresses at the point are usually the maximum tensile stress and/or the maximum shearing stress.

Maximum and minimum values of $\sigma_n$ occur at values of $\theta$ given by the expression

$$\tan 2\theta_p = \frac{2\tau_{xy}}{\sigma_x - \sigma_y} \tag{10-3}$$

When these values of $\theta$ are substituted into Eq. (10-1), maximum and minimum normal stresses $\sigma_{p1}$ and $\sigma_{p2}$ at the point are found to be

$$\sigma_{p1, \, p2} = \frac{\sigma_x + \sigma_y}{2} \pm \sqrt{\left(\frac{\sigma_x - \sigma_y}{2}\right)^2 + \tau_{xy}^2} \tag{10-4}$$

It is also found that the shearing stress $\tau_{nt}$ is zero on planes experiencing maximum and minimum values of normal stress. Planes free of shear stress are known as principal planes. Normal stresses occurring on principal planes are known as principal stresses.

In a similar manner, the maximum in-plane shearing stresses are found to be

$$\tau_p = \pm \sqrt{\left(\frac{\sigma_x - \sigma_y}{2}\right)^2 + \tau_{xy}^2} \tag{10-5}$$

A useful relation between the principal stresses $\sigma_{p1}$ and $\sigma_{p2}$ and the maximum in-plane shearing stress is

$$\tau_p = (\sigma_{p1} - \sigma_{p2})/2 \tag{10-6}$$

When stresses act in three directions it can be shown that there are, in general, three orthogonal planes on which the shearing stress is zero. These planes are principal planes, and the stresses acting on them (the principal stresses) will have three values: one maximum, one minimum, and a third stress between the other two. The maximum shearing stress $\tau_{\max}$ on any plane that could be passed through the point is

$$\tau_{\max} = \frac{\sigma_{\max} - \sigma_{\min}}{2} \tag{10-7}$$

Another useful relation between the principal stresses and the normal stresses on the orthogonal reference planes is

$$\sigma_{p1} + \sigma_{p2} = \sigma_x + \sigma_y \tag{10-8}$$

When a body is subjected to a system of loads, the body deforms. The dimensions of the deformed body can be expressed in terms of strains $\epsilon_x$, $\epsilon_y$, and $\gamma_{xy}$, which can be measured with strain gages. Relationships between strains, similar to Eqs. (10-1) and (10-2), are

$$\epsilon_n = \epsilon_x \cos^2 \theta + \epsilon_y \sin^2 \theta + \gamma_{xy} \sin \theta \cos \theta \tag{10-11a}$$

$$\gamma_{nt} = -2(\epsilon_x - \epsilon_y) \sin \theta \cos \theta + \gamma_{xy}(\cos^2 \theta - \sin^2 \theta) \tag{10-12a}$$

The similarity between Eqs. (10-11) and (10-12) for plane strain and Eqs. (10-1) and (10-2) for plane stress indicates that all of the equations developed for plane stress can be applied to plane strain by substituting $\epsilon_x$ for $\sigma_x$, $\epsilon_y$ for $\sigma_y$, and $\gamma_{xy}/2$ for $\tau_{xy}$. Thus,

$$\tan 2\theta_p = \frac{\gamma_{xy}}{\epsilon_x - \epsilon_y} \qquad (10\text{-}13)$$

$$\epsilon_{p1}, \epsilon_{p2} = \frac{\epsilon_x + \epsilon_y}{2} \pm \sqrt{\left(\frac{\epsilon_x - \epsilon_y}{2}\right)^2 + \left(\frac{\gamma_{xy}}{2}\right)^2} \qquad (10\text{-}14)$$

$$\gamma_p = 2\sqrt{\left(\frac{\epsilon_x - \epsilon_y}{2}\right)^2 + \left(\frac{\gamma_{xy}}{2}\right)^2} \qquad (10\text{-}15)$$

Hooke's law, Eq. (4-15), can be extended to biaxial and triaxial states of stress often encountered in engineering practice by using the principal of superposition. The results for a biaxial state of stress are

$$\epsilon_x = \frac{1}{E}(\sigma_x - v\sigma_y)$$

$$\epsilon_y = \frac{1}{E}(\sigma_y - v\sigma_x) \qquad (10\text{-}16)$$

$$\epsilon_z = -\frac{v}{E}(\sigma_x + \sigma_y)$$

When Eqs. (10-16) are solved for the stresses in terms of the strains,

$$\sigma_x = \frac{E}{1 - v^2}(\epsilon_x + v\epsilon_y)$$

$$\qquad (10\text{-}18)$$

$$\sigma_y = \frac{E}{1 - v^2}(\epsilon_y + v\epsilon_x)$$

Equations (10-18) can be used to calculate normal stresses from measured or computed normal strains.

Torsion test specimens are used to study material behavior under pure shear, and it is observed that a shearing stress produces only a single corresponding shear strain. Thus, Hooke's law extended to shearing stresses is simply

$$\tau = G\gamma \qquad (10\text{-}20)$$

The three elastic constants $E$, $v$, and $G$ for a material are not independent but are related by the equation

$$G = \frac{E}{2(1 + v)} \qquad (10\text{-}21)$$

A tension test of an axially loaded member is easy to conduct, and the results, for many types of materials, are well known. When such a member fails, the failure occurs at a specific principal (axial) stress $\sigma_f$. At this point, the principal (axial) strain is $\epsilon_f = \sigma_f/E$, the maximum shearing stress is $\tau_f = \sigma_f/2$, and the strain energy per unit volume of stressed material is $u = \sigma_f^2/2E$. For an element subjected to biaxial or triaxial loading, however, the situation is more complicated because the limits of normal stress, normal strain, shearing stress, and strain energy existing at failure for an axial load are not all reached simultaneously. In such cases, it becomes important to determine the best criterion for predicting failure. Several theories have been proposed for predicting failure of various types of material subjected to many combinations of loads. Unfortunately, none of the theories agree with test data for all types of materials and combinations of loads.

The maximum-normal-stress theory (also called Rankine's theory) predicts failure of a specimen subjected to any combination of loads when the maximum normal stress at any point in the specimen reaches the axial failure stress as determined by an axial tensile or compressive test of the same material, $\sigma_{max} = \sigma_f$. The maximum-shear-stress theory (also called Coulomb's theory or Guest's theory) predicts failure of a specimen subjected to any combination of loads when the maximum shear stress at any point reaches the failure stress as determined by an axial tension or axial compression test of the same material, $\tau_{max} = \tau_f = \sigma_f/2$. For ductile materials, the shearing elastic limit, as determined from a torsion test (pure shear), is about $0.57\sigma_f$. Therefore; the maximum-shear-stress theory errs on the conservative side by being based on the limit obtained from an axial test.

The maximum-distortion-energy theory (also called the Huber–Hencky–von Mises theory) predicts failure of a specimen subjected to any combination of loads when the distortion component of the strain energy intensity of any portion of the stressed member reaches the failure value of the distortion component of the strain energy intensity as determined from an axial tension or compression test of the same material. This theory assumes that the portion of the strain energy producing volume change is ineffective in causing failure by yielding. The strain energy of distortion is most readily computed by determining the total strain energy of the stressed material and subtracting the strain energy corresponding to the volume change.

Failure is predicted most closely by the maximum-distortion-energy theory. When the two largest principal normal stresses have the same sign, the maximum-normal-stress and the maximum-shear-stress theories are identical and both theories are conservative. When the two largest principal stresses have opposite signs, the maximum-shear-stress theory is conservative while the maximum-normal-stress theory over predicts failure stresses and is unsafe.

# REVIEW PROBLEMS

**10-181**  The stresses shown in Fig. P10-181 act at a point on the free surface of a structural member subjected to plane stress. Use the stress transformation equations for plane stress [Eqs. (10-1) and (10-2)] to solve for the normal and shear stresses on the inclined plane $A$–$B$.

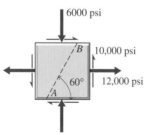

**Figure P10-181**

**10-182\***  The stresses shown in Fig. P10-182 act at a point on the free surface of a structural member subjected to plane stress. Use the stress transformation equations for plane stress [Eqs. (10-1) and (10-2)] to solve for the normal and shear stresses on the inclined plane $A$–$B$.

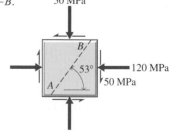

**Figure P10-182**

**10-183\*** The stresses shown in Fig. P10-183 act at a point on the free surface of a stressed body. The principal stresses at the point are 20 ksi C and 12 ksi T. Determine the unknown normal stresses on the horizontal and vertical planes and the angle $\theta_p$ between the $x$-axis and the maximum tensile stress at the point.

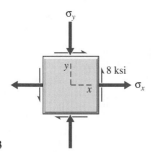

**Figure P10-183**

**10-184** At a point in a structural member subjected to plane stress there are normal and shear stresses on horizontal and vertical planes through the point, as shown in Fig. P10-184. Use Mohr's circle to determine
(a) The principal stresses and the maximum shear stress at the point.
(b) The normal and shear stresses on the inclined plane $A$–$B$ shown in the figure.

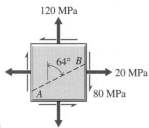

**Figure P10-184**

**10-185** The strain components at a point in a body under a state of plane strain are $\epsilon_x = +1000$ $\mu$in./in., $\epsilon_y = -800$ $\mu$in./in., and $\gamma_{xy} = -800$ $\mu$rad. Determine the principal strains and the maximum shearing strain at the point. Show the principal strain deformations and the maximum shearing strain distortion on a sketch.

**10-186\*** The strain components at a point in a body under a state of plane strain are $\epsilon_x = +1200$ $\mu$m/m, $\epsilon_y = +960$ $\mu$m/m, and $\gamma_{xy} = -960$ $\mu$rad. Determine the principal strains and the maximum shearing strain at the point. Show the principal strain deformations and the maximum shearing strain distortion on a sketch.

**10-187** At a point on the free surface of a steel ($E = 30,000$ ksi and $v = 0.30$) machine part, the strain rosette shown in Fig. P10-187 was used to obtain the following normal strain data: $\epsilon_a = +650$ $\mu$in./in., $\epsilon_b = +475$ $\mu$in./in., and $\epsilon_c = -250$ $\mu$in./in. Determine
(a) The stress components $\sigma_x$, $\sigma_y$, and $\tau_{xy}$ at the point.
(b) The principal strains and the maximum shearing strain at the point. Prepare a sketch showing all of these strains.

(c) The principal stresses and the maximum shearing stress at the point. Prepare a sketch showing all of these stresses.

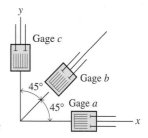

**Figure P10-187**

**10-188** The strain rosette shown in Fig. P10-188 was used to obtain normal strain data at a point on the free surface of an annealed titanium alloy machine part. The gage readings were $\epsilon_a = -306$ $\mu$m/m, $\epsilon_b = -456$ $\mu$m/m, and $\epsilon_c = +906$ $\mu$m/m. Determine
(a) The strain components $\epsilon_x$, $\epsilon_y$, and $\gamma_{xy}$ at the point.
(b) The stress components $\sigma_x$, $\sigma_y$, and $\tau_{xy}$ at the point.
(c) The principal stresses and the maximum shearing stress at the point. Express the stresses in SI units.

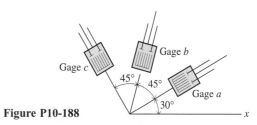

**Figure P10-188**

**10-189\*** The thin-walled pressure vessel shown in Fig. P10-189$a$ has an inside diameter of 20 in. and a wall thickness of ⅜ in. The vessel is subjected to an internal pressure of $p$ and a torque of $T$. At a point on the outside surface of the steel ($E = 30,000$ ksi and $v = 0.30$) vessel, the strain rosette shown in Fig. P10-189$b$ was used to obtain the following normal strain data: $\epsilon_a = +36$ $\mu$in./in., $\epsilon_b = +310$ $\mu$in./in., and $\epsilon_c = +150$ $\mu$in./in. Gages $a$ and $c$ are oriented in the axial and hoop directions of the vessel, respectively. Determine the pressure $p$ and the torque $T$.

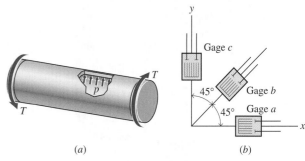

$(a)$          $(b)$

**Figure P10-189**

**10-190\*** A human femur is modeled as shown in Fig. P10-190. The abductor muscle force is $M = 4060$ N, and the femoral load is $J = 5210$ N.

(a) Determine the maximum tensile and compressive stresses at section $a$–$a$ if the section is modeled as a solid circular section 27 mm in diameter.

(b) Determine the maximum tensile and compressive stresses at section $a$–$a$ if the section is modeled as a hollow cylinder of outside diameter 27 mm and inside diameter 16 mm.

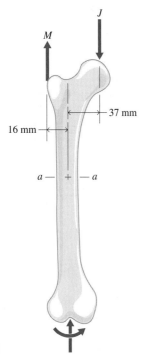

**Figure P10-190**

**10-191** The clamp of Fig. P10-191 is used to hold two boards. If the clamping force is 80 lb, determine the maximum tensile and compressive stresses at section $a$–$a$ if the clamp has a $\frac{1}{2}$- $\times$ $\frac{3}{16}$-in. cross section.

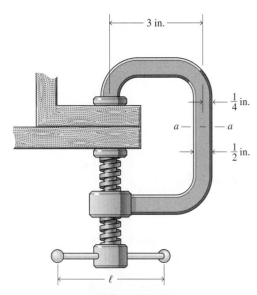

**Figure P10-191**

**10-192** A steel shaft with the left portion solid and the right portion hollow is loaded as shown in Fig. P10-192. If the maximum shearing stress in the shaft must not exceed 80 MPa and the maximum tensile stress in the shaft must not exceed 140 MPa, determine the maximum axial load $P$ that can be applied to the shaft.

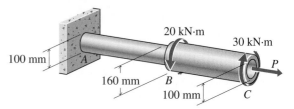

**Figure P10-192**

**10-193*** A hollow circular shaft has an outside diameter of 12 in. and an inside diameter of 4 in., as shown in Fig. P10-193. The shaft is in equilibrium when the indicated loads and torque are applied. If the maximum normal and shearing stresses at point A must be limited to 5000 psi T and 3000 psi, respectively, determine the maximum permissible value for the axial load $P$.

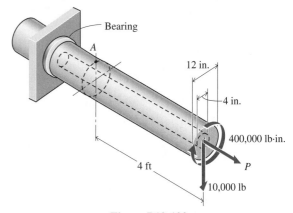

**Figure P10-193**

**10-194** The frame shown in Fig. P10-194 is subjected to a load of 100 kN. The material has a proportional limit of 220 MPa in tension and compression. Determine the factor of safety with respect to failure by yielding according to

(a) The maximum-shearing-stress theory of failure.
(b) The maximum-distortion-energy theory of failure.

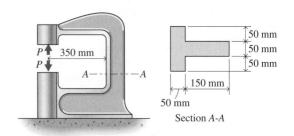

**Figure P10-194**

# COLUMNS

## 11-1 INTRODUCTION

In their simplest form, columns are long, straight, prismatic bars subjected to compressive, axial loads. As long as a column remains straight, it can be analyzed by the methods of Chapter 4; however, if a column begins to deform laterally, the deflection may become large and lead to catastrophic failure. This situation, called *buckling*, can be defined as the sudden large deformation of a structure due to a slight increase of an existing compressive load under which the structure had exhibited little, if any, deformation before the load was increased. For example, a yardstick will support a compressive load of several pounds without discernible lateral deformation, but once the load becomes large enough to cause the yardstick to "bow out" a slight amount, any further increase of load produces large lateral deflections.

Buckling of such a column is caused not by failure of the material of which the column is composed but by deterioration of what was a stable state of equilibrium to an unstable one. The three states of equilibrium can be illustrated with a ball at rest on a surface, as shown in Figure 11-1. The ball in Fig. 11-1a is in a stable equilibrium position at the bottom of the pit because gravity will cause it to return to its equilibrium position if perturbed. The ball in Fig. 11-1b is in a neutral equilibrium position on the horizontal plane because it will remain at any new position to which it is displaced, tending neither to return to nor move farther from its original position. The ball in Fig. 11-1c, however, is in an unstable equilibrium position at the top of a hill because, if it is perturbed, gravity will cause it to move even farther from its original location until it eventually finds a stable equilibrium position at the bottom of another pit.

As the compressive load on a column is gradually increased from zero, the column is at first in a state of stable equilibrium. During this state, if the column is perturbed by inducing small lateral deflections, it will return to its straight configuration when the lateral loads are removed. As the load is increased further, a critical value is reached at which the column is on the verge of experiencing a lateral deflection so, if it is perturbed, it will not return to its straight configuration. The load cannot be increased beyond this value unless the column is restrained laterally; should the lateral restraints be removed, the slightest perturbation will trigger large lateral deflections. For long, slender columns, the critical

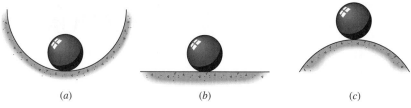

**Figure 11-1**

buckling load (the maximum load for which the column is in stable equilibrium) occurs at stress levels much less than the proportional limit for the material. This indicates that this type of buckling is an elastic phenomenon.

## 11-2 BUCKLING OF LONG STRAIGHT COLUMNS

The first solution for buckling of long slender columns was published in 1757 by the Swiss mathematician Leonhard Euler (1707–1783). Although the results of this section can be used only for long, slender columns, the analysis, similar to that used by Euler, is mathematically revealing and helps explain the behavior of columns.

The purpose of this analysis is to determine the minimum axial compressive load for which a column will experience lateral deflections. A straight, slender, pin-ended column of length $L$ centrically loaded by axial compressive forces $P$ at each end is shown in Fig. 11-2$a$. A pin-ended column is supported such that the bending moment and lateral movement are zero at the ends. In Fig. 11-2$b$, the load $P$ has been increased sufficiently to cause a lateral deflection $\delta$ at the midpoint of the span. Axes are selected with the origin at the center of the column for convenience. If the column is sectioned at an arbitrary position $x$, a free-

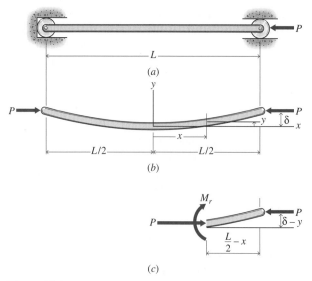

**Figure 11-2**

body diagram of the portion to the right of the section will appear as shown in Fig. 11-2c. The two forces constitute a couple of magnitude $P(\delta - y)$ that must equal the resisting moment $M_r$; thus, $M_r = P(\delta - y)$. The differential equation for the elastic curve, as given by Eq. (9-3), becomes

$$EI \frac{d^2y}{dx^2} = M_r = P(\delta - y)$$

or

$$\frac{d^2y}{dx^2} + \frac{P}{EI}y = \frac{P\delta}{EI} \tag{a}$$

Equation (a) is an ordinary second-order linear differential equation with constant coefficients and a constant right-hand side. Established methods for the solution of such equations show the solution to be of the form

$$y = A \sin px + B \cos px + C \tag{b}$$

where $A$, $B$, $C$, and $p$ are constants. By differentiating Eq. (b), substituting the results into Eq. (a), and collecting like terms, the following expression is obtained for evaluating $p$ and $C$:

$$\left(-p^2 + \frac{P}{EI}\right)(A \sin px + B \cos px) + \frac{PC}{EI} = \frac{P\delta}{EI}$$

from which it follows that

$$p^2 = \frac{P}{EI} \qquad \text{and} \qquad C = \delta$$

The constants $A$ and $B$ can be obtained from the two boundary conditions that the slope and deflection of the elastic curve are zero at the origin; that is, at $x = 0$, $y = 0$ and at $x = 0$, $dy/dx = 0$. The purpose of this analysis is to determine the minimum load for which lateral deflections occur; therefore, it is necessary to have a third boundary condition; namely, at $x = L/2$, $y = \delta > 0$. The midspan deflection $\delta$ is required to be nonzero; otherwise, $\delta = 0$, $y \equiv 0$ could be a solution of Eq. (a). When the boundary conditions are substituted into Eq. (b) and its first derivative, they give

$$0 = B + C$$
$$0 = Ap$$

from which

$$A = 0 \qquad \text{and} \qquad B = -C = -\delta$$

The solution to Eq. (a) thus becomes

$$y = \delta\left(1 - \cos\sqrt{\frac{P}{EI}}\, x\right) \tag{11-1}$$

To satisfy the condition that at $x = L/2$, $y = \delta$, the cosine term in Eq. (11-1) must vanish when $x = L/2$. Thus,

$$\cos \sqrt{\frac{P}{EI}} \frac{L}{2} = 0$$

This equation is satisfied when the argument of the cosine is an odd multiple of $\pi/2$ or

$$\sqrt{\frac{P}{EI}} \frac{L}{2} = \frac{\pi}{2}, \frac{3\pi}{2}, \frac{5\pi}{2}, \ldots$$

Only the first value has physical significance because it determines the minimum value of $P$ for a nontrivial solution. This value for $P$ is designated by $P_{cr}$ and is called the *critical buckling load* or the *Euler buckling load*,[1] in honor of Leonhard Euler. The Euler buckling load has the magnitude

$$P_{cr} = \frac{\pi^2 EI}{L^2} \tag{11-2}$$

The second moment of the cross-sectional area ($I$) is relative to the axis about which bending occurs. For a pin-ended, centrically loaded column with no intermediate bracing to restrain lateral motion, bending occurs about the "weak" axis—the axis of minimum second moment of area. When $I$ is replaced by $Ar^2$, where $A$ is the cross-sectional area and $r$ is the radius of gyration about the axis of bending, Eq. (11-2) becomes

$$\frac{P_{cr}}{A} = \frac{\pi^2 E}{(L/r)^2} = \sigma_{cr} \tag{11-3}$$

The dimensionless quantity $L/r$ is called the *slenderness ratio* and is determined for the axis about which bending tends to occur. Equation (11-3) is a particularly useful form if the critical stress $\sigma_{cr}$ is of concern in the design. Either Eq. (11-2) or (11-3) can be used if the critical force $P_{cr}$ is the primary concern in the design.

Since this analysis is based on simple beam bending theory, it is valid only as long as the stresses remain in the linearly elastic range. Since the proportional limit is difficult to measure, the yield strength is often used instead of the proportional limit as the limiting stress, and the smallest slenderness ratio for which the Euler buckling load equation is valid occurs when $\sigma_{cr} = \sigma_y$ (the yield strength).

The Euler buckling load as given by Eq. (11-2) or (11-3) agrees well with experimental data if the slenderness ratio is large ($L/r > 140$ for steel columns). Short compression members ($L/r < 40$ for steel columns) can be treated as compression blocks where yielding occurs before buckling. Many columns lie between these extremes, where neither solution is applicable. The intermediate length columns are analyzed by using empirical formulas described in Section 11-4.

---

[1] While the analysis predicts the buckling load, it does not determine the corresponding lateral deflection $\delta$. This deflection can assume any nonzero value small enough that the nonlinear factor $[1 + (dy/dx)^2]^{3/2}$ in the curvature expression is approximately unity.

## ■ Example Problem 11-1

An 8-ft-long pin-ended timber [$E = 1.9(10^6)$ psi and $\sigma_e = 6400$ psi] column has a 2- $\times$ 4-in. rectangular cross section. Determine

(a) The slenderness ratio.

(b) The Euler buckling load.

(c) The ratio of the axial stress under the action of the buckling load to the elastic strength $\sigma_e$ of the material.

## SOLUTION

(a) To determine the slenderness ratio, the minimum radius of gyration must be calculated. The second moment of area of a rectangular cross section is $bh^3/12$ and the area is $bh$; therefore, the radius of gyration is $\sqrt{I/A}$ or $h/2\sqrt{3}$. The minimum radius of gyration is found by using the centroidal axis parallel to the longer side of the rectangle. Thus, $b = 4$ in. and $h = 2$ in., so

$$r = h/2\sqrt{3} = 2/2\sqrt{3} = 0.5774 \text{ in.}$$

The slenderness ratio is then found to be

$$\frac{L}{r} = \frac{8(12)}{0.5774} = 166.3 \qquad \textbf{Ans.}$$

(b) The Euler buckling load is found by using Eq. (11-3). Thus,

$$P_{cr} = \frac{\pi^2 EA}{(L/r)^2} = \frac{\pi^2(1.9)(10^6)(2)(4)}{(166.3)^2} = 5424 \text{ lb} \cong 5420 \text{ lb} \qquad \textbf{Ans.}$$

The Euler buckling load could also have been found using Eq. (11-2);

$$P_{cr} = \frac{\pi^2 EI}{L^2} = \frac{\pi^2(1.9)(10^6)[4(2)^3/12]}{[8(12)]^2} = 5420 \text{ lb}$$

(c) The axial stress under the action of the buckling load is

$$\sigma_{cr} = \frac{P_{cr}}{A} = \frac{5424}{2(4)} = 678 \text{ psi}$$

and therefore

$$\frac{\sigma_{cr}}{\sigma_e} = \frac{678}{6400} = 0.1059 \qquad \textbf{Ans.}$$

This demonstrates that buckling can occur at stresses well below the elastic limit of a material for sufficiently slender columns! ■

## Example Problem 11-2

A 12-ft-long pin-ended column is made of 6061–T6 aluminum alloy. The column has a hollow circular cross section with an outside diameter of 5 in. and an inside diameter of 4 in. Determine

(a) The smallest slenderness ratio for which the Euler buckling load equation is valid.
(b) The critical buckling load.

### SOLUTION

(a) The smallest slenderness ratio for which the Euler buckling load equation is valid occurs when $\sigma_{cr} = \sigma_y$ (yield strength). Rewriting Eq. (11-3) as

$$\frac{L}{r} = \sqrt{\frac{\pi^2 E}{\sigma_{cr}}} = \sqrt{\frac{\pi^2 E}{\sigma_y}}$$

and substituting $E = 10(10^6)$ psi and $\sigma_y = 40{,}000$ psi (see Appendix A)

$$\left(\frac{L}{r}\right)_{min} = \sqrt{\frac{\pi^2 E}{\sigma_y}} = \sqrt{\frac{\pi^2(10)(10^6)}{40(10^3)}} = 49.67 \cong 49.7 \qquad \textbf{Ans.}$$

(b) The critical buckling load is found using Eq. (11-2).

$$P_{cr} = \frac{\pi^2 E I}{L^2}$$

First, check that the slenderness ratio of the column exceeds 49.7.

$$A = \frac{\pi}{4}(5^2 - 4^2) = 7.069 \text{ in}^2$$

$$I = \frac{\pi}{64}(5^4 - 4^4) = 18.113 \text{ in}^4$$

$$r = \sqrt{\frac{I}{A}} = \sqrt{\frac{18.113}{7.069}} = 1.6007$$

$$\frac{L}{r} = \frac{12(12)}{1.6007} = 89.96 > 49.7$$

Thus, the critical buckling load is

$$P_{cr} = \frac{\pi^2 E I}{L^2} = \frac{\pi^2(10)(10^6)(18.113)}{144^2} = 86{,}210 \text{ lb} \cong 86.2 \text{ kip} \qquad \textbf{Ans.}$$

While the theoretical limit for validity of Eqs. (11-2) or (11-3) is $\sigma_{cr} = \sigma_y$, experimental results indicate a limiting slenderness ratio at least 50% greater than the minimum given by $\sigma_{cr} = \sigma_y$. In this problem, the actual slenderness ratio (89.96) is 1.8 times the minimum ratio (49.7), so the Euler buckling load is expected to be valid. ■

## Example Problem 11-3

Two 51- × 51- × 3.2-mm structural steel angles 3 m long will be used as a pin-ended column. Determine the slenderness ratio and the Euler buckling load if

(a) The two angles are not connected and each acts as an independent member.
(b) The two angles are fastened together, as shown in Fig. 11-3, to act as a unit.

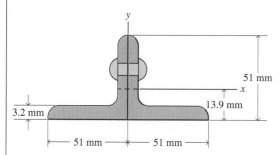

y

51 mm

x

13.9 mm

3.2 mm

51 mm    51 mm

**Figure 11-3**

## SOLUTION

(a) If the angles are not connected and each acts independently, the slenderness ratio is determined by using the minimum radius of gyration of the individual cross sections. From Appendix A, the minimum radius of gyration $r_{min}$ for this angle is 10.1 mm about the $ZZ$-axis. Thus, the slenderness ratio is

$$\frac{L}{r} = \frac{L}{r_{min}} = \frac{3000}{10.1} = 297 \qquad \textbf{Ans.}$$

The area for each angle is 312 mm²; therefore, the cross-sectional area for the column is 624 mm². The modulus of elasticity for structural steel is 200 GPa; therefore, the buckling load is [using Eq. (11-3)]

$$P_{cr} = \frac{\pi^2 EA}{(L/r)^2} = \frac{\pi^2 (200)(10^9)(624)(10^{-6})}{(297)^2}$$

$$= 13.964(10^3) \text{ N} \cong 13.96 \text{ kN} \qquad \textbf{Ans.}$$

(b) With the two angles connected as shown in Fig. 11-3, both $I_x$ and $I_y$ or $r_x$ and $r_y$ must be known in order to determine the minimum radius of gyration. Cross-sectional properties for the angles are given in Appendix A. The value of $I_y$ for the two angles is obtained by using the parallel axis theorem; thus,

$$I_y = 2(I_C + Ad^2) = 2(Ar_C^2 + Ad^2)$$

and

$$r_y = \sqrt{I_y/2A} = \sqrt{\frac{2A(r_C^2 + d^2)}{2A}} = \sqrt{r_C^2 + d^2}$$

This expression indicates that the radius of gyration for the two angles is the same as that for one angle, a fact obtained directly from the definition of radius of gyration. Then,

$$r_y = \sqrt{(15.9)^2 + (13.9)^2} = 21.12 \text{ mm}$$

In a similar fashion, the radius of gyration about the $x$-axis is the radius of gyration $r_x = 15.9$ mm for a single angle, since $d = 0$ for this axis. This means that the column tends to buckle about the $x$-axis where the slenderness ratio is

$$\frac{L}{r} = \frac{L}{r_{\min}} = \frac{3000}{15.9} = 188.68 \cong 188.7 \qquad \textbf{Ans.}$$

The corresponding Euler buckling load is

$$P_{cr} = \frac{\pi^2 EA}{(L/r)^2} = \frac{\pi^2 (200)(10^9)(624)(10^{-6})}{(188.68)^2}$$

$$= 34.599(10^3) \text{ N} \cong 34.6 \text{ kN} \quad \blacksquare \qquad \textbf{Ans.}$$

# PROBLEMS

## Introductory Problems

**11-1\*** A structural steel rod 1 in. in diameter and 50 in. long will be used to support an axial compressive load $P$. Determine
(a) The slenderness ratio.
(b) The smallest slenderness ratio for which the Euler buckling load equation is valid.
(c) The Euler buckling load.

**11-2\*** A hollow, circular structural steel column 6 m long has an outside diameter of 125 mm and an inside diameter of 100 mm. Determine
(a) The slenderness ratio.
(b) The smallest slenderness ratio for which the Euler buckling load equation is valid.
(c) The Euler buckling load.

**11-3** A wood column of air-dried Douglas fir is 10 ft long and has a 4- × 4-in. rectangular cross section. Determine
(a) The slenderness ratio.
(b) The smallest slenderness ratio for which the Euler buckling load equation is valid.
(c) The Euler buckling load.

**11-4** A 5-m-long column with the cross section shown in Fig. P11-4 is constructed from four pieces of timber.

The timbers are nailed together so that they act as a unit. Determine
(a) The slenderness ratio.
(b) The Euler buckling load. Use $E = 14$ GPa for the timber.
(c) The axial stress in the column when the Euler load is applied.

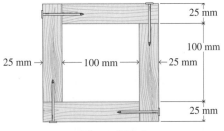

**Figure P11-4**

**11-5\*** Determine the allowable compressive load that a 5- × 5-in. air-dried red oak timber can support if it is 18 ft long and a factor of safety of 3 is specified.

## Intermediate Problems

**11-6** A 3-m-long column with the cross section shown in Fig. P11-6 is constructed from two pieces of air-dried

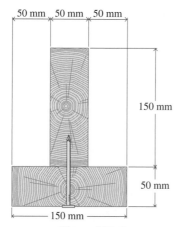

Figure P11-6

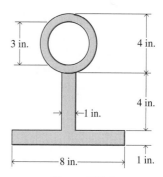

Figure P11-9

Douglas fir timber. The timbers are nailed together so that they act as a unit. Determine
(a) The slenderness ratio.
(b) The smallest slenderness ratio for which the Euler buckling load equation is valid.
(c) The Euler buckling load.

**11-7\*** A WT6 × 36 structural steel section (see Appendix A for cross-sectional properties) is used for an 18-ft-long column. Determine
(a) The slenderness ratio.
(b) The smallest slenderness ratio for which the Euler buckling load equation is valid.
(c) The Euler buckling load.

**11-8\*** Two L127 × 76 × 12.7-mm structural steel angles (see Appendix A for cross-sectional properties) are used for a column that is 4.5 m long. Determine the total compressive load required to buckle the two members if
(a) They act independently of each other.
(b) They are riveted together as shown in Fig. P11-8.

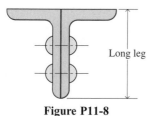

Figure P11-8

**11-9** Determine the maximum allowable compressive load for a 12-ft-long aluminum alloy ($E = 10,600$ ksi) column having the cross section shown in Fig. P11-9 if a factor of safety of 2.50 is specified.

**Challenging Problems**

**11-10\*** A 60-kN load is supported by a tie rod $AB$ and a pipe strut $BC$, as shown in Fig. P11-10. The tie rod has a diameter of 30 mm and is made of steel with a modulus of elasticity of 210 GPa and a yield strength of 360 MPa. The pipe strut has an inside diameter of 50 mm and a wall thickness of 15 mm, and is made of an aluminum alloy with a modulus of elasticity of 73 GPa and a yield strength of 280 MPa. Determine the factor of safety with respect to failure by yielding or buckling for the structure.

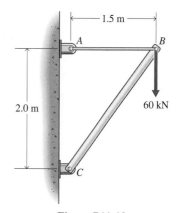

Figure P11-10

**11-11** The 2-in.-diameter solid circular column shown in Fig. P11-11 is made of an aluminum alloy [$E = 10,000$ ksi and $\alpha = 12.5(10^{-6})/°F$]. If the column is initially stress free, determine the temperature increase that will cause the column to buckle.

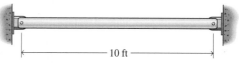

Figure P11-11

**11-12** Two C229 × 30 structural steel channels (see Appendix A for cross-sectional properties) are used for a column that is 12 m long. Determine the total compressive load required to buckle the two members if they are laced 150 mm back to back, as shown in Fig. P11-12.

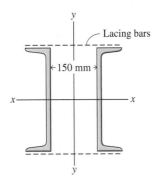

**Figure P11-12**

**11-13*** A simple pin-connected truss is loaded and supported as shown in Fig. P11-13. The members of the truss were fabricated by bolting two C10 × 30 channel sections (see Appendix A for cross-sectional properties) back to back to form an H-section. The channels are made of structural steel with a modulus of elasticity of 29,000 ksi and a yield strength of 36 ksi. Determine the maximum load $P$ that can be applied to the truss if a factor of safety of 1.75 with respect to failure by yielding and a factor of 4 with respect to failure by buckling is specified.

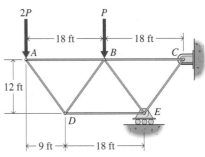

**Figure P11-13**

# 11-3 EFFECTS OF DIFFERENT IDEALIZED END CONDITIONS

The Euler buckling formula, as expressed by either Eq. (11-2) or Eq. (11-3), was derived for a column with pinned ends. The Euler equation changes for columns with different end conditions such as the four common ones shown in Fig. 11-4.

While it is possible to set up the differential equation with the appropriate boundary conditions to determine the Euler equation for each new case (see the Challenging Problems at the end of this section), a more common approach makes use of the concept of an effective length. The pin-ended column, by definition, has zero bending moments at each end. The length $L$ in the Euler equation, therefore, is the distance between successive points of zero bending moment. All that is needed to modify the Euler column formula for use with other end conditions is to replace $L$ by $L'$, where $L'$ is defined as the *effective length* of the column (the distance between two successive inflection points or points of zero moment).

The ends of the column in Fig. 11-4*b* are built in or fixed. Since the deflection curve is symmetrical, the distance between successive points of zero moment (inflection points) is half the length of the column. Thus, the effective length $L'$ of a fixed-end column for use in the Euler column formula is half the true length ($L' = 0.5L$). The column in Fig. 11-4*c*, being fixed at one end and free at the other end, has zero moment only at the free end. If a mirror image of this column is visualized below the fixed end, however, the effective length between points of zero moment is seen to be twice the actual length of the column ($L' = 2L$). The column in Fig. 11-4*d* is fixed at one end and pinned at the other end. The effective length of this column cannot be determined by inspection, as could be done in the previous two cases; therefore, it is necessary to

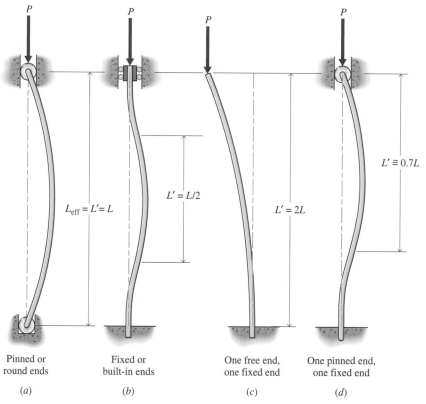

**Figure 11-4**

solve the differential equation to determine the effective length. This procedure yields $L' = 0.7L$.

A pin-ended column is usually loaded through a pin that, as a result of friction, is not completely free to rotate; hence, there will always be an indeterminate moment at the ends of a pin-connected column that will reduce the distance between the inflection points to a value less than $L$. Also, it is impossible to support a column so that all rotation is eliminated; therefore, the effective length of the column in Fig. 11-4$b$ will be somewhat greater than $L/2$. As a result, it is usually necessary to modify the effective column lengths indicated by the ideal end conditions. The amount of the corrections will depend on the individual application. In summary, the term $L/r$ in all column formulas in this book is interpreted to mean the effective slenderness ratio $L'/r$. In the problems in this book, the length given for a member is assumed to be the effective length unless otherwise noted.

## ■ Example Problem 11-4

A structural steel ($E = 29{,}000$ ksi) column 10 ft long must support an axial compressive load $P$, as shown in Fig. 11-5. The column has a 1- $\times$ 2-in. rectangular cross section. The left end of the column is fixed; the pin and bracket arrangement at the right end allows rotation about the pin, but prevents rotation about a vertical axis. Determine the maximum safe load for the column if a factor of safety of 2 with respect to failure by buckling is specified.

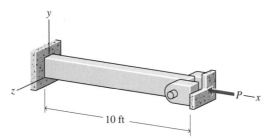

**Figure 11-5**

## SOLUTION

The relationship between factor of safety (FS) and loads is

$$FS = \frac{P_u}{P_{all}} \tag{a}$$

For failure by buckling, the ultimate load $P_u$ is the Euler buckling load $P_{cr}$, which is

$$P_{cr} = \frac{\pi^2 EI}{(L')^2} \tag{b}$$

Substituting Eq. (b) into Eq. (a) and rewriting yields

$$P_{all} = \frac{\pi^2 EI}{(L')^2(FS)} \tag{c}$$

The second moment of area $I$ and the effective length $L'$ depend on the plane in which the column buckles; either the $xy$- or $xz$-plane.

If buckling occurs in the $xy$-plane, the left end of the column is fixed and the right end is pinned. Thus, the effective length $L'$ is

$$L' = 0.7L = 0.7(10)(12) = 84 \text{ in.}$$

and the second moment of area $I$ is

$$I = \frac{1}{12} bh^3 = \frac{1}{12}(1)(2)^3 = 0.6667 \text{ in}^4$$

Therefore, for buckling in the $xy$-plane,

$$P_{all} = \frac{\pi^2 EI}{(L')^2(FS)} = \frac{\pi^2(29)(10^6)(0.6667)}{(84)^2(2)} = 13,522 \text{ lb}$$

For buckling in the $xz$-plane, both the left end of the column and the right end of the column are fixed. Thus, the effective length $L'$ is

$$L' = 0.5L = 0.5(10)(12) = 60 \text{ in.}$$

and the second moment of area $I$ is

$$I = \frac{1}{12} hb^3 = \frac{1}{12} (2)(1)^3 = 0.16667 \text{ in}^4$$

Therefore, for buckling in the $xz$-plane:

$$P_{all} = \frac{\pi^2 EI}{(L')^2 (FS)} = \frac{\pi^2 (29)(10^6)(0.16667)}{(60)^2 (2)} = 6626 \text{ lb}$$

For a column with this cross section and end conditions, buckling will occur in the $xz$-plane and the safe load that may be applied is

$$P_{all} = 6626 \text{ lb} \cong 6.63 \text{ kip} \quad \blacksquare \qquad \textbf{Ans.}$$

# PROBLEMS

## Introductory Problems

**11-14\*** An L102 × 76 × 6.4-mm aluminum alloy ($E = 70$ GPa) angle is used for a fixed-end, pinned-end column having an actual length of 2.5 m. Determine the maximum safe load for the column if a factor of safety of 1.75 with respect to failure by buckling is specified. See Appendix A for cross-sectional properties; they are the same as those for a steel angle of the same size.

**11-15\*** A W8 × 15 structural steel section (see Appendix A for cross-sectional properties) is used for a fixed-end, free-end column having an actual length of 14 ft. Determine the maximum safe load for the column if a factor of safety of 2 with respect to failure by buckling is specified. Use $E = 29,000$ ksi.

**11-16** A W254 × 33 structural steel ($E = 200$ GPa) section is used for a column with an actual length of 6 m. The column can be considered fixed at both ends for bending about the axis of the cross section with the smallest second moment of area and pinned at both ends for bending about the axis with the largest second moment of area. Determine the maximum axial compressive load $P$ that can be supported by the column if a factor of safety of 1.9 with respect to failure by buckling is specified.

## Intermediate Problems

**11-17** A WT7 × 24 structural-steel section (see Appendix A for cross-sectional properties) is used for a column

with an actual length of 45 ft. If the modulus of elasticity for the steel is 29,000 ksi and a factor of safety of 2 with respect to failure by buckling is specified, determine the maximum safe load for the column under the following support conditions:
(a) Pinned-pinned
(b) Fixed-free
(c) Pinned-fixed
(d) Fixed-fixed

**11-18\*** A solid circular rod with diameter $D$, length $L$, and modulus of elasticity $E$ will be used to support an axial compressive load $P$. The support system for the rod is shown in Fig. P11-18. Determine the critical buckling load in terms of $D$, $L$, and $E$ if buckling occurs in the plane of the page.

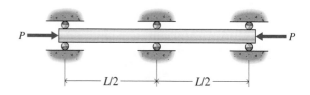

**Figure P11-18**

**11-19\*** The structural steel ($E = 29,000$ ksi) bar shown in Fig. P11-19 has a 1.0-in. diameter and is 6 ft long. The support at the top of the bar permits free vertical movement but no lateral movement or rotation. The bottom of the bar is free to move laterally but cannot rotate. Determine the maximum load $P$ that can be applied if a factor of safety of 2.5 with respect to failure by buckling is specified.

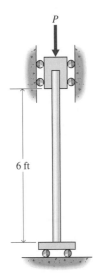

**Figure P11-19**

**11-20** Two 25-mm-diameter structural steel ($E = 200$ GPa) columns support a 400-kg mass, as shown in Fig. P11-20. The two short struts at the sides of the mass prevent horizontal movement of the mass but offer no constraint to vertical motion. Determine the factor of safety with respect to failure by buckling.

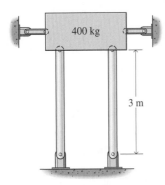

**Figure P11-20**

**Challenging Problems**

**11-21** Verify the effective length for the fixed-end, fixed-end column shown in Fig. 11-4b by solving the differential equation of the elastic curve and applying the appropriate boundary conditions. Place the origin of the

$xy$-coordinate system at the lower end of the column with the $x$-axis along the axis of the undeformed column.

**11-22** Verify the effective length for the fixed-end, free-end column shown in Fig. 11-4c by solving the differential equation of the elastic curve and applying the appropriate boundary conditions. Place the origin of the $xy$-coordinate system at the lower end of the column with the $x$-axis along the axis of the undeformed column.

**11-23** A free-body diagram for the fixed-end, pinned-end column shown in Fig. 11-4d is shown in Fig. P11-23. Use the free-body diagram to develop the differential equation of the elastic curve. Verify that the effective length for the fixed-end, pinned-end column is $L' = 0.7L$ by solving the differential equation of the elastic curve and applying the appropriate boundary conditions.

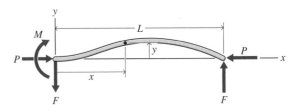

**Figure P11-23**

**11-24** A rigid block is supported by two fixed-end, fixed-end columns, as shown in Fig. P11-24a. Determine the effective lengths of the columns by solving the differential equation of the elastic curve and applying the appropriate boundary conditions. Assume that buckling occurs as shown in Fig. P11-24b.

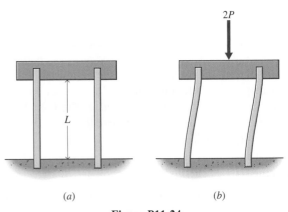

*(a)*      *(b)*

**Figure P11-24**

# 11-4 EMPIRICAL COLUMN FORMULAS—CENTRIC LOADING

Euler's formula, Eq. (11-2) or (11-3), is valid when the axial compressive stress for a column is less than the yield strength. The range of usefulness of the Euler formula is seen in Fig. 11-6, where the critical stress ($\sigma_{cr} = P_{cr}/A$) is plotted versus the slenderness ratio ($L/r$). The plot is for structural steel for which $E = 29{,}000$ ksi (200 GPa) and $\sigma_{\text{yield}} = 36$ ksi (250 MPa). The Euler curve is truncated at 36 ksi because the critical stress cannot exceed the yield strength. For structural steel, this occurs when $L/r = 89$. The dashed portion of the Euler curve indicates that the calculation of the critical stress is no longer useful because the stress exceeds the yield strength. For practical purposes, columns with large slenderness ratios are useless because they support only small loads. At the low end of the slenderness scale, the column would behave essentially as a short compression block and the critical stress would be the compressive strength of the material (for metals, usually the yield strength). The extent of this range, the *compression block* range, is a matter of judgment or is dictated by specifications. The range between the compression block and the slender ranges is known as the *intermediate* range. Neither compression-block theory nor the Euler formula gives results in the intermediate range that agree with test results.

Experimental results of many tests on axially loaded columns are shown in Fig. 11-7. The Euler curve agrees with experimental data for slenderness ratios in the slender range. Experimental data and compression-block theory agree in the compression-block range. Test results are not in agreement with compression-block theory or the Euler theory in the intermediate range. These intermediate-length columns may be analyzed by empirical formulas.

For design purposes, the entire range of stresses for a given material is covered by an appropriate set of specifications known as a column code. Depending on the code, the empirical formula may be specified for the intermediate range along with the limits of the intermediate range; the code may also specify the Euler formula in the slender range and the limits of the slender range; or the

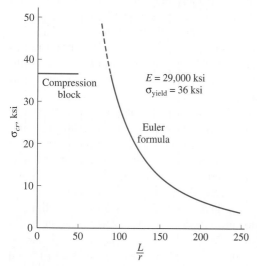

**Figure 11-6**

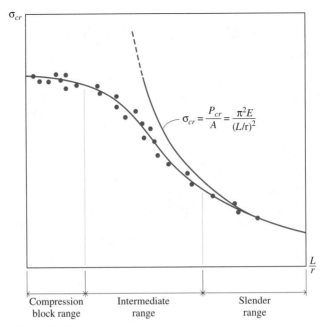

The equation shown in the figure:

$$\sigma_{cr} = \frac{P_{cr}}{A} = \frac{\pi^2 E}{(L/r)^2}$$

Axis label: $\sigma_{cr}$ (vertical), $\frac{L}{r}$ (horizontal)

Range labels: Compression block range | Intermediate range | Slender range

**Figure 11-7**

code may specify the critical stress and limits of the range for the compression-block range. If the code is written for allowable loads or stresses, the factor of safety will be either specified or included in the constants for the empirical formula. A few representative codes are listed in Table 11-1. Each code is written for allowable (or safe) stresses. A few representative codes for steel, aluminum, and timber will be discussed. The codes are taken from references listed in the footnotes of Table 11-1.

The formulas in Table 11-1 represent column equations that have been incorporated in various design codes. Slenderness ratios in this table are always the effective slenderness ratio $L'/r$. If no end conditions are specified in the problems presented later, the stated length is the effective length. Note that the use of high-strength materials will increase the allowable load ($P_{all} = \sigma_{all}A$) for short columns but will have little effect on the load carrying capacity of long columns, because the critical load (Euler load) depends on the modulus of elasticity, not on the elastic strength of the material. Note also that the use of fixed or restrained ends, which has the effect of reducing the length of the column, materially increases the load-carrying capacity of slender columns but has much less influence on short compression members.

The discussion so far has been concerned with primary instability, in which the column deflects as a whole into a smooth curve. No discussion of compression loading is complete without reference to local instability in which the member fails locally by crippling of thin sections. Thin open sections such as angles, channels, and H-sections are particularly sensitive to crippling failure. The design of such members to avoid crippling failure is usually governed by specifications controlling the width–thickness ratios of outstanding flanges. Closed-section members of thin material (thin-walled tubes, for example) must also be examined for crippling failure when the members are short. A discussion of these design issues may be found in references such as (a) and (b) listed in Table 11-1.

TABLE 11-1 Some Representative Column Codes for Centric Loading

| Code No. | Source | Material | Compression-Block and/or Intermediate-Range Formulas and Limitations ($L/r$ is the effective ratio $L'/r$) | | Slender Range |
|---|---|---|---|---|---|
| 1 | a | Structural steel with a yield point $\sigma_y$ | $0 \leq \dfrac{L}{r} \leq C_c$ | $\sigma_{all} = \dfrac{\sigma_y}{FS}\left[1 - \dfrac{1}{2}\left(\dfrac{L/r}{C_c}\right)^2\right]$ | $\dfrac{L}{r} \geq C_c$ |
| | | | | $C_c^2 = \dfrac{2\pi^2 E}{\sigma_y}$ | $\sigma_{all} = \dfrac{\pi^2 E}{1.92(L/r)^2}$ |
| | | | | $FS = \dfrac{5}{3} + \dfrac{3}{8}\left(\dfrac{L/r}{C_c}\right) - \dfrac{1}{8}\left(\dfrac{L/r}{C_c}\right)^3$ | |
| 2 | b | 2014-T6 (Alclad) Aluminum alloy | $\dfrac{L}{r} \leq 12$ | $\sigma_{all} = 28$ ksi $= 193$ MPa | $\dfrac{L}{r} \geq 55$ |
| | | | $12 \leq \dfrac{L}{r} \leq 55$ | $\sigma_{all} = \left[30.7 - 0.23\left(\dfrac{L}{r}\right)\right]$ ksi $= \left[212 - 1.585\left(\dfrac{L}{r}\right)\right]$ MPa | $\sigma_{all} = \dfrac{54{,}000}{(L/r)^2}$ ksi $= \dfrac{372(10^3)}{(L/r)^2}$ MPa |
| 3 | b | 6061-T6 Aluminum alloy | $\dfrac{L}{r} \leq 9.5$ | $\sigma_{all} = 19$ ksi $= 131$ MPa | $\dfrac{L}{r} \geq 66$ |
| | | | $9.5 \leq \dfrac{L}{r} \leq 66$ | $\sigma_{all} = \left[20.2 - 0.126\left(\dfrac{L}{r}\right)\right]$ ksi $= \left[139 - 0.868\left(\dfrac{L}{r}\right)\right]$ MPa | $\sigma_{all} = \dfrac{51{,}000}{(L/r)^2}$ ksi $= \dfrac{351(10^3)}{(L/r)^2}$ MPa |
| 4 | c | Timber with a rectangular cross section $b \times d$ where $d < b$ | $\dfrac{L}{d} \leq 11$ | $\sigma_{all} = F_c{}^*$ | $k \leq \dfrac{L}{d} \leq 50$ |
| | | | $11 \leq \dfrac{L}{d} \leq k$ | $\sigma_{all} = F_c\left[1 - \dfrac{1}{3}\left(\dfrac{L/d}{k}\right)^4\right]$ $k = 0.671\sqrt{E/F_c}$ | $\sigma_{all} = \dfrac{0.30E}{(L/d)^2}$ |

a. *Manual of Steel Construction*, 9th ed., American Institute of Steel Construction, New York, 1959.

b. *Specifications for Aluminum Structures*, Aluminum Association, Inc., Washington, D.C., 1986.

c. *Timber Construction Manual*, 3rd ed., American Institute of Timber Construction, John Wiley & Sons, Inc., New York, 1985.

*$F_c$ is the allowable stress for a short block in compression parallel to the grain.

## Example Problem 11-5

Two structural steel C10 × 25 channels are laced 5 in. back-to-back; as shown in Figs. 11-8a and b, to form a column. Determine the maximum allowable axial load for effective lengths of 25 ft and 40 ft. Use Code 1 for structural steel (see Appendix A for properties).

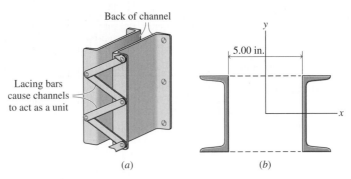

**Figure 11-8**

## SOLUTION

Both $I_x$ and $I_y$ (see Fig. 11-8*b*) or $r_x$ and $r_y$ must be known to determine the minimum radius of gyration. Properties of the channel section are given in Appendix A. The value of $I_y$ for two channels is obtained by using the parallel axis theorem: thus,

$$I_y = 2(I_C + Ad^2) = 2(Ar_C^2 + Ad^2)$$

and

$$r_y = \sqrt{I_y/(2A)} = \sqrt{\frac{2A(r_C^2 + d^2)}{2A}} = \sqrt{r_C^2 + d^2}$$

This expression indicates that the radius of gyration for the two channels is the same as that for one channel, a fact obtained directly from the definition of radius of gyration. Then,

$$r_y = \sqrt{(0.676)^2 + (2.50 + 0.617)^2} = 3.19 \text{ in.}$$

which is less than the tabular value of 3.52 in. for $r_x$. This means that the column tends to buckle with respect to the *y*-axis, and the slenderness ratios are

$$\frac{L}{r} = \frac{12(25)}{3.19} = 94.0 \quad \text{and} \quad \frac{12(40)}{3.19} = 150.5$$

From Code 1,

$$C_c^2 = \frac{2\pi^2 E}{\sigma_y} = \frac{2\pi^2(29{,}000)}{36} = 15{,}901 \qquad C_c = 126.1$$

The slenderness ratios above indicate that the 25-ft column is in the intermediate range because $L/r < C_c$ or $94.0 < 126.1$; hence, the factor of safety is

$$\text{FS} = \frac{5}{3} + \frac{3}{8}\left(\frac{94.0}{126.1}\right) - \frac{1}{8}\left(\frac{94.0}{126.1}\right)^3 = 1.894$$

The allowable stress is

$$\sigma_{\text{all}} = \frac{\sigma_y}{\text{FS}}\left[1 - \frac{1}{2}\left(\frac{L/r}{C_c}\right)^2\right] = \frac{36(10^3)}{1.894}\left[1 - \frac{1}{2}\left(\frac{94.0}{126.1}\right)^2\right] = 13{,}726 \text{ psi}$$

Hence, the safe load for the 25-ft column is

$$P = \sigma_{all}(A) = 13{,}726(2)(7.35) = 201.8(10^3) \text{ lb} \cong 202 \text{ kip} \qquad \textbf{Ans.}$$

The 40-ft column has a slenderness ratio of 150.5; therefore, it is in the slender range, since $L/r > C_c$ or $150.5 > 126.1$. Hence,

$$\sigma_{all} = \frac{\pi^2 E}{1.92(L/r)^2} = \frac{\pi^2 (29)(10^6)}{1.92(150.5)^2} = 6581 \text{ psi}$$

Hence, the safe load for the 40-ft column is

$$P = \sigma_{all}(A) = 6581(2)(7.35) = 96.74(10^3) \text{ lb} \cong 96.7 \text{ kip} \qquad \textbf{Ans.}$$

The Euler buckling load is inversely proportional to the length squared, $L^2$. Therefore, it might be expected that reducing the length of a column by a factor $n$ (in this case $n = 40/25 = 1.6$) would increase the buckling load by a factor of $n^2$ [in this case $n^2 = (1.6)^2 = 2.56$]. In this case, however, the safe load for the 25-ft-long column is only about 2.1 times that for the 40-ft-long column. ∎

## ▌Example Problem 11-6

A Douglas fir ($E = 11$ GPa and $F_c = 7.6$ MPa) timber column with an effective length of 3.5 m has a 150- $\times$ 200-mm rectangular cross section. Determine the maximum compressive load permitted by Code 4 of Table 11-1.

### SOLUTION

Code 4 of Table 11-1 indicates that the allowable stress (and hence, the allowable load) depends on the value of $L/d$, the ratio of the effective length of the column to the smallest cross-sectional dimension and a factor $k$ specified by the code. These values are

$$\frac{L}{d} = \frac{3.5(10^3)}{150} = 23.33$$

and

$$k = 0.671 \sqrt{\frac{E}{F_c}} = 0.671 \sqrt{\frac{11(10^9)}{7.6(10^6)}} = 25.53$$

Since $11 \leq L/d \leq k$ or $11 < 23.33 < 25.53$, the column is in the intermediate range and the allowable stress is given by the code as

$$\sigma_{all} = F_c \left[ 1 - \frac{1}{3} \left( \frac{L/d}{k} \right)^4 \right] = 7.6 \left[ 1 - \frac{1}{3} \left( \frac{23.33}{25.53} \right)^4 \right] = 5.833 \text{ MPa}$$

Thus, the allowable load is

$$P = \sigma_{all}(A) = 5.833(10^6)(150)(200)(10^{-6})$$
$$= 174.99(10^3) \text{ N} \cong 175.0 \text{ kN} \qquad \textbf{Ans.}$$

Note that $L/d = 23.33$ is the ratio of the effective length of the column to the smallest cross-sectional dimension and not the slenderness ratio

$$\frac{L}{r} = \frac{L}{\sqrt{I/A}} = \frac{3500}{\sqrt{(56.25 \times 10^6)/(30 \times 10^3)}} = 80.83$$

The corresponding Euler buckling load for this column would be

$$P_{cr} = \frac{\pi^2 EA}{(L/r)^2} = \frac{\pi^2 (11)(10^9)(30)(10^{-3})}{(80.83)^2} = 499(10^3) \text{ N}$$

which is 2.8 times larger than the allowable load permitted by Code 4 of Table 11-1. ■

## PROBLEMS

### Introductory Problems

**11-25\***  An air-dried red oak ($E = 1800$ ksi and $F_c = 4.6$ ksi) timber column with an effective length of 5 ft has a 3- × 3-in. rectangular cross section. Determine the maximum compressive load permitted by Code 4.

**11-26\***  A W254 × 89 structural steel ($E = 200$ GPa and $\sigma_y = 250$ MPa) column is pinned at both ends, is 4.5 m long, and supports an axial compressive load $P$. Determine the maximum load permitted by Code 1.

**11-27**  A 3-in.-diameter solid circular 6061–T6 aluminum alloy bar is to be used as a column with an effective length of 30 in. Determine the maximum axial compressive load $P$ permitted by Code 3.

### Intermediate Problems

**11-28\***  Three hollow circular structural steel tubes with inside diameters of 50 mm and outside diameters of 80 mm will be used for a 3.5-m-long pin-ended column. Determine the maximum compressive load permitted by Code 1 if
(a) The three tubes act as independent axially loaded members.
(b) The three tubes are welded together as shown in Fig. P11-28.

**Figure P11-28**

**11-29\***  Two C10 × 15.3 structural steel channels 12 ft long are used as a fixed-ended, pin-ended column. Determine the maximum load permitted by Code 1 if the channels are welded together to form a 10- × 5.2-in. box section, as shown in Fig. P11-29.

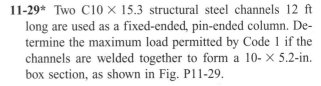
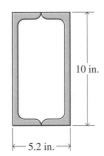

**Figure P11-29**

**11-30**  Four L76 × 76 × 12.7-mm structural steel angles 6 m long are used as a fixed-ended column. Determine the maximum load permitted by Code 1 if the angles are fastened together as shown in Fig. P11-30.

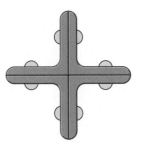

**Figure P11-30**

**11-31**  Four L4 × 3 × ³/₈-in. structural steel angles 13 ft long are used as a pin-ended column. Determine the maximum load permitted by Code 1 if the angles are welded together to form a 6- × 8-in. box section as shown in Fig. P11-31.

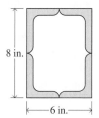

**Figure P11-31**

**11-32\*** A column of aluminum alloy 2014-T6 is composed of two L127 × 127 × 19.1-mm angles riveted together as shown in Fig. P11-32. The length between end connections is 3 m, and the end connections are such that there is no restraint to bending about the *y*-axis, but restraint to bending about the *x*-axis reduces the effective length to 2.1 m. Determine the maximum axial compressive load permitted by Code 2. The cross-sectional properties of aluminum and steel angles are the same (see Appendix A).

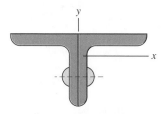

**Figure P11-32**

**11-33\*** A strut of aluminum alloy 6061-T6 having the cross section shown in Fig. P11-33 is to carry an axial compressive load of 175 kip. The strut is 4 ft long and is fixed at the bottom and pinned at the top. Determine the required dimension *d* of the cross section by using Code 3.

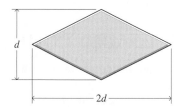

**Figure P11-33**

**11-34** A sand bin is supported by four 200- × 250-mm fixed-ended rectangular timber columns 6 m long. Assume that the load is equally divided among the columns and that the column load is axial. If the columns are made of timber with $E = 13$ GPa and $F_c = 9$ MPa, determine the maximum load permitted by Code 4.

**11-35** Three 2- × 4-in. timber studs are nailed together to form an 8-ft-long column with the cross section shown in Fig. P11-35. The columnn is fixed at the bottom and pinned at the top. If $E = 1600$ ksi and $F_c = 1100$ psi, use Code 4 to determine the maximum permissible axial compressive load that may be applied.

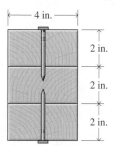

**Figure P11-35**

## Challenging Problems

**11-36\*** Four C178 × 22 structural steel channels 12 m long are used to fabricate a column with the cross section shown in Fig. P11-36. The column is fixed at the base and pinned at the top. The pin at the top offers no restraint to bending about the *y*-axis, but for bending about the *x*-axis, the pin provides restraint sufficient to reduce the effective length to 7.5 m. Determine the maximum axial compressive load permitted by Code 1.

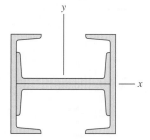

**Figure P11-36**

**11-37\*** A 25-ft-long plate and angle column consists of four L5 × 3½ × ½-in. structural steel angles riveted to a 10- × ½-in. structural steel plate, as shown in Fig. P11-37. Determine the maximum safe load permitted by Code 1
(a) If the column is pinned at both ends.
(b) If the column is fixed at the base and pinned at the top.

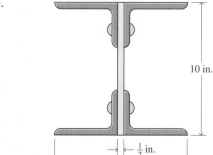

**Figure P11-37**

**11-38** A connecting rod made of SAE 4340 heat-treated steel has the cross section shown in Fig. P11-38. The pins at the end of the rod are parallel to the *x*-axis and are 1250 mm apart. Assume that the pins offer no restraint to bending about the *x*-axis but provide essentially complete fixity for bending about the *y*-axis. Determine the maximum axial compressive load permitted by Code 1.

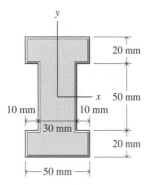

**Figure P11-38**

**11-39** The machine part shown in Fig. P11-39 is made of SAE 4340 heat-treated steel and carries an axial compressive load. The pins at the ends off no restraint to bending about the axis of the pin, but restraint about the perpendicular axis reduces the effective length to 36 in. Determine the maximum load permitted by Code 1.

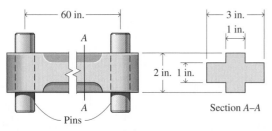

**Figure P11-39**

**Computer Problems**

**11-40** Two L152 × 89 × 12.7-mm angles of aluminum alloy 2014-T6 are welded together as shown in Fig. P11-40 to form a pin-ended column. The pins provide no restraint to bending about the *x*-axis but reduce the effective length to 0.7*L* for bending about the *y*-axis. The sectional properties for one angle are given on the figure where *C* is the centroid of one angle. Compute and plot the maximum compressive load permitted by Code 2 as a function of the column length *L* (0.2 m ≤ *L* ≤ 5.0 m).

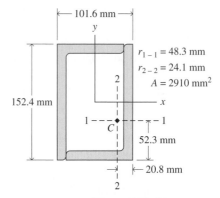

**Figure P11-40**

**11-41** Three 2- × 4-in. timber studs are nailed together to form a column with the cross section shown in Fig. P11-35. The column is fixed at the bottom and pinned at the top. If *E* = 1600 ksi and $F_c$ = 1100 psi, compute and plot the maximum permissible axial compressive load that may be applied as a function of the column length *L* (2 ft ≤ *L* ≤ 20 ft). Use Code 4.

# 11-5 ECCENTRICALLY LOADED COLUMNS

Although a given column will support the maximum load when the load is applied centrically, it is sometimes necessary to apply an eccentric load to a column. For example, a beam supporting a floor load in a building may in turn be supported by an angle riveted or welded to the side of a column, as shown in Fig. 11-9. Frequently, the floorbeam is framed into the column with a stiff connection, and the column is then subjected to a bending moment due to continuity of the floorbeam and column. The eccentricity of the load, or the bending moment, will increase the stress in the column or reduce its load-carrying capacity. Two methods will be presented for computing the allowable load on a column subjected to an eccentric load (or an axial load combined with a bending moment).

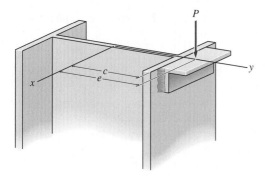

**Figure 11-9**

**Allowable Stress Method** This method is based on the specification that the sum of the direct and bending stresses $(P/A + Mc/I)$ shall not exceed the allowable stress prescribed for a centric load by the appropriate column formula,

$$\frac{P}{A} + \frac{Mc}{I} \leq \sigma_{\text{all}} \tag{11-4}$$

In Eq. (11-4), $\sigma_{\text{all}}$ is the allowable stress for a centric load and is calculated using the equations of Table 11-1 and using the largest value of the slenderness ratio for the cross section irrespective of the axis about which bending occurs. Values for $c$ and $I$ used in calculating the bending stress, however, depend on the axis about which the bending occurs. Equation (11-4), which is prescribed by most modern codes, usually produces a conservative design.

**Interaction Method** One of the modern expressions for treating combined loads is known as the *interaction formula*, of which several types are in use. The analysis for compression members subjected to bending and direct stress may be derived as follows. It is assumed that the stress in the column can be written as

$$\frac{P}{A} + \frac{Mc}{I} \leq \sigma_{\text{all}}$$

and, when this expression is divided by $\sigma_{\text{all}}$, it becomes

$$\frac{P/A}{\sigma_{\text{all}}} + \frac{Mc/I}{\sigma_{\text{all}}} \leq 1 \tag{11-5}$$

When considering eccentrically loaded columns, the value of $\sigma_{\text{all}}$ will, in general, be different for the two terms. In the first term $P/A$ represents an axial stress on a column; therefore, the value of $\sigma_{\text{all}}$ should be the average allowable stress on an axially loaded column as obtained by an empirical formula such as those presented in Table 11-1. In the second term, $Mc/I$ represents the flexural stress induced in the member as a result of the eccentricity of the load or an applied bending moment; therefore, the corresponding value of $\sigma_{\text{all}}$ should be the allowable flexural stress. Since the two values of $\sigma_{all}$ are different, one recommended form of Eq. (11-5) is the following interaction formula

$$\frac{P/A}{\sigma_a} + \frac{Mc/I}{\sigma_b} \leq 1 \qquad (11\text{-}6)$$

in which $P/A$ is the average axial stress on the eccentrically loaded column, $\sigma_a$ is the allowable average axial stress for an axially loaded column (note that the greatest value of $L/r$ should be used to calculate $\sigma_a$), $Mc/I$ is the flexural stress in the column, and $\sigma_b$ is the allowable flexural stress.

When Eq. (11-6) is used to select the most economical section, it will usually not be possible to obtain a section that will exactly satisfy the equation. Any section that makes the sum of the terms on the left side of the equation less than unity is considered safe, and the safe section giving the largest sum (less than unity) is the most efficient section.

## ▍ Example Problem 11-7

A W457 × 144 wide-flange section is used for the column shown in Fig. 11-10. The column is made of steel ($E$ = 200 GPa, $\sigma_y$ = 290 MPa, $\sigma_b$ = 190 MPa) and has an effective length of 6 m. An eccentric load $P$ ($e$ = 125 mm) is applied on the centerline of the web as shown in the figure. Determine the maximum safe load according to

(a) The allowable stress method.
(b) The interaction method.

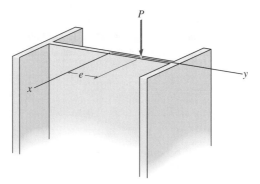

**Figure 11-10**

## SOLUTION

The cross-sectional properties for a W457 × 144 wide-flange section are

$$A = 18{,}365 \text{ mm}^2 \qquad c = 472/2 = 236 \text{ mm}$$

$$r_{min} = 67.3 \text{ mm} \qquad I_{min} = 83.7(10^6) \text{ mm}^4$$

$$r_{max} = 199 \text{ mm} \qquad I_{max} = 728(10^6) \text{ mm}^4$$

Since the column is made of steel, Code 1 of Table 11-1 will be used to calculate $\sigma_{all}$. The equation for determining $\sigma_{all}$ depends on the slenderness range.

$$C_c = \sqrt{\frac{2\pi^2 E}{\sigma_y}} = \sqrt{\frac{2\pi^2(200)(10^9)}{290(10^6)}} = 116.68$$

$$\frac{L}{r_{min}} = \frac{6(10^3)}{67.3} = 89.15 < C_c = 116.68$$

Since $L/r_{min} < C_c$, the intermediate column formula is applicable. The factor of safety FS for use in the intermediate column formula is

$$FS = \frac{5}{3} + \frac{3}{8}\left(\frac{L/r}{C_c}\right) - \frac{1}{8}\left(\frac{L/r}{C_c}\right)^3 = \frac{5}{3} + \frac{3}{8}\left(\frac{89.15}{116.68}\right) - \frac{1}{8}\left(\frac{89.15}{116.68}\right)^3 = 1.90$$

$$\sigma_{all} = \frac{\sigma_y}{FS}\left[1 - \frac{1}{2}\left(\frac{L/r}{C_c}\right)^2\right] = \frac{290(10^6)}{1.90}\left[1 - \frac{1}{2}\left(\frac{89.15}{116.68}\right)^2\right]$$

$$= 108.08(10^6) \text{ N/m}^2 = 108.08 \text{ MPa} = \sigma_a$$

(a) Using the allowable stress method, Eq. (11-4) gives

$$\frac{P}{A} + \frac{Mc}{I} = \frac{P}{A} + \frac{Pec}{I} \leq \sigma_{all}$$

$$\frac{P}{18,365(10^{-6})} + \frac{P(125)(236)(10^{-6})}{728(10^{-6})} \leq 108.08(10^6)$$

$$P \leq 1138(10^3) \text{ N}$$

Therefore, the maximum safe load that can be applied according to the allowable stress method is

$$P_{max} = 1138(10^3) \text{ N} = 1138 \text{ kN} \qquad \textbf{Ans.}$$

(b) Using the interaction method, Eq. (11-6) gives

$$\frac{P/A}{\sigma_a} + \frac{Mc/I}{\sigma_b} = \frac{P/A}{\sigma_a} + \frac{Pec/I}{\sigma_b} \leq 1$$

$$\frac{P/[18,365(10^{-6})]}{108.08(10^6)} + \frac{P(125)(236)(10^{-6})/[728(10^{-6})]}{190(10^6)} \leq 1$$

$$P \leq 1395(10^3) \text{ N}$$

Therefore, the maximum safe load that can be applied according to the interaction method is

$$P_{max} = 1395(10^3) \text{ N} = 1395 \text{ kN} \qquad \blacksquare \qquad \textbf{Ans.}$$

## PROBLEMS

### Introductory Problems

**11-42\*** A hollow circular steel member with outside and inside diameters of 150 mm and 120 mm, respectively, functions as a pin-ended column 4 m long. Determine the maximum load that the column can carry if the load is applied 20 mm from the axis of the member, as shown in Fig. P11-42. Use the allowable stress method and let $E = 200$ GPa and $\sigma_y = 250$ MPa.

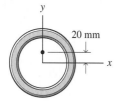

**Figure P11-42**

**11-43\*** A 2-in.-diameter steel strut is subjected to an eccentric compressive load, as shown in Fig. P11-43. The effective length for bending about the $x$-axis is 75 in.; but for bending about the $y$-axis, the end conditions reduce the effective length to 50 in. Determine the maximum load the strut can carry. Use the interaction method with $E = 29,000$ ksi, $\sigma_y = 36$ ksi, and $\sigma_b = 24$ ksi.

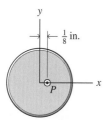

**Figure P11-43**

**11-44** Determine the maximum load $P$ that can be applied to the timber column shown in Fig. P11-44 if $E = 12$ GPa and the allowable stress for compression parallel to the grain is 9 MPa. Use the allowable stress method.

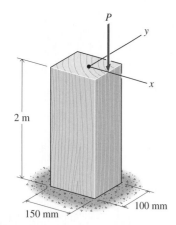

**Figure P11-44**

### Intermediate Problems

**11-45** Two C8 × 18.75 structural steel sections 25 ft long are laced 3 in. back to back as shown in Fig. P11-45 to form a pin-ended column. Determine the maximum load permitted by the interaction method if the load is applied at point $A$ of the cross section. Let $E = 29,000$ ksi, $\sigma_y = 36$ ksi, and $\sigma_b = 24$ ksi.

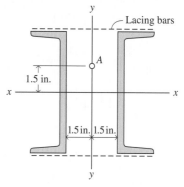

**Figure P11-45**

**11-46\*** A W356 × 64 structural steel section is used for a 12-m-long fixed-ended column. The load is applied at a point on the centerline of the web 150 mm from the axis of the column. If $E = 200$ GPa, $\sigma_y = 250$ MPa, and $\sigma_b = 160$ MPa, determine the maximum safe load according to
(a) The allowable stress method.
(b) The interaction method.

**11-47** A hollow square steel member with outside and inside dimensions of 4 in. and 2.8 in., (walls are 0.6 in. thick) functions as a pin-ended column 13 ft long. Determine the maximum load that the column can carry if the load is applied with a known eccentricity of 0.6 in. along a diagonal of the square as shown in Fig. P11-47. Use the allowable stress method and let $E = 29,000$ ksi and $\sigma_y = 36$ ksi.

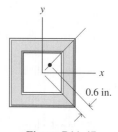

**Figure P11-47**

## Challenging Problems

**11-48\*** Two 50- × 150-mm structural steel bars 5 m long will be welded together as shown in Fig. P11-48 and used for a pin-ended column. Determine the maximum load permitted if the load is applied at the point on the cross section indicated in Fig. P11-48. Use the allowable stress method and let $E = 200$ GPa and $\sigma_y = 250$ MPa.

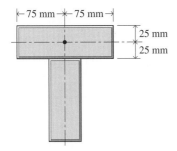

**Figure P11-48**

**11-49** A WT8 × 25 structural steel T-section is to be used as a compression member to transmit an eccentric load of 100 kip. The effective length of the member is 12 ft. Determine the maximum eccentricity $e$ permitted if the point of load application is on the centerline of the stem between the outer edge of the flange and the centroidal axis of the column. Use the interaction method with $E = 29{,}000$ ksi, $\sigma_y = 36$ ksi, and $\sigma_b = 24$ ksi.

**11-50** A 2014-T6 aluminum alloy compression member with an effective length of 1.25 m has the T-cross section shown in Fig. P11-50. Determine
(a) The maximum axial compressive load $P$ permitted by Code 2.
(b) Use the allowable stress method to determine the maximum bending moment $M$ in the $yz$-plane that can be applied as shown when the column is supporting a 175-kN axial load.

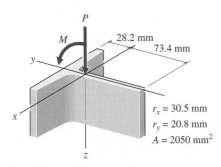

**Figure P11-50**

**11-51\*** Two L6 × $3\frac{1}{2}$ × $\frac{1}{2}$-in. structural steel ($E = 29{,}000$ ksi and $\sigma_y = 36$ ksi) angles are welded together, as shown in Fig. P11-51, to form a column with an effective length of 15 ft. Use the allowable stress method to determine
(a) The maximum axial compressive load $P$ that can be supported by the column.
(b) The maximum and minimum values for the distance $d$ when a 50-kip load is applied.

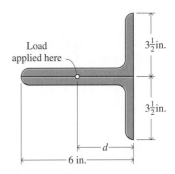

**Figure P11-51**

## Computer Problems

**11-52** A W457 × 144 wide-flange section is used for the column shown in Fig. P11-52. The column is made of steel ($E = 200$ GPa, $\sigma_y = 290$ MPa, $\sigma_b = 190$ MPa). An eccentric load $P$ is applied on the centerline of the web as shown in the figure.
(a) Calculate and plot the maximum safe load as a function of the eccentricity $e$ (0 mm ≤ $e$ ≤ 236 mm) as given by the allowable stress method for an effective length of 6 m.
(b) Calculate and plot (on the same graph) the maximum safe load as a function of the eccentricity $e$ (0 mm ≤ $e$ ≤ 236 mm) as given by the interaction method for an effective length of 6 m.
(c) Repeat parts $a$ and $b$ for an effective length of 9 m.

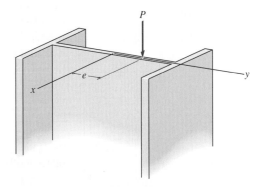

**Figure P11-52**

**11-53** Two C8 × 18.75 structural steel sections are laced 3 in. back to back, as shown in Fig. P11-53 to form a pin-ended column. An eccentric load $P$ is applied at point $A$ along the $y$-axis as shown in the figure. If $E =$ 29,000 ksi, $\sigma_y = 36$ ksi, and $\sigma_b = 24$ ksi,

(a) Calculate and plot the maximum safe load as a function of the eccentricity $e$ (0 in. $\leq e \leq 4$ in.) as given by the allowable stress method for an effective length of 15 ft.

(b) Calculate and plot (on the same graph) the maximum safe load as a function of the eccentricity $e$ (0 in. $\leq e \leq 4$ in.) as given by the interaction method for an effective length of 15 ft.

(c) Repeat parts $a$ and $b$ for an effective length of 25 ft.

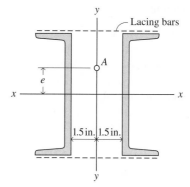

**Figure P11-53**

## 11-6 DESIGN

In previous chapters design usually involved strength as a controlling parameter. Since buckling is an elastic phenomenon, the modulus of elasticity (stiffness) is a more significant parameter than is strength (such as yield strength if failure is by yielding), if the column length is in the slender range. If the column is in the intermediate range, both yield strength and stiffness may be important parameters. When designing columns using the representative codes listed in Table 11-1, a designer must be aware of several factors. The codes are for specific materials, that is, materials with a specific value of yield strength and modulus of elasticity. In addition, some codes include a factor of safety, while others require that a factor of safety be introduced. Each of the codes has a range of applicability for the slenderness ratio. Finally, all codes in Table 11-1 are for axially loaded members.

For example, Code 2 is limited to a specific material, 2014-T6 aluminum alloy. The factor of safety (FS) is included in the Code. If the slenderness ratio lies between 12 and 55, the column is in the intermediate range and the empirical column formula is valid. If the slenderness ratio exceeds 55, a form of the Euler formula is used with a factor of safety of approximately 1.94 included. For a slenderness ratio less than 12, the axially loaded member is in the compression block range where buckling does not occur.

In this text, design will be limited to columns subjected to axial loads. Factors such as residual stress, out-of-straightness, and local buckling may be found in design codes established by professional organizations.

The following examples illustrate the use of the Code in Table 11-1.

## Example Problem 11-8

Select the lightest structural steel wide flange section listed in Appendix A to support an axial compressive load of 150 kip as a 15-ft column. Use Code 1.

## SOLUTION

When a rolled section is to be selected to support a specified load, it is usually necessary to make several trial solutions, since there is no direct relationship between areas and radii of gyration for different structural shapes. The best section is usually the section with the least area (smallest mass) that will support the load. A minimum area can be obtained by assuming $L/r = 0$. The load-carrying capacity of various sections with areas larger than this minimum can then be calculated, using the proper column formula, to determine the lightest one that will carry the specified load.

If $L/r$ is small,

$$\text{FS} \cong \frac{5}{3} = 1.667$$

$$\sigma_{\text{all}} = \frac{\sigma_y}{\text{FS}} = \frac{36,000}{1.667} = 21,596 \text{ psi}$$

$$A_{\min} = \frac{P}{\sigma_{\text{all}}} = \frac{150,000}{21,596} = 6.946 \text{ in}^2$$

A column should be selected from Appendix A with an area greater than 6.95 in² for the first trial. In this case, try a W8 × 24 section for which $A$ is 7.08 in² and $r_{\min}$ is 1.61 in. The value of $L/r$ for this column is

$$\frac{L}{r} = \frac{15(12)}{1.61} = 111.8$$

To determine which of the equations for the allowable stress is applicable, first determine the value of $C_c$.

$$C_c = \sqrt{\frac{2\pi^2 E}{\sigma_y}} = \sqrt{\frac{2\pi^2(29)(10^6)}{36,000}} = 126.1$$

Since $L/r = 111.8 < 126.1$, the column is in the intermediate range where the factor of safety is

$$\text{FS} = \frac{5}{3} + \frac{3}{8}\left(\frac{L/r}{C_c}\right) - \frac{1}{8}\left(\frac{L/r}{C_c}\right)^3 = \frac{5}{3} + \frac{3}{8}\left(\frac{111.8}{126.1}\right) - \frac{1}{8}\left(\frac{111.8}{126.1}\right)^3 = 1.912$$

and the allowable stress is

$$\sigma_{\text{all}} = \frac{\sigma_y}{\text{FS}}\left[1 - \frac{1}{2}\left(\frac{L/r}{C_c}\right)^2\right]$$

$$= \frac{36,000}{1.912}\left[1 - \frac{1}{2}\left(\frac{111.8}{126.1}\right)^2\right] = 11,428 \text{ psi}$$

The allowable load is

$$P_{\text{all}} = \sigma_{\text{all}}(A) = 11,428(7.08) = 80,910 \text{ lb} = 80.9 \text{ kip}$$

This load is less than the design load; therefore, a column with either a larger area, a larger radius of gyration, or both must be investigated. As a second trial value, use a W12 × 30 section for which $A$ is 8.79 in² and $r$ is 1.52 in. For this section $L/r$ is 118.4 (intermediate range), the factor of safety is 1.92, the allowable stress is 10,485 psi, and the load this column can support is

$$P_{all} = \sigma_{all}(A) = 10,485(8.79) = 92,200 \text{ lb} = 92.2 \text{ kip}$$

This load is also less than the design load of 150 kip. For the third trial, use a W8 × 40 section for which $A$ is 11.7 in² and $r$ is 2.04 in. For this section $L/r$ is 88.2 (intermediate range), the factor of safety is 1.89, the allowable stress is 14,542 psi, and the load this column can support is

$$P_{all} = \sigma_{all}(A) = 14,542(11.7) = 170,140 \text{ lb} = 170.1 \text{ kip}$$

Since the 40 lb/ft column (W8 × 40) is stronger than necessary and the 30 lb/ft column (W12 × 30) is not strong enough, any other section investigated should weigh between 30 and 40 lb/ft. The only other wide flange section is Appendix A that might satisfy the requirements is a W8 × 31 section for which $A$ is 9.13 in² and $r$ is 2.02 in. For this section $L/r$ is 89.1 (intermediate range), the factor of safety is 1.89, the allowable stress is 14,293 psi, and the load this column can support is 130,500 lb which is less than the design load of 150 kip.

Thus, a

W8 × 40 section should be used.          **Ans.**

This is not necessarily the best procedure. Different designers have different approaches to the trail-and-error procedure, and for certain problems one approach may be better than another. The important point to make here is that the problem of design involving rolled shapes (other than simple geometric shapes) is, in general, solved by trial and error.  ∎

## Example Problem 11-9

Determine the dimensions necessary for a 500-mm rectangular strut to carry an axial load of 6.75 kN. The material is aluminum alloy 2014-T6, and the width of the strut is to be twice the thickness. Use Code 2.

### SOLUTION

The code is represented by three different equations that depend on the value of $L/r$, which in turn will depend on the equation used. Thus, it will be necessary to assume that one of the equations applies and use it to obtain the dimension of the column, after which the value of $L/r$ must be calculated and used to check the validity of the equation used. Assume $L/r$ is less than 55 but greater than 12, in which case the straight-line equation is valid. The cross section is shown in Fig. 11-11, and the least second moment of area $I_x$ is equal to $bt^3/12$, the area $A$ is equal to $bt$, and the least radius of gyration is

$$r = \sqrt{\frac{bt^3/12}{bt}} = 0.2887t$$

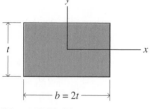

**Figure P11-11**

The slenderness ratio is

$$\frac{L}{r} = \frac{0.500}{0.2887t} = \frac{1.7319}{t}$$

and, when this value and the expression for the area are substituted in the straight-line formula of Code 2, it becomes

$$\sigma_{all} = \frac{P_{all}}{A} = [212 - 1.585(L/r)](10^6)$$

$$\frac{6.75(10^3)}{2t^2} = \left[212 - 1.585\left(\frac{1.7319}{t}\right)\right](10^6)$$

from which

$$t = 14.08(10^{-3}) \text{ m} = 14.08 \text{ mm}$$

The value of $L/r$ for this thickness is

$$\frac{L}{r} = \frac{1.7319}{0.01408} = 123.0$$

which is greater than 55 and indicates that the straight-line formula is not valid. The problem must be solved again using the slender range formula. Thus,

$$\frac{P}{A} = \frac{6.75(10^3)}{2t^2} = \frac{372(10^9)}{(1.7319/t)^2}$$

from which

$$t^4 = 27.22(10^{-9}) \text{ m}^4 \quad \text{and} \quad t = 12.84(10^{-3}) \text{ m} = 12.84 \text{ mm}$$

The value of $L/r$ is 134.9 for this thickness, which confirms the use of the slender range formula. The dimensions of the cross section are

$$t = 12.84 \text{ mm} \quad \text{and} \quad b = 25.7 \text{ mm} \quad \blacksquare \qquad \text{Ans.}$$

---

## PROBLEMS

### Introductory Problems

**11-54*** Select the lightest standard weight structural steel pipe that can be used to support an axial compressive load of 200 kN as a 4-m-long column. Use Code 1.

**11-55*** Select the lightest structural steel wide-flange section that can be used to support an axial compressive load of 200 kip as a 16-ft-long column. Use Code 1.

**11-56** A 7-m-long structural steel column will be used to support an axial compressive load of 400 kN. Select the lightest wide-flange section that can be used. Use Code 1.

### Intermediate Problems

**11-57*** A square 2014-T6 aluminum alloy member must support a 20,000-lb axial compressive load as a 12-ft-long column. Use Code 2 to determine the minimum cross-sectional area required.

**11-58*** A 2014-T6 aluminum alloy strut with a rectangular cross section and a length of 4 m will be used to support an axial compressive load of 15 kN. Determine the minimum dimensions required if the width of the strut is to be twice the thickness. Use Code 2.

**11-59** A Douglas fir ($E = 1800$ ksi and $F_c = 1350$ psi) timber column 16 ft long will be used to support an axial compressive load of 60 kip. Use Code 4 to determine the lightest structural timber that can be used.

**11-60** A Douglas fir ($E = 12$ GPa and $F_c = 9.3$ MPa) timber column 4 m long will be used to support an axial compressive load of 100 kN. Use Code 4 to determine the lightest structural timber that can be used.

**Challenging Problems**

**11-61*** The structure shown in Fig. P11-61 consists of a solid steel tie rod $BC$ and a standard weight structural steel pipe $AB$. The tie rod has been adequately designed. Using Code 1, determine the lightest pipe that can be used to support the load. The effective length of pipe $AB$ is 9 ft. Neglect the weight of the structure.

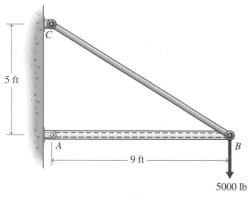

**Figure P11-61**

**11-62*** A structural steel standard weight pipe is used for a spreader bar, as shown in Fig. P11-62. If the structure is to support a load $P = 45$ kN and the cables have been adequately designed, determine the minimum size of pipe needed to support the load using Code 1. The effective length of the spreader bar is 1.5 m. Neglect the weight of the structure.

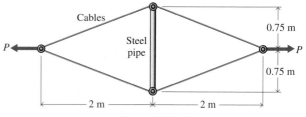

**Figure P11-62**

**11-63** Select the lightest structural steel wide-flange section that can be used for the compression members of the truss shown in Fig. P11-63. Assume that buckling is limited to the plane of the structure and that the tension members have been adequately designed. Use Code 1.

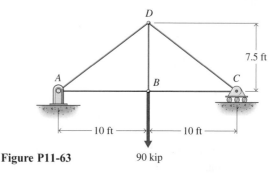

**Figure P11-63**

**11-64** Member $ABC$ of the structure shown in Fig. 11-64 supports a uniformly distributed load of 30 kN/m. Select the lightest standard weight structural steel pipe that can be used for member $BD$. Use Code 1 and consider $BD$ to be a pin-ended member. Neglect the weight of the structure.

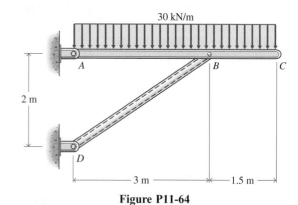

**Figure P11-64**

# 11-7 SUMMARY

Columns are long, straight, prismatic bars subjected to compressive loads. As long as a column remains straight, it can be analyzed as an axially loaded member, however, if a column begins to deform laterally, the deflection may become large and lead to catastrophic failure, called buckling. Buckling of a column is caused by deterioration of what was a stable state of equilibrium to an unstable one and not by failure of the material of which the column is composed. For long, slender columns, the maximum load for which the column is in stable equilibrium (the critical buckling load) occurs at stress levels much less than the proportional limit for the material.

For a straight, slender, pin-ended column that is centrically loaded by axial compressive forces $P$ at the ends and has experienced a small lateral deflection $\delta$ at the midpoint of the span, the differential equation for the elastic curve is

$$EI \frac{d^2y}{dx^2} = M_r = P(\delta - y)$$

which has the solution

$$y = \delta\left(1 - \cos\sqrt{\frac{P}{EI}}\, x\right) \qquad (11\text{-}1)$$

The minimum value of load $P$ for a nontrivial solution of Eq. (11-1) is

$$P_{cr} = \frac{\pi^2 EI}{L^2} \qquad (11\text{-}2)$$

The value given by Eq. (11-2) is called the critical buckling load or the Euler load. The second moment of the cross-sectional area $I$ in Eq. (11-2) refers to the axis about which bending occurs. When $I$ is replaced by $Ar^2$, where $r$ is the radius of gyration about the axis of bending. Eq. (11-2) becomes

$$\frac{P_{cr}}{A} = \frac{\pi^2 E}{(L/r)^2} \qquad (11\text{-}3)$$

The quantity $L/r$ is called the slenderness ratio and is determined for the axis about which bending tends to occur. For a pin-ended, centrically loaded column, bending occurs about the axis of minimum second moment of area (minimum radius of gyration).

Equation (11-2) agrees well with experiment if the slenderness ratio is large ($L/r > 140$ for steel columns). Short compression members ($L/r < 40$ for steel columns) can be treated as compression blocks where yielding occurs before buckling. Columns that lie between these extremes are analyzed by using empirical formulas (column design codes).

## REVIEW PROBLEMS

**11-65*** A 25-ft-long timber ($E = 1200$ ksi and $\sigma_e = 2.4$ ksi) column has the cross section shown in Fig. P11-65. The timbers are nailed together so that they act as a unit. Determine
(a) The slenderness ratio.

(b) The smallest slenderness ratio for which the Euler buckling load equation is valid.
(c) The Euler buckling load.
(d) The axial stress in the column when the Euler load is applied.

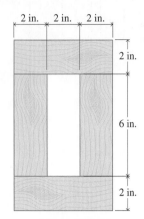

**Figure P11-65**

**11-66** A 3-m-long column with the cross section shown in Fig. P11-66 is fabricated from three pieces of timber ($E = 13$ GPa and $\sigma_e = 35$ MPa). The timbers are nailed together so that they act as a unit. Determine
(a) The slenderness ratio.
(b) The smallest slenderness ratio for which the Euler buckling load equation is valid.
(c) The Euler buckling load.
(d) The axial stress in the column when the Euler load is applied.

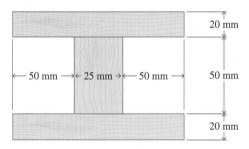

**Figure P11-66**

**11-67** Determine the maximum allowable axial compressive load for a 12-ft-long aluminum alloy ($E = 10,600$ ksi) column having the cross section shown in Fig. P11-67 if a factor of safety of 2.25 is specified.

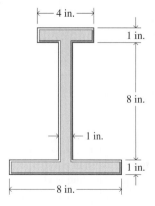

**Figure P11-67**

**11-68\*** A 25-mm-diameter tie rod and a pipe strut with an inside diameter of 100 mm and a wall thickness of 25 mm are used to support a 100-kN load as shown in Fig. P11-68. Both the tie rod and the pipe strut are made of structural steel with a modulus of elasticity of 200 GPa and a yield strength of 250 MPa. Determine
(a) The factor of safety with respect to failure by yielding.
(b) The factor of safety with respect to failure by buckling.

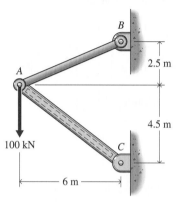

**Figure P11-68**

**11-69\*** Three S10 × 35 structural steel sections 40 ft long are used to fabricate a column with the cross section shown in Fig. P11-69. The column is fixed at the base and pinned at the top. The pin at the top offers no restraint to bending about the $y$-axis; but for bending about the $x$-axis, the pin provides restraint sufficient to reduce the effective length to 24 ft. Determine the maximum axial compressive load permitted by Code 1.

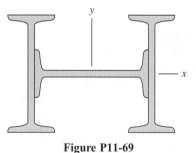

**Figure P11-69**

**11-70** The compression member $AB$ of the truss shown in Fig. P11-70 is a structural steel W254 × 67 wide-flange section with the $xx$-axis lying in the plane of the truss. The member is continuous from $A$ to $B$. Consider all connections to be the equivalent of pin ends. If Code 1 applies, determine
(a) The maximum safe load for member $AB$.
(b) The maximum safe load for member $AB$ if the bracing member $CD$ is removed.

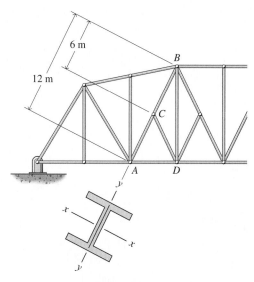

**Figure P11-70**

**11-71** A cold-rolled steel tension bar $AB$ and a structural steel ($E = 29,000$ ksi and $\sigma_y = 36$ ksi) compression strut $BC$ are used to support a load $P = 100$ kip, as shown in Fig. P11-71. Assume that the pins at $B$ and $C$ offer no restraint to bending about the $x$-axis but provide end conditions that are essentially fixed at $C$ and free at $B$ for bending about the $y$-axis. Select the lightest structural steel T-section listed in Appendix A that can be used for strut $BC$. Use Code 1.

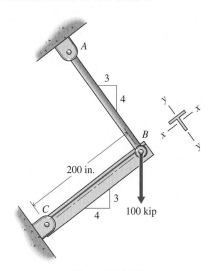

**Figure P11-71**

**11-72*** Two C229 × 30 structural steel channels are laced 120 mm back to back, as shown in Fig. P11-72, to form a column with an effective length of 10 m. The load $P$ will be applied at a point 50 mm from the axis of the column on the axis of symmetry parallel to the back of the channels. If $E = 200$ GPa and $\sigma_y = 250$ MPa, use the allowable stress method to determine the maximum load $P$ that can be supported by the column.

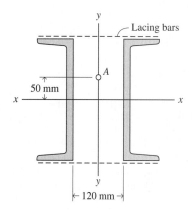

**Figure P11-72**

**11-73*** A W8 × 40 structural steel compression member has an effective length of 25 ft. The member is subjected to an axial load $P$ and a bending moment of 10 kip · ft about the $x$-axis of the cross section (see Appendix A). Use the interaction formula with an allowable flexural stress of $0.66\sigma_y$ to determine the maximum safe load $P$.

# TABLES OF PROPERTIES

**A**

**APPENDIX**

## PROPERTIES OF ROLLED STEEL STRUCTURAL SHAPES

## PROPERTIES OF STEEL PIPES AND STRUCTURAL TIMBERS

## AVERAGE PROPERTIES OF SELECTED MATERIALS

## TABLE OF BEAM DEFLECTIONS AND SLOPES

## TABLE A-1 Wide-Flange Beams (U.S. Customary Units)

| Desig-nation* | Area (in²) | Depth (in.) | FLANGE Width (in.) | FLANGE Thick-ness (in.) | Web Thick-ness (in.) | AXIS X–X $I$ (in⁴) | AXIS X–X $S$ (in³) | AXIS X–X $r$ (in) | AXIS Y–Y $I$ (in⁴) | AXIS Y–Y $S$ (in³) | AXIS Y–Y $r$ (in.) |
|---|---|---|---|---|---|---|---|---|---|---|---|
| W36 × 230 | 67.6 | 35.90 | 16.470 | 1.260 | 0.760 | 15000 | 837 | 14.9 | 940 | 114 | 3.73 |
| × 160 | 47.0 | 36.01 | 12.000 | 1.020 | 0.650 | 9750 | 542 | 14.4 | 295 | 49.1 | 2.50 |
| W33 × 201 | 59.1 | 33.68 | 15.745 | 1.150 | 0.715 | 11500 | 684 | 14.0 | 749 | 95.2 | 3.56 |
| × 152 | 44.7 | 33.49 | 11.565 | 1.055 | 0.635 | 8160 | 487 | 13.5 | 273 | 47.2 | 2.47 |
| × 130 | 38.3 | 33.09 | 11.510 | 0.855 | 0.580 | 6710 | 406 | 13.2 | 218 | 37.9 | 2.39 |
| W30 × 132 | 38.9 | 30.31 | 10.545 | 1.000 | 0.615 | 5770 | 380 | 12.2 | 196 | 37.2 | 2.25 |
| × 108 | 31.7 | 29.83 | 10.475 | 0.760 | 0.545 | 4470 | 299 | 11.9 | 146 | 27.9 | 2.15 |
| W27 × 146 | 42.9 | 27.38 | 13.965 | 0.975 | 0.605 | 5630 | 411 | 11.4 | 443 | 63.5 | 3.21 |
| × 94 | 27.7 | 26.92 | 9.990 | 0.745 | 0.490 | 3270 | 243 | 10.9 | 124 | 24.8 | 2.12 |
| W24 × 104 | 30.6 | 24.06 | 12.750 | 0.750 | 0.500 | 3100 | 258 | 10.1 | 259 | 40.7 | 2.91 |
| × 84 | 24.7 | 24.10 | 9.020 | 0.770 | 0.470 | 2370 | 196 | 9.79 | 94.4 | 20.9 | 1.95 |
| × 62 | 18.2 | 23.74 | 7.040 | 0.590 | 0.430 | 1550 | 131 | 9.23 | 34.5 | 9.80 | 1.38 |
| W21 × 101 | 29.8 | 21.36 | 12.290 | 0.800 | 0.500 | 2420 | 227 | 9.02 | 248 | 40.3 | 2.89 |
| × 83 | 24.3 | 21.43 | 8.355 | 0.835 | 0.515 | 1830 | 171 | 8.67 | 81.4 | 19.5 | 1.83 |
| × 62 | 18.3 | 20.99 | 8.240 | 0.615 | 0.400 | 1330 | 127 | 8.54 | 57.5 | 13.9 | 1.77 |
| W18 × 97 | 28.5 | 18.59 | 11.145 | 0.870 | 0.535 | 1750 | 188 | 7.82 | 201 | 36.1 | 2.65 |
| × 76 | 22.3 | 18.21 | 11.035 | 0.680 | 0.425 | 1330 | 146 | 7.73 | 152 | 27.6 | 2.61 |
| × 60 | 17.6 | 18.24 | 7.555 | 0.695 | 0.415 | 984 | 108 | 7.47 | 50.1 | 13.3 | 1.69 |
| W16 × 100 | 29.4 | 16.97 | 10.425 | 0.985 | 0.585 | 1490 | 175 | 7.10 | 186 | 35.7 | 2.52 |
| × 67 | 19.7 | 16.33 | 10.235 | 0.665 | 0.395 | 954 | 117 | 6.96 | 119 | 23.2 | 2.46 |
| × 40 | 11.8 | 16.01 | 6.995 | 0.505 | 0.305 | 518 | 64.7 | 6.63 | 28.9 | 8.25 | 1.57 |
| × 26 | 7.68 | 15.69 | 5.500 | 0.345 | 0.250 | 301 | 38.4 | 6.26 | 9.59 | 3.49 | 1.12 |
| W14 × 120 | 35.3 | 14.48 | 14.670 | 0.940 | 0.590 | 1380 | 190 | 6.24 | 495 | 67.5 | 3.74 |
| × 82 | 24.1 | 14.31 | 10.130 | 0.855 | 0.510 | 882 | 123 | 6.05 | 148 | 29.3 | 2.48 |
| × 43 | 12.6 | 13.66 | 7.995 | 0.530 | 0.305 | 428 | 62.7 | 5.82 | 45.2 | 11.3 | 1.89 |
| × 30 | 8.85 | 13.84 | 6.730 | 0.385 | 0.270 | 291 | 42.0 | 5.73 | 19.6 | 5.82 | 1.49 |
| W12 × 96 | 28.2 | 12.71 | 12.160 | 0.900 | 0.550 | 833 | 131 | 5.44 | 270 | 44.4 | 3.09 |
| × 65 | 19.1 | 12.12 | 12.000 | 0.605 | 0.390 | 533 | 87.9 | 5.28 | 174 | 29.1 | 3.02 |
| × 50 | 14.7 | 12.19 | 8.080 | 0.640 | 0.370 | 394 | 64.7 | 5.18 | 56.3 | 13.9 | 1.96 |
| × 30 | 8.79 | 12.34 | 6.520 | 0.440 | 0.260 | 238 | 38.6 | 5.21 | 20.3 | 6.24 | 1.52 |
| W10 × 60 | 17.6 | 10.22 | 10.080 | 0.680 | 0.420 | 341 | 66.7 | 4.39 | 116 | 23.0 | 2.57 |
| × 45 | 13.3 | 10.10 | 8.020 | 0.620 | 0.350 | 248 | 49.1 | 4.33 | 53.4 | 13.3 | 2.01 |
| × 30 | 8.84 | 10.47 | 5.810 | 0.510 | 0.300 | 170 | 32.4 | 4.38 | 16.7 | 5.75 | 1.37 |
| × 22 | 6.49 | 10.17 | 5.750 | 0.360 | 0.240 | 118 | 23.2 | 4.27 | 11.4 | 3.97 | 1.33 |
| W8 × 40 | 11.7 | 8.25 | 8.070 | 0.560 | 0.360 | 146 | 35.5 | 3.53 | 49.1 | 12.2 | 2.04 |
| × 31 | 9.13 | 8.00 | 7.995 | 0.435 | 0.285 | 110 | 27.5 | 3.47 | 37.1 | 9.27 | 2.02 |
| × 24 | 7.08 | 7.93 | 6.495 | 0.400 | 0.245 | 82.8 | 20.9 | 3.42 | 18.3 | 5.63 | 1.61 |
| × 15 | 4.44 | 8.11 | 4.015 | 0.315 | 0.245 | 48.0 | 11.8 | 3.29 | 3.41 | 1.70 | 0.876 |
| W6 × 25 | 7.34 | 6.38 | 6.080 | 0.455 | 0.320 | 53.4 | 16.7 | 2.70 | 17.1 | 5.61 | 1.52 |
| × 16 | 4.74 | 6.28 | 4.030 | 0.405 | 0.260 | 32.1 | 10.2 | 2.60 | 4.43 | 2.20 | 0.967 |
| W5 × 16 | 4.68 | 5.01 | 5.000 | 0.360 | 0.240 | 21.3 | 8.51 | 2.13 | 7.51 | 3.00 | 1.27 |
| W4 × 13 | 3.83 | 4.16 | 4.060 | 0.345 | 0.280 | 11.3 | 5.46 | 1.72 | 3.86 | 1.90 | 1.00 |

Courtesy of the American Institute of Steel Construction.

*W means wide-flange beam, followed by the nominal depth in inches, then the weight in pounds per foot of length.

## TABLE A-2 Wide-Flange Beams (SI Units)

| Desig-nation* | Area (mm²) | Depth (mm) | FLANGE Width (mm) | FLANGE Thick-ness (mm) | Web Thick-ness (mm) | AXIS X–X $I$ ($10^6$ mm⁴) | AXIS X–X $S$ ($10^3$ mm³) | $r$ (mm) | AXIS Y–Y $I$ ($10^6$ mm⁴) | AXIS Y–Y $S$ ($10^3$ mm³) | $r$ (mm) |
|---|---|---|---|---|---|---|---|---|---|---|---|
| W914 × 342 | 43610 | 912 | 418 | 32.0 | 19.3 | 6245 | 13715 | 378 | 391 | 1870 | 94.7 |
| × 238 | 30325 | 915 | 305 | 25.9 | 16.5 | 4060 | 8880 | 366 | 123 | 805 | 63.5 |
| W838 × 299 | 38130 | 855 | 400 | 29.2 | 18.2 | 4785 | 11210 | 356 | 312 | 1560 | 90.4 |
| × 226 | 28850 | 851 | 294 | 26.8 | 16.1 | 3395 | 7980 | 343 | 114 | 775 | 62.7 |
| × 193 | 24710 | 840 | 292 | 21.7 | 14.7 | 2795 | 6655 | 335 | 90.7 | 620 | 60.7 |
| W762 × 196 | 25100 | 770 | 268 | 25.4 | 15.6 | 2400 | 6225 | 310 | 81.6 | 610 | 57.2 |
| × 161 | 20450 | 758 | 266 | 19.3 | 13.8 | 1860 | 4900 | 302 | 60.8 | 457 | 54.6 |
| W686 × 217 | 27675 | 695 | 355 | 24.8 | 15.4 | 2345 | 6735 | 290 | 184 | 1040 | 81.5 |
| × 140 | 17870 | 684 | 254 | 18.9 | 12.4 | 1360 | 3980 | 277 | 51.6 | 406 | 53.8 |
| W610 × 155 | 19740 | 611 | 324 | 19.1 | 12.7 | 1290 | 4230 | 257 | 108 | 667 | 73.9 |
| × 125 | 15935 | 612 | 229 | 19.6 | 11.9 | 985 | 3210 | 249 | 39.3 | 342 | 49.5 |
| × 92 | 11750 | 603 | 179 | 15.0 | 10.9 | 645 | 2145 | 234 | 14.4 | 161 | 35.1 |
| W533 × 150 | 19225 | 543 | 312 | 20.3 | 12.7 | 1005 | 3720 | 229 | 103 | 660 | 73.4 |
| × 124 | 15675 | 544 | 212 | 21.2 | 13.1 | 762 | 2800 | 220 | 33.9 | 320 | 46.5 |
| × 92 | 11805 | 533 | 209 | 15.6 | 10.2 | 554 | 2080 | 217 | 23.9 | 228 | 45.0 |
| W457 × 144 | 18365 | 472 | 283 | 22.1 | 13.6 | 728 | 3080 | 199 | 83.7 | 592 | 67.3 |
| × 113 | 14385 | 463 | 280 | 17.3 | 10.8 | 554 | 2395 | 196 | 63.3 | 452 | 66.3 |
| × 89 | 11355 | 463 | 192 | 17.7 | 10.5 | 410 | 1770 | 190 | 20.9 | 218 | 42.9 |
| W406 × 149 | 18970 | 431 | 265 | 25.0 | 14.9 | 620 | 2870 | 180 | 77.4 | 585 | 64.0 |
| × 100 | 12710 | 415 | 260 | 16.9 | 10.0 | 397 | 1915 | 177 | 49.5 | 380 | 62.5 |
| × 60 | 7615 | 407 | 178 | 12.8 | 7.7 | 216 | 1060 | 168 | 12.0 | 135 | 39.9 |
| × 39 | 4950 | 399 | 140 | 8.8 | 6.4 | 125 | 629 | 159 | 3.99 | 57.2 | 28.4 |
| W356 × 179 | 22775 | 368 | 373 | 23.9 | 15.0 | 574 | 3115 | 158 | 206 | 1105 | 95.0 |
| × 122 | 15550 | 363 | 257 | 21.7 | 13.0 | 367 | 2015 | 154 | 61.6 | 480 | 63.0 |
| × 64 | 8130 | 347 | 203 | 13.5 | 7.7 | 178 | 1025 | 148 | 18.8 | 185 | 48.0 |
| × 45 | 5710 | 352 | 171 | 9.8 | 6.9 | 121 | 688 | 146 | 8.16 | 95.4 | 37.8 |
| W305 × 143 | 18195 | 323 | 309 | 22.9 | 14.0 | 347 | 2145 | 138 | 112 | 728 | 78.5 |
| × 97 | 12325 | 308 | 305 | 15.4 | 9.9 | 222 | 1440 | 134 | 72.4 | 477 | 76.7 |
| × 74 | 9485 | 310 | 205 | 16.3 | 9.4 | 164 | 1060 | 132 | 23.4 | 228 | 49.8 |
| × 45 | 5670 | 313 | 166 | 11.2 | 6.6 | 99.1 | 633 | 132 | 8.45 | 102 | 38.6 |
| W254 × 89 | 11355 | 260 | 256 | 17.3 | 10.7 | 142 | 1095 | 112 | 48.3 | 377 | 65.3 |
| × 67 | 8580 | 257 | 204 | 15.7 | 8.9 | 103 | 805 | 110 | 22.2 | 218 | 51.1 |
| × 45 | 5705 | 266 | 148 | 13.0 | 7.6 | 70.8 | 531 | 111 | 6.95 | 94.2 | 34.8 |
| × 33 | 4185 | 258 | 146 | 9.1 | 6.1 | 49.1 | 380 | 108 | 4.75 | 65.1 | 33.8 |
| W203 × 60 | 7550 | 210 | 205 | 14.2 | 9.1 | 60.8 | 582 | 89.7 | 20.4 | 200 | 51.8 |
| × 46 | 5890 | 203 | 203 | 11.0 | 7.2 | 45.8 | 451 | 88.1 | 15.4 | 152 | 51.3 |
| × 36 | 4570 | 201 | 165 | 10.2 | 6.2 | 34.5 | 342 | 86.7 | 7.61 | 92.3 | 40.9 |
| × 22 | 2865 | 206 | 102 | 8.0 | 6.2 | 20.0 | 193 | 83.6 | 1.42 | 27.9 | 22.3 |
| W152 × 37 | 4735 | 162 | 154 | 11.6 | 8.1 | 22.2 | 274 | 68.6 | 7.12 | 91.9 | 38.6 |
| × 24 | 3060 | 160 | 102 | 10.3 | 6.6 | 13.4 | 167 | 66.0 | 1.84 | 36.1 | 24.6 |
| W127 × 24 | 3020 | 127 | 127 | 9.1 | 6.1 | 8.87 | 139 | 54.1 | 3.13 | 49.2 | 32.3 |
| W102 × 19 | 2470 | 106 | 103 | 8.8 | 7.1 | 4.70 | 89.5 | 43.7 | 1.61 | 31.1 | 25.4 |

*W means wide-flange beam, followed by the nominal depth in mm, then the mass in kg per meter of length.

TABLE A-3 American Standard Beams (U.S. Customary Units)

| Desig-nation* | Area (in$^2$) | Depth (in.) | FLANGE Width (in.) | FLANGE Thick-ness (in.) | Web Thick-ness (in.) | AXIS X–X $I$ (in$^4$) | AXIS X–X $S$ (in$^3$) | AXIS X–X $r$ (in.) | AXIS Y–Y $I$ (in$^4$) | AXIS Y–Y $S$ (in$^3$) | AXIS Y–Y $r$ (in.) |
|---|---|---|---|---|---|---|---|---|---|---|---|
| S24 × 121 | 35.6 | 24.50 | 8.050 | 1.090 | 0.800 | 3160 | 258 | 9.43 | 83.3 | 20.7 | 1.53 |
| × 106 | 31.2 | 24.50 | 7.870 | 1.090 | 0.620 | 2940 | 240 | 9.71 | 77.1 | 19.6 | 1.57 |
| × 100 | 29.3 | 24.00 | 7.245 | 0.870 | 0.745 | 2390 | 199 | 9.02 | 47.7 | 13.2 | 1.27 |
| × 90 | 26.5 | 24.00 | 7.125 | 0.870 | 0.625 | 2250 | 187 | 9.21 | 44.9 | 12.6 | 1.30 |
| × 80 | 23.5 | 24.00 | 7.000 | 0.870 | 0.500 | 2100 | 175 | 9.47 | 42.2 | 12.1 | 1.34 |
| S20 × 96 | 28.2 | 20.30 | 7.200 | 0.920 | 0.800 | 1670 | 165 | 7.71 | 50.2 | 13.9 | 1.33 |
| × 86 | 25.3 | 20.30 | 7.060 | 0.920 | 0.660 | 1580 | 155 | 7.89 | 46.8 | 13.3 | 1.36 |
| × 75 | 22.0 | 20.00 | 6.385 | 0.795 | 0.635 | 1280 | 128 | 7.62 | 29.8 | 9.32 | 1.16 |
| × 66 | 19.4 | 20.00 | 6.255 | 0.795 | 0.505 | 1190 | 119 | 7.83 | 27.7 | 8.85 | 1.19 |
| S18 × 70 | 20.6 | 18.00 | 6.251 | 0.691 | 0.711 | 926 | 103 | 6.71 | 24.1 | 7.72 | 1.08 |
| × 54.7 | 16.1 | 18.00 | 6.001 | 0.691 | 0.461 | 804 | 89.4 | 7.07 | 20.8 | 6.94 | 1.14 |
| S15 × 50 | 14.7 | 15.00 | 5.640 | 0.622 | 0.550 | 486 | 64.8 | 5.75 | 15.7 | 5.57 | 1.03 |
| × 42.9 | 12.6 | 15.00 | 5.501 | 0.622 | 0.411 | 447 | 59.6 | 5.95 | 14.4 | 5.23 | 1.07 |
| S12 × 50 | 14.7 | 12.00 | 5.477 | 0.659 | 0.687 | 305 | 50.8 | 4.55 | 15.7 | 5.74 | 1.03 |
| × 40.8 | 12.0 | 12.00 | 5.252 | 0.659 | 0.462 | 272 | 45.4 | 4.77 | 13.6 | 5.16 | 1.06 |
| × 35 | 10.3 | 12.00 | 5.078 | 0.544 | 0.428 | 229 | 38.2 | 4.72 | 9.87 | 3.89 | 0.980 |
| × 31.8 | 9.35 | 12.00 | 5.000 | 0.544 | 0.350 | 218 | 36.4 | 4.83 | 9.36 | 3.74 | 1.00 |
| S10 × 35 | 10.3 | 10.00 | 4.944 | 0.491 | 0.594 | 147 | 29.4 | 3.78 | 8.36 | 3.38 | 0.901 |
| × 25.4 | 7.46 | 10.00 | 4.661 | 0.491 | 0.311 | 124 | 24.7 | 4.07 | 6.79 | 2.91 | 0.954 |
| S8 × 23 | 6.77 | 8.00 | 4.171 | 0.426 | 0.441 | 64.9 | 16.2 | 3.10 | 4.31 | 2.07 | 0.798 |
| × 18.4 | 5.41 | 8.00 | 4.001 | 0.426 | 0.271 | 57.6 | 14.4 | 3.26 | 3.73 | 1.86 | 0.831 |
| S7 × 20 | 5.88 | 7.00 | 3.860 | 0.392 | 0.450 | 42.4 | 12.1 | 2.69 | 3.17 | 1.64 | 0.734 |
| × 15.3 | 4.50 | 7.00 | 3.662 | 0.392 | 0.252 | 36.7 | 10.5 | 2.86 | 2.64 | 1.44 | 0.766 |
| S6 × 17.25 | 5.07 | 6.00 | 3.565 | 0.359 | 0.465 | 26.3 | 8.77 | 2.28 | 2.31 | 1.30 | 0.675 |
| × 12.5 | 3.67 | 6.00 | 3.332 | 0.359 | 0.232 | 22.1 | 7.37 | 2.45 | 1.82 | 1.09 | 0.705 |
| S5 × 14.75 | 4.34 | 5.00 | 3.284 | 0.326 | 0.494 | 15.2 | 6.09 | 1.87 | 1.67 | 1.01 | 0.620 |
| × 10 | 2.94 | 5.00 | 3.004 | 0.326 | 0.214 | 12.3 | 4.92 | 2.05 | 1.22 | 0.809 | 0.643 |
| S4 × 9.5 | 2.79 | 4.00 | 2.796 | 0.293 | 0.326 | 6.79 | 3.39 | 1.56 | 0.903 | 0.646 | 0.569 |
| × 7.7 | 2.26 | 4.00 | 2.663 | 0.293 | 0.193 | 6.08 | 3.04 | 1.64 | 0.764 | 0.574 | 0.581 |
| S3 × 7.5 | 2.21 | 3.00 | 2.509 | 0.260 | 0.349 | 2.93 | 1.95 | 1.15 | 0.586 | 0.468 | 0.516 |
| × 5.7 | 1.67 | 3.00 | 2.330 | 0.260 | 0.170 | 2.52 | 1.68 | 1.23 | 0.455 | 0.390 | 0.522 |

Courtesy of The American Institute of Steel Construction.

*S means standard beam, followed by the nominal depth in inches, then the weight in pounds per foot of length.

TABLE A-4 American Standard Beams (SI Units)

| Desig-nation* | Area (mm²) | Depth (mm) | FLANGE | | Web Thick-ness (mm) | AXIS X–X | | | AXIS Y–Y | | |
|---|---|---|---|---|---|---|---|---|---|---|---|
| | | | Width (mm) | Thick-ness (mm) | | $I$ ($10^6$ mm⁴) | $S$ ($10^3$ mm³) | $r$ (mm) | $I$ ($10^6$ mm⁴) | $S$ ($10^3$ mm³) | $r$ (mm) |
| S610 × 180 | 22970 | 622.3 | 204.5 | 27.7 | 20.3 | 1315 | 4225 | 240 | 34.7 | 339 | 38.9 |
| × 158 | 20130 | 622.3 | 199.9 | 27.7 | 15.7 | 1225 | 3935 | 247 | 32.1 | 321 | 39.9 |
| × 149 | 18900 | 609.6 | 184.0 | 22.1 | 18.9 | 995 | 3260 | 229 | 19.9 | 216 | 32.3 |
| × 134 | 17100 | 609.6 | 181.0 | 22.1 | 15.9 | 937 | 3065 | 234 | 18.7 | 206 | 33.0 |
| × 119 | 15160 | 609.6 | 177.8 | 22.1 | 12.7 | 874 | 2870 | 241 | 17.6 | 198 | 34.0 |
| S508 × 143 | 18190 | 515.6 | 182.9 | 23.4 | 20.3 | 695 | 2705 | 196 | 20.9 | 228 | 33.8 |
| × 128 | 16320 | 515.6 | 179.3 | 23.4 | 16.8 | 658 | 2540 | 200 | 19.5 | 218 | 34.5 |
| × 112 | 14190 | 508.0 | 162.2 | 20.2 | 16.1 | 533 | 2100 | 194 | 12.4 | 153 | 29.5 |
| × 98 | 12520 | 508.0 | 158.9 | 20.2 | 12.8 | 495 | 1950 | 199 | 11.5 | 145 | 30.2 |
| S457 × 104 | 13290 | 457.2 | 158.8 | 17.6 | 18.1 | 358 | 1690 | 170 | 10.0 | 127 | 27.4 |
| × 81 | 10390 | 457.2 | 152.4 | 17.6 | 11.7 | 335 | 1465 | 180 | 8.66 | 114 | 29.0 |
| S381 × 74 | 9485 | 381.0 | 143.3 | 15.8 | 14.0 | 202 | 1060 | 146 | 6.53 | 91.3 | 26.2 |
| × 64 | 8130 | 381.0 | 139.7 | 15.8 | 10.4 | 186 | 977 | 151 | 5.99 | 85.7 | 27.2 |
| S305 × 74 | 9485 | 304.8 | 139.1 | 16.7 | 17.4 | 127 | 832 | 116 | 6.53 | 94.1 | 26.2 |
| × 61 | 7740 | 304.8 | 133.4 | 16.7 | 11.7 | 113 | 744 | 121 | 5.66 | 84.6 | 26.9 |
| × 52 | 6645 | 304.8 | 129.0 | 13.8 | 10.9 | 95.3 | 626 | 120 | 4.11 | 63.7 | 24.1 |
| × 47 | 6030 | 304.8 | 127.0 | 13.8 | 8.9 | 90.7 | 596 | 123 | 3.90 | 61.3 | 25.4 |
| S254 × 52 | 6645 | 254.0 | 125.6 | 12.5 | 15.1 | 61.2 | 482 | 96.0 | 3.48 | 55.4 | 22.9 |
| × 38 | 4815 | 254.0 | 118.4 | 12.5 | 7.9 | 51.6 | 408 | 103 | 2.83 | 47.7 | 24.2 |
| S203 × 34 | 4370 | 203.2 | 105.9 | 10.8 | 11.2 | 27.0 | 265 | 78.7 | 1.79 | 33.9 | 20.3 |
| × 27 | 3490 | 203.2 | 101.6 | 10.8 | 6.9 | 24.0 | 236 | 82.8 | 1.55 | 30.5 | 21.1 |
| S178 × 30 | 3795 | 177.8 | 98.0 | 10.0 | 11.4 | 17.6 | 198 | 68.3 | 1.32 | 26.9 | 18.6 |
| × 23 | 2905 | 177.8 | 93.0 | 10.0 | 6.4 | 15.3 | 172 | 72.6 | 1.10 | 23.6 | 19.5 |
| S152 × 26 | 3270 | 152.4 | 90.6 | 9.1 | 11.8 | 10.9 | 144 | 57.9 | 0.961 | 21.3 | 17.1 |
| × 19 | 2370 | 152.4 | 84.6 | 9.1 | 5.9 | 9.20 | 121 | 62.2 | 0.758 | 17.9 | 17.9 |
| S127 × 22 | 2800 | 127.0 | 83.4 | 8.3 | 12.5 | 6.33 | 99.8 | 47.5 | 0.695 | 16.6 | 15.7 |
| × 15 | 1895 | 127.0 | 76.3 | 8.3 | 5.4 | 5.12 | 80.6 | 52.1 | 0.508 | 13.3 | 16.3 |
| S102 × 14 | 1800 | 101.6 | 71.0 | 7.4 | 8.3 | 2.83 | 55.6 | 39.6 | 0.376 | 10.6 | 14.5 |
| × 11 | 1460 | 101.6 | 67.6 | 7.4 | 4.9 | 2.53 | 49.8 | 41.7 | 0.318 | 9.41 | 14.8 |
| S76 × 11 | 1425 | 76.2 | 63.7 | 6.6 | 8.9 | 1.22 | 32.0 | 29.2 | 0.244 | 7.67 | 13.1 |
| × 8.5 | 1075 | 76.2 | 59.2 | 6.6 | 4.3 | 1.05 | 27.5 | 31.2 | 0.189 | 6.39 | 13.3 |

*S means standard beam, followed by the nominal depth in mm, then the mass in kg per meter of length.

## TABLE A-5 Standard Channels (U.S. Customary Units)

| Desig-nation† | Area (in²) | Depth (in.) | FLANGE Width (in.) | FLANGE Average Thick-ness (in.) | Web Thick-ness (in.) | AXIS X–X I (in⁴) | AXIS X–X S (in³) | AXIS X–X r (in.) | AXIS Y–Y I (in⁴) | AXIS Y–Y S (in³) | AXIS Y–Y r (in.) | $x_C$ (in.) |
|---|---|---|---|---|---|---|---|---|---|---|---|---|
| *C18 × 58 | 17.1 | 18.00 | 4.200 | 0.625 | 0.700 | 676 | 75.1 | 6.29 | 17.8 | 5.32 | 1.02 | 0.862 |
| × 51.9 | 15.3 | 18.00 | 4.100 | 0.625 | 0.600 | 627 | 69.7 | 6.41 | 16.4 | 5.07 | 1.04 | 0.858 |
| × 45.8 | 13.5 | 18.00 | 4.000 | 0.625 | 0.500 | 578 | 64.3 | 6.56 | 15.1 | 4.82 | 1.06 | 0.866 |
| × 42.7 | 12.6 | 18.00 | 3.950 | 0.625 | 0.450 | 554 | 61.6 | 6.64 | 14.4 | 4.69 | 1.07 | 0.877 |
| C15 × 50 | 14.7 | 15.00 | 3.716 | 0.650 | 0.716 | 404 | 53.8 | 5.24 | 11.0 | 3.78 | 0.867 | 0.798 |
| × 40 | 11.8 | 15.00 | 3.520 | 0.650 | 0.520 | 349 | 46.5 | 5.44 | 9.23 | 3.37 | 0.886 | 0.777 |
| × 33.9 | 9.96 | 15.00 | 3.400 | 0.650 | 0.400 | 315 | 42.0 | 5.62 | 8.13 | 3.11 | 0.904 | 0.787 |
| C12 × 30 | 8.82 | 12.00 | 3.170 | 0.501 | 0.510 | 162 | 27.0 | 4.29 | 5.14 | 2.06 | 0.763 | 0.674 |
| × 25 | 7.35 | 12.00 | 3.047 | 0.501 | 0.387 | 144 | 24.1 | 4.43 | 4.47 | 1.88 | 0.780 | 0.674 |
| × 20.7 | 6.09 | 12.00 | 2.942 | 0.501 | 0.282 | 129 | 21.5 | 4.61 | 3.88 | 1.73 | 0.799 | 0.698 |
| C10 × 30 | 8.82 | 10.00 | 3.033 | 0.436 | 0.673 | 103 | 20.7 | 3.42 | 3.94 | 1.65 | 0.669 | 0.649 |
| × 25 | 7.35 | 10.00 | 2.886 | 0.436 | 0.526 | 91.2 | 18.2 | 3.52 | 3.36 | 1.48 | 0.676 | 0.617 |
| × 20 | 5.88 | 10.00 | 2.739 | 0.436 | 0.379 | 78.9 | 15.8 | 3.66 | 2.81 | 1.32 | 0.692 | 0.606 |
| × 15.3 | 4.49 | 10.00 | 2.600 | 0.436 | 0.240 | 67.4 | 13.5 | 3.87 | 2.28 | 1.16 | 0.713 | 0.634 |
| C9 × 20 | 5.88 | 9.00 | 2.648 | 0.413 | 0.448 | 60.9 | 13.5 | 3.22 | 2.42 | 1.17 | 0.642 | 0.583 |
| × 15 | 4.41 | 9.00 | 2.485 | 0.413 | 0.285 | 51.0 | 11.3 | 3.40 | 1.93 | 1.01 | 0.661 | 0.586 |
| × 13.4 | 3.94 | 9.00 | 2.433 | 0.413 | 0.233 | 47.9 | 10.6 | 3.48 | 1.76 | 0.962 | 0.669 | 0.601 |
| C8 × 18.75 | 5.51 | 8.00 | 2.527 | 0.390 | 0.487 | 44.0 | 11.0 | 2.82 | 1.98 | 1.01 | 0.599 | 0.565 |
| × 13.75 | 4.04 | 8.00 | 2.343 | 0.390 | 0.303 | 36.1 | 9.03 | 2.99 | 1.53 | 0.854 | 0.615 | 0.553 |
| × 11.5 | 3.38 | 8.00 | 2.260 | 0.390 | 0.220 | 32.6 | 8.14 | 3.11 | 1.32 | 0.781 | 0.625 | 0.571 |
| C7 × 14.75 | 4.33 | 7.00 | 2.299 | 0.366 | 0.419 | 27.2 | 7.78 | 2.51 | 1.38 | 0.779 | 0.564 | 0.532 |
| × 12.25 | 3.60 | 7.00 | 2.194 | 0.366 | 0.314 | 24.2 | 6.93 | 2.60 | 1.17 | 0.703 | 0.571 | 0.525 |
| × 9.8 | 2.87 | 7.00 | 2.090 | 0.366 | 0.210 | 21.3 | 6.08 | 2.72 | 0.968 | 0.625 | 0.581 | 0.540 |
| C6 × 13 | 3.83 | 6.00 | 2.157 | 0.343 | 0.437 | 17.4 | 5.80 | 2.13 | 1.05 | 0.642 | 0.525 | 0.514 |
| × 10.5 | 3.09 | 6.00 | 2.034 | 0.343 | 0.314 | 15.2 | 5.06 | 2.22 | 0.866 | 0.564 | 0.529 | 0.499 |
| × 8.2 | 2.40 | 6.00 | 1.920 | 0.343 | 0.200 | 13.1 | 4.38 | 2.34 | 0.693 | 0.492 | 0.537 | 0.511 |
| C5 × 9 | 2.64 | 5.00 | 1.885 | 0.320 | 0.325 | 8.90 | 3.56 | 1.83 | 0.632 | 0.450 | 0.489 | 0.478 |
| × 6.7 | 1.97 | 5.00 | 1.750 | 0.320 | 0.190 | 7.49 | 3.00 | 1.95 | 0.479 | 0.378 | 0.493 | 0.484 |
| C4 × 7.25 | 2.13 | 4.00 | 1.721 | 0.296 | 0.321 | 4.59 | 2.29 | 1.47 | 0.433 | 0.343 | 0.450 | 0.459 |
| × 5.4 | 1.59 | 4.00 | 1.584 | 0.296 | 0.184 | 3.85 | 1.93 | 1.56 | 0.319 | 0.283 | 0.449 | 0.457 |
| C3 × 6 | 1.76 | 3.00 | 1.596 | 0.273 | 0.356 | 2.07 | 1.38 | 1.08 | 0.305 | 0.268 | 0.416 | 0.455 |
| × 5 | 1.47 | 3.00 | 1.498 | 0.273 | 0.258 | 1.85 | 1.24 | 1.12 | 0.247 | 0.233 | 0.410 | 0.438 |
| × 4.1 | 1.21 | 3.00 | 1.410 | 0.273 | 0.170 | 1.66 | 1.10 | 1.17 | 0.197 | 0.202 | 0.404 | 0.436 |

Courtesy of The American Institute of Steel Construction.

*Not part of the American Standard Series.

†C means channel, followed by the nominal depth in inches, then the weight in pounds per foot of length.

## TABLE A-6 Standard Channels (SI Units)

| Desig-nation* | Area (mm²) | Depth (mm) | FLANGE Width (mm) | FLANGE Thick-ness (mm) | Web Thick-ness (mm) | AXIS X–X I (10⁶ mm⁴) | AXIS X–X S (10³ mm³) | AXIS X–X r (mm) | AXIS Y–Y I (10⁶ mm⁴) | AXIS Y–Y S (10³ mm³) | AXIS Y–Y r (mm) | $x_C$ (mm) |
|---|---|---|---|---|---|---|---|---|---|---|---|---|
| C457 × 86 | 11030 | 457.2 | 106.7 | 15.9 | 17.8 | 281 | 1230 | 160 | 7.41 | 87.2 | 25.9 | 21.9 |
| × 77 | 9870 | 457.2 | 104.1 | 15.9 | 15.2 | 261 | 1140 | 163 | 6.83 | 83.1 | 26.4 | 21.8 |
| × 68 | 8710 | 457.2 | 101.6 | 15.9 | 12.7 | 241 | 1055 | 167 | 6.29 | 79.0 | 26.9 | 22.0 |
| × 64 | 8130 | 457.2 | 100.3 | 15.9 | 11.4 | 231 | 1010 | 169 | 5.99 | 76.9 | 27.2 | 22.3 |
| C381 × 74 | 9485 | 381.0 | 94.4 | 16.5 | 18.2 | 168 | 882 | 133 | 4.58 | 61.9 | 22.0 | 20.3 |
| × 60 | 7615 | 381.0 | 89.4 | 16.5 | 13.2 | 145 | 762 | 138 | 3.84 | 55.2 | 22.5 | 19.7 |
| × 50 | 6425 | 381.0 | 86.4 | 16.5 | 10.2 | 131 | 688 | 143 | 3.38 | 51.0 | 23.0 | 20.0 |
| C305 × 45 | 5690 | 304.8 | 80.5 | 12.7 | 13.0 | 67.4 | 442 | 109 | 2.14 | 33.8 | 19.4 | 17.1 |
| × 37 | 4740 | 304.8 | 77.4 | 12.7 | 9.8 | 59.9 | 395 | 113 | 1.86 | 30.8 | 19.8 | 17.1 |
| × 31 | 3930 | 304.8 | 74.7 | 12.7 | 7.2 | 53.7 | 352 | 117 | 1.61 | 28.3 | 20.3 | 17.7 |
| C254 × 45 | 5690 | 254.0 | 77.0 | 11.1 | 17.1 | 42.9 | 339 | 86.9 | 1.64 | 27.0 | 17.0 | 16.5 |
| × 37 | 4740 | 254.0 | 73.3 | 11.1 | 13.4 | 38.0 | 298 | 89.4 | 1.40 | 24.3 | 17.2 | 15.7 |
| × 30 | 3795 | 254.0 | 69.6 | 11.1 | 9.6 | 32.8 | 259 | 93.0 | 1.17 | 21.6 | 17.6 | 15.4 |
| × 23 | 2895 | 254.0 | 66.0 | 11.1 | 6.1 | 28.1 | 221 | 98.3 | 0.949 | 19.0 | 18.1 | 16.1 |
| C229 × 30 | 3795 | 228.6 | 67.3 | 10.5 | 11.4 | 25.3 | 221 | 81.8 | 1.01 | 19.2 | 16.3 | 14.8 |
| × 22 | 2845 | 228.6 | 63.1 | 10.5 | 7.2 | 21.2 | 185 | 86.4 | 0.803 | 16.6 | 16.8 | 14.9 |
| × 20 | 2540 | 228.6 | 61.8 | 10.5 | 5.9 | 19.9 | 174 | 88.4 | 0.733 | 15.7 | 17.0 | 15.3 |
| C203 × 28 | 3555 | 203.2 | 64.2 | 9.9 | 12.4 | 18.3 | 180 | 71.6 | 0.824 | 16.6 | 15.2 | 14.4 |
| × 20 | 2605 | 203.2 | 59.5 | 9.9 | 7.7 | 15.0 | 148 | 75.9 | 0.637 | 14.0 | 15.6 | 14.0 |
| × 17 | 2180 | 203.2 | 57.4 | 9.9 | 5.6 | 13.6 | 133 | 79.0 | 0.549 | 12.8 | 15.9 | 14.5 |
| C178 × 22 | 2795 | 177.8 | 58.4 | 9.3 | 10.6 | 11.3 | 127 | 63.8 | 0.574 | 12.8 | 14.3 | 13.5 |
| × 18 | 2320 | 177.8 | 55.7 | 9.3 | 8.0 | 10.1 | 114 | 66.0 | 0.487 | 11.5 | 14.5 | 13.3 |
| × 15 | 1850 | 177.8 | 53.1 | 9.3 | 5.3 | 8.87 | 99.6 | 69.1 | 0.403 | 10.2 | 14.8 | 13.7 |
| C152 × 19 | 2470 | 152.4 | 54.8 | 8.7 | 11.1 | 7.24 | 95.0 | 54.1 | 0.437 | 10.5 | 13.3 | 13.1 |
| × 16 | 1995 | 152.4 | 51.7 | 8.7 | 8.0 | 6.33 | 82.9 | 56.4 | 0.360 | 9.24 | 13.4 | 12.7 |
| × 12 | 1550 | 152.4 | 48.8 | 8.7 | 5.1 | 5.45 | 71.8 | 59.4 | 0.288 | 8.06 | 13.6 | 13.0 |
| C127 × 13 | 1705 | 127.0 | 47.9 | 8.1 | 8.3 | 3.70 | 58.3 | 46.5 | 0.263 | 7.37 | 12.4 | 12.1 |
| × 10 | 1270 | 127.0 | 44.5 | 8.1 | 4.8 | 3.12 | 49.2 | 49.5 | 0.199 | 6.19 | 12.5 | 12.3 |
| C102 × 11 | 1375 | 101.6 | 43.7 | 7.5 | 8.2 | 1.91 | 37.5 | 37.3 | 0.180 | 5.62 | 11.4 | 11.7 |
| × 8 | 1025 | 101.6 | 40.2 | 7.5 | 4.7 | 1.60 | 31.6 | 39.6 | 0.133 | 4.64 | 11.4 | 11.6 |
| C76 × 9 | 1135 | 76.2 | 40.5 | 6.9 | 9.0 | 0.862 | 22.6 | 27.4 | 0.127 | 4.39 | 10.6 | 11.6 |
| × 7 | 948 | 76.2 | 38.0 | 6.9 | 6.6 | 0.770 | 20.3 | 28.4 | 0.103 | 3.82 | 10.4 | 11.1 |
| × 6 | 781 | 76.2 | 35.8 | 6.9 | 4.6 | 0.691 | 18.0 | 29.7 | 0.082 | 3.31 | 10.3 | 11.1 |

*C means channel, followed by the nominal depth in mm, then the mass in kg per meter of length.

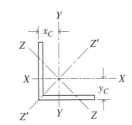

TABLE A-7 Equal Leg Angles (U.S. Customary Units)

| Size and Thickness (in.) | Weight (lb/ft) | Area (in²) | AXIS X–X OR Y–Y | | | | *AXIS Z–Z |
|---|---|---|---|---|---|---|---|
| | | | $I$ (in⁴) | $S$ (in³) | $r$ (in.) | $x_C$ or $y_C$ (in.) | $r$ (in.) |
| L8 × 8 × 1 | 51.0 | 15.0 | 89.0 | 15.8 | 2.44 | 2.37 | 1.56 |
| × 7/8 | 45.0 | 13.2 | 79.6 | 14.0 | 2.45 | 2.32 | 1.57 |
| × 3/4 | 38.9 | 11.4 | 69.7 | 12.2 | 2.47 | 2.28 | 1.58 |
| × 5/8 | 32.7 | 9.61 | 59.4 | 10.3 | 2.49 | 2.23 | 1.58 |
| × 1/2 | 26.4 | 7.75 | 48.6 | 8.36 | 2.50 | 2.19 | 1.59 |
| L6 × 6 × 1 | 37.4 | 11.0 | 35.5 | 8.57 | 1.80 | 1.86 | 1.17 |
| × 7/8 | 33.1 | 9.73 | 31.9 | 7.63 | 1.81 | 1.82 | 1.17 |
| × 3/4 | 28.7 | 8.44 | 28.2 | 6.66 | 1.83 | 1.78 | 1.17 |
| × 5/8 | 24.2 | 7.11 | 24.2 | 5.66 | 1.84 | 1.73 | 1.18 |
| × 1/2 | 19.6 | 5.75 | 19.9 | 4.61 | 1.86 | 1.68 | 1.18 |
| × 3/8 | 14.9 | 4.36 | 15.4 | 3.53 | 1.88 | 1.64 | 1.19 |
| L5 × 5 × 7/8 | 27.2 | 7.98 | 17.8 | 5.17 | 1.49 | 1.57 | 0.973 |
| × 3/4 | 23.6 | 6.94 | 15.7 | 4.53 | 1.51 | 1.52 | 0.975 |
| × 5/8 | 20.0 | 5.86 | 13.6 | 3.86 | 1.52 | 1.48 | 0.978 |
| × 1/2 | 16.2 | 4.75 | 11.3 | 3.16 | 1.54 | 1.43 | 0.983 |
| × 3/8 | 12.3 | 3.61 | 8.74 | 2.42 | 1.56 | 1.39 | 0.990 |
| L4 × 4 × 3/4 | 18.5 | 5.44 | 7.67 | 2.81 | 1.19 | 1.27 | 0.778 |
| × 5/8 | 15.7 | 4.61 | 6.66 | 2.40 | 1.20 | 1.23 | 0.779 |
| × 1/2 | 12.8 | 3.75 | 5.56 | 1.97 | 1.22 | 1.18 | 0.782 |
| × 3/8 | 9.8 | 2.86 | 4.36 | 1.52 | 1.23 | 1.14 | 0.788 |
| × 1/4 | 6.6 | 1.94 | 3.04 | 1.05 | 1.25 | 1.09 | 0.795 |
| L3½ × 3½ × 1/2 | 11.1 | 3.25 | 3.64 | 1.49 | 1.06 | 1.06 | 0.683 |
| × 3/8 | 8.5 | 2.48 | 2.87 | 1.15 | 1.07 | 1.01 | 0.687 |
| × 1/4 | 5.8 | 1.69 | 2.01 | 0.794 | 1.09 | 0.968 | 0.694 |
| L3 × 3 × 1/2 | 9.4 | 2.75 | 2.22 | 1.07 | 0.898 | 0.932 | 0.584 |
| × 3/8 | 7.2 | 2.11 | 1.76 | 0.833 | 0.913 | 0.888 | 0.587 |
| × 1/4 | 4.9 | 1.44 | 1.24 | 0.577 | 0.930 | 0.842 | 0.592 |
| L2½ × 2½ × 1/2 | 7.7 | 2.25 | 1.23 | 0.724 | 0.739 | 0.806 | 0.487 |
| × 3/8 | 5.9 | 1.73 | 0.984 | 0.566 | 0.753 | 0.762 | 0.487 |
| × 1/4 | 4.1 | 1.19 | 0.703 | 0.394 | 0.769 | 0.717 | 0.491 |
| L2 × 2 × 3/8 | 4.7 | 1.36 | 0.479 | 0.351 | 0.594 | 0.636 | 0.389 |
| × 1/4 | 3.19 | 0.938 | 0.348 | 0.247 | 0.609 | 0.592 | 0.391 |
| × 1/8 | 1.65 | 0.484 | 0.190 | 0.131 | 0.626 | 0.546 | 0.398 |

Courtesy of The American Institute of Steel Construction.

*Axes $Z$–$Z$ and $Z'$–$Z'$ are principal centroidal axes. The second moment of area $I_{ZZ}$ is $I_{min}$ and may be determined from $I_{ZZ} = r^2 A$. The second moment of area $I_{Z'Z'}$ is $I_{max}$ and may be determined from $I_{ZZ} + I_{Z'Z'} = I_{XX} + I_{YY}$.

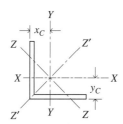

## TABLE A-8 Equal Leg Angles (SI Units)

| Size and Thickness (mm) | Mass (kg/m) | Area (mm²) | AXIS $X$–$X$ OR $Y$–$Y$ | | | | *AXIS $Z$–$Z$ |
|---|---|---|---|---|---|---|---|
| | | | $I$ ($10^6$ mm⁴) | $S$ ($10^3$ mm³) | $r$ (mm) | $x_C$ or $y_C$ (mm) | $r$ (mm) |
| L203 × 203 × 25.4 | 75.9 | 9675 | 37.0 | 259 | 62.0 | 60.2 | 39.6 |
| × 22.2 | 67.0 | 8515 | 33.1 | 229 | 62.2 | 58.9 | 39.9 |
| × 19.1 | 57.9 | 7355 | 29.0 | 200 | 62.7 | 57.9 | 40.1 |
| × 15.9 | 48.7 | 6200 | 24.7 | 169 | 63.2 | 56.6 | 40.1 |
| × 12.7 | 39.3 | 5000 | 20.2 | 137 | 63.5 | 55.6 | 40.4 |
| L152 × 152 × 25.4 | 55.7 | 7095 | 14.8 | 140 | 45.7 | 47.2 | 29.7 |
| × 22.2 | 49.3 | 6275 | 13.3 | 125 | 46.0 | 46.2 | 29.7 |
| × 19.1 | 42.7 | 5445 | 11.7 | 109 | 46.5 | 45.2 | 29.7 |
| × 15.9 | 36.0 | 4585 | 10.1 | 92.8 | 46.7 | 43.9 | 30.0 |
| × 12.7 | 29.2 | 3710 | 8.28 | 75.5 | 47.2 | 42.7 | 30.0 |
| × 9.5 | 22.2 | 2815 | 6.61 | 57.8 | 47.8 | 41.7 | 30.2 |
| L127 × 127 × 22.2 | 40.5 | 5150 | 7.41 | 84.7 | 37.8 | 39.9 | 24.7 |
| × 19.1 | 35.1 | 4475 | 6.53 | 74.2 | 38.4 | 38.6 | 24.8 |
| × 15.9 | 29.8 | 3780 | 5.66 | 63.3 | 38.6 | 37.6 | 24.8 |
| × 12.7 | 24.1 | 3065 | 4.70 | 51.8 | 39.1 | 36.3 | 25.0 |
| × 9.5 | 18.3 | 2330 | 3.64 | 39.7 | 39.6 | 35.3 | 25.1 |
| L102 × 102 × 19.1 | 27.5 | 3510 | 3.19 | 46.0 | 30.2 | 32.3 | 19.8 |
| × 15.9 | 23.4 | 2975 | 2.77 | 39.3 | 30.5 | 31.2 | 19.8 |
| × 12.7 | 19.0 | 2420 | 2.31 | 32.3 | 31.0 | 30.0 | 19.9 |
| × 9.5 | 14.6 | 1845 | 1.81 | 24.9 | 31.2 | 29.0 | 20.0 |
| × 6.4 | 9.8 | 1250 | 1.27 | 17.2 | 31.8 | 27.7 | 20.2 |
| L89 × 89 × 12.7 | 16.5 | 2095 | 1.52 | 24.4 | 26.9 | 26.9 | 17.3 |
| × 9.5 | 12.6 | 1600 | 1.19 | 18.8 | 27.2 | 25.7 | 17.4 |
| × 6.4 | 8.6 | 1090 | 0.837 | 13.0 | 27.7 | 24.6 | 17.6 |
| L76 × 76 × 12.7 | 14.0 | 1775 | 0.924 | 17.5 | 22.8 | 23.7 | 14.8 |
| × 9.5 | 10.7 | 1360 | 0.732 | 13.7 | 23.2 | 22.6 | 14.9 |
| × 6.4 | 7.3 | 929 | 0.516 | 9.46 | 23.6 | 21.4 | 15.0 |
| L64 × 64 × 12.7 | 11.5 | 1450 | 0.512 | 11.9 | 18.8 | 20.5 | 12.4 |
| × 9.5 | 8.8 | 1115 | 0.410 | 9.28 | 19.1 | 19.4 | 12.4 |
| × 6.4 | 6.1 | 768 | 0.293 | 6.46 | 19.5 | 18.2 | 12.5 |
| L51 × 51 × 9.5 | 7.0 | 877 | 0.199 | 5.75 | 15.1 | 16.2 | 9.88 |
| × 6.4 | 4.75 | 605 | 0.145 | 4.05 | 15.5 | 15.0 | 9.93 |
| × 3.2 | 2.46 | 312 | 0.079 | 2.15 | 15.9 | 13.9 | 10.1 |

*Axes $Z$–$Z$ and $Z'$–$Z'$ are principal centroidal axes. The second moment of area $I_{ZZ}$ is $I_{min}$ and may be determined from $I_{ZZ} = r^2 A$. The second moment of area $I_{Z'Z'}$ is $I_{max}$ and may be determined from $I_{ZZ} + I_{Z'Z'} = I_{XX} + I_{YY}$.

TABLE A-9 Unequal Leg Angles (U.S. Customary Units)

| Size and Thickness | Weight (lb/ft) | Area (in²) | AXIS X–X | | | | AXIS Y–Y | | | | *AXIS Z–Z | |
|---|---|---|---|---|---|---|---|---|---|---|---|---|
| | | | $I$ (in⁴) | $S$ (in³) | $r$ (in.) | $y_C$ (in.) | $I$ (in⁴) | $S$ (in³) | $r$ (in.) | $x_C$ (in.) | $r$ (in.) | Tan $\alpha$ |
| L9 × 4 × 5/8 | 26.3 | 7.73 | 64.9 | 11.5 | 2.90 | 3.36 | 8.32 | 2.65 | 1.04 | 0.858 | 0.847 | 0.216 |
| × 1/2 | 21.3 | 6.25 | 53.2 | 9.34 | 2.92 | 3.31 | 6.92 | 2.17 | 1.05 | 0.810 | 0.854 | 0.220 |
| L8 × 6 × 1 | 44.2 | 13.0 | 80.8 | 15.1 | 2.49 | 2.65 | 38.8 | 8.92 | 1.73 | 1.65 | 1.28 | 0.543 |
| × 3/4 | 33.8 | 9.94 | 63.4 | 11.7 | 2.53 | 2.56 | 30.7 | 6.92 | 1.76 | 1.56 | 1.29 | 0.551 |
| × 1/2 | 23.0 | 6.75 | 44.3 | 8.02 | 2.56 | 2.47 | 21.7 | 4.79 | 1.79 | 1.47 | 1.30 | 0.558 |
| L8 × 4 × 1 | 37.4 | 11.0 | 69.6 | 14.1 | 2.52 | 3.05 | 11.6 | 3.94 | 1.03 | 1.05 | 0.846 | 0.247 |
| × 3/4 | 28.7 | 8.44 | 54.9 | 10.9 | 2.55 | 2.95 | 9.36 | 3.07 | 1.05 | 0.953 | 0.852 | 0.258 |
| × 1/2 | 19.6 | 5.75 | 38.5 | 7.49 | 2.59 | 2.86 | 6.74 | 2.15 | 1.08 | 0.859 | 0.865 | 0.267 |
| L7 × 4 × 3/4 | 26.2 | 7.69 | 37.8 | 8.42 | 2.22 | 2.51 | 9.05 | 3.03 | 1.09 | 1.01 | 0.860 | 0.324 |
| × 1/2 | 17.9 | 5.25 | 26.7 | 5.81 | 2.25 | 2.42 | 6.53 | 2.12 | 1.11 | 0.917 | 0.872 | 0.335 |
| × 3/8 | 13.6 | 3.98 | 20.6 | 4.44 | 2.27 | 2.37 | 5.10 | 1.63 | 1.13 | 0.870 | 0.880 | 0.340 |
| L6 × 4 × 3/4 | 23.6 | 6.94 | 24.5 | 6.25 | 1.88 | 2.08 | 8.68 | 2.97 | 1.12 | 1.08 | 0.860 | 0.428 |
| × 1/2 | 16.2 | 4.75 | 17.4 | 4.33 | 1.91 | 1.99 | 6.27 | 2.08 | 1.15 | 0.987 | 0.870 | 0.440 |
| × 3/8 | 12.3 | 3.61 | 13.5 | 3.32 | 1.93 | 1.94 | 4.90 | 1.60 | 1.17 | 0.941 | 0.877 | 0.446 |
| L6 × 3½ × 1/2 | 15.3 | 4.50 | 16.6 | 4.24 | 1.92 | 2.08 | 4.25 | 1.59 | 0.972 | 0.833 | 0.759 | 0.344 |
| × 3/8 | 11.7 | 3.42 | 12.9 | 3.24 | 1.94 | 2.04 | 3.34 | 1.23 | 0.988 | 0.787 | 0.767 | 0.350 |
| L5 × 3½ × 3/4 | 19.8 | 5.81 | 13.9 | 4.28 | 1.55 | 1.75 | 5.55 | 2.22 | 0.977 | 0.996 | 0.748 | 0.464 |
| × 1/2 | 13.6 | 4.00 | 9.99 | 2.99 | 1.58 | 1.66 | 4.05 | 1.56 | 1.01 | 0.906 | 0.755 | 0.479 |
| × 3/8 | 10.4 | 3.05 | 7.78 | 2.29 | 1.60 | 1.61 | 3.18 | 1.21 | 1.02 | 0.861 | 0.762 | 0.486 |
| × 1/4 | 7.0 | 2.06 | 5.39 | 1.57 | 1.62 | 1.56 | 2.23 | 0.830 | 1.04 | 0.814 | 0.770 | 0.492 |
| L5 × 3 × 1/2 | 12.8 | 3.75 | 9.45 | 2.91 | 1.59 | 1.75 | 2.58 | 1.15 | 0.829 | 0.750 | 0.648 | 0.357 |
| × 3/8 | 9.8 | 2.86 | 7.37 | 2.24 | 1.61 | 1.70 | 2.04 | 0.888 | 0.845 | 0.704 | 0.654 | 0.364 |
| × 1/4 | 6.6 | 1.94 | 5.11 | 1.53 | 1.62 | 1.66 | 1.44 | 0.614 | 0.861 | 0.657 | 0.663 | 0.371 |
| L4 × 3½ × 1/2 | 11.9 | 3.50 | 5.32 | 1.94 | 1.23 | 1.25 | 3.79 | 1.52 | 1.04 | 1.00 | 0.722 | 0.750 |
| × 3/8 | 9.1 | 2.67 | 4.18 | 1.49 | 1.25 | 1.21 | 2.95 | 1.17 | 1.06 | 0.955 | 0.727 | 0.755 |
| × 1/4 | 6.2 | 1.81 | 2.91 | 1.03 | 1.27 | 1.16 | 2.09 | 0.808 | 1.07 | 0.909 | 0.734 | 0.759 |
| L4 × 3 × 1/2 | 11.1 | 3.25 | 5.05 | 1.89 | 1.25 | 1.33 | 2.42 | 1.12 | 0.864 | 0.827 | 0.639 | 0.543 |
| × 3/8 | 8.5 | 2.48 | 3.96 | 1.46 | 1.26 | 1.28 | 1.92 | 0.866 | 0.879 | 0.782 | 0.644 | 0.551 |
| × 1/4 | 5.8 | 1.69 | 2.77 | 1.00 | 1.28 | 1.24 | 1.36 | 0.599 | 0.896 | 0.736 | 0.651 | 0.558 |
| L3½ × 3 × 1/2 | 10.2 | 3.00 | 3.45 | 1.45 | 1.07 | 1.13 | 2.33 | 1.10 | 0.881 | 0.875 | 0.621 | 0.714 |
| × 3/8 | 7.9 | 2.30 | 2.72 | 1.13 | 1.09 | 1.08 | 1.85 | 0.851 | 0.897 | 0.830 | 0.625 | 0.721 |
| × 1/4 | 5.4 | 1.56 | 1.91 | 0.776 | 1.11 | 1.04 | 1.30 | 0.589 | 0.914 | 0.785 | 0.631 | 0.727 |
| L3½ × 2½ × 1/2 | 9.4 | 2.75 | 3.24 | 1.41 | 1.09 | 1.20 | 1.36 | 0.760 | 0.704 | 0.705 | 0.534 | 0.486 |
| × 3/8 | 7.2 | 2.11 | 2.56 | 1.09 | 1.10 | 1.16 | 1.09 | 0.592 | 0.719 | 0.660 | 0.537 | 0.496 |
| × 1/4 | 4.9 | 1.44 | 1.80 | 0.755 | 1.12 | 1.11 | 0.777 | 0.412 | 0.735 | 0.614 | 0.544 | 0.506 |
| L3 × 2½ × 1/2 | 8.5 | 2.50 | 2.08 | 1.04 | 0.913 | 1.00 | 1.30 | 0.744 | 0.722 | 0.750 | 0.520 | 0.667 |
| × 3/8 | 6.6 | 1.92 | 1.66 | 0.810 | 0.928 | 0.956 | 1.04 | 0.581 | 0.736 | 0.706 | 0.522 | 0.676 |
| × 1/4 | 4.5 | 1.31 | 1.17 | 0.561 | 0.945 | 0.911 | 0.743 | 0.404 | 0.753 | 0.661 | 0.528 | 0.684 |
| L3 × 2 × 1/2 | 7.7 | 2.25 | 1.92 | 1.00 | 0.924 | 1.08 | 0.672 | 0.474 | 0.546 | 0.583 | 0.428 | 0.414 |
| × 3/8 | 5.9 | 1.73 | 1.53 | 0.781 | 0.940 | 1.04 | 0.543 | 0.371 | 0.559 | 0.539 | 0.430 | 0.428 |
| × 1/4 | 4.1 | 1.19 | 1.09 | 0.542 | 0.957 | 0.993 | 0.392 | 0.260 | 0.574 | 0.493 | 0.435 | 0.440 |
| L2½ × 2 × 3/8 | 5.3 | 1.55 | 0.912 | 0.547 | 0.768 | 0.813 | 0.514 | 0.363 | 0.577 | 0.581 | 0.420 | 0.614 |
| × 1/4 | 3.62 | 1.06 | 0.654 | 0.381 | 0.784 | 0.787 | 0.372 | 0.254 | 0.592 | 0.537 | 0.424 | 0.626 |

*Axes $Z$–$Z$ and $Z'$–$Z'$ are principal centroidal axes. The second moment of area $I_{ZZ}$ is $I_{min}$ and may be determined from $I_{ZZ} = r^2 A$. The second moment of area $I_{Z'Z'}$ is $I_{max}$ and may be determined from $I_{ZZ} + I_{Z'Z'} = I_{XX} + I_{YY}$. The angle $\alpha$ locates the principal centroidal axes.

## TABLE A-10 Unequal Leg Angles (SI Units)

| Size and Thickness | Mass (kg/m) | Area (mm²) | AXIS $X-X$ $I$ ($10^6$ mm⁴) | $S$ ($10^3$ mm³) | $r$ (mm) | $y_C$ (mm) | AXIS $Y-Y$ $I$ ($10^6$ mm⁴) | $S$ ($10^3$ mm³) | $r$ (mm) | $x_C$ (mm) | *AXIS $Z-Z$ $r$ (mm) | Tan $\alpha$ |
|---|---|---|---|---|---|---|---|---|---|---|---|---|
| L229 × 102 × 15.9 | 39.1 | 4985 | 27.0 | 188 | 73.7 | 85.3 | 3.46 | 43.4 | 26.4 | 21.8 | 21.5 | 0.216 |
| × 12.7 | 31.7 | 4030 | 22.1 | 153 | 74.2 | 84.1 | 2.88 | 35.6 | 26.7 | 20.6 | 21.7 | 0.220 |
| L203 × 152 × 25.4 | 65.8 | 8385 | 33.6 | 247 | 63.2 | 67.3 | 16.1 | 146 | 43.9 | 41.9 | 32.5 | 0.543 |
| × 19.1 | 50.3 | 6415 | 26.4 | 192 | 64.3 | 65.0 | 12.8 | 113 | 44.7 | 39.6 | 32.8 | 0.551 |
| × 12.7 | 34.2 | 4355 | 18.4 | 131 | 65.0 | 62.7 | 9.03 | 78.5 | 45.5 | 37.3 | 33.0 | 0.558 |
| L203 × 102 × 25.4 | 55.7 | 7095 | 29.0 | 231 | 64.0 | 77.5 | 4.83 | 64.6 | 26.2 | 26.7 | 21.5 | 0.247 |
| × 19.1 | 42.7 | 5445 | 22.9 | 179 | 64.8 | 74.9 | 3.90 | 50.3 | 26.7 | 24.2 | 21.6 | 0.258 |
| × 12.7 | 29.2 | 3710 | 16.0 | 123 | 65.8 | 72.6 | 2.81 | 35.2 | 27.4 | 21.8 | 22.0 | 0.267 |
| L178 × 102 × 19.1 | 39.0 | 4960 | 15.7 | 138 | 56.4 | 63.8 | 3.77 | 49.7 | 27.7 | 25.7 | 21.8 | 0.324 |
| × 12.7 | 26.6 | 3385 | 11.1 | 95.2 | 57.2 | 61.5 | 2.72 | 34.7 | 28.2 | 23.3 | 22.1 | 0.335 |
| × 9.5 | 20.2 | 2570 | 8.57 | 72.8 | 57.7 | 60.2 | 2.12 | 26.7 | 28.7 | 22.1 | 22.4 | 0.340 |
| L152 × 102 × 19.1 | 35.1 | 4475 | 10.2 | 102 | 47.8 | 52.8 | 3.61 | 48.7 | 28.4 | 27.4 | 21.8 | 0.428 |
| × 12.7 | 24.1 | 3065 | 7.24 | 71.0 | 48.5 | 50.5 | 2.61 | 34.1 | 29.2 | 25.1 | 22.1 | 0.440 |
| × 9.5 | 18.3 | 3230 | 5.62 | 54.4 | 49.0 | 49.3 | 2.04 | 26.2 | 29.7 | 23.9 | 22.3 | 0.446 |
| L152 × 89 × 12.7 | 22.8 | 2905 | 6.91 | 69.5 | 48.8 | 52.8 | 1.77 | 26.1 | 24.7 | 21.2 | 19.3 | 0.344 |
| × 9.5 | 17.4 | 2205 | 5.37 | 53.1 | 49.3 | 51.8 | 1.39 | 20.2 | 25.1 | 20.0 | 19.5 | 0.350 |
| L127 × 89 × 19.1 | 29.5 | 3750 | 5.79 | 70.1 | 39.4 | 44.5 | 2.31 | 36.4 | 24.8 | 25.3 | 19.0 | 0.464 |
| × 12.7 | 20.2 | 2580 | 4.16 | 49.0 | 40.1 | 42.2 | 1.69 | 25.6 | 25.7 | 23.0 | 19.2 | 0.479 |
| × 9.5 | 15.5 | 1970 | 3.24 | 37.5 | 40.6 | 40.9 | 1.32 | 19.8 | 25.9 | 21.9 | 19.4 | 0.486 |
| × 6.4 | 10.4 | 1330 | 2.24 | 25.7 | 41.1 | 39.6 | 0.928 | 13.6 | 26.4 | 20.7 | 19.6 | 0.492 |
| L127 × 76 × 12.7 | 19.0 | 2420 | 3.93 | 47.7 | 40.4 | 44.5 | 1.07 | 18.8 | 21.1 | 19.1 | 16.5 | 0.357 |
| × 9.5 | 14.6 | 1845 | 3.07 | 36.7 | 40.9 | 43.2 | 0.849 | 14.6 | 21.5 | 17.9 | 16.6 | 0.364 |
| × 6.4 | 9.82 | 1250 | 2.13 | 25.1 | 41.1 | 42.2 | 0.599 | 10.1 | 21.9 | 16.7 | 16.8 | 0.371 |
| L102 × 89 × 12.7 | 17.7 | 2260 | 2.21 | 31.8 | 31.2 | 31.8 | 1.58 | 24.9 | 26.4 | 25.4 | 18.3 | 0.750 |
| × 9.5 | 13.5 | 1725 | 1.74 | 24.4 | 31.8 | 30.7 | 1.23 | 19.2 | 26.9 | 24.3 | 18.5 | 0.755 |
| × 6.4 | 9.22 | 1170 | 1.21 | 16.9 | 32.3 | 29.5 | 0.870 | 13.2 | 27.2 | 23.1 | 18.6 | 0.759 |
| L102 × 76 × 12.7 | 16.5 | 2095 | 2.10 | 31.0 | 31.8 | 33.8 | 1.01 | 18.8 | 21.9 | 21.0 | 16.2 | 0.543 |
| × 9.5 | 12.6 | 1600 | 1.65 | 23.9 | 32.0 | 32.5 | 0.799 | 14.2 | 22.3 | 19.9 | 16.4 | 0.551 |
| × 6.4 | 8.63 | 1090 | 1.15 | 16.4 | 32.5 | 31.5 | 0.566 | 9.82 | 22.8 | 18.7 | 16.5 | 0.558 |
| L89 × 76 × 12.7 | 15.2 | 1935 | 1.44 | 23.8 | 27.2 | 28.7 | 0.970 | 18.0 | 22.4 | 22.2 | 15.8 | 0.714 |
| × 9.5 | 11.8 | 1485 | 1.13 | 18.5 | 27.7 | 27.4 | 0.770 | 13.9 | 22.8 | 21.1 | 15.9 | 0.721 |
| × 6.4 | 8.04 | 1005 | 0.795 | 12.7 | 28.2 | 26.4 | 0.541 | 9.65 | 23.2 | 19.9 | 16.0 | 0.727 |
| L89 × 64 × 12.7 | 14.0 | 1775 | 1.35 | 23.1 | 27.7 | 30.5 | 0.566 | 12.5 | 17.9 | 17.9 | 13.6 | 0.486 |
| × 9.5 | 10.7 | 1360 | 1.07 | 17.9 | 27.9 | 29.5 | 0.454 | 8.70 | 18.3 | 16.8 | 13.6 | 0.496 |
| × 6.4 | 7.29 | 929 | 0.749 | 12.4 | 28.4 | 28.2 | 0.323 | 6.75 | 18.7 | 15.6 | 13.8 | 0.506 |
| L76 × 64 × 12.7 | 12.6 | 1615 | 0.866 | 17.0 | 23.2 | 25.4 | 0.541 | 12.2 | 18.3 | 19.1 | 13.2 | 0.667 |
| × 9.5 | 9.82 | 1240 | 0.691 | 13.3 | 23.6 | 24.3 | 0.433 | 9.52 | 18.7 | 17.9 | 13.3 | 0.676 |
| × 6.4 | 6.70 | 845 | 0.487 | 9.19 | 24.0 | 23.1 | 0.309 | 6.62 | 19.1 | 16.8 | 13.4 | 0.684 |
| L76 × 51 × 12.7 | 11.5 | 1450 | 0.799 | 16.4 | 23.5 | 27.4 | 0.280 | 7.77 | 13.9 | 14.8 | 10.9 | 0.414 |
| × 9.5 | 8.78 | 1115 | 0.637 | 12.8 | 23.9 | 26.4 | 0.226 | 6.08 | 14.2 | 13.7 | 10.9 | 0.428 |
| × 6.4 | 6.10 | 768 | 0.454 | 8.88 | 24.3 | 25.2 | 0.163 | 4.26 | 14.6 | 12.5 | 11.0 | 0.440 |
| L64 × 51 × 9.5 | 7.89 | 1000 | 0.380 | 8.96 | 19.5 | 20.7 | 0.214 | 5.95 | 14.7 | 14.8 | 10.7 | 0.614 |
| × 6.4 | 5.39 | 684 | 0.272 | 6.24 | 19.9 | 20.0 | 0.155 | 4.16 | 15.0 | 13.6 | 10.8 | 0.626 |

*Axes $Z-Z$ and $Z'-Z'$ are principal centroidal axes. The second moment of area $I_{ZZ}$ is $I_{min}$ and may be determined from $I_{ZZ} = r^2 A$. The second moment of area $I_{Z'Z'}$ is $I_{max}$ and may be determined from $I_{ZZ} + I_{Z'Z'} = I_{XX} + I_{YY}$. The angle $\alpha$ locates the principal centroidal axes.

T

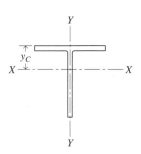

TABLE A-11 Structural Tees (U.S. Customary Units)

| Desig-nation* | Area (in²) | Depth of Tee (in.) | FLANGE Width (in.) | Thick-ness (in.) | Stem Thick-ness (in.) | AXIS X–X $I$ (in⁴) | $S$ (in³) | $r$ (in.) | $y_C$ (in.) | AXIS Y–Y $I$ (in⁴) | $S$ (in³) | $r$ (in.) |
|---|---|---|---|---|---|---|---|---|---|---|---|---|
| WT18 × 115 | 33.8 | 17.950 | 16.470 | 1.260 | 0.760 | 934 | 67.0 | 5.25 | 4.01 | 470 | 57.1 | 3.73 |
| × 80 | 23.5 | 18.005 | 12.000 | 1.020 | 0.650 | 740 | 55.8 | 5.61 | 4.74 | 147 | 24.6 | 2.50 |
| WT15 × 66 | 19.4 | 15.155 | 10.545 | 1.000 | 0.615 | 421 | 37.4 | 4.66 | 3.90 | 98.0 | 18.6 | 2.25 |
| × 54 | 15.9 | 14.915 | 10.475 | 0.760 | 0.545 | 349 | 32.0 | 4.69 | 4.01 | 73.0 | 13.9 | 2.15 |
| WT12 × 52 | 15.3 | 12.030 | 12.750 | 0.750 | 0.500 | 189 | 20.0 | 3.51 | 2.59 | 130 | 20.3 | 2.91 |
| × 47 | 13.8 | 12.155 | 9.065 | 0.875 | 0.515 | 186 | 20.3 | 3.67 | 2.99 | 54.5 | 12.0 | 1.98 |
| × 42 | 12.4 | 12.050 | 9.020 | 0.770 | 0.470 | 166 | 18.3 | 3.67 | 2.97 | 47.2 | 10.5 | 1.95 |
| × 31 | 9.11 | 11.870 | 7.040 | 0.590 | 0.430 | 131 | 15.6 | 3.79 | 3.46 | 17.2 | 4.90 | 1.38 |
| WT9 × 38 | 11.2 | 9.105 | 11.035 | 0.680 | 0.425 | 71.8 | 9.83 | 2.54 | 1.80 | 76.2 | 13.8 | 2.61 |
| × 30 | 8.82 | 9.120 | 7.555 | 0.695 | 0.415 | 64.7 | 9.29 | 2.71 | 2.16 | 25.0 | 6.63 | 1.69 |
| × 25 | 7.33 | 8.995 | 7.495 | 0.570 | 0.355 | 53.5 | 7.79 | 2.70 | 2.12 | 20.0 | 5.35 | 1.65 |
| × 20 | 5.88 | 8.950 | 6.015 | 0.525 | 0.315 | 44.8 | 6.73 | 2.76 | 2.29 | 9.55 | 3.17 | 1.27 |
| WT8 × 50 | 14.7 | 8.485 | 10.425 | 0.985 | 0.585 | 76.8 | 11.4 | 2.28 | 1.76 | 93.1 | 17.9 | 2.51 |
| × 25 | 7.37 | 8.130 | 7.070 | 0.630 | 0.380 | 42.3 | 6.78 | 2.40 | 1.89 | 18.6 | 5.26 | 1.59 |
| × 20 | 5.89 | 8.005 | 6.995 | 0.505 | 0.305 | 33.1 | 5.35 | 2.37 | 1.81 | 14.4 | 4.12 | 1.57 |
| × 13 | 3.84 | 7.845 | 5.500 | 0.345 | 0.250 | 23.5 | 4.09 | 2.47 | 2.09 | 4.80 | 1.74 | 1.12 |
| WT7 × 60 | 17.7 | 7.240 | 14.670 | 0.940 | 0.590 | 51.7 | 8.61 | 1.71 | 1.24 | 247 | 33.7 | 3.74 |
| × 41 | 12.0 | 7.155 | 10.130 | 0.855 | 0.510 | 41.2 | 7.14 | 1.85 | 1.39 | 74.2 | 14.6 | 2.48 |
| × 34 | 9.99 | 7.020 | 10.035 | 0.720 | 0.415 | 32.6 | 5.69 | 1.81 | 1.29 | 60.7 | 12.1 | 2.46 |
| × 24 | 7.07 | 6.985 | 8.030 | 0.595 | 0.340 | 24.9 | 4.48 | 1.87 | 1.35 | 25.7 | 6.40 | 1.91 |
| × 15 | 4.42 | 6.920 | 6.730 | 0.385 | 0.270 | 19.0 | 3.55 | 2.07 | 1.58 | 9.79 | 2.91 | 1.49 |
| × 11 | 3.25 | 6.870 | 5.000 | 0.335 | 0.230 | 14.8 | 2.91 | 2.14 | 1.76 | 3.50 | 1.40 | 1.04 |
| WT6 × 60 | 17.6 | 6.560 | 12.320 | 1.105 | 0.710 | 43.4 | 8.22 | 1.57 | 1.28 | 172 | 28.0 | 3.13 |
| × 48 | 14.1 | 6.355 | 12.160 | 0.900 | 0.550 | 32.0 | 6.12 | 1.51 | 1.13 | 135 | 22.2 | 3.09 |
| × 36 | 10.6 | 6.125 | 12.040 | 0.670 | 0.430 | 23.2 | 4.54 | 1.48 | 1.02 | 97.5 | 16.2 | 3.04 |
| × 25 | 7.34 | 6.095 | 8.080 | 0.640 | 0.370 | 18.7 | 3.79 | 1.60 | 1.17 | 28.2 | 6.97 | 1.96 |
| × 15 | 4.40 | 6.170 | 6.520 | 0.440 | 0.260 | 13.5 | 2.75 | 1.75 | 1.27 | 10.2 | 3.12 | 1.52 |
| × 8 | 2.36 | 5.995 | 3.990 | 0.265 | 0.220 | 8.70 | 2.04 | 1.92 | 1.74 | 1.41 | 0.706 | 0.773 |
| WT5 × 56 | 16.5 | 5.680 | 10.415 | 1.250 | 0.755 | 28.6 | 6.40 | 1.32 | 1.21 | 118 | 22.6 | 2.68 |
| × 44 | 12.9 | 5.420 | 10.265 | 0.990 | 0.605 | 20.8 | 4.77 | 1.27 | 1.06 | 89.3 | 17.4 | 2.63 |
| × 30 | 8.82 | 5.110 | 10.080 | 0.680 | 0.420 | 12.9 | 3.04 | 1.21 | 0.884 | 58.1 | 11.5 | 2.57 |
| × 15 | 4.42 | 5.235 | 5.810 | 0.510 | 0.300 | 9.28 | 2.24 | 1.45 | 1.10 | 8.35 | 2.87 | 1.37 |
| × 6 | 1.77 | 4.935 | 3.960 | 0.210 | 0.190 | 4.35 | 1.22 | 1.57 | 1.36 | 1.09 | 0.551 | 0.785 |
| WT4 × 29 | 8.55 | 4.375 | 8.220 | 0.810 | 0.510 | 9.12 | 2.61 | 1.03 | 0.874 | 37.5 | 9.13 | 2.10 |
| × 20 | 5.87 | 4.125 | 8.070 | 0.560 | 0.360 | 5.73 | 1.69 | 0.988 | 0.735 | 24.5 | 6.08 | 2.04 |
| × 12 | 3.54 | 3.965 | 6.495 | 0.400 | 0.245 | 3.53 | 1.08 | 0.999 | 0.695 | 9.14 | 2.81 | 1.61 |
| × 9 | 2.63 | 4.070 | 5.250 | 0.330 | 0.230 | 3.41 | 1.05 | 1.14 | 0.834 | 3.98 | 1.52 | 1.23 |
| × 5 | 1.48 | 3.945 | 3.940 | 0.205 | 0.170 | 2.15 | 0.717 | 1.20 | 0.953 | 1.05 | 0.532 | 0.841 |
| WT3 × 10 | 2.94 | 3.100 | 6.020 | 0.365 | 0.260 | 1.76 | 0.693 | 0.774 | 0.560 | 6.64 | 2.21 | 1.50 |
| × 6 | 1.78 | 3.015 | 4.000 | 0.280 | 0.230 | 1.32 | 0.564 | 0.861 | 0.677 | 1.50 | 0.748 | 0.918 |
| WT2 × 6.5 | 1.91 | 2.080 | 4.060 | 0.345 | 0.280 | 0.526 | 0.321 | 0.524 | 0.440 | 1.93 | 0.950 | 1.00 |

Courtesy of The American Institute of Steel Construction.

*WT means structural T-section (cut from a W-section), followed by the nominal depth in inches, then the weight in pounds per foot of length.

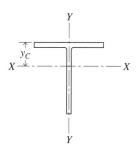

## TABLE A-12 Structural Tees (SI Units)

| Desig-nation* | Area (mm²) | Depth of Tee (mm) | FLANGE Width (mm) | Thick-ness (mm) | Stem Thick-ness (mm) | AXIS X–X $I$ (10⁶ mm⁴) | $S$ (10³ mm³) | $r$ (mm) | $y_C$ (mm) | AXIS Y–Y $I$ (10⁶ mm⁴) | $S$ (10³ mm³) | $r$ (mm) |
|---|---|---|---|---|---|---|---|---|---|---|---|---|
| WT457 × 171 | 21805 | 455.9 | 418.3 | 32.0 | 19.3 | 389 | 1098 | 133 | 102 | 196 | 936 | 94.7 |
| × 119 | 15160 | 457.3 | 304.8 | 25.9 | 16.5 | 308 | 914 | 142 | 120 | 61.2 | 403 | 63.5 |
| WT381 × 98 | 12515 | 384.9 | 267.8 | 25.4 | 15.6 | 175 | 613 | 118 | 99.1 | 40.8 | 305 | 57.2 |
| × 80 | 10260 | 378.8 | 266.1 | 19.3 | 13.8 | 145 | 524 | 119 | 102 | 30.4 | 228 | 54.6 |
| WT305 × 77 | 9870 | 305.6 | 323.9 | 19.1 | 12.7 | 78.7 | 328 | 89.2 | 65.8 | 54.1 | 333 | 73.9 |
| × 70 | 8905 | 308.7 | 230.3 | 22.2 | 13.1 | 77.4 | 333 | 93.2 | 75.9 | 22.7 | 197 | 50.3 |
| × 63 | 8000 | 306.1 | 229.1 | 19.6 | 11.9 | 69.1 | 300 | 93.2 | 75.4 | 19.6 | 172 | 49.5 |
| × 46 | 5875 | 301.5 | 178.8 | 15.0 | 10.9 | 54.5 | 256 | 96.3 | 87.9 | 7.16 | 80.3 | 35.1 |
| WT229 × 57 | 7225 | 231.3 | 280.3 | 17.3 | 10.8 | 29.9 | 161 | 64.5 | 45.7 | 31.7 | 226 | 66.3 |
| × 45 | 5690 | 231.6 | 191.9 | 17.7 | 10.5 | 26.9 | 152 | 68.8 | 54.9 | 10.4 | 109 | 42.9 |
| × 37 | 4730 | 228.5 | 190.4 | 14.5 | 9.0 | 22.3 | 128 | 68.6 | 53.8 | 8.32 | 87.7 | 41.9 |
| × 30 | 3795 | 227.3 | 152.8 | 13.3 | 8.0 | 18.6 | 110 | 70.1 | 58.2 | 3.98 | 51.9 | 32.3 |
| WT203 × 74 | 9485 | 215.5 | 264.8 | 25.0 | 14.9 | 32.0 | 187 | 57.9 | 44.7 | 38.8 | 293 | 63.8 |
| × 37 | 4755 | 206.5 | 179.6 | 16.0 | 9.7 | 17.6 | 111 | 61.0 | 48.0 | 7.74 | 86.2 | 40.4 |
| × 30 | 3800 | 203.3 | 177.7 | 12.8 | 7.7 | 13.8 | 87.7 | 60.2 | 46.0 | 5.99 | 67.5 | 39.9 |
| × 19 | 2475 | 199.3 | 139.7 | 8.8 | 6.4 | 9.78 | 67.0 | 62.7 | 53.1 | 2.00 | 28.5 | 28.4 |
| WT178 × 89 | 11420 | 183.9 | 372.6 | 23.9 | 15.0 | 21.5 | 141 | 43.4 | 31.5 | 103 | 552 | 95.0 |
| × 61 | 7740 | 181.7 | 257.3 | 21.7 | 13.0 | 17.1 | 117 | 47.0 | 35.3 | 30.9 | 239 | 63.0 |
| × 51 | 6445 | 178.3 | 254.9 | 18.3 | 10.5 | 13.6 | 93.2 | 46.0 | 32.8 | 25.3 | 198 | 62.5 |
| × 36 | 4560 | 177.4 | 204.0 | 15.1 | 8.6 | 10.4 | 73.4 | 47.5 | 34.3 | 10.7 | 105 | 48.5 |
| × 22 | 2850 | 175.8 | 170.9 | 9.8 | 6.9 | 7.91 | 58.2 | 52.6 | 40.1 | 4.07 | 47.7 | 37.8 |
| × 16 | 2095 | 174.5 | 127.0 | 8.5 | 5.8 | 6.16 | 47.7 | 54.4 | 44.7 | 1.46 | 22.9 | 26.4 |
| WT152 × 89 | 11355 | 166.6 | 312.9 | 28.1 | 18.0 | 18.1 | 135 | 39.9 | 32.5 | 71.6 | 459 | 79.5 |
| × 71 | 9095 | 161.4 | 308.9 | 22.9 | 14.0 | 13.3 | 100 | 38.4 | 28.7 | 56.2 | 364 | 78.5 |
| × 54 | 6840 | 155.6 | 305.8 | 17.0 | 10.9 | 9.66 | 74.4 | 37.6 | 25.9 | 40.6 | 265 | 77.2 |
| × 37 | 4735 | 154.8 | 205.2 | 16.2 | 9.4 | 7.78 | 62.1 | 40.6 | 29.7 | 11.7 | 114 | 49.8 |
| × 22 | 2840 | 156.7 | 165.6 | 11.2 | 6.6 | 5.62 | 45.1 | 44.5 | 32.3 | 4.25 | 51.1 | 38.6 |
| × 12 | 1525 | 152.3 | 101.3 | 6.7 | 5.6 | 3.62 | 33.4 | 48.8 | 44.2 | 0.587 | 11.6 | 19.6 |
| WT127 × 83 | 10645 | 144.3 | 264.5 | 31.8 | 19.2 | 11.9 | 105 | 33.5 | 30.7 | 49.1 | 370 | 68.1 |
| × 65 | 8325 | 137.7 | 260.7 | 25.1 | 15.4 | 8.66 | 78.2 | 32.3 | 26.9 | 37.2 | 285 | 66.8 |
| × 45 | 5690 | 129.8 | 256.0 | 17.3 | 10.7 | 5.37 | 49.8 | 30.7 | 22.5 | 24.2 | 188 | 65.3 |
| × 22 | 2850 | 133.0 | 147.6 | 13.0 | 7.6 | 3.86 | 36.7 | 36.8 | 27.9 | 3.48 | 47.0 | 34.8 |
| × 9 | 1140 | 125.3 | 100.6 | 5.3 | 4.8 | 1.81 | 20.0 | 39.9 | 34.5 | 0.454 | 9.03 | 19.9 |
| WT102 × 43 | 5515 | 111.1 | 208.8 | 20.6 | 13.0 | 3.80 | 42.8 | 26.2 | 22.2 | 15.6 | 150 | 53.3 |
| × 30 | 3785 | 104.8 | 205.0 | 14.2 | 9.1 | 2.39 | 27.7 | 25.1 | 18.7 | 10.2 | 99.6 | 51.8 |
| × 18 | 2285 | 100.7 | 165.0 | 10.2 | 6.2 | 1.47 | 17.7 | 25.4 | 17.7 | 3.80 | 46.0 | 40.9 |
| × 13 | 1695 | 103.4 | 133.4 | 8.4 | 5.8 | 1.42 | 17.2 | 29.0 | 21.2 | 1.66 | 24.9 | 31.2 |
| × 7 | 955 | 100.2 | 100.1 | 5.2 | 4.3 | 0.895 | 11.7 | 30.5 | 24.2 | 0.437 | 8.72 | 21.4 |
| WT76 × 15 | 1895 | 78.7 | 152.9 | 9.3 | 6.6 | 0.733 | 11.4 | 19.7 | 14.2 | 2.76 | 36.2 | 38.1 |
| × 9 | 1150 | 76.6 | 101.6 | 7.1 | 5.8 | 0.549 | 9.24 | 21.9 | 17.2 | 0.624 | 12.3 | 23.3 |
| WT51 × 10 | 1230 | 52.8 | 103.1 | 8.8 | 7.1 | 0.219 | 5.26 | 13.3 | 11.2 | 0.803 | 15.6 | 25.4 |

*WT means structural T-section (cut from a W-section), followed by the nominal depth in mm, then the mass in kg per meter of length.

TABLE A-13 Properties of Standard Steel Pipe (U.S. Customary Units)

| DIMENSIONS | | | | PROPERTIES | | | | |
|---|---|---|---|---|---|---|---|---|
| Nominal Diam. $d$ (in.) | Outside Diam. $d_o$ (in.) | Inside Diam. $d_i$ (in.) | Wall Thickness $t$ (in.) | Weight $w$ (lb/ft) | $A$ (in$^2$) | $I$ (in$^4$) | $S$ (in$^3$) | $r$ (in.) |
| **Standard Weight** | | | | | | | | |
| $^1/_2$ | 0.840 | 0.622 | 0.109 | 0.85 | 0.250 | 0.017 | 0.041 | 0.26 |
| $^3/_4$ | 1.050 | 0.824 | 0.113 | 0.13 | 0.333 | 0.037 | 0.071 | 0.33 |
| 1 | 1.315 | 1.049 | 0.133 | 1.68 | 0.494 | 0.087 | 0.133 | 0.42 |
| $1^1/_4$ | 1.660 | 1.380 | 0.140 | 2.27 | 0.669 | 0.195 | 0.235 | 0.54 |
| $1^1/_2$ | 1.900 | 1.610 | 0.145 | 2.72 | 0.799 | 0.310 | 0.326 | 0.62 |
| 2 | 2.375 | 2.067 | 0.154 | 3.65 | 1.075 | 0.666 | 0.561 | 0.79 |
| $2^1/_2$ | 2.875 | 2.469 | 0.203 | 5.79 | 1.704 | 1.530 | 1.064 | 0.95 |
| 3 | 3.500 | 3.068 | 0.216 | 7.58 | 2.228 | 3.017 | 1.724 | 1.16 |
| $3^1/_2$ | 4.000 | 3.548 | 0.226 | 9.11 | 2.680 | 4.787 | 2.39 | 1.34 |
| 4 | 4.500 | 4.026 | 0.237 | 10.79 | 3.174 | 7.233 | 3.21 | 1.51 |
| 5 | 5.563 | 5.047 | 0.258 | 14.62 | 4.300 | 15.16 | 5.45 | 1.88 |
| 6 | 6.625 | 6.065 | 0.280 | 18.97 | 5.581 | 28.14 | 8.50 | 2.25 |
| 8 | 8.625 | 7.981 | 0.322 | 28.55 | 8.399 | 72.49 | 16.81 | 2.94 |
| 10 | 10.750 | 10.020 | 0.365 | 40.48 | 11.91 | 160.7 | 29.9 | 3.67 |
| 12 | 12.750 | 12.000 | 0.375 | 49.56 | 14.58 | 279.3 | 43.8 | 4.38 |
| **Extra Strong** | | | | | | | | |
| $1^1/_2$ | 1.900 | 1.500 | 0.200 | 3.63 | 1.068 | 0.391 | 0.412 | 0.61 |
| 2 | 2.375 | 1.939 | 0.218 | 5.02 | 1.477 | 0.868 | 0.731 | 0.77 |
| $2^1/_2$ | 2.875 | 2.323 | 0.276 | 7.66 | 2.254 | 1.924 | 1.338 | 0.92 |
| 3 | 3.500 | 2.900 | 0.300 | 10.25 | 3.016 | 3.894 | 2.23 | 1.14 |
| 4 | 4.500 | 3.826 | 0.337 | 14.98 | 4.407 | 9.610 | 4.27 | 1.48 |
| 6 | 6.625 | 5.761 | 0.432 | 28.57 | 8.405 | 40.49 | 12.22 | 2.20 |
| **Double Extra Strong** | | | | | | | | |
| $1^1/_2$ | 1.900 | 1.100 | 0.400 | 6.41 | 1.885 | 0.568 | 0.564 | 0.55 |
| 2 | 2.375 | 1.503 | 0.436 | 9.03 | 2.656 | 1.311 | 1.104 | 0.70 |
| $2^1/_2$ | 2.875 | 1.771 | 0.552 | 13.69 | 4.028 | 2.871 | 1.997 | 0.84 |
| 3 | 3.500 | 2.300 | 0.600 | 18.58 | 5.466 | 5.993 | 3.42 | 1.05 |
| 4 | 4.500 | 3.152 | 0.674 | 27.54 | 8.101 | 15.28 | 6.79 | 1.37 |
| 6 | 6.625 | 4.897 | 0.864 | 53.16 | 15.64 | 66.33 | 20.0 | 2.06 |

TABLE A-14 Properties of Standard Steel Pipe (SI Units)

| Nominal Diam. $d$ (mm) | Outside Diam. $d_o$ (mm) | Inside Diam. $d_i$ (mm) | Wall Thickness $t$ (mm) | Mass $m$ (kg/m) | $A$ (mm$^2$) | $I$ ($10^6$ mm$^4$) | $S$ ($10^3$ mm$^3$) | $r$ (mm) |
|---|---|---|---|---|---|---|---|---|
| **Standard Weight** | | | | | | | | |
| 13 | 21.3 | 15.8 | 2.77 | 1.264 | 161.3 | 0.007 | 0.672 | 6.6 |
| 19 | 26.7 | 20.9 | 2.87 | 1.681 | 214.8 | 0.015 | 1.163 | 8.5 |
| 25 | 33.4 | 26.6 | 3.38 | 2.499 | 318.7 | 0.036 | 2.179 | 10.7 |
| 32 | 42.2 | 35.1 | 3.56 | 3.376 | 431.6 | 0.081 | 3.851 | 13.7 |
| 38 | 48.1 | 40.9 | 3.68 | 4.045 | 515.5 | 0.129 | 5.342 | 15.8 |
| 51 | 60.3 | 52.5 | 3.91 | 5.428 | 693.5 | 0.277 | 9.193 | 20.0 |
| 64 | 73.0 | 62.7 | 5.16 | 8.611 | 1099 | 0.637 | 17.44 | 24.1 |
| 76 | 88.9 | 77.9 | 5.49 | 11.27 | 1437 | 1.256 | 28.25 | 29.5 |
| 89 | 101.6 | 90.1 | 5.74 | 13.55 | 1729 | 1.992 | 39.17 | 34.0 |
| 102 | 114.3 | 102.3 | 6.02 | 16.05 | 2048 | 3.011 | 52.60 | 38.4 |
| 127 | 141.3 | 128.2 | 6.55 | 21.74 | 2774 | 6.310 | 89.31 | 47.8 |
| 152 | 168.3 | 154.1 | 7.11 | 28.21 | 3600 | 11.71 | 139.3 | 57.2 |
| 203 | 219.1 | 202.7 | 8.18 | 42.46 | 5419 | 30.2 | 275.5 | 74.7 |
| 254 | 273.1 | 254.5 | 9.27 | 60.20 | 7684 | 66.9 | 490 | 93.2 |
| 305 | 323.9 | 304.8 | 9.53 | 73.71 | 9406 | 116.3 | 718 | 111.3 |
| **Extra Strong** | | | | | | | | |
| 38 | 48.3 | 38.1 | 5.08 | 5.399 | 689 | 0.163 | 6.75 | 15.4 |
| 51 | 60.3 | 49.3 | 5.54 | 7.466 | 953 | 0.361 | 11.98 | 19.5 |
| 64 | 70.0 | 59.0 | 7.01 | 11.39 | 1454 | 0.801 | 21.93 | 23.5 |
| 76 | 88.9 | 73.7 | 7.62 | 15.24 | 1946 | 1.621 | 36.54 | 29.0 |
| 102 | 114.3 | 97.2 | 8.56 | 22.28 | 2843 | 4.000 | 69.67 | 37.6 |
| 152 | 168.3 | 146.3 | 10.97 | 42.49 | 5423 | 16.85 | 200 | 55.9 |
| **Double Extra Strong** | | | | | | | | |
| 38 | 48.3 | 27.9 | 10.16 | 9.53 | 1216 | 0.236 | 0.564 | 13.9 |
| 51 | 60.3 | 38.2 | 11.07 | 13.43 | 1714 | 0.546 | 1.104 | 17.9 |
| 64 | 70.0 | 45.0 | 14.02 | 20.36 | 2600 | 1.195 | 1.997 | 21.4 |
| 76 | 88.9 | 58.4 | 15.24 | 27.63 | 3526 | 2.494 | 3.42 | 26.7 |
| 102 | 114.3 | 80.1 | 17.12 | 40.96 | 5226 | 6.360 | 6.79 | 34.8 |
| 152 | 168.3 | 124.4 | 21.95 | 79.06 | 10090 | 27.61 | 20.0 | 52.3 |

TABLE A-15 Properties of Standard Structural Timber (U.S. Customary Units)

| DIMENSIONS* | | PROPERTIES | | | |
|---|---|---|---|---|---|
| Nominal Size $b \times h$ (in.) | Dressed Size (in.) | Weight $w$ (lb/ft) | Area $A$ (in$^2$) | Second Moment of Area $I$ (in$^4$) | Section Modulus $S$ (in$^3$) |
| $2 \times 4$ | $1\frac{5}{8} \times 3\frac{5}{8}$ | 1.64 | 5.89 | 6.45 | 3.56 |
| 6 | $5\frac{5}{8}$ | 2.54 | 9.14 | 24.1 | 8.57 |
| 8 | $7\frac{1}{2}$ | 3.39 | 12.2 | 57.1 | 15.3 |
| 10 | $9\frac{1}{2}$ | 4.29 | 15.4 | 116 | 24.4 |
| 12 | $11\frac{1}{2}$ | 5.19 | 18.7 | 206 | 35.8 |
| $4 \times 4$ | $3\frac{5}{8} \times 3\frac{5}{8}$ | 3.65 | 13.1 | 14.4 | 7.94 |
| 6 | $5\frac{5}{8}$ | 5.66 | 20.4 | 53.8 | 19.1 |
| 8 | $7\frac{1}{2}$ | 7.55 | 27.2 | 127 | 34.0 |
| 10 | $9\frac{1}{2}$ | 9.57 | 34.4 | 259 | 54.5 |
| 12 | $11\frac{1}{2}$ | 11.6 | 41.7 | 459 | 79.9 |
| $6 \times 6$ | $5\frac{1}{2} \times 5\frac{1}{2}$ | 8.40 | 30.3 | 76.3 | 27.7 |
| 8 | $7\frac{1}{2}$ | 11.4 | 41.3 | 193 | 51.6 |
| 10 | $9\frac{1}{2}$ | 14.5 | 52.3 | 393 | 82.7 |
| 12 | $11\frac{1}{2}$ | 17.5 | 63.3 | 697 | 121 |
| 14 | $13\frac{1}{2}$ | 20.6 | 74.3 | 1128 | 167 |
| $8 \times 8$ | $7\frac{1}{2} \times 7\frac{1}{2}$ | 15.6 | 56.3 | 264 | 70.3 |
| 10 | $9\frac{1}{2}$ | 19.8 | 71.3 | 536 | 113 |
| 12 | $11\frac{1}{2}$ | 23.9 | 86.3 | 951 | 165 |
| 14 | $13\frac{1}{2}$ | 28.0 | 101 | 1538 | 228 |
| 16 | $15\frac{1}{2}$ | 32.0 | 116 | 2327 | 300 |
| $10 \times 10$ | $9\frac{1}{2} \times 9\frac{1}{2}$ | 25.0 | 90.3 | 679 | 143 |
| 12 | $11\frac{1}{2}$ | 30.3 | 109 | 1204 | 209 |
| 14 | $13\frac{1}{2}$ | 35.6 | 128 | 1948 | 289 |
| 16 | $15\frac{1}{2}$ | 40.9 | 147 | 2948 | 380 |
| 18 | $17\frac{1}{2}$ | 46.1 | 166 | 4243 | 485 |
| $12 \times 12$ | $11\frac{1}{2} \times 11\frac{1}{2}$ | 36.7 | 132 | 1458 | 253 |
| 14 | $13\frac{1}{2}$ | 43.1 | 155 | 2358 | 349 |
| 16 | $15\frac{1}{2}$ | 49.5 | 178 | 3569 | 460 |
| 18 | $17\frac{1}{2}$ | 55.9 | 201 | 5136 | 587 |
| 20 | $19\frac{1}{2}$ | 62.3 | 224 | 7106 | 729 |

*Properties and weights are for dressed sizes.

TABLE A-16 Properties of Standard Structural Timber (SI Units)

| Nominal Size $b \times h$ (mm) | Dressed Size (mm) | Mass $m$ (kg/m) | Area $A$ ($10^3$ mm$^2$) | Second Moment of Area $I$ ($10^6$ mm$^4$) | Section Modulus $S$ ($10^3$ mm$^3$) |
|---|---|---|---|---|---|
| $51 \times 102$ | $41 \times 92$ | 2.44 | 3.77 | 2.68 | 58.3 |
| 152 | 140 | 3.78 | 5.86 | 10.03 | 140 |
| 203 | 191 | 5.04 | 7.83 | 23.8 | 251 |
| 254 | 241 | 6.38 | 9.88 | 48.3 | 400 |
| 305 | 292 | 7.72 | 12.0 | 85.7 | 587 |
| $102 \times 102$ | $92 \times 92$ | 5.43 | 8.46 | 5.99 | 130 |
| 152 | 140 | 8.42 | 13.2 | 22.4 | 313 |
| 203 | 191 | 11.2 | 17.6 | 52.9 | 557 |
| 254 | 241 | 14.2 | 22.2 | 107.8 | 893 |
| 305 | 292 | 17.3 | 26.9 | 191.1 | 1310 |
| $152 \times 152$ | $140 \times 140$ | 12.5 | 19.6 | 31.8 | 454 |
| 203 | 191 | 17.0 | 26.7 | 80.3 | 846 |
| 254 | 241 | 21.6 | 33.7 | 163.6 | 1350 |
| 305 | 292 | 26.0 | 40.9 | 290 | 1980 |
| 356 | 343 | 30.6 | 48.0 | 470 | 2740 |
| $203 \times 203$ | $191 \times 191$ | 23.2 | 36.5 | 110 | 1150 |
| 254 | 241 | 29.4 | 46.0 | 223 | 1850 |
| 305 | 292 | 35.5 | 55.8 | 396 | 2700 |
| 356 | 343 | 41.6 | 65.5 | 640 | 3740 |
| 406 | 394 | 47.6 | 75.3 | 969 | 4920 |
| $254 \times 254$ | $241 \times 241$ | 37.2 | 58.1 | 283 | 2340 |
| 305 | 292 | 45.1 | 70.4 | 501 | 3420 |
| 356 | 343 | 52.9 | 82.7 | 810 | 4740 |
| 406 | 394 | 60.8 | 95.0 | 1227 | 6230 |
| 457 | 445 | 68.6 | 107 | 1766 | 7950 |
| $305 \times 305$ | $292 \times 292$ | 54.6 | 85.3 | 607 | 4150 |
| 356 | 343 | 64.1 | 100 | 981 | 5720 |
| 406 | 394 | 73.6 | 115 | 1486 | 7540 |
| 457 | 445 | 83.1 | 130 | 2138 | 5620 |
| 508 | 495 | 92.7 | 145 | 2958 | 11950 |

*Properties and masses are for dressed sizes.

TABLE A-17 Properties of Selected Materials (U.S. Customary Units)

Exact values may vary widely with changes in composition, heat treatment, and mechanical working. More precise information can be obtained from manufacturers.

| Materials | Spe-cific Weight (lb/in$^3$) | ELASTIC STRENGTH[a] | | | ULTIMATE STRENGTH | | | Endur-ance Limit[c] (ksi) | Modu-lus of Elas-ticity (1000 ksi) | Modu-lus of Rigid-ity (1000 ksi) | Coeffi-cient of Thermal Expan-sion (10$^{-6}$/°F) |
|---|---|---|---|---|---|---|---|---|---|---|---|
| | | Ten-sion (ksi) | Comp. (ksi) | Shear (ksi) | Ten-sion (ksi) | Comp. (ksi) | Shear (ksi) | | | | |
| **Ferrous metals** | | | | | | | | | | | |
| Wrought iron | 0.278 | 30 | b | | 48 | b | 25 | 23 | 28 | | 6.7 |
| Structural steel | 0.284 | 36 | b | | 66 | b | | 28 | 29 | 11.0 | 6.6 |
| Steel, 0.2% C hardened | 0.284 | 62 | b | | 90 | b | | . | 30 | 11.6 | 6.6 |
| Steel, 0.4% C hot-rolled | 0.284 | 53 | b | | 84 | b | | 38 | 30 | 11.6 | |
| Steel, 0.8% C hot-rolled | 0.284 | 76 | b | | 122 | b | | | 30 | 11.6 | |
| Cast iron—gray | 0.260 | | | | 25 | 100 | | 12 | 15 | | 6.7 |
| Cast iron—malleable | 0.266 | 32 | b | | 50 | b | | | 25 | | 6.6 |
| Cast iron—nodular | 0.266 | 70 | | | 100 | | | | 25 | | 6.6 |
| Stainless steel (18–8) annealed | 0.286 | 36 | b | | 85 | b | | 40 | 28 | 12.5 | 9.6 |
| Stainless steel (18–8) cold-rolled | 0.286 | 165 | b | | 190 | b | | 90 | 28 | 12.5 | 9.6 |
| Steel, SAE 4340, heat-treated | 0.283 | 132 | 145 | | 150 | b | 95 | 76 | 29 | 11.0 | |
| **Nonferrous metal alloys** | | | | | | | | | | | |
| Aluminum, cast, 195-T6 | 0.100 | 24 | 25 | | 36 | | 30 | 7 | 10.3 | 3.8 | |
| Aluminum, wrought, 2014-T4 | 0.101 | 41 | 41 | 24 | 62 | b | 38 | 18 | 10.6 | 4.0 | 12.5 |
| Aluminum, wrought, 2024-T4 | 0.100 | 48 | 48 | 28 | 68 | b | 41 | 18 | 10.6 | 4.0 | 12.5 |
| Aluminum, wrought, 6061-T6 | 0.098 | 40 | 40 | 26 | 45 | b | 30 | 13.5 | 10.0 | 3.8 | 12.5 |
| Magnesium, extrusion, AZ80X | 0.066 | 35 | 26 | | 49 | b | 21 | 19 | 6.5 | 2.4 | 14.4 |
| Magnesium, sand cast, AZ63-HT | 0.066 | 14 | 14 | | 40 | b | 19 | 14 | 6.5 | 2.4 | 14.4 |
| Monel, wrought, hot-rolled | 0.319 | 50 | b | | 90 | b | | 40 | 26 | 9.5 | 7.8 |
| Red brass, cold-rolled | 0.316 | 60 | | | 75 | | | | 15 | 5.6 | 9.8 |
| Red brass, annealed | 0.316 | 15 | b | | 40 | b | | | 15 | 5.6 | 9.8 |
| Bronze, cold-rolled | 0.320 | 75 | | | 100 | | | | 15 | 6.5 | 9.4 |
| Bronze, annealed | 0.320 | 20 | b | | 50 | b | | | 15 | 6.5 | 9.4 |
| Titanium alloy, annealed | 0.167 | 135 | b | | 155 | b | | | 14 | 5.3 | |
| Invar, annealed | 0.292 | 42 | b | | 70 | b | | | 21 | 8.1 | 0.6 |
| **Nonmetallic materials** | | | | | | | | | | | |
| Douglas fir, green[d] | 0.022 | 4.8 | 3.4 | | | 3.9 | 0.9 | | 1.6 | | |
| Douglas fir, air dry[d] | 0.020 | 8.1 | 6.4 | | | 7.4 | 1.1 | | 1.9 | | |
| Red oak, green[d] | 0.037 | 4.4 | 2.6 | | | 3.5 | 1.2 | | 1.4 | | 1.9 |
| Red oak, air dry[d] | 0.025 | 8.4 | 4.6 | | | 6.9 | 1.8 | | 1.8 | | |
| Concrete, medium strength | 0.087 | | 1.2 | | | 3.0 | | | 3.0 | | 6.0 |
| Concrete, fairly high strength | 0.087 | | 2.0 | | | 5.0 | | | 4.5 | | 6.0 |

[a]Elastic strength may be represented by proportional limit, yield point, or yield strength at a specified offset (usually 0.2 percent for ductile metals).

[b]For ductile metals (those with an appreciable ultimate elongation), it is customary to assume the properties in compression have the same values as those in tension.

[c]Rotating beam.

[d]All timber properties are parallel to the grain.

**TABLE A-18** Properties of Selected Materials (SI Units)

Exact values may vary widely with changes in composition, heat treatment, and mechanical working. More precise information can be obtained from manufacturers.

| Materials | Density (Mg/m³) | ELASTIC STRENGTH[a] | | | ULTIMATE STRENGTH | | | Endurance Limit[c] (MPa) | Modulus of Elasticity (GPa) | Modulus of Rigidity (GPa) | Coefficient of Thermal Expansion (10⁻⁶/°C) |
|---|---|---|---|---|---|---|---|---|---|---|---|
| | | Tension (MPa) | Comp. (MPa) | Shear (MPa) | Tension (MPa) | Comp. (MPa) | Shear (MPa) | | | | |
| **Ferrous metals** | | | | | | | | | | | |
| Wrought iron | 7.70 | 210 | [b] | | 330 | [b] | 170 | 160 | 190 | | 12.1 |
| Structural steel | 7.87 | 250 | [b] | | 450 | [b] | | 190 | 200 | 76 | 11.9 |
| Steel, 0.2% C hardened | 7.87 | 430 | [b] | | 620 | [b] | | | 210 | 80 | 11.9 |
| Steel, 0.4% C hot-rolled | 7.87 | 360 | [b] | | 580 | [b] | | 260 | 210 | 80 | |
| Steel, 0.8% C hot-rolled | 7.87 | 520 | [b] | | 840 | [b] | | | 210 | 80 | |
| Cast iron—gray | 7.20 | | | | 170 | 690 | | 80 | 100 | | 12.1 |
| Cast iron—malleable | 7.37 | 220 | [b] | | 340 | [b] | | | 170 | | 11.9 |
| Cast iron—nodular | 7.37 | 480 | | | 690 | | | | 170 | | 11.9 |
| Stainless steel (18–8) annealed | 7.92 | 250 | [b] | | 590 | [b] | | 270 | 190 | 86 | 17.3 |
| Stainless steel (18–8) cold-rolled | 7.92 | 1140 | [b] | | 1310 | [b] | | 620 | 190 | 86 | 17.3 |
| Steel, SAE 4340, heat-treated | 7.84 | 910 | 1000 | | 1030 | [b] | 650 | 520 | 200 | 76 | |
| **Nonferrous metal alloys** | | | | | | | | | | | |
| Aluminum, cast, 195-T6 | 2.77 | 160 | 170 | | 250 | | 210 | 50 | 71 | 26 | |
| Aluminum, wrought, 2014-T4 | 2.80 | 280 | 280 | 160 | 430 | [b] | 260 | 120 | 73 | 28 | 22.5 |
| Aluminum, wrought, 2024-T4 | 2.77 | 330 | 330 | 190 | 470 | [b] | 280 | 120 | 73 | 28 | 22.5 |
| Aluminum, wrought, 6061-T6 | 2.71 | 270 | 270 | 180 | 310 | [b] | 210 | 93 | 70 | 26 | 22.5 |
| Magnesium, extrusion, AZ80X | 1.83 | 240 | 180 | | 340 | [b] | 140 | 130 | 45 | 16 | 25.9 |
| Magnesium, sand cast, AZ63-HT | 1.83 | 100 | 96 | | 270 | [b] | 130 | 100 | 45 | 16 | 25.9 |
| Monel, wrought, hot-rolled | 8.84 | 340 | [b] | | 620 | [b] | | 270 | 180 | 65 | 14.0 |
| Red brass, cold-rolled | 8.75 | 410 | | | 520 | | | | 100 | 39 | 17.6 |
| Red brass, annealed | 8.75 | 100 | [b] | | 270 | [b] | | | 100 | 39 | 17.6 |
| Bronze, cold-rolled | 8.86 | 520 | | | 690 | | | | 100 | 45 | 16.9 |
| Bronze, annealed | 8.86 | 140 | [b] | | 340 | [b] | | | 100 | 45 | 16.9 |
| Titanium alloy, annealed | 4.63 | 930 | [b] | | 1070 | [b] | | | 96 | 36 | |
| Invar, annealed | 8.09 | 290 | [b] | | 480 | [b] | | | 140 | 56 | 1.1 |
| **Nonmetallic materials** | | | | | | | | | | | |
| Douglas fir, green[d] | 0.61 | 33 | 23 | | | 27 | 6.2 | | 11 | | |
| Douglas fir, air dry[d] | 0.55 | 56 | 44 | | | 51 | 7.6 | | 13 | | |
| Red oak, green[d] | 1.02 | 30 | 18 | | | 24 | 8.3 | | 10 | | 3.4 |
| Red oak, air dry[d] | 0.69 | 58 | 32 | | | 48 | 12.4 | | 12 | | |
| Concrete, medium strength | 2.41 | | 8 | | | 21 | | | 21 | | 10.8 |
| Concrete, fairly high strength | 2.41 | | 14 | | | 34 | | | 31 | | 10.8 |

[a]Elastic strength may be represented by proportional limit, yield point, or yield strength at a specified offset (usually 0.2 percent for ductile metals).

[b]For ductile metals (those with an appreciable ultimate elongation), it is customary to assume the properties in compression have the same values as those in tension.

[c]Rotating beam.

[d]All timber properties are parallel to the grain.

TABLE A-19 Beam Deflections and Slopes

| Case | Load and Support (Length $L$) | Slope at End ($+\triangle$) | Maximum Deflection (+ upward) | Equation of Elastic Curve (+ upward) |
|------|------------------------------|------------------------------|-------------------------------|---------------------------------------|
| 1 | | $\theta = -\dfrac{PL^2}{2EI}$ at $x = L$ | $y_{max} = -\dfrac{PL^3}{3EI}$ at $x = L$ | $y = -\dfrac{Px^2}{6EI}(3L - x)$ |
| 2 | | $\theta = -\dfrac{wL^3}{6EI}$ at $x = L$ | $y_{max} = -\dfrac{wL^4}{8EI}$ at $x = L$ | $y = -\dfrac{wx^2}{24EI}(x^2 - 4Lx + 6L^2)$ |
| 3 | | $\theta = -\dfrac{wL^3}{24EI}$ at $x = L$ | $y_{max} = -\dfrac{wL^4}{30EI}$ at $x = L$ | $y = -\dfrac{wx^2}{120EIL}(10L^3 - 10L^2x + 5Lx^2 - x^3)$ |
| 4 | | $\theta = +\dfrac{ML}{EI}$ at $x = L$ | $y_{max} = +\dfrac{ML^2}{2EI}$ at $x = L$ | $y = \dfrac{Mx^2}{2EI}$ |
| 5 | | $\theta_1 = -\dfrac{Pb(L^2 - b^2)}{6LEI}$ at $x = 0$ $\theta_2 = +\dfrac{Pa(L^2 - a^2)}{6LEI}$ at $x = L$ | $y_{max} = -\dfrac{Pb(L^2 - b^2)^{3/2}}{9\sqrt{3}\,LEI}$ at $x = \sqrt{(L^2 - b^2)/3}$ $y_{\underset{\text{not max}}{\text{center}}} = -\dfrac{Pb(3L^2 - 4b^2)}{48EI}$ | $y = -\dfrac{Pbx}{6EIL}(L^2 - b^2 - x^2)$ $0 \le x \le a$ |
| 6 | | $\theta_1 = -\dfrac{PL^2}{16EI}$ at $x = 0$ $\theta_2 = +\dfrac{PL^2}{16EI}$ at $x = L$ | $y_{max} = -\dfrac{PL^3}{48EI}$ at $x = L/2$ | $y = -\dfrac{Px}{48EI}(3L^2 - 4x^2)$ $0 \le x \le \dfrac{L}{2}$ |
| 7 | | $\theta_1 = -\dfrac{wL^3}{24EI}$ at $x = 0$ $\theta_2 = +\dfrac{wL^3}{24EI}$ at $x = L$ | $y_{max} = -\dfrac{5wL^4}{384EI}$ at $x = L/2$ | $y = -\dfrac{wx}{24EI}(x^3 - 2Lx^2 + L^3)$ |
| 8 | | $\theta_1 = -\dfrac{ML}{6EI}$ at $x = 0$ $\theta_2 = +\dfrac{ML}{3EI}$ at $x = L$ | $y_{max} = -\dfrac{ML^2}{9\sqrt{3}\,EI}$ at $x = L/\sqrt{3}$ $y_{\underset{\text{not max}}{\text{center}}} = -\dfrac{ML^2}{16EI}$ | $y = -\dfrac{Mx}{6EIL}(L^2 - x^2)$ |

# ANSWERS TO SELECTED PROBLEMS

## Chapter 1

**1-1** $m = 18.63$ slug
**1-2** $W = 6620$ N
**1-4** $W = 560$ N
**1-7** $W = 768$ lb
**1-8** $W = 486$ N
**1-9** $W = 22.33$ lb
**1-13** (a) $g = 7.71$ ft/s$^2$
     (b) $g = 81.7$ ft/s$^2$
**1-14** (a) $W = 39.8$ N
     (b) $W = 4810$ N
**1-17** $W = 165.4$ lb
**1-18** $m = 227$ kg
**1-19** $V = 7.21$ L
**1-21** $\rho = 691$ kg/m$^3$
**1-22** $\mu = 0.1489\,(10^{-3})$ lb $\cdot$ s/ft$^2$
**1-26** Specific heat =
     6000 lb $\cdot$ft/slug $\cdot$ °R
**1-27** $G = L^3/MT^2$
**1-28** $E = F/L^2$
**1-31** $P = F$; $I = L^4$
**1-32** $P = F$; $I = ML^2$
**1-35** $x = L$; $b = T$; $a = 1/T$,
     $\alpha$ (dimensionless)
**1-36** $w = L^3$; $a = L$; $b = L^2$
**1-39** (a) 0.015 (%$D = -2.36$%)
     (b) 55 (%$D = -0.609$%)
     (c) 64,000 (%$D = +0.398$%)
**1-42** (a) 0.473 (%$D = +0.1776$%)
     (b) 826 (%$D = -0.0585$%)
     (c) 340,000 (%$D = +0.0374$%)
**1-43** (a) 0.05662 [%$D =$
     $-5.30\,(10^{-3})$%]
     (b) 74.83 [%$D =$
     $+1.109\,(10^{-3})$%]
     (c) 27,380 [%$D =$
     $-10.37(10^{-3})$%]
**1-45** (a) $F = 170.9$ lb
     (b) $F = 138.9$ lb

**1-46** $F = 2.10(10^{18})$ N
**1-47** $A = 259$ hectares

## Chapter 2

**2-1** $\mathbf{R} = 150$ lb $\measuredangle\ 36.87°$
**2-2** $\mathbf{R} = 408$ N $\measuredangle\ 36.03°$
**2-5** $\mathbf{F}_B = 291$ lb $\measuredangle\ 25.28°$
**2-6** $\mathbf{R} = 10.97$ kN $\measuredangle\ 5.23°$
**2-9** $\theta = 67.60°$ or $112.40°$
**2-10** $\mathbf{R}_3 = 307$ N $\measuredangle\ 42.41°$
**2-11** (a) $\mathbf{R} = 223$ lb $\measuredangle\ 22.45°$
     (b) $\mathbf{R}_x = 207$ lb $\rightarrow$ $\mathbf{R}_y = 85.5$ lb $\downarrow$
**2-12** $\mathbf{R} = 2.71$ kN $\measuredangle\ 3.09°$
**2-17** $F_u = 582$ lb
     $F_v = 718$ lb
**2-18** $F_{AB} = 776$ N
     $F_{BC} = 2900$ N
**2-21** $F_u = 2800$ lb
     $F_v = 2830$ lb
**2-22** $F_a = 4330$ N
     $F_p = 2500$ N
**2-23** $F_u = 707$ lb
     $F_v = 948$ lb
**2-26** $F_u = 383$ N
     $F_v = 273$ N
**2-27** $F_2 = 42.6$ kip
     $F_3 = 12.13$ kip
**2-28** $N_1 = 610$ N
     $\theta = 55.42°$
**2-33** $F_x = 866$ lb
     $F_y = 500$ lb
**2-34** $F_x = 338$ N
     $F_y = 725$ N
**2-37** (a) $F_{1x} = 171.0$ lb
     $F_{1y} = 470$ lb
     $F_{2x} = 650$ lb
     $F_{2y} = -375$ lb
     (b) $F_{1x'} = -211.3$ lb
     $F_{1y'} = 453$ lb

     $F_{2x'} = 724$ lb
     $F_{2y'} = 194.1$ lb
**2-38** (a) $F_x = 369$ N
     $F_y = 638$ N
     $F_z = 516$ N
     (b) $\mathbf{F} = (369\,\mathbf{i} + 638\,\mathbf{j} + 516\,\mathbf{k})$ N
**2-41** (a) $F_{1x} = -514$ lb
     $F_{1y} = 613$ lb
     $F_{2x} = 940$ lb
     $F_{2y} = 342$ lb
     (b) $F_{1x'} = -752$ lb
     $F_{1y'} = 274$ lb
     $F_{2x'} = 643$ lb
     $F_{2y'} = 766$ lb
**2-42** (a) $F_x = 25.4$ kN
     $F_y = -15.21$ kN
     $F_z = 5.07$ kN
     (b) $\mathbf{F} = (25.4\,\mathbf{i} - 15.21\,\mathbf{j} + 5.07\,\mathbf{k})$ kN
     (c) $F_{12} = 25.8$ kN
     (d) $\alpha = 30.57°$
**2-43** (a) $F_{AD} = 69.3$ lb
     (b) $F_{CD} = 96.8$ lb
     (c) $\alpha = 41.91°$
**2-44** (a) $\mathbf{T}_A = (603\,\mathbf{i} + 905\,\mathbf{j} - 1509\,\mathbf{k})$ N
     (b) $T_{A/BD} = 1074$ N
     (c) $\alpha = 54.72°$
**2-47** $\mathbf{R} = 10.14$ lb $\measuredangle\ 88.63°$
**2-48** $\mathbf{R} = 846$ N $\measuredangle\ 47.19°$
**2-51** $\mathbf{R} = (113.1\,\mathbf{i} + 113.1\,\mathbf{j} + 229\,\mathbf{k})$ lb
     $\theta_x = 66.13°$
     $\theta_y = 66.13°$
     $\theta_z = 34.91°$
**2-52** $R = 28.6$ kN
     $\theta_x = 82.15°$
     $\theta_y = 69.55°$

$\theta_z = 22.03°$

**2-55** $\mathbf{R} = (-105.7 \mathbf{i} - 83.4 \mathbf{j} - 569 \mathbf{k})$ lb

$\theta_x = 100.41°$

$\theta_y = 98.20°$

$\theta_z = 166.69°$

**2-56** $R = 34.0$ kN

$\theta_x = 77.48°$

$\theta_y = 94.55°$

$\theta_z = 13.34°$

**2-59** $\mathbf{R} = 5220$ lb ⤢ $59.04°$

**2-60** (a) $\mathbf{R} = 52.5$ N ⤢ $63.59°$

(b) $F_u = 41.9$ N

$F_v = 18.46$ N

**2-62** $\mathbf{R} = (-39.5 \mathbf{i} + 51.5 \mathbf{j} + 24.9 \mathbf{k})$ kN

$\theta_x = 124.58°$

$\theta_y = 42.17°$

$\theta_z = 68.98°$

**2-65** (a) $R = 215$ lb

$\theta_x = 72.25°$

$\theta_y = 71.06°$

$\theta_z = 26.44°$

(b) $F_{1/2} = 38.2$ lb

(c) $\alpha = 75.25°$

## Chapter 3

**3-1** $F_2 = 220$ lb

$F_3 = 269$ lb

**3-2** $F_1 = 339$ N

$F_2 = 400$ N

**3-3** $N = 23.4$ lb

$P = 8.78$ lb

**3-6** $T_1 = 1799$ N

$T_2 = 317$ N

**3-9** $T_{AB} = 12.73$ lb

$T_{BC} = 9.06$ lb

$T_{CD} = 12.73$ lb

$\alpha = 6.34°$

**3-10** $N_{AB} = 1870$ N

$N_D = 1432$ N

$N_E = 3610$ N

**3-11** $\theta_{min} = 10.89°$

**3-14** $P = 201$ N

**3-17** (a) $N_1 = 160.0$ lb

$N_2 = 90.0$ lb

(b) $T = 120.0$ lb

(c) $\theta = 36.87°$

**3-18** (a) $N_A = 866$ N

$N_B = 500$ N

(b) $T = 436$ N

(c) $\theta = 6.59°$

**3-19** $T_A = 1116$ lb

$T_B = 1116$ lb

$T_C = 1116$ lb

**3-22** $T_A = 603$ N

$T_B = 439$ N

$T_C = 925$ N

**3-23** $T_{AD} = -225$ lb

$T_{BD} = 621$ lb

$T_{CD} = -484$ lb

**3-32** $\mathbf{F}_4 = 1377$ N ⤢ $1.89°$

**3-33** $N_A = 539$ lb

$N_B = 340$ lb

$N_C = 879$ lb

**3-36** $F_L = 24.6$ kN

$F_D = 5.66$ kN

**3-37** $T_{AB} = 267$ lb

$T_{AC} = 141.3$ lb

$T_{AD} = 162.7$ lb

**3-38** $T_{AD} = 1677$ N T

$F_{AB} = 995$ N C

$F_{AC} = 700$ N C

## Chapter 4

**4-1** $t = 0.1207$ in.

**4-2** $\sigma_A = 106.7$ MPa

$\sigma_B = 133.3$ MPa

$\sigma_C = 53.3$ MPa

**4-3** $\sigma_A = 5.74$ ksi

$\sigma_B = 5.96$ ksi

**4-4** $A_A = 4330$ mm$^2$

$A_B = 10,000$ mm$^2$

$A_C = 20,000$ mm$^2$

**4-8** (a) $P = 113.7$ kN

(b) $P = 156.0$ kN

(c) $P = 54.3$ kN

(d) $P = 216$ kN

**4-9** $d = 3.18$ in.

**4-13** $\tau = 1050$ psi

**4-14** $d = 2.35$ mm

**4-15** (a) $d_{AB} = 0.304$ in.

$d_{BC} = 0.293$ in.

(b) $d_A = 0.304$ in.

**4-22** $\sigma = 75.6$ MPa

$\tau = 57.0$ MPa

**4-23** $\sigma = 18.29$ psi

$\tau = 73.4$ psi

**4-26** $P_{max} = 291$ kN

**4-27** $\theta = 36.87°$

$P = 75.0$ kip

**4-30** (a) $\phi = 59.04°$

(b) $P_{max} = 34.0$ kN

**4-31** $P_{max} = 5440$ lb

**4-36** $\delta = 0.480$ mm

**4-37** $\gamma = 0.0200$ rad

**4-40** $\gamma = 27.9(10^{-3})$ rad

**4-41** $\epsilon = 75.0(10^{-3})$ in./in.

$\epsilon = 75.0(10^{-3})$ in./in.

**4-44** (a) $\delta = 5.00$ mm

(b) $\epsilon_{avg} = 1667$ $\mu$m/m

(c) $\epsilon_{max} = 5000$ $\mu$m/m

**4-45** (a) $\delta = \dfrac{\gamma L^2}{6E}$

(b) $\epsilon_{avg} = \dfrac{\gamma L}{6E}$

(c) $\epsilon_{max} = \dfrac{\gamma L}{3E}$

**4-48** $E = 78.7$ GPa

$\nu = 0.326$

$\sigma_p = 354$ MPa

**4-49** $E = 15,000$ ksi

$\nu = 0.333$

$G = 5630$ ksi

**4-50** (a) $E = 70.0$ GPa

(b) $\sigma_p = 150$ MPa

(c) $\sigma_{ult} = 450$ MPa

(d) $\sigma_{0.05} = 220$ MPa

(e) $\sigma_{0.2} = 275$ MPa

(f) $\sigma_f = 450$ MPa

(g) $\sigma_{tf} = 483$ MPa

(h) $E_t = 1.000$ GPa

(i) $E_s = 8.89$ GPa

**4-53** (a) $E = 26,700$ ksi

(b) $\sigma_p = 40$ ksi

(c) $\sigma_{ult} = 73$ ksi

(d) $\sigma_{0.05} = 44$ ksi

(e) $\sigma_{0.2} = 47$ ksi

(f) $\sigma_f = 65$ ksi

(g) $\sigma_{tf} = 91.6$ ksi

(h) $E_t = 1875$ ksi

(i) $E = 14,370$ ksi

**4-54** (a) $\delta_L = 2.12$ mm

$\delta_o = 0.0889$ mm

$\delta_i = 0.0593$ mm

(b) $\delta_L = -2.57$ mm

$\delta_o = -0.108$ mm

$\delta_i = -0.072$ mm

**4-55** $\delta_L = 4.39$ in.

$\delta_d = 0.234$ in.

**4-57** $\delta_p = 0.0472$ in. ↑

**4-60** (a) $\sigma = 50.0$ MPa

(b) $\epsilon_a = -441$ $\mu$m/m

(c) $\delta_d = -0.0338$ mm

**4-61** (a) $\sigma_{max} = 20.5$ ksi

(b) $\delta_{AB} = 65.6(10^{-3})$ in.

(c) $\delta_{BC} = -6.00(10^{-3})$ in.

(d) $\delta_{total} = 77.6(10^{-3})$ in.

**4-62** $P = 187.5$ kN

**4-65** (a) $\delta = 0.0231$ in.

(b) $\sigma = 1389$ psi

(c) $\sigma_b = 1389$ psi

(d) $\tau_{max} = 694$ psi

**4-66** (a) $\delta_1 = 3.16$ mm

(b) $\delta_r = 4.33$ mm

**4-69** $\delta = 0.0638$ in.

**4-72** (a) $\sigma_A = 56.6$ MPa

$\sigma_B = 84.9$ MPa

(b) $\delta_C = 0.849$ mm ↓

**4-73** (a) $P = 737$ kip

(b) $\delta = 0.01517$ in. ↓

**4-75** (a) $\sigma_A = 0$ ksi

$\sigma_B = 12.73$ ksi T

$\sigma_C = 12.73$ ksi C

(b) $\delta_{aa} = 0.0306$ in. →

**4-76** (a) $\sigma_p = 1.074$ MPa T

$\sigma_b = 11.93$ MPa T

(b) $x_p = 200.1023$ mm

**4-79** $\sigma_{st} = 48.8$ ksi T

$\sigma_{br} = 25.6$ ksi C

**4-83** $\sigma = 13.25$ ksi T

**4-84** (a) $\sigma = 75.0$ MPa T

(b) $\delta_d = -0.0804$ mm

**4-86** $\sigma = 110.2$ MPa

**4-88** (a) $\sigma_A = 218$ MPa T

$\sigma_B = 79.1$ MPa C

(b) $\delta_C = 0.331$ mm →

**4-89** $\sigma_{st} = 1839$ psi T

$\sigma_{br} = 817$ psi C

**4-94** $d = 51$ mm

**4-95** W8 × 24

**4-97** $d_{AB} = 0.874$ in.

$d_{BC} = 1.070$ in.

**4-98** Concrete: 133.7 mm ×

133.7 mm

Steel: 140.2 mm × 140.2 mm

**4-100** $d_{min} = 8.24$ mm

**4-102** (a) $\sigma_{AB} = 25.5$ MPa

$\sigma_{BC} = 62.2$ MPa

(b) $\delta = 0.709$ mm

**4-103** $\sigma_{AB} = 4.07$ ksi C

$\sigma_{BC} = 33.9$ ksi T

$\sigma_{CD} = 12.73$ ksi T

$\delta = 0.0213$ in.

**4-104** (a) $\sigma_n = 6.67$ MPa

(b) $\sigma_b = 6.67$ MPa

(c) $\sigma_b = 3.33$ MPa

(d) $\sigma_b = 385$ kPa

**4-107** (a) $\sigma_A = 8.42$ ksi T

$\sigma_B = 14.73$ ksi C

(b) $\delta_C = 0.0236$ in. ↓

**4-108** $P_{max} = 90.3$ N

## Chapter 5

**5-1** (a) $\mathbf{M}_1 = 100$ lb · in. ↰

(b) $\mathbf{M}_2 = 525$ lb · in. ↲

**5-2** $\mathbf{M}_A = 135$ N · m ↰

$\mathbf{M}_B = 90$ N · m ↰

$\mathbf{M}_C = 180$ N · m ↰

**5-5** $\mathbf{M}_O = 1047$ lb · in. ↲

$\mathbf{M}_A = 1286$ lb · in. ↲

**5-6** $\mathbf{M}_A = 158.4$ N · m ↰

$\mathbf{M}_B = 33.9$ N · m ↰

**5-9** $P = 150.6$ lb

**5-10** (a) $\mathbf{M}_A = 6.93$ kN · m ↰

(b) $\mathbf{M}_D = 10.39$ kN · m ↰

**5-13** (a) $\mathbf{M}_A = (11.87 \mathbf{k})$ kip · in.

(b) $\mathbf{M}_A = 11.87$ kip · in. ↰

**5-14** (a) $\mathbf{M}_O = (-215 \mathbf{k})$ N · m

(b) $\mathbf{M}_O = 215$ N · m ↲

**5-17** $\mathbf{M}_A = (-74.0 \mathbf{k})$ lb · ft

**5-18** $\mathbf{F} = 107.8$ N ⦨ 21.80°

**5-21** (a) $\mathbf{M}_A = (-13.40 \mathbf{i} + 8.37 \mathbf{j} - 0.893 \mathbf{k})$ kip · in.

(b) $\theta_x = 147.85°$

$\theta_y = 58.05°$

$\theta_z = 93.23°$

**5-23** $\mathbf{M}_B = (1709 \mathbf{i} - 621 \mathbf{j} - 2660 \mathbf{k})$ lb · in.

**5-24** $\mathbf{M}_A = (-253 \mathbf{i} + 13.18 \mathbf{j} + 209 \mathbf{k})$ N · m

**5-25** $\mathbf{M}_C = (324 \mathbf{i} + 1029 \mathbf{j} - 662 \mathbf{k})$ lb · in.

**5-29** $M_{BC} = 1216$ lb · in.

**5-30** $M_{Ox} = -830$ N · m

**5-33** $M_{Ox} = 320$ lb · in.

**5-34** $M_{CD} = -2620$ N · m

**5-37** $M_{Ox} = 2190$ lb · in.

**5-38** (a) $M_{Oy} = -105.0$ N · m

(b) $M_{Ox} = 75.0$ N · m

**5-40** (a) $\mathbf{M}_{CE} = (46.1 \mathbf{i} - 87.8 \mathbf{j})$ N · m

(b) $\mathbf{M}_{\perp CE} = (81.9 \mathbf{i} + 43.0 \mathbf{j} + 196.0 \mathbf{k})$ N · m

$\theta_x = 67.80°$

$\theta_y = 78.56°$

$\theta_z = 25.26°$

**5-42** $\mathbf{M} = 168.1$ N · m ↰

$d = 221$ mm

**5-43** $\mathbf{M} = (-300 \mathbf{i})$ lb · in.

**5-45** $\mathbf{M} = (2500 \mathbf{k})$ lb · in.

**5-46** $T = 842$ N · m

$\theta_x = 27.08°$

$\theta_y = 68.70°$

$\theta_z = 74.08°$

**5-48** (a) $C = 140.2$ N · m

$\theta_x = 55.20°$

$\theta_y = 62.37°$

$\theta_z = 47.34°$

(b) $C_{OA} = 56.7$ N · m

**5-50** $\mathbf{F}_A = 3$ kN ↓

$\mathbf{C}_A = 450$ N · m ↲

**5-51** $\mathbf{F}_A = (50.0 \mathbf{j})$ lb

$\mathbf{C}_A = (-300 \mathbf{k})$ lb · in.

**5-54** $\mathbf{R}_A = 12.37$ kN ⦩ 75.96°

$d = 3.23$ m

**5-55** $\mathbf{R} = 154.4$ lb ⦫ 13.20°

$d = 0.648$ ft

**5-56** $\mathbf{R} = 597$ N ⦨ 39.90°

$d = 94.0$ mm, left of $O$

**5-60** $\mathbf{F} = (462 \mathbf{i} + 615 \mathbf{j} + 1846 \mathbf{k})$ N

$\mathbf{C} = (415.4 \mathbf{i} + 34.6 \mathbf{j} - 115.4 \mathbf{k})$ N · m

**5-61** $\mathbf{F} = (50.0 \mathbf{i} + 50.0 \mathbf{j} - 200 \mathbf{k})$ lb

$\mathbf{C} = (-650 \mathbf{i} + 250 \mathbf{j} - 100.0 \mathbf{k})$ lb · in.

**5-64** $\mathbf{R} = 39.1$ N ⦨ 39.81°

$x_R = -2.6$ m

**5-66** $\mathbf{P} = (-342 \mathbf{j} - 940 \mathbf{k})$ N

$\mathbf{C} = (-132.5 \mathbf{i} - 211 \mathbf{j} + 77.0 \mathbf{k})$ N · m

**5-67** $\mathbf{P} = (-10.00 \mathbf{j} - 20.00 \mathbf{k})$ lb

$\mathbf{C}_D = (346 \mathbf{i} + 2420 \mathbf{j} - 610 \mathbf{k})$ lb · in.

**5-70** $x_G = 11.60$ mm

$y_G = 29.0$ mm

**5-71** $x_G = 2.09$ in.

$y_G = 7.83$ in.

$z_G = 1.739$ in.

**5-73** $x_G = 7.39$ in.

$y_G = 9.02$ in.

$z_G = 4.07$ in.

**5-74** $x_G = 225$ mm

$y_G = 263$ mm

$z_G = 173.6$ mm

**5-76** $x_C = 133.3$ mm

$y_C = 100.0$ mm

**5-77** $y_C = 3b/20$

**5-80** $y_C = 3b/4$

**5-81** $x_C = 2a/5$

$y_C = a/2$

**5-82** $y_C = 4.69$ mm

**5-85** $x_C = 1.250$ in.

$y_C = z_C = 0$

**5-88** $x_C = 26.5$ mm

$y_C = 9.15$ mm

**5-89** $x_C = y_C = r/\pi$

**5-90** $z_G = 18r/25$

$x_G = y_G = 0$

**5-93** $x_C = 1.500$ in.

$y_C = 2.00$ in.

**5-94** $x_C = 22.0$ mm

$y_C = 60.0$ mm

**5-96** $x_C = 65.5$ mm

$y_C = 190.2$ mm

**5-98** $x_C = 4.23$ mm

$y_C = 194.9$ mm

**5-99** $x_C = y_C = 0.431$ in.

**5-101** $x_C = 4.23$ in.

$y_C = 8.33$ in.

**5-103** $x_G = 10.86$ in

$y_G = 9.95$ in.

$z_G = 3.5$ in.

**5-104** $x_G = z_G = 0$

$y_G = 165.0$ mm

**5-106** (a) $x_C = y_C = 0$
$z_C = 231$ mm
(b) $x_G = y_G = 0$
$z_G = 190.4$ mm

**5-108** $R = 2250$ N $\downarrow$
$d = 2.67$ m

**5-109** $R = 1350$ lb $\downarrow$
$d = 3.83$ ft

**5-111** $R = 2500$ lb $\downarrow$
$d = 10.5$ ft

**5-112** $R = 667$ N $\downarrow$
$d = 2.20$ m

**5-114** $R = 794$ kN $\leftarrow$
$d = 3.00$ m

**5-115** $R = 33.9$ lb $\downarrow$
$x_R = 9.03$ ft
$y_R = 8.15$ ft

**5-118** $\mathbf{M}_O = (-326\,\mathbf{i} - 623\,\mathbf{j} + 668\,\mathbf{k})$ N $\cdot$ m

**5-119** $R = 750$ lb $\rightarrow$
$y = 9.60$ in. (from ground)

**5-123** $R = 40.0$ lb $\downarrow$
$\mathbf{M}_A = 350$ lb $\cdot$ in. $\downarrow$

**5-124** $R = (-130\,\mathbf{k})$ N
$x = 1.385$ m
$y = 8.42$ m

**5-128** $y_C = -64.3$ mm

**5-130** $x_C = z_C = 0$
$y_C = 122.0$ mm
$x_G = z_G = 0$
$y_G = 206$ mm

**5-131** $R = 3470$ lb $\downarrow$
$x = 5.66$ ft

## Chapter 6

**6-13** $\mathbf{A} = 1240$ lb $\uparrow$
$\mathbf{B} = 1160$ lb $\uparrow$

**6-14** $\mathbf{A} = 2.00$ kN $\uparrow$
$\mathbf{M}_A = 11.00$ kN $\cdot$ m $\uparrow$

**6-15** $Q = 62.9$ lb
$\mathbf{A} = 39.8$ lb $\measuredangle$ 22.13°

**6-18** $T = 828$ N

**6-21** (a) $\mathbf{C} = 134.2$ lb $\searrow$ 63.43°
(b) $\mathbf{A} = 61.2$ lb $\searrow$ 11.31°
$\mathbf{B} = 108.0$ lb $\uparrow$
(c) $\tau_C = 1367$ psi

**6-22** (a) $T_{AB} = 600$ N
$\mathbf{C} = 960$ N $\measuredangle$ 51.34°
(b) $\tau_C = 6.11$ MPa
(c) $\delta_{AB} = 0.465$ mm

**6-25** (a) $\mathbf{A} = 72.0$ lb $\nearrow$ 5.31°
$T_{CD} = 59.6$ lb C
(b) $\sigma_{CD} = 304$ psi

(c) $\tau_A = 183.3$ psi
(d) $\delta_{CD} = 0.1358(10^{-3})$ in.

**6-28** $P = 3410$ N

**6-30** $\theta = 20.10°$

**6-31** $P = 12.86$ kip

**6-32** $F = 221$ N

**6-35** (a) $T_D = 92.5$ lb
$\mathbf{B} = 46.1$ lb $\leftarrow$
$\mathbf{C} = 108.1$ lb $\uparrow$
(b) $\tau_B = 417$ psi
$\tau_C = 979$ psi

**6-44** $\mathbf{A} = 212$ N $\nearrow$ 45°
$\mathbf{B} = 335$ N $\nearrow$ 63.43°
$\mathbf{E} = 541$ N $\measuredangle$ 56.31°

**6-45** $\mathbf{T}_{AC} = 85.7$ lb $\nearrow$ 45°
$\mathbf{B} = 48.2$ lb $\measuredangle$ 57.40°

**6-46** $\mathbf{F}_D = 8.75$ N $\downarrow$
$\mathbf{B} = 13.75$ N $\downarrow$

**6-49** $\mathbf{A} = 224$ lb $\searrow$ 63.43°
$\mathbf{B} = 566$ lb $\searrow$ 45°
$\mathbf{C} = 447$ lb $\searrow$ 26.57°

**6-51** (a) $F_A = 5890$ lb
(b) $F_{CD} = 6020$ lb C

**6-52** $F_s = 5650$ N C
$\mathbf{A} = 3200$ N $\measuredangle$ 67.09°
$\mathbf{B} = 2970$ N $\searrow$ 65.22°
$F_{CD} = 1246$ N C

**6-55** (a) $\mathbf{F}_B = 14.14$ lb $\measuredangle$ 45°
$\mathbf{F}_C = 32.5$ lb $\leftarrow$
$\mathbf{F}_D = 24.6$ lb $\searrow$ 23.96°
(b) $\tau = 1152$ psi
(c) $\delta_{AB} = 1.42(10^{-5})$ in.

**6-58** $F = 426$ N

**6-59** (a) $\mathbf{B} = 1050$ lb $\downarrow$
$\mathbf{C} = 1000$ lb $\uparrow$
(b) $E = 1500$ lb
(c) $\sigma_D = 13.33$ ksi
(d) $\delta_D = 1.839(10^{-3})$ in.

**6-60** (a) $\mathbf{A} = 15.36$ kN $\searrow$ 26.57°
$\mathbf{B} = 15.42$ kN $\measuredangle$ 82.12°
$\mathbf{C} = 15.01$ kN $\downarrow$
$\mathbf{D} = 17.18$ kN $\searrow$ 22.62°
(b) $\mathbf{F} = 32.2$ kN $\uparrow$
$\mathbf{M}_F = 12.68$ kN $\cdot$ m $\uparrow$
(c) $\tau_A = 43.5$ MPa

**6-67** $\sigma_B = 6020$ psi
$\sigma_C = 7950$ psi

**6-68** (a) $\sigma_{AD} = 238$ MPa
$\sigma_{BE} = 44.1$ MPa
(b) $\delta_{AD} = 0.511$ mm

**6-71** (a) $\sigma_A = 26.8$ ksi
$\sigma_B = 21.4$ ksi
(b) $d_C = 0.0357$ in. $\downarrow$
(c) $\tau_D = 63.1$ ksi

**6-72** (a) $\sigma_A = 227$ MPa

$\sigma_B = 87.3$ MPa
(b) $d_D = 1.794$ mm $\downarrow$

**6-74** (a) $d_D = 6.51$ mm $\downarrow$
(b) $\tau_C = 143.8$ MPa

**6-75** (a) $\sigma_A = 3170$ psi T
$\sigma_B = 508$ psi C
(b) $\delta_A = 15.85(10^{-3})$ in.
$\delta_B = -0.508(10^{-3})$ in.

**6-82** $AB$: 2.50 kN C
$AD$: 2.17 kN T
$BC$: 4.33 kN C
$BD$: 5.00 kN T
$CD$: 2.17 kN T

**6-83** $AB$: 5000 lb T
$AE$: 1000 lb T
$BC$: 3330 lb T
$BD$: 2000 lb T
$BE$: 1667 lb C
$CD$: 2670 lb C
$DE$: 2670 lb C

**6-86** $CJ$: 4.28 kN C
$JK$: 9.40 kN T

**6-88** $BC$: 6.19 kN C
$BG$: 0.809 kN T
$DE$: 12.38 kN C

**6-89** (a) $\sigma_{CD} = 3.82$ ksi
(b) $\delta_{CF} = -0.0527$ in.
(c) $\delta_{EF} = -0.0263$ in.

**6-91** $\sigma_{CD} = 4.73$ ksi C
$\sigma_{CF} = 1.478$ ksi C
$\sigma_{CG} = 1.773$ ksi T
$\sigma_{FG} = 5.91$ ksi T
$\epsilon_{CD} = -163.0$ $\mu$in./in.
$\epsilon_{CF} = -50.9$ $\mu$in./in.
$\epsilon_{CG} = +61.1$ $\mu$in./in.
$\epsilon_{FG} = +204$ $\mu$in./in.

**6-96** $AB$: 2.35 kN C
$AC$: 5.89 kN C
$AH$: 3.53 kN T
$CD$: 3.53 kN C
$DF$: 5.89 kN C
$EF$: 2.35 kN C
$FG$: 3.53 kN T
$GH$: 3.53 kN T
$BC, CG, CH, DE, DG$: 0 kN

**6-98** $CD$: 20.2 kN T
$DE$: 10.45 kN C
$DF$: 23.2 kN T

**6-99** $DE$: 10.00 kip C
$DJ$: 8.00 kip C
$JK$: 7.00 kip T

**6-102** $T_A = 3340$ N
$T_B = 3390$ N
$T_C = 4060$ N

**6-103** $\mathbf{A} = (50\,\mathbf{k})$ lb

**6-105** $M_A = (1150 \text{ i} + 350 \text{ j}) \text{ lb} \cdot \text{ft}$
$P = 150$ lb
A $(928 \text{ i} - 116.7 \text{ k})$ lb
B $= (-278 \text{ i} - 200 \text{ j} + 267 \text{ k})$ lb

**6-106** C $= (85.0 \text{ j} + 147.2 \text{ k})$ N
D $= (294 \text{ k})$ N
$F_{AB} = (-85.0 \text{ j} + 147.2 \text{ k})$ N

**6-108** $P = 2130$ N
$A = B = 1064$ N

**6-111** (a) $P = 21.1$ lb
(b) $P = 12.37$ lb

**6-112** $\mu_s = 0.0875$

**6-115** $\mu_s = 0.333$

**6-116** $\mu_s = 0.556$

**6-118** In equilibrium
$N = 14.93$ N
$F = 4$ N

**6-119** $P = 211$ lb

**6-121** $P = 0.889$ lb

**6-124** $\theta = 46.40°$

**6-125** (a) $P = 173.1$ lb $\rightarrow$
(b) Not in equilibrium
(c) $P = 22.6$ lb $\rightarrow$

**6-126** (a) $\theta_{min} = 19.92°$
$\theta_{max} = 37.82°$
(b) $\mu_{smin} = 0.567$
$\mu_{smax} = 0.570$

**6-130** (a) $P = 144.4$ N
(b) $\tau_A = 1.855$ MPa
$\tau_B = 4.24$ MPa

**6-134** $T = 59.1$ N

**6-135** (a) $P = 103.9$ lb
(b) $P = 2410$ lb

**6-137** (a) $C = 766$ lb $\cdot$ in.
(b) $C = 3150$ lb $\cdot$ in.

**6-138** (a) $T = 128.8$ N $\cdot$ m
(b) $T = 58.7$ N $\cdot$ m

**6-140** (a) $T = 21.1$ kN
$C_A = 1.460$ kN $\cdot$ m
$C_B = 0.450$ kN $\cdot$ m
(b) $\tau_A = 26.6$ MPa
$\tau_B = 8.22$ MPa

**6-141** $a = 4.52$ in.

**6-144** $d = 1.114$ mm

**6-146** $(d_{AB})_{min} = 19.65$ mm
$(d_A)_{min} = 14.98$ mm
$(d_B)_{min} = 14.98$ mm
$(d_C)_{min} = 13.44$ mm

**6-147** $b_{min} = {}^5\!/_{64}$ in.

**6-150** $(d_B)_{min} = 3$ mm
$(d_C)_{min} = 3$ mm
$(d_D)_{min} = 3$ mm

**6-152** $d_{min} = 10$ mm

**6-155** $P = 168.7$ lb

**6-156** $T = 17.68$ N
$B = 57.7$ N $\uparrow$

**6-157** A $= (-350 \text{ j} + 25 \text{ k})$ lb
B $= (225 \text{ k})$ lb
$T_C = 450$ lb

**6-160** (a) $\delta_{AB} = -1.053$ mm
$\delta_{EF} = 0.526$ mm
(b) $\tau_D = 93.0$ MPa

**6-163** (a) $\sigma_{AC} = 1331$ psi
$\sigma_{DE} = 3550$ psi
(b) $\delta_{AC} = -5.01(10^{-3})$ in.
$\delta_{DE} = 7.51(10^{-3})$ in.
(c) $\tau_B = 1136$ psi

**6-164** $\mu_s = 0.395$

## Chapter 7

**7-1** (a) $T_{AB} = 80$ kip $\cdot$ ft $\curvearrowright$
$T_{BC} = 20$ kip $\cdot$ ft $\curvearrowleft$
$T_{CD} = 20$ kip $\cdot$ ft $\curvearrowright$
$T_{DE} = 45$ kip $\cdot$ ft $\curvearrowright$

**7-2** (a) $T_{max} = 30$ kN $\cdot$ m $\curvearrowright$

**7-5** (a) $\tau_{max} = 45.8$ ksi
(b) $\theta = 0.367$ rad

**7-6** (a) $\tau = 65.8$ MPa
(b) $\tau = 43.9$ MPa
(c) $\theta = 0.0274$ rad
(d) $\theta = 0.0713$ rad

**7-9** $T_{max} = 2.60$ kip $\cdot$ in.

**7-10** (a) $\tau_{max} = 139.3$ MPa
(b) $\theta_{D/B} = 0.00389$ rad $\curvearrowright$
(c) $\theta_{E/A} = 0.0894$ rad $\curvearrowright$

**7-13** (a) $\tau_{max} = 11.21$ ksi
(b) $\theta_{D/A} = 0.00611$ rad $\curvearrowright$

**7-16** (a) $\tau_3 = 127.3$ MPa
(b) $\theta_{E/A} = 0.0850$ rad $\curvearrowleft$

**7-17** (a) $T_{max} = 9.94$ kip $\cdot$ in.
(b) $\theta_A = 0.1559$ rad

**7-18** (a) $d_{AB} = 68.8$ mm
$d_{BC} = 93.4$ mm
(b) $d = 74.9$ mm

**7-19** (a) $d_{AB} = 0.546$ in.
(b) $d_{CD} = 0.402$ in.
(c) $L = 2.12$ ft

**7-22** $\theta = \dfrac{3TL}{16Gt\pi r^3}$

**7-23** $\theta = \dfrac{qL^2}{\pi Gc^4}$

**7-28** $T_{max} = 9.71$ kN $\cdot$ m

**7-29** (a) $\sigma_n = 4.12$ ksi T
$\tau_{nt} = 1.501$ ksi
(b) $\sigma_{max} = 4.39$ ksi T & C

**7-31** (a) $T_{BC} = 500$ lb $\cdot$ ft
(b) $T_{CD} = 250$ lb $\cdot$ ft
(c) $\sigma_{max}$(motor) $= 11.46$ ksi
T & C

$\sigma_{max}$(power) $= 15.64$ ksi
T & C

**7-33** (a) $T_1 = 32.0$ kip $\cdot$ in.
(b) $\sigma_{max} = 20.4$ ksi T
(c) $\sigma_{max} = 6.05$ ksi C

**7-35** (a) $U_M = 2100$ lb $\cdot$ ft
(b) $U_G = -2100$ lb $\cdot$ ft

**7-36** (a) $U_C = 30.0$ kJ
(b) $U_G = -23.5$ kJ

**7-37** (a) $U_p = 927$ lb $\cdot$ ft
(b) $U_G = 0$ lb $\cdot$ ft
(c) $U_F = -927$ lb $\cdot$ ft

**7-39** $U = 28.0$ lb $\cdot$ ft

**7-41** $P = 1615$ hp

**7-42** $d = 102.0$ mm

**7-44** (a) $P = 286$ kW
(b) $\theta = 0.0600$ rad

**7-45** (a) $\tau_{max} = 3.96$ ksi
(b) $\tau_{max} = 4.95$ ksi

**7-47** (a) $d = 1.951$ in.
(b) $d = 1.074$ in.

**7-48** (a) $\tau_{max} = 138.5$ MPa
(b) $\tau_{max} = 27.7$ MPa
(c) $\theta_{C/A} = -0.0386$ rad

**7-53** $\tau_B = 7.13$ ksi

**7-54** (a) $\tau_s = 8.99$ MPa
$\tau_m = 10.23$ MPa
(b) $\theta = 3.60(10^{-3})$ rad

**7-57** (a) $T = 10.18$ kip $\cdot$ ft
(b) $\theta_{B/A} = 0.01241$ rad $\curvearrowleft$

**7-58** $d = 77.5$ mm

**7-59** (a) $T = 31.8$ kip $\cdot$ in.
(b) $\tau_{max} = 500$ psi
(c) $T = 668$ kip $\cdot$ in.

**7-62** (a) $\tau_{AB} = 70.0$ MPa
$\tau_{BC} = 62.2$ MPa
(b) $\theta_{B/A} = 0.01437$ rad
(c) $\sigma_{max} = 70.0$ MPa T & C

**7-63** $d_C = 0.0564$ in.
$d_D = 0.219$ in.

**7-64** (a) $\tau_s = 94.5$ MPa
(b) $\tau_b = 78.7$ MPa
(c) $\sigma_{max} = 78.7$ MPa C
(d) $\theta_{C/A} = 0.01518$ rad

**7-72** $d_{60} = 110$ mm
$d_{6000} = 25$ mm
$N = 6000$ rpm

**7-73** $d = 2.50$ in.

**7-76** $d = 50$ mm

**7-78** $d = 60$ mm

**7-81** (a) $\tau_{max} = 9.05$ ksi
(b) $\theta_{B/A} = 0.01811$ rad $\curvearrowleft$
(c) $\theta_{C/A} = 0.0487$ rad $\curvearrowleft$

**7-82** $\theta_{A/C} = -0.0483$ rad

**7-85** (a) $T = 3410$ lb $\cdot$ in.

(b) $\tau_{max} = 462$ psi
(c) $\theta = 0.00418$ rad

**7-86** $\tau_{CD} = 50.7$ MPa

## Chapter 8

**8-1** (a) $I_x = 736$ in$^4$
(b) $I_y = 256$ in$^4$

**8-2** $I_x = 116.5(10^6)$ mm$^4$
$I_y = 318(10^6)$ m$^4$

**8-4** $I_{xC} = 390(10^6)$ mm$^4$
$I_{yC} = 142.4(10^6)$ mm$^4$

**8-5** $I_{xC} = 2730$ in$^4$
$I_{yC} = 190.8$ in$^4$

**8-7** (a) $I_x = 5510$ in$^4$
$I_y = 20,900$ in$^4$
(b) $I_{xC} = 1572$ in$^4$
$I_{yC} = 4810$ in$^4$

**8-10** $M_r = 33.8$ kN · m
**8-11** $M_r = 64.0$ kip · in.
**8-12** $\sigma = 17.65$ MPa T
**8-15** $M_r = 102.7$ kip · in.
**8-18** (a) $\sigma = 252$ MPa
(b) $M_r = -5.25$ kN · m
**8-19** $h = 0.0249$ in.
**8-22** $M_r = 8.08$ kN · m
**8-23** $\sigma_{max} = 19.62$ ksi T & C
**8-24** $\sigma_{maxT} = 5.58$ MPa T
$\sigma_{maxC} = 3.88$ MPa C
**8-27** (a) $\sigma_{maxT} = 13.91$ ksi T
(b) $\sigma_{maxC} = 7.36$ ksi C
**8-28** $\sigma_{bot} = 87.5$ MPa T
**8-31** (a) $\sigma_A = 10.48$ ksi C
(b) $\sigma_B = 7.86$ ksi T
**8-37** $V = -1000$ lb
$M = -1000x$ lb · ft
**8-38** $V = -2x$ kN
$M = -x^2$ kN · m
**8-42** $V = w(L + x)$ N
$M = \dfrac{w}{2}(x^2 + 2Lx - 2L^2)$ N · m
**8-43** $\sigma = 3000$ psi T
**8-45** (a) $V = -2000$ lb
$M = -2000x - 6000$ lb · ft
(b) $V = -4000$ lb
$M = -4000x - 6000$ lb · ft
(c) $V = -1000x + 6000$ lb
$M = -500x^2 + 6000x - 8000$ lb · ft
(d) $V = -4000$ lb
$M = -4000x + 40,000$ lb · ft
**8-46** (a) $V = -12(x + 2)$ kN
$M = -6x^2 - 24x - 30$ kN · m
(b) $V = -12x + 51$ kN
$M = -6x^2 + 51x - 30$ kN · m

(c) $V = 3$ kN
$M = 3x + 66$ kN · m
(d) $V = -21$ kN
$M = -21x + 210$ kN · m

**8-50** (a) $V = -15x + 25.5$ kN
$M = -7.5x^2 + 25.5x - 10.88$ kN · m
(b) $\sigma = 44.1$ MPa T
(c) $\sigma = 53.0$ MPa C, top; T, bot

**8-51** $\sigma_{top} = 3.27$ ksi C
$\sigma_{bot} = 2.08$ ksi T

**8-53** (a) $V = 3.18 \cos\left(\dfrac{\pi x}{10}\right)$ kip
$M = 10.13 \sin\left(\dfrac{\pi x}{10}\right)$ kip · ft
(b) $V_{max} = 3.18$ kip
$M_{max} = 10.13$ kip · ft

**8-54** (a) $V = \dfrac{400}{\pi^2} - \dfrac{200}{\pi} \sin\left(\dfrac{\pi x}{8}\right)$ kN
$M = \dfrac{400}{\pi^2}(x - 4)$
$- \dfrac{1600}{\pi^2} \cos\left(\dfrac{\pi x}{8}\right)$ kN · m
(b) $V_{max} = 40.5$ kN at $x = 0$
$M_{max} = 34.1$ kN · m at $x = 1.757$ m

**8-57** $P = 9.40$ kip
**8-61** $V_0 = 5000$ lb
$V_4 = 5000$ lb & 3000 lb
$V_8 = 3000$ lb & $-1000$ lb
$V_{12} = -1000$ lb & $-7000$ lb
$V_{16} = -7000$ lb
$M_4 = 20,000$ lb · ft
$M_8 = 32,000$ lb · ft
$M_{12} = 28,000$ lb · ft

**8-62** $V_0 = -15.00$ kN
$V_{2.5} = -15.00$ kN & 15.50 kN
$V_{5.5} = 15.50$ kN & $-4.50$ kN
$V_{7.5} = -4.50$ kN
$M_{2.5} = 37.5$ kN · m
$M_{5.5} = -9.00$ kN · m

**8-66** $V_0 = 45.0$ kN
$V_2 = 45.0$ kN & 25.0 kN
$V_4 = 25.0$ kN
$V_8 = -95.0$ kN
$M_2 = 90.0$ kN · m
$M_4 = 140.0$ kN · m
$M_{max} = 150.4$ kN · m

**8-67** $V_0 = 1800$ lb
$V_5 = 1800$ lb & 300 lb
$V_{10} = 300$ lb
$V_{20} = -2200$ lb
$M_5 = 9000$ lb · ft & 8000 lb · ft
$M_{10} = 9500$ lb · ft
$M_{max} = 9680$ lb · ft

**8-69** $\sigma_{maxT} = 1588$ psi T
$\sigma_{maxC} = 2650$ psi C
**8-70** $\sigma_{max} = 194.2$ MPa T & C
**8-73** $\sigma_{max} = 19.30$ ksi C
$\sigma_{max} = 63.7$ ksi T
**8-75** $V_{-2} = -1000$ lb
$V_0 = -1000$ lb & 2333 lb
$V_4 = 333$ lb & $-1667$ lb
$V_6 = -1667$ lb & 1000 lb
$V_8 = 1000$ lb
$M_{-2} = 2000$ lb · ft
$M_0 = 0$ lb · ft
$M_4 = 5333$ lb · ft & 1333 lb · ft
$M_6 = -2000$ lb · ft
**8-76** (a) $V_{1.5} = -120$ kN & 95.0 kN
$V_{3.5} = 15.00$ kN & $-20.0$ kN
$V_6 = -20.0$ kN & 30.0 kN
$M_{1.5} = -90.0$ kN · m
$M_{3.5} = 20.0$ kN · m
$M_6 = -30.0$ kN · m
(b) $\sigma_{max} = 61.4$ MPa T & C
**8-79** (a) $V_0 = -2550$ lb
$V_8 = -654$ lb & 1092 lb
$V_{12} = -2690$ lb
$M_{max} = 8100$ lb · ft
$M_8 = 7570$ lb · ft
(b) $\sigma_{max} = 21.6$ ksi T & C
(c) $\sigma_{BE} = 730$ psi T
(d) $\tau_A = 5.52$ ksi
**8-80** (b) $x = 5.00$ m
**8-84** $\tau_a = 0$
$\tau_b = 187.5$ kPa
$\tau_c = 250$ kPa
**8-85** $\tau = 152.9$ psi
$\tau = 179.6$ psi
$\tau = 214$ psi at NA
**8-88** (a) $\tau_J = 2.31$ MPa
(b) $\tau_{max} = 2.60$ MPa
**8-90** $\sigma_{max} = 128.9$ MPa T & C
$\tau_{max} = 17.49$ MPa
**8-91** (a) $\sigma_{max} = 1600$ psi
(b) $\tau_{max} = 83.3$ psi
**8-95** (a) $\tau_J = 84.5$ psi
(b) $\tau_{NA} = 111.5$ psi
(c) $\sigma_{max} = 876$ psi T & C
**8-96** (a) $F_B = 28.5$ kN
(b) $d_{min} = 24.6$ mm
**8-97** (a) $F_J = 519$ lb
(b) $n = 6$ nails
$x = 2$ in.
**8-105** 4- × 12-in. timber
**8-106** 203- × 254-mm timber
**8-108** S305 × 47
**8-111** 2- × 8-in. timber
**8-114** (a) $\sigma_{max} = 70$ MPa C

(b) $M_r = 59.5$ kN $\cdot$ m

**8-115** $\sigma_b = 28.5$ ksi T

$\sigma_t = 19.03$ ksi C

**8-119** (a) $V_0 = 1200$ lb & 1960 lb

$V_{10} = -1040$ lb &

$-1640$ lb

$V_{15} = -1640$ lb

$M_0 = 3600$ lb $\cdot$ ft

$M_{max} = 10{,}003$ lb $\cdot$ ft

$M_{10} = 8200$ lb $\cdot$ ft

(b) $V = -300x + 1960$ lb

$M = -150x^2 + 1960x$

$+ 3600$ lb $\cdot$ ft

**8-120** two L178 $\times$ 102 $\times$ 9.5

**8-123** $\sigma_{max} = 6.38$ ksi C

$\sigma_{max} = 10.03$ ksi T

**8-124** (a) $\sigma_{max} = 70.2$ MPa T

(b) $\sigma_{max} = 30.1$ MPa C

(c) $\tau_{max} = 7.00$ MPa

(d) $\tau_J = 6.53$ MPa

# Chapter 9

**9-1** (a) $y = \dfrac{P}{6EI}(-x^3 + 3L^2x - 2L^3)$

(b) $\delta_A = \dfrac{PL^3}{3EI}$ ↓

(c) $\theta_A = \dfrac{PL^2}{2EI}$ ⟋

**9-2** (a) $y = \dfrac{w}{24EI}(-x^4 + 4Lx^3$

$-6L^2x^2)$

(b) $\delta_B = \dfrac{wL^4}{8EI}$ ↓

(c) $\theta_B = \dfrac{wL^3}{6EI}$ ⟍

**9-6** $\delta_M = 0.781$ mm ↑

**9-7** $\delta_M = 0.242$ in. ↓

**9-9** (a) $y = \dfrac{P}{48EI}(4x^3 - 3L^2x)$

(b) $\theta_A = \dfrac{PL^2}{16EI}$ ⟍

(c) $\delta_M = \dfrac{PL^3}{48EI}$ ↓

**9-10** (a) $\delta_B = 10.37$ mm ↓

(b) $\delta_c = 22.0$ mm ↓

**9-12** (a) $\delta_p = 7.48$ mm ↑

(b) $\delta_{max} = 9.14$ mm ↓

**9-15** $\delta_M = \dfrac{11wL^4}{384EI}$ ↓

**9-17** (a) $y = \dfrac{P}{6EI}(x^3 - 3ax^2)$

(b) $a = 3L/4$

(c) $a = L/3$

**9-18** (a) $\delta_B = \dfrac{7PL^3}{12EI}$ ↓

(b) $\delta_C = \dfrac{23PL^3}{12EI}$ ↓

**9-21** $\delta_M = \dfrac{205wa^4}{384EI}$ ↓

**9-24** (a) $\delta_B = \dfrac{5wL^4}{24EI}$ ↓

(b) $\delta_c = \dfrac{3wL^4}{2EI}$ ↓

**9-27** (a) $y = \dfrac{w}{24EI}(-x^4 +$

$4L^3x - 3L^4)$

(b) $\delta_A = \dfrac{wL^4}{8EI}$ ↓

**9-29** (a) $y = \dfrac{w}{840EIL^3}(-x^7 +$

$7L^6x - 6L^7)$

(b) $y_0 = \dfrac{wL^4}{140EI}$ ↓

(c) $V_B = \dfrac{wL}{4}$ ↑

$M_B = \dfrac{wL^2}{20}$ ⟲

**9-31** (a) $y = \dfrac{w}{2\pi^4 EI}[-32L^4 \cos\dfrac{\pi x}{2L}$

$- 4\pi^2 L^2 x^2 + 8(\pi - 2)\pi L^3 x$

$+ 4\pi(4 - \pi)L^4]$

(b) $y_0 = 0.1089\dfrac{wL^4}{EI}$ ↓

(c) $V_B = \dfrac{2wL}{\pi}$ ↑

$M_B = \dfrac{4wL^2}{\pi^2}$ ⟲

**9-33** (a) $\delta_B = \dfrac{PL^3}{2EI}$ ↓

(b) $\delta_C = \dfrac{11PL^3}{6EI}$ ↓

**9-34** (a) $\delta_B = \dfrac{2PL^3}{3EI}$ ↓

(b) $\delta_C = \dfrac{2PL^3}{EI}$ ↓

**9-38** (a) $\delta_0 = \dfrac{19PL^3}{324EI}$ ↓

(b) $\delta_{5L/6} = \dfrac{PL^3}{48EI}$ ↑

**9-39** $\delta_0 = \dfrac{29wL^4}{24EI}$ ↓

**9-40** (a) $\delta_A = \dfrac{2wL^4}{3EI}$ ↓

(b) $\delta_M = \dfrac{305wL^4}{384EI}$ ↓

**9-43** $\delta_C = \dfrac{wL^4}{24EI}$ ↑

**9-44** (a) $\delta_a = \dfrac{11wa^4}{24EI}$ ↓

(b) $\delta_M = \dfrac{205wa^4}{384EI}$ ↓

**9-51** $\delta_C = \dfrac{37wL^4}{3EI}$ ↓

**9-52** $\delta_B = \dfrac{11wL^4}{24EI}$ ↑

**9-56** (a) $\delta_B = \dfrac{13wL^4}{384EI}$ ↓

(b) $\delta_D = \dfrac{5wL^4}{144EI}$ ↑

**9-57** $\delta_C = 0.0333$ in. ↓

**9-58** $\delta = \dfrac{13wL^4}{384EI}$ ↓

**9-60** $\delta = 5.42$ mm ↓

**9-63** $\delta_B = \dfrac{5PL^3}{6EI}$ ↓

**9-64** $\delta_D = \dfrac{wL^4}{4EI}$ ↓

**9-67** $\delta_C = \dfrac{11wL^4}{192EI}$ ↓

**9-69** (a) $R_A = \dfrac{3wL}{8}$ ↑

$R_B = \dfrac{5wL}{8}$ ↑

$M_B = \dfrac{wL^2}{8}$ ⟲

(b) $\delta_{L/2} = \dfrac{wL^4}{192EI}$ ↓

**9-70** $M = \dfrac{wL^2}{12}$ ⟳

**9-73** $P = \dfrac{3wL}{8}$ ↑

**9-74** (a) $R_A = \dfrac{wL}{10}$ ↑

$R_B = \dfrac{2wL}{5}$ ↑

$M_B = \dfrac{wL^2}{15}$ ⟲

(b) $\delta_{max} = 0.00239\dfrac{wL^4}{EI}$ ↓

**9-75** (a) $R_A = \dfrac{wL}{24}$ ↑

$R_B = \dfrac{7wL}{24}$ ↑

$M_B = \dfrac{wL^2}{24}$ ⟲

(b) $\delta_{max} = 0.001267\dfrac{wL^4}{EI}$ ↓

**9-78** (a) $R_A = \dfrac{P}{2}$ ↑

$R_B = \dfrac{P}{2}$ ↑

$M_A = \dfrac{PL}{8}$ ⟳

$$M_B = \frac{PL}{8} \downarrow$$

(b) $\delta_{L/2} = \frac{PL^3}{192EI} \downarrow$

**9-80** (a) $R_A = \frac{7wL}{16} \uparrow$

$$R_B = \frac{5wL}{8} \uparrow$$

$$R_C = \frac{wL}{16} \downarrow$$

(b) $M_B = -\frac{wL^2}{16}$

(c) $\delta_{3L/2} = \frac{wL^4}{256EI} \uparrow$

**9-81** (a) $R_A = \frac{4P}{11} \uparrow$

$$R_C = \frac{23P}{22} \uparrow$$

$$R_D = \frac{9P}{22} \downarrow$$

$$M_D = \frac{3PL}{22} \curvearrowright$$

(b) $\delta_B = \frac{13PL^3}{132EI} \downarrow$

**9-84** (a) $R_A = 26.7$ kN $\uparrow$
$R_D = 45.3$ kN $\uparrow$
$M_D = 111.4$ kN · m $\downarrow$

(b) $\delta_B = 8.99$ mm $\downarrow$

**9-92** $P = \frac{wL}{3}$

**9-93** $R_A = \frac{9M}{16L} \uparrow$

$$R_B = \frac{9M}{16L} \downarrow$$

$$M_D = \frac{M}{8} \curvearrowright$$

**9-96** $R_A = \frac{13wL}{16} \uparrow$

$$R_B = \frac{33wL}{16} \uparrow$$

$$R_C = \frac{wL}{8} \uparrow$$

**9-97** (a) $P = \frac{2wL}{9} \downarrow$

(b) $R_A = \frac{13wL}{24} \uparrow$

$$M_A = \frac{7wL^2}{72} \curvearrowright$$

$$R_B = \frac{49wL}{72} \uparrow$$

**9-98** (a) $R_A = \frac{57wL}{64} \uparrow$

$$M_A = \frac{9wL^2}{32} \curvearrowright$$

$$R_C = \frac{7wL}{64} \uparrow$$

(b) $\delta_B = \frac{13wL^4}{384EI} \downarrow$

**9-102** $R_A = \frac{5P}{32} \uparrow$

$$M_A = \frac{5PL}{32} \curvearrowright$$

$$R_B = \frac{5P}{32} \uparrow$$

$$R_D = \frac{27P}{32} \uparrow$$

$$M_D = \frac{11PL}{32} \downarrow$$

**9-103** $R_B = 6.29$ kip
**9-106** $\sigma = 69.6$ MPa T
**9-107** (a) $\sigma_{max\ AB} = 8.50$ ksi T & C
$\sigma_{max\ CD} = 6.77$ ksi T & C
(b) $\tau_{max\ AB} = 1440$ psi
$\tau_{max\ CD} = 559$ psi
**9-110** 51- × 254-mm timber
**9-112** $d = 50$ mm
**9-114** W533 × 92
**9-116** $t = 15.00$ mm
**9-117** $\sigma_{max} = 6.60$ ksi T & C
**9-120** (a) $\delta_C = 11.65$ mm $\downarrow$
(b) $\sigma_{max} = 80.9$ MPa T

**9-123** (a) $R_C = \frac{wL}{4} \uparrow$

(b) $R_A = \frac{wL}{4} \uparrow$

$$M_A = \frac{wL^2}{8} \curvearrowright$$

$$R_B = \frac{3wL}{4} \uparrow$$

$$M_B = \frac{wL^2}{4} \downarrow$$

**9-126** (a) $R_A = \frac{wL}{8} \uparrow$

$$R_B = \frac{33wL}{16} \uparrow$$

$$R_C = \frac{13wL}{16} \uparrow$$

(b) $\delta_{2L} = \frac{11wL^4}{96EI} \downarrow$

## Chapter 10
**10-1** $\sigma = 14.23$ ksi T
$\tau = 11.82$ ksi
**10-2** $\sigma = 121.5$ MPa T
$\tau = 9.64$ MPa
**10-4** $\sigma = 48.3$ MPa C
$\tau = 103.8$ MPa
**10-7** (a) $\tau_{xy} = -5.10$ ksi

(b) $\sigma_y = 0.150$ ksi C
**10-12** $\sigma_n = 41.6$ MPa T
$\tau_{nt} = +28.8$ MPa
**10-13** $\sigma_n = 30.9$ ksi T
$\tau_{nt} = -1.794$ ksi
**10-16** $\sigma_n = 163.8$ MPa T
$\tau_{nt} = +9.85$ MPa
**10-17** (a) $\sigma_n = 13.78$ ksi T
$\tau_{nt} = +6.27$ ksi
(b) $\sigma_n = 21.8$ ksi T
$\sigma_t = 9.23$ ksi T
$\tau_{nt} = +1.724$ ksi
**10-18** (a) $\sigma_n = 56.4$ MPa C
$\tau_{nt} = +47.2$ MPa
(b) $\sigma_n = 5.98$ MPa T
$\sigma_t = 86.0$ MPa C
$\tau_{nt} = +19.64$ MPa
**10-21** (a) $|\tau_h| = |\tau_v| = 5.33$ ksi
(b) $\tau_{nt} = +5.33$ ksi
**10-22** $\sigma_x = 221$ MPa T
$\tau_{nt} = +61.0$ MPa
**10-28** $\sigma_{p1} = 77.7$ MPa T
$\sigma_{p2} = 7.72$ MPa C
$\sigma_{p3} = 0$ MPa
$\tau_p = \tau_{max} = 42.7$ MPa
$\theta_p = 34.72°$
**10-29** $\sigma_{p1} = 14.00$ ksi T
$\sigma_{p2} = 6.00$ ksi C
$\sigma_{p3} = 0$ ksi
$\tau_p = \tau_{max} = 10.00$ ksi
$\theta_p = -18.43°$
**10-31** $\sigma_{p1} = 12.34$ ksi T
$\sigma_{p2} = 17.34$ ksi C
$\sigma_{p3} = 0$ ksi
$\tau_p = \tau_{max} = 14.84$ ksi
$\theta_p = -16.31°$
**10-33** $\sigma_{p1} = 31.0$ ksi T
$\sigma_{p2} = 1.000$ ksi T
$\sigma_{p3} = 0$ ksi
$\tau_p = 15.00$ ksi
$\tau_{max} = 15.50$ ksi
$\theta_p = 26.57°$
**10-34** $\sigma_{p1} = 44.0$ MPa T
$\sigma_{p2} = 18.00$ MPa T
$\sigma_{p3} = 0$ MPa
$\tau_p = 13.00$ MPa
$\tau_{max} = 22.0$ MPa
$\theta_p = 33.69°$
**10-36** $\sigma_{p1} = 84.0$ MPa T
$\sigma_{p2} = 24.0$ MPa T
$\sigma_{p3} = 0$ MPa
$\tau_p = 30.0$ MPa
$\tau_{max} = 42.0$ MPa
$\theta_p = -26.57°$
**10-38** $\sigma_{p1} = 105.0$ MPa T
$\sigma_{p2} = 145.0$ MPa C

$\sigma_{p3} = 0$ MPa
$\tau_{xy} = -75.0$ MPa
$\theta_p = 18.43°$

**10-39** $\sigma_c = 92.3$ psi C

**10-43** $\sigma_n = 6.62$ ksi T
$\tau_{nt} = +11.03$ ksi

**10-45** $\sigma_{p1} = 31.0$ ksi T
$\sigma_{p2} = 1.000$ ksi T
$\sigma_{p3} = 0$ ksi
$\tau_p = 15.00$ ksi
$\tau_{max} = 15.50$ ksi
$\theta_p = 26.57°$

**10-46** $\sigma_{p1} = 77.7$ MPa T
$\sigma_{p2} = 7.72$ MPa C
$\sigma_{p3} = 0$ MPa
$\tau_p = \tau_{max} = 42.7$ MPa
$\theta_p = 34.72°$

**10-50** (a) $\sigma_{p1} = 98.2$ MPa T
$\sigma_{p2} = 118.2$ MPa C
$\sigma_{p3} = 0$ MPa
$\tau_p = \tau_{max} = 108.2$ MPa
(b) $\sigma_n = 60.3$ MPa C
$\tau_{nt} = -95.8$ MPa

**10-52** $\epsilon_n = -2410$ $\mu$m/m

**10-53** $\epsilon_{AC} = +596$ $\mu$in./in.

**10-55** $\epsilon_n = -633$ $\mu$in./in.
$\epsilon_t = +473$ $\mu$in./in.
$\gamma_{nt} = +1332$ $\mu$rad

**10-56** $\epsilon_n = +264$ $\mu$m/m
$\epsilon_t = -24.1$ $\mu$m/m
$\gamma_{nt} = +1219$ $\mu$rad

**10-59** $\epsilon_n = +362$ $\mu$in./in.
$\epsilon_t = +748$ $\mu$in./in.
$\gamma_{nt} = -306$ $\mu$rad

**10-60** $\epsilon_n = -398$ $\mu$m/m
$\epsilon_t = -402$ $\mu$m/m
$\gamma_{nt} = +721$ $\mu$rad

**10-63** (a) $\gamma_{nt} = +425$ $\mu$rad
(b) $\epsilon_y = +1675$ $\mu$in./in.

**10-65** $\epsilon_{p1} = +672$ $\mu$in./in.
$\epsilon_{p2} = -272$ $\mu$in./in.
$\epsilon_{p3} = 0$ $\mu$in./in.
$\gamma_p = \gamma_{max} = 943$ $\mu$rad
$\theta_p = 16.00°$

**10-66** $\epsilon_{p1} = +1004$ $\mu$m/m
$\epsilon_{p2} = -364$ $\mu$m/m
$\epsilon_{p3} = 0$ $\mu$m/m
$\gamma_p = \gamma_{max} = 1367$ $\mu$rad
$\theta_p = -10.28°$

**10-69** $\epsilon_{p1} = +1009$ $\mu$in./in.
$\epsilon_{p2} = -759$ $\mu$in./in.
$\epsilon_{p3} = 0$ $\mu$in./in.
$\gamma_p = \gamma_{max} = 1768$ $\mu$rad
$\theta_p = 4.07°$

**10-70** $\epsilon_{p1} = +777$ $\mu$m/m
$\epsilon_{p2} = -437$ $\mu$m/m

$\epsilon_{p3} = 0$ $\mu$m/m
$\gamma_p = \gamma_{max} = 1215$ $\mu$rad
$\theta_p = +8.62°$

**10-73** $\epsilon_{p1} = +908$ $\mu$in./in.
$\epsilon_{p2} = +388$ $\mu$in./in.
$\epsilon_{p3} = 0$ $\mu$in./in.
$\gamma_p = 519$ $\mu$rad
$\gamma_{max} = 908$ $\mu$rad
$\theta_p = -16.85°$

**10-74** $\epsilon_{p1} = +970$ $\mu$m/m
$\epsilon_{p2} = +580$ $\mu$m/m
$\epsilon_{p3} = 0$ $\mu$m/m
$\gamma_p = 391$ $\mu$rad
$\gamma_{max} = 970$ $\mu$rad
$\theta_p = -25.09°$

**10-77** $\epsilon_x = +500$ $\mu$in./in.
$\epsilon_y = -300$ $\mu$in./in.
$\gamma_{xy} = +600$ $\mu$rad
$\gamma_p = \gamma_{max} = 1000$ $\mu$rad

**10-78** $\epsilon_x = +640$ $\mu$m/m
$\epsilon_y = -800$ $\mu$m/m
$\gamma_{xy} = +960$ $\mu$rad
$\gamma_p = \gamma_{max} = 1730$ $\mu$rad

**10-81** $\epsilon_{p1} = +966$ $\mu$in./in.
$\epsilon_{p2} = -241$ $\mu$in./in.
$\epsilon_{p3} = 0$ $\mu$in./in.
$\gamma_p = \gamma_{max} = 1207$ $\mu$rad
$\theta_p = 6.59°$

**10-82** $\epsilon_{p1} = +1072$ $\mu$m/m
$\epsilon_{p2} = -505$ $\mu$m/m
$\epsilon_{p3} = 0$ $\mu$m/m
$\gamma_p = \gamma_{max} = 1576$ $\mu$rad
$\theta_p = +19.27°$

**10-85** $\gamma_{xy} = \pm641$ $\mu$rad
$\epsilon_{p2} = -774$ $\mu$in./in.
$\epsilon_{p3} = 0$ $\mu$in./in.
$\gamma_p = \gamma_{max} = 1188$ $\mu$rad
$\theta_p = \pm16.35°$

**10-86** $\gamma_{xy} = \pm912$ $\mu$rad
$\epsilon_{p2} = -180$ $\mu$m/m
$\epsilon_{p3} = 0$ $\mu$m/m
$\gamma_p = \gamma_{max} = 960$ $\mu$rad
$\theta_p = \pm35.90°$

**10-89** $\sigma_x = 8.94$ ksi T
$\sigma_y = 0.1753$ ksi C
$\tau_{xy} = -1.520$ ksi

**10-90** $\sigma_x = 191.2$ MPa T
$\sigma_y = 130.4$ MPa T
$\tau_{xy} = -42.6$ MPa

**10-92** $\sigma_x = 174.8$ MPa T
$\sigma_y = 193.8$ MPa C
$\tau_{xy} = 64.6$ MPa

**10-95** $\sigma_{p1} = 9.15$ ksi T
$\sigma_{p2} = 0.579$ ksi C
$\sigma_{p3} = 0$ ksi
$\tau_p = \tau_{max} = 4.86$ ksi

$\theta_p = -9.22°$

**10-98** (a) $\epsilon_x = +780$ $\mu$m/m
$\epsilon_y = -251$ $\mu$m/m
$\gamma_{xy} = -782$ $\mu$rad
(b) $\epsilon_{p1} = +912$ $\mu$m/m
$\epsilon_{p2} = -383$ $\mu$m/m
$\epsilon_{p3} = -261$ $\mu$m/m
$\gamma_p = \gamma_{max} = 1294$ $\mu$rad
$\theta_p = -18.59°$

**10-99** (a) $\epsilon_x = +750$ $\mu$in./in.
$\epsilon_y = -250$ $\mu$in./in.
$\gamma_{xy} = -750$ $\mu$rad
(b) $\epsilon_{p1} = +875$ $\mu$in./in.
$\epsilon_{p2} = -375$ $\mu$in./in.
$\epsilon_{p3} = -214$ $\mu$in./in.
$\gamma_p = \gamma_{max} = 1250$ $\mu$rad
$\theta_p = -18.43°$

**10-101** (a) $\epsilon_x = -350$ $\mu$in./in.
$\epsilon_y = +1150$ $\mu$in./in.
$\gamma_{xy} = +520$ $\mu$rad
(b) $\sigma_x = 0.514$ ksi T
$\sigma_y = 33.5$ ksi T
$\tau_{xy} = +5.72$ ksi
(c) $\sigma_{p1} = 34.5$ ksi T
$\sigma_{p2} = 0.451$ ksi C
$\sigma_{p3} = 0$ ksi
$\tau_p = \tau_{max} = 17.46$ ksi
$\theta_p = -9.56°$

**10-102** (a) $\epsilon_x = +525$ $\mu$m/m
$\epsilon_y = +1350$ $\mu$m/m
$\gamma_{xy} = -975$ $\mu$rad
(b) $\sigma_x = 75.2$ MPa T
$\sigma_y = 121.4$ MPa T
$\tau_{xy} = -27.3$ MPa
(c) $\sigma_{p1} = 134.0$ MPa T
$\sigma_{p2} = 62.5$ MPa T
$\sigma_{p3} = 0$ MPa
$\tau_p = 35.8$ MPa
$\tau_{max} = 67.0$ MPa
$\theta_p = 24.88°$

**10-104** (a) $\sigma_x = 64.7$ MPa T
$\sigma_y = 2.52$ MPa T
$\tau_{xy} = -42.8$ MPa
(b) $\epsilon_{p1} = +1272$ $\mu$m/m
$\epsilon_{p2} = -655$ $\mu$m/m
$\epsilon_{p3} = -304$ $\mu$m/m
$\gamma_p = \gamma_{max} = 1927$ $\mu$rad
$\theta_p = -27.00°$
(c) $\sigma_{p1} = 86.5$ MPa T
$\sigma_{p2} = 19.28$ MPa C
$\sigma_{p3} = 0$ MPa
$\tau_p = \tau_{max} = 52.9$ MPa

**10-107** $\sigma = 11.95$ ksi

**10-108** $\sigma = 3.70$ MPa

**10-111** $t = 0.474$ in.

**10-112** $t = 6.25$ mm

**10-116** (a) $t = 3.57$ mm
(b) $\epsilon = 350$ $\mu$m/m
(c) $\Delta D = 1.050$ mm

**10-117** (a) $\sigma_a = 4.73$ ksi
$\sigma_h = 9.45$ ksi

**10-119** $\sigma_{p1} = 5.10$ ksi T
$\sigma_{p2} = 1.119$ ksi C
$\sigma_{p3} = 0$ ksi
$\tau_p = \tau_{\max} = 3.11$ ksi
$\theta_p = 25.10°$

**10-120** $\sigma_{p1} = 56.6$ MPa T
$\sigma_{p2} = 29.3$ MPa C
$\sigma_{p3} = 0$ MPa
$\tau_p = \tau_{\max} = 43.0$ MPa
$\theta_p = 35.74°$

**10-123** $\sigma_C = 4.78$ ksi T
$\sigma_D = 12.87$ ksi C

**10-124** $\sigma_{CD} = 89.8$ MPa T
$\sigma_{DE} = 126.1$ MPa C

**10-126** $\sigma_{p1} = 59.0$ MPa T
$\sigma_{p2} = 14.84$ MPa C
$\sigma_{p3} = 0$ MPa
$\tau_p = \tau_{\max} = 36.9$ MPa
$\theta_p = 26.64°$

**10-127** $\sigma_{\max} = 8670$ psi T
$\tau_{\max} = 4340$ psi

**10-130** $\sigma_{\text{top}} = 11.60$ MPa T
$\sigma_{\text{bot}} = 10.42$ MPa C

**10-131** $\sigma_A = 1000$ psi T
$\sigma_B = 200$ psi C
$\sigma_C = 1800$ psi C
$\sigma_D = 600$ psi C

**10-135** $\sigma_{p1} = 12.62$ ksi T
$\sigma_{p2} = 1.463$ ksi C
$\sigma_{p3} = 0$ ksi
$\tau_p = \tau_{\max} = 7.04$ ksi
$\theta_p = 11.43°$

**10-136** (a) $\sigma_n = 87.6$ MPa T
$\tau_{nt} = 31.2$ MPa
(b) $\sigma_{p1} = 107.6$ MPa T
$\sigma_{p2} = 40.2$ MPa T
$\sigma_{p3} = 2.50$ MPa C
$\tau_{\max} = 55.0$ MPa
$\theta_p = -19.61°$

**10-137** $P = 11.11$ kip
$T = 4.16$ kip $\cdot$ in.

**10-140** $\sigma_{p1} = 5.07$ MPa T
$\sigma_{p2} = 75.8$ MPa C
$\sigma_{p3} = 0$ MPa
$\tau_{\max} = \tau_p = 40.4$ MPa
$\theta_p = -14.50°$

**10-141** $\sigma_{\text{top}} = 175.6$ psi C
$\sigma_{\text{bot}} = 170.2$ psi T

**10-149** Failure predicted by

maximum shear stress and
maximum distortion energy

**10-152** Maximum normal stress:
$\sigma_{PL} > 190.0$ MPa
Maximum shear stress:
$\sigma_{PL} > 315$ MPa
Maximum distortion energy:
$\sigma_{PL} > 275$ MPa

**10-153** Maximum normal stress:
$FS = 2.00$
Maximum shear stress:
$FS = 2.00$
Maximum distortion energy:
$FS = 2.11$

**10-156** Maximum normal stress:
$FS = 1.246$
Maximum shear stress:
$FS = 1.246$
Maximum distortion energy:
$FS = 1.346$

**10-157** (a) $p = 1714$ psi
(b) $p = 1979$ psi

**10-158** (a) $T = 39.3$ kN $\cdot$ m
(b) $T = 45.3$ kN $\cdot$ m

**10-162** $d_i = 108.7$ mm

**10-163** $d_{\min} = 1.580$ in.

**10-167** (a) $R = 4.38$ kip
(b) $R = 4.58$ kip

**10-168** (a) $D = 94.9$ mm
(b) $D = 92.0$ mm

**10-170** $t = 14.47$ mm

**10-172** (a) Safe
(b) Safe

**10-173** Safe

**10-175** $d = 3.68$ in.

**10-176** $d = 57.1$ mm

**10-178** $P = 2840$ N

**10-182** $\sigma_n = 46.6$ MPa C
$\tau_{nt} = -47.4$ MPa

**10-183** $\sigma_x = 9.86$ ksi T
$\sigma_y = 17.86$ ksi C
$\theta_p = 15.00°$

**10-186** $\epsilon_{p1} = +1575$ $\mu$m/m
$\epsilon_{p2} = +585$ $\mu$m/m
$\epsilon_{p3} = 0$ $\mu$m/m
$\gamma_p = 990$ $\mu$rad
$\gamma_{\max} = 1575$ $\mu$rad
$\theta_p = -37.98°$

**10-189** $p = 198.4$ psi
$T = 1206$ kip $\cdot$ in.

**10-190** (a) $\sigma_{\max} = 131.4$ MPa T
$\sigma_{\max} = 135.4$ MPa C
(b) $\sigma_{\max} = 149.0$ MPa T
$\sigma_{\max} = 155.2$ MPa C

**10-193** $P = 186.0$ kip

# Chapter 11

**11-1** (a) $L/r = 200$
(b) $L/r = 89.2$
(c) $P_{cr} = 5.62$ kip

**11-2** (a) $L/r = 149.9$
(b) $L/r = 88.9$
(c) $P_{cr} = 388$ kN

**11-5** $P = 6.61$ kip

**11-7** (a) $L/r = 145.9$
(b) $L/r = 89.2$
(c) $P_{cr} = 142.4$ kip

**11-8** (a) $P_{cr} = 128.5$ kN
(b) $P_{cr} = 382$ kN

**11-10** $FS = 2.62$

**11-13** $P = 3.86$ kip

**11-14** $P = 38.3$ kN

**11-15** $P = 4.32$ kip

**11-18** $P_{cr} = \dfrac{\pi^3 E D^4}{16 L^2}$

**11-19** $P = 1.084$ kip

**11-25** $P = 12.15$ kip

**11-26** $P = 1302$ kN

**11-28** (a) $P = 429$ kN
(b) $P = 928$ kN

**11-29** $P = 166.2$ kip

**11-32** $P = 1097$ kN

**11-33** $d = 3.50$ in.

**11-36** $P = 455$ kN

**11-37** (a) $P = 163.0$ kip
(b) $P = 280$ kip

**11-42** $P = 400$ kN

**11-43** $P = 18.29$ kip

**11-46** (a) $P = 245$ kN
(b) $P = 360$ kN

**11-48** $P = 312$ kN

**11-51** (a) $P = 67.8$ kip
(b) $1.447$ in. $< d < 2.42$ in.

**11-54** $d = 127$ mm

**11-55** W10 $\times$ 60

**11-57** $A = 9.61$ in$^2$

**11-58** $44.4 \times 88.8$ mm

**11-61** $d = 2.5$ in.

**11-62** $d = 32$ mm

**11-65** (a) $L/r = 156.7$
(b) $L/r = 70.3$
(c) $P_{cr} = 23.2$ kip
(d) $\sigma = 483$ psi

**11-68** (a) $FS_{AB} = 1.322$
$FS_{AC} = 22.9$
(b) $FS_{AC} = 6.53$

**11-69** $P = 431$ kip

**11-72** $P = 247$ kN

**11-73** $P = 69.1$ kip

# INDEX